QUANTITATIVE UNDERSTANDING
OF BIOSYSTEMS

Foundations of Biochemistry and Biophysics

This textbook series focuses on foundational principles and experimental approaches across all areas of biological physics, covering core subjects in a modern biophysics curriculum. Individual titles address such topics as molecular biophysics, statistical biophysics, molecular modeling, single-molecule biophysics, and chemical biophysics. It is aimed at advanced undergraduate- and graduate-level curricula at the intersection of biological and physical sciences. The goal of the series is to facilitate interdisciplinary research by training biologists and biochemists in quantitative aspects of modern biomedical research and to teach key biological principles to students in physical sciences and engineering.

New books in the series are commissioned by invitation. Authors are also welcome to contact the publisher (Lou Chosen, Executive Editor: lou.chosen@taylorandfrancis.com) to discuss new title ideas.

Light Harvesting in Photosynthesis
Roberta Croce, Rienk van Grondelle, Herbert van Amerongen, Ivo van Stokkum (Eds.)

An Introduction to Single Molecule Biophysics
Yuri L. Lyubchenko (Ed.)

Biomolecular Kinetics: A Step-by-Step Guide
Clive R. Bagshaw

Introduction to Experimental Biophysics: Biological Methods for Physical Scientists, Second Edition
Jay L. Nadeau

Biomolecular Thermodynamics: From Theory to Application
Douglas E. Barrick

Quantitative Understanding of Biosystems: An Introduction to Biophysics, Second Edition
Thomas M. Nordlund, Peter M. Hoffmann

https://www.crcpress.com/Foundations-of-Biochemistry-and-Biophysics/book-series/CRCFOUBIOPHY

QUANTITATIVE UNDERSTANDING OF BIOSYSTEMS
AN INTRODUCTION TO BIOPHYSICS
SECOND EDITION

Thomas M. Nordlund
Peter M. Hoffmann

CRC Press
Taylor & Francis Group
Boca Raton London New York

CRC Press is an imprint of the
Taylor & Francis Group, an **informa** business

CRC Press
Taylor & Francis Group
6000 Broken Sound Parkway NW, Suite 300
Boca Raton, FL 33487-2742

First issued in paperback 2020

© 2019 by Taylor & Francis Group, LLC
CRC Press is an imprint of Taylor & Francis Group, an Informa business

No claim to original U.S. Government works

ISBN-13: 978-1-138-63341-4 (hbk)
ISBN-13: 978-0-367-77991-7 (pbk)

Library of Congress Cataloging-in-Publication Data

Names: Nordlund, Thomas M., author. | Hoffmann, Peter M., author.
Title: Quantitative understanding of biosystems : an introduction to biophysics / Thomas M. Nordlund, Peter M. Hoffmann.
Description: Second edition. | Boca Raton : CRC Press, Taylor & Francis Group, [2019] | Series: Foundations of biochemistry and biophysics ; 6 | Includes bibliographical references and index.
Identifiers: LCCN 2018044931 | ISBN 9781138633414 (hardback : alk. paper)
Subjects: LCSH: Biophysics--Textbooks. | Molecular biology--Textbooks.
Classification: LCC QH505 .N67 2019 | DDC 612/.014--dc23
LC record available at https://lccn.loc.gov/2018044931

Visit the Taylor & Francis Web site at
http://www.taylorandfrancis.com

and the CRC Press Web site at
http://www.crcpress.com

Contents

PART II   Structure and Function

4   Water

11 Direct Ultraviolet Effects on Biological Systems 317

PART IV   Biological Activity: (Classical) Microworld

Preface

This book is an introductory textbook/learning tool for biological physics, not a survey of current knowledge in biophysics. When faced with competing descriptions that are (1) simple and didactic or (2) more complex, accurate and correct, the authors have chosen the former. The book addresses a common need of university students enrolled in physical and biological science and engineering programs to learn the physical and quantitative aspects of biological systems, especially on a micro- and nanoscale, as a counterpart to descriptive and genetic approaches developed in other courses. A task typically assigned to the "biophysics" faculty in a physics department is to teach introductory biophysics to a single class with both undergraduate and graduate students. Among these students may be sophomore to senior physics students; senior biology, chemistry, and engineering students; and a mix of chemistry, biochemistry, and physics grad students. Preparations in math, physics, and biochemistry vary widely. This book is for students who have taken at least a year of calculus-based physics. Readers will be aided by supplementary resources (see www.crcpress.com/9781138633414), including "active" versions of graphs and diagrams in the text, along with links to mathematical and biological/biochemical source data.

FOCUS AND APPROACH

This book contains material for an introductory undergraduate course covering the biological physics of macromolecules, subcellular structures, and whole cells, including interactions with light. This quantitative, intermediate-level textbook emphasizes the following:

1. Mathematical and computational tools—graphing, calculus, simple differential equations, diagrammatic analysis, and visualization tools
2. Randomness, variation, statistical mechanics, distributions, and spectra
3. The biological micro- and nanoworld: structures, processes, and the physical laws
4. Quantum effects—photosynthesis, UV damage, electron and energy transfer, and spectroscopic characterization of biological structures
5. Experimental methods—an overview of important methods in biophysics, plus a chapter on atomic force microscopy (AFM). This is a new addition to the second edition.

Besides the treatment of quantum effects, lacking in most general biophysics textbooks, this text includes online supplementary materials, with multimedia learning tools such as graphs, video clips, and animations that illustrate intrinsically dynamic processes. This multimedia component serves as a guide to reading and understanding the written text. The video clips and active graphs emphasize what is most obvious about biological systems: *living things move.*

Quantitative Understanding of Biosystems is a learning tool, not a repository of current research knowledge, and does not attempt to be comprehensive. Appendix materials point students to reviews of subject areas not covered in the text. This text also uses some of the "effective learning" ideas incorporated into most first-year general physics textbooks. These include graph- and diagram-centered physics and math, frequent checks of understanding and "tool training," and repetition of important ideas at higher levels or from different points of view. The multimedia aspect described above supports this learning approach.

Throughout the text, we focus on the behavior and properties of microscopic structures that underlie living systems. Students should gain significant computational and project experience and become competent at mastering higher-level software and quantitatively characterizing biosystems.

Biophysics and genetic engineering? The impact and relation of the genetic engineering and biophysics, addressed at the beginning of Chapter 1, should be discussed early in the course. "CRISPR/cas9," a gene-editing tool largely unknown to bioscientists in 2011, is now familiar to many high school science students. Because CRISPR gives detailed control of the biological synthetic apparatus to the investigator at relatively low cost, it would appear most structure/function and cause/effect questions—the soul of the biophysics enterprise—could be ceded to genetic engineers. Is there no longer a need for the detailed and sometimes expensive nano-structural investigations (e.g., atomic force microscopy) carried out by the biophysicist? Several chapters in this second edition will address the role of CRISPR in biophysical investigations.

TARGET AUDIENCE

The target audience of this book is the advanced *undergraduate student*, including physics, chemistry, biology/biochemistry, and biomedical engineering majors who have had one full year of physics, a year of general chemistry, plus an additional two to three semesters of science courses including coverage of the following areas: modern physics or physical chemistry, basic statistical mechanics, biochemistry, or biomolecular structure. Math or applied math majors with biological interests will also find a useful learning experience in this text. The level of math will be the equivalent of concurrent enrollment in an ordinary differential equations course. For the physics major or *graduate student* fascinated by but ignorant of the physics of biological systems, and for the chemistry, biochemistry, or bioengineering major who wants to learn about quantum and nanometer effects in biology, this book will provide an introduction sufficient to enter graduate programs in the biosciences or for employment in many biotech companies. This course is often cross-listed as an undergraduate/graduate course, with graduate students typically expected to complete more advanced problems and projects.

OMITTED TOPICS

This book is designed for the large majority of undergraduate physics programs that can offer, at most, one semester of biophysics. (As noted below in "Course Features and Suggestions to Instructors," it can be turned into a two-semester course by an experienced biophysics instructor.) Hard choices among important topics must be made. Our philosophy has been to choose topics and coverage (1) that constitute a self-contained, coherent *one*-semester course, (2) that give the "non-biophysics career" student a picture of the power and use of physics in this multidisciplinary field, as well as some serious computational experience in covered areas, and (3) that gives the biophysics career-bound student a solid foundation to enter bioscience graduate programs or gain employment in biotech firms. Finally, this work encourages active, long-term learning; i.e., students are asked to perform computations, simulations, and/or miniprojects in the areas covered, with the goal that they remember it for years to come.

Following is a list of important topics in biological physics that have been slighted.

1. Electrostatics, ion flow and nerve function. The contribution of electrostatic terms to free energy is introduced in Chapter 14. While some homework problems highlight the electric work done in moving ions across a typical membrane potential, as well as control of flow in ion channels, this second edition does not seriously deal with this extremely complex topic that is central to neuroscience.
2. Evolution, genetics, systems biology. Though Chapters 1 and 7 address recent developments in computational evolution and their integration with experimental genetic engineering, these three topics are largely left to other courses.
3. Human vision. Ray optics of the eye is covered in first-year physics textbooks. Chapter 15 on light-driven reactions describes the fundamental quantum-detection aspect of vision.
4. Brain function. This topic is much too large, multidisciplinary, and difficult to provide more than surface-level treatment, if given only a short time.

COURSE FEATURES AND SUGGESTIONS TO INSTRUCTORS

1. Know yourself and your students. The majority of your 2017 students probably do not address learning and academic challenges the same way as those from 2007. While the approach of physics is to work out answers and applications from basic principles, today's students have been trained from childhood to search for answers using online tools. Online tools are extremely valuable in biophysics and are used frequently in this text, but they are, indeed, *tools*, not a way of thinking.

2. Form small "working groups" of students with differing backgrounds (e.g., physics and biochemistry majors). Require them to collaborate on certain course projects. The goal is to exploit the strengths and strengthen the weaknesses of each. If possible, set aside a weekly period for collaborative computational, experimental (e.g., video microscopy), and presentational work.

3. Assign homework and projects. Three types of assignments are available: normal and advanced Problems, and Miniprojects. Advanced calculations are typically for graduate students (extra credit for undergraduate) and Miniprojects demanding 1–4 hours of research (open-ended, typically requiring reading of recent articles, development of a spreadsheet calculation, or simulations) can be assigned to pairs or groups of students with mixed backgrounds. Problems added to the second edition are often open-ended, with students and instructor determining the proper scope of the answers.

4. Be aware of student level. The text can be used for students ranging from bright sophomores to graduate students. It is designed to be appropriate for physics, as well as chemistry/biochemistry and biology students. The reading level is directed to undergraduates, with referrals to the supplementary CD for further explanations and practice in difficult areas. Chapter 3, on mathematical methods, should not normally be covered in its entirety. Graduate students in physics, math, or engineering may not need any of it; undergraduate or graduate students in biology may need much of it.[1] Use it selectively to address individual student needs.

5. Teach students to read the text and check that they do. Students must go through (1) preparing to read a section (knowing what to expect, what to understand if everything else escapes you, where to go if the math is overwhelming); (2) reading the section; (3) completing a guided exercise; (4) checking understanding through short-answer questions with solutions provided; (5) reading again, if necessary; and (6) completing assigned problems and projects, at graded levels (sophomore to graduate).

6. Focus on quantitative understanding. Quantitative analysis is what physics is good for. Written, qualitative descriptions are forgotten within a few months and cannot take the place of quantitative tasks and understanding. Calculating the effectiveness of sunscreen X from its spectrum and absorption coefficient will lead to better understanding than reading that sunscreen X has an SPF of 30. Encourage students to reason, calculate, and connect, not just remember.

7. Use computational and other computer-based problems and projects. Frequent use of common software such as Microsoft Excel is made, because most students know the basics, most employers expect such expertise, and such software reinforces understanding of data points and distribution functions. Molecular structure software, freely available on the Web and easy to learn, will facilitate dealing quantitatively with complex biomolecular structures. Miniprojects at the ends of many chapters often employ such methods. Online software and its interaction with web browsers and plugins change frequently. Before the course begins, the instructor should check how files in the Supplementary Resources (available online at http://textbooksonline.tandf.co.uk/), as well as the online software in Chapter 5, behave on both Windows and Mac computers with the latest and with older operating systems and software. If necessary, arrange for student access to institutional computers preloaded with appropriate operating systems and software.

The author believes this textbook demands an "expert" biophysics instructor less than some other texts, owing to its structure and resources. The coverage—physics, chemistry, biology, and math—is unusually broad for a "physics" textbook, but none of it is particularly difficult.

The software is basic and typically available to all students: NGL, Protein Workshop, JsMol/Jmol, PV, Simple Viewer, Ligand Explorer, 3DNA, RNAView, and other online visualization tools. Common software such as Microsoft Excel will be used, with "live" spreadsheets rather than static graphs (Mathematica, Maple, SigmaPlot, etc. can *also* be used) to exploit common computer expertise and to encourage understanding of connections between a function or distribution and the values on the graph axes.

---

[1] The student who is unfamiliar with more than half of Chapter 3 should be encouraged to learn more math before enrolling in the biophysics course. In extreme cases, offer a separate course section that excludes Chapter 8 and quantum-intensive sections of Chapters 9 through 11. Be aware, however, that this course constitutes a rare opportunity for a student to learn the relevance of quantum mechanics to biology.

8. Embrace quantum effects. This text takes a diagrammatic approach to understanding quantum effects in biological systems, with differential equation extensions for advanced students. Physical aspects of photobiophysics and photosynthesis, neglected topics of our day with connections to renewable energy, can then be handled with quantum treatments that have applications to other biophysical and biomedical phenomena.

9. Adjust coverage to the particular goals of your course. A typical application is a one-semester core course for physics majors in a biophysics track, a recommended elective for biomedical engineering and biochemistry majors, and a recommended elective for biology majors in a quantitative track. The text could serve as the basis for a two-semester course: Part III or IV could be placed into a second semester, possibly with additional material. A second semester might be appropriate, for example, for students who need more time to learn quantum physics and experimental methods, and who want to more thoroughly explore nanobiophysics and its relation to nanotechnology.

Suggested chapter coverage: one-semester course, well-prepared students

Week	1	2	3	4	5	6	7	8	9	10	11	12	13	14	15
Chapters	1, 2	4	5	6	7	8	9	10 or 11	12	13	14	15	16	17 or 18	19

See the book's dedicated webpage at the publisher's site for online supplements, including full-color images available to download: www.crcpress.com/9781138633414.

MATLAB® is a registered trademark of The MathWorks, Inc. For product information, please contact:

The MathWorks, Inc.
3 Apple Hill Drive
Natick, MA 01760-2098 USA
Tel: 508 647 7000
Fax: 508-647-7001
E-mail: info@mathworks.com
Web: www.mathworks.com

Acknowledgments

The contributions of Thomas M. Nordlund to this second edition would not have been possible without the involvement and support of

- editor Lou Chosen, who guided the planning and composition of both text editions;
- Drs. Neal Kay and Takumi Yamada, who restrained Thomas M. Nordlund's chaotic cardiac electrical activity;
- Dr. Moya Nordlund, who said, "Do it", and watched that a planned two-hour work session did not turn into five hours;
- advisor/mentors Hans Frauenfelder, Robert Knox, Paul LaCelle and Gerhard Schwarz, who fueled and directed my scientific development.

Authors

Thomas M. Nordlund grew up learning to care for a variety of animals and crops on a small farm maintained by his veterinarian father and administrator/homemaker mother. After spending considerable time in the forests of the northwest United States, chasing rainbow trout and the elusive steelhead, as well as occasionally putting out fires, he decided to become a physics major, or perhaps a biology major. Eventually, he received a PhD degree in physics from the University of Illinois, Urbana–Champaign, with Hans Frauenfelder and his remarkable group of graduate students. Postdoctoral positions at the Biozentrum der Universität Basel ("Learn to speak another language!") and the University of Rochester led to a faculty position at the latter. He moved to the University of Alabama at Birmingham in 1990, where he continues as an Emeritus faculty of the Department of Physics. His course in life took on a clearer direction and fulfillment when he met his wife, Moya, a pianist and music educator, who has taught him that learning music and physics are not so different. A Fellow of the American Physical Society, he performed research in biomolecular dynamics for 30 years and has recently realized that the *grand challenge* in physics is now to teach physics to (i) the new American student and (ii) the future theologian/pastor.

Peter M. Hoffman is a professor in the Department of Physics and Astronomy at Wayne State University in Detroit, MI. The founder of the biomedical physics program at Wayne State University and the author of *Life's Ratchet*, Dr. Hoffmann has been involved in soft matter and biophysics research for 25 years. He earned his PhD in materials science and engineering from Johns Hopkins University.

PART I

INTRODUCTION, APPROACH, AND TOOLS

CHAPTER 1

CONTENTS

Introduction to a New World

The biological world, we firmly believe, is subject to the normal laws of physics, contrary to opinions expressed by many mainstream scientists and philosophers 100 years ago and by some more recently. At least, we have no compelling data suggesting that physics does not apply to or cannot properly describe biosystems. On the other hand, we have a multitude of observations and data confirming that life is indeed complicated and that it may be impractical, if not impossible, to satisfactorily describe these complications using physical methods we are accustomed to using with inert matter. So why study bio*physics* at all? Why not leave it to biology, biochemistry, and molecular biology? Are physicists just trying to get a piece of the action? Or do biologists need physicists to do calculations and some difficult experiments for them?

Physicists, indeed, want a "piece of the action," in that one of the primary remaining frontiers of science lies in the study of biosystems. We could convincingly argue that the progress in biomedical science during the past 40 years is primarily due to advances in molecular biology and genetic analysis and manipulation. However, when we look at the core discoveries enabling these "molecular biology" and other major advances in the biosciences, we find physicists or scientists who had a good understanding of physics: Watson, Crick (physicist), Wilkins (physicist), and Franklin and their determination of DNA's double-helical structure and base complementarity; Kendrew and Perutz and their determinations from X-ray crystallography of the first protein structures, those of myoglobin and hemoglobin, in atomic detail; theoretical physicist Max Delbrück's deduction, with Hershey and Luria, of a hereditary molecule and virus replication mechanisms; Berg, Gilbert (physicist), and Sanger (recombinant DNA); Deisenhofer (physicist), Huber, and Michel (the photosynthetic reaction center); Neher and Sakmann (both biophysicists, ion channels in cells); Lauterbur and Mansfield (both physicists, magnetic resonance imaging [MRI]); Kornberg (chemical physicist, DNA $\rightarrow$ RNA transcription); Betzig, Hell, and Moerner (physicists/chemical physicists, super-resolved fluorescence microscopy)[1]… the Nobel list goes on. The point is that physical approaches have historically been responsible for new ideas and great advances in biological understanding, as well as practical biomedical diagnoses and treatments.

I will use "biophysics" and "biological physics" interchangeably, except for clarity and emphasis in certain situations.

1.1 BIOPHYSICS AND GENETIC ENGINEERING

Since the first edition of this textbook appeared, major advances in most areas of the biosciences have been made. The weightiest of these advances seems, to the general public, to have been in genetic engineering. "CRISPR/cas9," a gene-editing tool largely unknown to bioscientists in 2011, is now familiar to many high school science students. Because CRISPR gives detailed control of the biological synthetic apparatus to the investigator at relatively low cost, it would appear most structure/function, mechanism, and cause/effect questions—the soul of the biophysics enterprise—could be ceded to genetic engineers. If one wants to know whether a missing or modified protein is the cause of a disease or unexpected effect, one can simply locate the gene coding for the protein in a normal organism, delete or modify it, and test for the effect. Does this not avoid the detailed and sometimes tedious and expensive nano-structural investigations (e.g., atomic force microscopy) carried out by the biophysicist? The answer is an emphatic "No." The type of question CRISPR can essentially answer is a *correlation* question: Is gene A connected to effect X? Without the assistance of other bioscientists with quite different trainings, the genetic engineer cannot hope to explain the physical mechanism of a biological activity. The biophysicist, for example, wants to know not only whether protein A can carry out process X, but also how the protein does its work: how much energy and other materials are required, what the structure and structural changes of protein A are and how they allow process X to occur. What CRISPR has essentially done for the field of biophysics is to allow a much more efficient search for (i) the specific biological macromolecules involved in a process and (ii) the parts of each macromolecule that are crucial for its natural functioning. Even the field of medicine, which today often relies solely on the genetic correlations afforded by CRISPR technology, will encounter the limitations of an "information/correlation" solution to the complex systems, structures and processes of biology.

1.2 BIOLOGICAL AND NONLIVING WORLDS CONTRASTED

Living organisms move; nonliving things don't. While not completely true, this simple statement encompasses much of the surprising nature of biological systems. Most of us are familiar with motion of living things (including ourselves). Close examination of some familiar types of motion usually hint at physical mechanisms involved. Figure 1.1a and b shows two insects with somewhat different types of wings. Some of you recall the historic declaration, supposedly by German engineers in the 1930s,[2] that bees could not possibly fly, based on then-current physical understanding. The truth is, aerodynamics is not simple and there is a good reason why aircraft are designed by engineers who spend more than a week or two learning only Bernoulli's principle. A more common type of motion, swimming, is similarly not simple but nicely separates into two regimes: large scale, corresponding to the fish's motion in Figure 1.1c, and small scale, applicable to most cells, bacteria, and smaller structures. For several reasons, motion in water offers more hints at mechanisms. The shape, motion, and symmetry of the fish's tail, as well as the swirling of debris in the water, indicate important features of the motion. The movement of water skippers on top of a pond (Figure 1.1d) offers an example of motion involving surface tension, often addressed in first-year physics texts. Hints of underlying physics again abound in this diagram and the accompanying web-support video clip: (1) the size and shape of the depression in the water below the skipper's feet, (2) the water rippling (or lack thereof) as the skipper moves, (3) the speed at which the skipper moves, and (4) the size and mass of the skipper. In contrast to the rather rare macroscopic case of the water skipper, surface tension and interfacial forces play an important role on a microscopic scale. Indeed, the huge magnitude (to microscopic organisms) of the air–water surface tension force can be seen in the video clip corresponding to Figure 1.2f.

Figure 1.2 shows a variety of moving creatures on a microscopic scale. Because much of this course will deal with such motion, we will not dwell on it now, but comparisons with Figure 1.1 can quickly lead to revealing questions. For example, does swirling of debris occur on a microscopic scale? Are the motile elements (paddles, fins, cilia, etc.) similar? Can microorganisms create waves? Do bacteria coast in the water? How do things move inside of cells? Do microorganisms avoid sinking to the bottom the same way fish do? The answers to these questions point to the heart of the unique dynamics—the causes and mechanisms of motion—on a microscopic scale.

A final important type of motion was identified by botanist Robert Brown in 1827 while studying the fertilization process in the plant *Clarkia pulchella*.[3] At the time, Brownian motion was thought to occur only in living organisms and to show the most elemental form of the motion of living things. The former was found to be wrong in fairly short order, because other, clearly nonliving small particles also showed such microscopic motion. As often happens, wrong conclusions can nonetheless point in the right direction. Unlike the macroscopic systems engineered by humans, cells and subcellular systems must constantly cope with, even exploit, this random Brownian motion while accomplishing major life processes such as transport and movement, enzymatic catalysis, replication, and energy production. Indeed, Brownian motion, illustrated in Figure 1.3, is a major focus of this text and modern biophysics.

"Life will find a way." While quotes from science fiction movies such as *Jurassic Park* are rarely reliable guides to scientific truth, in this case the quote correctly declares an important characteristic of living organisms. Living systems constantly adapt to their surroundings, thus demonstrating that all organisms have the equivalent of a nervous system—i.e., sensors that relay information about surroundings with some sort of feedback/response mechanism (see Figure 1.4). While not unique to

Aquarium or fish experts may recall air-filled fish bladders and the fact that your dead aquarium fish typically float or sink.

(a)

(b)

(c)

(d)

FIGURE 1.1

Motion and change. The motion of "large" living things constitutes a legitimate segment of "biophysics," but we will ignore "large" for the most part. Examination of the pictures suggests what physical principles might be involved. (a, b) Bumblebee and dragonfly: aerodynamics, Bernoulli's principle; (c) fish: fluid dynamics, Archimedes' principle; (d) water skipper: interfaces and surface tension. Flying is rare in living systems and will not be considered in this text; swimming in water, however, is ubiquitous, especially on a microscopic scale. The video clips corresponding to panels c and d show many important features of the motion that will be considered later in the text: swirling of gunk in the water, the shape, motion, and symmetry of the fish's fins, and the size of the depression below the water skipper's feet.

living systems—modern automobile engines have dozens of sensors monitored by a computer that commands adaptations to specified changes—biosystems have amazingly complex and compliant systems that can often adapt to thousands of types of change, as long as the change is not too great. The majority of these adaptive systems cannot usefully be called "nervous systems," because they do not involve a myriad of nerve cells and a brain. A partial chart of the metabolic pathways shown in Figure 1.5, corresponding to sensing and feedback within a single cell, suggests that this sense/respond system is indeed complex and is built in at the biochemical level. These "biochemical" elements refer to sensor systems that consist of a chemical reaction involving one or several molecules (organic, inorganic, and macromolecules) that are converted to other forms, as well as a macromolecular enzyme that not only catalyzes the reaction, but whose catalytic activity is affected by yet other molecules in the environment. By examining the chart carefully in the online materials, we see that virtually all of the reactions involve cycles that have feedback (control) from several or many other reactions. Solving two or three simultaneous (coupled) algebraic and differential equations is manageable. However, for

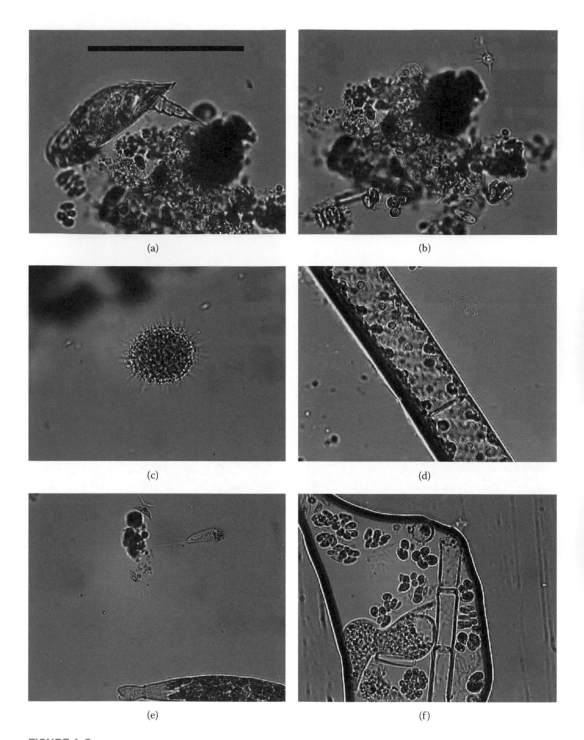

FIGURE 1.2

Moving objects in pond water, ranging in size from about 500 to 1 µm. The larger motile organisms in the left column (a, c, e) can easily be identified through their drive elements (tail, spines, cilia). Smaller moving objects, such as those inside the rodlike algae (d), can only be identified through movement in the accompanying web resource files (f) Motion of organisms in a shrinking drop of water is shown. Micrographs recorded with 40 × microscope objective. The 100-µm scale bar in panel a applies to all figures. The differing fluid dynamics of these organisms and those in Figure 1.1 will be discussed in Chapter 12.

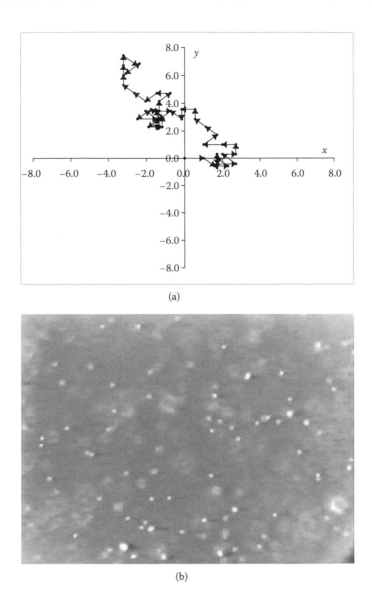

(a)

(b)

FIGURE 1.3
Random motion. (a) Static plot of a random walk simulation. (b) Movie frame of 0.8-μm polymer spheres in water; 30 frames per second. See text web resources for video clip of real Brownian motion.

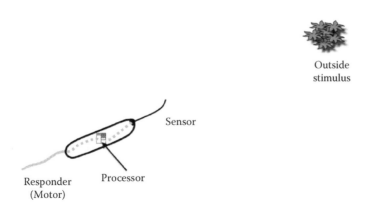

Outside
stimulus

Sensor

Responder   Processor
(Motor)

FIGURE 1.4
Living systems, at least at the cellular level, can sense their surroundings and adapt to changes. They must have sensors, processors of sensed information, and response systems.

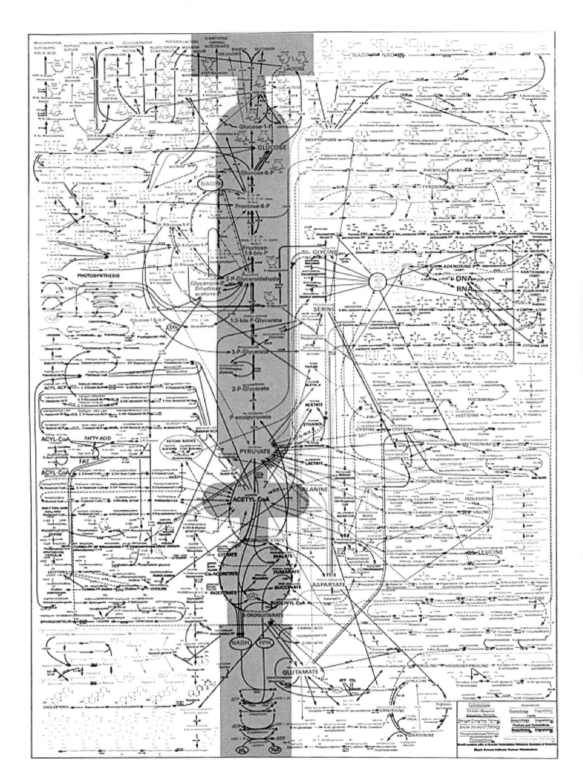

FIGURE 1.5
Partial chart of the metabolic pathways. This diagram, not meant to be read here, can be found in interactive form at http://www.sigmaaldrich.com/technical-documents/articles/biology/interactive-metabolic-pathways-map.html. Alternatively, see http://www.expasy.ch/cgi-bin/show_thumbnails.pl. The myriad of intersecting, correlated, coalescing, and crossing arrows shows that, by almost any definition, complexity plays a key role in living systems. We cannot formally address the issue of complexity in this book.

four or more equations you need a computer. Even if you can solve the system of equations, your ability to determine all unknown parameters is severely limited. Such solutions may show instabilities, with large variations in outcome caused by small changes in initial conditions. On top of these difficulties, add the necessity to account for both predictable and random variations in several or many of the parameters, and the task of the "mathematical biologist" becomes formidable. This area of complexity is being addressed seriously but is mostly beyond the scope of this text.

1.3 HIERARCHICAL STRUCTURE AND FUNCTION

One of the general features of biological systems is the intricate structure that can be observed on virtually any length scale. Starting from an organismal scale (~1 m), where obvious "interesting" structure can be observed, increasing the magnification by an order of magnitude over and over again brings new, interesting, often unexpected structure on every scale until the atomic (~10^{-10} m) is reached. This *structural hierarchy* over 10 orders of magnitude contrasts sharply with most nonliving materials, where structural intricacy can be observed over 1–2 orders, occasionally a few more (see Figure 1.6).

What are the purposes of life's hierarchical structures? They cannot solely be movement or ability to do work, because something like an old-fashioned tractor can move and do work but has hierarchical structure on scales from perhaps 1 m to 1 mm. A glimpse at one reason for life's structural organization comes from considering a modern tractor, which often has computerized controls. Computers have intricate structures on scales from about a meter to tens of nanometers. We could then postulate that one reason for life's hierarchy of structure is the need for an "intelligent" (sense, respond) system—a system that can be programmed to carry out certain tasks, monitor how systems and subsystems are operating, and make needed adjustments. A modern robot, still crude compared to simple organisms, mimics the structural hierarchy of a living system even more, partly by design (biomimicry) and partly by necessity: smaller parts and sensors must enable fine control of larger appendages; even smaller circuitry must communicate signals from a control center to the parts; the control center (brain) and sensors must be made of components smaller still, which allows for application to a variety of sensing needs.

Microscopic examination of biological structures, along with some understanding from colloid chemistry, suggests a practical reason for biological hierarchical organization: *self-assembly*. Biological structures are mostly assembled on-site.

Unlike a Sony™ robot, there are no large central factories, where suppliers ship raw materials, highly trained technicians or programmed robots do assembly and testing, and transport agents reship to final sites of use. Consider a simple two-dimensional biological membrane, for example (Figure 1.7; more complex examples occur in real cells). This example membrane is made up of two components: *lipid* molecules that form the bulk of the layered structure and protein molecules that create a hole through the membrane. The lipid molecules are highly asymmetric in terms of their polarity and hydrophobicity.

The *polar head group* prefers to be surrounded by water and other polar molecules than by hydrophobic molecules. (In quantitative terms, it has lower free energy in water than in hydrophobic solvents.) In contrast, the double *hydrocarbon tail* has high free energy in water. A solution to the energy minimization principle in physical systems is to have head groups aggregate together in a coordinated, directional manner, leaving a plane of lipid head groups facing the same direction. The hydrophobic tails will achieve a relative energy minimum if they are separated from the surrounding water by a second layer of lipids facing the opposite direction. This type of *bilayer* structure will form spontaneously if lipids of appropriate structure are placed in water in sufficient quantity. Thus, we see that one of the primary needs of a biological organism, to separate itself from its environment, can be achieved by a structure, the membrane, which will self-assemble. This self-assembly phenomenon underlies virtually all forms of life on Earth: in principle, the biological cell will assemble itself, given water and lipid molecules in appropriate amounts. Because an organism

---

Because an organism could not long survive wild variations in its metabolism or energetics, such mathematical instability should not correspond to real life. However, physicists know that a mathematical instability or infinity sometimes corresponds to a real, new phenomenon or particle.

Biomimicry (n.), biomimetic (adj.): copying or imitating biological designs in a nonliving system.

There are indeed sites where certain specialized agents and structures, e.g., hormones and red blood cells, are produced in a fully developed organism. Hormones, however, are small and simple organic molecules in a biological context (molecular mass < 1,000), and red blood cells are rather an exception than a rule in cell assembly. We will study later chaperonins, local environments, and agents designed to assist assembly of certain nanometer-sized structures, in Chapters 6 and 17.

Polarity (degree of charge separation) and hydrophobicity (water-avoiding character) are usually correlated.

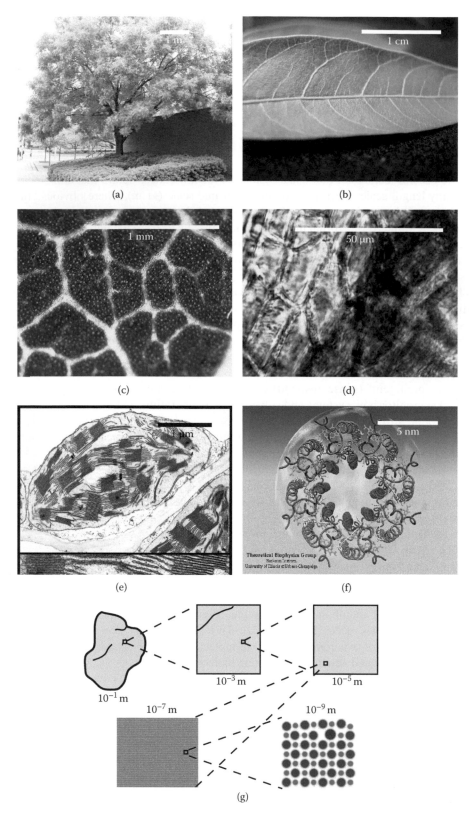

FIGURE 1.6

Hierarchical and nonhierarchical structure. (a–f) Interesting, intricate structure in biosystems occurs over many scales. (g) Inanimate matter typically shows structure over 1 or 2 orders of magnitude in size scale. A crystal may have cracks and defects, but these have no functional role other than strength and perhaps conductivity. Image (e) courtesy of Radboud University Nijmegen. (Image Gallery, www.vcbio.science.ru.nl/en/); image (f), Theoretical Biophysics Group, University of Illinois, Urbana-Champaign.

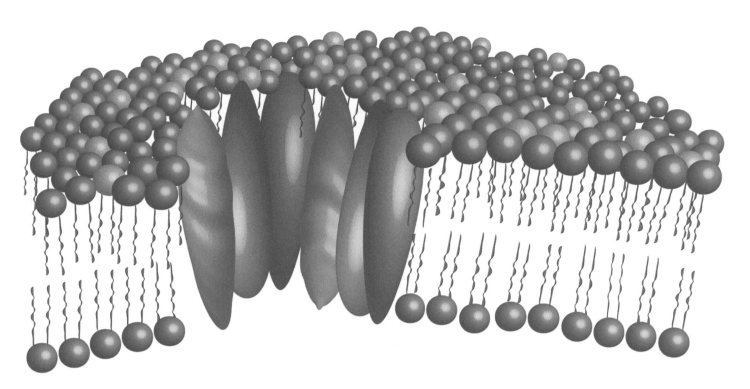

FIGURE 1.7
A pore structure in a biological membrane. The bulk membrane constituents, lipid molecules, spontaneously self-assemble to form a bilayered, two-dimensional structure in an aqueous environment. To enable this assembly, lipid molecules must be elongated, with ends differing in polarity or *hydrophobicity*. The lower (free) energy structure will form with the more polar, less hydrophobic end (*polar head group*) associated with water and with other polar head groups, and the hydrophobic hydrocarbon "tail" associated other "tail" groups. Similarly, a pore can spontaneously form when pore molecules of sufficient hydrophobicity are introduced after the membrane is formed. Lower energy results when pore molecules move from water to membrane. Favorable interactions between pore molecules further lower the energy when they aggregate together. Biological membranes are typically 5–10 nm thick.

completely separated from its environment could not adapt to or exploit its environment (e.g., eat), some passage through the membrane must be provided. The *gating protein* does just that. In Figure 1.7, the gating protein molecules aggregate to form a ring in and a hole through the membrane. From the principle of energy minimization, what type of structural features must the protein have? Think about how polarity and hydrophobicity of the ends and center of the protein, protein-protein interactions compared to protein-lipid interactions, and the molecular contents of the hole, among other details, might allow protein molecules placed into the surrounding water to spontaneously incorporate themselves into the membrane and form a ring. Would it help if the protein molecules were produced or transported adjacent to (or even in) the membrane? What properties of the protein would determine the diameter of the ring? Clearly, even this more complex structure could also assemble itself.

Does self-assembly tend to produce structural hierarchy? In the membrane example, the needed polarity-hydrophobicity properties of the lipids and the proteins demand "intricate" protein and lipid structure on a 0.5- to 5-nm scale to allow them to properly orient in a 5–10-nm-thick membrane. So, we already see structural hierarchy on the 10^{-10} to 10^{-7} m (0.1–100 nm) scale. What about the transverse, planar dimensions of the membrane? It could, of course, extend endlessly, but the structural stability would suffer.

We have suggested some reasons for biological structures that must have "interesting" structure on scales from nearly 1 nm to 100 μm—5 orders of magnitude. Furthermore, these structures can create themselves spontaneously, given the right conditions and materials. What about the size regime from 0.1 mm to 1 m? You may wish to create larger structures by assembling cells (e.g., as in a developing embryo) until useful organelles, organs, and support structures are created, limited by strength and elasticity needs.

EXERCISE 1.1 SIZE OF A CELL MEMBRANE

Knowing almost nothing about biological membranes, can we estimate limits on its dimensions? First, recalling the cell's need to separate its interior from the exterior, we see the membrane must eventually reconnect to itself. Forming a bilayer sphere or spheroid would solve this problem; so would a parallelepiped, but sharp corners are always high-energy structures in nature and are unstable.

1. Estimate a lower limit on the size (diameter) of a sphere formed by a 5.0-nm-thick bilayer membrane.
2. How big could these spheroids get before they were too unstable to hold their shape?

Answers

1. The sphere diameter could not be lower than about 50 nm (0.05 μm), if the bilayer is 5.0 nm thick, and the sphere needs to contain a decent amount of an organism's innards.
2. Though we really should consider the water-membrane interface, the bending rigidity of the membrane, and typical forces applied when the sphere moves through water, we could make a first guess that the membrane is a bit less stable than a plastic or rubber (polymer) sheet—the membrane bilayer we construct does not have transverse covalent bonds throughout the membrane sheet, while a polymer does. Plastic sheets generally range from 0.5 to 5 mils thick (1 mil = 10^{-3} in ~2.5×10^4 nm). Heavy-duty sandwich bags that are about 20 × 20 cm and hold their shape reasonably well are nearly the thickest, 5 mils (~10^5 nm). If filled to approximate a sphere, the diameter would be about 0.1 m. Simple scaling to a 50-nm thickness would suggest a membrane sphere maximum diameter of about (50/100,000) × 0.1 m ~50 μm. While this kind of simplistic scaling is dangerous, it provides a reasonable estimate of the size of a "large" cell membrane.

The bottom line: biological structural hierarchy is in part generated by the need for self-assembly of organisms. Is this hierarchical organization advantageous to an organism? This book will show some of the many ways in which this hierarchy works very effectively on a microscopic scale. To learn more about the macroscopic structure of organisms, consult your biology, anatomy, and physiology texts.

1.4 SOME IMPORTANT QUANTITIES TO GET STARTED

In this textbook, hundreds of important quantities and properties of light and matter are presented and used to perform quantitative computations. Although we will deal with most as we come to them, a reminder of some common quantities of mass, size, energy, and power helps to set the stage for faster comprehension of the context and scales of biosystems (Tables 1.1 and 1.2). For a more complete listing of constants, see nist.gov; search "physical constants."

Take special note of the combinations of constants that are listed both in standard SI units and in slightly modified SI units. When nanometers, piconewtons, and unit electric charges are used, many characteristic physical quantities take on values not far from 1. Though there is *a priori* no deep significance to this scale, you have already learned something important about the scales that are important in biological systems. Especially notice the value of $k_B T_R$ at 20°C—4.0 pN nm, the amount of thermal energy contained in each mode (degree of freedom) of a system at equilibrium (at 20°C). The value 4.0 pN nm is the energy all other energies must be compared with on a microscopic or molecular scale. If, for example, a molecule of fuel such as adenosine triphosphate

TABLE 1.1 Masses and Sizes

Object	Size (Radius)	Mass (u, Da)[a]	Comments
Electron	<1 fm, 0.0529 nm	5.486×10^{-4}	Elementary electron radius, Bohr radius
Proton	0.8 fm	1.007	
Neutron	1.3 fm	1.009	I. I. Rabi, *Physical Review* 43 (1933):838
H	0.0529 nm	1.008	
H bond	0.27 nm	—	
C	0.067 nm	12.011	Abundance average
O	0.048 nm	16.00	
H_2O	0.135 nm	18.02	
Amino acid unit	—	110	Average mass of a unit in protein polymers
Nucleotide unit	—	309/324	Average mass of a unit in the DNA/RNA polymer
Cells: prokaryotic eukaryotic	0.3–0.6 µm	5–50×10^{10}	Shapes, especially eukaryotic, vary dramatically
	2–8 µm	2–100×10^{13}	

[a] To obtain values in kilograms, divide u, the atomic mass unit, by 6.022×10^{26}.

TABLE 1.2 Short List of Useful Physical Constants

Quantity	Symbol	Value	Comments
Elementary charge	e	1.602×10^{-19}°C	Coulombs
Speed of light	c	2.998×10^8 m/s	In vacuum
Planck constant	h	6.626×10^{-34} J s	
		4.136×10^{-15} eV s	
	$\hbar$	1.055×10^{-34} J s	
		6.582×10^{-16} eV s	
Avogadro constant	N_A	6.022×10^{23} mol^{-1}	
Boltzmann constant	k_B	1.381×10^{-23} J/K	
Gravitational constant	G	6.674×10^{-11} N m^2 kg^{-2}	
Combination constants for calculation and estimation			
Thermal energy (per mode)	$k_B T_R$	$4.045 \text{ pN nm} \approx \frac{1}{40} \text{ eV}$	At 293 K
Gravitational energy of protein in cell	$(m_{protein}\, g)\,(1\ \mu m)$	2×10^{-6} pN nm	Rough; top to bottom of eukaryotic cell
	hc	1240 eV nm	
Electric proportionality constant	$\dfrac{1}{4\pi\varepsilon_0}$	8.988×10^9 N m^2/C^{-2} 144.0 pN (nm)^2/(e)^2	
Electric potential	$\dfrac{e^2}{4\pi\varepsilon_0}$	2.307×10^{-28} J m 1.440 eV nm	
Bohr radius	$a_0 = \dfrac{4\pi\varepsilon_0 \hbar^2}{m_e e^2} = \dfrac{(hc)^2}{m_e e^2 (e^2/4\pi\varepsilon_0)}$	5.292×10^{-11} m 0.05292 nm	
H ground-state energy	$E_0 = \dfrac{e^2}{4\pi\varepsilon_0}\dfrac{1}{2a_0}$	2.180×10^{-18} J 13.61 eV	

(ATP) is to provide energy to drive a molecular motor forward, that energy must be at least several times 4.0 pN nm. Furthermore, if we find that typical forces applied during motor-driven motion are on the piconewton scale, we can guess that a typical "step lengths" taken by a motor when one fuel molecule is "burned" would be a few nanometers. And we would be right! Already we would know something about the nanotechnology of living cells.

NASA's Phoenix Mars Lander had actually *touched and analyzed* water on Mars—"a feat never before accomplished outside of Earth." Most of us grew up thinking scientists had told us years before that water was at the poles of Mars, but the 2008 data provided the needed direct physical data for water in Martian soil. https://www.nasa.gov/mission_pages/phoenix/images/index.html

1.5 BIOPHYSICS AND BIOCHEMISTRY OPERATE IN WATER (WATER 1)

On July 31, 2008, a NASA Phoenix Mars Lander (Figure 1.8) team announced the confirmation of water on Mars:

> *We have water said William Boynton, University of Arizona, lead scientist for the Thermal and Evolved Gas Analyzer (TEGA). We've now finally touched it and tasted it—and from my standpoint it tastes very fine.*

NASA Release 08–195.

We all know why Martian water might be an important discovery, but it is important to remind ourselves why water is so crucial to life as we know it. In their 2000 book entitled *Rare Earth—Why Complex Life Is Uncommon in the Universe*, Peter Ward and Donald Brownlee note five principal reasons why complex life was able to develop on Earth:

1. The presence of carbon and other important life-forming elements
2. Water at or near the surface
3. An appropriate atmosphere
4. A long period of thermal stability, stabilizing liquid water
5. An abundance of heavy elements

FIGURE 1.8
NASA's Phoenix Mars Lander and the ground at the landing area. NASA announced on July 31, 2008, that the craft had touched and analyzed liquid water on Mars—a feat never before accomplished outside of Earth: "We have water," said William Boynton, University of Arizona, lead scientist for the Thermal and Evolved Gas Analyzer (TEGA). "We've now finally touched it and tasted it—and from my standpoint it tastes very fine." Image courtesy of NASA/JPL-Caltech/University Arizona/Texas A&M University.

Two of the above relate to water. Why is water so important? You should be able to quickly list a few reasons, perhaps from introductory biology, chemistry, and physics courses: large heat capacity (stabilizes temperature), good solvent properties (many compounds can dissolve), unusual freezing-density properties (ice floats), etc. We deal with water in more detail in Chapter 4, but throughout the rest of this book you should note the involvement of water in life's processes, sometimes directly as a chemical reactant, more often as a hospitable environment.

For those of you intrigued by astrobiology generally, or by the possibility of life on Mars in particular, make sure you keep reading both the next section and current science news. NASA analysis of Curiosity Rover data in 2017 suggests that though some images (e.g., NASA image PIA17595.jpg; see jpl.nasa.gov/spaceimages/details.php?id=PIA17595) have been interpreted as lake beds, suggesting liquid water, there seems to be no carbonate, an indicator of atmospheric CO_2, the greenhouse gas believed necessary to allow liquid water. While the case for liquid water and life on Earth is clear, that for our nearest planetary neighbor is still murky.

A recurring theme in biosystems is that life processes take place in water and the (bio)physics you apply to life must take this into account.

Still, the same general physical laws apply in biophysics, in particular those dealing with topics such as ideal gases, statistical physics, and fluid dynamics. These are central to dealing with biophysical systems. You may also discover that some "facts" are rather approximations. For example, did you accept the "fact" that the surface frictional force depended only on the magnitude of the normal force (Equation 1.1)?

$$f = \mu N \tag{1.1}$$

Perhaps you suspected that the friction should depend on the surface area. Or perhaps you thought about what would happen if the surface roughness, which determines the magnitude of μ, was a bit *too* much, and some surface bumps and peaks were large enough to abruptly bring the sliding object to a halt, or nudge it into the air. Do the surfaces scratch each other? What about the air-resistance case?

You may have seen the air-friction equation

$$f = b_1 v \tag{1.2}$$

where v is the speed of an object passing through a fluid (air) and b_1 is a coefficient of friction. Some texts may have added

$$f = b_2 v^2 \tag{1.3}$$

stating that Equation 1.2 applies to small objects at low speeds and Equation 1.3 to larger objects at high speeds. The description of a force in terms of successive powers of a variable is an example of a physical approximation, which is crucial to the quantitative description of physical systems, perhaps most crucial in the field of biophysics. Physics is not an *exact* science; it is a quantitative science that is as exact as can be or as exact as needed. This fluid friction example is important to recall because it reminds you that series and summations are approximations that can be as exact as needed. The "exact" friction/speed relation would be written

$$f = b_0 + b_1 v + b_2 v^2 + \ldots$$
$$= \sum_{n=0}^{\infty} b_n v^n \tag{1.4}$$

where the $n = 0$ term is important in cases of surface friction. For most cases of reasonably symmetric objects, only one or two terms in this summation need be kept to adequately describe observed behavior.

What is the physical significance of these expansion parameters? To be clear, this expansion is just a convenient way to characterize and parameterize the interaction between a moving object and

It may also be useful to refer to an introductory calculus text, in anticipation of reviewing series and integrals.

its fluid environment; there are no fundamental meanings as that of m in $F = ma$. However, given that biosystems are generally complicated, finding meaning behind phenomenological or empirical parameters can be very helpful in understanding at a detailed microscopic level. We later find that the $n = 1$ parameter can be tied to two simple, major properties of the system: the average radius of the object and the viscosity of the fluid. Radius seems fundamental, but what about viscosity? It is a very useful and needed parameter for describing motion in fluids when we do not know molecular interaction details (or do not want to deal with them).

This is our first pass at dealing with the physical involvement of water in biosystems. A large part of the apparent difference between life in air and life in water relates to the relative importance of the first three terms in Equation 1.4. This focus on terms in equations is not to imply that the secrets of life lie in values of mathematical expansion parameters; it is rather an assertion that when one wants to *quantitatively* describe biosystems and their behaviors, one should look to relatively *simple* principles and associated *equations*. The wonder of life lies in the incredible variety of unexpected behaviors that can be *observed* and *measured*. It is also underscored by our unexpected ability to describe many of them *quantitatively* with simple principles.

> It is useful to occasionally remind yourself of the answer to questions like, "Why do some things have mass?" The answer is likely, "I have no idea. My physics books don't even explain it."

1.6 IMPORTANT OR "HOT" ISSUES IN BIOPHYSICS, OR HOW TO BE OUT-OF-DATE QUICKLY

During the past 50 years new aspects and ways to study biosystems ("advances") have come and, sometimes, gone. Some are truly fundamental—they explain an important feature or quality of a biosystem—and stand the test of time. Examples include the existence and structure of fundamental and active biomolecules such as proteins[4] and DNA.[5] Many discoveries of the last 100 years resulted from improvements in imaging technology on scales of 10^{-6} to 10^{-10} m. We therefore devote a new chapter to atomic force microscopy, a major new imaging technique. Some advances stand the test of time, not because they offer fundamental understanding, but because they provide important and standard descriptors. An example is the Michaelis–Menten equation for enzyme-catalyzed reactions, first developed by Adrian Brown and Victor Henri in 1902–1903 and explained microscopically by Michaelis and Menten in 1913 in terms of reaction rates and quasi-equilibrium. This treatment for catalyzed reactions has provided a standard way for biochemists to parameterize and compare catalyzed reactions, even though it may not correctly describe the details of the reactions. When two different investigators work to isolate and characterize a new enzyme, for example, it is important to ensure that when they publish results, the enzyme they describe is similar in purity and activity. Michaelis–Menten parameters provide a measure of this purity and activity. Biochemistry students reading this text can likely attest to the importance of the Michaelis–Menten equation, but researchers today, 100 years after the original development, are still trying to understand its details and correct its flaws.[6]

> The terms *biomolecules*, *biological molecules*, *biological macromolecules*, and *biopolymers* are mostly used interchangeably, often by different subgroups of bioscientists, and refer to large molecules, molecular mass about 1 kDa (1,000 MW) or larger, made up of distinct subunits usually joined into a linear polymer. "Polymer" should not be taken here to imply that all subunits must be identical.

> An enzyme is a protein molecule, usually of mass >5 kDa, that catalyzes a specific reaction.

Current technological developments and the needs of society often drive interest in science. Perhaps the most recent driver of interest in the biological and biomedical sciences has been genetic engineering, with CRISPR methodologies promising ways to correct and improve genes that govern abilities and health. Promises of health and long life are hard to displace, so careers in genetics will probably be attractive for the foreseeable future. Unlike advances that "merely" provide better understanding of biological mechanisms or that lead to profitable biomimetic devices, however, the attractions of gene editing for a privileged individual's benefit promise considerable social and political controversy.

Photosynthesis has been a popular topic among biologists/botanists, photobiologists, microbiologists, photochemists, biophysicists, and others for hundreds of years. The 1980s witnessed a renewed interest in its study by physicists because of the perfection of ultrafast light excitation and detection techniques. Measurement of photons and photon-induced electron flow is a traditional area of physics. The realization that improvements in pulsed lasers and photodetectors (photomultipliers, streak cameras), along with progress in X-ray crystallography of difficult-to-crystallize macromolecules, allowed measurement of the very rapid processes of photosynthesis and determination of microscopic structures responsible for photosynthetic energy *harvesting* and charge separation, producing a boom in basic photobiophysics research. Serious attempts to mimic nature in harvesting the sun's energy also began at this time (or began again). The oil/energy "crisis" of the late 1970s almost certainly drove popularity and funding of photosynthetic research in the 1980s. The fading of the crisis in the 1990s had a corresponding opposite effect. The twenty-first century has begun with great attention to renewable energy. Photosynthesis relates to this energy effort, but researchers are also focusing on a more direct goal: making natural photosynthesis more efficient in order to increase crop yields for an ever-increasing population.

Some biophysical "hot" topics flare up and burn out quickly because of difficult measurements, experimental mistakes, vested interests, unjustified interpretations, and simply the complicatedness of biology. From this chapter you already should expect that the role of water in biosystems is major. Expectations, encouraged by some *unexpected* results, have at times led to interesting assertions about water. Felix Franks, one of the true water experts, wrote a 1981 book entitled *Polywater,* describing the rise of an amazing new form of water.[7] As reviewed by *Discover* magazine:

> *Polywater was supposed to be an alternate form of ordinary H_2O in which the molecules were linked to produce a strange new substance, denser and far more viscous than water, which remained a liquid all the way from −70 degrees Fahrenheit to almost 500 degrees.... from 1968 to 1972, when hundreds of researchers participated in the Polywater boom, the world's supposed total supply was measured in drops.... that made polywater hard to study and delayed the emergence of the boring, bitter truth: that polywater was nothing more than a solution of impurities....*
>
> *There have been many good books about successful scientific research, but Polywater is especially valuable as a reminder of all the research that leads nowhere.... For most laymen, however, the message of Polywater may be encouraging: there is, after all, something wonderful about an enterprise in which people can feel so much remorse and chagrin, not over cruelty, treachery, or hypocrisy, but simply over being enthusiastically and publicly wrong.*
>
> *Discovery, June, 1981, pp. 88–89*

What is the lesson here? Scientists do make mistakes, but they occasionally should, in a field as complicated as biophysics.

1.7 READ APPENDIX A

This textbook focuses on learning how to do (understand) biological physics. It is not a survey of current knowledge in biophysics; comprehensiveness and currency are not the goal. *Quantitative Understanding of Biosystems* attempts to encourage the reader to think quantitatively about physically complex systems, to understand them in terms of basic scientific principles, and to support this understanding with calculations. Physical understanding and calculations for biological systems appear very different from physics and biology, however. These disparate areas are all crucial to understanding, however, and we must work to patch up our areas of weakness or discomfort. To help patch up areas of weakness, Appendix A points those from both "physics" (see Appendix A.1) and "biology" (see Appendix A.2) to books and tools that may be useful for review.

1.8 PROBLEM-SOLVING

PROBLEM 1.1 FINDING BIOCHEMICAL INFORMATION

Go to the website http://www.expasy.ch/cgi-bin/search-biochem-index, where metabolic pathways and cellular processes charts are posted. (In case the site has changed, search for the "Roche Applied Science Biochemical Pathways" chart.) Click on the "Cellular and Molecular Processes" link, which takes you to a thumbnail view of the chart, details of which will be revealed by a click. The chart is organized like a map, with horizontal and vertical coordinates, e.g., (N,5), which takes you to entries like "30S initiation complex." Pick five entries, find out what they are, and describe them in two to three sentences each. Also, state where you found the information.

PROBLEM 1.2 DESCRIBING ORGANISMS

Examine the video clips on the online version of Figures 1.1 and 1.2. (1) Estimate the speed and approximate size of a fish in Figure 1.1c and of two microorganisms from Figure 1.2c and e. In each case, describe the organism in words, or copy and paste a frame showing the organism. (2) State what appears to be the drive mechanism (motor) for the organism's movement.

PROBLEM 1.3 MEASURING IN A FLUID ENVIRONMENT

Air and water are both fluids, but common objects behave quite differently in them. Locate the following: a tall, clear glass (a tall graduated cylinder is good, if you have access to it), a steel ball bearing, a glass marble and a plastic sphere or pellet, and an 8.5 × 11 in sheet of paper. Crumple the paper into a 2- to 4-cm-diameter ball. (1) Drop the ball bearing, marble, plastic pellet, and paper from a height of about 1 m. Estimate (measure) the fall time of the fastest-falling object and estimate how much (in percent) additional time it takes for the slower objects to fall. Consider dropping objects in pairs. (2) Repeat in water. Fill the tall glass with water and repeat the fall-time measurement with the objects. (3) Discuss the behavior in 1 and 2 in terms of Newton's laws and estimate the percent deviation from the Physics 101 treatment of fall time, $h = \frac{1}{2} gt^2$. What are the major causes of the deviations?

PROBLEM 1.4 COMPARING THERMAL AND GRAVITATIONAL ENERGIES

Find values for thermal and gravitational energies from Table 1.2. (1) Discuss the significance and use of this thermal energy. (Consult a freshman physics or chemistry text, if necessary.) (2) What is the significance of the size of the gravitational energy compared to thermal? Do not simply state that one can always neglect E_{grav} compared to $E_{thermal}$.

PROBLEM 1.5 LIVING IN WATER

List five physical properties of water (besides those noted in Section 1.4) that life relies on to carry out biological processes. Explain the relationship between biological process and water property in one or two sentences.

PROBLEM 1.6 QUANTIFYING FORCE AND SPEED

Use a first-year physics text, online search, or other resource to find values for b_1 and b_2 for a "small" spherical object traveling at low speed in air. Using Excel or another graphing tool, plot resistive force versus speed and determine the speed at which the quadratic term equals the linear term. This crossover point is approximately the speed at which an object becomes "large and fast."

MINIPROJECT 1.7 ANALYZING FRICTION

Investigate applicability of Equations 1.1 through 1.4 to systems in air and water. Consider issues such as which terms in Equation 1.4 are needed for which systems, air versus water, size and shape of object, speed. You might concentrate on one issue, such as ballistics of cannonballs (an important military/historical issue), submarine friction in water, space-vehicle atmospheric reentry, etc., but focus on the issue of the relationship between speed and resistive force.

MINIPROJECT 1.8 VIDEO EXPERIMENTS (PROJECT FOR ONE OR TWO PERSONS)

Much of the bioscientific research data, as well as communications of everyday life in 2019, is recorded in video format. The objective of this project is to become familiar with the basic nature of video records and to record video clips that can be analyzed as in Problem 1.2. Cameras may range from those associated with research-grade fluorescence microscopes, to hobby microscopes (e.g., Celestron Digital Microscope Kit), to cellphone cameras. Research cameras may simply record sequences of high-resolution still pictures, at 10–60 frames per second, with file sizes in the 10's of MB for a second of video. Most cameras, however, apply video compression on-the-fly to the raw video, such that parts of each frame's image are constructed, in effect, from adjoining images making file sizes 10–100 times smaller. Choose your camera and decide whether you will record video of microscopic or macroscopic objects.

1. Write down the model of your camera and its resolution (e.g., 3,000 × 2,000 camera pixels [not cell phone screen pixels or points]). Does each video frame have this same resolution? Explain.
2. Determine the compression format of your chosen camera's video, e.g., uncompressed, .avi, wmv, QuickTime, mp4, m4v, 3gp. Determine the structure of recorded files and discuss, in about 200 words. Include information on the spatial 2D resolution, the time between frames, and the compressor's affect on resolution.
3. Put an appropriate distance calibration device (ruler) in the imaging area and record a short video of this calibration. Locate and install a video-display program that is capable of single-frame selection and display of pixel location (x, y) of points on the screen. One of your computer's video-play or -edit programs may be able to do this. (Several freeware programs such as ImageJ are designed to do this. ImageJ can also perform sophisticated image processing, such as outlining of objects and location of their centers.)
4. Record a video of a small (relative to screen size), moving object, whose center or a characteristic point can be located, moving over at least 10% of the screen. Divide the object's trajectory into 40 or more distinct x_i, y_i pixels corresponding to times t_i, where the first point corresponds to time = 0. Record these coordinates in a spreadsheet.
5. Convert the x, y pixels to physical coordinates in meters or other appropriate units using spreadsheet computations. Calculate the average velocity of the object between each set of coordinates. Similarly, calculate the average accelerations.
6. Estimate the uncertainties in position, velocity, and acceleration. Justify your estimate.

REFERENCES

1. Nobel Foundation, The Nobel Prize in Physiology or Medicine (1962, 1969, 1991, 2003); The Nobel Prize in Chemistry (1962, 1980, 1988, 2006, 2014). http://nobelprize.org/prizes/medicine/, or see https://en.wikipedia.org/wiki/Nobel_Prize_in_Physiology_or_Medicine and https://en.wikipedia.org/wiki/Nobel_Prize_in_Chemistry.
2. McMasters, J. H. 1989. The flight of the bumblebee and related myths of entomological engineering. *Am Sci* 77:164–168.
3. Brown, R. 1828. A brief account of microscopical observations made in the months of June, July and August, 1827, on the particles contained in the pollen of plants. *Philos Mag* 4:161–173.
4. Kendrew, J. C., G. Bodo, H. M. Dintzis, R. G. Parrish, H. Wyckoff, and D. C. Phillips. 1958. A three-dimensional model of the myoglobin molecule obtained by x-ray analysis. *Nature* 181:662–666; Muirhead, H. and M. F. Perutz. 1963. Structure of haemoglobin. A three-dimensional Fourier synthesis of reduced human haemoglobin at 5–5 A resolution. *Nature* 199:633–638.
5. Watson, J. D., and F. H. C. Crick. 1953. Molecular structure of nucleic acids: A structure for deoxyribose nucleic acid. *Nature* 171:737–738; Watson, J. D., and F. H. C. Crick. 1953. Genetical implications of the structure of deoxyribonucleic acid. *Nature* 171:964–967.
6. Maiwald, T., H. Hass, B. Steiert, J. Vanlier, R. Engesser, et al. 2016. Driving the model to its limit: Profile likelihood based model reduction. *PLOS One* 11(9):e0162366. doi:10.1371/journal.pone.0162366.
7. Franks, F. 1983. *Polywater.* Cambridge, MA: The MIT Press.

CHAPTER 2

CONTENTS

How (Most) Physicists Approach Biophysics

2

This chapter contains little biophysics content, except in the homework. The purpose of this brief chapter is to point out the exceedingly different approach a physicist may take when grappling with a new sort of problem or phenomenon. This different approach to the study of a biological system is the cause of much of the difficulty "physicists" and "biologists" (sometimes) have in understanding each other: the biologist thinks the physicist *knows nothing* about real biology; the physicist thinks the biologist *understands nothing* at all. Unfortunately, both are right to some degree, which makes the need for collaboration and communication most urgent. In the physicist's mind, there may be a simple, elegant physical model for understanding a biological system but one that explains little the biologist is interested in. In the biologist's mind, there may be a compendium of information about a system that has few or no general, unifying principles. Both desperately need each other, but even better is a hybrid of the two—a biophysicist knows some of what the biologist knows and thinks like the physicist.

The claim to understanding by a physicist is not just a matter of self-conceit. It is the focus of the different drive a physicist has: the drive to simplify and relegate all new observations to a subcategory of a general physical principle, or in rare cases, to a new physical principle. The extreme version of this drive is the quest for the "Theory of Everything" in particle physics and cosmology.

2.1 DEALING WITH NONSPHERICAL COWS: DRIVE FOR SIMPLICITY

The old biophysics joke, ending with the physicist's didactic instruction, "Consider a spherical cow," is likely unknown to readers under the age of 30. The joke illustrates the tendency of the physicist to oversimplify complicated matters. The biologist describes complicated behavior of an organism. The physicist proceeds to propose a simplified model that can explain certain features using a few parameters—perhaps the volume and mass of the cow. While the biologist rolls his eyes at the physicist's naïveté and ignorance, the physicist is smug in knowing that he/she can calculate at least two properties of the cow by knowing only one parameter (the cow's radius, the density being approximately that of water) and does not have to rely on asking the farmer or state fair auctioneer. The feebleness of this spherical model is, of course, that its one or two parameters explain almost nothing of importance or interest to the biologist or farmer. This text tries to avoid the spherical physicist's failure to distinguish issues of importance from calculations easily accomplished but to retain the principle of attempting, whenever possible, to quantify from simple models. This approach is perhaps the most important distinguishing scientific characteristic of a physicist: the drive to quantify, generalize, and simplify. Homework Miniproject 2.7, Energetics of the Cow, provides an exercise in calculating something of a bit more importance to society than a cow's volume. Though this problem is only half serious, it provides an example of how a physicist may differ from a biologist or other scientist: the biologist focuses on all of the features of the cow that allow it to be distinguished from, or classified with, other organisms; the physicist will focus on how the cow is like many other things, even automobiles, in terms of universal physical (and biological!) principles like energy usage and efficiency, and proceeds to describe and explain the cow or its behavior quantitatively. Both of these approaches are of great benefit in the study of biophysical systems: the first keeps track of the known facts and their importance; the second begins the process of quantitative description and understanding in terms of fundamental principles. Either approach is ineffective in the absence of the other.

> Because physical principles are clearest for small systems, it may be expected that the primary contribution of physics will be to subcellular biosystems. The physics of complex systems is making strides, however, in the description of larger systems.

2.2 TWO APPROACHES TO BIOSYSTEMS

2.2.1 Approach 1: Use Principles to Explain and Predict the Phenomena

Starting from general principles such as conservation laws, a physicist, chemist, or biologist can begin to explain and quantify a large number of phenomena. Conservation of energy, $E_i = E_f$, is the most widely applicable and useful conservation law and can be effectively applied, for example, to photosynthesis (Chapter 10). Because the basic process is the conversion of light energy to chemical energy, using conservation of energy is virtually a "no-brainer." Coupled with principles from quantum mechanics, biology, and biochemistry, it has explained and predicted energy and electron transfer processes in a wide variety of photosynthetic organisms; more on this issue later.

How do we treat biosystems and processes that are less obviously "physical" or "quantum"? Conservation of energy is built into all reaction theory, including biochemical equations such as the Michaelis–Menten equation (Chapter 15). However, reaction theory uses quantities of free energy, usually Gibbs free energy, not simply energy. Why is this? The primary element passed over is the entropy and its critical role in systems of many particles. Biosystems are clearly dominated by very many particles, so entropy must be crucial. We learn that even in systems of single particles, entropy can play a dominant role when the process is one that has multiple possible outcomes. So, the principle of conservation of energy must be expanded to include entropy and many particles. This expansion of the concept of energy quickly leads to the principle that a *system* tends toward a minimum in free energy, but nagging questions remain when one is interested in the behavior of one particular molecule within the system.

The title of this section is not "Energy, Entropy, and Free Energy," but "Approach 1: Use Principles to Explain and Predict the Phenomena." We should not try to tackle the free energy issue at this point but rather understand the approach of studying a biosystem from basic principles. Notice what we have been doing: starting from the principle of conservation of energy, we expanded the principle after acknowledging a crude fact of biology, that it involves many particles. To the biologist, acknowledging multiple particles would be an almost irrelevant observation about, for example, a paramecium or fruit fly. To the physicist and chemist, another tool has been added to the resource store that can be applied to biology. This is the physics "principle" approach: start from a law of physics, try to apply it to a biosystem, expand the principle when it falls short, and try to apply it again.

2.2.2 Approach 2: Describe the System as Completely as Possible and Infer the Organizing Principles

To the biologist, the previous approach is the long way 'round: if you want to know a system properly, observe it in all its complexity, put together an organized description of how it behaves, and then try to digest the description and classify the organism or behavior in terms of known systems. Whether the description is initially organized and viewed through the lens of "laws of physics" is a secondary matter: the main focus is the particular bacterium or biosystem. Approach 2 must be kept at hand to keep us on track toward a proper understanding of a biosystem. Approach 2 has also been used in well-established fields of physics far from biology. Superconductivity is one of many examples.[1]

A comparison of the two approaches, their strengths and weaknesses, is summarized in Table 2.1. Stated crudely, Approach 1 views the object (organism, system) by trying to observe important characteristics that can be quantitatively measured and subjected to physical analysis. Approach 2 views the object as one to be compared and filed with similar objects already known and studied.

Let us take a concrete example of the two approaches and begin to describe and analyze a real biological system. Figure 1.2e shows a microorganism from pond water. If you examine the organism carefully, you will see a pear-shaped organism with a "feeler" sticking out the forward end. (This easier to discern in the animated pdf/mp4 file.)

TABLE 2.1   Example Approaches to the Study of Biosystems

How We See the System	Primary Features	Notes
Approach 1: View by looking for quantifiable physical properties	Observe and measure the obvious features	May miss important features that do not obviously fit
	Fit the features (dimensions, mass, speed, acceleration, energies) into physical law: Newton's Laws, energy conservation, fluid laws...	
	Compute important properties of system: energetics, generated forces, elastic properties, quantum properties	
	Compare properties with those found in nonliving and similar biosystems	
	Generate a model to explain the features you have quantified	Physical modeling of one or more features is a main goal
Approach 2: View by comparing the new with known biosystems	Observe as many features and behaviors as possible	A "complete" description is desired
	Look for features that fit into known biological classification schemes: structure, habitat, behaviors...	
	Classify new system with similar known systems: bacterium, algae, fungus...	
	Compare with similar systems	
	Update classification schemes: evolutionary/genetic, ecological...	Fitting into known schemes is a primary goal

2.3 COMPARISON OF "PHYSICS" AND "BIOLOGY" APPROACHES TO ORGANISM 1.2e

2.3.1 Approach 1

In this section physical parameters that fit into physical laws are indicated in italics. The two most obvious physical features have to do with the motion (*speed, direction changes*) and the *structure* of the microorganism. From the length scale and the video frame rate (15 frames/s), we can estimate the speed. Knowing the speed, we can calculate the fluid *friction force* and thus know the force needed to drive the organism through the water. Force times distance gives work, which can be compared with the *energy* (fuel) the organism consumes. (What is this fuel? We should try to find out.) We might also note that the speed of this organism is comparatively slow. Other, smaller microorganisms visible in the video clip move much faster. The microorganism is pear-shaped, with a bent, rod-shaped appendage projecting in the forward direction, like a "feeler." Does this bent rod drive the motion (a "*motor*") or is it a *sensor*? If it were to drive the motion, the drive would have to be connected to the forward, bent part of the rod. The end of the flagellum, indeed, seems to flail around, and the body does seem to follow the tip of the flagellum, rather than vice versa, but is the motion energetic enough to drive the organism? It seems unusual: most pictures show flagella at the rear, like man-made submarines. Let's assume this forward flagellum provides the driving force, at least until we calculate the needed force; otherwise, we have to look for other, invisible, small cilia or flagella. What is the purpose of the pear shape? As the organism seems not to move quickly, the pear shape is not likely connected to streamlining the shape for motion. Alternatively, if the microorganism needs to support the long rod pointing in the forward

This is a close call because the motion is relatively slow, probably demanding little force. A quick search of *Peranema* on the Internet results in an answer: the end of the forward-facing flagellum rotates rapidly and pulls the body forward through the water. In fact, use of the word *flagellum* implies that it is a motor organ.

direction, the narrower shape where the rod connects to the main body may have structural support advantages. Does the *supporting force* depend on the radius of curvature at the support point? We could keep going from here, but you see the driving force of Approach 1: focusing on quantities that readily fit into physical modeling and computations. We have not tried to classify the organism or compare it to others, except to note that the speed was comparatively slow to moderate.

2.3.2 Approach 2

Is it animal, plant, bacterium, fungus, or protist? What is this microorganism called? Where did it come from? It came from a freshwater pond, so that eliminates all the marine (saltwater) organisms. Some possibilities found in a biology reference book include bacterium, diatom, protozoa, algae, rotifer, gastrotrich, worm, bryozoa, hydra, water bear (Tartigrade), and arthropod. Is it a single cell? It's not green, so it's probably not photosynthetic. What is the size of the organism? It's fairly large, so it cannot be a prokaryote, which range from 0.5–10 μm in diameter. Eukaryotic cells are 10–100 μm, so it could be one. What structures can we identify inside the organism? We need higher magnification. There is a flagellum at the front, not the rear. The front of the flagellum seems to wave or rotate. Looking up descriptions of freshwater organisms (e.g., www.microscopy-uk.org.uk/pond/index.html), protozoa or rotifer looks the closest. Ah-ha! Found it on the Web: it looks like a flagellated *Peranema*. See Figure 2.1. Now we can look up this name and do more detailed "research."

These two approaches to the (now-known) *Peranema* wildly differ from each other. The first starts with physical structures and motions and jumps right into important physical questions and calculations that are elementary to a physicist but advanced detail to a biologist. Approach 1 has its greatest weakness in knowing little about the organism and the context in which it exists. While it can be inferred that motion, speed, forces, and energy use are important in any biosystem, it could be that the observed activity is unusual, or it could be ubiquitous for microscopic pond life. It could be that the great, unsolved question about *Peranema* has nothing to do with forces and motions. The scientist who used Approach 1 only would have the well-deserved scorn of a good biologist: she doesn't even know what it is! Approach 2, in contrast, illuminates little of the physical mechanisms involved in the organism's behavior. It is focused primarily on identifying and classifying the organism, so that known facts can be looked up in published works. It is unlikely that important known facts about *Peranema* will be missed in Approach 2. This approach is what students entering college usually consider "research" on a topic, but in the absence of an approach like 1, few (potentially) new questions are uncovered. Though this text focuses more on Approach 1, the serious student and researcher must always use both approaches, either personally or with help from a "biology-minded" colleague.

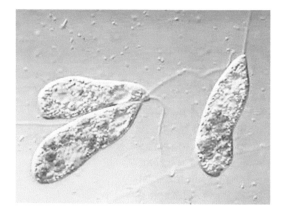

FIGURE 2.1
Freshwater pond microorganism, *Peranema*, apparently the same as that in Figure 1.2e. The long flagellum at the front flicks back and forth and around to pull the organism forward. Image courtesy of Niles Biologicals, www.nilesbio.com/prod149.html. (This organism can be purchased!).

2.4 MEMORIZATION: ITS ADVANTAGES AND DANGERS

The "biology" approach 2 described above clearly has the danger of focusing one's brainpower and work on looking up and remembering facts about an organism or biosystem that have been found by others—memorization. It can lead to a dead-end of knowledge and little curiosity about how things work, except for the fact that there are so many different varieties of microscopic life forms in ponds. On the other hand, most biosystems, unlike the usual systems encountered by a physicist, are incredibly complex. There aren't a lot of facts to be remembered about a field-effect transistor; most of the issues are computational or experimental. Considering even the simplest cell will quickly reveal an almost countless number of facts that could or should be known.

A focus on memorization in physics frequently results in the complaint, "There are so many equations to remember!"

The first response of the physics instructor is usually, "There are many more things to remember in biology. Why complain about physics?" A second response is more pertinent, however. Equations of physics are not "facts to remember" but rather "tools to use." Even memorizing the results of an application of a physical equation, like $F_{net} = ma$, to a particular system, like a spherical object moving at constant speed in a fluid, resulting in frictional force $f = 6\pi\eta rv$, should be done with a clear understanding that the result may not apply to anything else. There are some advantages to remembering this result for a closed-book exam, but remembering when and how to use the various physics tools is the hard part of physics. Remembering the bending force relation and how to calculate the bending force of a *Peranema* flagellum is far more mentally taxing than remembering that *Peranema* is a flagellated protozoa and has a flagellum in the front that bends. Focus on the bending properties of the protozoa's flagellum, in the absence of a more wide-ranging Approach 2 knowledge, carries the "irrelevancy" danger with it, like knowing your spouse likes to eat Italian food with garlic but knowing little else—not the basis for a lasting relationship.

2.5 PROBLEM-SOLVING

For each problem or project, draw a diagram or briefly explain why one is not needed. Use SI units, unless otherwise indicated.

PROBLEM 2.1 CALCULATING ELEMENTARY PHYSICAL DISTANCES RELATED TO SOLID MATTER

1. From the density and atomic mass of silicon, calculate the average distance between atoms in solid silicon.
2. Repeat for carbon in graphite and in diamond.
3. For important applications of these materials, what material, electric, or other physical properties are these distances related to?

PROBLEM 2.2 CALCULATING ELEMENTARY PHYSICAL DISTANCES RELATED TO LIQUID MATTER

1. Calculate the average distance between water molecules in liquid water.
2. Calculate the average distance between sodium ions in a 1 molar (1 M), aqueous solution of NaCl.
3. Plot a graph of $\log_{10}$ distance versus $\log_{10}$ concentration for NaCl concentrations from 1 pM (10^{-12} M) to 10 M.
4. What is the slope of the graph line in part 3? What would change if concentrations were converted to SI units, particles/m^3?

PROBLEM 2.3 BIOLOGICAL IMPORTANCE OF PROBLEM 2.2

The distance calculations in the previous problem might have been done after it was learned that a cell with some intracellular (interior) aqueous NaCl concentration might encounter an aqueous solution of a different NaCl concentration. What elementary physical properties and processes related to biological activity might be related to these distances. Explain. Because distances involve individual particles and microscopic distances, your explanation should also involve those.

PROBLEM 2.4   DNA

You learn that the long, linear polymer, DNA, the biological information molecule, is sometimes stored by being wrapped around spherical protein particles (histones, about 10 nm in diameter), roughly 2½ times per sphere. The information in DNA is primarily in the form of the sequence and identity of the units. You also learn that DNA has, on average, one negative charge per polymer unit. Draw a diagram of this structure. Discuss some physical issues related to this structure and its biological role. Include discussions sizes of structures, elastic properties, likely electrical properties of the spheres, access of information-reading molecules to the information, forces, and stability. (Note: this form of DNA is associated with the terms *chromatin* and *nucleosomes*.)

  DNA is made up of four types of units (A, T, G, and C) in various orders and is not technically a polymer, which has a single type of repeating unit.

PROBLEM 2.5   PHOTOSYNTHESIS: PHOTON TO CHEMICAL ENERGY CONVERSION

This is not the way photosynthesis actually works. You may have to remind yourself of focal properties of mirrors from your first-year physics text.

Suppose a Web site claimed that a photosynthetic plant had been discovered deep in a Costa Rican jungle. It had large leaves (about 20 cm across) covered with many small, closely packed, almost spherically curved mirrors. The backs of the mirrors were about 1.0 mm from the front of the leaf. The leaf was transparent, except for a dark, square lattice of veins at the top surface, diameter 0.1 mm, intersecting directly above each small mirror. The claim was that heated fluid in the veins circulated to another part of the plant, where the heat drove a chemical reaction producing glucose.

1. Draw a diagram of this purported leaf.
2. Determine the dimensions, arrangement, and focal length the mirrors must have if they are to focus light energy onto the dark veins. Is this physically possible?
3. Knowing that the noonday sun provides an intensity of about 1000 W/m² to the surface of the Earth, calculate approximately how much power each vein would have to conduct away from the leaf. Is this energy transport physically possible?

MINIPROJECT 2.6   SIMPLE CELL

Understanding the simplest form of life is one of the great, unsolved issues in the biosciences. You should not get too deep into this at present, but facing big questions early may produce lifelong interest. At this point, do not get more advanced than a first-year college biology or physics level.

Read about the fundamental properties of a biological cell on a reliable Web site (e.g., universities, national laboratories) or from a high school or college biology text.

1. From your reading, draw a diagram of a very simple living cell. Use your best artistic talent, but more importantly, identify critical structures and their roles.
2. Pick two or three of the parts of the cell to which you can readily relate major laws or equations of physics or physical chemistry. Write down the equation(s) that apply to the structure you have selected. Do not actually apply the equation to your cell model, unless the result is exceedingly simple. For example, the cell has a wall separating its inside from the outside. It must have certain dimensional

characteristics: diameter, thickness, etc. If you decide your cell must be rigid (fixed shape) or semirigid, some restrictions on the elastic properties (bending rigidity, etc.) of the cell are implied.

3. Briefly discuss any specific environmental conditions (e.g., external forces/ pressures, materials, temperature and other thermodynamic parameters, energy source) your simple cell requires.

MINIPROJECT 2.7   ENERGETICS OF THE COW

The information needed to work this problem can be found on the web, but use a reliable Web site. Two sites quoting similar values are more reliable. Look up a dairy cow's average weight (mass) and daily consumption of grass and other food, in kilograms, and its average discharge of methane and similar volatile gases in cubic meters at 1 atm pressure.

1. Determine the average available energy, in joules, per kilogram of the food. You could obtain an upper limit estimate of this energy from looking at a box of unsweetened cereal or oatmeal and finding the calories per serving. Make sure you understand what a nutritionist means by a "calorie."
2. Calculate, in joules, the cow's average intake of food energy, its average production of volatile gas energy (the energy we could obtain by burning the gas), and the energy conversion efficiency, food → gas.
3. Compare a cow's "mileage" to an automobile's. Convert a car's miles per gallon (mpg) to meters per joule (mpJ) of fuel energy, normalized to a mass of 1000 kg (multiply the car's mpJ by 1000 kg/car mass). Estimate a cow's normalized mpJ (per joule of food energy). To make this mileage more memorable, convert the cow's mpJ to an equivalent mpg rating. (A "gallon" of grass is that mass of grass containing the energy equivalent of a gallon of gasoline.) Present all these ratings in a table. If you need help here, ask among the physics majors: in a group of 30, there is usually one who grew up on a farm. The sophisticated student farmer may calculate both the cow's free-pasture mileage and its barnyard mileage. (Just kidding…but why not?)
4. Find the number of dairy cows in the United States and determine how many (average) automobiles could be operated from the methane and related gases produced by the cows.

> While this project may seem a bit of a joke, resource use and methane production by dairy cattle has been shown to be an important environmental issue. In a later chapter a similar problem of energy use and efficiency will be encountered on a nanometer to micrometer distance. Some of the very same issues will be contended with: how much energy, exactly, can you get from one molecule of adenosine triphosphate (ATP)?

MINIPROJECT 2.8   ANALYSIS OF FIGURE 2.2

Taking the drawings in Figure 2.2 as examples of organisms you have seen in a microscope, outline the analysis you would do using Approaches 1 and 2 discussed in this chapter. Use Table 1.1 as a guide. If you need to know sizes of the organisms, either estimate or search the Web.

Notes: (1) The names in the diagram provide information for Approach 2. (2) For Approach 1, energy and/or force considerations are useful here. Surface effects are also clearly involved in Figure 2.2. The surface tension Σ, a constant for a given gas-liquid interface, equals the energy per unit surface area (J/m^2), so if an object is going to increase the liquid surface area (e.g., form a depression in a previously flat surface), it must do work. We explore surface effects more in Chapters 4 and 6.

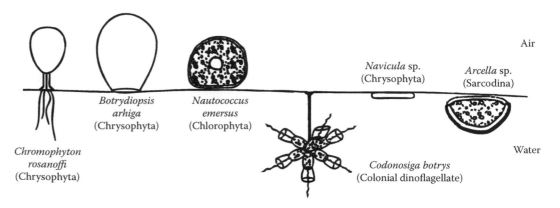

FIGURE 2.2

Diagram of biological structures observed at the interface between water and air. Besides a locale for interesting and unique biological organisms, this boundary-layer environment presents obvious physical parameters the physicist normally does not encounter: surface tension, simultaneous high (water) and low (air) dielectric constants, gravity with and without buoyant force—a bonanza for both biologist and physicist. The text discusses two contrasting approaches to the study of such systems. (Drawing follows Valkanov, A., *Limnologica Berlin* 6, 381–403, 1968, Also discussed in Atlas, R. M. and Bartha, R., *Microbial Ecology*, Benjamin Cummings, Upper Saddle River, NJ, 342–342, 1998.[2])

PROJECT 2.9   HOW DO "OTHER PEOPLE" APPROACH THE BIOSCIENCES: INTERVIEW THE EXPERT(S)

The goal is to gain experience, early on, with a different viewpoint on questions that will be addressed in the course. Material submitted should be electronic, with a master file (e.g., MS Word, html, pdf) containing a written summary of the interview(s) and links to video, audio, and/or other interview media.)

- Decide on the bioscience topic and best candidates for your interview(s). The person(s) should be from a department other than that of your major and should have at least 2 years of research experience. Investigate your institution's guidelines for recording and posting of interview media. Expect to find different rules for media recordings of your institution's professional employees and their enrolled students.
- Write down a draft of the questions you plan to ask, estimate the time needed for the interview, and make your first contacts with potential interviewees. Offer to provide a list of questions.
- Decide on logistics: personal meeting (recommended) with video or audio recording; online meeting, using Facetime, Skype, your institution's learning management software (e.g., Blackboard, Canvas).
- Obtain any needed permissions; prepare and edit audio or video recording, as well as a summary transcript of questions and answers and your own conclusions and insights gained from the interview.

Suggested questions for the interviewer:

- Describe your field of scientific interest and what led you to it.
- What do you think are the two most important unsolved scientific problems in your field of bioscience?
- How does your research relate to these problems?

- How do expose yourself to important discoveries that may relate to your field, but that are made by persons outside your normal research area? For instance, an engineer or astronomer may publish a new method for analyzing images in an astronomy journal.
- What are a few of the most important sources of information you use? How and how often do you access these sources? How do you organize this information (e.g., bibliographic database)?
- How are physics, biology, and chemistry involved in your own research?
- What is your strategy for investigating a new research problem?
- What other fields of science do you follow, in order to keep a sufficiently broad scientific awareness? Hobbies?
- How do "political" issues, from those at the federal government level, to those in your own university, to those in your scientific societies, affect your research and career?
- Do you make presentations of your research to nonscientists?

REFERENCES

1. Parinov, I. A. 2007. *Microstructure and Properties of High-Temperature Super-conductors*. New York: Springer. An abstract of this book reads, "The main features of high-temperature superconductors...layered anisotropic structure and the supershort coherence length.... The 'composition-technique-experiment-theory-model,' employed here...helps to draw a comprehensive picture of modern representations of the microstructure, strength and the related structure-sensitive properties of the materials considered," an embodiment of our Approach 2 to complex systems.
2. Valkanov, A. 1968. Das Neuston. *Limnologica Berlin* 6:381–403. Also discussed in Atlas, R. M., and R. Bartha. 1998. *Microbial Ecology*. Upper Saddle River, NJ: Benjamin Cummings, 342–342.

CHAPTER 3

CONTENTS

Math Tools
First Pass

3.1 WHAT MATH DO WE NEED?

The math needed to deal with biophysical systems can be vast and is limited only by the systems one chooses to deal with and the aspects one chooses to describe. This chapter highlights some essential math.

3.2 NOTATION: MATHEMATICS VERSUS PHYSICS NOTATIONS

3.2.1 Derivatives and Differentials

Physics texts generally develop notations that directly suit the common equations used in physics, such as Newton's law. Derivatives are mostly with respect to time. Shorthand, these derivatives are often written with superposed dots standing for a time derivative. Engineering dynamics texts that commonly delve into finite changes in motional variables and finite integrations thereof sometimes introduce other variations that are of use in engineering applications. Table 3.1 summarizes some of the notations. This text uses the "normal" notation for derivatives.

Although not absolutely necessary, derivatives should be written as $\frac{dx}{dt}$, rather than dx/dt, to facilitate checking of physical units and making changes of variables.

> This chapter is intended as a survey of the more advanced math that is needed for biophysics.

> The dot notation referring to derivatives with respect to a variable besides t (time) should be avoided.

3.3 APPROXIMATIONS

Biosystems are complex and impossible to treat exactly. One of the primary tasks of the biophysicist is to determine how to extract quantitative information that is meaningful and as close as possible to exact. Most of this chapter deals with mathematical approximations, after a discussion of useful material for chemical approximations. Physics has a long history of making approximations. In fact, one could say that all of physics consists of modeling parts of the real universe approximately so that simple equations can be used. One always has to ignore matter or events that exist far (enough) away from the system under consideration. Some impurities must be ignored if progress is to be made, even though experience has taught that small amounts of foreign substances sometimes produce dramatic and useful behavior. Impurities in semiconductors and enzymes in cells are good examples. Nevertheless, when biological life presents us with such a myriad of processes and structures, large and small, some simplification is needed for any progress in understanding to be made, even when important phenomena may be missed. Those missed materials or behaviors will hopefully be understood later.

3.3.1 Complexity: Approximations That Eliminate the "Butterfly Effect"

Originally, this effect referred to a butterfly's wings causing air currents that resulted, ultimately, in a hurricane on the other side of the world. In science the term *butterfly effect* normally is employed

TABLE 3.1 Derivatives and Differentials: Normal and Shorthand Notation[a]

Normal Notation	Short	Typical Integral or Differential Equations	Notes
dx/dt	$\dot{x}$	$v = \dot{x} \quad a = \ddot{x}$	Physics definitions
ds/dt	$\dot{s}$	$v = \dot{s} \quad a = \ddot{s}$	Engineering texts: "s"
dv/dt	$\dot{v}$	$v\,dv = a\,dx,$ or $\dot{x}\,d\dot{x} = \ddot{x}\,dx$	Useful when $a = a(x) =$ function of x
		$\int v\,dv = \int a\,dx = \int -\frac{k}{m}x\,dx$	Harmonic oscillator: integral form
d^2x/dt^2	$\ddot{x}$	$F_{net} = ma = md^2x/dt^2$	Newton's second law
		$m\ddot{x} + kx - 0$	Harmonic oscillator differential equation
$f(x, y)$	—	Many different	Common math version of function of two independent variables
$\partial f/\partial x$	f_x		Partial derivatives of a function of two variables;
$\partial f/\partial y$	f_y		Short form should be avoided

[a] Here, x, s, position (one-dimensional); v, velocity; a, acceleration; k, spring constant; m, mass; F, force.

in complex systems, where infinitesimal changes in initial conditions produce large changes in long-term behavior. Complex systems to which this applies behave chaotically, at least under certain conditions. They may not be complex in the sense of having many parts but generally have special types of interactions between parts of the system. A quantum hockey puck bouncing in a small, properly shaped rink, resulting in unpredictable paths, is an example.[1] Approximations to these systems must be done extremely carefully, for they often eliminate the entire effect of interest. Chaos is now known to have important consequences in biosystems. We touch on some of these effects in various chapters, but the interested reader should do further reading.[2-4] It should still be noted that approximations can be made even in chaotic systems. It takes a certain amount of time for the unpredictability to develop. If times are short compared to this time-to-chaos, approximations may be quite successful at determining quantitative values of parameters of interest.

3.3.2 Experimental: Ignoring Chemical Components in the Biosystem

Simplifying biosystems by ignoring or removing certain components has been the basis for thousands of arguments between physical and biological scientists. The physicist, engineer, or chemist says to take apart and simplify the system until some simplified version of the system can be described mathematically. The biologist says such dismantling amounts to eliminating the "life" of the system. Nevertheless, thousands of biochemists and biologists have spent their lives purifying specific components of organisms and determining how they behave in pure form. Such purification (simplification of the system) is pursued most vigorously when a chemical reaction that should not take place does. The search is then on for the rare impurity component responsible for the reaction—the enzyme. The trick in eliminating components found in natural systems is to keep (or eliminate) those that have strong effects. For example, much of the structure and behavior of the DNA molecule has been studied with purified DNA dissolved in 0.15 M NaCl (or KCl), with small concentrations of buffers (weak acids/bases) added to keep the pH ($-\log_{10}$ of the H^+ concentration) between 7 and 8 in water. No DNA-containing cell has only these chemicals, of course, and even these simplified systems have varying, often uncontrolled amounts of other chemical components, like dissolved O_2, N_2, and other gases. However, DNA takes on its "usual" double-stranded form, can make double to single-strand transitions (ds → ss), can be cut by adding natural enzymes (endo- and exo-nucleases), and carry on other biologically important activities in this simplified system.

Note that there are often exotic, unexplained components of the "simplified" system, such as ethylene diamine tetraacetic acid (EDTA). This addition of a chemical component to "simplify"

We will see that the DNA double strand is not at all the only biologically active form of DNA.

the system is interesting because its presence may be to prevent the activity of low-concentration components that are difficult to remove from an actual sample. In the case of EDTA, the undesired, low-concentration components are heavy metal ions that bind to DNA and strongly affect its structure and action. EDTA does its job of system simplification by "caging" these metal ions, so they cannot bind to the DNA. Biochemists have learned that certain rare impurities cannot be tolerated in simplified systems. There are many other examples, such as different buffers that cannot be used for certain purified molecules, enzymes and nucleic acids. In a healthy, living cell, these "bad" impurities, buffers, ions, and other agents are generally not present, having been filtered out or not inserted by the cellular systems that create the cell and replenish its contents.

3.3.3 Mathematical: Approximations to Biological Structure and Dynamics

> This book will not be very careful about use of accepted names for mathematical terms. Other books may use different names for expansions and approximations.

These approximations are needed when first dealing with a biosystem so that mathematical complexity does not overwhelm. Researchers who become interested in certain problems that have initially been approximated usually find computer methods that improve some of the approximations. We focus on these methods for the rest of this chapter.

3.3.3.1 Expansions: Taylor series

The value of function, $f(x)$, near a point, x_0, can be approximated, if it has *proper* mathematical behavior, by a Taylor series expansion:

$$f(x) = f(x_0) + \left(\frac{df}{dx}\right)_{x=x_0} (x - x_0) + \frac{1}{2!}\left(\frac{d^2 f}{dx^2}\right)_{x=x_0} (x - x_0)^2 + \frac{1}{3!}\left(\frac{d^3 f}{dx^3}\right)_{x=x_0} (x - x_0)^3 + \ldots \quad (3.1)$$

In the majority of applications to biosystems, *proper* mathematical behavior is also demanded by nature: physical systems do not generally have infinities or abrupt changes in values of physical parameters at certain points. In the present case, one can see that if the function or any of its derivatives takes on an infinite or undefined value for a value of x of physical interest, there is generally something wrong with the mathematical model.

Physics is, however, a pragmatic business. Many of you have seen the example of quantum tunneling through a square potential barrier. Square potential energy functions, $V(x)$, that change values upon infinitesimal changes in the value of x, are unphysical because they imply infinite forces at a point. (Nature abhors infinities.) However, the unphysicality is manageable. Many quantities of interest, such as the wave function, energy, probabilities, and transmission coefficients can be calculated just fine; simply avoid asking certain questions, like what is the force at the edge of the potential barrier. The mathematical difficulty needed to smooth the sharp edges of the barrier is often not worth the gain in understanding or accuracy: the interesting quantities are often changed very little (usually not enough to justify an additional publication in a peer-reviewed journal, for example).

3.3.3.2 Example 1: Oscillation in an asymmetric potential well

The harmonic (symmetric) oscillator is well known to you all. The force exerted by the "spring" on the mass and the potential energy functions can be written

$$F(x) = -kx$$

$$V(x) = \frac{1}{2}k(x - x_0)^2 \quad (3.2)$$

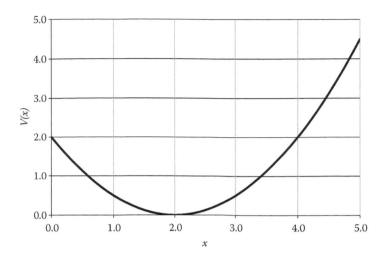

FIGURE 3.1
Harmonic potential function, with minimum at $x = 2.0$ m and $k = 1.0$ N/m.

Here we have no reason to apply the Taylor expansion, because the function is already in the form of the expansion, with only the second term nonzero. A graph of such a potential is shown in Figure 3.1, with $x_0 = 2.0$ m and $k = 1.0$ N/m. The oscillation angular frequency and frequency of a mass m are, respectively,

$$\omega = \sqrt{\frac{k}{m}} \quad \text{and} \quad f = \frac{1}{2\pi}\sqrt{\frac{k}{m}} \tag{3.3}$$

When the application is to spectroscopy, the potential is often rewritten

$$V(x) = \frac{1}{2}m\omega^2(x - x_0)^2 \tag{3.4}$$

What if the oscillator is asymmetric? If the potential is extremely asymmetric (anharmonic), it is not of much use to try to use to try to model it as an oscillator, so we consider cases where the potential is not too far from harmonic, at least near some position. A common case in bio- and other systems is when the potential energy is more accurately expressed by completely different functions (not polynomials). A not so simple example is the potential sometimes used to model vibrations in three-dimensional solids,

$$V(r) = -\frac{\alpha e^2}{4\pi\varepsilon_0 r} + \lambda e^{-r/\rho} \tag{3.5}$$

You should recognize one term as an attractive coulomb potential, multiplied by an extra constant (presently of no interest) and an added decaying exponential, all as a function of r, the radial distance. As the two terms are opposite in sign, there is probably a minimum in the potential. If there weren't, there would be no sense in trying to approximate the potential as almost harmonic. The tasks we have to accomplish:

1. Find the value of r, call it r_0, where the potential energy is a minimum.
2. Calculate the second, third, fourth, etc., derivatives of the function; evaluate them at $r = r_0$. (The first derivative is zero at r_0.)
3. Use Equation 3.1 to write the approximation for $V(r)$, valid near $r = r_0$.
4. After confirming that the values of the $(r - r_0)^3$, $(r - r_0)^4$, etc. terms are small compared to the $(r - r_0)^2$ term, use the coefficients of the squared term to determine approximate oscillation force constant, frequency, etc.

The first task is to find where that minimum in potential is located. How do we do that? Differentiate the function and set the first derivative to zero; the resulting values of r locate minima and maxima.

$$\frac{dV}{dr} = 0 = \frac{\alpha e^2}{4\pi\varepsilon_0 r^2} - \frac{\lambda e^{-r/\rho}}{\rho} \tag{3.6}$$

Though SI units are standard in this book, e.g., joules (J) for energies, eV energy units (1 eV = 1.602×10^{-19} J) will be found to be quite useful for energies of single particles, because the eV values often take on values near 1 for position ranges of interest in biosystems, 0.1–10 nm.

Most of you will realize this equation is not simple to solve. What do we do? Sometimes the unknown r_0 can be determined exactly in terms of another unknown, such as $V(r = r_0)$, the minimum energy, where the new unknown can more easily be estimated or measured from experiments or other data. In Figure 3.2 this value can be seen to be about −3.0 eV. You can also try a computerized math program (e.g., Mathematica®, MATLAB®, or Maple®) or else solve using the "guess-and-try" method, after putting in numerical values for all constants. Usually, three or four guesses are enough to arrive at a quite accurate answer. Using numerical values does not provide a general, analytical answer, but it is an extremely powerful and fast way to get answers for particular cases. Practicality can often win out over mathematical elegance. We leave the rest of this solution for the homework.

Let's look at some mathematically simpler cases. Many potentials with a minimum can be approximately written as the sum of two polynomials or inverse polynomials of opposite sign:

$$V_1(r) = A(r - r_0)^n - B(r - r_0)^m \tag{3.7}$$

$$V_2(r) = \frac{A}{r^n} - \frac{B}{r^m} \tag{3.8}$$

> Play around with parameter values in "live" Microsoft Excel graph in the online supplement to see how the function shape depends on these parameters.

The reader should convince herself that n is usually greater than m in Equation 3.7 and the reverse in Equation 3.8; otherwise, the functions would not have a proper minimum. These functions are plotted in Figures 3.3 and 3.4 for values near the minimum. Since Equation 3.7 is already in the form of a polynomial, let's approximate Equation 3.8 near the minimum with the Taylor expansion quadratic term (the linear term is zero near the minimum), taking $n = 3$ and $m = 2$. First, determine units for A and B. In the SI system, A must have units $J \cdot m^3$ and B, $J \cdot m^2$; otherwise, the two terms would not have overall units of energy. Next, find the minimum:

$$\frac{dV_2}{dr} = 0 = -\frac{nA}{r^{n+1}} + \frac{mB}{r^{m+1}} = -\frac{3A}{r^4} + \frac{2B}{r^3}, \text{ or } 0 = nA + mBr; \text{ so } r_0 = \frac{3A}{2B}$$

Then

$$\left(\frac{d^2V_2}{dr^2}\right)_{r=r_0} = \frac{12A}{r^5} - \frac{6B}{r^4} = 4\frac{B^5}{A^4}.$$

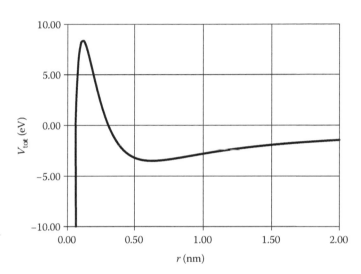

FIGURE 3.2
Anharmonic potential with minimum. A short-range, exponential repulsive core is added to a long-range, coulomb-type potential. For this sort of potential, positions below about 0.1 nm would not be allowed.

So the potential near r_0 can be (harmonically) approximated

$$V_2(r) = \frac{1}{2}\left(4\frac{B^5}{A^4}\right)(r - r_0)^2 + \cdots \simeq \frac{81B}{4}\frac{(r - r_0)^2}{r_0^4} \tag{3.9}$$

The "spring constant" for motion near $r = r_0$ is then $k_2 = \frac{81B}{2r_0^4}$. Let us check the units of k_2:

$$\frac{B}{r_0^4} : \frac{J \cdot m^2}{m^4} = \frac{N \cdot m \cdot m^2}{m^4} = \frac{N}{m}$$

which is correct. The oscillation angular frequency is

$$\omega_2 = \sqrt{\frac{k}{m}} = \left(\frac{81B}{2mr_0^4}\right)^{1/2}$$

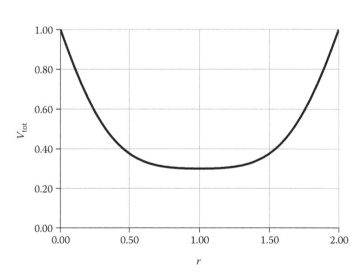

FIGURE 3.3
Anharmonic potential $V(r) = A(r - r_0)^n - B(r - r_0)^m$, with $n = 2, m = 3$, $A = 1.0, B = 0.3, r_0 = 1.0$. The potential is reasonable for only narrow ranges of A and B and for values of r not too far from r_0.

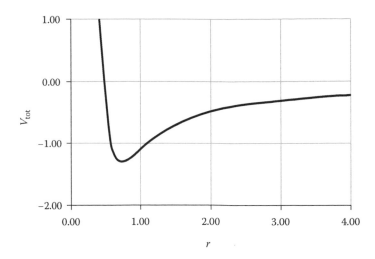

FIGURE 3.4
Anharmonic potential $V(r) = \frac{A}{r^n} - \frac{B}{r^m}$, with $n = 3$, $m = 2$, $A = 1$, $B = 2$. The potential near the minimum can be approximated a quadratic function. See text.

Though the approximation of potential functions that have a minimum are one of the most common applications of Taylor series expansions in physics, readers should keep this approximation as one of the tools in their problem-solving "toolkits" for other aspects of biosystems.

3.3.3.3 Example 2: Small angles

Trigonometric functions commonly occur as mathematical solutions to equations found in physics, chemistry, biology, microscopy (imaging), and other fields. If the angles in these functions are small, about 10° (~0.17 rad) or less, the common trig functions can be approximated:

$$\sin \theta \simeq \tan \theta \simeq \theta \text{ (small } \theta, \text{ in radians)} \tag{3.10}$$

$$\cos \theta \simeq \frac{1}{2}\theta^2 \text{(small } \theta, \text{ in radians)} \tag{3.11}$$

These results come straight from the Taylor expansions of the functions. The form of these solutions in real applications is typically not $\sin \theta$ or $\cos \theta$, but rather

$$F(x) = A \sin(kx) + B \cos(kx) = C \sin(kx + \delta), \text{ or} \tag{3.12}$$

$$F(x) = A\sin\left(\frac{n\pi x}{L}\right) + B\cos\left(\frac{n\pi x}{L}\right) = C\sin\left(\frac{n\pi(x - x_0)}{L}\right) \tag{3.13}$$

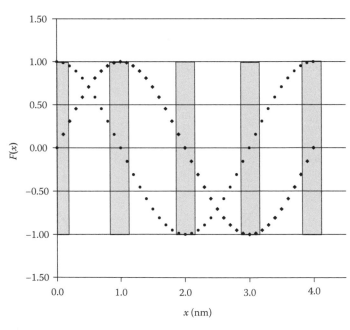

FIGURE 3.5
Sine and cosine functions for $n = 4$. Other values of n can be explored in the associated online Microsoft Excel file. In the shaded regions the functions can be approximated by linear and quadratic functions: $\sin\left(\frac{4\pi x}{L}\right) \simeq \frac{4\pi(x - x_0)}{L}$ and $\cos\left(\frac{4\pi x}{L}\right) \simeq 1 - \frac{1}{2}\left(\frac{4\pi(x - x_0)}{L}\right)^2$ for $x_0 = 0$ nm. The reader should be able to write down modified approximations valid near $x_0 = 2$ nm, 4 nm, etc.

In these functions k has units of m⁻¹ and L units of meters. Graphs of these functions are shown in Figure 3.5, along with regions in which the Taylor series approximation can be simply made. Sine and cosine functions have familiar "oscillatory" shapes and represent familiar solutions to the time dependence of the position of a harmonic oscillator. However, oscillation is not the only application. A sine or cosine function, appropriately modified, could represent a reasonable potential energy function, for example, for a small particle constrained to move along a polymer track, where maxima or minima are located at the center of the polymer unit. In this case, the particle could be trapped in a local minimum, where small oscillations could be parameterized using a quadratic potential, $V(x) \sim \frac{1}{2}k(x - x_0)^2$, as shown in Figure 3.6.

3.3.3.4 Example 3: Exponentials

Exponential functions are also readily approximated using the Taylor expansion:

$$\begin{aligned} e^{ax} &= 1 + ax + \frac{(ax)^2}{2!} + \frac{(ax)^3}{3!} + \cdots \\ e^{-kt} &= 1 - kt + \frac{(kt)^2}{2!} - \frac{(kt)^3}{3!} + \cdots \end{aligned} \tag{3.14}$$

The first form of the equation is common in math and physics applications; the second is a common solution to reaction equations. See Section 3.3.4. Equation 3.14 is valid for both positive and negative values of a or k. Exponential functions are common solutions to equations arising in physics, chemistry, and biology, but it is less common that the approximation in Equation 3.14 is useful in describing a reaction or other process, as chemical reactions must usually be described out to long times.

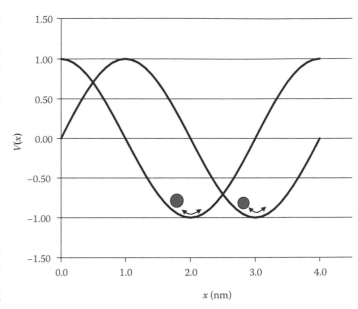

3.3.4 Approximations in Differential Equations

We encounter differential equations quite often in biophysics. When dealing with actual biosystems, many of the equations encountered are standard and require little sophisticated differential-equation solving skills. In the majority of cases, the guess-and-try method, the sister of that recommended earlier when faced with a difficult algebraic equation, works quite nicely after learning the usual functions to guess. Rather than trying to introduce these functions all at once, we bring them in as necessary.

FIGURE 3.6
Oscillations near a minimum of a sine or cosine potential energy function. Readers should be prepared to approximate the potential function near its minimum as a quadratic, of form $V(x) \sim \frac{1}{2} k(x - x_0)^2$.

3.3.4.1 Small and large: Compared to what?

Approximations of use in differential equations usually consist of cases where small or large values of a quantity or parameter are of interest. The approximation may be introduced as an "exact" steady-state or infinite-time solution (neither of which exist exactly in the real world). When you hear, "For large values of x, we see that…," it makes sense to ask "Large compared to what?" In real applications, nothing is ever "large" or "small" by itself. Only in comparison to something else is it large or small.

In differential equations occurring in biosystems, long- and short-time solutions are often desired. If the solution to the equation is known, determining the measure of short and long is usually straightforward. If the exponential function

$$A(t) = A_0 e^{-k_1 t} \tag{3.15}$$

where A refers to a number or concentration of a component A, is the solution, then short and long times simply refer to the parameter defining how rapidly the function changes. In this case, times must be compared to the time $\tau_1 = 1/k_1$, the inverse of the reaction rate. Short time means $t \ll \tau_1$ and long time means $t \gg \tau_1$. Suppose a solution is

$$\frac{1}{A(t)} = \frac{1}{A_0} + k_2 t \tag{3.16}$$

A first response might be to again say $t \ll 1/k_2$ or $t \gg 1/k_2$, but following the admonition, "Check the units!" belies this conjecture. Time cannot be compared to $1/k_2$ because $(k_2 t)$ must have units of (number of particles/m³)⁻¹ or M⁻¹ (inverse molar); so k_2 has units of m³/s or M⁻¹s⁻¹. The next, and correct, response is to compare $k_2 t$ to $1/A_0$: short time means $k_2 t \ll 1/A_0$ and long time means $k_2 t \gg 1/A_0$. The same approach can be taken if the independent variable is position rather than time.

What if you do not yet have the analytical form of the solution to the differential equation? Suppose you have a reversible reaction,

$$A \underset{k_r}{\overset{k_f}{\rightleftharpoons}} B \tag{3.17}$$

and you know the reaction proceeds via mass action, or

$$\frac{dA}{dt} = -k_f A + k_r B$$

$$\frac{dB}{dt} = -\frac{dA}{dt} \quad \text{if no particles are lost} \tag{3.18}$$

with $A(0) = A_0$ and $B(0) = 0$. We also note, by particle conservation, $A(t) + B(t) = A_0$.

3.3.4.2 Short time

The shortest time is $t = 0$. At $t = 0$, $B(t) = B(0) = 0$. So our first stab at "short time" lets us approximate Equation 3.18 by

$$\frac{dA(t)}{dt} \simeq -k_f A(t) \quad (\text{at } t = 0) \tag{3.19}$$

> Exponential functions should be among your first guesses. Other guesses at differential equation solutions should include sine and cosine functions and polynomial functions.

The solution of this equation, found by experience or guess, is $A(t) = A_0 e^{-k_f t}$. However, it is pretty useless to have a functional solution that is valid only at $t = 0$. How long is this approximate solution valid? It is valid as long as $k_f A(t) \gg k_r B(t)$. We could leave it at this inequality, but can we get a characteristic time τ that marks the border between "short" and "long"? We might try rewriting the differential equation using $B(t) = A_0 - A(t)$:

$$\frac{dA(t)}{dt} = -k_f A(t) + k_r (A_0 - A(t))$$

$$= -(k_f + k_r)A + k_r A_0 \tag{3.20}$$

There seem to be two possible characteristic times in Equation 3.20: $1/k_r$ and $1/(k_f + k_r)$. We address this problem later in more detail (chemical reactions), but we have discovered a fundamental feature of dynamic processes, that characteristic *time constants are determined by the sum of forward and backward rates.*

3.3.4.3 Long time

In Equation 3.18 or 3.20, how can we define the long-time behavior? We do not know what the value of $A(t)$ or $B(t)$ is at any "long" time, so we cannot try to neglect A or B. Let's use another feature of functions that usually apply to real, spontaneous physical, chemical, and biological phenomena: unless a dynamic process is oscillatory, rapid events take place at short times and slow events at long times. Stated another way, things stop changing if you wait long enough. In Equation 3.18 or 3.19, this slowing of change translates to $dA/dt \sim dB/dt \sim 0$. Using this long-time/steady-state/equilibrium approximation in Equation 3.20 leads to $\frac{dA}{dt} \cong 0 = -\left(k_f + k_r\right)A + k_r A_0$, which gives the steady-state or long-time solution

> We will eventually have to be more careful to distinguish steady state from equilibrium and long time.

$$A(t \to \infty) = A_\infty = \frac{k_r}{k_f + k_r} A_0 \tag{3.21}$$

How do we decide when t is midway between "short" and "long"? When A is about halfway between A_0 and $\frac{k_r}{k_f + k_r} A_0$, which should be at about $t = (k_f + k_r)^{-1}$.

Now that we know what the long-time concentration of A is, A_∞, can we now go back and guess the solution to the differential equation? Exponential decay of $A(t)$ is always a good guess, but A should only decay to a value of A_∞, not zero. So a good guess would be $A - A_\infty = Ce^{-(k_f + k_r)t}$, where C is not yet known. (We previously argued that the overall rate should be the sum of rates.) With this guess and the differential equation, we very quickly find that $C = A_0 - A_\infty$, so the solution to Equation 3.20, the reversible $A \Leftrightarrow B$ reaction, is

$$A(t) - A_\infty = (A_0 - A_\infty)e^{-(k_f + k_r)t} \tag{3.22}$$

3.4 VECTORS

This section is brief—just a reminder that in dealing with directional physical processes, vectors are most effectively treated by the component method. In three dimensions,

$$
\begin{aligned}
\vec{A} &= \hat{x}A_x + \hat{y}A_y + \hat{z}A_z \quad &\text{(rectangular)} \\
&= \hat{r}A_r + \hat{\theta}A_\theta + \hat{z}A_z \quad &\text{(cylindrical)} \\
&= \hat{r}A_r + \hat{\theta}A_\theta + \hat{\phi}A_\phi \quad &\text{(spherical)}
\end{aligned}
\tag{3.23}
$$

Note r and θ do not have the same meaning in cylindrical and spherical coordinates. The cylindrical and spherical versions are much less common than the rectangular. In biological applications the system of coordinates chosen is often dictated by the measurements that can be made. For example, transfer of electronic energy between two transition dipoles (absorbers of light) is described in terms of angles and the radial distance between the dipoles, because those two quantities (especially the latter) are amenable to experimental determination.

3.5 TWO- AND THREE-DIMENSIONAL GEOMETRY

Two-dimensional geometry and trigonometry are straightforward. The angle in two dimensions, θ, is the angle as measured from the $+x$ axis. Relations between rectangular (x, y) and polar (r, θ) coordinates are simple, with corresponding relations for the x and y components of a vector of length A making an angle θ with the x axis:

$$
\begin{aligned}
&x = r\cos\theta \;\text{ or }\; A_x = A\cos\theta \\
&y = r\sin\theta \;\text{ or }\; A_y = A\sin\theta \\
&\tan\theta = \frac{y}{x} \;\text{ or }\; \tan\theta = \frac{A_y}{A_x} \\
&r^2 = x^2 + y^2 \;\text{ or }\; A^2 = A_x^2 + A_y^2
\end{aligned}
\tag{3.24}
$$

Cylindrical coordinates (r, θ, z) use r and θ variables as in polar coordinates, with z the distance along the z axis (Figure 3.7). Conventions for angles in three-dimensional diagrams in spherical coordinates often differ in physics and math textbooks. For spherical coordinates (r, θ, ϕ) this text uses the convention that the angle θ stands for the angle a vector makes with the $+z$ axis; ϕ measures the angle "around" the z axis of the vector's projection in the x–y plane, the $+x$ axis being $\phi = 0$, shown in Figure 3.7. In the subsections below we review relations between the coordinate systems.

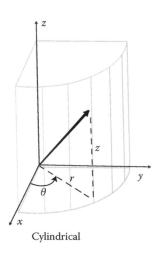
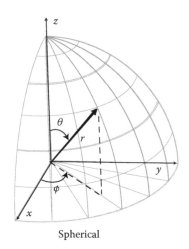

Cylindrical   Spherical

FIGURE 3.7
Coordinates in three dimensions. Cylindrical coordinates (r, θ, z): the r and θ coordinates are the same as in two-dimensional polar coordinates, while z is the same as in rectangular. Spherical coordinates (r, θ, ϕ): the angle θ takes on values from 0 to π; ϕ, from 0 to 2π; r is the length of the vector. We do not try to avoid the inconsistencies in use of r and θ in cylindrical and spherical coordinates. A well-drawn diagram will avoid confusion. Note that if a physical function contains the angle ϕ, the value of the function at $\phi = 2\pi$ should be the same as the value at $\phi = 0$, except in unusual cases.

3.5.1 Rectangular Coordinates: $\vec{r}$, or (x, y, z)

In most vector applications rectangular coordinates are the first to be used, because of their mathematical simplicity and familiarity. For position vectors (whose tails are at the origin of the coordinate system),

Length: $r = \sqrt{x^2 + y^2 + z^2}$

Displacement (difference): $\vec{r}_2 - \vec{r}_1 = (\Delta x, \Delta y, \Delta z) = (x_2 - x_1, \ y_2 - y_1, \ z_2 - z_1)$

Distance between points: $\sqrt{(x_2 - x_1)^2 + (y_2 - y_1)^2 (z_2 - z_1)^2}$

Vector sum: $\vec{r}_1 + \vec{r}_2 = (x_1 + x_2, y_1 + y_2, z_1 + z_2) = \hat{x}(x_1 + x_2) + \hat{y}(y_1 + y_2) + \hat{z}(z_1 + z_2).$

For vectors other than position or displacement vectors, $\vec{F} = (F_x, F_y, F_z)$,

Length: $\left|\vec{F}\right| = F = \sqrt{F_x^2 + F_y^2 + F_z^2}$

Vector sum: $\vec{F}_1 + \vec{F}_2 = (F_{1x} + F_{2x}, F_{1y} + F_{2y}, F_{1z} + F_{2z}) = \hat{x}(F_{1x} + F_{2x}) + \hat{y}(F_{1y} + F_{2y}) + \hat{z}(F_{1z} + F_{2z})$

3.5.2 Cylindrical Coordinates

Usually, only used in cases of cylindrical symmetry, cylindrical coordinates (r, θ, z) use the polar coordinates from the two-dimensional x–y plane, with the rectangular coordinate z describing the third dimension. The length of a vector is $\sqrt{r^2 + z^2}$ ($\sqrt{A_r^2 + A_z^2}$ for nonposition vectors). Distances between points cannot be written so easily. The relation between cylindrical and rectangular coordinates is the same as for two-dimensional polar coordinates, with an extra rectangular coordinate z added:

$$x = r\cos\theta$$
$$y = r\sin\theta$$
$$z = z$$
$$r^2 = x^2 + y^2$$
$$\tan\theta = \frac{y}{x} \quad \text{cylindrical} \Leftrightarrow \text{rectangular}$$

(3.25)

A reason to use cylindrical coordinates in a biosystem might be treatment of transport down a cylindrical neuron.

3.5.3 Spherical Coordinates

Most commonly used in cases of spherical symmetry, spherical coordinates (r, θ, ϕ) may also be used when analysis in terms of spherical functions, e.g., spherical harmonics, is desired. Relationships between rectangular and spherical coordinates:

$$x = r\sin\theta\cos\phi$$
$$y = r\sin\theta\sin\phi$$
$$z = r\cos\theta$$
$$\cos\theta = \frac{z}{r}$$
$$\tan\theta = \frac{y}{x}$$
$$r^2 = x^2 + y^2 + z^2 \quad \text{spherical} \Leftrightarrow \text{rectangular}$$

(3.26)

$$\vec{A} = \hat{x}A_x + \hat{y}A_y + \hat{z}A_z$$
$$A_x = |\vec{A}|\sin\theta\cos\phi$$
$$A_y = |\vec{A}|\sin\theta\sin\phi$$
$$A_z = |\vec{A}|\cos\theta$$
$$|\vec{A}|^2 = A^2 = A_x^2 + A_y^2 + A_z^2$$

Spherical coordinates might be used in cases of diffusion of molecules though the membrane of a spherical cell or diffusion from a point-like source or to a point-like sink of molecules (Figure 3.8, lower).

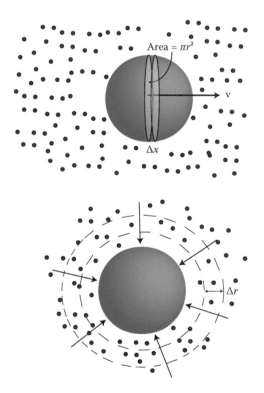

FIGURE 3.8
Diagram for flux of particles encountered by a sphere. (Upper) If the sphere is moving it will encounter a flux of particles over its transverse cross-sectional area, πr^2. (Lower) Simultaneously, a flux of diffusing particles will impinge on the entire sphere surface area, $4\pi r^2$.

3.6 CALCULUS

Calculus is needed in biophysics because biological systems are not generally uniform, homogeneous, and constant and because the fundamental equations describing dynamic processes are usually differential.

Equations for position, velocity, and acceleration often cannot assume constant force for more than a short time or distance because, as shown in Figure 1.2, the environment of a moving biological object generally changes rapidly. In this situation, an expression like $vdv = adx$ (Table 3.1), especially useful when force or acceleration can be written as a function of position, $a = a(x)$, is ideally set up to integrate directly, either analytically or numerically.

You might notice that this has violated the rule that there are no short times, short distances, or "rapid" except in comparison to some standard. The reference in the previous paragraph was to a microorganism that encountered obstacles on a regular basis and had to change direction. An even more obvious case of a rapidly changing surrounding would seem to be the inside of a cell, which is far from a uniform intracellular solution. What are the standards by which short time and short distance should be judged? We find in Chapter 16 that distances and times to which observational times should be compared are often surprisingly small, as small as fractions of an angstrom and a nanosecond or less. These short times and distances for change would imply that much motion can be treated as having constant, even zero, acceleration, and no integration need be done. In fact, integration is often unnecessary for analyzing the movement of microorganisms. Even though the driving force generated by a bacterium's flagellum may change significantly during an observation, the motion can be treated as a series of constant-velocity movements, connected by (comparatively) short periods of acceleration.

We do need calculus, however. Often, we do not track the movement of individual objects but rather the flux created by movement of many objects. The need for comparative scales is seen as soon as one realizes that the motion of one biological "particle"—a microorganism or even a molecular motor—may be driven by the chemical conversion of large numbers of molecules to a lower-energy form (e.g., ATP to ADP). The organism may consume hundreds of millions of fuel molecules during a movement of a few hundred microns. These fuel molecules must diffuse from the surrounding into the organism: an integral must be done.

3.6.1 Drawing Diagrams That Help You Do the Math

Physics is done with diagrams. The art of physics does not consist of selecting a correct equation from a vast number of possibilities. The equations are not the thing; the diagrams are. Diagrams tell you how to think about a problem, how to determine the important principles that apply, how to set it up for solution, and how to find the equations to use. The equations are tools you use; they come near the end of the analysis, though they sometimes demand time and effort to solve.

Consider the example of a microorganism, either moving or motionless, immersed in a surrounding solution of fuel molecules needed for motion or metabolism. In order to balance energy use with energy input, a certain number of fuel molecules must be taken into the cell per unit time. Assuming that every molecule that hits the cell is absorbed, determine the rate at which molecules are absorbed by the cell. How do we set up a problem like this if we haven't seen it done before? Start with a diagram. The diagram needs to show the spherical cell, the surrounding molecules, possible movement of the cell, and a flux of molecules hitting the cell. What do we do then? We don't yet know, but let's draw the diagram and work from there. Figure 3.8 shows a first try at a diagram. The upper panel shows the cell moving through the medium with molecules randomly distributed. Some brief thought should convince the reader that the moving sphere sweeps out an area of πr^2 as it moves through the fluid. If the concentration of molecules is c molecules/m^3 and the sphere speed is v m/s, the volume swept out by the sphere after it has moved a distance Δx is $dV = (\pi r^2)dx = (\pi r^2)vdt$. The rate the volume is swept out (m^3/s) is $dV/dt = (\pi r^2)v$. For c molecules/m^3, the number of molecules swept up per second is $dN_1/dt = c(dV/dt) = (\pi r^2)vc$. This dN_1/dt is the "active" rate at which "fuel" molecules are swept up by the moving sphere. Note the frequent use of differentials in this sort of diagram. The fact that we did not need to perform any actual integrations resulted from the intuitive insight that the sphere cross-sectional area was sufficient to give us the number of molecules swept out. (We did not need to consider differential areas of the sphere surface and integrate over the sphere area. We are ignoring fluid swirling.)

In addition, there is a "passive" rate, dN_2/dt, given by the rate at which the randomly diffusing molecules hit the surface of the sphere. Writing the value of this rate is more difficult and the upper part of Figure 3.8 does not help much. How can we combine a diagram and intuition to give us direction in calculation of dN_2/dt? We presume that the concentration of fuel molecules is c far from the sphere, but because molecules near the sphere get absorbed, the concentration near the sphere is smaller. This gradient of concentration would set up a flux of molecules moving inward toward the sphere. The molecules would probably move, on average, radially toward the center of the sphere. We have drawn this radial flux in the lower panel of Figure 3.8. This panel also shows additional spherical surfaces perpendicular to the inward moving molecule. If there is inward, radial motion of molecules, the same number of molecules/s should pass through every surface; otherwise, molecules would build up or disappear at one of the spheres. Because the surface areas are $4\pi r^2$, where r is the radius of any of these (imaginary) surfaces, the flux (particles per unit surface area per second) should decrease as $1/r^2$. The diagram also suggests one might want to calculate the rate at which molecules enter and leave the spherical shell volume between two surfaces—they should be equal. We are going to leave this diffusion calculation for a later chapter, but the point here is that diagrams drive the calculation: there is no magic formula that is to be sought for this case, though the utility of the final equation makes it worth recording. Knowing only the final formula for a spherical cell is of little help when the next cell encountered is rod shaped. Knowing the drawing tools to use to figure out the next case is more difficult than formula memorization, but it is, in contrast, of great use.

3.6.2 Differentials in Two and Three Dimensions

In physics we often see diagrams of large objects that are broken up into small pieces—differential length (dx), area (dA) or volume elements (dV). Figure 3.9 shows such volume elements in cylindrical and spherical coordinates; corresponding elements in Cartesian coordinates are, of course, simple cubes of volume $dxdydz$. Forces and torques, for example, can often be written in terms of some functional form (alternatively, the numerical value) valid at the location of the element. To determine the net effect on the entire object, the differential force elements dF are integrated over the entire object. Fluxes or intensities of particles or radiation hitting a surface must also often be analyzed in terms of differential fluxes hitting differential area elements. The power per unit surface area of sunlight hitting the Earth's surface as a function of latitude and local slope of the land or of light hitting the leaves on a tree are examples. Figure 3.10a shows how a diagram for this example might start out: sphere for Earth, parallel rays of light coming from the sun, small differential area marked on the surface of the sphere. The next step is to set up the mathematical strategy for calculating the power/area. The approach using differentials allows the gradual determination of the correct mathematical expression. First, differential power dP hitting the differential area dA must be proportional to dA, with a proportionality factor that depends on the angles of longitude and latitude: $dP = f(\theta,\phi)\,dA$, where $f(\theta,\phi)$ is a function that describes (as a function of θ and ϕ) the angle at which light rays strike

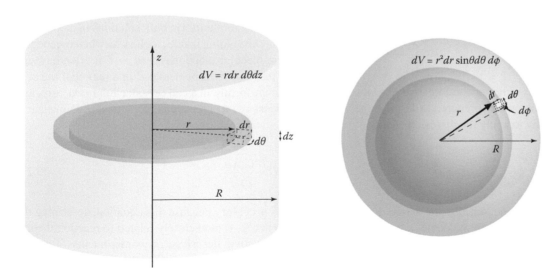

FIGURE 3.9
Differential volume elements in cylindrical and spherical coordinates.

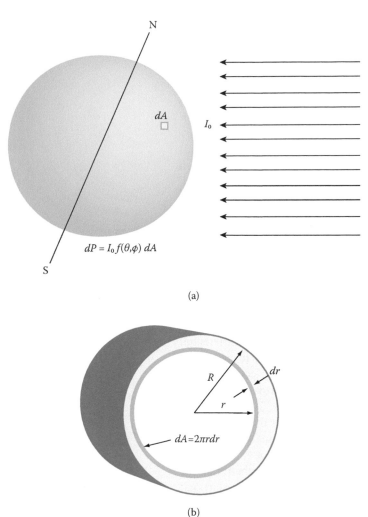

FIGURE 3.10
(a) Diagram for sunlight hitting the Earth. Light strikes a differential area dA located on the surface of the sphere. The corresponding differential power dP depends on the incident light intensity and the angle of incidence, which depends on the angular coordinates (θ,ϕ) of dA, and dA. (b) Differential area diagram appropriate for transport down a cylinder. The shown differential area assumes that flow is cylindrically symmetric, independent of θ.

An exception to this is when the reaction volume being measured is limited by the measuring device. A laser beam used to measure a reaction may, for example, be focused to a volume of a few cubic microns, throughout which reactant concentrations may be constant.

the surface, or equivalently, the effective area the local surface presents perpendicular to the light rays. Three more problems then remain: (1) determination of $f(\theta,\phi)$, (2) writing dA in terms of $d\theta d\phi$ and sphere radius, and, if needed, (3) integration of dP over the desired portion of the sphere surface, which might be the state of Iowa or the farmland of Ukraine during the middle of July. We will not do this integration here, as our purpose is to develop diagrammatic and mathematical strategies and tools, not catalog results for particular cases.

The previous example of sunlight hitting the Earth's surface seems a bit out of the area of "biophysics," until one realizes this is a fundamental issue of photosynthesis, which underlies all life on Earth. Nevertheless, an example of the use of differentials for a case more immediately tied to the physics of a biological system might be the flow down a cylindrical tube. In fact, such flow has great biophysical significance, as it applies to macroscopic blood flow in vessels, to movements of fluids and solutes along microscopic, cylindrically shaped cells, and to molecular motion along the outside and inside of microtubules (excluding molecular motors). Figure 3.10b shows how differential geometry might be set up for this case. In the case of cylindrical symmetry, the fundamental differentials dr and $d\theta$ can be reduced to a differential area, $dA = \int_0^{2\pi} d\theta \, rdr = 2\pi rdr$, because there is no dependence on angle. The flow then needs to be expressed as a function of distance from the cylinder's center, as do the boundary conditions. See Section 3.7.

3.6.3 Derivatives and Partial Derivatives

Equations with a single, simple derivative often occur in chemical reaction theory, as in Equation 3.18. In this and subsequent reaction equations, dependence only on time is expected, so only a derivative with respect to time occurs: $\frac{d[C]}{dt} = -k[A]\cdot[B]$. The symbol d is a full derivative and reflects that there is only one independent variable. The full derivative might be appropriate for a reaction taking place in a uniform, bulk solution. However, when does one encounter "uniform bulk solution" in biophysics? As we have seen, the volume inside a cell is very heterogeneous. For a typical bimolecular reaction $A + B \rightarrow C$ it often happens that one of the reactants will be depleted in a region of the cell.

It is rather uncommon that reactants are distributed uniformly in a cell or subcellular structure, and we must be prepared to describe the spatial dependence of most reactions. We will spend time learning to apply the diffusion equation:

$$\frac{\partial C(x,t)}{\partial t} = -D\frac{\partial^2 C(x,t)}{\partial x^2} \tag{3.27}$$

The appearance of the partial derivative symbol ∂ is necessary because this equation, describing the movement of molecules whose local concentration is $C(x, t)$ (and closely related to reaction theory) depends on both position and time. The partial derivative, described in words, means "Take the time derivative of the function $C(x, t)$, holding the value of x constant." Though even Equation 3.27 applies only to a large ensemble of molecules, whose concentration does not vary *too* rapidly with position or time, the presence of two partial derivatives is a first attempt to deal with real, restricted populations of molecules, which behave differently in different parts of their "container."

3.6.4 Change of Variables and Jacobians

The two most common changes of variables that occur in physics courses are position, $x \to$ (position)$^{-1}$, k, and wavelength, $\lambda \to$ frequency, f, with relations

$$k = 1/x \tag{3.28}$$

$$v = \lambda f \tag{3.29}$$

What do we do when given a continuous function of one of the variables, say, $I(f)$, but we must rewrite the function in terms of the variable λ? The original function might be, for example, a theoretical expression for light intensity as a function of frequency, $I(f)$, while we have a measured spectrum as a function of wavelength, $F(\lambda)$. How do we compare these two spectra? If we choose to convert the measured spectrum to a function of f, we first simply associate the intensity for each wavelength with the corresponding frequency, $f = v/\lambda$. Now we have an F as a function of f. But this substitution is not enough; it is not even correct, if we intend to plot the value of f on the horizontal axis (abscissa). We have to multiply this "reassociated" function by the differential factor $d\lambda/df$. Why? The fundamental *physical* reason is conservation of photons or energy: you cannot end up with a different number of photons or different energy when you change variables. This conservation is mathematically expressed as

$$F(\lambda)d\lambda = F(\lambda(f))\frac{d\lambda}{df}df = I(f)df \tag{3.30}$$

The relation between $I(\lambda)$ and $F(f)$ is then

$$I(f) = F(\lambda(f))\frac{d\lambda}{df} = F(\lambda(f))\frac{d}{df}\left(\frac{v}{f}\right) = -\frac{v}{f^2}F(\lambda(f)) \tag{3.31}$$

If $F(\lambda) = F[\lambda(f)]$ has been measured by a spectrometer, the amplitude at each measured point should be multiplied by the factor $v/f^2 = \lambda^2/v$ to convert to a spectrum as a function of frequency. If, on the other hand, you decided to convert an analytic expression for $I(f)$ to a function of λ, to be compared directly with the measured spectrum, you would multiply $I(f)$ by $df/d\lambda$. The expressions above can also be integrated over all values of f and λ (the area under the curves), corresponding to the fact that the total number of photons (or total energy) is the same, whether you express quantities as a function of frequency or wavelength.

What if the function is of two or three variables? The most familiar examples are functions of spatial coordinates. How do we convert from functions of (x, y) to functions of (r,θ)? The most general form is

$$\int dx \int dy F(x, y) = \int dr \int d\theta\, F(x(r,\theta), y(r,\theta))\frac{\partial(x, y)}{\partial(r,\theta)} = \int dr \int d\theta\, G(r,\theta) \tag{3.32}$$

The expression in the middle of Equation 3.32 is the mathematical way of expressing your desire to rewrite $F(x, y)$ in terms of the variables (r, θ), but you cannot merely change every x you see to $x = r \cos \theta$ and $y = r \sin \theta$. The rightmost term (the "Jacobian") in the middle integral may be unfamiliar to some of you. For the rectangular–polar coordinate relation it is calculated as follows:

$$\frac{\partial(x, y)}{\partial(r,\theta)} = \begin{vmatrix} dx/dr & dx/d\theta \\ dy/dr & dy/d\theta \end{vmatrix} = \begin{vmatrix} \cos\theta & -r\sin\theta \\ \sin\theta & r\cos\theta \end{vmatrix} = r, \tag{3.33}$$

The vertical bars indicate the determinant. The result is the well-known factor for integrals in polar coordinates. Three-dimensional functions can be treated likewise, but with a 3×3 determinant, resulting in an $r^2 \sin \theta$ factor.

Do we really need this mathematical horsepower to handle biological systems? Why not just remember how to do integrals properly in polar and spherical coordinates? You should, in fact, remember integral formulas, but the issue here is a framework for dealing with any situation where you need to change from one set of (continuous) independent variables to another. As just shown, this conversion happens all the time in spectroscopy, which is one of the most important classes of experimental methods in the biological and biomedical sciences today, as well as in the sciences in general. You should also expect such mathematics to occur when you need to express a probability (population) distribution in terms of speed, $P(v)$, or a variable related to speed, such as $E = \frac{1}{2}mv^2$ or $p = mv$.

3.7 DIFFERENTIAL EQUATIONS

Why do we need them? Differential equations are ubiquitous in the quantitative sciences, but, in principle, should be even more common in biosystems. Why? Because the primary quality exhibited by living things is change with time and position: life moves. Differentials and differential equations are then a natural way to describe these changes. Though an important feature of the biological microworld is that changes are not always continuous—molecular motors, for example, may occupy discrete positions—the connection to the macroscopic world, or at least the extrapolation thereto, is often through continuous, differential approximations. We consider below some of the differential equations that commonly apply to biosystems, focusing on linear equations, where derivatives and terms involving the function occur linearly.

The following short discussions of differential equations focuses on our "guess the solution and try it" method, rather than general mathematical rigor. One should also look for familiar differential equations, such as occurs for the harmonic oscillator. These approaches usually suffice in biophysics.

3.7.1 First Order, Linear

$$\frac{df(t)}{dt} + bf(t) = g(t) \tag{3.34}$$

A more general case would replace constant b with a f unction, $b(t)$. Functional dependence on t may also be replaced with dependence on x, and d/dt replaced by d/dx.

If $g(t) = 0$, the equation is referred to as the homogeneous equation. The general approach to the inhomogeneous equation is to find the solution to the homogeneous, then add a particular solution to the inhomogeneous equation. Knowing that the derivative of an exponential is a constant times the exponential, a natural guess for the homogeneous solution is $f(t) = Ae^{kt}$. Because $\frac{d}{dt}\left(Ae^{kt}\right) = Ake^{kt}$, we have

$$kAe^{kt} + bAe^{kt} = kf + bf = 0 \tag{3.35}$$

implying $k = -b$ and

$$f(t) = Ae^{-bt} \tag{3.36}$$

There are varieties of ways of "guessing" a particular solution to the case where $g(t) \neq 0$, but let's take a constant $g(t) = a$. The simplest particular solution is when $df/dt = 0$, so $f(\text{particular}) = a/b$. So the general solution to Equation 3.34 with $g(t) = a$, is

$$f(t) = \alpha e^{-bt} + \frac{a}{b} \tag{3.37}$$

Note here one undetermined constant, α, occurs in the solution, so the value of f or df/dt must be known at some time t_0 in order to determine α. If, for example,

$$f(0) = 0: \quad f(t) = \frac{a}{b}\left(1 - e^{-bt}\right) \tag{3.38}$$

If

$$\left(\frac{df}{dt}\right)_{t=0} = -R: \quad f(t) = \frac{a}{b}\left(\frac{R}{a}e^{-bt} + 1\right) \tag{3.39}$$

Equation 3.38 corresponds to a function whose initial value is known; Equation 3.39 corresponds to a function whose initial derivative—a rate—is known.

What biophysical cases might these equations and solutions apply to? We first look to classical chemical reaction theory for the simplest reaction, $A \rightarrow B$, with corresponding equation $\frac{dA(t)}{dt} = -kA(t)$. This expression is Equation 3.34 with $g(t) = a = 0$ and the solution is $A(t) = A_0 e^{-kt}$. $B(t) = 1 - A(t) = A_0\left(1 - e^{-kt}\right)$ then corresponds to the solution in Equation 3.38.

3.7.2 First Order, Nonlinear (but Not the Most General Nonlinear)

$$\frac{df}{dt} + bf + cf^2 = g(t) \tag{3.40}$$

Let's consider the form in which this type of differential equation commonly occurs in chemistry and biophysics: the so-called bimolecular reaction, $A + B \rightarrow C$ considered earlier. If this reaction occurs only to the right and is governed by "mass action," then $\frac{dC(t)}{dt} = kA(t) \cdot B(t)$. This equation is difficult to treat in general, so let's put in some of the necessary initial conditions: suppose there is less of B than A, and call the initial values of each A_0 and B_0. The only time an additional C appears is when an A reacts, so $C(t) = A_0 - A(t)$: $\frac{dC}{dt} = \frac{d}{dt}(A_0 - A) = -\frac{dA}{dt} = kA \cdot B$. When a C appears, a B must also have disappeared, so $B = B_0 - (A_0 - A)$, and

> A more general mathematical case would have $b(t)$ and $c(t)$.

$$\frac{dA}{dt} = -k(B_0 - A_0)A - kA^2 \tag{3.41}$$

This form now looks like Equation 3.40, with $g(t) = 0$, $b = -k(B_0 - A_0)$ and $c = -k$.

Let's make things simple at first and assume $A_0 = B_0$, which implies $A(t) = B(t)$. Equation 3.41 becomes

$$\frac{dA}{dt} = -kA^2 \tag{3.42}$$

Can we guess a solution to this equation? Probably not, but a bit of rearrangement leads to

> Recognized because we can get all terms with $A(t)$ on the left side, and all terms with t on the right.

$$\frac{dA}{A^2} = -kdt, \tag{3.43}$$

which can be directly integrated, leading to $-\frac{1}{A} = -kt - \frac{1}{A_0}$, where we have chosen the integration constant such that the solution gives $A(0) = A_0$. The solution to Equation 3.42 can then be written

$$\frac{1}{A} - \frac{1}{A_0} = kt. \tag{3.44}$$

We could likewise write $\frac{1}{B} - \frac{1}{B_0} = kt$.

In biosystems nature is unlikely to oblige our mathematical desires and set $A_0 = B_0$. What if $A_0 \neq B_0$, say $A_0 > B_0$? We expect both A and B to decrease as time proceeds, but B should go to zero before A does. This relative rate of decrease suggests we might guess a solution for the ratio, $A(t)/B(t)$,

$$\frac{B}{A} = \alpha F(t), \tag{3.45}$$

where one expects $F(t)$ to decrease with time. One can start trying simple functions for $F(t)$, e.g., $F(t) = 1/(\beta t)$, but this does not work. What should the next try be? The ratio B/A must decrease rapidly with time, as $B \rightarrow 0$. What simple function decreases rapidly with time? Exponential functions do, so let's try

$$\frac{B}{A} = \alpha e^{-\beta t} \tag{3.46}$$

Substituting into Equation 3.41 is a success and leads to

$$\frac{B(t)}{A(t)} = \frac{B_0}{A_0} e^{-k(A_0 - B_0)t} \tag{3.47}$$

What if the back reaction $A + B \leftarrow C$ must be considered? Then $\frac{dC(t)}{dt} = kA(t) \cdot B(t) - k'C(t)$ and

$$\frac{dA}{dt} = k'A_0 - [k(B_0 - A_0) + k']A - kA^2 \tag{3.48}$$

www.wolfram.com/
mathematica/;
www.maplesoft.com/.

The form is again that of Equation 3.40. We see that the difference between Equations 3.41 and 3.48 is just a constant term $k'A_0$ ($g(t) = $ constant in Equation 3.40. We might try again to simplify to the (rather unbiophysical) case, $A_0 = B_0$, but soon realize this conjecture does not simplify the math because of the $k'A$ term. Likewise, rearrangement cannot lead an easy integration. We should probably, then, make use of a computerized differential equation solver, like those contained in the programs Mathematica or Maple.

These differential equations are easy to write down, and they clearly may apply to reasonable chemical reactions, but one sees the solutions are not always easy to find for common applications in biophysics.

More generally, $b = b(t)$ and $c = c(t)$.

3.7.3 Second Order, Linear

$$\frac{d^2 f}{dt^2} + b\frac{df}{dt} + cf = g(t) \tag{3.49}$$

First, note when a second-order differential equation is encountered that two characteristic time constants can occur. Second, this equation is a more general form of the most familiar differential equation from first-year physics, the harmonic oscillator, if $b = 0$ and $c = \omega_0^2$, where $\omega_0^2 = \frac{k}{m}$, m is mass and k is force constant. The value $g(t)$, if nonzero, is a driving force. Third, important examples occur with functional, rather than constant, coefficients multiplying the derivative terms.

What are some good guesses at solutions to constant coefficient cases? Exponentials, polynomials, and sinusoidal functions might be considered. If there could be two characteristic time constants, one should consider sums and products of such functions. Reasonable guesses might then be

$$f(t) = Ae^{\alpha t} + Be^{\beta t}, \ (A + Bt + Ct^2), \ (A + Bt)e^{\alpha t}, \ A\sin \omega_1 t + B\cos \omega_2 t, \ e^{\alpha t}(A\sin \omega t + B\cos \omega t) \tag{3.50}$$

The polynomial is quickly ruled out, but all of the other functions are possible solutions for various values of b and c, with $g(t) = 0$ or $g_0 \cos(\omega t)$. The reader is invited to check these.

An important case of Equation 3.49, with time derivatives replaced by spatial derivatives, $b = 0$ and $c \rightarrow c(x)$ occurs in the time-independent quantum Schrödinger equation, with the time derivative replaced by a spatial derivative:

$$\frac{d^2 f(x)}{dx^2} + c(x)f(x) = 0 \qquad (3.51)$$

Those who have studied modern physics or quantum theory recognize Equation 3.51 as the time-independent form of the general Schrödinger equation, $\frac{-\hbar^2}{2m}\frac{\partial^2 \Psi(x,t)}{\partial x^2} + V(x)\Psi(x,t) = i\hbar \frac{\partial \Psi(x,t)}{\partial t}$. We deal with this equation more seriously in Chapter 8.

3.7.4 Boundary (Spatial) and Initial (Time) Conditions

Boundary conditions on differential equations refer to spatial positions where values of functions are specified, usually by experimental arrangement or the physical structure involved. Initial conditions refer to function values assumed at time $t = 0$ or at any specified time, often $t = \infty$. Most often, initial conditions are irrelevant to a biosystem, as it is usually in a steady state. Unfortunately, steady state often hides intrinsic system response times, which can be of great importance. Initial conditions occur in a living biophysical system when it encounters some external agent that causes a sudden change. Important examples include (1) response to a stimulus, causing a sudden release of signaling molecules, further causing Ca^{2+} release and sudden voltage changes across cell membranes; (2) a rapid external temperature change; and (3) injury or sudden failure of a structure (e.g., blood vessel blockage). Of course, *in vitro* (in a test tube) experiments designed by researchers frequently employ sudden time changes in some parameter in order to measure important system response times.

First-order differential equations need one boundary (or initial) condition, corresponding to one integration constant; second-order equations need two. Two spatial and two time derivatives apply to the quantitative treatment of classical waves (e.g., light or water waves). Two spatial derivatives and one time derivative occur in diffusion and in quantum systems. Two time derivatives occur in systems governed by Newton's second law, $F_{net} = md^2x(t)/dt^2$, but as previously noted, F_{net} is often zero in biosystems, and the second derivative is 0.

3.7.5 Sketching and Guessing Solutions

Boundary conditions should first be treated to an appropriate diagram. For our level of math, this diagram often readily leads to a solution to the differential equation. Figure 3.11 shows a sketch of a one-dimensional (in space) system where values of the solution to a second-order differential equation are known at two positions, $x = 0$ and $x = x_1$. The function values are assumed to be 0 at $x = 0$ and at $x = x_1$. In sketching a possible solution to a differential equation over this interval one must start drawing lines (curves) from these known points [0, 0] and [x_1, 0]. The goal is then to determine, from the sketch and from other information known about the system and the differential equation, whether the correct connections between known points is $a1$, $b1$, $c1$, $d1$, $a2$, $b2$, $c2$, or $d2$. If, for example, you know that the function's second derivative is negative ($d^2f/dx^2 \leq 0$) everywhere between $x = 0$ and $x = x_1$, then $a1$ is the proper choice; if $d^2f/dx^2 \geq 0$, then $d2$ is correct. Curve $a1$ would also be the likely solution if one knew that the solution being sought had the lowest curvature possible. If one knew that the solution must be symmetric (antisymmetric) about the point $x = x_1/2$, choice $a1$ ($a2$) would be correct. If we knew that the slope, df/dx, was everywhere positive, we would know that what "we knew" is wrong—the diagram rules it out. If we were to then guess mathematical solutions corresponding to (one-dimensional) Figure 3.11, it is often fairly quick to round up the usual

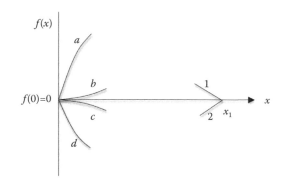

FIGURE 3.11
Sketches of possible solutions to a (second-order) differential equation where two boundary conditions apply. The goal of the sketch is to properly connect possible curves, a, b, c, d to curves 1 and 2.

suspects. Sine/cosine, polynomial, exponential functions, sums, or products thereof are all reasonable trial functions. If $x_1 = \infty$ and the function value were known to go to zero there, a great reduction in possible function choices results. These "usual suspects" are often adequate in biophysics. In Chapter 8 we will see examples where various drawn curves correspond to different allowed values of coefficients (energies) in the differential equation.

3.7.6 Steady State and Equilibrium

The difference between the steady state and equilibrium is, as noted earlier, both subtle and obvious. A system in equilibrium is a system that is also dead; a system in a steady state can be very alive. Steady state means that the first time derivative of some descriptor of a system is zero: $df/dt = 0$. There may be much going on to keep this steady state, however. In biosystems, this steady state generally corresponds to energy input and energy and work output. Just as $F_{net} = 0$ in Newton's law does not imply that all forces are zero, $df/dt = 0$ does not imply that all contributions to the value of f are zero. In Figure 3.8 (lower), for example, the density of particles may be constant in time everywhere in space, $dN(r, t)/dt = 0$, but there is still a constant flow of particles inward toward the spherical cell. Of course, we then know that some external agent or large "bath" must be constantly supplying new particles at $r = \infty$. Sometimes, as in Michaelis–Menten chemical kinetics, there may be an *experimental* steady state for one particular quantity that persists only over a certain length of time. Over a longer time scale an uninteresting equilibrium may develop. The existence of two different sorts of "steady states" does not mean our mathematical steady-state assumption, $df/dt = 0$, is incorrect, just that we have to pay attention to time scales and not measure too soon or too long.

A typical invocation of the steady-state assumption occurs in chemical reaction kinetic equations like Equation 3.18:

$$\frac{dA}{dt} = -k_f A + k_r B = 0, \tag{3.52}$$

which allows for easy calculation of the steady-state ratio of B/A:

$$\left(\frac{B}{A}\right)_{ss} = \frac{k_f}{k_r} \tag{3.53}$$

Don't let this easy math fool you, however. The steady-state result may be straightforward to emulate in the test tube but may not correspond to any relevant state of a biosystem.

Equilibrium, in contrast to steady state, is not really a mathematical term, even though it implies $df/dt = 0$. A system in equilibrium is assumed to have started from some initial condition, with given fixed values of matter and energy, and then evolved under governing physical and chemical law to an equilibrium state at $t = \infty$. Though no human can measure the value of $f(t)$ until $t = \infty$, it is usually only necessary to measure over a time long compared to some system response time. As we will see later, equilibrium is defined by a free energy minimum, not by $df/dt = 0$. Having said this, does the steady-state ratio in Equation 3.53 correspond to equilibrium? The answer is normally "Yes," if this situation is set up by placing initial amounts of A and B in a sealed container under some fixed external conditions, where they are free to move and encounter other particles, and allowing sufficient time for equilibrium to be reached. There are many other ways to achieve a steady state, $dA/dt = 0$, which is not equilibrium. An example common in biology is where a steady external supply of A is provided to the "container," while B has a route to disappear from the system:

$$\frac{dA}{dt} = -k_f A + k_r B + R_A$$
$$\frac{dB}{dt} = k_f A - (k_r + k_b)B \tag{3.54}$$

This system can also have steady states—$dA/dt = 0$ and $dB/dt = 0$—but it is neither isolated nor at equilibrium.

3.8 DISTRIBUTIONS

Why do we need them? You may have seen before that a gas-filled container will have molecules with velocities and energies spread according to the Maxwell–Boltzmann distribution (Figure 3.12, corresponding kinetic energy distribution in Figure 3.12b). Distributions are a fact of life dictated by fundamental statistical mechanics. We see in Chapter 15 that an ensemble of molecules undergoes chemical reactions at a rate governed primarily by their energies—more energetic molecules react more rapidly. In most cases of bulk chemical reactions, this dependence of reaction rate on an individual molecule's energy—the distribution—can be ignored. The measured rate is an average over the entire molecular population and can be described by one temperature-dependent rate "constant" (e.g., k values in the immediately preceding equation). This one-parameter description is usually justified because when the most energetic molecules react, they are rapidly replaced by other molecules that gain additional energy from interactions (collisions) with solvent (or other) molecules, governed by the temperature. So, when dealing with reactions, can we discard distributions and work with rate constants only? Absolutely not. Rate constants are not constant at all but rather dependent on a variety of environmental parameters. Chief among these parameters is temperature. The temperature dependence can only be properly described in reference to the distribution of (usually) the accessible energy states of the particles. See Arrhenius rate.

Spectroscopy is perhaps the most widely used experimental method in science today. Four of spectroscopy's biggest advantages are (1) its noninvasive nature—a living organism, cell or semiconductor nanocircuit can be probed without damage; (2) its sensitivity—a single photon, reflecting the properties of a single particle, can sometimes be detected; (3) its tunability—by choosing the wavelength a wide variety of system properties can be probed; and (4) its response time—from 10^{-15} s to almost infinite, limited by the apparatus. An absorption *spectrum* is fundamentally a distribution function representing the *probability* that a photon of wavelength between λ and $\lambda + d\lambda$, will be absorbed by a system. Similar statements can be made about emission and activation spectra.

Probability and its associated distribution functions are fundamental features of matter, whether a "perfect" crystal of silicon or the molecular motors creating force and motion in muscles and flagella. We ignore distributions at the danger of misunderstanding fundamental properties of the systems we desire to study.

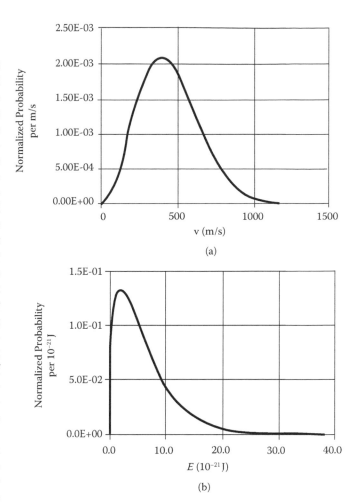

FIGURE 3.12
(a) Maxwell–Boltzmann speed distribution, $P(v)v^2dv = A\exp-\left(\frac{mv^2}{2k_BT}\right)v^2dv$, the probability of observing a speed between v and $v + dv$. The exponential is the Boltzmann factor and v^2dv comes because the direction-independent speed can have its corresponding velocity vector tip anywhere within the spherical shell volume $4\pi v^2dv$ (4π is absorbed into the constant A). In chemical reactions the most energetic molecules react fastest but are rapidly replaced by molecules that gain energy through collisions with solvent or gas molecules. (b) Corresponding kinetic energy distribution.

3.8.1 Large Populations and Single Particles

The previous section would seem to imply that distributions apply when the number of particles is large, perhaps approaching Avogadro's number, 6.02×10^{23}. The idea that a *smooth* distribution accurately describes the population of particles does require a large number of particles, or else a smaller number, even as small as 1, as long as adequate time is allowed by the experiment or biological process for the smaller number of particles to explore (pass through) the majority of the allowed and accessible states. These two extremes—ensembles of large numbers of particles whose properties are measured during a brief snapshot in time or one particle observed for an infinite length of

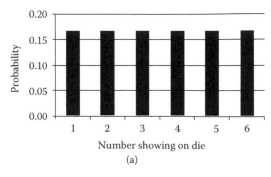

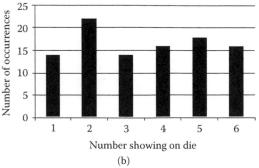

FIGURE 3.13

Intrinsic and measured probability distributions: (a) A uniform, discrete probability distribution describing the intrinsic probability of the roll of a die. All probabilities equal 1/6. (b) A measured distribution of die throws resulting from 100 throws. The measured distribution is clearly different from the "intrinsic," showing that experimental determination of the intrinsic probability distribution demands a large number of measurements (compared to 100, in this case).

time—constitute the *ergodic* hypothesis. A system is ergodic with respect to a measured property p if the distribution function describing that property is the same whether the ensemble average or the time average is measured:

$$\text{Ergodic hypothesis: } \langle p \rangle_{\text{time}} = \langle p \rangle_{\text{ensemble}} \tag{3.55}$$

We will see that biological systems are frequently non-ergodic, because the number of particles (e.g., protein molecules) is too small or there is insufficient time for a molecule to explore even a small fraction of the accessible states. This complication is often not a limitation imposed by our experimental methods, but by nature herself. Furthermore, the lack of ergodicity is not a limitation at all: nature has no need to explore all possibilities. If it did, the time needed to do so would not allow organisms to live.

3.8.2 Intrinsic versus Measured Probabilities

A die (singular of dice) is rolled once. Possible values showing on the top of the die are 1, 2, 3, 4, 5, and 6. Assuming the die is honest, we would say the result of a throw is equally likely to be any of the possible numbers—the probability of each is 1/6. This "intrinsic" probability distribution would be uniform. See Figure 3.13a. Intrinsic probability distributions are sometimes calculated theoretically, as above, but when theory is absent, measurement of probabilities needs to be done. In Figure 3.13b, 100 throws of a die have been done in order to estimate the probability distribution. Significant differences are obvious between a and b. Figure 3.13b is indeed a probability distribution (or would be if the ordinate were divided by $N = 100$ throws), but it is quite far from the intrinsic distribution. Which is closer to reality? In a biosystem with 100 "identical" particles, Figure 3.13b, or rather a remeasurement of Figure 3.13b each time a system is observed, is more appropriate. There would always be a somewhat different distribution of results each time an observation was made. If 10^{23} particles were in the system, Figure 3.13a would be more appropriate. We will see later that an estimate of the number of particles (or repeated measurements) needed for the intrinsic probability to accurately describe an experiment is when $\sqrt{N}/N \ll 1$. In the example given $\sqrt{100}/100 = 0.1$ is, depending on desired accuracy, barely adequate.

3.8.3 Discrete Distributions: Averages and Deviations

Suppose a system can be described by a quantity E, which can take on discrete, exact values, possibilities indicated by integral indices $i = 1,2,3...$, and that no intermediate values are possible or allowed. Suppose that E is then measured N times and that result E_i is obtained n_i times. The experimental probability P_i of result i is then

$$\text{Probability}: \quad P_i = \frac{n_i}{N} \tag{3.56}$$

and

$$N = \sum_{i=1}^{m} n_i, \tag{3.57}$$

where m is the number of possible values of E_i,

$$\text{Normalization}: \quad \sum_{i=1}^{m} P_i = 1 \tag{3.58}$$

Equation 3.58 represents the *normalization condition*: probabilities should add to 1. The function P_i, $i = 1, 2...m$ is the discrete distribution function and P_i is a probability: it has no units (dimensions).

The (weighted) average (mean, expectation value) of any quantity subject to a discrete probability distribution is calculated as

$$\langle f_i \rangle = \frac{n_1 f_1 + n_2 f_2 + \cdots}{n_1 + n_2 + \cdots} = \frac{\displaystyle\sum_{i=1}^{m} n_i f_i}{\displaystyle\sum_{i=1}^{m} n_i} = \frac{\displaystyle\sum_{i=1}^{m} n_i f_i}{N} = \sum_{i=1}^{m} P_i f_i \tag{3.59}$$

(Note that P is assumed to be normalized and the triangular brackets indicate the average of f_i is taken over all values of i and that there is no longer any dependence on i.) The variance, which describes the average deviation squared of the quantity from its mean, is

$$\text{Variance} \equiv \left\langle \left(f_i - \langle f_i \rangle \right)^2 \right\rangle = \left\langle \left(f_i \right)^2 \right\rangle - \langle f_i \rangle^2 \tag{3.60}$$

The standard deviation is the square root of the variance:

$$\text{Standard deviation} \equiv \sqrt{\text{Variance}} \equiv \sqrt{\left\langle \left(f_i - \langle f_i \rangle \right)^2 \right\rangle} = \sqrt{\left\langle \left(f_i \right)^2 \right\rangle - \langle f_i \rangle^2} \tag{3.61}$$

These definitions of deviations from the average are equally weighted for values of f_i higher or lower than the average, and they are weighted by the square of the difference from the mean. There are other defined deviations from the mean, such as the absolute value of the difference from the mean, but the above is generally the best deviation to use in physical situations. The reader is advised to become familiar with both calculator- and spreadsheet-based (e.g., Microsoft Excel) computations of averages and deviations. The calculator will prove useful for checking spreadsheet computations and the latter provides convenient graphing capability.

3.8.4 Continuous Distributions

In continuous distributions a quantity f is a continuous function of some variable x, $f = f(x)$, with x values varying form some minimum to a maximum, often $-\infty$ to $+\infty$ or 0 to ∞. However, a crucial difference from the discrete distribution is that the probability of a result is given by $P(x)dx$, and that $P(x)$ has units of probability per unit x (probability per meter, if $x = $ distance). The summations used in Equations 3.59 through 3.61 must be replaced by integrals:

$$\langle f(x) \rangle = \frac{\displaystyle\int_{x_{min}}^{x_{max}} P(x) f(x) dx}{\displaystyle\int_{x_{min}}^{x_{max}} P(x) dx} \tag{3.62}$$

Do not mistake significant figures or experimental rounding for a discrete distribution. For example, spectrophotometer measurement of absorption versus photon wavelength may record data points only every nanometer, but this simply means that each recorded data point $A(\lambda)$ represents an average over the interval $\lambda \rightarrow \lambda + \delta\lambda = \lambda + 1$ nm. The spectrum is still, in principle, a continuous function of λ, unless we are measuring the discrete (line) spectra of atoms or rotational or vibrational line spectra of molecules, where only discrete values of wavelength may be allowed.

The normalization condition for $P(x)$ is

$$\int_{x_{min}}^{x_{max}} P(x)dx = 1 \tag{3.63}$$

If P is a function of two or more variables, two or three integrals must be done. See Sections 3.5 and 3.6.

It is crucial to reiterate that $P(x)$ is not a probability: the probability that P takes some value at any particular $x = x_1$ is always zero for a continuous distribution. The probability that P takes on an average value of $P(x)$ for values of x between x and $x + dx$ is what is defined by the continuous probability distribution.

3.8.5 Distribution Functions

3.8.5.1 Uniform

A uniform, discrete distribution has been presented in the previous section. Probabilities for each possible result of a measurement, $f_1, f_2 \ldots f_m$, are equal. The average value of f would be

$$\langle f_i \rangle = \frac{nf_1 + nf_2 + \cdots + nf_m}{n + n + \cdots + n} = \frac{n\sum_{i=1}^{m} f_i}{n\sum_{i=1}^{m} 1} = \frac{\sum_{i=1}^{m} f_i}{m} \text{ and the average value of } f^2,$$

$$\langle f_i^2 \rangle = \frac{nf_1^2 + nf_2^2 + \cdots + nf_m^2}{n + n + \cdots + n} = \frac{n\sum_{i=1}^{m} f_i^2}{n\sum_{i=1}^{m} 1} = \sum_{i=1}^{m} \frac{1}{m} f_i^2 = \sum_{i=1}^{m} P_i f_i^2.$$

Remember two important facts: (1) m is the number of different possible values of f ($m = 6$, in the case of the die) and (2) these average and variance values only apply if a very large number of measurements are done. In the case of 100 rolls of the die, each possible result is not necessarily rolled $n_i = n$ times, and Equation 3.59 must be used directly.

For a uniform, continuous distribution, $P(x) = A = $ constant, over the interval $x = a$ to $x = b$, the average value of a function

$$\langle f(x) \rangle = \frac{\int_{a}^{b} P(x)f(x)dx}{\int_{a}^{b} P(x)dx} = \frac{\int_{a}^{b} Af(x)dx}{\int_{a}^{b} Adx} = \frac{\int_{a}^{b} f(x)dx}{(b-a)} \tag{3.64}$$

The average value of $f(x) = x$, for instance, is $(b + a)/2$.

3.8.5.2 Binomial

The binomial distribution in physics is usually applied to a system of magnetic spins (spin ½), where spin can be either of two states, up or down. The general binomial distribution can be applied to any two-state system, however. We denote the (intrinsic) probabilities as $P(\text{up}) = p$ and $P(\text{down}) = q$, with normalization condition

$$p + q = 1. \tag{3.65}$$

If we now take an ensemble of N spins, we may want to know, for example, the probability that n of the spins are "up" and $N - n$ are "down," with the difference

$$(\text{number up}) - (\text{number down}) = 2n - N \equiv m \tag{3.66}$$

Note the use of m differs here from previous sections.

For the case $p = q = \frac{1}{2}$, the binomial distribution is

$$P(m) = \frac{N!}{\left(\dfrac{N+m}{2}\right)!\left(\dfrac{N-m}{2}\right)!}\left(\frac{1}{2}\right)^N \tag{3.67}$$

For equal numbers up and down,

$$P(0) = \frac{N!}{\left(\dfrac{N}{2}\right)!\left(\dfrac{N}{2}\right)!2^N} \tag{3.68}$$

We see from Equation 3.68 that for large values of N, these calculations produce very small values of $P(0)$, as well as all $P(m)$ values, because the latter are even smaller:

N	$P(0)$
2	0.500
4	0.375
6	0.312
8	0.273
10^4	0.008

(You will find that calculators and spreadsheets are often incapable of calculating factorials of numbers higher than about 170, so we will need to find alternate ways to calculate values from Equations 3.67 and 3.68.)

If p and q are not equal,

$$P_{\text{binomial}}(m) = \frac{N!}{\left(\dfrac{N+m}{2}\right)!\left(\dfrac{N-m}{2}\right)!}\, p^{\frac{N+m}{2}} q^{\frac{N-m}{2}} \tag{3.69}$$

Figure 3.14 shows a graph of a binomial distribution for $N = 10$ and $p = 0.65$.

3.8.5.3 Gaussian

The (continuous) normalized Gaussian distribution in one dimension is

$$P_{\text{Gauss}}(x)dx = \frac{1}{\sigma\sqrt{2\pi}}e^{-(x-x_0)^2/2\sigma^2}dx \tag{3.70}$$

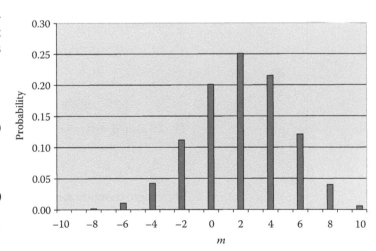

FIGURE 3.14

Binomial distribution. The graph plotted is for $N = 10$ and $p = 0.60$. For $N = 10$, that maximum possible value of m is 10, which has a probability of 0.006. The center of the distribution is displaced to the right of 0 because $p > 1/2$. Note also that certain values of m are not allowed, odd values of m in the case of $N =$ even. This graph would also apply to one-dimensional random walks if $p =$ probability for a right step, $q =$ left probability, and $p + q = 1$.

This distribution is symmetric about its maximum at $x = x_0$. At $x = x_0 \pm \sigma$ the value of $P(x)$ is $e^{-1/2} = 0.6065 \times P(x_0)$. The value σ is also the standard deviation of the Gaussian distribution. In two dimensions,

$$P_{\text{Gauss}}(x, y)dxdy = \frac{1}{\sigma_x \sigma_y 2\pi} e^{-(x-x_0)^2/2\sigma_x^2} e^{-(y-y_0)^2/2\sigma_y^2} dxdy \tag{3.71}$$

If $\sigma_x = \sigma_y$ and $x_0 = y_0 = 0$, this distribution can be written in polar coordinates, independent of angle θ, using Jacobian methods from an earlier section:

$$P_{\text{Gauss}}(x, y)dxdy = P_{\text{Gauss}}(r)2\pi rdr = \frac{r}{\sigma^2} e^{-r^2/2\sigma^2} dr$$

The Gaussian distribution results from the binomial distribution if N, the number of spins, is very large[5]:

$$P_{\text{binom}}^{N\ \text{large}}(m) = \frac{N!}{\left(\dfrac{N+m}{2}\right)!\left(\dfrac{N-m}{2}\right)!} p^{\frac{N+m}{2}} q^{\frac{N-m}{2}} \cong \frac{1}{\sqrt{2\pi Npq}} e^{-\left(\frac{m}{2}-(p-1/2)N\right)^2 / 2Npq} \tag{3.72}$$

An alternate version of Equation 3.72 can be written in terms of the number n of spins in the "up" position:

$$P_{\text{binom}}^{N\ \text{large}}(n) = \frac{N!}{(n)!(N-n)!} p^n q^{N-n} \cong \frac{1}{\sqrt{2\pi Npq}} e^{-(n-Np)^2/2Npq} \tag{3.73}$$

It can be seen from Equation 3.73 that the maximum probability for the number of "up" spins is at $n = Np$. The average value of n is the same, $\langle n \rangle = Np$. Both distributions show that $\sigma^2 = Npq$. If we now consider the probability for having a value of n anywhere between n and $n + \Delta n$, $P(n + \Delta n) = P(n_1) + P(n_2) + P(n_3) + \ldots = P(n)\Delta n$, where the interval Δn contains a large number of n values but is small enough such that the value of $P(n)$ does not change much, we see that the continuous form of Equation 3.73 can be written as

$$P_{\text{binom}}^{N\ \text{large}}(n)dn \cong \frac{1}{\sqrt{2\pi Npq}} e^{-(n-Np)^2/2Npq} dn.$$

3.8.5.4 Poisson

The Poisson distribution can also be derived from the binomial in the limit

$$p \ll 1 \text{ or } n \ll N \tag{3.74}$$

This approximation applies when one of the two states is intrinsically very improbable. The result is[5]

$$P(n) = \frac{\lambda^n}{n!} e^{-\lambda}, \quad \lambda \equiv Np \tag{3.75}$$

3.8.5.5 Maxwell–Boltzmann velocity distribution

One-dimensional velocity:

$$P(v_x)dv_x = \left(\frac{m}{2\pi k_B T}\right)^{1/2} e^{-mv_x^2/2k_B T} dv_x \tag{3.76}$$

Three-dimensional velocity:

$$P(v_x, v_y, v_z)dv_x dv_y dv_z = \left(\frac{m}{2\pi k_B T}\right)^{3/2} e^{-m(v_x^2+v_y^2+v_z^2)/2k_B T} dv_x dv_y dv_z \tag{3.77}$$

These velocity distributions are Gaussian and centered about 0, so clearly the average velocity is $\langle \bar{v} \rangle = \langle (v_x, v_y, v_z) \rangle = \langle (0,0,0) \rangle$.

3.8.5.6 From velocity, to speed, to momentum, to kinetic energy distributions

The normalized Maxwell–Boltzmann *speed distribution* in three dimensions has already been presented in Figure 3.12a. The equation behind this graph is

$$P(v)dv = 4\pi \left(\frac{m}{2\pi k_B T}\right)^{3/2} e^{-mv^2/2k_B T} v^2 dv \tag{3.78}$$

which can be derived using methods discussed in Section 3.5. The distribution of (absolute) momentum, $p = mv$, can be found

$$P(p) = P(v)\frac{dv}{dp} = \frac{4\pi}{m^3}\left(\frac{m}{2\pi k_B T}\right)^{3/2} e^{-p^2/2mk_B T} p^2 \tag{3.79}$$

The kinetic energy, $K = 1/2\, mv^2$, distribution function is

$$P(K) = P(v)\frac{dv}{dK} = \frac{4\pi\sqrt{2K}}{m}\left(\frac{1}{2\pi k_B T}\right)^{3/2} e^{-K/k_B T} \tag{3.80}$$

3.9 PROBLEM-SOLVING

PROBLEM 3.1 TAYLOR SERIES

Write a three-term (nonzero terms) Taylor series expansion for cos (*kx*), where $k = 10^7$ m^{-1}, around the point $kx = 2\pi$. Determine the % error of the approximation at $kx = 1.7\pi$, 1.9π, 2.1π and 2.3π.

PROBLEM 3.2 DIFFERENTIAL EQUATIONS AND RATES

Assume the differential Equation 3.20 and its solutions, Equations 3.21 and 3.22, apply to a reaction $A \underset{k_r}{\overset{k_f}{\rightleftharpoons}} B$, and that $A(0) = 2.00$ mM (millimolar, 10^{-3} molar), $B(0) = 0$, $k_f = 5.00 \times 10^5 \, s^{-1}$ and that A is measured to be 0.60 mM at $t = 30 \, \mu s$.

1. Determine the values of k_r and A_∞. Feel free to use the "guess and try" technique. (Hint: start with values of k_r not far from that of k_f.)
2. What is the characteristic time constant of the exponentially decaying part of the solution (τ in the term $e^{-t/\tau}$).

PROBLEM 3.3 BINOMIALS AND GAUSSIANS

Check the approximation in Equation 3.73. Use Microsoft Excel or a graphing program to plot the binomial and corresponding Gaussian distribution for $N = 10$ and for $N = 100$. Use $p = 0.60$ and put plots on the same graph.

PROBLEM 3.4 OPTICAL SPECTRA

The Planck radiation law for blackbody radiation can be written $F(\lambda, T) = \frac{2\pi c^2 h}{\lambda^5} \frac{1}{e^{hc/\lambda kT} - 1}$, where c is the speed of light, h is Planck's constant, λ is the wavelength, k is the Boltzmann constant, and T is the absolute temperature in kelvin.

1. What are the units of this expression? Show that the units are consistent with intensity (power per m^2) per unit wavelength interval.
2. Calculate the intensity at $T = 310$ K in the wavelength range $\lambda = 10 \pm 0.01 \, \mu m$.
3. Use Microsoft Excel or another spreadsheet graphing program to graph $F(\lambda, 310 \, K)$. Choose a wavelength range that shows the peak of the distribution and the falloff on both sides to one-tenth of the peak value or less.
4. Convert the Planck spectral distribution to a function $I(f, T)$, intensity per unit frequency interval, where frequency $f = c/\lambda$.
5. Plot the spectrum in 4 using your graphing software, using starting and ending frequencies corresponding to those in c.

PROBLEM 3.5 DIFFUSION EQUATION

1. Determine whether the functions $c_1(x,t) = A e^{-x^2/4Dt}$, $c_2(x,t) = At^{-1} e^{-x^2/4Dt}$ and $c_3(x,t) = At^{-1/2} e^{-x^2/4Dt}$, where A is a constant, are solutions to the one-dimensional diffusion Equation 3.27.
2. What must the units of D be if x and t are expressed in SI units?

PROBLEM 3.6 BIMOLECULAR REACTION EQUATION

1. Show that Equation 3.47 is a solution to Equation 3.41.
2. Draw a graph of Equation 3.47 for $A_0 = 20B_0$ and for $A_0 = 1.5B_0$. Choose a fixed value for kB_0.

PROBLEM 3.7 DISTANCE DISTRIBUTIONS

The positions of 200 particles were measured to be in the following ranges in one dimension:

x (m)	Number of Particles
0.0–0.5	2
0.5–1.0	4
1.0–1.5	14
1.5–2.0	26
2.0–2.5	34
2.5–3.0	42
3.0–3.5	35
3.5–4.0	25
4.0–4.5	13
4.5–5.0	4
5.0–5.5	0
5.5–6.0	1

1. Plot this position distribution using spreadsheet/graphing software.
2. Determine the average and standard deviation of the distribution.
3. Write down the Gaussian distribution that most accurately describes this distribution and plot this Gaussian on a graph with the measured values.

PROBLEM 3.8 RADIAL DISTANCE DISTRIBUTION

A Gaussian radial probability distribution in two dimensions is $P(r)dr = \frac{1}{2\pi\sigma^2}e^{-r^2/2\sigma^2}(2\pi r)dr$.

1. Plot this distribution for $\sigma = 0.5$ (with proper units).
2. If 100 particles have radial positions Gaussian-distributed according to the plot above, determine the most likely numbers of particles with radial positions in the ranges 0.10 ± 0.10, 0.30 ± 0.10, $0.50 \pm 0.10 \ldots 1.90 \pm 0.10$.

PROBLEM 3.9   DIFFUSIVE SPREADING

Distribution functions that spread with time from an initially infinitely narrow distribution are $c(x,t) = \frac{N}{(4\pi Dt)^{1/2}} e^{-x^2/4Dt}$ in one dimension and $c(r,t) = \frac{N}{(4\pi Dt)^{3/2}} e^{-r^2/4Dt}$ in three. N is the number of particles, D is a diffusion constant, and t is time.

1. What are the SI units of D?
2. Plot the 1D function as a function of x over the range 0–5 μm using spreadsheet/graphing software for $N = 1$, $D = 10^{-12}$ (SI units). Leave room in your spreadsheet for one more column to calculate part 3. Choose three times: one showing a narrow distribution on your graph, one broad, and one in between.
3. For the same values of parameters in 2 above, plot the three-dimensional function.

PROBLEM 3.10   POISSON DISTRIBUTION

1. Using Excel or another graphing program, make four separate plots of the Poisson distribution, Equation 3.75, for $N = 10^4$ and $p = 10^{-3}$, 10^{-2}, 10^{-1}, 0.5. Make sure each graph is easy to read.
2. For which of the graphs is the Poisson approximation valid? Do not simply assert that a value of p or n is not sufficiently small. Argue from the shapes of the graphs or other objective criteria.
3. How does the shape of the Poisson differ from the Gaussian distribution? You may want to plot a comparable Gaussian graph.

PROBLEM 3.11   DIMERIZATION REACTION EQUATION

1. Write the differential equation that applies to the reaction of two identical molecules to form a dimer: $A + A \xrightarrow{k} A_2$. The forward reaction rate is k. Neglect any back reaction. Hint: Read Section 3.7.2 carefully.
2. Write the solution to the equation.
3. Graph the concentrations of A and A_2 as a function of time if the initial concentration of A is 0.100 M (molar) and the reaction rate is 10^6 M^{-1} s^{-1}. Pay attention to a factor of 2 and explain why the units of k are what they are.
4. Qualitatively explain how the graph would differ if there were a "slow" reverse reaction. Note that a reverse reaction rate, k_r, would have units s^{-1}, so that k and k_r cannot directly be compared.

REFERENCES

1. Fleming, G. R., and M. A. Ratner. 2008. Grand challenges in basic energy sciences. *Phys Today* 61:28–33.
2. Alon, U. 2006. *An Introduction to Systems Biology: Design Principles of Biological Circuits*. Boca Raton, FL: Chapman & Hall/CRC.
3. Klafter, J., M. F. Shlesinger, and G. Zumofen. 1996. Beyond brownian motion. *Phys Today* 49:33–39.
4. West, G. B., and J. H. Brown. 2004. Life's universal scaling laws. *Phys Today* 57:36–42.
5. Reif, F. 1967. *Statistical Physics*. New York: McGraw-Hill, 350–357.

PART II

STRUCTURE AND FUNCTION

CHAPTER 4

CONTENTS

Water

4

4.1 INTRODUCTION

It is all but inconceivable that the complex structural organization and chemistry of living systems could exist in other than an aqueous medium.[1]

This quote from a biochemistry textbook reflects a profound truth, though it would likely never be made by a physicist, who may be less familiar with the broad scope of (earthly) biological life. Life as we know it is based largely on the properties of carbon and water. Even the search for extraterrestrial life on Mars has largely focused on evidence for water as the most certain indicator of life.[2] This chapter ignores other possible bases for life and focuses on the unusual features of liquid and, to a lesser extent, solid and gaseous water that we can readily implicate in biological structure and function.

Many of the properties of water that are important to life are evident in everyday life: its abilities to dissolve many substances, to cool, to change phase (solid, liquid, gas), to wet common surfaces, to keep your plants from dying. Many, however, are hidden because they involve microscopic structures and processes. As we will see, some of the common properties we associate with water are quite different on a microscopic scale. We may recall from introductory physics the difficulty Isaac Newton had convincing some people of his first law: *A body persists in its state...of uniform motion unless acted upon by an external unbalanced force.* This difficulty largely had to do with the common experience that wagons tended not to move if the draft horse did not pull. Conditions in which, as we now know, the ubiquitous frictional force was low were not terribly difficult to find, however, as when a thrown object or cannonball followed its approximately parabolic path through the air. If we humans were of micron size and lived in water, however, it is highly doubtful that Newton ever could have convinced anyone of his first law, as the frictional force virtually always rises to exactly match any applied or generated motive force—we would never observe "inertial" motion. But, then, this speculation is rather irrelevant, because a micron-sized human could not have a brain to think up such laws of motion.

The subject of water structure and properties has a long and continuing publication history, and the interested student can find volumes of information by searching for works edited by Felix Franks,[3,4] as well as the book by Hasted.[5] A useful, though not peer-reviewed, online compilation can be found at www1.lsbu.ac.uk/water/index.html. This chapter provides an initial overview of water structure and properties. Many more properties will be dealt with in later chapters as they become relevant. Properties we now introduce include water's molecular geometry, hydrogen bonding, and macroscopic properties, including flow, heat capacity and conductivity, phase changes, surface tension, and dielectric properties. Table 4.1 lists a few of the common properties of water needed in many biophysical computations.

4.2 STRUCTURE

We separate discussion of the structure of water into two, somewhat arbitrary pieces: the structure of the water molecule (i.e., water in the gaseous state) and the structure of the water molecules in liquid water or ice. This separation is not clean, as the structure of individual molecules is changed slightly by the phase.

TABLE 4.1 Common Properties of Water

Molecular Formula	H_2O
Mass	18.0153 amu (or g/mol)
	2.99072×10^{-26} kg/molecule
Molecular volume (liquid)	14.6 $Å^3$/molecule (van der Waals)
Concentration	55.345 M (mol/L)
Number density	3.34×10^{28} m^{-3}
Density	997.05 kg/m^3 (liquid, 25°C, 1 atm)
	999.97 kg/m^3 (liquid, 3.984°C, 1 atm)
	916.72 kg/m^3 (solid, 0°C, 1 atm)
Melting point	0°C (273.15 K)
Boiling point	100.0°C
Specific heat (constant pressure)	4182 J/(kg·K)
Dielectric constant	78.5 ε_0
Viscosity (dynamic)	0.8909×10^{-3} Pa s (25°C)
	1.0016×10^{-3} Pa s (20°C)
Surface tension	0.07198 J/m^2

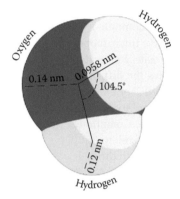

FIGURE 4.1
Molecular structure of water. Covalent bond lengths, angles and the van der Waals radii of the atoms are indicated. The atoms are colored according to the Corey, Pauling, Kultin (CPK) color scheme. A version of this color scheme can be found at www.rasmol.org/software/RasMol_Latest_Manual.html#chcolours.

Anion is a negatively charged atom or molecule; cation is a positively charged atom or molecule.

4.2.1 Molecular Structure

Water, H_2O, has the simple and familiar molecular structure shown in Figure 4.1. It is planar, with an H-O-H angle of 104.5° in the gaseous and also, approximately, in the liquid state. This angle and the interatomic distances were initially determined by analysis of infrared absorption experiments, which are sensitive to bond bending and stretching.[6] Two of the most important properties of this molecule are (1) the electronegativity difference between O and H and (2) the presence of two H atoms, in equivalent positions. The electronegativity difference means that much of each hydrogen's electron density resides on the oxygen, leaving the hydrogen with a fractional positive charge. That fractional positive charge ($H^{+\delta}$) means that the proton can more easily be attracted to another electronegative atom, like N or O, forming hydrogen bonds. Another O may be part of an adjacent water molecule, for instance. The presence of two equivalent hydrogen atoms means that water can then form multi-molecule networks or lattices and not just dimers.

The hydrogen-bonding and networking ability of water provide the basis for many of its biological and biochemical roles in life. Because most biological structures are surrounded by water, an electron or a proton can, for instance, move over long distances through water to a site where it is needed, i.e., where its free energy is lower. An example of this is charge movement and control by "buffers"—molecules such as Na_3PO_4 (known generically as phosphate buffer) and $Na_3COO(CH_2–COO)_2COH$ (known generically as citrate buffer). By dissociating varying numbers of sodium ions these buffer molecules can attach varying numbers of protons from water and thereby adjust the pH ($=-\log_{10}[H^+]$, where $[H^+]$ is the molar free hydrogen ion concentration in the solution and $[H^+][OH^-] = 10^{-14}$). What this really means is that though the net charge of a solution is always zero, the amount of charge in the form of protons, which can be bound to many anionic sites, can be controlled. We later see that the structure of water allows the movement of protons (not water reorientation, as discussed below) from one to another water molecule on a time scale as fast as 10^{-14} to 10^{-10} s.

What is the strength of a typical hydrogen bond in water? This is critical to our understanding not only of water and biomolecular structure but fluctuations and reactions. A typical water–water H bond has strength (free energy) ~19 kJ/mol in liquid and 54 kJ/mol in ice. (These are actually measured water reorientation "activation energies.") As bound water molecules are somewhat immobile, it may be expected that their reorientation would demand an activation energy somewhere between 19 and 54 kJ/mol.

4.2.2 Water: Liquid and Solid

Liquids, by definition, do not have a fixed structure on a macroscopic scale, though their densities are comparable to those of solids. Their molecules are in constant motion and do not oscillate about any fixed point. As you may recall from other courses, X-rays scattered from atoms in a solid macroscopic sample (e.g., a "crystal") interfere with each other and form patterns (diffract) that reflect the arrangement of the atoms in the crystal, if the atoms have some ordered arrangement over the size of the sample.

Does this diffraction require the atoms to be absolutely motionless? It had better not because we know that motion only ceases when the temperature is absolute zero. Even when atoms in a solid move, if they oscillate about some fixed point, diffraction still occurs; the brightness of the spots is simply reduced. This is described by the *temperature factor*. We will focus more on this later when we discuss protein structure. What about liquids, where atoms generally do not oscillate about a fixed point, at least on time scales of typical X-ray scattering experiments, minutes or longer? Do scientists measure diffraction from ice and then extrapolate to liquid? Yes, as a matter of fact they do, but direct measurements of liquid water are also performed.

4.2.2.1 Solid water (ice)

X-ray and neutron scattering experiments on ice have led to the "tridymite ice I" structure shown in Figure 4.2. Water molecules are approximately tetrahedrally arranged, with each water involved (potentially) in four hydrogen bonds and the H–O–H angle 109.4°. Oxygen atoms show approximate sp^3 orbital hybridization, with two "lone-pair" orbitals and two bonds to H atoms. This ice structure, with its rigid lattice, is more open

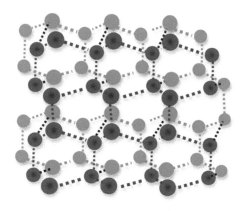

FIGURE 4.2
Model of ice I structure inferred from X-ray scattering data, showing only oxygen atoms. Hydrogen atoms form the (hydrogen) bonds and are located on the lines between oxygen atoms. Only two layers are shown in the dimension perpendicular to the page. The approximate tetrahedral arrangement of water molecules reflects oxygen's approximate sp^3 hybridized bonding, with two hydrogens and two oxygen lone-pair orbitals. The average H–O–H bond angle in ice I is 109.4° compared to gaseous water's 104.5°. The latter value applies approximately to liquid water, also.

than the liquid water structure, resulting in the familiar observation that ice floats on water. We will come back to this in a moment. The "rigid lattice" does not imply that ice cubes from your freezer are single crystals: this would be rare indeed. Many defects occur, owing to impurities, air bubbles, cracks, and other molecule misorientations, and X-ray or neutron diffraction measurements on ice generally imply polycrystalline samples.

How can an extrapolation of the solid data to a liquid state be done? If we take the accepted H–O–H angle for liquid water as 104.5°, we could simply distort the ice I structure and insert defects and interstitials as necessary. Such distorted structures have been developed that agree with diffraction data, but the approach is not very satisfying, because a distorted crystal would not have any obvious liquid character. There is a second approach whose main features are to keep the basic ice structure but surround ice microcrystals with disordered liquid water, with continual breaking and reforming of the hydrogen bonds at the crystal periphery. This model is also perhaps reasonable, but the skeptic (each one of us) should begin asking, "Where are the data?" If we propose a theoretical structural model by changing angles, bending bonds, and introducing variable amounts of fluctuating, disordered regions, a single scattering experiment that may be sufficient for a single, uniform crystal will not suffice. We don't solve this problem here, as the structure and dynamics of water are still under active investigation.[7,8] Sometimes, simple systems can be astonishingly difficult to explain completely. Because water underlies the astonishing complexity of life on Earth, perhaps this difficulty should not surprise us.

4.2.2.2 Liquid water experiments

Scientists have also been recording X-ray as well as neutron diffraction directly from liquids for many years in order to determine the liquid "structure." There are several possible reasons for these measurements. First, even for a liquid of spherical particles the packing of the spheres in the first layer around a given particle may be different than farther out. Second, there may be some localized structures throughout the macroscopic liquid that may drift and constantly assemble and reassemble at the periphery of the structure. Third, there may be (should be) preferred sites for (nonspherical) water molecules in the neighborhood of any particular water molecule. The movement of one molecule away from a preferred site would be followed quickly by motion of another molecule into that site. This preference for specific sites for many or most water molecules, should, even though that site may itself be moving, produce measurable effects in the scattering pattern. At the very least, a diffraction measurement would reflect the distances of the nearest few neighbors to each water

X-ray "scattering" usually refers to measurements of scattered intensity as a function of one (axial) angle from the incident beam, averaged over the angle about the beam axis, whereas "diffraction" often indicates intensity was measured as a function of both angles, with diffracted *spots* corresponding to Bragg scattering. Polycrystalline (powder) samples give scattered rings rather than spots.

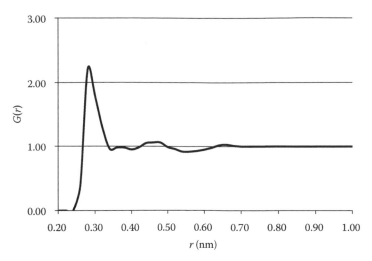

FIGURE 4.3

Radial distribution describing the distribution of water molecules around any given water molecule, as derived from X-ray diffraction measurements. The radial distance is the distance between oxygen atoms. (From Hasted, J.B., *Aqueous Dielectrics*, Chapman and Hall, London, UK, 1973.)

molecule. We should expect this pattern of preferred sites for water molecules to reveal itself again when we study the structure of biomolecules dissolved in water.

The time average of molecular positions in liquid water does indeed show structure. In the 1930s, Morgan and Warren[9] observed that the sharp scattering pattern of ice gave way to a blurred pattern when the ice melted. This "blurring" sounds reasonable, but keep in mind there are several possible outcomes for a crystal that melts. Diffracted spots/rings could move or disappear, spots could become rings, or rings could broaden. The latter applies in this case. Some position information can be described by the X-ray or neutron scattering radial distribution function:

$$G(r) = \rho(r)/\rho_0 \qquad (4.1)$$

where $\rho(r)$ is the (average) number of scattering particles in a differential volume element located a distance r from the primary particle and ρ_0 is the bulk molecular density of the particles. A sketch of such a distribution function for water is shown in Figure 4.3. Because water is made of H and O, care must be taken to understand the meaning of r. The data shown reflect the O–O distance between the central and secondary water molecules. (Hydrogen is often not seen in X-ray scattering.) The "molecular density" then refers to the number of O atoms (number of water molecules) per unit volume in water located a distance r from the O atom of the primary water molecule. (For a more accurate distribution function, see p. 69 in the book by Hasted.)

What do these density distribution functions such as Figure 4.3 mean? Because we are looking at the O–O radial probability density, orientation of water molecules directly adjacent to the primary water molecule may imply more (or fewer) O atoms per unit volume than for random orientations because H atoms also take up space. Suppose, for example, there were some microscopic (but bigger than Angstrom-size) icelike structures that maintained their integrity for a sufficiently long time: X-rays would scatter differently than if there were no such structure, if water molecules were completely independent and randomly positioned and moving. In the past, such structures were referred to as "flickering clusters." This scattering is similar to polycrystalline samples. If a sample truly had random molecular positions, with little or no space between molecules, $G(r)$ would be zero from $r = 0$ to $r \sim r_{mol}$, where r_{mol} is approximately the van der Waals radius of the molecule and $G(r) = 1$ for $r > r_{mol}$. The peaks and troughs in the data indicate preferred and nonpreferred distances for water molecules. Does this imply that a snapshot of molecular locations at a given time is truly spherically symmetric? If the sample were truly amorphous, it would. However, thinking of the polycrystalline sample model again, microcrystals of varying, but nonspherically symmetric structures, could be scattering the X-ray or neutron beam. We cannot tell, from scattering data alone, what the true, instantaneous structure is.

The best way out of the conundrum of "several models fit the data" is not to make increasingly complex theoretical models to analyze the existing data but rather to apply other experimental techniques—get more data. At the beginning of Section 4.2 we noted that the H–O–H angle in liquid water was 104.5°, determined from infrared absorption measurements. This suggests that liquid water is not quite in the tetrahedral structure characteristic of ice I (and crystalline quartz). Can we get more information out of similar measurements? Yes. X-ray and neutron scattering are governed by the positions of atomic nuclei. Other measurements such as X-ray absorption and X-ray Raman scattering give other sorts of structural information, as they excite core (tightly bound) electrons in H_2O and are sensitive to the environment of the H_2O, including hydrogen bonding. The conclusions made from one such study are as follows:

Figure 4.3 is already "processed" data, not experimental data. The original experimental data are normally a graph of scattered intensity versus $2\sin(\theta/2)/\lambda$, where θ is the scattering angle and λ is the X-ray wavelength.

Even infrared data have recently been shown to produce somewhat ambiguous structural information when applied to liquid water. However, a difference in H–O–H angle between liquid and solid seems correct.

Most molecules in liquid water are in two hydrogen-bonded configurations with one strong donor and one strong acceptor hydrogen bond in contrast to the four hydrogen-bonded tetrahedral structure in ice. Upon heating from 25°C to 90°C, 5 to 10% of the molecules change from tetrahedral environments to two hydrogen-bonded configurations. Our findings are consistent with neutron and X-ray diffraction data, and combining the results sets a strong limit for possible local structure distributions in liquid water. Serious discrepancies with structures based on current molecular dynamics simulations are observed.[10]

We note that the conclusion involves a "strong limit on possible local structure distributions" and "serious discrepancies with structures," so ambiguity in the liquid water structure remains, though some models seem to be ruled out.

The structure of liquid water seems, then, to be similar to that of ice, with the following exceptions: (1) it is fluid; (2) only half the possible hydrogen bonds are in place at any given instant; (3) the H–O–H angle is, on average, slightly smaller; and (4) molecules in liquid water fill in spaces that are present in the ice structure. It is left to homework to further discuss features of the distance dependence of water's molecular density.

4.2.2.3 Fluctuating structure

Water molecules are quite free to move; that is evident from the fact that water is a liquid of quite low viscosity, even if we don't try to decipher the details of the X-ray and other data discussed in previous sections. How fast and how far do water molecules move? Do individual molecules move independently, or do large aggregates move as microcrystals? This topic will be addressed in several later sections of the book, after we have prepared ourselves a bit more for quantum and fluid mechanics. It may be unsatisfying, but many physical properties of water that are still incompletely understood are central to biological life; we can only do our best to fit these properties in with our observations of biosystems and marvel at the inventiveness of life.

Computer simulations of the molecular motion of water molecules in the condensed, liquid state show rotations on a picosecond (10^{-12} s) time scale. The *rotational correlation time* determined from dielectric and NMR (nuclear magnetic resonance) measurements is about 8 ps and measures approximately how fast the molecular motions are that govern the viscosity and rotational adjustment to perturbations in the liquid.[11] We will later treat translational motion of water molecules in water and rotations of H_2O in the gaseous state, but it is clear that water rotates on a very short time scale compared to most cellular processes, even when we take into account that, on average, a water molecule is H-bonded to two others at any given time. Some measurements suggest discreet trimeric (3-water), tetrameric (4-water), pentameric (5-water) and hexameric (6-water) H-bonded complexes in liquid water, though rotational vibrational spectroscopy of a hexameric form shows "geared" motions of H_2O pairs, facilitated by quantum mechanical tunneling, within the hexamer on times scales of about 10^{-10} s.[12]

> As a rule of thumb, most cellular processes occur on a time scale of 10^{-3} s or longer; photosynthetic energy transport and some other electronic processes are exceptions to this rule.

4.3 UNUSUAL PHYSICAL PROPERTIES

4.3.1 Macroscopic Flow Properties: Viscosity

Viscosity (SI units $N \cdot s/m^2$, or $Pa \cdot s$) measures the ease at which macroscopic objects can move through a fluid. The ease of movement can be quantified, for example, by the time it takes a standard ball bearing to drop a certain vertical distance in a cylinder of the fluid (with various corrections). For a liquid fluid, water has a low viscosity, though it is only slightly lower than alcohols and some other common organic solvents. See Table 4.2. What does this mean? Macroscopic objects move through water comparatively easily. This statement leaves many questions: (1) Is motion of microscopic objects governed by this same viscosity? (2) Does the speed of movement in the fluid matter? (3) What are the fluid molecules doing as objects move through the fluid? It will take time to address these questions, but we should consider them at the outset. The viscosity most commonly occurs in equations relating the fluid frictional force to an object's velocity. In Equation 1.2, viscosity is part of the coefficient of friction, b_1 (which is simplified to b for later chapters).

> We speak of the "dynamic viscosity," η, SI units $Pa \cdot s$ or cgs units "poise." The "kinematic viscosity" ν is related to dynamic viscosity by $\nu = \eta/\rho$, SI units m^2 s or cgs units "stokes" and is often used when viscous and inertial forces are being compared. This book uses dynamic viscosity with SI units.

TABLE 4.2   Resistance to Movement: Viscosities of Fluids (25°C)

Fluid	Viscosity (Pa·s)
Air	1.8×10^{-5}
Acetone	3.06×10^{-4}
Methanol	5.44×10^{-4}
Benzene	6.04×10^{-4}
Water	1.0×10^{-3}
Ethanol	1.07×10^{-3}
Methanol	0.544×10^{-3}
Blood	2.3×10^{-3}
Sulfuric acid	2.42×10^{-2}
Motor oil (No. 10)	0.2
Glycerol	1.5
Glucose (solid)	$6.6 \times 10^{+10}$

TABLE 4.3   Heat Storage:[a] $Q = mc_p \Delta T$

Substance	c_p (kcal/kg·K)
Lead	0.031
Mercury	0.033
Copper	0.092
Glass	0.20
Marble	0.21
Aluminum	0.214
Ice (−10°C)	0.53
Ethanol	0.581
Sea water	0.93
Water	1.00

[a] Specific heats of substances at constant pressure.

TABLE 4.4   Heat Transport: Thermal Conductivities of Selected Substances

Substance	Thermal Conductivity (W m^{-1} K^{-1})
Air	0.0256
Nitrogen (300 K/80 K)	0.026/0.13
Carbon tetrachloride	0.10
Diphenyl	0.14
Toluene	0.13
Water	0.60
Styrofoam	0.010
Ice (0°C)[a]	2.2
Copper	390
Diamond	2450

[a] "Low-density ice," which we normally experience in nature, has thermal conductivity up to 10 times less than the listed, standard value. (Coles, W.D., Experimental determination of thermal conductivity of low-density ice, Technical Note 3143, Lewis Flight Propulsion Laboratory, Cleveland, OH, 1954.)

4.3.2 Heat Storage: Specific Heat *c*

Most of us are familiar with the high heat-absorbing power of water from everyday life, where we use water to rapidly cool hot objects (cooling steel from a forge or an egg from boiling water), as well as from introductory physics and chemistry. Table 4.3 shows the numbers behind this ability to absorb and contain heat. Water has a significantly greater capacity to absorb heat for a given temperature rise than all common solids and liquids. The implication is that if heat is generated in an aqueous environment, the temperature will not rise very much. As indicated in Table 4.3 caption, the most common occurrence of specific heat is in calculation of temperature rise upon introduction of a quantity of heat, Q:

$$Q = mc\Delta T \tag{4.2}$$

In biological applications the specific heat at constant pressure c_p is normally used because that is the usual environment of biosystems.

4.3.3 Heat Conduction: Thermal Conductivity

Thermal conductivity measures the rate at which a material can transport heat, measured as power per unit distance transported per unit temperature difference. We think of copper as a good heat conductor and materials like Styrofoam poor conductors (good insulators). Table 4.4 bears this out, and it also shows that liquids have conductivities intermediate between solids and gases. Note that Styrofoam's low conductivity is due primarily to the high volume fraction of air. Water clearly has a high thermal conductivity for liquids, about 4 times the average of the other listed liquids. Because heat transport involves movement of thermal energy (molecular kinetic energy) from one atom or molecule of a substance to neighboring atoms/molecules, it would follow that tight coupling between neighboring atoms would facilitate transport. Coupling efficiency is clear in the cases of diamond (tight coupling of atoms) and air (loose coupling), with liquids in the middle. However, we should be careful of comparisons between materials, because heat transport mechanisms can be very different—molecular collisions in the case of gases and free conduction electrons in the case of metals. We might correlate the unusually high thermal conductivity of water for a liquid with the extensive hydrogen bonding between molecules, coupled with the relative freedom molecules have to move, when compared to ice. We must, however, be careful what we mean when we discuss thermal conductivity of "ice."

The thermal conductivity of ice, listed in Table 4.4, is 2.2 W m^{-1} K^{-1}—considerably *larger* than that of liquid water, 0.602 W m^{-1} K^{-1}. Many who have lived above a latitude of 45° N claim that a layer of ice on a pond (Figure 4.4) helps to insulate and keep the water beneath in a liquid state, saving the fish. Another claimed example of ice as insulator is the igloo or snow cave, where contact of occupant with the frozen walls is not a problem, whereas immersion in melt water can be deadly. The tabulated conductivity values, however, imply that ice would more quickly conduct heat out of the water or a body than would liquid water itself. The ice-as-insulator report might then be considered incorrect, except that so many lives have relied on this presumption. Yet another insight into water's surprises can be gained by accepting the experience of high-latitude residents and more carefully examining the data of the

FIGURE 4.4
A layer of snow and ice on Lake George in south-central Alaska hides active life beneath, where the water temperature is above 0°C. December-to-February high (low) temperatures in nearby Fairbanks average about −5°C (−26°C), with record average low of −41°C. The layer of lower-density, snowy ice, up to 1.5 m thick, both insulates the liquid water below and inhibits convective heat loss. Dense, blue ice from the Colony Glacier can also be seen on the shore. Importantly, lake ice generally allows Alaska's residents to land planes and walk on the ice. 2017. (Photo courtesy of Kirsten McDaniel.)

scientist in Table 4.4. First, the tabulated conductivity is at a temperature of 0°C, where water can exist as both liquid and solid. Every northerner knows that ice does not form on a pond when the temperature is 0°C. It must be at least several degrees below 0. Only the careful scientist could produce solid ice at exactly 0°C, and this ice would look like the lenses in a $1,000 pair of eyeglasses—crystal clear and colorless. The ice one sees on a frozen body of water (Figure 4.4) never appears perfectly clear. In fact, the ice is usually covered with a layer of snow or snowy ice. But what exactly is snow? Isn't snow just a bunch of ice crystals interspersed with small air pockets? What if you compressed the snow to get rid of the air and to fuse the ice crystals together? Perhaps then the ice would be clear. But unless you compressed the snow at 0°C, where repetitive thawing and refreezing could occur, the packed ice would appear, at best, like the ice cubes one often gets from a freezer, with cracks and air bubbles, giving the ice a milky appearance.

The experiment of compressing snow until it became ice has been done by nature for millions of years, in glaciers and in Antarctica. (Denser, glacial ice, characterized by its bluish color, can be seen on the slope in Figure 4.4, covered by a layer of recent snow.) More formal research was also done in the 1950s, by W.D. Coles for the National Advisory Committee for Aeronautics, who found that when snow gets packed to a higher and higher density, thermal conductivity increases by a factor of 30 or more, to a maximum of about 20 $Btu/hr \cdot ft^2 \cdot (°F/in)$.[13] (The reader should do the conversion to SI units and find that this value is about 3 $W/m \cdot K$.) Conductivity of ice in real life can therefore easily be less than that of liquid water. Additional issues include (1) natural selection: the higher-density, higher-conductivity ice would conduct more heat from the water, tending to melt itself, until less-dense, more-insulating ice and snow formed on top, after which ice of higher density could form; and (2) convection: a cold wind blowing on liquid water will remove heat even more efficiently because of liquid convection, whereas no convection can occur in solid ice. To bring the pond-ice question to a close, the thermal conductivity of the frozen material (including "snow") on top of liquid lake water in northern latitudes has an effective thermal conductivity that can vary considerably, but will often be less than that of liquid water.

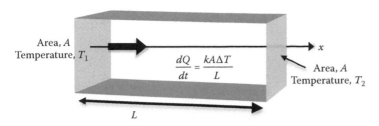

FIGURE 4.5
Heat conduction in a one-dimensional geometry.

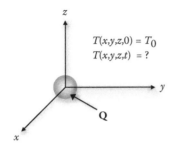

FIGURE 4.6
A more common geometry for heat dissipation in a biosystem. An amount of heat Q is suddenly deposited in a spherical volume that, together with its surroundings, is initially at a temperature T_0. What is the time and position dependence of the temperature?

The geometry used when thermal conductivity is introduced in first-year physics is typically linear, as in Figure 4.5, where the two ends of the volume are held a constant temperatures differing by ΔT, with accompanying equation:

$$\frac{dQ}{dt} = \frac{kA\Delta T}{L} \qquad (4.3)$$

Excluding the geometrical factors, the thermal conductivity k determines how fast heat flows. In this simple case, the temperature profile, if the material is uniform and isotropic, depends linearly on position, and is independent of time:

$$T(x, y, z, t) = T_1 + \frac{x(T_2 - T_1)}{L} \qquad (4.4)$$

This geometry occasionally occurs in a biosystem. A more common biosystem calculation involves heat quickly deposited in a small volume immersed in an aqueous environment initially at a temperature T_0. See Figure 4.6. This small volume may be a molecule that has just absorbed a photon and converted the energy to heat, or it may be a molecular motor that has just moved a load (done work) by "burning" adenosine triphosphate (ATP) energy at 60% efficiency. How do you describe this situation? In contrast to the case of Figure 4.5, there is clearly time and three-dimensional spatial dependence of the temperature. We must learn to deal with such situations.

4.3.4 Phase Changes of Water

The latent heats of fusion and vaporization of water and several other materials are shown in Table 4.5. The amount of heat needed to change water from one phase to the next is about 3 times larger than for the other listed liquids.

Water has an additional unusual characteristic that appears when changing from a liquid to a solid. Figure 4.7 shows the inverse of water's density as a function of temperature between 8°C and 0°C. Normally, materials contract (density increases) monotonically as temperature decreases. Water, however, reverses this process of contraction between 4°C and 0°C. Even though the magnitude of this effect is small, about 0.03%, the implications are extensive. One of these implications is shown in Figure 4.4.

TABLE 4.5 Latent Heats of Fusion, Vaporization, and Corresponding Temperatures for Water and Several Other Substances

Substance	Melting Point (°C)	Latent Heat of Fusion, L_f (J/kg)	Boiling Point (°C)	Latent Heat of Vaporization, L_v (J/kg)
Ethanol	−114.4	10.8×10^4	78.3	8.55×10^5
Methanol	−96	9.92×10^4	64.7	10.7×10^5
Acetone	−94.7	9.80×10^4	56.5	5.51×10^5
Mercury	−38.9	1.14×10^4	356.6	2.96×10^5
Water	0.0	33.5×10^4	100.0	22.6×10^5
Benzene	5.5	12.6×10^4	80.1	13.7×10^5
Glycerol	17	20.6×10^4	290	9.57×10^5

Note: Pressure is 1 atm.

4.3.5 Water Has a High Surface Tension

Table 4.6 shows that water is again at the extreme end of the scale for liquids, as measured by the air/water surface tension. What does surface tension measure? First, look at the units—they are N/m or, equivalently, J/m². Surface tension is the work needed to create an additional square meter of surface. For example, when a water skipper (the insects that scamper across the surface of ponds and streams) steps onto water, it makes a depression in the water, which creates a larger surface area. Water requires a relatively large amount of work to be done (force) to make this depression, which allows the water skipper to walk on the water. A more generally important implication, however, is that water will have more difficulty leaking through small holes than the other liquids. We will see that this is what makes all life that is based on cells possible. Note also, at the other extreme, the small value of lung-surfactant's surface tension in contact with air. This issue will be probed further in a biomedical homework miniproject.

When two materials such as water and air, water and oil, or water and glass contact each other, water molecules at the interface have to decide whether they prefer to be in contact with other water molecules or with the foreign material (in anthropomorphic terms). The surface tension involves both substances: one cannot speak of "water's surface tension," but of "water–air" or "water–oil" or "water–insect leg" surface tension. The surface tension of water (as well as that of other liquids) in contact with air arises from the fact that water molecules have a lower free energy when surrounded on all sides by water. See Figure 4.8. If, for example, the surface area of the water in Figure 4.8a increases owing to either a bulge or indentation in the surface, the water surface tension will resist because the larger surface area increases the (free) energy. Figure 4.8b shows a falling water drop (or water drop in zero gravity), on which the net external force is approximately zero. Because no other significant free energies exist that can affect the shape, the surface tension energy will find a minimum. This minimum, where the surface area is smallest, corresponds to a spherical shape because the water volume must remain fixed. (A conical, cubical, ellipsoidal, etc., shape will have a greater surface/volume ratio.) These examples involve the surface tension of a water–air interface, but they do not have great biological significance on the cellular level.

Another familiar example occurs when a water skipper (water strider; Figure 1.1d) stands on the surface of the water. Several forces on the contact point come into play: the gravitational force, the water–air surface tension, and the water–insect leg surface tension (or insect leg hydrophobic "force"). The insect could not walk on water if its leg surfaces were "wettable," i.e., not hydrophobic, having low surface tension with water. Analysis of the water skipper case is simple in principle. The increase in energy due to the surface area increase equals the gravitational energy decrease owing to the lowering of the insect's center of gravity: $n\Sigma\Delta A = mg\Delta h$, where n is the number of legs or appendages contacting the water, assuming all contribute equally. However, the complication is in the geometrical relationship between the depression's depth Δh and its shape. As this is

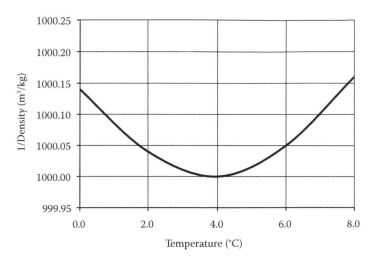

FIGURE 4.7
Inverse of water density as a function of temperature. The minimum (maximum of density) occurs at about 4°C. As water in a pond cools below 4°C, it rises to the top, where ice is formed. Surface ice, commonly with lower thermal conductivity than water, insulates the liquid below it from further cooling.

TABLE 4.6	Surface Tensions and Surface Tension: Viscosity Ratio	
Substance	**Surface Tension[a] (w/air) Σ, (N/m or J/m2)**	**Σ/η (m/s)**
Ethanol	0.0221	20.6
Methanol	0.0227	41.7
Acetone	0.0252	82.4
Benzene	0.0288	47.7
Pyridine	0.0380	—
Sulfuric acid	0.0551	2.28
Glycerol	0.0640	0.0445
Hydrazine	0.0667	—
Water	0.0728 (25°C)	72.8
	0.0588 (100°C)	
Tissue fluids	0.050	—
Soapy water	0.037	—
Lung surfactant	0.001	—
Water (with butanol)	0.0018	—

[a] Interface with air, except as noted.

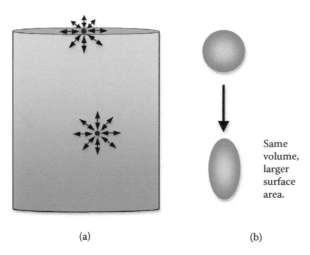

(a)   (b)

Same volume, larger surface area.

FIGURE 4.8
Surface tension of water at the interface with another substance (e.g., air). (a) A water molecule at the surface is attracted less by the air above than by the water beneath, so the water will try to minimize the surface area, which will minimize the free energy of the system, subject to any other forces in the system. Other forces may arise from gravity, the forces from container walls, and the (near) incompressibility of water. (b) A falling water drop is subject to near-zero net external force and will adopt a spherical shape to minimize, subject to its fixed volume, its internal pressure, and surface energy (surface area).

not a general issue in biophysical systems, we do not spend time trying to solve it.

If no force acts normal to a surface, it will remain flat, but if the pressures on two sides of a surface differ, the surface will distort until the surface tension force cancels the pressure force, $F_p = \Delta p A$. The relationship between area and the shape of the surface must be known. In the case of an ellipsoidal surface, the result of the analysis is the Young–Laplace equation

$$\Delta p = \Sigma \left(\frac{1}{R_x} + \frac{1}{R_y} \right) \tag{4.5}$$

where Δp is the pressure difference between the sides of the surface, Σ is the surface tension, and R_x and R_y are the surface radii of curvature in the two dimensions (equal for a sphere). This equation would apply to a balloon, where someone has manually placed a higher pressure inside the balloon, but it also applies to water droplets and liposomes in aqueous solution, because any surface curvature produces a surface tension force that must be balanced by a pressure difference.

EXAMPLE: PRESSURE INSIDE WATER DROPS

Calculate the internal pressure of spherical water droplets of radii 1 mm and 1 μm. We use Equation 4.5 for water droplets in 1 atm air at 25°C. The Δp we calculate will be the pressure increase due to the bending of the surface of the droplet above that of a flat ($r = \infty$) surface, which we presume to be 1 atm in this case. Setting the values of R_x and R_y equal in Equation 4.5 gives $\Delta p = 2\Sigma/R$. The value of Σ from Table 4.6 is 0.0728 N/m. Inserting radius values of 10^{-3} and 10^{-6} m gives $\Delta p = 146$ and 1.46×10^5 N/m². Since 1 atm = 1.01×10^5 N/m² (sea level), the pressure increases are 0.0014 atm for the 1 mm drop and 1.4 atm for the 1 μm drop.

4.3.6 Surface Tension: Viscosity Ratio

Of the common liquids that are biofriendly (exclude hydrazine and sulfuric acid), glycerol is closest to water in terms of surface tension. However, when we compare the viscosity, which measures the lossiness of an object's motion through water (how much force is needed to move an object through the liquid) with the surface tension, which measures how much force the surface can maintain without breaking, we see that water is even more unusual. The other high-surface-tension liquids acquire their high-Σ values along with high viscosity. See Table 4.6. Water stands out among all liquids with high surface tension and low viscosity.

4.3.7 Dielectric Properties

The dielectric constant ε that appears in the most familiar equation, Coulomb's law,

$$F(r) = \frac{q_1 q_2}{4\pi \varepsilon r^2} \tag{4.6}$$

describes how effectively a medium can reduce the electric force between charges. It is, in principle, a macroscopic parameter and applies when macroscopic charged bodies are immersed in a uniform, isotropic medium with no structure on the scale of the charged objects. Before we get into the complexities of some dielectric theory, let's look at the numbers. Table 4.7 shows that water has one of the highest static dielectric constants of common liquids (of all materials, in fact)—78.5, which clearly points to water's ability to attenuate electric forces by a large factor.

> This book follows the convention that the dielectric constant of vacuum is $\varepsilon_0 = 8.8542 \times 10^{-12}$ C^2/(N·m^2) and that of materials is $\varepsilon = \kappa\varepsilon_0$, where κ is unitless. However, in this book, as in the speech of most scientists and engineers, ε is often spoken in terms of its κ, i.e., "The dielectric constant of water is 78.5." Because any calculation you make is off by a factor of about 10^{11} if you make a mistake, you will not make this mistake if you *think* about the magnitude of any force or energy you calculate.

The microscopic theory of the dielectric constant, found in any first-year physics textbook, supposes that the uniform medium is actually composed of small particles (atoms or molecules) that have a permanent or induced dipole moment. The permanent dipoles align when an electric field is applied or the induced dipoles are induced, by application of the electric field. (For larger dielectric constants, the permanent dipole factor is generally larger.) This macroscopic view has been adequate for most engineering purposes (until recently) because charged objects and media were mostly of millimeter size or larger. In biophysics, as in most basic sciences, this is not adequate because we are usually trying to understand objects of 10 μm or smaller—cells to molecules. We want to know, for instance, the average force between a DNA phosphate group (average charge $-e$) and a charged aspartic acid amino acid on a protein. There is no way this biosystem, that of a protein, DNA, and water with dissolved ions, with all relevant particles of size a few nanometers or smaller and separated by the same distances, can be treated as charges in a uniform dielectric.

> Readers may get irritated with the frequent repetition of "average." Its use is, however, necessary and correct. It is a prelude to a major physical phenomenon that dominates microscopic biosystems: random motion and forces and their role in maintaining life.

Shall we give up on understanding in terms of fundamental physical law and just try to measure the end result? We propose not to do that, but to do what physicists, chemists, and engineers have always done: use the existing theory and see if it works. If it does, try to understand why it does; if it does not, try to correct the theory.

Correcting dielectric theory on a microscopic scale is not new. Most senior-level books on electromagnetism (or see Hasted[5]) will describe Debye's (Nobel Prize in Chemistry, 1936), Onsager's (Nobel Prize in Chemistry, 1968) and Fröhlich's attempts to correct the continuum dielectric model for microscopic effects in liquids.[14–16] The results are

$$\frac{\varepsilon - 1}{\varepsilon + 2} \frac{M}{\rho} = \frac{4\pi N_A}{3}\left(\alpha_a + \alpha_e + \frac{p^2}{3k_B T}\right) \quad \text{(Debye)} \tag{4.7}$$

$$\frac{(\varepsilon - n^2)(2\varepsilon + n^2)}{\varepsilon(n^2 + 2)^2} = \frac{4\pi N p^2}{9 k_B T} \quad \text{(Onsager)} \tag{4.8}$$

$$\frac{(\varepsilon - n^2)(2\varepsilon + n^2)}{\varepsilon(n^2 + 2)^2} = \frac{4\pi N_A g p^2}{9 k_B T V}\left(\text{Kirkwood} - \text{Fröhlich}\right) \tag{4.9}$$

In these equations, N_A is Avogadro's number, ρ is density, M is molecular mass, α_a and α_e are the atomic and electronic polarizabilities, p is electric dipole moment (of water), N is the number of molecules per unit volume, n is the index of refraction, and g is the molecular correlation parameter (for water, numerical value between about 1.5 and 3.0), V is the molar volume (for water, 18 mL, or 18×10^{-6} m^3).

What are these equations? On the left side of each equation is the dielectric constant; on the right, the molecular dipole moment. Without taking a detailed look at these equations, we would expect that "typical" liquids with similar molecular dipole moments would have similar dielectric constants: aligned dipoles counteract an applied electric field according to their dipole moments, not

TABLE 4.7 Dielectric Properties of Fluids[a]

Substance	Static Dielectric Constant, ε	Dipole Moment (Debye)
Hexane	1.9	0.00
Benzene	2.3	0.00
Diethyl ether	4.3	1.15
Chloroform	4.8	1.15
Ammonia	16.9	1.47
Acetone	20.7	2.72
Ethanol	24.3	1.68
Methanol	32.6	1.66
Dimethyl sulfoxide	48.9	3.96
Water (298 K)	78.5	1.84 (gas)
(Liquid, 273 K)	88	2.5–3 (in liquid and solid?)
(Ice, 273 K)	91.5	
Formamide	110.0	3.37

[a] Water has a high dielectric constant and relatively low permanent electric dipole moment.

according to what atoms they are made up of, yet water's ($p = 1.84$ D) dielectric constant of 78.5 is significantly larger than those of methanol ($p = 1.66$ D), ethanol ($p = 1.68$ D) and even acetone ($p = 2.72$ D). This has been the problem with dielectric theory applied to water.

Onsager[12] gave an answer to this puzzle:

> *The formation of a "hydrogen bond" increases the electric moment of the group which carries the hydrogen.... We can estimate that a water molecule in the liquid has a dipole moment of the order $\mu \sim 3 \times 10^{-18}$ esu, with $\mu^*/\mu \sim 3/2$. The increment from $\mu_0 = 1.8 \times 10^{-18}$ esu in the vapor could hardly be accounted for by ordinary homogeneous induction.*

Note: Dipole moment units. Recall that a dipole moment has units of charge times distance. The most commonly used units for the dipole moment of a molecule is the debye (D), especially in chemistry. Textbooks of electromagnetic theory tend to use cgs units: esu-cm. SI units are coulomb-m. We use D, with frequent footnotes on the conversion to SI units. The dipole moment of water is generally thought to be 1.84 ± 0.02 D in the gaseous form and dispersed in nonpolar solvents like benzene. Equivalencies:

$$1 \text{ D} = 10^{-18} \text{ esu-cm} = 10^{-29}/2.998 \text{ coulomb-m.}$$

Onsager is stating that the 1.8-debye dipole moment of H_2O in the gaseous form or dispersed in a solvent like benzene, is not appropriate for H_2O surrounded by other H_2O molecules, because of hydrogen bonds; rather, a value of about 3 D applies. Why? We recall the initial discussion of the strong effects of H bonds on the structure of liquid water, as well as ice. An applied electric field aligns electric dipoles, whose field attenuates the applied field. Except at a temperature of absolute zero, these dipoles are always, however, fluctuating in and out of alignment because of thermal fluctuations. If a dipole is physically attached to other dipoles via hydrogen bonds, thermal fluctuations will not be as effective at misaligning this much larger object.

So, where do we stand with the dielectric properties of water? We do not have a nice equation that accurately predicts the dielectric constant—water is still a bit of a mystery—but we have useful equations and a decent explanation for why hydrogen bonding might significantly increase water's dielectric effects and ability to reduce electric forces. We do not, however, have a good picture of how to proceed if we want to calculate the electric force between DNA phosphate and protein aspartic acid. Depending on the geometry, there may be mostly water, mostly protein structure, or a mixture of the two surrounding and separating the charged groups. This problem is beyond the scope of this textbook, though significant progress has been made in accurately calculating electrostatic effects in proteins.[17]

This is not really the end of the "unusual" water properties we will encounter, but we will wait until we face them in specific examples.

4.4 SUMMARY OF IMPORTANT PHYSICAL PROPERTIES

Which of the "unusual" properties of water are important to the understanding of biosystems? Because this textbook is not intended to be simply a treatise on the behavior and properties of water, all of the properties mentioned are important to biology. To make this section concise, the biological/biophysical relevance of the eight or so unusual water properties are presented below in Table 4.8.

4.5 BULK VERSUS LOCAL STRUCTURES

Roughly 70% of the weight of a cell is water. Because cells range in size from about 1 to 100 μm, their volumes will range from about 0.5 to 500,000 μm^3. Because water molecules are roughly 0.3 nm in length, the volume is about 10^{-11} μm^3; so there are about 10^{10}–10^{16} water molecules in the cell. These numbers would normally qualify as being "large," and most water in cells would qualify

TABLE 4.8 Unusual Properties of Water and the Biological Processes and Systems They Affect

Water Property	Biological Importance
Hydrogen bonding	Extensive H-bond networks: produce water's strong solvent ability, reduce electrostatic forces, allow water's relatively small dipole moment to produce a large dielectric constant, govern most biological processes inside of cells
Viscosity	Low viscosity (for a liquid): produces relatively small frictional losses; bacterial motility, molecular motors
Specific heat	High specific heat: results in smaller temperature changes upon heat introduction; temperature regulation of all organisms (heat and cold protection)
Thermal conductivity	High thermal conductivity: allows fast temperature equilibration and dispersion of excess heat; temperature regulation of all organisms, protection of molecules from destruction upon photon absorption
Heat of fusion	High heat of fusion: helps prevent freezing; cold protection of all organisms
Heat of vaporization	High heat of vaporization aids cooling: thermal protection of mammals, plants in hot weather
Density versus T	Decreasing density from 4°C to 0°C: makes ice form on top of water; survival of organisms in ponds and lakes outside of the tropics (and indirectly, resulting survival of even tropical organisms)
Surface tension	High surface tension: prevents water from leaking out of cells; resulting large capillary effect allows plants and trees to survive, which allows all other organisms on Earth to survive; allows formation of cell membranes
Dielectric constant	High dielectric constant reduces electric forces: governs protein folding and interactions between proteins, nucleic acids, and cell surfaces; virtually all cellular processes in all organisms are affected

as "bulk" water with corresponding properties. Most properties of water—bond lengths and angles, thermal properties, etc.—studied over the past 100 years with volumes of 1 mL or more would then apply. (We will deal with the fact that the cell is filled with an aqueous solution, not pure water, with slightly altered freezing points, viscosity, etc., in Section 4.7.) There are two major fallacies in the seemingly natural assumption that we can treat water in the cell as normal water, however. First, the cell is not a hollow container filled with water and dissolved molecules. Second, virtually all of the action in cells takes place at surfaces or boundaries between two different materials.

4.5.1 Cytoskeleton Fills and Organizes the Interior of Eukaryotic Cells

This section is slightly out of place, in that it does not deal with properties of water per se. However, because we naturally think of water in terms of a glass of water, a pond, or an ocean, it is important to disavow ourselves of the image of water as an extended body, a bulk medium when we think of fundamental biophysics. Though fish swim in lakes, the fundamental processes of life take place within a cell, where water predominates, but where it often cannot be treated as "bulk" or "extended." Figure 4.9 shows a micrograph that uses fluorescent dyes to stain specific types of cell materials, including the cell nucleus (the organized storage site for DNA), actin (a protein central to the muscle system), and microtubules (hollow protein that acts as structural supports and tracks for molecular motors). Though the entire cell may be of size 100 μm, some spaces within the cell have less than a micron separating adjacent filaments. Figure 4.10 shows an electron micrograph of structure involving the packing near the nucleus. The scale here is much smaller than in Figure 4.9, as seen by the 5- and 10-nm diameter beads in the figure: free space may even be small on a nanometer scale.

The restriction of spaces inside a cell makes detailed analyses of many biological reactions *in situ* very difficult. Large molecules and objects may be greatly restricted in movement and diffusion may not be close to random, while smaller molecules may deviate only slightly from classical, diffusive reaction theory. What is the best approach to take? First, there are cases where diffusion is obviously restricted, both theoretically

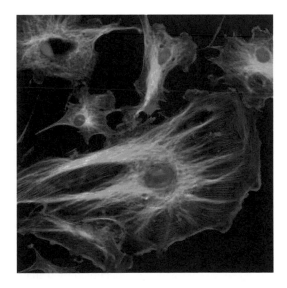

FIGURE 4.9

Cell cytoskeleton. Cells are not empty containers filled with an aqueous solution. The shape of eukaryotic cells is controlled by the cytoskeleton, microtubules, and fibers that permeate the cell interior. Cell nuclei are stained blue, microtubules green, and actin filaments red. Bovine pulmonary artery endothelial cells. (Courtesy of the National Institutes of Health, rsb.info.nih.gov/ij/images/.)

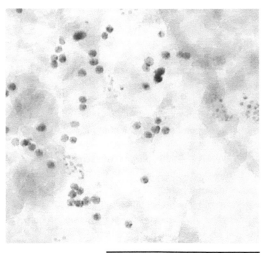

200 nm

FIGURE 4.10
Nuclear actin and protein 4.1 SABD (spectrin-actin binding domain) imaged by electron microscopy in human fibroblasts (a type of cell that synthesizes the extracellular matrix and collagen), showing dense packing in the nuclear matrix. The spherical structures are 5- and 10-nm beads. (By permission, from Krauss, S.W. et al., *Proc. Natl Acad. Sci. USA*, 100, 10752–10757, 2003.)

and experimentally. We will encounter an example of this in a method for DNA purification, which makes use of the restricted diffusion of DNA in the presence of a matrix of polyethylene glycol (PEG). A large fraction of the DNA purification done for the human genome project has been based on this phenomenon; successful high-tech companies are selling products for this purpose. Another obvious restricted-diffusion example we will explore is the motion of a large molecular complex, in the presence of a nearby cell membrane. On the other hand, there appear to be many examples where the worry about restricted and anomalous diffusion and reactions is overblown, where experiments show that potentially large deviations from diffusive motion are not observed. Dix and Verkman[18] summarized the following four points concerning molecular crowding effects inside of cells:

1. Crowding can slow the diffusion of solutes in aqueous-phase compartments and in membranes without leading to anomalous diffusion.
2. Large reductions in solute diffusion and/or anomalous diffusion are probably indicators of interactions between the solute and cellular or membrane components or of fixed barriers to diffusion.
3. Crowding reduces the diffusion of small solutes and many macromolecules in cytoplasm by only a few-fold compared to their diffusion in water.
4. Discrepancies between simulations and experiments on crowding effects on solute diffusion require further investigation.

We will not dwell now on the distinction between "slowed" and "anomalous" diffusion (point 1 above), except to state that there is, indeed, some effect of crowding, but slowed but nonanomalous diffusion does not change the form of the time dependence, $(\Delta r)^2 \sim t$. Point 2 refers mostly to the obvious cases of restricted motion but adds that immobility of a barrier or large macromolecule is an important consideration.

4.5.2 "Bound" Water Molecules

In his chemical physics book *Aqueous Dielectrics*, J. B. Hasted presents an interesting view of bound water molecules in terms of "physically and economically important…moist materials," the first examples of which are animal and plant tissue. Animal and plant tissues have been the most important materials to the human economy throughout recorded history. One might even extend this statement to the entire ecosystem's "economy." Our study of biosystems could then be thought of as a study of the two most economically important materials in the Earth's economy.

There are two prime aspects of bound water: (1) Biological molecules and structures usually have water molecules semipermanently attached to their surfaces and interiors, some of which take on specific chemical roles. Some water molecules, for example, are found in X-ray crystal structures of proteins to be bound to specific sites on the protein. More generally, since a globular protein is not entirely a closest-packed crystal, we find that extra spaces in the interior of the protein are usually occupied by water molecules that may not be specifically bound at one site. These water molecules may be surrounded, not by other water molecules, but by amino acid side chains (part of the protein structure) that can have widely varying properties—charge, hydrophobicity, H-bonding predisposition, chemical reactivity. We will deal with specific examples as we encounter them. (2) Bound water molecules often have different properties than H_2O in gas or bulk water.

What are some of the effects of binding on water properties? As the dielectric constant of water appears to have direct bearing on the stability of biomolecules, and since it has been studied in some detail, we consider that property first. Figure 4.11 shows a graph of ε of (bound) water as a function of its distance from a phosphate charge in polyadenine, a polymer like DNA, but with only adenine (A) bases. (This graph is calculated from Onsager/Debye theory, but is based on measurements.[19]) Though the ε values in this graph may not be exactly right, they illustrate the strong effect that binding of water to a charged site can have: ε varies from about 4 at short distances to 65 or so at large. What does this mean for us, in terms of biological structures or stability? Consider a protein that binds DNA, such as the "CRO" protein shown in Figure 4.12. Because phosphate charges of $-1e$ occur at every unit of the DNA polymer, the protein will certainly have to bind near a negative charge, and the protein is likely to have a positive charge in its structure near the site where it binds to the DNA phosphate. That negative charge will almost certainly have water molecules bound near it. Any electric force between the two biomolecules will then depend not only directly on the distance, as for example with a Coulomb force, $F = q_1 q_2 / 4\pi\varepsilon r^2$, but also on the distance-dependent dielectric constant. If the effect in Figure 4.11 is correct, this predicts that the force between protein and DNA will increase rapidly as charges on the protein and DNA become closer than about 4Å. The water distance-dependent dielectric effect will reinforce the effect of any solution counterions (Na$^+$, K$^+$, Mg^{2+}, Cl$^-$, OH$^-$, SO$_4^{2-}$, etc.).

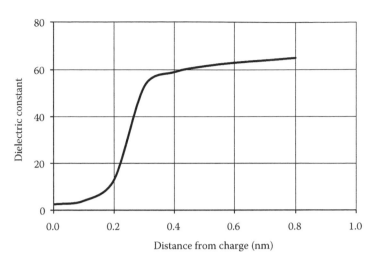

FIGURE 4.11
Dependence of dielectric constant of water, bound to polyadenine (a nucleic acid consisting of only adenine bases). Similar measurements on protein solutions could not be done because bound water manifests itself there only as a small departure from the principal dispersions from protein and bulk water dipoles. See text. (After Takashima, S. et al., *Biophys. J.*, 49, 1003–1008, 1986.)

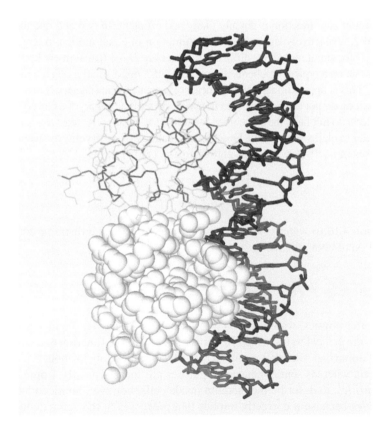

FIGURE 4.12
CRO protein bound to DNA. Before CRO binds, it must interact with the DNA from a varying distance, separated by water and ions. Electric forces will be distance dependent, e.g., Coulomb force $F = q_1 q_2 / 4\pi\varepsilon r^2$, but not only r, but ε, the dielectric constant of the intervening solution will vary (decrease) as the distance from DNA phosphate charges decreases. (From Wikipedia Commons author P99am; commons.wikimedia.org/wiki/File:Cro_protein_complex_with_DNA.png; creativecommons.org/licenses/by-sa/3.0/deed.en.)

4.6 DIFFUSION AND CHEMICAL REACTIONS IN WATER

We postpone primary consideration of molecular diffusion in water until later chapters, but we should note at the beginning of our study that molecular diffusion—normal, anomalous, driven—is central to any description and understanding of biosystems. Where the volume of bulk water is large (where the number of water molecules at surfaces is much smaller than the number in bulk, surrounded by other water molecules), where spaces created by cytoskeleton are open (judged by the size of the species that diffuses), or where for other reasons isotropic movement of a diffusing molecule is allowed (even if the motion is slower than in bulk water), diffusion lies either at the heart of biochemical reactions, or in a supporting role, providing sufficient numbers of molecules needed for reactions. Nondiffusive mechanisms are important, even critical to life, as the pumping of ions across membranes is needed for generation of nerve impulses, production of energy, etc., or the motor-driven transport of hydrophobic, non-water-soluble materials is needed for construction of membranes or for waste disposal, but diffusion of supporting molecules like ATP and Na^+Cl^- is required for all. Diffusion also underlies chemical reactions, even when the time for diffusion of molecules is small compared to times for other processes. When we come to chemical reactions we will deal with many processes that are "not diffusion controlled," but this usually means that diffusion brings the reacting molecules together, but that they take several (or many) "attempts" at reacting before they actually react.

> We should perhaps exclude biomedical issues such as macroscopic bone and muscle mechanics from this generalization, but even those have diffusive processes underlying the maintenance of their living state.

The simplest mathematical descriptions of diffusive processes in water require a molecule's diffusion constant D, which appears in diffusion schemes in the form

$$\langle r^2 \rangle = 6Dt \tag{4.10}$$

in three dimensions, where r is the distance from the starting point at $t = 0$, t is the time and angle brackets indicates the average, usually the average over many independently diffusing molecules. This equation implies that motion in any direction is equally likely and efficient. In two and one dimensions, the 6 is replaced by 4 and 2, respectively, again with the assumption of equal motion in any allowed direction. Typical values of D for small molecules diffusing in water are $D \sim 1 \, (\mu m)^2/ms = 1 \times 10^{-9} \, m^2/s$. The former value tells us that on a typical biochemical time scale of 1 ms, a small molecule moves 2–3 μm from the starting point. This is approximately the diameter of a prokaryotic (bacterial) cell, which suggests to us already that bacteria do not need specialized transport systems as eukaryotic cells do; this point further suggests that bacteria do not have the same need for cell cytoskeletons. At least to a small extent, the rate of diffusion in water explains some of the basic design features of eukaryotic versus prokaryotic cells.

Water diffuses in itself, of course; its self-diffusion coefficient is

$$D(H_2O) = 2.272 \times 10^{-9} \, m^2/s$$
$$D(D_2O) = 2.109 \times 10^{-9} \, m^2/s \tag{4.11}$$

Applying Equation 4.10 to water gives a time for water to diffuse a distance corresponding to its own diameter (3 Å) of about 6.6 ps.

4.7 SOLUTES AND THE SOLVENT POWER OF WATER

In condensed matter physics, water in cells would be described as H_2O molecules plus impurities. These impurities are critical to life, as phosphorus might be to the function of a silicon semiconductor. The Na^+Cl^- "impurities" in a cell are about 0.15 M, corresponding to about $0.15/55.4 = 2700 \, ppm$ (parts per million), whereas semiconductor impurities are more typically 1 ppm. We use the term *solutes*, not *impurities*, to describe the medium inside cells, however. Semiconductor scientists use the term *impurities* because it correctly implies that processes in the semiconductor are radically different with the impurity and further, that almost any uncontrolled impurities (e.g., metal atoms) completely disrupt the desired function. Biological cells, in contrast, contain many solutes and are much more resistant to extraneous impurities, unless the impurities happen to be poisons, such as CO, arsenic, or singlet oxygen (O_2 in the electronic spin = 0 state); or "signaling" molecules, such as Ca^{2+}, hormones, or NO (nitric oxide). In the latter cases, consult a textbook discussing biological signal transduction (e.g., Voet and Voet, *Biochemistry*, Chapter 19) to read about the wide variety of cell responses to these specific cellular signaling molecules.

The most abundant solute in the "cytosol"—the liquid contents of a cell—is NaCl. The most well-known physical effect of the NaCl is the depression in the freezing point of water. The depression of the freezing point of water can be described approximately by

$$\Delta T_f = -K_f\, m_{\text{solute}}\, i \tag{4.12}$$

where K_f is the cryoscopic constant (about $1.86°C\cdot kg/mole$ for water), m_{solute} is the "molality" (moles of solute/kg of solvent), and i is the number of separate particles (usually ions) formed by the dissolved solute: $i = 2$ for NaCl and 3 for $CaCl_2$. This depression does not depend, to a first approximation, on the specific solute, only on its concentration and ionic number. For 0.15 M (~0.15 mole/kg) NaCl, $\Delta T_f \sim -0.6°C$.

More important biological functions of NaCl, however, include acting as counterions to reduce electrical forces and energies between charged groups, carrying electrical current and allowing charge neutralization when cells require time-dependent increases in specific ions, such as Ca^{2+} or H^+. In this sense, Na^+Cl^- provides a bath of charges of either sign that can be used to neutralize charge. There are specific roles Na^+ plays, however, such as in the "Na^+-glucose symport," which transports glucose and Na^+ from intestinal cells to the bloodstream. (Cl^- maintains charge neutrality.) The transport of Na^+ to the bloodstream increases the resorption of water. Administration of glucose, salt, and water has acted to save the lives of many persons suffering from severe diarrhea.

4.7.1 Solubility in Water

Water dissolves more substances than any other known solvent. Water is sometimes called a universal solvent. Though this solvent power can be correlated with many properties already discussed—hydrogen bonding, the dielectric constant (reduces electric interactions), and the related orienting of water dipoles alongside solute dipoles—no one knows a general way to calculate substance solubility based on these properties of water. Predicting solubility is generally not attempted.

Important types of supramolecular structures can be dissolved by water because of its strong dipolar and dielectric interactions. Hydrophobic ("oily") molecules are generally insoluble in water, but if a molecule is constructed with a hydrophobic end and a dipolar or charged end, important large structures, lipid bilayer membrane structures such as *liposomes*, can be formed which are dissolved by water. Membranes were discussed briefly in Chapter 1 (see Figure 1.7). They play a central role in biology. Because membranes are an integral part of the cell surface, we rarely speak of the "solubility" of lipid structures, but the strongly favorable (negative free energy of) interaction between water and the polar or charged heads of lipid molecules would qualify liposomes as soluble entities.

> How large a structure can be described as "dissolved" is a matter of judgment. If a chunk of material is placed in water and stirred then gradually disappears, it is usually described as dissolved. But this is a judgment call, based on ignorance. What if one could observe particles with a magnifying glass? Or with a microscope? What if particles sink to the container bottom in 10 s? In 10 days? We apply the term *dissolved* when discreet structures are stable and dispersed in water, with little tendency to aggregate or adhere to container surfaces.

> Hydrophobic means "water fearing." Hydrophobic interaction does not represent a truly fundamental intermolecular force. Hydrophobic reflects a more favorable interaction (lower free energy) of a molecule with air or with glass or plastic container surfaces than with water. Nevertheless, we occasionally follow the tradition in biochemistry in speaking of hydrophobic interactions.

4.7.2 Water Ionization

Water is said to readily ionize, described by the equilibrium chemical equation

$$H_2O \Leftrightarrow H^+ + OH^-, \text{ with } K_w \equiv [H^+][OH^-] = 10^{-14}\,M^2 \text{ (at 25°C)} \tag{4.13}$$

where the brackets indicate the molar concentrations of the species. This notation is used to define the pH of aqueous systems,

$$pH \equiv -\log_{10}[H^+] \tag{4.14}$$

keeping in mind that additional H^+ or OH^- ions can come from other molecules added to the water. (Strong acids rapidly transfer all their protons to water, as strong bases do their OH^- groups.) In actuality, the H^+ ion does not exist as a distinct physical/chemical species, but rather bound to water molecules in the form of H_3O^+, $H_5O_2^+$, etc. While pH 7 is commonly interpreted as "neutral," $[H^+] = [OH^-]$ in pure water, if the value of K_w changes due to a temperature or pressure change, the pH corresponding to "neutral" will also change. The "ionization constant" K_w describes the formation of separated ions, not the ejection of an electron, as in physics, and is generally treated as a universal constant, though the ionization is, in fact, temperature dependent, and the process can be described as normal physicochemical processes of equilibrium:

$$\frac{[H^+][OH^-]}{H_2O} = 1.821 \times 10^{-16} M \quad (4.15)$$

with $[H_2O] = 55.345$ M at 25°C, and

$$2H_2O \rightarrow H_3O^+ + OH^-, \Delta G = 79.907 \text{ kJ/mol} \quad (4.16)$$

We treat these topics in more detail in Chapter 14. It should be noted that in any living or model biosystem, pure water, where all H^+ and OH^- come from water molecules, does not exist. Nevertheless, Equation 4.13 still applies, because the concentration of water is at least hundreds of times greater that concentrations of other ionizable molecules.

Finally, note that in a macroscopic volume of water, even one as small as a bacterial cell, overall charge neutrality holds. Nature may give a large macromolecule a net charge, but compensating opposite charges can be found within several water layers around the molecule.

4.8 POINTS TO REMEMBER

In this chapter we begin an important chapter afterword intended for those *rare* times when you are overwhelmed. My intent is not to minimize other issues, but some things are more important than others if our goal is to describe the dynamics of biosystems quantitatively.

So, if you forget everything else from this chapter, *don't* forget the following:

- Water consists of H_2O, H_3O^+, and OH^-. Know how to work with pH.
- Water is dynamic, moving rapidly, and interconverting between the forms above.
- Water in cells is "local"; its altered structure near micro-objects is important.
- Overall charge neutrality generally holds on distance scales of 1 μm or larger.
- Know how to calculate electric forces and energies with dielectric constants.
- Understand log–log plots.

4.9 PROBLEM-SOLVING

PROBLEM 4.1 WATER AND MOLECULAR DENSITY DISTRIBUTION FUNCTIONS

Figure 4.3 shows the radial distance dependence of the radial probability density as a function of (O–O) distance between primary and peripheral water molecules.

1. Calculate the value of ρ_0, the bulk molecular density of water, in molecules/m³ and in molecules/(nm)³.
2. Why is $G(r) = 0$ below about 0.25 nm?
3. Explain the physical meaning of the value of $G(r)$ at large distances.
4. Explain the physical meaning of the maximum of the function $G(r)$.
5. Determine the number of water molecule O atoms located a distance r_{max}, corresponding to where G is maximum. Don't forget the geometry is a spherical shell. Take the shell thickness Δr to be 0.04 nm (shell between $r_{max} \pm \Delta r/2$).
6. Repeat step 5 for $r = 0.6$ nm.

PROBLEM 4.2 WATER: TETRAHEDRAL ANGLES AND H BONDS

Section 4.2 stated that water molecules in ice are arranged tetrahedrally.

1. Draw this structure, showing enough water molecules for at least one tetrahedral unit. Diagrams such as this are well known (e.g., see Voet and Voet, *Biochemistry*), but try to draw the diagram without consulting such textbooks.
2. Determine the tetrahedral angle (the one that is slightly more than 100°) from basic geometrical considerations. Write the angle, precise to three decimal places.
3. A water molecule has only two H atoms. Explain how each water molecule can (potentially) form four hydrogen bonds, and not just two.

PROBLEM 4.3 WATER AND STRUCTURAL FLUCTUATIONS

Section 4.2 stated that the 8-ps rotational motion of water molecules means that any extended, H-bonded complexes of water molecules cannot endure for long. Develop this argument more convincingly (or disprove it!) by applying it to the following.

1. A planar trimeric ring of water molecules, each of which is H-bonded to the neighbor on each side. Draw the structure of the ring, showing positions of O and H atoms and H bonds.
2. A tetrahedral arrangement of water molecules (like in ice), where each water is H bonded to four neighbors. First, draw the oxygen atoms of four water molecules at the corners of a tetrahedron, place the fifth O in the center of the tetrahedron, draw in H atoms, show the four H bonds to the central water, and discuss what happens when the central water molecule rotates.

PROBLEM 4.4 POND IN THE NORTH

Assume a pond has maximum depth of about 1 m, with temperature of 6.0°C at the bottom. Assume that the air above the pond in the winter is at a constant –10.0°C, and that such conditions apply nearly all winter (~4 months). The pond bottom is, of course, connected to the Earth, which can transfer heat to or from the pond. As things work now, the pond bottom normally stays at about the same temperature, because the water above it is not frozen (0°C or higher) and cold weather creates an ice sheet on top of the pond, which insulates it. Qualitatively discuss what would happen, including biological implications, if ice did not form on the surface, if water kept getting more dense as it went from 4°C to 0°C and froze. Consider time scales of both a few days and a few thousand years. Also discuss what difference the variable thermal conductivity of the ice would make (Section 4.3.3).

PROBLEM 4.5   FREEZING POND

Refer to your first-year physics text if you need to. Remember that for a two-layer insulation problem, the top and bottom temperatures are known, but the temperature at the interface between the two materials is not, initially. The solution is to equate dQ/dt through both layers.

Assume a 1.0-m-deep pond has constant depth, with temperature at the bottom constant at 6.0°C. Assume the air above the pond is at –15.0°C, kept constant by light wind blowing over the pond. Describe any assumptions you need to make in the following.

1. Discuss what will happen to the water near the surface as it cools but before it freezes.
2. Calculate the rate, in J/s, at which heat is transported out of the pond per square meter of surface area, if there is no ice. Ignore convection in the water—a bad approximation to make, but convection rapidly makes things difficult.
3. Assume there is now a sheet of ice of thickness d on the surface of the pond. Calculate the rate/m² at which heat is now transported out of the pond and find the ice thickness. First use the thermal conductivity of ice from Table 4.4, then a conductivity five times less. (Hint: What should the temperature at the bottom of the ice be?)

PROBLEM 4.6   ELECTRIC FORCES AND BIOMOLECULAR REACTIONS

Biochemical reactions are typically governed by an "activation energy," an amount of energy that must be provided in order for the reaction to proceed. A typical activation energy may be about $25k_BT_R$ per molecule, where k_B is the Boltzmann constant and T_R is room temperature in kelvin. The activation energy is often the energy needed to produce a conformational (structural) change in a biomolecule or complex between two biomolecules. Suppose this conformational change must decrease the distance between two charges of +e on a protein, in order to enable a chemical reaction. Assume the initial separation of the charges is 0.50 nm. If the $20k_BT$ activation energy goes into pressing the charges closer together, calculate the decrease in charge separation if

1. The medium between the charges has a constant dielectric constant of 6.0.
2. The medium has a constant dielectric constant of 78.5.
3. The medium is water, with dielectric constant described as in Figure 4.11. (You may approximate the values from the graph, as necessary.)

PROBLEM 4.7   DIFFUSION OF WATER IN WATER: LOG–LOG PLOTS

1. Use Microsoft Excel or a graphing program to make a log–log plot of the time dependence of the root-mean-square (RMS) distance from a starting point ($r_{RMS} = \sqrt{\langle r^2 \rangle}$) of a water molecule diffusing in water. Plot $\log(r_{RMS})$ versus $\log(t)$. What is the slope of this graph?
2. How long will it take to diffuse 2 Å? 2 nm? 2 μm? 2 mm?

PROBLEM 4.8 SURFACE MOLECULES AND SURFACE: VOLUME RATIOS FOR SMALL VOLUMES

1. Calculate the surface area to volume ratios of spheres of radius 30 nm, 3.0 μm, and 300 μm. Find biological objects that correspond, roughly, to each sphere.
2. Assume these spheres are filled with liquid water. Calculate the number of molecules in each volume.
3. Calculate the number of water molecules on the surface of each sphere. Draw a diagram and describe how you determine the surface area occupied by a water molecule. If you like, consult www1.lsbu.ac.uk/water/water_properties.html or another reliable source for any information you may want. (However, you can get an acceptable answer from information in this chapter.)

PROBLEM 4.9 DIELECTRIC THEORY

1. Use the Debye, Onsager, and Kirkwood–Fröhlich equations to calculate the dielectric constant of water at 20°C. Be careful to use consistent units and think about use of °C versus K temperatures in the calculations. (Use these "hand-calculated" values to check values in the graphs in step 2.) You will need to do some digging to find needed values of parameters.
2. On one graph, plot the value of each theoretical ε versus temperature, from 0°C to 100°C.

PROBLEM 4.10 DROPS AND MINIDROPS

A water drop of radius 2 mm is floating in the weightless environment of the space station, where the ambient temperature and pressure are 25°C and 1 atm. Suppose the drop hits some objects and breaks up into 100 smaller droplets, of equal radii.

1. Calculate the change in PV energy of the drop. Recall from first-year physics that PV represents an energy, that changes in either P or V produce energy changes, and that the change in PV in this case is the interior pressure change of the drop-to-100-droplets transition times the total volume, not the Δp between inside and outside of a drop times volume. Indicate whether the energy increases or decreases.
2. Calculate the change in surface tension energy ΣA when the drop breaks up into 100 minidrops. Indicate whether the energy increases or decreases.
3. Find the total energy change, 1 drop → 100 drops. If two minidrops collide at very low relative velocity, will they tend to combine and form a larger drop? Explain, and explain why formation of small droplets from large (not vice versa) generally happens in the weightlessness of space.
4. Graduate computation. Assume two identical mini-drops traveling with equal and opposite velocities collide head on. If they collide with a low enough velocity, they may combine into one drop, which may oscillate in shape and heat up slightly but which will have zero translational velocity. If the velocity is larger, the two drops will collide and form two or even more separate drops. Estimate the value of the velocity below which the drops may combine.

MINIPROJECT 4.11 SURFACE TENSION AND LUNG FUNCTION

A biomedical miniproject on lung surface tension. Research and write a ~250- to 300-word report on the small surface tension value for the surfactant found in lungs. Include (at a minimum) discussion of what this surfactant is and why it is needed. Discuss one or two primary biomedical disorders involving the surfactant, the severity of the disorder, and how the disorder is treated. Include at least one calculation of the effect of the surfactant and its absence.

PROBLEM 4.12 FREEZING-POINT DEPRESSION

Graph the freezing-point depression (temperature reduction) of water with NaCl varying from 0 to 5.0 M. Look up the "salt" concentration of sea water, as well as its freezing point. Mark these on the graph and discuss. Consult https://chem.libretexts.org/Core/Physical_and_Theoretical_Chemistry/Acids_and_Bases/Acids_and_Bases_in_Aqueous_Solutions/The_pH_Scale/Temperature_Dependence_of_the_pH_of_pure_Water if you need further clarification.

PROBLEM 4.13 WATER IONIZATION AND TEMPERATURE

The ionization constant K_w of water is 10^{-14} M² at a temperature of 25°C. At 100°C, K_w decreases to about $10^{-12.26}$ M². What should the pH of pure water be at 100°C? Neglect absorption of atmospheric CO_2 or other materials. Hint: Must $[H^+]=[OH^-]$? (Note that this is not the same as asking the pH of a 100°C buffer solution that has a pH of 7.00 at 25°C, or the pH reading of a pH-meter's electrode at 100°C.)

REFERENCES

1. Voet, D., and J. G. Voet, 2004. *Biochemistry*, 39. New York: John Wiley & Sons.
2. Catling, D. C., et al. 2005. Martian vistas (Editor's summary and following articles). *Nature* 436:42–69.
3. Franks, F., Ed. 1972–1982. *Water, A Comprehensive Treatise*, vol 7. New York: Plenum Press.
4. Franks, F., Ed. 2009. *Water Science Reviews 5: Volume 5: The Molecules of Life.* (Reissue edition). Cambridge, UK: Cambridge University Press.
5. Hasted, J. B. 1973. *Aqueous Dielectrics*. London, UK: Chapman and Hall.
6. Eisenberg, D., and W. Kauzmann. 1969. *The Structure and Properties of Water.* London, UK: Oxford University Press.
7. Finney, J. L., A. Halibrucker, I. Kohl, A. K. Soper, and D. T. Bowron. 2002. Structures of high and low density amorphous ice by neutron diffraction. *Phys Rev Lett* 88:225503-1 to 225503-4.

8. Wernet, P., D. Nordlund, U. Bergmann, M. Cavalleri, M. Odelius., et al. 2004. The structure of the first coordination shell in liquid water. *Science* 304:995–999.

9. Morgan, J., and B. E. Warren. 1938. X-ray analysis of the structure of water. *J Chem Phys* 6:666–673.

10. Soper, A. K. 2000. The radial distribution functions of water and ice from 200 to 673 K and at pressures up to 400 MPa. *Chem Phys* 258:121–137.

11. Franks, F. 2000. *Water: A Matrix of Life*. Cambridge, UK: Royal Society of Chemistry, 27–28.

12. Richardson, J. O., C. Pérez, S. Lobsiger, A. A. Reid, T. Shields, G. C. Berhane, Z. Kisiel, D. J. Wales, B. H. Pate, and S. C. Althorpe. 2016. Concerted hydrogen-bond breaking by quantum tunneling in the water hexamer prism. *Science* 351(6279):1310–1313.

13. Coles, W. D. 1954. Experimental determination of thermal conductivity of low-density ice. Technical Note 3143. Cleveland, OH: Lewis Flight Propulsion Laboratory.

14. Debye, P. J. W. 1912. Polar molecules. *Phys Z* 13:97; can be found in *The Collected Papers of Peter J. W. Debye*. 1954. New York: Interscience, 173.

15. Onsager, L. 1936. Electric moments of molecules in liquids. *J Am ChemSoc* 58:1486–1493.

16. Fröhlich, H. 1987. *Theory of Dielectrics (2nd Ed.)*. Oxford, UK: Oxford University Press, 1–200.

17. Sheinerman, F. B., R. Norel, and B. Honig. 2000. Electrostatic aspects of protein–protein interactions. *Curr Opin Struct Biol* 10:153–159. See also Ritchie, A. R. and L. J. Webb. 2015. Understanding and manipulating electrostatic fields at the protein-protein interface using vibrational spectroscopy and continuum electrostatics calculations. *J Phys Chem B* 119:13945–13957.

18. Dix, J. A., and A. S. Verkman. 2008. Crowding effects on diffusion in solutions and cells. *Annu Rev Biophys* 37:247–263.

19. Takashima, S., A. Casaleggio, F. Giuliano, M. Morando, P. Arrigo, and S. Ridella. 1986. Study of bound water of poly-adenine using high frequency dielectric measurements. *Biophys J* 49:1003–1008.

20. Krauss, S. W., Chen, C., Penman, S., and Heald, R. 2003. Nuclear actin and protein 4.1: Essential interactions during nuclear assembly *in vitro*. Proc Natl Acad Sci USA 100, 10752–10757.

CHAPTER 5

CONTENTS

Structures

From 0.1 to 10 nm and Larger

This chapter is an extended introduction to the molecules that underlie all biological structure and function. The section headings may look inconsequential, equations are few and far between, and the detail is less than what is typically found in advanced biochemistry textbooks. Nevertheless, the contents provide a fundamental vocabulary and set of working tools needed for a successful grasp of the physics of biosystems. The omission of the "theoretical-chemical" foundations of amino acid structure (precursors, chirality, metabolites, etc.), chromatography, cloning, and many other topics essential to the future bioscientist is mostly intentional, as it would divert attention away from our main goal—to describe quantitatively the *dynamics* of biosystems.

Variation in molecular mass over many orders of magnitude—hierarchical structural design— is a feature that separates biosystems from nonliving systems. Living cells are made up of atoms, of course, but molecules of up to millions of dalton (Da) are also common. These large molecules are not formally "polymers," because they are not made up of a single type of monomer. Such large molecules have always been a challenge to deal with because of their complexity. Even displaying them has been a perennial challenge. One of the greatest early contributions to the understanding of biomolecules was made by an artist, Irving Geis, who drew pictures of the myoglobin (Figure 5.1a) and hemoglobin structures determined by Nobel Prize winners John Kendrew[1] and Max Perutz.[2] Although Kendrew and Perutz certainly deserved their prizes, Geis brought visualization to a set of hundreds of atomic coordinates that could not be displayed by computer at that time, the 1960s. Geis gave us the pictures that let scientists sit around a table in a bar to argue about structure and mechanisms: hard to do with a sheaf of papers covered with numbers. A later development in the display of biomolecular structure was made by Jane Richardson, whose *The Anatomy and Taxonomy of Protein Structure*[3] was a standard bookshelf occupant in biophysics and molecular biology research laboratories for 20 years, until desktop computers could routinely display structures from online databases. Richardson brought the routine use of protein "ribbon" displays of structure, sometimes called cartoon structures, though they are based on the atomic coordinates, not on careless sketches. Such a display of myoglobin is in Figure 5.1b.

> We should also take seriously the work of coauthors of these articles, without whom the Nobel prizes would have been delayed or nonexistent.

The study of biomolecules has always been associated with human disease, its causes and cures. Disease can, in fact, often be connected to the specific structures of biomolecules and their modifications, not just an invasion of foreign microorganisms or viruses. Two examples include prions (mad cow disease) and cancer. In 1984, Stanley Prusiner and group purified and isolated the specific protein that one prion was composed of, the Prion Protein (PrP). This protein occurs in infectious PrP^{Sc} (Sc = "Scrapie") and noninfectious PrP^{C} (C = "cellular") forms. The latter has 210 amino acids and molecular mass 36,000 D, making it about 30% larger than the myoglobin molecule shown in Figure 5.1. Like myoglobin, it also has considerable α-helical structure (Section 5.6.1.3). The infectious PrP^{Sc} form has a higher proportion of β sheets (Section 5.6.1.5), rather than α helices, and apparently stimulates PrP^{C} molecules to convert to PrP^{Sc} and aggregate into *amyloid plaques* that accumulate outside of cells. Human prion diseases are presently untreatable and generally fatal when the disease is fully active, but amyloid plaques can sometimes be present without any obvious ill effects. Like cancer, prion disease is not a closed subject. Prusiner won the 1997 Nobel Prize in Physiology or Medicine for his prion research.

Cancer comprises an enormously complex set of diseases, with thousands of research articles appearing every year. A few recent discoveries relate to an unanticipated cancer-DNA structural issue. Flavahan, Drier Y. et al., in a paper entitled, "Insulator dysfunction and oncogene activation in IDH mutant gliomas" (Nature Online December 23, 2015; see www.broadinstitute.org/news/7706 for a nontechnical account), found unusual changes in the way a genome folds up in brain tumors.

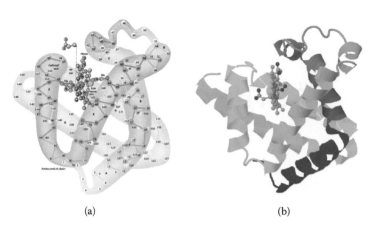

(a)

(b)

FIGURE 5.1

(a) Early drawing of sperm whale oxy-myoglobin atomic structure by Irving Geis. (Reproduced by permission: Irving Geiss Collection, Howard Hughes Medical Institute. Rights owned by HHMI; not to be reproduced without permission.) (b) "Cartoon" display of oxy-myoglobin structure, popularized by Jane Richardson, but displayed here from the PDB using Jmol software. (1A6M, based on X-ray crystal structure by Vojtechovsky, J. et al. See http://www.wwpdb.org or http://www.pdb.org.)

The primary work, however, has been done by thousands of scientists around the world who have voluntarily, or because of journal publication requirements, submitted sets of atomic coordinates to the databases. Most scientific journals now require structure submission to these databases prior to publication.

"Cookies" and "pop-ups" must sometimes be enabled. Special attention must also be paid to the version of java installed on the computer used to display structures. See Help/Troubleshooting at www.rcsb.org/pdb and discuss with local IT authorities.

The changes relate to key parts of the genome called insulators, which physically prevent genes in one region from interacting with the control switches in neighboring regions. When the insulators lose control in tumors, a growth factor gene falls under the control of an always-on gene switch. Such "always-on" growth is the primary problem of cancer. While gene folding is larger-scale than the 0.1–10 nm structures of the present chapter, the fact that small sections of a DNA strand can trigger structural effects on a scale of an entire genome points again to the importance of hierarchical structure and function in biology.

5.1 SOFTWARE TO DISPLAY AND ANALYZE BIOLOGICAL STRUCTURES

Since the 1990s, virtually all displays of complex biomolecules, as well as even larger, supramolecular structures like viruses and ribosomes, have been via desktop or laptop computers and online databases. The basic source data for these structures are primarily X-ray and neutron crystallography, nuclear magnetic resonance (NMR), and electron microscopy, combined with various structural refinement computations, including molecular dynamics. Two of the primary biomolecular structural databases are the Protein Data Bank (PDB) and Nucleic Acid Database (NDB). These are U.S. government–supported databases with counterparts funded internationally. Currently, databases may be found at

PDB: www.rcsb.org/pdb/home/home.do; see also www.wwpdb.org/ for worldwide sites;

NDB: http://ndbserver.rutgers.edu/

The PDB presently archives about 130,000 structures, the NDB, about 9,000. Most of the NDB nucleic acid structures are also located in the PDB, where a variety of different software packages can be used to display and manipulate the structures. Most of this software has user-friendly graphical interfaces, as well as "Instructions for Dummies" and extended user manuals.

Prior to about 2000, software had to be installed as separate programs on your computer, was not compatible with some operating systems, and ran slowly on the average personal computer. Many packages now operate via most Web browsers and Java scripts or applets, though some must still install browser plug-in viewers or stand-alone applications. Common apps in 2017 include NGL, Jmol/JSmol, Protein Workshop, Kiosk Viewer. The plug-in and stand-alone viewers are quite powerful, capable of sophisticated analyses and computations, but some software issues (esp. Java) may be noted. The indexing and search engines in these databases are becoming user-friendly, also, and do not require prior knowledge of structure codes (e.g., "1la8" stands for the solution structure of the DNA hairpin 13-mer CGCGGTGTCCGCG) to display a structure successfully. Codes may normally be found in published journal articles, however, offering a quick way to locate structures. If a structure is found in the database, links to the original publication, as well as to related articles, can also be found. Figure 5.2 shows a typical page displayed from the PDB. The reader should browse through the PDB and NDB databases in preparation for what is to come.

Before we consider the molecules important to biosystems, we should survey the collection. Table 5.1 summarizes, in terms of approximate mass ranges and amounts in cells, various categories of molecules and other subcellular components. Most single lipid molecules have mass less than 1000 Da, but because lipid molecules most commonly occur in two-dimensional aggregates, they are placed in the 10^3–10^7 Da range. Similarly, proteins, DNA, and polysaccharides are polymers with variable numbers of (nonidentical) units. Their masses range from as small as 1 kDa to 10 MDa.

FIGURE 5.2

Page displayed from PDB Web site http://www.rcsb.org/pdb/explore/explore.do?structureId=1la8 on May 11, 2017, showing part of the main page for structure 1la8, the DNA hairpin 13-mer, CGCGGTGTCCGCG. Links to original articles, as well as software viewers like NGL, JSmol, Simple Viewer, Protein Workshop, Ligand Explorer, and Kiosk Viewercan be found. Sets of atomic coordinates may also be downloaded for independent analysis.

Many molecules critical to life are not listed in Table 5.1 because their amounts are so tiny. We will not dwell on this subject, but it is interesting to note that some atoms and molecules occurring in humans in trace quantities are both required nutrients and poisons, depending on the amount.[4] Examples among the transition metals include chromium, copper, manganese, zinc, and even iron. These elements are often needed because they take part in oxidation-reduction reactions involved in cellular respiration, chemical detoxification, metabolism, and neurotransmitter synthesis. We will also see zinc playing a central role in DNA-binding proteins, in a role referred to as *zinc fingers*. However, elements such as lead and cadmium may be metabolized like iron, cannot play the same biochemical role as iron, and act strictly as poisons.

TABLE 5.1 Masses of Structures in Cells

Molecular Mass Range (Da)	Molecule or Structure				Cell Mass(%)
$1-10^2$	H_2O				70
	Ions: $H_3O^+,Na^+,K^+,Mg^{2+},Ca^{2+},NH_4^+$				1.2
	$OH^-,Cl^-,SO_4^{2-},PO_4^{3-},CH_3COO^-$				
	Sugars: glyceraldehyde, erythrose, ribose, glucose, fructose, sucrose, lactose...				1
10^2-10^3					
	Amino acids: glycine, leucine, tryptophan, serine...				0.4
	Nucleotides: adenine, thymine, cytosine, guanine				0.4
	Fatty acids: myristoleic, palmitic, stearic, linoleic, etc.				1
10^3-10^5	Proteins	DNA	↑Lipids	Polysaccharides	Proteins: 15
			↓		DNA: 6
10^5-10^7			Membranes		Lipids: 2
					Polysaccharides: 2

5.2 SOLVENTS

Chapter 4 discussed water, the solvent of life, in some detail. Protein and DNA structures in the previous section did not include water in any of the structures. Go back to the PDB and look for the structure *1ply*, which is the structure of Poly d(A)·Poly d(T), a synthetic nucleic acid polymer made up of strand of adenine (A) paired with a strand of thymine (T). (The "d" stands for "deoxy"; more about that later.) A display of the resulting structure, using Protein Workshop software with "balls and sticks" style, is shown in Figure 5.3a. Figure 5.3b shows another view of the structure, but this time with bound water molecules (actually, just the oxygen; X-ray diffraction is weakly sensitive to hydrogen atoms) displayed.

> This particular structure was determined from a polycrystalline fiber of Poly (dA)·Poly (dT).

Why can't we display the ubiquitous solvent water molecules? You learned about X-ray diffraction through scattering off a crystal, where atoms had fixed locations. Solvent water molecules, as we just learned, move around rapidly and do not oscillate around a fixed location. This is true even in a protein crystal, except when the waters are bound to fixed locations. X-rays scattered from

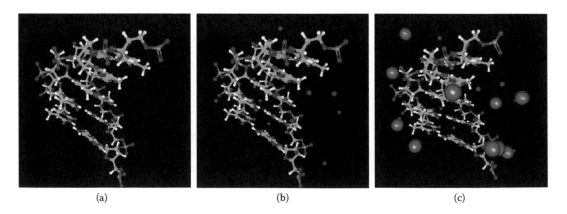

(a)　　　　　(b)　　　　　(c)

FIGURE 5.3

Structure of Poly d(A)·Poly (d(T) (PDB code *1ply*) displayed using Protein Workshop software (a) without bound water molecules (O atoms of water) displayed; (b) with water displayed. Structures of biomolecules determined from crystallography or NMR methods may show bound water molecules but will have no other solvent molecules, because the methods are incapable of locating atoms that have no fixed location. Note the water oxygen atoms are colored red like the phosphate group oxygens and; (c) 1ply with localized water and Na^+ ions displayed.

nonfixed solvent waters cannot interfere and form diffraction spots, so they do not appear in structures. This is a weak aspect of the apparently powerful structural software we have just examined: they do not see solvent because the experimental methods used to obtain their atomic coordinates do not see them. Methods other than diffraction or NMR must generally be used to study solvent molecules in biosystems.

5.3 SMALL MOLECULES

5.3.1 Important Ions

The ions H_3O^+, Na^+, K^+, Mg^{2+}, Ca^{2+}, NH_4^+, OH^-, Cl^-, SO_4^{2-}, PO_4^{3-}, CH_3COO^-, and others are fundamentally important to biosystems because of water and the pervasive presence of the electric force. The electric force is the basis of all molecular electronic structure and interactions. Water, with its high, but variable, dielectric constant modulates and controls the magnitude of the force. Some of the general uses of ions are described below.

5.3.1.1 Counterions

The most general use of small ions such as Na^+, K^+, Mg^{2+}, Ca^{2+}, NH_4^+, OH^-, and Cl^- in biosystems is as counterions. Given the freedom to do so, ions naturally move toward opposite charges, greatly reducing the electric field extending from the charge. This electric field reduction in most of space lowers the overall energy. Most often the counterions form a mobile layer of countercharge, each ion having no fixed position. The spatial dependence of the concentration of such layers of charge can be described in general thermodynamic terms. We will briefly sketch this approach later.

Counterions may also be bound to particular charged sites of biomolecules. ("Counter" in counterions implies the existence of another charge to "counter.") We just looked at structure *1ply*, a short Poly d(A)Poly d(T) polymer in the PDB, turning on the display of bound waters to visualize them. Many of you may have noticed localized Na^+ ions, the display of which has been turned off in Figure 5.3a and b, but on in Figure 5.3c. These sodium ions and water molecules are localized, since data was collected using X-ray diffraction; so they are truly a part of the structure. However, structures in the PDB are usually from crystals of DNA, with large numbers of identical DNA polymers bound together in an array, often with high concentrations of unusual ionic additives that stimulate formation of the crystal, rather than the environment in a living cell, where DNA molecules are surrounded by water, ions, and perhaps other DNA and protein molecules in a noncrystalline structure. Could the localized water molecules and Na^+ ions be an artifact of the sample preparation? Proof of an answer to this question is rare, but the description of the structure suggests the H_2O's and Na^+'s are not all artifacts of the sample preparation:

> *The helix contrasts itself from B-DNA in terms of a very narrow minor groove. Difference electron-density maps have revealed that a continuous spine of water molecules, two per base pair, propagates along this groove with the same symmetry as the DNA and establishes new links between the two strands. In addition to this hydrated DNA helix, the monoclinic unit cell... accommodates about 20 sodium ions and 12 water molecules in the vicinity of phosphate groups. These structured guest molecules provide an intricate network of bridges, ranging in size from a single sodium ion to a multiple sodium–water–water–sodium unit, connecting phosphate groups belonging to adjacent DNA helices.*
>
> **Chandrasekaran et al.**
> *http://www.rcsb.org/pdb/explore/explore.do?structureId=1PLY*

There are several important statements in this abstract, as well as some terms that must be read carefully. First, there is a "spine" of water molecules (sometimes called a *spine of hydration*) located in the minor groove between the two "strands." These are water molecules bound very much in the interior part of *each* DNA molecule, not on the outside of the structure. Carefully examine Figure 5.3b to locate this spine of water molecules. Contrast, the statement, "These structured guest molecules provide an intricate network of bridges, ranging in size from a single sodium ion to a multiple sodium–water–water–sodium unit, connecting phosphate groups belonging to adjacent DNA helices." *Adjacent helices* refers to those of an entire DNA double helix and those of its neighbors. Counterions located between DNA molecules in a crystal are not surprising, of course. If DNA is a negatively charged molecule, positive counterions must be present somewhere to neutralize that charge of the crystal. Where else would they be, other than between adjacent DNA molecules?

So, what have we found with water molecules and small ions as they actually behave when associated with biomolecules?

1. Water molecules and ions will form relatively unstructured, but localized, mobile layers around charged groups on biomolecules like DNA.

2. Water molecules and ions like Na^+ can bind specifically near charged sites on biomolecules and remain relatively immobile, as measured by successful X-ray crystallographic location determinations. In DNA, these sites may constitute a spine of hydration and counterions to DNA phosphate charges.

3. Some ions and water molecules may be immobilized at sites governed by interactions between adjacent molecules in the (closely packed) crystal used for X-ray crystallography. These bound ions would likely not be there in the absence of the close proximity enforced by the crystal structure. Does this mean these localized ions are irrelevant to biology? Probably not entirely. The true action in biology occurs when molecules like DNA directly interact (come in contact) with other molecules. If intervening Na^+ ions stabilize (reduce the energy of) the structure when two DNA molecules interact in a crystal, a similar effect is likely to occur in a cell when DNA comes into contact with another DNA molecule or with a protein molecule. If Na^+ ions are present in the cell interior when intermolecular contact occurs, they will bind, if doing so reduces the energy.

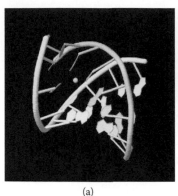 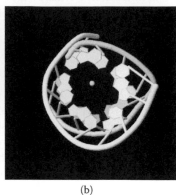

(a)   (b)

FIGURE 5.4
Structure of *1dnt* (RNA/DNA dodecamer [12 nucleotide units]) with magnesium binding sites using Jmol Java script software. The structure is viewed (a) from the side and (b) from the end of the helix. A magnesium ion is bound in the center of the helix.

FIGURE 5.5
Structure of *3icb*, the vitamin D-dependent calcium-binding protein from bovine intestine, displayed using Protein Workshop software. Bound calcium ions are evident in the structure. Calcium ions are frequently employed as signaling molecules and are bound to specific enzymes, stimulating or inhibiting their activity. On the left side of the protein structure is also a localized sulfate ion. Thirty-six localized water molecules are not shown.

What about doubly charged ions, like Ca^{2+} and Mg^{2+}? Figure 5.4 shows a typical location for a Mg^{2+} ion in a nucleic acid structure: bound to DNA. Doubly charged positive ions are often bound more tightly because of the larger electric force. (This tighter binding does not necessarily apply to other systems, such as protein-bound DNA, where size and space issues may be important.) Two Ca^{2+} ions and one SO_4^{2-} ion bound to a protein are shown in Figure 5.5.

Are these bound ions simply stumbled upon in the structure by crystallographers while they are reconstructing the protein or DNA electron density map? Usually not. Crystallographers select well-characterized biomolecules to measure, ones that have been purified by biochemists and whose activity is known. It is rare that a biochemist will pass a protein to a crystallographer without knowing all the small molecules and ions essential to its structure or action. The biochemist may not initially understand how an enzyme works, but you can rely on them to check the dependence of enzyme activity on pH, temperature, five or six standard ions, and a variety of other molecules—the usual suspects. The important ions will then be included in the soup in which the crystallographer forms the crystal.

Though the details of the above examples of water and counterions apply to DNA, or even just to Poly (dA)·Poly (dT) and/or the RNA/DNA dodecamer, the general principles of water and ions forming local structure, sometimes in the form of binding at specific sites, in order to minimize the energy of the system, is a general characteristic of biomolecular structure.

5.3.1.2 Ionic equilibrium

"Ionic" solids like NaCl dissociate completely when placed in water under almost all circumstances: the equilibrium in $NaCl \rightleftharpoons Na^+ + Cl^-$ is far to the right. In other words, the free energy of the system on the right side of the chemical equation is much lower than the left. Other molecules, like H_2O, H_2SO_4, and H_3PO_4 (or Na_3PO_4 or Na_2HPO_4 or NaH_2PO_4—phosphate "buffer") dissociate to varying degrees, depending on the amounts of other ions in the solution:

> This section is designed for readers who are not chemistry experts. Those who are will find this treatment unsophisticated.

$$Na_3PO_4 \rightleftharpoons Na_2PO_4^- + Na^+ \rightleftharpoons NaPO_4^{2-} + 2Na^+ \rightleftharpoons PO_4^{3-} + 3Na^+ \tag{5.1}$$

Readers may have learned to deal with such equilibria in first-year chemistry; see acid–base equilibrium and weak acids. Such ionic equilibria are of general importance to biology because they affect the pH, and pH affects most biochemical reactions and biomolecular structures. Consider the rightmost state in Equation 5.1 when water is included:

> Water in all its forms, H_2O, H_3O^+, OH^-, should always be included in this equilibrium.

$$PO_4^{3-} + 3Na^+ + 3H_2O \rightleftharpoons HPO_4^{2-} + 3Na^+ + OH^- + 2H_2O \rightleftharpoons$$
$$H_2PO_4^- + 3Na^+ + 2OH^- + H_2O \rightleftharpoons H_3PO_4 + 3Na^+ + 3OH^- \tag{5.2}$$

Because H_2O can dissociate, forming H^+ and OH^-, when it encounters the variety of phosphate ions, the isolated water equilibrium with $[H^+] = [OH^-] = 10^{-7}$ M will be changed. In other words, the pH will not be 7. Thus phosphate can act as a buffer, adjusting or stabilizing the pH. If the phosphate concentration is much larger, say 10^{-4} M, than the 10^{-7} M concentration of water's ions, the pH of the solution will not change much if some process generates 10^{-6} M H^+ or OH^- (hydroxide), because the generated ions will interact primarily with the phosphate in its various ionic forms. Near pH 7, phosphate is predominantly in the $H_2PO_4^-$ and HPO_4^{2-} ionic forms.

> The "phosphate concentration" normally means the total concentration of all ionic forms of phosphate. Buffers are typically present in concentrations between 10^{-3} and 10^{-1} M, or 1–100 mM.

5.3.2 Carriers of Energy and Genetic Information: Nucleotides

The primary carriers of energy, in their ready-to-use form, in biological systems are adenosine triphosphate (ATP) and guanosine triphosphate (GTP). Structures of both are shown in Figure 5.6. Energy is extracted from ATP or GTP by removal of the outermost phosphate group, releasing about $20k_BT_r \sim 8 \times 10^{-20}$ J or 50 kJ/mol. We have just noted that independent phosphate groups exist in four forms, with charge ranging from –3 to 0: PO_4^{3-}, HPO_4^{2-}, $H_2PO_4^-$, and H_3PO_4. In ATP and GTP, two of the phosphates have bonding that prevents ionization of two of the phosphate OH groups. The end phosphate group has two OH groups that can ionize. Thus an ATP or GTP can exist in four forms, with charge ranging from 0 to –4: H_4ATP, H_3ATP^-, H_2ATP^{2-}, $HATP^{3-}$, and ATP^{4-}. Between pH 6.5 and 7.5, the H_2ATP^{2-}, $HATP^{3-}$, and ATP^{4-} species are most numerous. In cells, ATP is often associated with Mg^{++}, $MgATP^{2-}$, because of stronger binding of Mg^{++}. Mg^{++} turns out to be an important factor governing the structure of DNA.

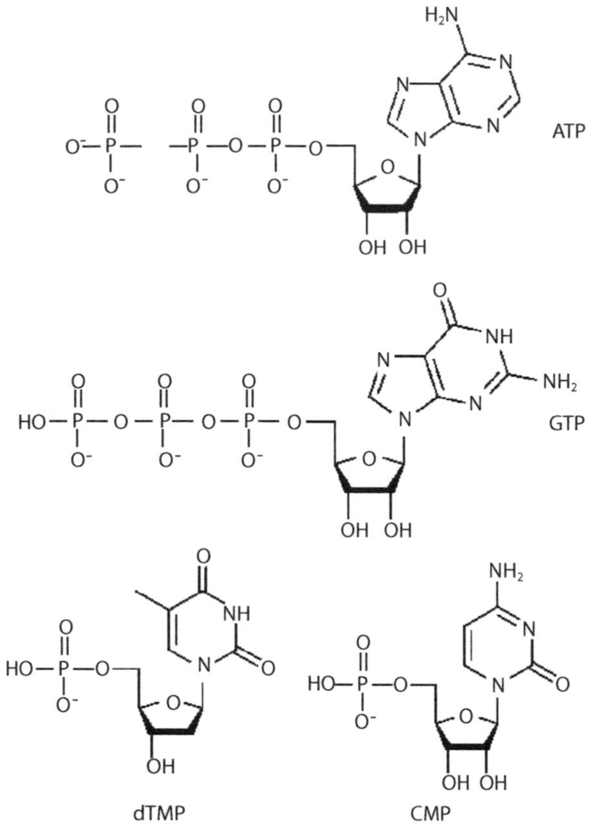

FIGURE 5.6
ATP, GTP, dTMP, and CMP structures. The two triphosphate molecules are shown in their most common biological structure, adenosine-5′-triphosphate and guanosine-5′-triphosphate. The 5′ indicates that the nearest phosphate group is bound to the 5′ carbon of the adenosine or guanine ribose. GTP is less universal as an energy source than ATP but is used for energy in protein synthesis. GTP is essential to signal transduction, particularly with G proteins, where it is converted to guanosine diphosphate (GDP) through the action of GTPases. The "d" in dTMP indicates "deoxy"—thymine-5′-monophosphate with one O (from the 2′OH) group removed from the ribose ring; CMP has both OH groups. (The H left in dTMP's 2′ site is often not shown in chemical structures. In general, H atoms bound on ringlike carbon groups are also not usually shown.) Deoxynucleotides occur in DNA, (oxy) nucleotides in RNA. The suffix "-tide" indicates the molecule has one or more phosphates; "-side," as in nucleoside, indicates no phosphates. All four nucleosides, A, G, T, and C, are used in the genetic codes in DNA.

5.3.3 Carriers of Energy and Electron-Donating Capability: NADH and NADPH

The Earth's atmosphere is about 21% oxygen. Oxygen causes iron to rust, to oxidize: O_2 is a good *oxidizing agent*. On the other hand, many biological molecules must be in their *reduced* state to maintain biological activity. For example, your blood is red in the presence of O_2 because it contains hemoglobin in its reduced form. The iron atom at the active site of hemoglobin is in its reduced, more electron-dense, charge $+2e$ form: $Fe^{3+} + e \rightleftharpoons Fe^{2+}$. A living cell must constantly employ reducing agents, electron donors, to keep the cell in a sufficiently reduced state.

Electron donors in biosystems may change the charge state of a molecule, but often the donated electron carries a proton along with it. Thus the net reaction may be to add a hydrogen atom, H, to a molecule. Sometimes two electrons plus a proton are transferred.

Two of the common *reducing agents* in biosystems are NADH and NADPH, shown in Figure 5.7. These molecules are *dinucleotides* (note again the presence of the adenine base) because they are made of two nucleotides joined through their phosphate groups. One nucleotide has an

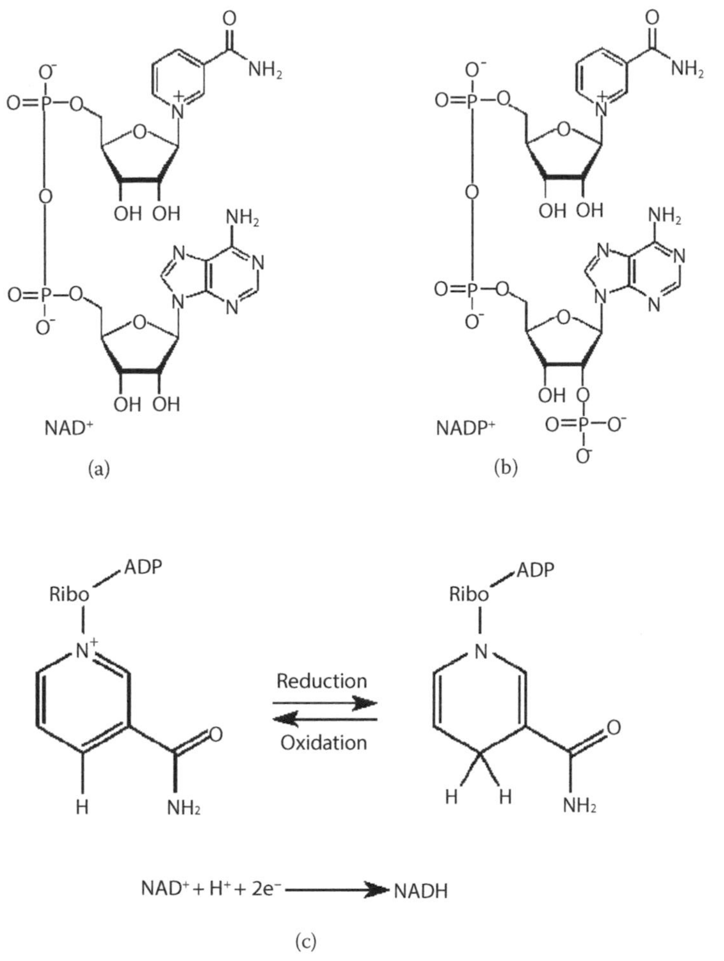

FIGURE 5.7
(a) NAD$^+$; (b) NADP$^+$; (c) the reaction converting these to their reduced forms, NADH and NADPH. Two electrons, along with a proton, are added to the oxidized form in this reduction reaction.

adenine base, the other nicotinamide, which is not a normal base used in DNA. NAD$^+$ accepts two electrons and a proton from other molecules and becomes reduced NADH, which can then be used as a reducing agent to donate electrons. Such electron transfer reactions are the main function of NAD$^+$/NADH. However, it is also used as a *substrate* of enzymes that add or remove chemical groups from proteins, immediately after the proteins are synthesized. The NAD$^+$/NADH ratio in living cells is an important component of what is called the "redox state" of a cell, a quantity that reflects the activity and the health of cells. In plant chloroplasts, NADP$^+$ is reduced by the enzyme ferredoxin-NADP$^+$ reductase during photosynthesis. The NADPH produced is then used as reducing power for the *biosynthetic reactions* in the *Calvin cycle* of photosynthesis. In animals, NADPH provides the reducing equivalents for biosynthetic reactions (lipid and cholesterol synthesis and fatty acid chain elongation) and for oxidation-reduction involved in protection against the toxicity of reactive oxygen species. Both NADH and NADPH are termed *coenzymes* because they bind to enzymes and help catalyze reactions. The molecular masses of NADH and NADPH are 663 and 745 Da (without counterions), respectively, so they are borderline "small." Note also that the net charge of NAD$^+$ (NADH) is −1 (−2); that of NADP$^+$ (NADPH) is −3 (−4).

Hydrogen is the smallest reducing agent in biosystems, though it is rarely in the form of molecular hydrogen H_2. Often the hydrogen is in the form of a single H atom, attached to another molecule, as in NADH, or a metastable product of another reaction. After hydrogen donates an electron (or an entire H atom, sometimes referred to by the symbol H·, the hydrogen radical) to the electron acceptor, H$^+$ or H_2O are common leftover, oxidized forms of hydrogen.

NADH and NADPH are also net carriers of energy. Either can act as an electron-donating energy source in biosystems, moving other components from their lower-energy oxidized state to a reduced state. ATP, in comparison, donates a phosphate group when it gives energy to another molecule in the system. In both cases, the component accepting the electron or phosphate moves to a state that is needed for biological activity to progress. Among many examples, we will see NADH/NADPH used in photosynthetic systems and ATP in molecular motors.

5.4 MEDIUM-SIZED MOLECULES: COMPONENTS OF LARGE BIOMOLECULES

5.4.1 Ringlike Structures and Delocalized Electrons: Prosthetic Groups and Catalysis

We begin discussion of medium-sized molecules by focusing on two ringlike structures, heme and chlorophyll, which lie at the center of three major biological processes: (1) oxidation reduction, or electron transfer reactions, (2) oxygen binding and transport of oxygen, and (3) photosynthesis. These three processes all involve reactions in which electrons move and/or chemical bonds are broken. Electron movement, chemical bond breakage—we are speaking of quantum mechanical processes. Whether we like it or not, quantum processes lie at the heart of biology. This does not mean, however, that we must wait until we have finished an advanced quantum theory course before we can attack significant biological processes. As you will see, the molecules are complicated, but the quantum theory we need to apply is not. The most advanced quantum theory we need—De Broglie wavelength, Heisenberg uncertainty principle, particle-in-a-box, guessing wave functions, Pauli Exclusion Principle—are not much beyond what is covered in first-year physics and chemistry texts. Here we introduce the structures, postponing the treatment of processes until later, when we have reviewed some quantum mechanics.

Heme and chlorophyll illustrate one basic attribute of molecules involved in catalyzing or assisting chemical reactions: the presence of several or many loosely bound electrons. The simplest chemical reaction is the transfer of an electron from one molecule to another, so it makes sense that a catalyst should have many electrons available to assist the process. Metals such as platinum, which have conduction electrons, are used in automobile combustion. Biological processes are also often assisted by metal atoms or ions, bound at the *active site* of an enzyme. The two examples below have active metal ions at their center, but these ions lie in the center of a large, carbon-rich, ringlike structure, which predisposes the metal for certain general types of reactions. The final "tweaking" of the activity of the metal is done by alterations at the edge of the ring, or an attractive force directly on the metal ion, by parts of the protein structure in which the metal is embedded, forming the *enzyme*.

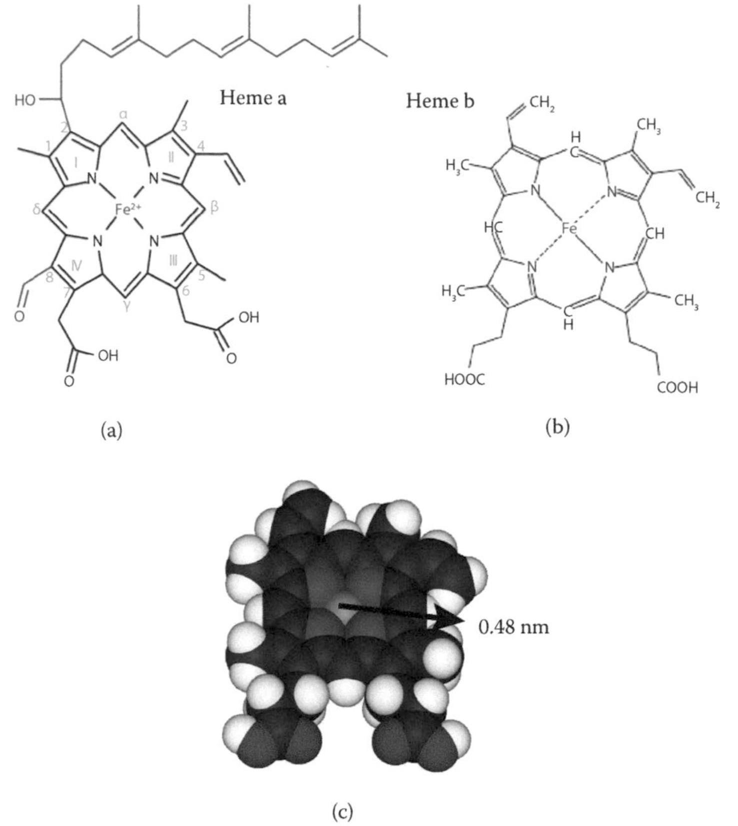

FIGURE 5.8
(a) Heme a; (b) Heme b; (c) Space-filling model of heme b. The radius of the heme ring is about 0.48 nm. The five-carbon subrings are numbered I–IV.

5.4.1.1 Heme

Heme is a *prosthetic group*, a group at the center of activity, of proteins and enzymes (see Figure 5.8). The molecular mass of heme is between 600 and 700 Da, depending on the particular heme. Proteins with heme prosthetic groups are called *hemoproteins* or *heme proteins*. Hemoproteins have biological functions including the transport of O_2, chemical catalysis, diatomic gas detection, and electron transfer. The heme iron serves as a source or sink of electrons during electron transfer or *redox* (oxidation–reduction) chemistry, and as a binding site for molecules such as O_2 (good) and CO (not good). Hemoproteins achieve their variety of functions by modifying the environment of the heme, sometimes forming covalent bonds to groups on the edge of the heme or providing a different charge or dielectric environment. We now note a few obvious features of the heme group, leaving the details for later.

Iron in the heme group is commonly in the reduced Fe^{2+} (II) electronic state but can cycle readily to the oxidized Fe^{3+} (III) state. The change of electronic state makes a profound difference in the activity of the protein to which heme is (noncovalently) bound. O_2 and other diatomic molecules reversibly bind to Fe (II) hemoglobin and myoglobin, but not to the Fe (III) form. In the cytochromes, Fe (II) heme can donate an electron, whereas Fe (III) accepts electrons.

Common forms of heme, shown in Figure 5.8, are heme a and heme b, named for the cytochrome enzymes, cytochrome a and cytochrome b, in which they are found. The overall *porphyrin* ring is made up of four 5-carbon subrings, numbered I through IV. Figure 5.8c, a space-filling version of the structure, reminds us that these structures do not have empty spaces in them. The cytochromes catalyze electron transfer reactions. Heme b, also called *iron protoporphyrin IX*, is also present in hemoglobin and myoglobin, both of which bind oxygen. *Porphyrin* refers to the basic ring structure of heme as well as chlorophyll.

5.4.1.2 Characteristics common to heme and chlorophyll

The structures of the porphyrins have an outer hydrocarbon ring that can be identified by alternating single and double bonds. These alternating bond notations indicate not that the double and single bonds are located exactly where they are drawn but rather that the electrons in the ring are *delocalized*. This implies that porphyrin constitutes a circle, of radius about 0.48 nm, in which electrons are free to wander over the circumference. We will use this property in a simple quantum calculation later in the text. When a molecule has a large number of electrons that are loosely bound, that molecule will be efficient at absorbing light. This is especially so when many electrons are confined to the same ring system. We will show that for heme, the energies of photons that can be absorbed correspond to the visible region of the spectrum. From this we predict that heme and chlorophyll should be highly colored, which they are.

The four nitrogens and the central metal ion of the porphyrin ring of heme and chlorophyll appear to be drawn differently in Figures 5.8 and 5.9. Are there really four N–Fe bonds in heme and just two N–Mg bonds in chlorophyll? No, though the details are a bit complicated. The electron densities that constitute the N-metal bonds depend on the details of the porphyrin ring, and lines drawn to indicate "bonds" are usually a matter of the drawer's preference. To a first approximation, consider the equivalent of about two N-to-metal covalent bonds but distributed between the four nitrogens. Motion of the Fe or Mg perpendicular to the porphyrin plane will be only weakly restricted because it does not directly stretch a covalent bond.

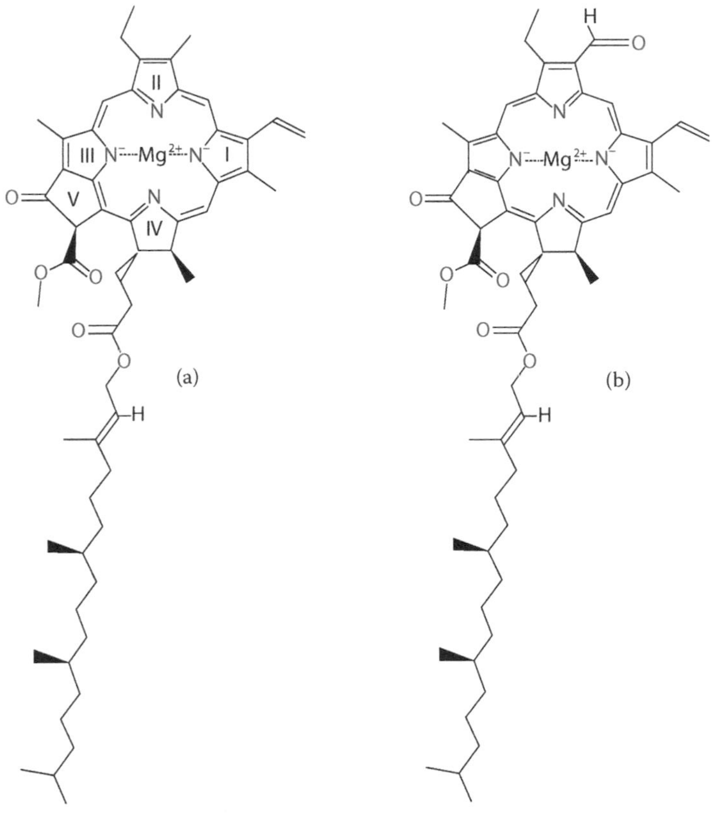

FIGURE 5.9
(a) Chlorophyll a; (b) Chlorophyll b. Chlorophylls have an additional ring V, a hydrocarbon "tail" (the phytyl side chain), a central Mg^{2+}, and different bonding in subring IV, compared to the heme group. The triangular symbol at the edge of the chlorophyll ring stands for a CH_3 (methyl) group. By convention, H atoms bound to carbon rings are not displayed.

5.4.1.3 Chlorophyll

Chlorophyll molecules are the primary molecular entities responsible for light absorption, electron transfer, and energy conversion in green plants and photosynthetic microorganisms. Like heme, chlorophyll exists in many variants. Chlorophyll a and b are shown in Figure 5.9. The most obvious differences between heme and chlorophyll are (1) a magnesium ion in place of the iron and (2) a long hydrocarbon *tail*. Some differences less obvious to the physicist reader are (3) the extra 5-carbon ring and (4) the CH_3 group in ring. The Fe $\rightarrow$ Mg replacement suggests the electronic processes governed by chlorophyll will be quite different from those of heme, which is true. The chlorophyll hydrocarbon tail, which you should catalog as a *hydrophobic* part of the molecule, suggests that chlorophyll should be found near hydrophobic regions. Because water constitutes 70% of the cell mass and is not hydrophobic (of course), chlorophyll should be found either in hydrophobic regions of proteins or in membranes. Both guesses are correct.

5.4.2 Multipurpose Molecules: Building Blocks, Energy Carriers, Control/Signaling Molecules

This section primarily introduces the building blocks of the major four biopolymers: proteins, nucleic acids, membranes, and carbohydrates. As we already discovered, two of the DNA building blocks, ATP and GTP, are also carriers of energy. Study of the biochemistry of the cell will reveal that most of the polymer building blocks can act as energy carriers and/or controls on reactions in the cell. Sometimes this control is simple thermodynamics: additional ATP will speed up (slow down) reactions that consume (produce) ATP. In other cases the building block may specifically bind to an enzyme that triggers or inhibits production of yet another enzyme. The amino acid glutamic acid (Glu) has a function as a neurotransmitter, in addition to being a building block of proteins.

5.4.2.1 Amino acids

There are 20 "standard" amino acids used as building blocks of proteins. The basic structure of the "α-amino" acids is shown in Figure 5.10. The "front" of the amino acid has an $-NH_2$ amino group, while the "rear" has a $-COOH$ carboxyl. In aqueous solution at pH 7, COOH will spontaneously lose a proton and NH_2 will gain one, forming the so-called zwitterion. A central alpha carbon ($\alpha-C$) has an "R group" covalently bound to it. The R group identifies the particular amino acid. Table 5.2 summarizes the structure, some key properties, and nomenclature of the 20 amino acids. Proline (Pro) is unique among the side chains in its unusual bonding to the core amino acid structure. Pro bonds to the amino-terminus of the amino acid, as well as to the α-C, making those two substructures part of a ring structure. Pro residues in polyamino acids tend to induce sharp turns in the polymer.

Though the dielectric character of an R group is often used to predict whether the group will be exposed to water or buried inside the protein in which it resides, such prediction is inexact. Many amino acids have been found in both environments, depending on the details of nearby amino acids. It should also be noted that among the charged amino acid side chains, the Histidine (His) charge depends on pH in the physiological range (pH 6–8). The pK_a of the His R group is 6.0: at pH 6.0, half the His residues will have charge +1, half 0. At pH 7.0, the majority will be uncharged.

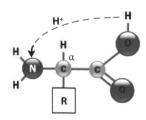

FIGURE 5.10
Design of an amino acid. The central carbon, to which the R group is bound, is termed the *alpha carbon*: α-C. The –COOH carboxyl group can easily lose a proton, while the –NH_2 amino group can gain one, making the dipolar "zwitterions."

TABLE 5.2 Twenty Standard Amino Acids

Name Abbreviation	Molecular Mass (D)[a]	R Group (Side Chain)	Percent[b]	Dielectric Character	DNA Codon(s)[c]
Glycine Gly	57.0	$-H$	7.08	Nonpolar	GGU, GGC, GGA, GGG
Alanine Ala	71.1	$-CH_3$	8.26	Nonpolar	GCU, GCC, GCA, GCG
Valine Val	99.1	$-CH(CH_3)_2$	6.86	Nonpolar	GUU, GUC, GUA, GUG
Leucine Leu	113.2	$-CH_2-CH(CH_3)_2$	9.65	Nonpolar	UUA, UUG, CUU, CUC, CUA, CUG
Isoleucine Ile	113.2	$-CH(CH_3)-CH_2-CH_3$	5.93	Nonpolar	AUU, AUC, AUA
Methionine Met	131.2	$-CH_2-CH_2-S-CH_3$	2.41	Nonpolar	AUG
Proline Pro	97.1	(pyrrolidine ring: $-CH-CH_2-CH_2-CH_2-N-$)	4.72	Nonpolar	CCU, CCC, CCA, CCG
Phenylalanine Phe	147.2	$-CH_2-$ (phenyl)	3.86	Nonpolar	UUU, UUC
Tryptophan Trp	186.2	$-CH_2-$ (indole)	1.09	Nonpolar	UGG
Serine Ser	87.1	$-CH_2-OH$	6.60	Dipolar	UCU, UCC, UCA, UCG, AGU, AGC
Threonine Thr	101.1	$-CH(OH)-CH_3$	5.35	Dipolar	ACU, ACC, ACA, ACG
Asparagine Asn	114.1	$-CH_2-C(=O)-O$	4.06	Dipolar	AAU, AAC
Glutamine Gln	128.1	$-CH_2-CH_2-C(=O)-NH_2$	3.93	Dipolar	CAA, CAG
Tyrosine Tyr	163.2	$-CH_2-$ (phenyl)$-OH$	2.92	Dipolar	UAU, UAC
Cysteine Cys	103.1	$-CH_2-SH$	1.37	Dipolar	UGU, UGC
Lysine Lys	128.2	$-CH_2-CH_2-CH_2-NH_3^+$	5.82	Charged (+)	AAA, AAG
Arginine Arg	156.2	$-CH_2-CH_2-CH_2-NH-C(NH_2)(=NH_2^+)$	5.53	Charged (+)	CGU, CGC, CGA, CGG, AGA, AGG

(Continued)

TABLE 5.2 (*Continued*) Twenty Standard Amino Acids

Name Abbreviation	Molecular Mass (D)[a]	R Group (Side Chain)	Percent[b]	Dielectric Character	DNA Codon(s)[c]
Histidine His	137.1	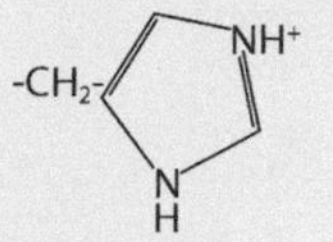	2.27	Charged (+)	CAU, CAC
Aspartic acid Asp	115.1	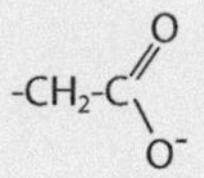	5.46	Charged (−)	GAU, GAC
Glutamic acid Glu	129.1	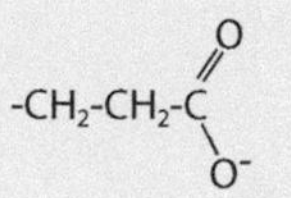	6.74	Charged (−)	GAA, GAG
Average amino acid	111.0				

[a] Mass of an amino acid unit in a polypeptide. Add 18.0 Da, the mass of H_2O, to get the mass of the parent amino acid. Subtract 56.0 Da to get the mass of the R group.

[b] Probability of amino acid's occurrence in an average protein, measured by average occurrence in a protein sequence database of about 560,000 sequence entries, www.expasy.org/sprot/relnotes/relstat.html as of May 20, 2017; numbers will change slightly as database is updated.

[c] 3-RNA-base triplet coding for the corresponding amino acid.

5.4.2.2 Nucleotides

Note again that the convention, not always followed, is that carbon members of a hydrocarbon ring system are implied at unlabeled corners where bonds originate. If there are fewer than four bond lines extending from such a corner, there is an implied H atom bound to that C atom.

We have already seen two of the four standard nucleotides when we introduced ATP and GTP as energy carriers. All four nucleotides are shown in Figure 5.11. The smallest structure associated with a nucleic acid is called a base. Two categories of bases exist, *pyrimidines* and *purines*. The pyrimidine bases, cytosine and thymine (thymine is replaced by uracil in RNA), are six-sided rings made of four carbon and two nitrogen atoms. The purine bases, adenine and guanine, are double-ring systems made from carbon and nitrogen atoms. The four bases are abbreviated A or Ade, G or Gua, C or Cyt, and T or Thy. These same abbreviations are used for the bases with ribose and phosphates attached. Some nomenclature and abbreviations are found in Table 5.3.

Important molecular features of the nucleotides to note include all three parts, the base, the sugar, and the phosphate. First, every base ring has two or three groups that can form hydrogen bonds. This forms the basis of the well-known complementarity of base pairs. The rings of guanine

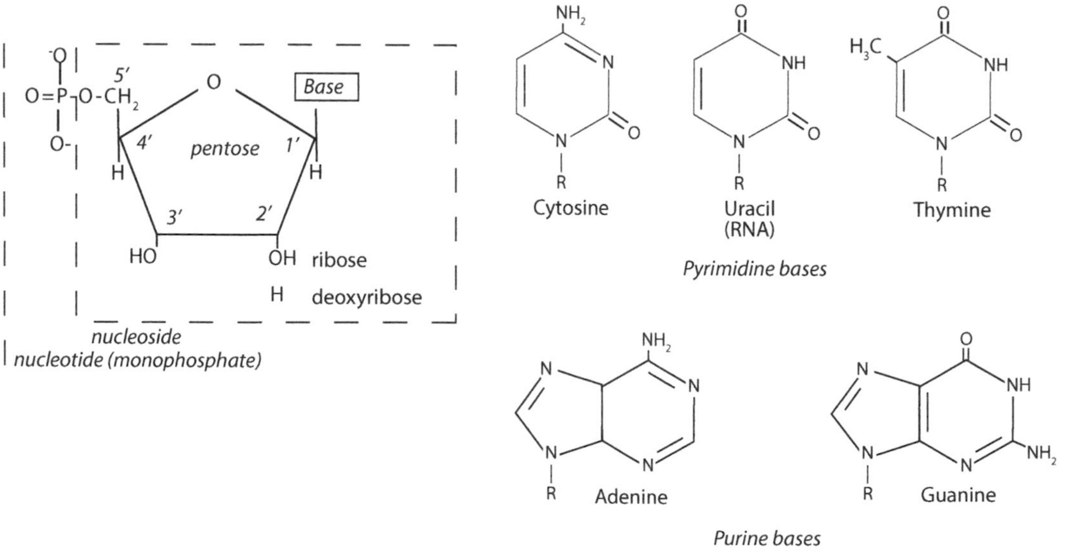

FIGURE 5.11
Design of a nucleotide (monophosphate).

TABLE 5.3   Nucleic Acid Component Notation and Abbreviations

Term	Meaning	Names	Abbreviation			Mass (D)
Base	The bare ring structure bonded to the ribose ring[a]	Adenine	A			
		Guanine	G			
		Cytosine	C			
		Thymine	T			
		Uracil	U			
Nucleoside	Base bonded to ribose	Adenosine	Ado			
		Guanosine	Guo			
		Cytidine	Cyd			
		Thymidine	Thd			
		Uridine	Urd			
Nucleotide	Nucleoside bonded to 1, 2, or 3 phosphates[b]	Adenylic acid	AMP	ADP	ATP	331.2 (AMP)[d]
		Guanylic acid	GMP	GDP	GTP	347.2 (GMP)
		Cytidylic acid	CMP	CDP	CTP	307.2 (CMP)
		Thymidylic acid	TMP	TDP	TTP	322.2 (TMP)
		Uridylic acid	UMP	UDP	UTP	324.2 (UMP)
Base pairs	Base pairs as bonded in DNA (RNA)	(deoxy)A-(deoxy)T	dA-dT			613.4
		(deoxy)G-(deoxy)C	dG-dC			616.4
		(ribo)A-(ribo)U	rA-rU			615.4
		Average base pair[c]	—			614.6
Sugar phosphate	(deoxy) Ribose plus phosphate, as bonded in DNA (RNA)	Deoxyribose phosphate	—			176.1
		Ribose phosphate	—			192.1

[a] The covalent bond between a base and the 1′ ribose carbon atom is called the glycosidic bond.

[b] The terminal phosphate has two O– groups; intermediate phosphates have one. A, G, C, and T nucleotides sometimes notated, e.g., deoxyadenylic acid. Thymidylic acid occurs in the ribo-form, but only rarely.

[c] Average in the human genome (DNA) (from http://rh.healthsciences.purdue.edu/vc/theory/dna/).

[d] Deoxyribose form, except for UMP (http://en.wikipedia.org/wiki/Uridine_monophosphate).

have an NH_2 and an NH group, which can each donate a proton, and a double-bonded O, which can accept a proton. (There is another ring nitrogen that could accept a proton, but it normally is not in a position to do so, except in unusual DNA structures.) Adenine has an NH_2 proton donor and an adjacent N acceptor on its six-membered ring. Cytosine has a NH_2 donor group and N and O acceptor groups. Finally, thymine has NH donor and O acceptor groups. The five-membered sugar ring can be either ribose (in RNA) or deoxyribose (in DNA). Ribose has a 2′-OH group (primed numbers like 2′ indicate number on the ribose ring system), whereas deoxyribose has 2′H. A base bonded to the ribose 1′-C constitutes a ribonucleoside or deoxyribonucleoside. If a single phosphate group is bonded to the 5′-C of the ribose, the result is a (deoxy)ribonucleotide.

5.4.2.3 Sugars

We have seen one of the sugars, ribose, as part of the nucleosides and nucleotides. We might say that DNA and RNA are a mixed polymer of sugar, nucleic acid bases, and phosphate. In contrast, carbohydrates, or saccharides, are polymers made up of purely of sugar units. Besides ribose, there are many such units and many variants in which groups attached to the sugar ring are modified. An example is glucose and glucosamine, in which the OH in the glucose 2 position is changed to an amino NH_2 group in glucosamine. A few of the most common sugars are shown in Figure 5.12. Simple sugars are called *monosaccharides*, but note that some sugars commonly known by name are actually disaccharides. Among these are sucrose (glucose–fructose), lactose (galactose–glucose) and maltose (glucose–glucose). Sugars are, of course, important nutrients. Perhaps the most familiar nutrient, sucrose, is digested to glucose and fructose, its two components, before absorption into the body.

In DNA, the ribose 2 position was called 2′ to distinguish it from the 2 position in the base ring. The prime is not included when describing sugars by themselves.

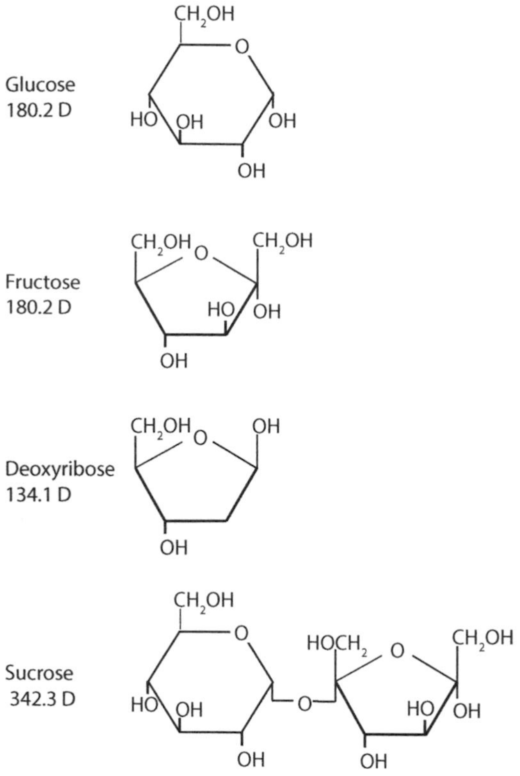

FIGURE 5.12
A few mono- and disaccharides, with molecular masses noted. Sucrose is a glucose–fructose disaccharide. Lactose and maltose are galactose–glucose and glucose–glucose disaccharides, respectively (not shown). Cellulose is a polymer of glucose (not shown).

5.5 FORCES AND FREE ENERGIES

In the following sections we will frequently refer to "interactions" that govern the structures of biopolymers. Though we will discuss "interactions" in more detail in Chapters 12–15, it is advisable to briefly remind ourselves of what we mean by interactions.

Interactions can be described by forces or free energies. The physicist is usually more comfortable with forces, the chemist with free energies. Physics speaks of the force on *an object* and perhaps the potential energy from which the force derives. Chemistry commonly uses free energies to describe chemical reactions involving *many particles*. What is the relation between the two? A brief sketch of the relation follows.

We learned in Physics 201 that forces and potential energies are related, in one dimension, by

$$F(x) = -\frac{dV(x)}{dx}, \qquad (5.3)$$

where $V(x)$ is the potential energy function. We learned that the force and potential energy might be caused by a spring with a spring constant: $V(x) = 1/2\,kx^2$, $F(x) = -d/dx(1/2\,kx^2) = -kx$. The spring force is understood to derive ultimately from potential energies or forces between atoms in the spring, as bonds are stretched. In thermodynamics, we learn to deal with systems of many (very many) particles and describe interactions in terms of free energies:

$$G = H - TS \cong U - TS \cong (K + V) - TS \qquad (5.4)$$

where G is Gibbs free energy, H is enthalpy, T is (absolute) temperature, S is entropy, K is kinetic energy, V is potential energy, and we note that potential energy is one term in the free energy. Is potential energy the only term from which a force can be derived? The bottom line is that any term in the free energy that depends on the variable x will produce a force in the x direction. In first-year physics, S is normally irrelevant because one particle is involved (or S is simply constant), and K does not explicitly depend on x, leaving only V to be responsible for $F(x)$. A linear polymer in solution, however, will spontaneously form a more or less compact ball, not a straight line, because there are many possible ways to form a compact ball and very few ways to get a straight line. This is entropy. If you manage to grab hold of the two ends of the balled-up polymer and pull, it will resist, because you are trying to force the polymer into a line—a very low-entropy state. In this case, entropy does depend on the extension, x.

$$
\begin{aligned}
F(x) &= -\frac{d}{dx}G(x) = -\frac{d}{dx}\big(H(x) - TS(x)\big) \\[1ex]
&= -\frac{d}{dx}H(x) + T\frac{d}{dx}S(x) \\[1ex]
&\cong -\frac{d}{dx}V(x) + T\frac{d}{dx}S(x) \\[1ex]
&\cong -\frac{V(x_2) - V(x_1)}{x_2 - x_1} + T\frac{S(x_2) - S(x_1)}{x_2 - x_1} \\[1ex]
&= -\frac{\Delta V}{\Delta x} + T\frac{\Delta S}{\Delta x}
\end{aligned}
\qquad (5.5)
$$

So, assuming temperature is independent of x. Both of the force terms, *enthalpic* and *entropic*, are ubiquitous in biosystems. We see the familiar spring force term $-dV/dx$ coming from the enthalpy, but we neglect the entropic force at our peril.

Suppose we measure free energy change ΔG and want to determine the contributions ΔH and $T\Delta S$. How do we separate ΔH and ΔS? By varying temperature. If H and S are intrinsically temperature independent (not always the case, but we desperately hope so), then a plot of $\Delta G = \Delta H - T\Delta S$ versus T will yield a straight line with slope $-\Delta S$ and $T = 0$ intercept ΔH—not so difficult, though we always have to extrapolate to $T = 0$.

What if we apply a force, measure the opposing force, $F(x)$, and want to know if $V(x)$, $TS(x)$, or both are responsible for the force? First, we have to integrate

$$G(x) = -\int F(x)dx \tag{5.6}$$

or

$$\Delta G = \Delta H - T\Delta S = G(x_2) - G(x_1) = -\int_{x_1}^{x_2} F(x)\,dx$$

For cases of constant force we get

$$\Delta G = -F(x_2 - x_1) = -F\Delta x \tag{5.7}$$

Again, we can do measurements as a function of temperature and determine the components that are temperature independent and dependent. In many cases, as with stretching of polymers, the entropic force is smaller than the enthalpic force and/or the enthalpic force occurs only at large extensions of the polymer. We will cite one such case involving DNA.

There is an additional issue when discussing states and forces in biopolymers. From our thermodynamics experience (see Chapter 14, if necessary) we know that the Gibbs free energy difference between two states, $S_1 \rightleftharpoons S_2$, governs their numerical equilibrium at constant pressure:

$$\frac{N_1}{N_2} = \exp\left(-\frac{\Delta G}{k_B T}\right) = \exp\left(\frac{\Delta S}{k_B} - \frac{\Delta H}{k_B T}\right) \tag{5.8}$$

but this leaves several questions. First, what are the two states, S_1 and S_2? For a polymer structure like a protein α helix, we would presume from the following sections that S_1 is the polypeptide, stabilized by internal interactions, surrounded by water, with α-helical structure intact. What is S_2? Is S_2 an intact α helix without internal hydrogen bonds, surrounded by water? (This has not been observed.) Is S_2 a random structure surrounded by water? What does "random" mean? Second, what about other interactions? If the structure becomes randomized, different parts of the structure, including hydrophobic ("water-avoiding") amino acids, become exposed to water. Third, does temperature have an effect? Depending on values of ΔS and ΔH, a moderate variation in temperature may change the structure from helical to (more) random, which brings up another question: Is the "hydrogen bond energy" the value of ΔG, ΔH, or something else? (The short answer is that when you read about "bond energies," the reference usually refers to ΔH or ΔV, a change in potential energy, which is often not far from ΔH.)

Disentangling the contributions of various interactions to protein stability is based, in the end, on experimental measurements with proper theoretical analysis. In the previous paragraph we saw an outline of one basic approach.

1. Find a way to measure the state ratio N_1/N_2 on the complete protein in the relevant solvent and measure this ratio as a function of temperature. This gives values of ΔH_{tot} and ΔS_{tot}.
2. Devise experiments to measure individual contributions, like ΔH_{HB} and ΔS_{HB} for H bonding of each individual pair. This usually involves smaller polypeptides and measurements in several solvents.
3. Simultaneously, calculate the contribution of each type of interaction. Quite advanced methods have been developed for computing contributions to $\Delta V \sim \Delta H$. ΔS is difficult.
4. Determine whether $\Delta H_{tot} = \Delta H_1 + \Delta H_2 + \Delta H_3 + \ldots$ and $\Delta S_{tot} = \Delta S_1 + \Delta H_2 + \Delta S_3 + \ldots$. If equality is obtained and theory agrees with experiment, you are done.
5. Now, repeat all the above procedures, but measure forces instead.

Why is calculating ΔS so difficult compared to ΔH? We will postpone a serious attack on this problem until we have some basic statistical mechanics under our belts in Chapter 14. A summary of the issue is as follows. To calculate a potential energy of interaction, you need only know the two interacting particles (entities), their relative spatial positions, and the forces between them, usually electrostatic. There are a few difficulties: the calculations must sometimes be quantum mechanical and solving the Schrödinger equation for multielectron systems must use approximations. As we saw in Chapter 4, the surrounding water can also be problematic, if we have to include position-dependent dielectric constants or the actual interactions with local water molecules. This can be managed.

The entropy difference is another story. Potential energy involves the particles, where they actually are, and the forces between them. To compare two states, do two calculations. Entropy, by contrast, involves the interacting particles, where they are and how many different ways they can be there, how many other places they can be and how many ways they can be at those other places, and the same for all the *other* particles (water, other amino acids) that *could* be where your particles of interest are located. We wait until Chapter 14 to do a few simple entropy calculations.

5.6 BIOPOLYMERS

The vast majority of structural and active molecules in biosystems are polymers, made up primarily of the "medium-sized" molecules we have just discussed, with a sprinkling of small molecules or ions to enable specific interactions with other molecules. The four biopolymers we consider here and their primary roles are as follows:

> A membrane technically is not a polymer because the components are not covalently linked together.

- *Proteins*: provide structure and activity (catalysis, force generation)
- *DNA/RNA*: maintain the information needed for life
- *Membranes*: separate the inside of cells from the outside
- *Polysaccharides*: reinforce cell walls of plants, insects, bacteria; provide nutrients; regulate intercellular recognition

5.6.1 Proteins

Proteins are referred to as *polyamino acids* or *polypeptides*. The former term refers to the makeup of proteins as a linear polymer of the 20 amino acids. The latter refers to the type of covalent bond between amino acids units, the peptide bond. In this section we lightly explore several levels of structural classification that have been traditionally been used for proteins.

The structure of proteins has traditionally been broken down into four categories:

1. Primary structure: the amino acid sequence, e.g., Gly-Pro-Ala-Tyr-Ala-Leu-Leu…; an ordered list of the amino acids, starting from the "amino terminus." See Section 5.6.1.1 below.
2. Secondary structure: a more global level of structural organization describing the relative orientations of adjacent groups of about 5–20 amino acids. Standard secondary structures include the α and other helices, β sheets, the collagen helix, and several other observed structures, including pseudorandom coils.
3. Tertiary structure: a yet more global descriptor of structure describing the relative orientations and packing of all secondary structural components in a polypeptide chain, including disordered regions.
4. Quaternary structure: a description of the relative orientation and packing of two or more tertiary structures (polypeptides), if the protein of interest tends to form aggregates.

Experimental and theoretical approaches to these levels of structural organization tend to be opposite. Theorists start with the primary structure and try to deduce the next higher levels of organization. Experimentalists who use electron microscopy, X-ray and neutron crystallography, two-dimensional NMR spectroscopy (nuclear Overhauser effect), sedimentation equilibria, mass spectrometry, chemical footprinting, etc., tend to first produce grosser levels of structural organization and work to finer detail. Because the experimental methods require an enormous investment in equipment, training, and person hours of work, some think it would be better to have a computer program that predicts protein structure and function from the amino acid sequence (primary structure). For others, the investment in personnel and laboratory facilities provides the needed checks (the "real" answer) on computers and theory, as well as the infrastructure base for future advances in biology and should be a primary focus of effort and finances. As we will see, these two approaches have, in fact, become unified in the "database" approach to structural prediction.

Three primary problems in protein structure have faced the biophysics community for the past 50 years or so: (1) prediction of the three-dimensional structure of proteins from a knowledge of the amino acid sequence of the protein, (2) quantitatively describing the *protein folding* process, from a linear polymer to a well-defined, closely packed three-dimensional structure, and (3) understanding the function of a protein enzyme from its structure. These goals remain primary in biophysics, though the rise of molecular genetics over the past 20 years has emphasized a *site-directed mutagenesis* approach to function. This approach to protein function is to change the amino acid sequence and observe how a measurable function changes. In the traditional site-directed mutagenesis approach, several amino acids thought to be most directly involved in functional activity are changed one by one, and the activity measured. The changes made to each amino acid site are usually to another amino acid having similar properties, such as a replacement of Glu by Asp, or to a contrasting amino acid, which may have different charge properties, if local charge is believed to play an important role. In each case the mutant protein is first screened for a basically intact structure because many amino acid substitutions completely disrupt the protein structure, leading to insolubility or other major changes.

Site-directed mutagenesis is a genetic approach to protein structure, in which the DNA sequence (the *gene*) that codes for the protein is altered, the new DNA is placed into an organism (e.g., yeast or *Escherichia coli* bacteria) that *expresses* the gene (synthesizes the protein) on a large scale, and the mutant protein is purified and tested. The changes made are usually carefully planned and done one at a time. In a more recent approach, mutant protein sequences are made randomly and *en masse*. Thousands of mutant proteins may simultaneously be created, the vast majority of which are nonfunctional. However, if even 0.1% have some activity, you may have several viable mutated proteins that reveal important information. This approach is similar to modern drug-design approaches.

Directed Evolution. The above approaches to studying protein structure and function center on the isolation and purification of the native protein, which is subjected to painstaking and expensive structural analyses based on X-ray crystallography, nuclear magnetic resonance, molecular dynamics computer simulations, etc. The analysis usually resulted in the identification of important parts of the structure, such as active or substrate-binding sites. Researchers then proposed what specific amino acids were primarily responsible for the structure and activity at those sites and replaced, one by one, those identified amino acids using genetic engineering techniques. This approach to investigating the relation of structure to function has been termed *protein engineering*. Even in the 1990s, such modified proteins could number in the hundreds, as techniques like CRISPR-cas9 were developed.

In the twenty-first century, investigators sometimes require not only thousands, but even 10^{11}, or 10^{50}, or 10^{96}... different variations of a given enzyme (protein) sequence, in order to find sequences that result in enzymatic activity. The reason for the huge numbers of desired variations is that, contrary to an old assumption that a protein had to have a unique sequence to enable its enzymatic activity,

Enzymes with the same fold, catalyzing the same reaction, and sharing the same ancestor typically share less than 20 percent of their amino acids

(Wagner, Arrival of the Fittest.)

Note that this does not mean that one can *randomly* change 80% of the amino acids in an enzyme and still retain activity. The interested reader should explore these two assertions in the cited book and in a problem at chapter's end.

The *genotype* of a protein refers to its amino acid sequence, specified by the corresponding DNA sequence. The *phenotype* refers to its enzymatic activity(ies). Overall structures within a phenotype can be similar, but the biochemical activities under a given set of conditions is primary. More generally, a protein or proteins might allow a cell to survive on a specific sugar—a phenotype— whether or not the protein(s) catalyze the same reaction or set of reactions as another protein(s) that also allow such survival.

What kind of method could *physically manage* 10^{50} variants of a protein sequence? The answer is none, at least during the 4 billion years of life on earth. The approach that Wagner and others have taken is to map genotypic networks of protein sequences using large-scale computing, analyze those networks for viable prospects for activity, and ultimately identify a few thousand for experimental study. All 10^{50} network points (proteins) need not explicitly be catalogued by the computer; only the organization rules of the network, i.e., of "protein space," plus an intelligent search strategy for specific phenotypes of proteins (or protein systems) are required.

At some point in this twenty-first-century protein-space search, hundreds or thousands of specific proteins must be physically produced and separated from other molecules that would interfere with proper measurements of enzymatic activity. We cannot address details of the techniques for the synthesis of thousands of variants of a given protein sequence, but they have been described in several recent reviews.

The directed-evolution approach has proven so successful that commercial services can be contracted, ThermoFisher Scientific and SeSaM-Biotech GmbH, for example. Figure 5.13 shows an overview of one general approach to directed evolution. Starting from the natural gene for a protein with known activity, a library of genetic variants is generated through random or user-managed mutations. Cellular, e.g., *E. coli*, or in vitro microdroplet methods are then used to express the corresponding library of protein molecules. Proteins must then be screened for the desired activity or other property, and one or more proteins are selected as viable members of the next generation. The gene(s) for this (these) selected protein(s) are then used in place of the natural gene for the next generation of directed evolution.

Since determining 3D structure is considerably more laborious and expensive than determining the sequence of a protein, the PDB database contains about 560,000 entries as of May, 2017, whereas the UniProt entries number nearly 86,000,000.

Informatics is often incorporated into protein-engineering or directed-evolution approaches. Informatics can best be understood by thinking of the online databases of protein structure and sequence. PDB, www.rcsb.org/pdb/home/home.do, and UniProtKB/Swiss-Prot www.expasy.org/sprot/, for example, both contain sequence information, but the former focuses more on measured 3D structure and the latter on sequence, classification, predicted local structure, and similarity to other proteins. Large numbers of sequences and structures, along with activity information, reside in such databases. Computer analysis of the data can reveal correlations between structure and sequence, for example, though no prior suspicion of such a correlation existed. Now the researcher who discovers a new protein needs to only purify, sequence, and characterize the biochemical activity of the protein, then compare the new sequence to existing entries in the database, and by analogy postulate the new structure. The Basic Local Alignment Search Tool (BLAST) in UniProt, for example, finds regions of local sequence similarity between a new and a known protein, which can be used to guess functional and evolutionary relationships between sequences.

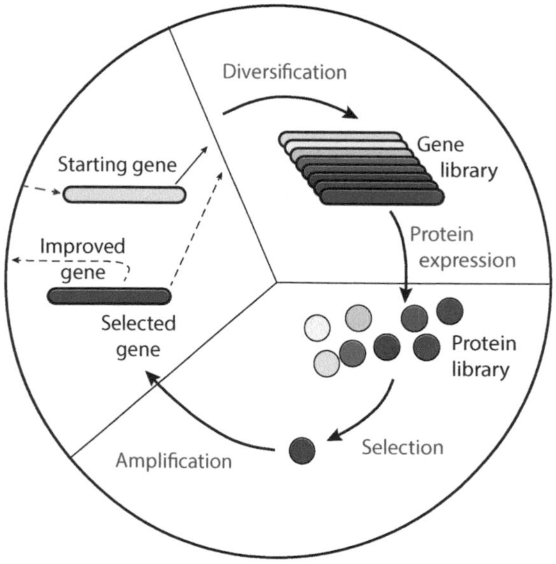

The above approaches to protein structure and activity do not focus on the forces and energies—the physical laws—that govern protein structure and function. Biochemistry and molecular genetics have made great strides in producing modified proteins that have modified activity. It is then the job of the physical bioscientist to explain, at a fundamental level, how function is related to structure. We will focus on energetic and statistical descriptions of structure.

FIGURE 5.13
A few mono- and disaccharides, with molecular masses noted. Sucrose is a glucose–fructose disaccharide. Lactose and maltose are galactose–glucose and glucose–glucose disaccharides, respectively (not shown). Cellulose is a polymer of glucose (not shown).

5.6.1.1 Polypeptides

Polypeptides are linear polymers made up of amino acids bound together by *peptide bonds*. The peptide bond, Figure 5.14, is formed when two amino acids come together, amino end of one adjacent to the carboxyl end of the other, H_2O is removed, and a covalent C–N bond formed. In order to successfully "come together" in a cell, the amino acids must be provided with the appropriate biological infrastructure—a protected space, specialized catalytic enzymes (existing polypeptides), coding nucleic acids, and materials resources.

5.6.1.1.1 Rigidity of the peptide bond

An important characteristic of the O = C–N–H peptide bond is its planarity and resistance to distortion. Rotations can occur about most other bonds, subject to side groups encountering the van der Waals radii of other side groups, but rotation about the peptide C–N bond is hindered in general. The planarity is an experimental result of X-ray crystallography, first noted by Pauling and Corey in the 1930s and 1940s, but Pauling, the world's expert in chemical bonding, then explained the planarity as resulting from a partial double-bond character of the C–N. The bonding structure is actually a mix of O = C–N–H and O–C = N–H, which lowers the energy of the system. This "resonance" energy reduction, about 85 kJ/mol or $35k_BT_r$ (compared to 0 when twisted 90°), turns out to be maximal when the four atoms are in a plane. The value $35k_BT_r$ indicates that fluctuations at room temperature are unlikely to significantly bend the bond.

FIGURE 5.14
Formation of a peptide bond. H_2O atoms removed during the reaction are shown in red. Though this "formula" structure does not accurately indicate three-dimensional structure, successive R groups are shown on opposite sides of the dipeptide to indicate that they generally avoid contact with each other.

5.6.1.1.2 Steric avoidance of side chains and (ϕ, ψ) angles

Amino acid side chains (R groups) will tend to avoid each other, which is indicated by placing successive R groups on opposite sides of the formula structures of Figure 5.14. This is called *steric avoidance* or *steric hindrance*. In contrast to the rigid peptide C–N bond, rotations can occur around the N–C_α bond with relatively small energy rises, until R groups are brought into contact with each other. The angle around the N–C_α bond is labeled ϕ. The angle around the C_α–$C_{peptide}$ bond is referred to as ψ.

The traditional structural formula diagrams of Figure 5.14 are easy to decipher, but they do not reflect the actual orientations in three dimensions. A tetrapeptide Ala-Leu-Ala-Leu structure taken from the PDB better illustrates the placement of atoms in a short polypeptide (an "oligopeptide"). The identity and locations of the individual amino acids can be deciphered after a brief study of the ball-and-stick model of the tetrapeptide, shown at the left in Figure 5.15. A CPK model to the right better shows the space-filling qualities of the atoms but is more difficult to interpret.

5.6.1.2 Structural rules and protein folding

5.6.1.2.1 Ramachandran diagrams

The tendency of nonbonding atoms to avoid each other (steric repulsion or avoidance) was quantified in the 1960s in the form of a graph by Gopalasamudram Narayana Ramachandran. These graphs are now called Ramachandran diagrams or Ramachandran plots. Computations of allowed (ϕ,ψ) angles are based on van der Waals minimal distances shown in Table 5.4.

An example of a traditional Ramachandran plot is shown in Figure 5.16. This diagram is for polyalanine but is often taken as a generic Ramachandran plot for protein structure. There is, in fact, a Ramachandran diagram for each amino acid. The "allowed" diagram for glycine, for example, which has only an H atom for a side chain, covers 2/3 of the entire (ϕ,ψ)

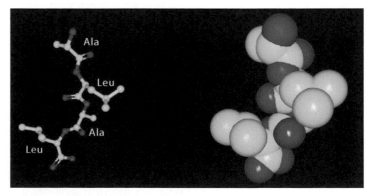

FIGURE 5.15
Three-dimensional structure of tetrapeptide Ala-Leu-Ala-Leu. Tetrapeptide is that bound to a protein subtilisin BPN (not shown). Colors: C, green; O, red; N, blue; H, not shown. From structure 1sua, http://www.rcsb.org/pdb/explore.do?structureId=1SUA, displayed using Protein Workshop software.

TABLE 5.4 van der Waals Distances Used for Ramachandran Diagrams

Atomic Pair	Allowed Distance (nm)	Marginally Allowed Distance (nm)
H····H	0.20	0.19
H····O	0.24	0.22
H····N	0.24	0.22
H····C	0.24	0.22
O····O	0.27	0.26
O····N	0.27	0.26
O····C	0.28	0.27
N····N	0.27	0.26
N····C	0.29	0.28
C····C	0.30	0.29
C····CH$_2$	0.32	0.30
CH$_2$····CH$_2$	0.32	0.30

Source: Ramachandran, G. N. and Sasisekharan, V., *Adv. Protein Chem.,* 23, 283–437, 1968.

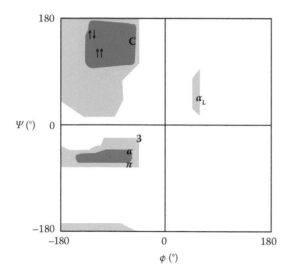

FIGURE 5.16

Traditional Ramachandran diagram for poly-L-alanine. Blue areas indicate "allowed" values of (ϕ,ψ), green areas where atom van der Waals radii begin to overlap. Real proteins have distinct values for each pair of amino acids in their structure, but certain classes of structure tend to have similar (ϕ,ψ) values. Values of (ϕ,ψ) for various uniform structures are indicated: α, right-handed alpha helix (normal); α_L, left-handed alpha helix; ↑↑, parallel β pleated sheet; ↑↓, antiparallel β pleated sheet; C, collagen helix; 3, right-handed 3$_{10}$ helix; π, right-handed π helix. Such diagrams were developed in the 1960s.

space: values of $|\phi|$ and $|\psi|$ must only be in the range of about 50°–180°. Several comparisons of measured angles from protein crystallography data have been made, usually with Gly and Pro excluded, the former because of the small size of the side chain and the latter because of the unusual ring structure. The most well known of these comparisons was performed in 1989.[6] The similarity to the Ala diagram is striking. The large majority of angle pairs lie well within the two largest allowed areas on the left side of Figure 5.16. On the other hand, detailed comparison of experimentally observed angles for specific amino acids with those predicted from corresponding steric-allowed Ramachandran plots for the same amino acid show some clear differences. One of the more important is that the edges of the allowed (and marginally allowed) areas tend to be diagonally sloped from top left to bottom right in the measured diagram. The edges in the Figure 5.16 are mostly horizontal or vertical. See www.fos.su.se/~svenh/index.html for a more complete discussion.

More recently, statistical-distribution Ramachandran diagrams have been developed through analysis of large structural datasets now available for proteins. These analyses can be represented by a probability density, $P(\phi,\psi)$, that the backbone angles have values ϕ and ψ. An example is shown in Figure 5.17 for 12 specific amino acids. The label, "ALA.all" means that datasets for alanine with each of the other amino acids located to its right, were averaged together. Other distributions in the original paper examine the effect of residues to the left or right of the residue of interest. Such analyses are complicated by the choice of structures to include in the dataset. A large number of protein structures is desirable, but different proteins will include different numbers of amino acid residues in various secondary conformations—α and other helices, β sheets, loops, etc. Averaging all amino acids of all proteins together will not produce any structural insight, of course, so restriction to a local structure type (e.g., α-helical regions) must be made. If specific amino acid information is desired, then restriction to the desired amino acid (e.g., alanine) must be made. If the influence of nearest neighbor is of interest, then restriction to specific pairs or triplets must be done. For example, *x-ala-y*, where *ala* is the residue of interest and *gly* and *x* and *y* are neighbors; *x* and *y* may be specified, unspecified, or averaged over a set of amino acids. Unfortunately, a simple restriction to peptides of interest can reduce an apparently large set of protein structural data to statistically small sets of data on a specified amino acid pair or triplet from a specified structure type. Ting et al., using a database of about 3,000 proteins, identified 62,000 amino acid residues located in "loop" regions that connect other more highly structured regions and determined both neighbor-dependent and neighbor-independent Ramachandran distributions. Intelligent statistical analyses gave $P(\phi,\psi)$ distributions like those in Figure 5.17. Comparisons must be done carefully. Figure 5.16, the older Ramachandran diagram, reflects a determination whether alanine, surrounded by other alanines, can exist in a given (ϕ,ψ) conformation without van der Waals radii overlapping. The conformations were then classified by structural type—helix, sheet, etc. In contrast, the upper left distribution of Figure 5.17 reflects the probability that an alanine from a "loop" region in a real protein structure, with any of 20 amino acids to its right, takes on a (ϕ,ψ) conformation. (An individual amino acid in a "loop" region can take on α helix-like (ϕ,ψ) angles even though the region is not completely α helical.) The latter

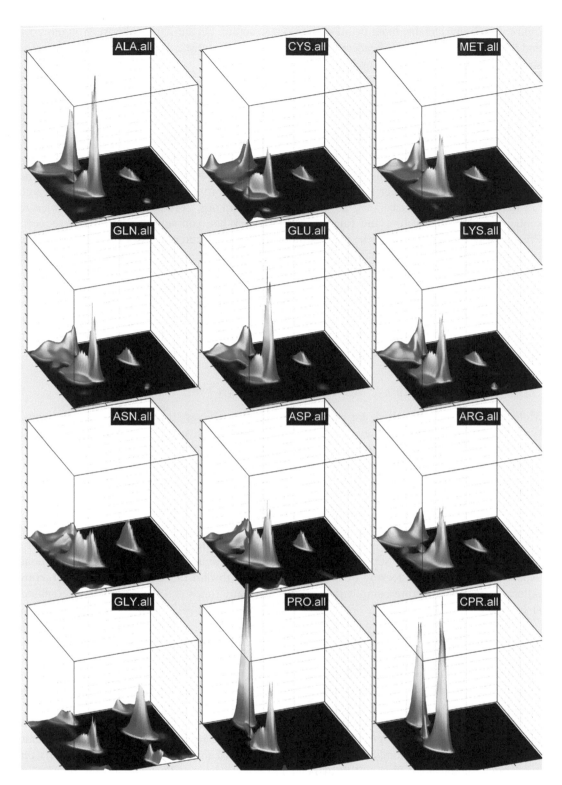

FIGURE 5.17

Neighbor-independent Ramachandran probability densities for 12 amino acid types. Axes: *x*-axis (horizontal)=φ from −180° to 180°; *y*-axis=ψ from −180° to 180°; *z*-axis (vertical)=probability density functions. These probability distributions show considerably more and different detail than the older diagram in Figure 5.15. "CPR" means cis proline. (Reproduced from Ting, D. et al., *PLoS Comput. Biol.*, 6, e1000763, 2010. https://doi.org/10.1371/journal.pcbi.1000763). "This is an open-access article distributed under the terms of the Creative Commons Attribution License, which permits unrestricted use, distribution, and reproduction in any medium, provided the original author and source are credited."

figure shows considerably more detail on allowed structure, but similarities between Figures 5.16 and 5.17 (upper left) show up. Readers interested in computer statistical analysis of protein structure should put a note in Chapter 13 or 14 to refer back to this topic.

5.6.1.2.2 Other rules for protein structure prediction

Dozens of more sophisticated rules for protein structure have been devised since the 1980s. Such sets of rules lie in the domain of biophysics called protein structure prediction, usually based on the amino acid sequence. These sets of rules are sometimes computational, invoking energy minimization techniques, molecular dynamics (Newton's laws of motion), database guidelines for local structure, and other tools. Besides purchasing software tools, students and researchers may use online services such as www.predictprotein.org. "PredictProtein is a service for sequence analysis, structure and function prediction. When you submit any protein sequence PredictProtein retrieves similar sequences in the database and predicts aspects of protein structure and function." This service is based on software developed under the auspices of the National Library of Medicine, National Institutes of Health.[7] There are many other such services: https://cmm.cit.nih.gov/, http://www.expasy.org, https://bioinformatics.ca/links_directory/tool/9479/psa, http://predictioncenter.org/, though some may not offer support. Do an Internet search on "protein structure prediction" to find others. While "PredictProtein" uses database methods to predict structure, other institutions compute structures from force fields, some even using quantum techniques for smaller structures.

> The flavor of the database approach is this: if your amino acid sequence is similar to that of 20 other known, similar structures in the database, your structure is probably similar to them.

5.6.1.2.3 Computational approaches (molecular mechanics)

Molecular mechanical methods for protein and other biopolymer structure prediction are sometimes termed *ab initio* when they attempt to compute the relative positions of atoms in a structure from knowledge of only the sequence of units in the polymer. The predicted structure is determined as that having lowest total energy. Through extensive modeling of the interaction of atoms in model pairs of amino acids, force fields have been constructed, as well as constraints on possible relative orientations of the amino acids. In practice, such structure prediction is never truly *ab initio*, because likely positions and velocities of atoms must be chosen as a starting point for computation. In fact, the initial positions of atoms often must be fairly close to the final computed positions, or the computer will converge to only a metastable structure or will not converge at all. Since a unique final set of 1,000 or so atomic coordinates is never found, an additional aspect is the intelligent steering of computed structure toward the more likely final structure. Alternatively, the structure computation can be run many times and the ensemble of computed structures analyzed.

Distance geometry. Usually applied to data measured by two-dimensional NMR (nuclear magnetic resonance) Overhauser effects spectroscopy, this method is based on a calculation of matrices of distance constraints on pairs of atoms, bond and torsion angles, and van der Waals radius constraints. The set of distances is then converted into three-dimensional space, the coordinates of all atoms of the proteins.

Simulated annealing. This is a *molecular dynamics* method, based on positions, velocities, and forces in real space. A "guessed" starting structure is heated to a high temperature in a simulation, to give all atoms a large kinetic energy that can overcome any local barriers that might inhibit atoms from getting to their equilibrium locations. During many discrete cooling steps the starting structure then evolves (hopefully) toward the energetically favorable final structure.

Molecular dynamics simulations. Introduced in the previous paragraph, molecular dynamics simulators assign initial starting positions and velocities of all (or the important) atoms, and then use force fields, energy constraints, and other rules to calculate the positions and velocities of all atoms as a function of time, as well as the total energy of the system. Ideally, the total energy will decrease as the positions of atoms approach their equilibrium positions, which ideally will approach the real positions in space. The appeal of this method to the physicist is that it is based on fundamental laws of nature (Newton's), occasionally supplemented by quantum mechanical laws. Dozens of software packages and model force fields exist for molecular dynamics simulations: Abalone, ADUN, AMBER, CHARMM, GROMOS, MOSCITO, and more. See https://en.wikipedia.org/wiki/Comparison_of_software_for_molecular_mechanics_modeling for a more extended list and links to software sources. Many of the software packages are truly "packages," which use force fields (distance-dependent potentials), originally developed for the programs AMBER, OPLS, CHARMM, and GROMOS, in different computational codes and graphic displays.

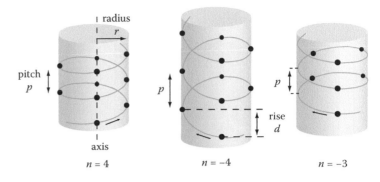

5.6.1.3 Alpha and other helices

Helices made up of identical units attached together in a linear array are cylindrical coils described by four numbers: (1) helix radius r, (2) helix pitch p, (3) the number of repeat units per turn, n, and (4) the total number of units in the helix, N. An additional quantity commonly used is the helix rise, defined as $d = p/n$. As part of a three-dimensional structure, the direction of the helix axis, which may vary for a long helix, is also a descriptor for the helix. Figure 5.18 shows the geometry of such a helix. A right-handed helix has positive values of n; a left-handed helix has negative n. A right-handed helix is defined by a right-hand rule. Curl the fingers of your right hand along the path of the helix, and your thumb points in the direction of travel along the axis of the helix. The leftmost helix in Figure 5.18 is right-handed, while the other two are left-handed.

The "length" of the helix can refer to several different quantities. The most common is the distance *along the helix axis* from end to end. This is what the "length of the helix" will refer to in this text. A second sort of length may be called the "contour length" or path length you travel moving *along the backbone* of the helical strand from end to end. You will find the relation between these two lengths in the homework. Unfortunately, the term *contour length* becomes ambiguous when we deal with real biological structures, especially the double helix, DNA. Long helices always change directions, introducing a third "length": the straight-line distance between the two ends of the strand, which may be much smaller than the helical "length" or "path length." In this context, contour length often refers to the length of the helix, if the axis were straight. The best advice is to not be too fussy about definitions. If a situation seems ambiguous, decide which meaning of length makes most sense and proceed, or determine both lengths and explain them.

The α helix is one of the most common local arrangements of amino acids. As you have seen from protein structures in this chapter, α helices tend to be local, consisting usually of 5–20 amino acids. The average number of amino acids in globular protein α helices is 12, though numbers higher than 50 have been observed.[8] The length of an α helix of 12 units is about 1.8 nm and about three helical turns. These numbers can be determined from the data in Table 5.5, which also contains helical parameters for several less common helices found in nature.

Different amino acid sequences have different tendencies to form α helices. Met, Ala, Leu, uncharged Glu, and Lys have high helix-forming propensities, whereas Pro, Gly, and negatively charged Asp have low propensities.[9] Pro tends to disrupt helices because it cannot donate an amide hydrogen bond. The Pro ring structure also restricts its backbone φ dihedral angle to about −70°, which is not conducive to α-helix formation. However, Pro sometimes acts as the first residue preceding a helix. Gly also tends to disrupt helices: its high conformational flexibility makes it *entropically unfavorable* to adopt the relatively constrained α-helical structure. (More on entropy, energy, and stability in Section 5.6.1.4, and in Chapters 14 and 15.) The (real) α helix of Figure 5.19 tends to follow all these rules of thumb.

FIGURE 5.18

Helices made from discreet units. The pitch is the distance (in the axial direction) between equivalent points on the helix separated by travel of one complete rotation around the axis. The number of units per turn is n, which does not have to be an integer. The helical rise is the distance (in the axial direction) between two adjacent units. Note that whether the structure is a true helix depends on the structure of a unit. If units were small spheres connected by straight lines, $n = 2$ and $n = 1$ would correspond to a flat zigzag and straight line, respectively.

TABLE 5.5	Polypeptide Helical Parameters			
Type of Helix[a]	**Pitch p (nm)**	**n (units/turn)**	**ϕ (°)**	**Ψ (°)**
α helix (RH)	0.54	3.6	−57	−47
α helix (LH)	0.54	−3.6	57	47
3_{10} helix (RH)	0.60	3.0	−49	−26
π helix (RH)	0.52	4.4	−57	−70
Collagen helix[b]	1.00	−3.3	−51	153

[a] RH, right-handed; LH, left-handed.
[b] Parameters for one strand of the triple helix.

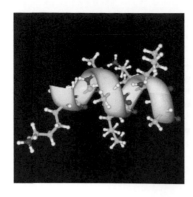

FIGURE 5.19
Hydrogen bonds in an alpha helix. An α-helical section of the 1DJF oligopeptide. Each of the red double-bonded oxygen atoms is H-bonded to an NH group of an amino acid four units away. The N$\cdots$O distance is about 0.28 nm, considered optimum for a hydrogen bond. Some amino acids and atoms are not shown, for clarity. The sequence of this peptide is (amino end)-Gln-Ala-Pro-Ala-Tyr-Lys-Lys-Ala-Ala-Lys-Lys-Leu-Ala-Glu-Ser-(carboxyl end), with display only from residue 4 (Ala) to residue 13 (Ala). The α-helical section is preceded on the amino side by Pro. (From Montserret, R. et al., *Biochemistry*, 39, 8362–8373, 2000. PDB ID 1DJF from the Protein Data Bank, displayed using Protein Workshop software.) (From Jmol: An open-source Java viewer for chemical structures in 3D. http://www.jmol.org/; Protein Workshop: Moreland, J.L. et al., *BMC Bioinformatics*, 6, 21, 2005; NGL Viewer: Rose, A.S. et al., *Bioinformatics*, 2018.)

Can we connect the α helix to the Ramachandran angles discussed earlier? The answer is yes. Those angles are $\phi = -57°$ and $\psi = -47°$, as shown in Table 5.5, along with angles for the other listed structures.

5.6.1.3.1 Alpha helix conformation stabilized by H bonds

The right-handed α helix is fairly common in proteins, which implies that it is a low-energy conformation of polypeptides, almost independent of the component amino acids (excepting, of course, the disruptive bonding character of Pro). The reason for this is hydrogen bonding: the peptide unit (O = C–N–H) hydrogen of the nth amino acid in the sequence is hydrogen bonded to the peptide unit oxygen of the $(n-4)$th amino acid. Figure 5.19 shows an example of a helical section from a short protein from the PDB. For clarity, the H bonds are not drawn in, and some amino acids and side groups are not displayed. Each red O atom is H bonded to an NH group of an amino acid four units away, giving the structure additional stability above that of a "random" coiled structure.

Alpha helix H bonds have energies of only about -2 to -6 kJ/mol (-0.8 to $2.5 k_B T_r$) in aqueous solution—not a significant energy unless more than a few amino acids contribute.[10] Part of the reason for this small energy is water. The high dielectric constant will tend to reduce all types of electrostatic energies. Indeed, α-helix H bond strengths are found to be higher in solvents with lower dielectric constants, illustrating one of the key roles of water in protein structure. Water itself can also form H bonds to amino acids, in effect competing with the α-helix H bonding. Both the dielectric and competitive effect of water tend to reduce the strength of intra-protein H bonds, and you would predict that α helices might be more stable if shielded from water, for example, on the interior of a large protein. The issue of H bond stabilization of α helices in proteins in one sense is simple: they are commonly observed, so they must be stable. The issue of whether the hydrogen bonds are primarily responsible for the stability is complex, however. Entropies must also be considered; entropies not only of the α-helix amino acids but also of the water.

5.6.1.4 Interactions that govern protein structure

What are the various forces that we should consider when analyzing the stability of a protein helix?

1. Hydrogen bonding, energies about -2 to -6 kJ/mol (-0.8 to $2.5 k_B T_r$). H bonds can involve not only N–H$\cdots$O = C of the backbone but also atoms of the side chains (R groups) with water or with each other. The optimal distance for the N$\cdots$O hydrogen bond is about 0.28 nm. The latter is more important in a large protein, where a side chain of an amino acid from another part of the protein approaches the α-helix of interest. H bonds are usually considered significant when the atoms of a donor–acceptor pair are closer than about 0.35 nm. There can be many of these bonds, up to about two per amino acid.

2. Ionic interactions, often called "salt bridges" in the literature, energies about -4 kJ/mol ($-2 k_B T_r$) for opposite charges of magnitude $1e$ separated by 0.4 nm in a dielectric medium of $\varepsilon = 78$. The force can vary considerably but involves only charged amino acids, Lys, Arg, Asp, Glu and sometimes His and solvent counterions, if the charged group is exposed to solvent. Though these interactions can be locally important, there are not many of them and they are not expected to significantly stabilize a protein.

3. Dipole–dipole interactions, energies about -10 kJ/mol ($-4 k_B T_r$) for dipoles of magnitude 5×10^{-30} C-m (1.5 Debye), separated by 0.4 nm in a dielectric medium $\varepsilon \sim 4$. Why should we use $\varepsilon \sim 4$ here, whereas we used $\varepsilon \sim 78$ for charge–charge interactions? Sometimes we *should*

TABLE 5.6 Energies of Transfer from Water to Nonpolar Solvent at $T_r = 25°C$

MolecularGroup	Nonpolar Solvent	ΔH (kJ/mol)	$-T_r\Delta S$ (kJ/mol)	ΔG (kJ/mol)
CH_4	Benzene	11.7	−22.6	−10.9
CH_4	CCl_4	10.5	−22.6	−12.1
C_2H_6	Benzene	9.2	−25.1	−15.9
C_2H_4	Benzene	6.7	−18.8	−12.1
C_2H_2	Benzene	0.8	−8.8	−8.0

Source: Kauzmann, W., *Adv. Protein Chem.*, 14, 1–63, 1959.

use the latter value, such as when amino acids with polar side chains are surrounded by water, but the most important and numerous dipoles are often those associated with the peptide bonds, and these may be in the protein interior. These dipoles recur every 0.54 nm or so in an alpha helix, but importantly, they are all rigidly aligned in the same direction.

4. Hydrophobic interactions. This interaction is complex, partly because it does not involve a force, per se. The force is based on electrostatic interactions and entropy but cannot readily be calculated. Measurements of some relevance are straightforward, however: a molecular group similar to that in a protein is transferred from a nonpolar to a polar solvent and the free energy measured. More precisely, the partitioning of a molecular group between the two solvents is made, determining ΔG. The temperature dependence of the partitioning determines both ΔH and ΔS, via Equation 5.8. Some results of measurements made 50 years ago are in Table 5.6.

Table 5.6 molecules are "nonpolar" CH_4, C_2H_6, C_2H_4, and C_2H_2, models for the nonpolar amino acids side chains like $-CH_3$ in Ala, $-CH-C_2H_6$ in Val, and $-CH_2-CH-C_2H_6$ in Leu. Values for ΔH indicate immediately that the hydrophobic effect is not like any of the other protein-stabilizing interactions, which involve some negative potential energy of interaction. The positive value of ΔH shows that the molecular groups CH_4, C_2H_6, C_2H_4, and C_2H_2 do not enthalpically "prefer" to be (have lower potential energy) in benzene than in water. In terms of ΔH they would prefer to be in water. The key is in the entropy, which at room temperature is positive, indicating the system has greater freedom (more distinct ways to be in the benzene-surrounded state), and the magnitude of $-T\Delta S$ is negative and twice the value of ΔH. So, we conclude that the hydrophobic interaction is somewhat larger than k_BT_r, since its average value is about −10 kJ/mol ($-4k_BT_r$), but the effect will be important because one-third of the amino acids will have side chains with some degree of hydrophobicity.

Free energies of transfer of amino acid side chains, including both hydrophobic and nonhydrophobic parts have been collated by Engelman et al.[11] See Table 5.7. These numbers make clear that some amino acids have side chains with polar and nonpolar parts. The transfer energies have been used primarily to predict which parts of α-helical membrane protein sequences will span a membrane. Running sums (or averages) of these free energies are used to evaluate the hydrophobicity of the polypeptide. The total free energy of transfer of amino acids 1–20 is calculated and assigned to position 1; the total of 2–21 is assigned to position 2, 3–22 to position 4, etc. The summing is done because a minimum of about 20 amino acids is needed to span the hydrophobic part of a membrane. (Many prefer to change the sign of ΔG_{total} to reflect the transfer from inside the membrane to the water.) A graph of this summed free energy of transfer (from low ε to water) versus amino acid position is then made. Regions that rise above 85 kJ/mol are predicted to be fixed inside the membrane. An example of such a graph is shown in Figure 5.20.

A "hydropathy score" has also been proposed to describe the degree of hydrophobicity of the 20 amino acid side chains,[12] shown in Table 5.7, last column. Hydropathy scores reflect primarily hydrophobic effects, but they have been designed to predict overall structures of proteins, so there is an admixture of effects besides "hydrophobicity." The basic idea is that sections of the polypeptide chain that are, on average, hydrophobic will tend to be on the inside of a protein, away from water. To perform a structural prediction, a window size is set. The window size is the number of amino acids whose hydropathy scores will be averaged and assigned to the first amino acid in the window. (It makes no sense to consider a 1-amino acid section to be hydrophobic if it is surrounded by 10 hydrophilic amino acids.)

TABLE 5.7 Transfer Free Energies for Amino Acid Side Chains in α-Helical Polypeptides from Water to an $\varepsilon = 2$ Dielectric Environment[a] and Hydropathy Index

Amino Acid	$\Delta G_{hydrophobic}$ (kJ/mol)	$\Delta G_{hydrophilic}$ (kJ/mol)	ΔG_{total} (kJ/mol)	KD Hydropathy Index[b]
Phe	−15.5		−15.5	2.8
Met	−14.2		−14.2	1.9
Ile	−13.0		−13.0	4.5
Leu	−11.7		−11.7	3.8
Val	−10.9		−10.9	4.2
Cys	−8.4		−8.4	2.5
Trp	−20.5	12.5	−8.0	−0.9
Ala	−6.7		−6.7	1.8
Thr	−9.2	4.2	−5.0	−0.7
Gly	−4.2		−4.2	−0.4
Ser	−6.7	4.2	−2.5	−0.8
Pro	−7.5	8.4	−0.9	−1.6
Tyr	−15.5	16.7	1.2	−1.3
His	−12.5	25.1	12.6	−3.2
Gln	−12.2	29.3	17.1	−3.5
Asn	−9.2	29.3	20.1	−3.5
Glu	−10.9	45.1	34.2	−3.5
Lys	−15.5	52.2	36.7	−3.9
Asp	−8.8	47.2	38.4	−3.5
Arg	−18.4	69.8	51.4	−4.5

[a] Values are given for the hydrophobic and hydrophilic components of the transfer of amino acid side chains from water to a dielectric environment of $\varepsilon = 2$. (Adapted from Engelman et al., *Annu. Rev. Biophys. Biophys. Chem.*, 15, 321–353, 1986.)

[b] The Kyte–Doolittle (KD) score ranges from −4.6 (most water-loving) to +4.6. See text.

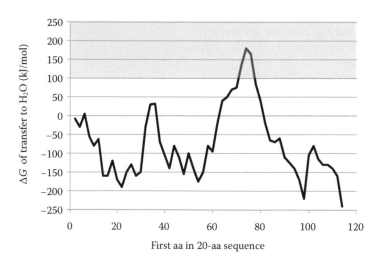

FIGURE 5.20

Application of the Goldman, Engelman, Steitz (GES) hydrophobicity scale to the protein glycophorin. Regions of the 20-peptide running sum rising above 85 kJ/mol are considered likely candidates for intramembrane regions of the protein. Direct calculations of fractions in and out of the membrane can be calculated from these plots. (Adapted from Engelman, D. M. et al., *Annu. Rev. Biophys. Biophys. Chem.*, 15, 321–353, 1986.)

The default window size is 9. A computer program starts with the first window of amino acids, calculates the average window hydropathy score, and proceeds to the end of the protein, computing the average score for each window. The averages are then plotted on a graph. The y axis represents the hydrophobicity scores, and the x axis represents the window number. An example is shown in Figure 5.21. This graph was obtained after copying the sequence of the cystic fibrosis transmembrane conductance regulator, a 1480-amino acid protein, from the https://www.ncbi.nlm.nih.gov/protein/ database and pasting it into the Kyte–Doolittle hydropathy online calculator at http://gcat.davidson.edu/rakarnik/kyte-doolittle-background.htm#ref. This score and the calculator do not, of course, reflect any fundamental law of protein conformation. Application to a new, unknown protein takes a good amount of trial and error. Conclusions reached must always be viewed as hypotheses to be checked by measurement.

How did Kyte–Doolittle determine the values of individual hydropathy amino acid scores? They used proposed values on proteins whose structures were known and found the parameters that predicted protein structure the best. The structural prediction is not of the positions of all atoms but of regions of the protein exposed at the surface (water-exposed) and regions on

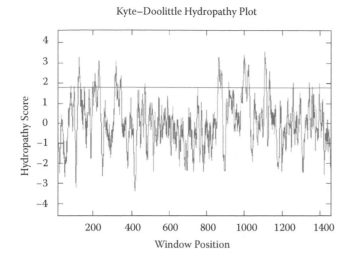

FIGURE 5.21

Kyte–Doolittle hydropathy score plot for the cystic fibrosis transmembrane conductance regulator, a 1480-amino acid protein. Score above 1.6 indicates regions that are likely inside the membrane. Detailed information on this protein can be found in the NIH database: http://www.ncbi.nlm.nih. gov/sites/entrez?cmd=Retrieve&db=protein&dopt=GenPept&list_uids=180332.

the protein interior. When looking for surface regions in a globular protein, for example, surface regions were identified as peaks below the zero line. The method can also be applied to proteins that extend across membranes. For membrane protein applications, a window size of 19 worked best, similar to the transfer free energy method. Intramembrane regions were identified by peaks with scores greater than 1.6. Refer again to Figure 5.21. Online problems can be tried out at http://gcat.davidson.edu/rakarnik/ kyte-doolittle-background.htm#ref, but any protein sequence, introduced into the online calculator in a standard format, can be analyzed. Sequences can be copied and pasted, directly from National Institutes of Health (NIH) sequence databases, for example. See also http://gcat.davidson.edu/rakarnik/kyte-doolittle-background.htm for more information.

The hydropathy index approach to intramembrane insertion of protein regions has a significant weakness. The weakness relates to major areas of cell function, including selective ion transport across membranes. See Section 14.13.4, Membrane Potentials and Work. Most proteins extending across a membrane must be oriented in a certain direction in order to allow directional transport and signaling. Membrane *channel proteins* may also contain charged amino acids located in the interior of the membrane. Neither of these properties is readily dealt with by the hydropathy index, which prefers charges to be located near counter-charges and surrounded by water and has little or no sensitivity to the reversal of polypeptide chain direction in the membrane. Elazar et al. have developed a method that follows the "positive-inside" rule—positive charges of the protein prefer to be located inside a volume enclosed by a membrane, where the prevailing ionic environment is negatively charged—and predicts that this interaction can drive membrane insertion even of marginally polar segments predicted by hydropathy alone to be external to the membrane.[13] See http://topgraph.weizmann. ac.il/bin/steps or search for "TopGraph." New protein/membrane prediction tools frequently appear on the Web, due to the importance of transmembrane proteins in cellular function and recognition.

Why should ΔS be so negative (favorable) for transfer of CH_4 from water to a hydrocarbon-rich environment? The explanation may be found in Chapter 4, where we learned that unlike benzene and common organic solvents, water (1) can form extended structures and (2) often has different properties and structure near other molecules and surfaces.

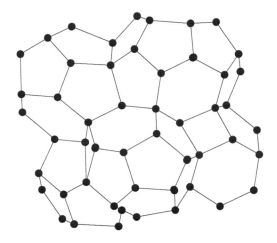

FIGURE 5.22

Hypothetical (partial) ordered shell of water molecules that could surround a nonpolar protein side chain. Black dots indicate where water O atoms are located. The ordering could extend more than one layer from the shell. The array of O atoms is located such that water H atoms can form H bonds with minimal distortions from the tetrahedral arrangement described in Chapter 4.

The inside of cells is generally a reducing environment, except for the endoplasmic reticulum where lipids and membrane proteins are synthesized.

Virtually intact hair has been found in 4,000-year-old Egyptian mummies.

We usually think of adjacent water molecules assuming specific, fixed orientations near a charge or object it is attracted to. This ordering of water indeed happens and would lower the entropy. A similar thing happens when water molecules surround a molecule or surface it is *not* attracted to. *Some* water molecules *have* to be adjacent to the hydrophobic object; they may move as far away as possible and turn one particular side toward the object, but this, too, is ordering and lowers the entropy. Any replacement of the water by organic molecules that don't particularly care if the hydrophobic object is there or not (i.e., don't orient) will be a more favored situation (higher entropy, lower free energy).

These properties of the hydrophobic interaction, as well as the other interactions, which depend on dielectric constant, should convince us again that water is critical to biology in the form of polypeptide conformational stability. Now we also know some numerical estimates for relevant energies, and we know that both ΔH and ΔS are important, the latter because water molecules dislike being organized. The details of the structure explaining the entropic effects, including extended shells like Figure 5.22, are still being debated, but some sort of extended water ordering from a hydrophobic or mixed hydrophobic/hydrophilic protein surface is likely.[14]

5. Disulfide bonds. Table 5.2 indicates about 1.4% of protein amino acids are cysteines (Cys). A typical 300-amino acid protein might have four Cys residues. These Cys amino acids have –SH groups at the far end of their side chains. In an *oxidizing environment*, these –SH groups are oxidized to form a disulfide group, if two sulfur atoms are close enough together: $-SH + -SH + O \rightarrow -S-S- + H_2O$. The oxygen atom comes from some oxidizing agent. Experience with the reducing environment inside cells and oxidizing environment outside is common. We know our blood is red when it is inside the red blood cells. When we cut our finger and bleed, we see a brownish scab form. The brown is primarily from the hemoglobin iron atom, which becomes oxidized outside the reducing environment of the live red blood cell.

Some proteins have an unusual number of cysteine amino acids. Disulfide bonds between appropriately aligned helices stabilize the structure of *alpha keratin*, a primary component of hair and nails in mammals. (Vulcanized rubber has similar stabilization.) These disulfides give hair and nails strength, stability, and insolubility, as well as the pungent aroma when burned. Hair and other α keratins consist of α-helically coiled single protein strands (with a slightly smaller helical pitch of 0.51 nm), which are further twisted into superhelical ropes that are further coiled. A cartoon of the smallest-scale supramolecular structure of α keratin is shown in Figure 5.23. Some of you may have experienced the reduction (breaking) of hair disulfide bonds by chemical means (ammonium thioglycolate or glycerol monothioglycolate) and re-oxidation (reforming of disulfide bonds) by hydrogen peroxide. Some keywords in this sequence of chemical names—*ammonium* gives the familiar, strong smell; *thio* refers to a compound with an –SH group—may give away this process of a salon hair "permanent." Though this is not a chemistry book, it is good for the physicist reader to identify important phenomena encountered in daily life.

Disulfide bonds can play an important role in managing protein structure. Some of the structural effects can be guessed at from toys used by young children. Figure 5.24 shows a colored beads-on-a-string model for a polypeptide chain, initially stretched straight in Figure 5.24(a). We have argued that this structure would collapse for entropic reasons into some sort of random coil, if interactions between units in the chain are negligible. As a bit of playtime suggests, the positional freedom of all units in the chain allows a large number of different collapsed structures, one of them shown in (b). Suppose, however, that the green beads (arrows) represent cysteine amino acids that can form covalent disulfide bonds. The number of possible structures is drastically reduced by disulfide formation. In (c) and (d), two such structures are shown, with disulfide bonds between different pairs of the four cysteines. Though (c) and (d) superficially appear like (b), a bit more playtime shows that the ends of the chain have structural freedom, but the loops between the bonded green beads are restricted to more coordinated breathing motions. Even more insight into protein assembly and structure can be discovered. Since we know, or will find out, that single polypeptide chains are produced in the cell from one end to the other, we might guess that disulfide bonds would preferentially form between the first two cysteines that came out of the production site, if that site had a an oxidative nature that facilitated formation of disulfides. This would rule out structure (d) and favor structures like (e). If, however, the protein production factory (the ribosome) places some type of chain-bending limiter or reaction inhibitor on cysteines as they come off the assembly line, structure (d) would not be precluded and the cell could control disulfide bond formation in another location. For the most part, this latter situation holds in eukaryotic cells (Chapter 7), where the generally reducing environment inside cells moves disulfide formation over to specialized sites like the endoplasmic reticulum.

Figure 5.25 shows a small but heavily disulfide-bond stabilized protein of another sort, a potassium channel blocking toxin from the venom of scorpions. Toxins of this class typically have a Cys content 10 times that of the average protein—one in eight amino acids may be a Cys. The scorpion toxin is a potent blocker of K(+) channels. The protein structure is stabilized by four pairs of Cys residues that form disulfide bonds. Miniproject 5.10 investigates some implications of the stability of this sort of protein.

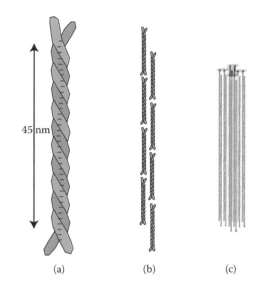

(a) (b) (c)

FIGURE 5.23
(a) Structure of α keratin dimer. The two strands are each slightly distorted α helices of about 310 amino acids, wound around each other in a left-handed manner. (b) In hair, such coiled coils further associate into larger protofilaments, which further associate into protofibrils, which further associate into (c) microfibrils, which further associate into macrofibrils, which further associate into hair. (Think of hierarchical structure!) The small dashes indicate disulfide cross-links, which strengthen the structure. (Hand drawings of these and other proteins by Irving Geis can be seen in Dickerson, R. E., and Geis, I., *The Structure and Action of Proteins*, W. A. Benjamin, Menlo Park CA, 1969.)[39]

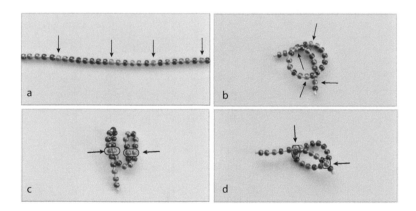

FIGURE 5.24
(a) Structure of α keratin dimer. The two strands are each slightly distorted α helices of about 310 amino acids, wound around each other in a left-handed manner. (b) In hair, such coiled coils further associate into larger protofilaments, which further associate into protofibrils, which further associate into (c) microfibrils, which further associate into macrofibrils, which further associate into hair. (Think of hierarchical structure!) (d) The small dashes indicate disulfide cross-links, which strengthen the structure. (Hand drawings of these and other proteins by Irving Geis can be seen in Dickerson, R. E., and Geis, I., *The Structure and Action of Proteins, W. A. Benjamin*, Menlo Park, CA, 1969.)

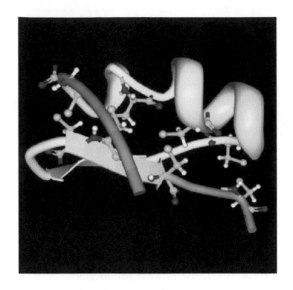

FIGURE 5.25
Disulfide bonds in the small potassium channel blocking toxin 4 (PDB 1n8m) of the venom of *Pandinus imperator* scorpions. (From Guijarro, J.I. et al., *Protein Sci.*, 12, 1844–1854, 2003.) The toxin is a potent blocker of K(+) channels. Four pairs of Cys residues positioned for disulfide bonds can be seen. Sulfur atoms are in yellow. Displayed from PDB using Protein Workshop software; only Cys residues have side group atoms drawn. (From Jmol: An open-source Java viewer for chemical structures in 3D. http://www.jmol.org/; Protein Workshop: Moreland, J.L. et al., *BMC Bioinformatics* 6, 21, 2005; NGL Viewer: Rose, A.S. et al., *Bioinformatics*, 2018.)

5.6.1.5 β Sheet

The interactions addressed above also apply to other protein secondary structures. We have seen α helices in protein structures viewed in the PDB. We have also encountered another structure cartooned as a flattened arrow, such as that visible in Figure 5.25. Such arrows indicate a structure called the β sheet. The Ramachandran diagram (Figure 5.16, arrow symbols in upper left) suggests that the structure is not far from that of an α helix, but the details differ considerably. Beta sheet structures have sections of the polypeptide chain aligned parallel or anti-parallel to each other. There must, of course, be breaks in the β structure to allow the chain to curl back on itself. The hydrogen bonds that stabilize the β sheet are between amino acids on these (anti) aligned strands of polypeptide, shown in Figure 5.26. In typical globular proteins a sheet may be made

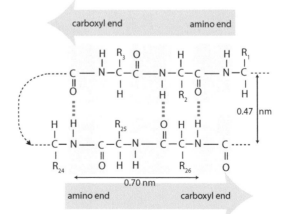

FIGURE 5.26
Antiparallel β sheet, showing H bonds (heavy dotted lines) between O and N atoms of the peptide group on adjacent strands. In cartoon diagrams of β-sheet structure, sheets are shown as flat arrows, pointed in the amino $\rightarrow$ carboxyl direction. Repeat distances are noted in the diagram. Sheets are referred to as *pleated* because adjacent C_α atoms (attached to R groups) are offset into and out of the plane of the paper.

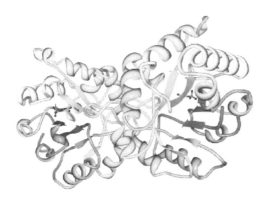

FIGURE 5.27
Interior β sheets in triose phosphate isomerase from yeast have a pronounced twist. (PDB structure 2YPI, from Lolis, E. and Petsko, G. A. *Biochemistry*, 29, 6619–6625, 1990, displayed using Protein Workshop software.) (From Jmol: An open-source Java viewer for chemical structures in 3D. http://www.jmol.org/; Protein Workshop: Moreland, J.L. et al., *BMC Bioinformatics* 6, 21, 2005; NGL Viewer: Rose, A.S. et al., *Bioinformatics*, 2018.)

up of 2–15 strands, each strand of length about 2 nm. Beta sheets in larger globular proteins are often twisted in a right-handed manner, especially when they occur in the center of the protein. See Figure 5.27.

5.6.2 Hints of Function: Structural vs. Catalytic Proteins

We have just examined the first of several important structural proteins found in living organisms, keratin (Figure 5.23). Keratin is a protein that spontaneously forms fibers and can thus be used for many structural purposes. Myoglobin (Figure 5.1), on the other hand, is shaped like a globule and has a role in transport. These two are examples of the two major classes of polypeptides in biosystems: *fibrous* and *globular* proteins.

5.6.2.1 Fibrous proteins

Fibrous proteins are the ropes, cables, and frames of biological systems. As we saw in Chapters 1 and 4, cells and cell compartments are not empty spaces filled with water and dissolved molecules but rather closely packed three-dimensional mazes. An important capability of most cells is the ability to *maintain or change shape* in response to changes in the environment. We have seen microorganisms change shape as they moved and maneuvered around objects. Our first associations with this cellular shape change were no doubt similar to something like a person pulling in his stomach when trying to squeeze through a narrow space. This association is not completely off the mark, as the muscles used to contain a waistline are made up of fibrous proteins.

Fibrous proteins have two major roles: maintaining a basic framework (shape) and moving that framework. The structural role is uncomplicated, at least compared to enzymatic activity (chemical reactions) and movement. As a consequence, their structures are rather simple, consisting primarily of α helices, β sheets, and triple helices, interspersed by small regions of relatively unordered structure. (We address the triple helix when we get to collagen.)

5.6.2.1.1 Keratin

The keratins form a group of structural proteins that make up the major part of skin, hair, nails, horns, and feathers. Keratins are members of what are referred to as *intermediate filaments* with diameters of 7–15 nm. The α keratins are found in mammals, β keratins in birds and reptiles. We focus on the former, which were introduced in Section 5.6.1.4.

The structure of keratin has been described as a helix of helices, organized in a hierarchical fashion, as shown in Figure 5.23a–c. The central part of one unit of the keratin "dimer" in Figure 5.23a is made up of about 310 amino acids and is about 45 nm long. The entire unit contains about 400 amino acids. Molecular masses of mammalian α keratins vary from 40 to 50 kDa. The details of keratin structure, especially above the level of the α helix and at the atomic level, are not known very well. The reason for this applies to most fibrous proteins—X-ray and neutron diffraction do not work as well because crystals do not form. The fibers from which diffraction has been measured have only one-dimensional order. We will find a similar problem faced Watson and Crick when they performed diffraction from fibers of DNA, but unlike DNA, crystals of keratin still cannot be formed.

One of the reasons keratin crystals do not readily form is an interaction we have discussed in general but not applied specifically to keratin. The lowest-level structure of an α keratin consists of α-helical domains that consist of amino acid repeats of a pattern of seven, ABCDEFG. A and D are hydrophobic amino acids. This "heptad" repeat is characteristic of sequences able to form coiled-coil structures.[15] What structural consequences do these heptads have? One basic feature of the α helix is that there are about 3.6 residues per turn. This means that every fourth amino acid will be on the same side of the helix as the first but offset a bit. This offset will create a hydrophobic line that will wind around the outside of the helix. When two such α helices come near each other, the hydrophobic interaction will tend to make the two helices wind around each other—a helix of helices.

The mechanical properties of keratin and other fibers will be addressed in Chapter 12, but for now we note that keratin is a quite stiff spring because of the multiple disulfide bonds, in addition to the hydrophobic forces that help form the coiled coil. On the other hand, if the disulfide bonds are broken by a reducing agent ($-S-S- \rightarrow -SH + HS-$), keratin can be stretched to almost double its normal length, in the process changing from an α helix to a β sheet structure.

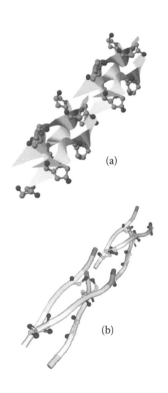

FIGURE 5.28

Triple-helical structure of a collagen-like oligopeptide (Pro-Hyp-Gly)$_n$ ($n = 10, 11$). (a) Ribbon structure with side groups displayed, using Jmol software. (From Jmol: An open-source Java viewer for chemical structures in 3D. http://www.jmol.org/; Protein Workshop: Moreland, J.L. et al., *BMC Bioinformatics* 6, 21, 2005; NGL Viewer: Rose, A.S. et al., *Bioinformatics*, 2018.) (b) Strand structure with Gly residues (only) displayed, using Protein Workshop. Gly side chains are on the interior of the triple helix, where the close packing would not permit a larger amino acid. In vivo, collagen polypeptides are much longer, but cannot be crystallized. (PDB structure 1V7H, from Okuyama, K. et al., *Biopolymers*, 76, 367–377, 2004.)

5.6.2.1.2 Collagen

Collagen is the most abundant protein of vertebrate animals. The amino acid sequence is repetitive, usually with one modified Pro or Lys amino acid—(Pro-Hyp-Gly)$_n$, where Hyp is hydroxyproline—but also contains a lesser amount of other amino acids. Three major types of collagen are found in humans: Type I, in skin, bones, tendon, blood vessels, and the cornea; Type II, in cartilage; Type III, in blood vessels and fetal skin. A "molecule" of type I collagen is composed of three polypeptides, with total molecular mass about 280 kDa. The three polypeptides are wrapped around each other to form a triple helix of length about 300 nm and diameter 1.4 nm.

Each individual polypeptide chain is a left-handed helix, 3.3 amino acids per turn, and a pitch of 1.00 nm. The left-handed helices wrap around each other to form a right-handed triple helix of 10 Pro-Hyp-Gly residues per turn and pitch 8.6 nm. Figure 5.28 shows the atomic structure of a collagen-like oligomer (Pro-Hyp-Gly)$_{10}$. Figure 5.28a shows helical structures, while Figure 5.28b illustrates why collagen triple helices need Gly in the third position. Crowding in the interior of the triple helix is such that there is little room for a side group larger than Gly's H atom. The three polypeptide chains are offset by one unit in the longitudinal direction so that Gly N–H groups can form a strong hydrogen bond with a Pro O on each of the other two chains. This, in addition to the relative inflexibility of the Pro amino acids, gives collagen great tensile strength and rigidity. The ropelike structure helps to convert a longitudinal stress into a lateral compression force. The close packing of the triple helix prevents significant compression. Note also that the left-handed twists of the individual polypeptides and the right-handed triple helical twists tend to mutually prevent the other from unraveling—a design that ropes and cables also use. Keratin and muscle fibers are similarly arranged.

The collagen triple helices are either packed into higher-order fibrils, much like their keratin cousins, or form complex networks. The collagen fibrils can have diameters of 10–200 nm and typically consist of different types of collagen. As we should expect, these fibrils are further organized into higher-order structures, summarized in Table 5.8. The structures generally correspond to the role of the various tissues. Tendons must withstand great longitudinal stress, so bundles of fibrils are employed. Skin must flex in two dimensions; fibril sheets oriented at various angles assist this flexing. The cornea of the eye must be transparent, so yet another sheet structure is used. (Problem 5.12 addresses this transparency issue.) Finally, cartilage must support stresses in three dimensions (the structure must be able to resist shearing forces along any axis), so no sheet- or bundle-like structure will suffice. A more random arrangement of fibrils apparently suffices.

TABLE 5.8	Higher-Order Structure of Collagen Fibrils in Various Tissues
Tissue	**Structure**
Tendon	Parallel bundles
Skin	Layered sheets, oriented at random angles
Cornea	Sheets stacked crossways to reduce light scattering
Cartilage	No uniform arrangement

When two fibrils encounter each other, covalent bonds are formed that further strengthen the network. These are not the disulfide –S–S– bonds of keratin but typically involve Lys and His amino acids, which tend to occur at the amino and carboxyl ends of the polypeptide chains. However, other unusual amino acids appear to be involved.[16] The extent of networking in some tissues increases with age. We will address the issue of effects of ultraviolet (UV) light on collagen in Chapter 11.

5.6.2.1.3 Elastin

Covalent networking of collagen fibrils clearly increases the strength and rigidity of tissues, but the issue of flexibility is less clear. Skin should ideally be strong yet flexible, so a pure collagen network is not optimal. For this reason, another fibrous protein, *elastin*, is mixed in the matrix with collagen. Elastin is a polypeptide made primarily from Ala, Gly, and Val. As with collagen and keratin, its structure and properties are not completely understood.

5.6.2.1.4 Muscle fibers and microtubules

This category of fibrous protein includes actin and myosin of muscle and microtubules of various sorts. Muscle and related fibers are the fibrous proteins that have not only important structural properties but also active, *motile* properties. While collagen and keratin primarily react to forces applied from the outside, muscle fibers and associated proteins generate forces. These fibers may also have connections between adjacent fibers that give collective, coordinated properties, like collagen and keratin. However, *motor proteins* associated with muscle fibers and microtubules generate forces that move materials along fibrous or tubular tracks—a major new feature not seen in our previous fibers. We will therefore postpone details of this final class of fibrous proteins to Chapter 15.

5.6.2.2 Globular proteins and active sites

Fibrous proteins are simple compared to globular proteins. Except for the motor-associated fibers just mentioned, fibrous proteins' primary role is as a construction material. The material properties are somewhat unique and will be dealt with in later chapters, but compared to the multiple jobs carried out by globular proteins, fibers are simplicity indeed. Globular proteins carry out the vast majority of chemical reactions that allow organisms to extract energy from the outside world, convert energy to usable form, and construct new proteins and DNA—new cells and organisms. Globular proteins are the primary catalysts of the thousands of chemical reactions that take place in organisms. Nevertheless, this section on globular protein properties will be rather brief. Though similarities exist between the thousands of globular proteins found in the PDB, the activities of these enzymes cannot be generalized into three or four types, associated with a small number of structural sorts. The most useful classification of globular proteins usually involves details of active sites and structures designed for specific reactions or motions. There are many of these reactions. Wikipedia tells us that there are more than 2×10^6 proteins in the human body, most of which have catalytic activity. Cataloging this large collection of enzymatic activity has been attempted in biochemistry texts and books devoted solely to enzymes, but even these barely scratch the surface. We will, rather, discuss some of the major features of all globular proteins—their overall structure, packing, active sites, and motional properties—and deal with specific enzymes involved in major phenomena we treat later in the book. Examples include heme, photosynthetic, and molecular motor proteins.

To a large extent new proteins are added to the PDB because of their similarity to others already in the bank, so similarities in PDB structures will be greater than in actual organisms.

5.6.2.2.1 Globular protein atomic packing

Globular proteins involved in catalysis do not form extended arrays like the fibrous proteins do. The α-helical and β-sheet sections of a polypeptide do form structures similar to those in fibrous proteins, but in globular proteins those structures are local, rarely extending more than 50 amino acids. The reason for the relative confinement of repetitive structure in globular proteins is the amino acid sequence. Keratin and collagen have repetitive sequences. While some globular proteins have small sections that may repeat several times, these sections are ultimately interrupted by an amino acid that prevents the extension of the same structure further. As noted earlier, a Pro residue often ends a normal α-helical section in a protein because it enforces a certain bend in the chain. (While this bend may be a regular part of the collagen helix, it disrupts a standard α helix.) From that point, another helix, sheet, or unstructured section may begin. Because peptide bonds do not enforce a strict direction on the next section of the helix, the direction will have a good deal of randomness, unless some other structure is encountered. The net result is a globular shape to the protein.

Globular proteins are closely packed structures. Figure 5.29 illustrates the actual packing of atoms in the protein myoglobin and the impression one may get from a stick model structure. Proteins such as myoglobin do have interior cavities, which often will be occupied by water molecules, but interior water molecules generally number in the dozens, compared to the thousands of amino acids atoms.

5.6.2.2.2 Active sites and catalytic groups

Enzymes usually catalyze very specific reactions. Catalysis of chemical reactions requires strong interactions and electrons and/or protons that can be moved easily. Certain of the charged and dipolar amino acid side chains minimally satisfy these requirements, but catalysis often involves a metal atom. Metal atoms or ions do not naturally fit into the polypeptide covalent structure; they must be included in an enzyme by other means. A polypeptide must have some way of attaching the metal atom to its structure. The protein cannot rely on ionic interactions, e.g., the COO^- of Glu attracting Mg^{2+} through Coulomb attraction, because there will generally be many counter ions in the solution and water will reduce the strength of the Coulomb force. We have seen one approach to the incorporation of Fe and Mg atoms into heme and chlorophyll proteins. In both of these examples the metal ion is bound in the center of a large ringlike structure and the ring is bound to the protein. The porphyrin ring in this case also contains many electrons that can easily be moved to adjust interactions, allowing a great deal of electronic control over the catalyzed reaction. Heme, in particular, is involved in reactions that range

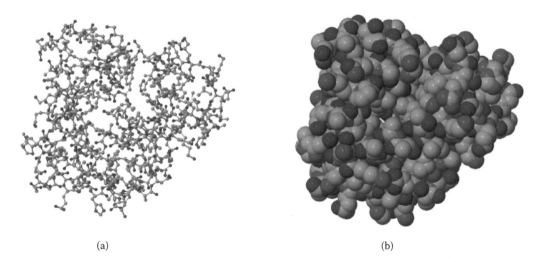

(a)   (b)

FIGURE 5.29
Neutron-diffraction structure of oxymyoglobin: (a) Ball-and-stick model of structure; (b) CPK space-filling model. The ball-and-stick and other displays do not give the correct picture of the closely packed structure of a globular protein. (PDB ID 1MBD, by Phillips, S. E. and Schoenborn, B. P. X-ray structure of sperm whale deoxymoglobin refined at 1.4A resolution, 1982. Displayed using Jmol software.) (From Jmol: An open-source Java viewer for chemical structures in 3D. http://www.jmol.org/; Protein Workshop: Moreland, J.L. et al., *BMC Bioinformatics* 6, 21, 2005; NGL Viewer: Rose, A.S. et al., *Bioinformatics*, 2018.)

TABLE 5.9 Catalytic Groups, Example Enzymes, and Reactions Catalyzed

Catalytic Group/Cofactor[a]	Enzyme	Reactions or Processes[b]
Heme (Fe-porphyrin)	Myoglobin, hemoglobin	O_2 transport and storage
	Cytochrome c	Electron transport
Chlorophyll (Mg-porphyrin)	Antenna protein	Photosynthetic energy transfer
	Chlorophyll a/b protein	Photosynthetic electron transfer
Bacteriochlorophyll a	Bacteriochlorophyll a protein	Reduction, energy production
Flavin	Phototropin	Self-phosphorylation, causes plant stems to bend toward light
Flavin adenine dinucleotide	D-amino acid oxidase	Oxidizes amino acids
Zinc ($+3$ His $+OH^-$)	Carbonic anhydrase	$CO_2 + H_2O \longrightarrow CH_3^- + H^+$
Zinc finger		DNA binding
Fe-S clusters	Ferredoxin	Electron transport
Asp (ionized), Glu (un-ionized)	Lysozyme	Dissolves mucopolysaccharide cell walls
Leu zipper (α helix with 7th Leu)	AP1 transcription factor	DNA binding, protein production, cancer

[a] There are slight variations of each catalytic group, depending on the enzyme.
[b] O_2 binding is arguably a reaction.

from reversible binding of O_2 and other molecules (the "reaction" changes a free O_2 to a bound O_2) to electron transfer reactions that generate needed electron-donating compounds to the cell to detoxifying reactions of organic compounds. This flexibility of action is enabled by the many adjustments that can be made on the iron and its interactions by changes on the porphyrin ring.

We still have the question of how the protein holds onto the metal atoms. The secret is again in the porphyrin ring. It has a large number and variety of atoms on its periphery—atoms that can be attracted to a polypeptide site through ionic, hydrophobic, and even van der Waals forces. No single interaction is large, but there are often 5–10 of these protein-porphyrin interactions, which add up to a significant binding free energy. In a cytochrome b enzyme, for example, the free energy of heme binding to protein is about 50 kJ/mol.[17] This interaction includes two interactions of the polypeptide chain directly with the Fe atom in the heme group. Some other common catalytic groups found at enzyme active sites are shown in Table 5.9.

The active site of an enzyme often consists of the nonprotein catalytic group and a local polypeptide structure designed to hold and position the group. This catalytic group is not synthesized by the same cellular apparatus that creates the polypeptide chain, so the polypeptide must allow for later addition of the group. Table 5.9 lists another possibility: an active site can consist entirely of amino acids, carefully positioned. The Leu zipper, with Leu incorporated into an α helix at every 7th position, is such a case. The Leu zipper is used in DNA-binding proteins. The *transcription factor* noted in Table 5.9 helps control the *transcription* of the DNA code for proteins to RNA and thus controls protein production and cell proliferation.

The polypeptide shape allows for the inclusion of the catalytic group, but the shape is not, and cannot be, rigid. The same noncovalent interactions that give the protein its shape—H bonds, hydrophobic interactions, ionic forces—also attach the catalytic group to the protein. Since magnitudes of these forces are comparable, the protein must adapt its shape to accommodate the group. Heme can be removed from myoglobin (Mb), for example, by addition of organic solvents, and the heme-less protein redissolved in aqueous solution.[18] In this manner, structures of normal Mb and heme-free Mb can be compared. X-ray or other crystallography could be used to determine which atoms move how far when the catalytic group is added.

However, a funny thing sometimes happens on the way to the crystallography lab. When catalytic groups are removed, some enzymes lose their stability. They may aggregate and sink to the bottom of the container or they may take on a structure that cannot be crystallized. The catalytic groups sometimes interact strongly enough with the polypeptide to stabilize the entire structure. In cases like myoglobin, the heme-less structure is less stable and several structures exist in equilibrium, but the protein is structured nonetheless.[19]

A more refined way of studying the attachment of molecular groups to proteins is *site-directed mutagenesis* (SDM). When the structure of a protein is known enough to allow a guess of which amino acids interact with the catalytic group, biochemists trained in SDM can insert the DNA sequence coding for that protein into *vectors* that, when inserted into certain microorganisms, will produce that *native protein*. The DNA sequence can then be altered to change any particular amino acid to any other, producing a *mutant protein*. If an Asp side chain, for example, is believed to ionically interact with another group, that Asp can be changed to an uncharged Asn or another amino acid to test the hypothesis. If the region of the interaction appears crowded, the Asp could be changed to a slightly larger Glu to check. Such site-directed mutagenesis methods have been used thousands of times over the past 40 years to test detailed hypotheses of enzyme structure and function. A PDB search for myoglobin structures done in 2009 or later will turn up more structures of mutant proteins than native proteins, reflecting the usefulness of the SDM technique.

The bound nonamino acid catalytic or prosthetic groups discussed above are often generically referred to as enzyme *cofactors* and are chemically unchanged after the enzymatic reaction. There are also groups such as NADH and folic acid that are needed by enzymes to catalyze reactions but that are more loosely bound. They are called coenzymes and are chemically changed during the reaction. Because coenzymes are changed, they technically could be referred to as *substrates*, but because coenzymes generally deliver a small molecular group like H or CH_3 to the primary substrate and are regenerated by other systems, the term *coenzyme* is used.

5.6.2.2.3 Atoms in a protein are in constant motion

The many X-ray crystal structures presented in this chapter give the impression that protein molecules are rigid structures, except perhaps for small vibrations that characterize atoms in all crystals at finite temperature. A 1988 review of the subject by Frauenfelder, Parak, and Young summarized the understanding of the fluctuational/motional character of proteins as follows:

> *"A protein cannot be said to have 'a' secondary structure but exists mainly as a group of structures not too different from one another in free energy, but frequently differing considerably in energy and entropy. In fact the molecule must be conceived as trying out every possible structure."*
>
> *This remarkably contemporary statement made by Linderstøm-Lang & Schellman in 1959 was based mainly on early studies of deuterium/hydrogen exchange in proteins. Perutz & Mathews realized in 1966 that hemoglobin and myoglobin must have flexible elements that permit dioxygen to reach the binding site at the heme iron. Klotz and Weber suggested that proteins with a given primary sequence exist in many conformational states and are dynamic systems. The view of the protein as a strongly fluctuating system is, however, just one extreme. At the other extreme is the picture of the protein as a rigid aperiodic crystal. The picture in favor at any given time depends strongly on the dominant experimental techniques and on the power of the theoretical tools. Deuterium exchange and NMR techniques, for instance, favor a flexible protein, whereas X-ray and neutron diffraction techniques tend to emphasize a rigid macromolecule. The early theoretical models, based on simple calculations, tended to treat the protein as a relatively rigid system, whereas recent computer simulations provide convincing demonstrations of protein motions. Thus the textbook view of a protein has oscillated between the two extremes, rigidity and flexibility.[20]*

In the 1970s, Frauenfelder, Petsko, and Tsernoglou set out to determine if temperature-dependent X-ray diffraction data could measure mobilities of individual atoms in complex protein molecules.[21] The idea was that as temperature increased, increased vibrations of atoms would give rise to decreased diffraction spot intensities because the equivalent atom in different parts of a protein crystal would not oscillate in phase and constructive X-ray interference would therefore not be as complete. They successfully demonstrated this idea, and the structures of many proteins in the PDB now show individual atom mobilities. Going back to the "hydrogen atom of protein physics," myoglobin, we can display the atomic mobilities, as in Figure 5.30. The displays in these graphics have set the atom coloring to "temperature (relative)," sometimes referred to as *B factors*. We will expand on the functional implications of these mobilities in Chapter 15. Figure 5.30 shows that differing atomic mobilities are evident even in small oligopeptides.

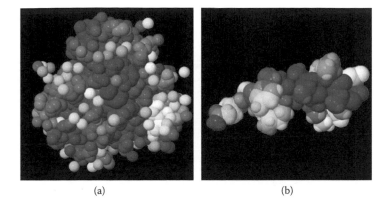

(a) (b)

FIGURE 5.30
Structures of proteins, showing atomic mobility. Displays done with Jmol software, coloring atoms by temperature (relative). Dark blue atoms move the least; red, the most. (From Jmol: An open-source Java viewer for chemical structures in 3D. http://www.jmol.org/; Protein Workshop: Moreland, J.L. et al., *BMC Bioinformatics* 6, 21, 2005; NGL Viewer: Rose, A.S. et al., *Bioinformatics*, 2018.) (a) Myoglobin (with bound carbon monoxide, PDB ID 1MBC, by Kuriyan, J. et al., *J. Mol. Biol.*, 192, 133–154, 1986.) (b) An amino acid 15-mer. (PDB ID 1DJF, as in Figure 5.17.)

So what should our view of protein structure be: A well-defined structure with each atom in its place, or a collection of atoms in chaotic motion? Because coordinated motions of groups of atoms in enzymes are prerequisites for most catalytic activity (Chapter 15), we would expect atomic mobility to be an integral part of protein structure. However, if coordinated motion is required, chaotic motion must be limited. If you look into the details of the atomic motions, you will find that most have amplitudes less than 0.2 nm, significantly less than the 20 nm or more diameter of typical proteins. So atoms indeed move, but structures of proteins at the atomic level are well defined. Proteins must have the flexibility to move, but coordinated motion requires a framework in which random motions are constrained.

> RNA is known to have some catalytic activity, but we will not investigate that here. See *ribozymes*.

5.6.3 Polynucleotides

The topic of polynucleotides deserves a chapter entirely to itself because of the importance of the genetic code and its governance of the characteristics of developing organisms. However, from a dynamic, physical point of view, DNA and RNA do not do much—move, exert forces, catalyze reactions, etc.—at least compared to the polypeptides that make up enzymes. The genetic code is information. This information enables all known forms of life. The information is being exploited to treat disease and develop new medicines. The information perhaps explains the evolution of various life forms. In the late twentieth and early twenty-first centuries bioscience research has indeed largely shifted from reactions, processes, and motions to the genetic codes governing production of the controls for these processes. We should, of course, be interested in this information and control, but our primary focus in this biophysics course is on the processes, component design, governing physical laws, and quantitative description of the processes. Our interest in nucleic acids will center not on the information they encode but how this information is managed and read. These are jobs of the proteins.

> When Francis Crick used the term *dogma* in 1958, he intended the term to mean an idea for which there was no reasonable evidence (at the time).

Two major forms of nucleic acids or polynucleotides exist: DNA and RNA. Hybrids of the two occur and are a normal part of the production of proteins. We will not elaborate on this production process here, but some terminology is advisable. Any time you look for references on DNA and proteins, you are liable to come upon the following terms:

1. The *Central Dogma of Molecular Biology* states that DNA directs its own replication and its transcription to RNA, which directs its translation to proteins. See Figure 5.31. The dogma is not the whole story, but it suffices for now. See any recent molecular biology or biochemistry book for information on other side paths.
2. *Replication*, the basis for biological inheritance, occurs in all living organisms and involves copying their DNA. The process is "semiconservative": one strand of the original

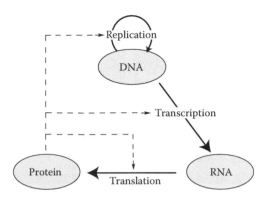

FIGURE 5.31

The Central Dogma of Molecular Biology: DNA directs its own replication, as well as its transcription to RNA, which directs its translation to proteins (solid arrows). The dashed arrows indicate that specific proteins are the active agents in each process (not part of the Central Dogma).

double-stranded (ds) DNA molecule serves as template for the reproduction of the *complementary* strand. Following DNA replication, two (normally) identical DNA molecules have been produced from one ds DNA molecule. Cellular proofreading and repair ensure near-perfect fidelity for DNA replication. In the lab, the polymerase chain reaction (PCR) artificially replicates a specific target DNA fragment from a pool of DNA in vitro (in a test tube).

3. *Transcription*, the synthesis of an RNA copy of DNA, is the process of converting the DNA nucleotide sequence into an equivalent RNA sequence. The DNA sequence is copied by the RNA polymerase enzyme (with helpers) to produce *messenger RNA* (mRNA) molecules, which carry a genetic message from the DNA to the protein-synthesizing machinery of the cell.

4. *Translation* is the production of protein amino acids coded for by mRNA. mRNA is a series of three-nucleotide segments (*codons*) that specify an amino acid. mRNA travels to the ribosome, where each mRNA codon binds to a *complementary* tRNA *anticodon*, to which the amino acid specified by the codon is covalently bound. The amino acid is transferred by the ribosome to the carboxyl terminus of the growing amino acid chain.

5. *Recombination* is the process by which a strand of DNA (or RNA) is broken and joined to a different DNA molecule. In eukaryotes, recombination commonly occurs during *meiosis* as chromosomal crossover between paired chromosomes. This process leads to offspring having different combinations of genes than their parents.

6. *Repair* is the process by which damaged DNA (DNA with broken chains or improper covalent links from bases to other molecules), DNA with incorrectly paired bases (e.g., A–C, G–T, A–U), DNA with bulges caused by unpaired bases, and other abnormal DNA structures are identified and fixed. (Transcription by RNA polymerase in *E. coli* has an error rate of about 1 in 10^4 bases, while DNA replicated by DNA polymerases has about 1 error in 10^8–10^{10}.) Damaged DNA is recognized by a constant cellular proofreading process. A properly paired and structured DNA segment that may have been inserted into a cell's DNA by a virus cannot generally be repaired, through enzymes can sometimes recognize and destroy such foreign DNA in bacteria. In Chapter 16 we will examine the Eco RI repair system, which simply tries to eliminate any nonnative DNA, but the reader should also be aware of the more complex CRISPR-based editing system, which has been employed by some bacteria for millions of years to resist viral infection from a DNA library of previous encounters with viruses.[22]

5.6.3.1 DNA: Deoxyribonucleic acids

A single strand of DNA is a polymer of the four individual deoxynucleotides dA, dT, dG, and dC formed by the covalent linkage of the phosphate group of one base to the ribose group of the preceding nucleotide. "Preceding" in this case means toward the 5′ end, which corresponds to the fact that DNA polymerases that replicate DNA add bases on the 3′ side of the chain being synthesized.

Double-stranded DNA consists of a strand of DNA wound around an axis, usually in the form of a helix and hydrogen bonded to its complementary strand, which winds around the axis in the opposite direction. The covalent structure of such a chain with its H-bonded complement is shown in Figure 5.32.

The complement of a DNA strand consists of a strand made up of the complementary base of each base in the first strand, read from last base to first. The complement to a base X is the base that best hydrogen bonds to X. This means that for every base T, there is a base A and for every C there is a G:

$$N_A = N_T \text{ and } N_G = N_C \tag{5.9}$$

These equalities are called Chargaff's rules, after Erwin Chargaff, who experimentally discovered them in the 1940s, before anything was known about the double helix or base pairing. Chargaff also noted that the overall base composition was characteristic of the organism and did not depend on which tissue the DNA was extracted from—crucial pieces of evidence that guided later researchers, including Watson and Crick, the discoverers of the DNA double helix.

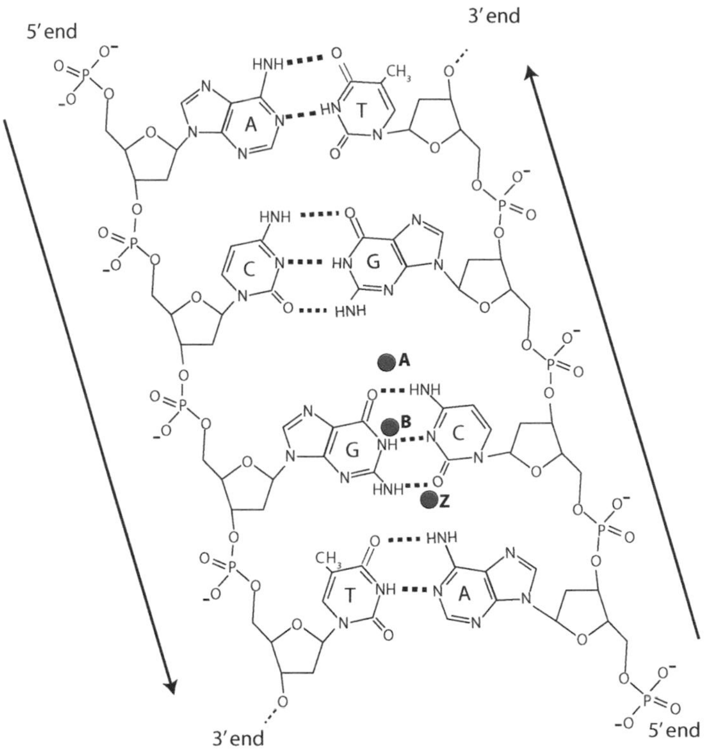

5.6.3.1.1 Overall properties of DNA double helices

Nucleic acids can take on some interesting and complex three-dimensional structures, as we observe in the ribosome pictures. These complex structures occur primarily in ribonucleic acids, not deoxyribonucleic acids. We will leave others the task of explaining why a small change, like having an –OH group rather than an –H atom at the ribose 2′ position, should result in very different stable structures. Nevertheless, compared to protein structures, nucleic acids appear limited indeed. The main source of this limitation on DNA structural possibilities is the limited number of bases, four, as compared to amino acids, 20, plus a few variants. If we were to suppose that each polymer unit can take on just one possible orientation but that different units allow different orientations, then an N-mer would have n^{N-1} possible conformations, where N is the number of units in the polymer and $n = 4$ (20) for nucleic acids (proteins). For just a 10-mer, proteins would have $10^{13}/10^6 \sim 10^7$ times more possible structures than DNA. This ratio goes up rapidly as the number of units increases. Though this is not a terribly accurate model of real conformational possibilities, it gives a clear message that DNA is not designed to take on very active and diverse roles like proteins must. However, nucleic acids primarily store information. If the proteins charged with "reading" the nucleic acid "books" had to deal with books as varied as a modern ebook and Egyptian hieroglyphs, life would be in great difficulty.

So, let's first look at some common DNA structures: double helices. The most commonly observed DNA helix in cells is the B helix, sometimes referred to as the *Watson-Crick* structure. We will not go into the history of James Watson's and Francis Crick's 1953 deduction that the X-ray diffraction patterns of spools of DNA fibers measured by Maurice Wilkins and Rosalind Franklin, as well as Chargaff's rules and other observations on the composition of DNA, could be explained by a double-helical structure, with two strands wound around each other in a right-handed manner, complementary bases paired together in the center of the helix.

FIGURE 5.32
Diagrammatic structure of double-stranded DNA. Each strand is read in the 5′ → 3′ direction. The complementary strand has the complementary base (T/A, C/G) H bonded to each base on the primary strand. The solid dots labeled A, B, and Z indicate the approximate position of the helix axis relative to a G–C base pair for A, B, and Z DNA. (In Z DNA, the bases are flipped 180° from those in B and A DNA.) In three dimensions, the base pairs would be rotated almost 90° out of the page, with the helix axis nearly normal. The double-stranded (ds) tetranucleotide shown can be abbreviated d(ACGT)·d(ACGT) or d(ACGT)$_2$. This particular d(ACGT) strand is *self-complementary*: the strand and its complement are identical.

1962 Nobel Prize in Physiology and Medicine to Watson, Crick, and Wilkins. A maximum of three persons could be named for one discovery. Franklin died of cancer at the age of 37 in 1958, probably from X-ray exposure. Consult "Contributions of 20th Century Women to Physics" http:www. physics.ucla.edu/~cwp for more information.

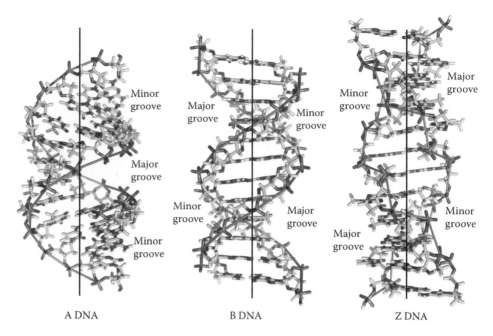

A DNA   B DNA   Z DNA

FIGURE 5.33

Models of A, B, and Z DNA. A DNA: right-handed helix, bases are inclined 20° with respect to and moved away from the helix axis, larger helix diameter leaves space along the axis. B DNA: right-handed helix, helix axis is perpendicular to base planes and intersects base-pair H bonds; no space along axis. Z DNA: left-handed helix, major groove almost flat (base pairs near side of helix), sugar-phosphate backbone follows zigzag path, bases flipped 180° from those in A and B DNA. Diagram by Richard Wheeler, http://en.wikipedia.org/wiki/File:A-DNA,_B-DNA_and_Z-DNA.png, reproduced by permission of the artist.

Figure 5.33 shows structural models of A, B, and Z DNA. Think of these structures as categories of DNA helix types, since there is no "A DNA" or "B DNA" in a living cell. Natural DNA will have a varied sequence, and the structure along the helix will vary accordingly. Examination of even short DNA helices found in structural databases will reveal variations in values of helix parameters listed in Table 5.10. Nevertheless, the B class of double helix more closely resembles the majority of DNA found in cells. Several other types of DNA helix have been defined, but we will not discuss them.

Of the three helix types, B DNA is right-handed and most nearly symmetric. Its base pairs are nearly centered about the helix axis (see Figure 5.32) and base planes are nearly normal to the axis.

TABLE 5.10	Properties of DNA Double Helices[a]		
Property	**A DNA**	**B DNA**	**Z DNA**
Sense	Right-handed	Right-handed	Left-handed
Repeating unit	1 base pair	1 base pair	2 base pairs
Base pairs/helix turn	11	10–10.5	12
Helix pitch (rise per turn)	2.8 nm	3.4 nm	4.4 nm
Rise per base pair	0.26 nm	0.34 nm	0.37 nm
Diameter	2.6 nm	2.0 nm	1.8 nm
Rotation about axis per base pair	33°	36°	9°, 51° (60° for two)
Angle (base normal to helix axis)	19° to 20°	–1° to 6°	–9° to 7°
Minor groove	Wide, shallow	Narrow, deep	Narrow, deep
Major groove	Narrow, deep	Wide, deep	Flat

[a] All of these parameters can have variations. In the end *measurements* matter more than *definitions* of A, B, and Z.

There are about 10 base pairs per helical turn, with adjacent base pairs separated by 0.34 nm. (We will use 10 bps/turn in further discussions.) The two DNA strands are not equidistantly separated in the axial direction. (Think of the extreme of two strands touching each other as they wrap around the helix axis.) The result is a *major groove* and a *minor groove* in the helix. The major groove is the groove along which the base edges closest to the ribose-base bond are placed; in B DNA it is the wider groove. A major difference of A and Z DNA is that bases are displaced toward the side of the helix, in Z forming almost a flat major groove at the exterior of the helix. Z DNA, the left-handed of the helices, can be formed only when G-C content is high and ionic conditions favorable, usually high salt concentrations (e.g., 0.2 M KCl and 20 mM $MgCl_2$).

5.6.3.1.2 *Interactions stabilizing nucleic acid structures*

Interactions between complementary strands do not involve any new fundamental forces beyond those described for polypeptides. However, it is useful to classify the forces somewhat differently.

Base-pair H bonds are of the type N···H···N and N···H···O. The T–A hydrogen bond is not as strong at the G–C bond. Note in Figure 5.32 that the G–C bond is threefold, while the A–T is twofold. Base-pairing H bonds generally have free energies of –2 to –8 kJ/mol per base pair. This energy is not large enough to be the primary stabilizing energy for the double helix. Nevertheless, two complementary strands, where all bases are H-bonded according to the A–T/G–C pairing rule, can form a relatively uniform double helix. If one base on one strand is changed, making a single base pair *mismatch*, a mismatched base will generally bulge out of the helix (look for base mismatches in the NDB) or the double helix may completely come apart, depending on the initial helix stability.

Base stacking interactions involve a particular type of hydrophobic interaction, which, as we discussed, is no fundamental force at all but rather reflects a lower free energy when molecular groups that prefer a low-dielectric environment are next to each other and not surrounded by water. Use of the term *hydrophobic interaction* is, in fact, not very useful when applied to DNA. We will not elaborate on the debate concerning the exact nature of the stacking interaction but will refer the interested reader to the classic measurements by Davis and Tinoco[23] and will tabulate theoretical results by Friedman and Honig in Table 5.11. These latter researchers find stacking interaction free energies ranging from about –12 to –30 kJ/mol, depending on the computation methodology and the stacking bases. We will make nothing of the small differences between the base stacks other than to say that the general consensus is that A/A and G/G stacking is stronger than C/C and T/T, more so than indicated by the theory used in Table 5.11.

Table 5.11 free energy data is from 1995. More recently, direct methods to measure base stacking forces (not free energies) were devised, using single-stranded DNA (ssDNA), which has no competing H-bonding forces. The base A was studied because it is observed that ssPoly(dA) can form an ordered helix-like structure in solution, whereas poly (dT) cannot. Marszalek and colleagues covalently attached ssDNA, Poly(dA), consisting only of adenine base units to a gold surface, and then pulled the other end with an AFM tip. The force thus measured will contain both entropic and enthalpic contributions. They detected one plateau in the elasticity of the ssDNA at around 23 pN, which was expected, and then a second plateau around 113 pN.[24] What do these plateaus mean? We will show soon that any (soluble) polymer molecule that does not interact with itself will form a loosely packed globule in solution. When the two ends are pulled, the polymer will resist with an entropic force that behaves like an entropic spring: force $\propto \Delta x$, starting from 0. If there is an additional fixed stacking force of 23 pN between adjacent units in the polymer, this force will not show up until an applied force of 23 pN is reached, when the bases become unstacked (Figure 5.34). The polymer will continue extending indefinitely at that force until the limits of the structure are reached. If we multiply that force by the total extension at that force and divide by the number of bases, we get the work done to disrupt one base stack, about 15 kJ/mol. This is close to (minus) the calculated stacking energies in Table 5.11 and not far from earlier measurements for A/T base pairs in double-helix experiments complicated by hydrogen bonding. Similar early experiments gave stacking energies for G–C base pairs in DNA of about –60 kJ/mol. The conclusion: base stacking contributes more free energy than base-pairing hydrogen bonds, but a clear separation of the two (and other) contributions is difficult.

TABLE 5.11	Free Energy Changes for the Stacking of DNA Bases[a]
Dinucleotide	**ΔG (kJ/mol)**
AA	–14 to –31
AT	–12 to –27
TA	–14 to –24
TT	–13 to –23
GC	–14 to –27
CG	–13 to –23
GG	–15 to –27
UU	–12 to –21

[a] Theoretical results from Friedman, R.A. and Honig, B., *Biophys.* 69, 1528–1535, 1995.

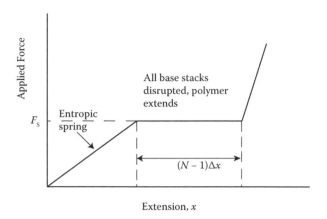

FIGURE 5.34

Expected ideal graph of force versus extension for a polymer like Poly(dA), which randomly coils into a globule in solution but which also forms stacks of adjacent bases, which will shorten the length of the polymer. When the ends of the coiled polymer are pulled, entropy resists approximately like a linear spring. When the applied force is equal to the stacking force F_s, bases will all unstack and the length will extend (approximately) $(N - 1)\Delta x$, where Δx is the length extension due to unstacking one base and N is the number of bases in the polymer. The work or free energy done to unstack one base is $F_s\Delta x$.

5.6.3.1.3 Ionic interactions

DNA and RNA are negatively charged polymers, with a charge of $-e$ per phosphate group. Whether single or double stranded, this high charge might be expected to have a dominant effect on structure. We might even expect the double helix to be unstable because of the repulsion of phosphate charges on opposite strands, as well as between neighboring phosphates on the same strand. This is, in fact, the case if DNA at low concentration is dissolved in pure water. Because DNA is a polymer that can be of widely varying length, the DNA concentration is usually indicated in terms of the base or base-pair concentration rather than the concentration of the entire polymer molecule. This makes sense when the DNA (or RNA) contains more than about 100 bases, because polymers of this length behave as "long" polymers, with negligible end effects. The double-strandedness, for example, is relatively independent of DNA concentration. Shorter DNA is more complicated. We will address this in more detail in Chapter 15. A 10^{-6} M DNA concentration (meaning [base] = 10^{-6} M) in pure water will also contain 10^{-6} M Na^+ (or K^+ or whatever the counterion to the DNA phosphate ion is). Though this is still one Na^+ per DNA PO_3^-, these sodium ions are free to wander throughout the solution. Because of the huge volume occupied by 55 M water compared to 10^{-6} M DNA, the Na^+ ions will not spend much time near the DNA, in spite of the Coulomb attraction, reduced as it is by water's dielectric constant. The net result is that the phosphate–phosphate charge repulsion does, in fact, destabilize the double helix: most DNA will not be in the form of a B helix. The destabilization gets stronger as the DNA concentration gets lower. In cells or under "normal" salt concentrations, however, the concentrations of counterions are much higher, about 0.15 M. This provides sufficient numbers of positive ions to act as countercharges to phosphates. The phosphate–phosphate repulsion is greatly reduced and the double helix becomes more stable.

> 10^{-6} M Na^+ corresponds to a 10^{-6} M *ionic strength*, in biochemistry terms. See glossary.

Many doubly charged metal ions like Mg^{2+}, Mn^{2+}, and Co^{2+} bind more strongly than expected to phosphate groups. We will let the chemists explain this binding, but the net result is that one of these ions counts for 10^2–10^3 Na^+ or K^+ ions, not just the expected factor of 4 in ionic strength. For this reason, we must take special note when we read of an experiment that uses 10 mM $MgCl_2$. Such a concentration may seem low but has an effect 100 times larger, in terms of stabilization of the double helix or facilitation of protein binding to DNA.

Ionic strength is an important factor in the stability of the double helix, but ions are typically not an issue in a cell, where ionic strength is controlled. When the Na^+ concentration is under experimental control, the stability of the double helix may be an issue. An approximation for the "melting temperature," the temperature at which half of the helices come apart, of long DNA (DNA with more than about 100 bases) in a solution of NaCl, near pH 7.0, is

$$T_m \approx 41.1 X_{G+C} + 16.6\log[Na^+] + 81.5 \qquad (5.10)$$

where X_{G+C} is the fraction of the bases that are G or C.[25] Note that a higher T_m means a more stable helix. Increasing $[Na^+]$ from 1 to 100 mM raises the temperature at which the double helix comes apart by about 30°C.

Equation 5.10 reflects another observation about DNA duplex (double-helix) stability: more G–C base pairs increases stability. Because this equation is purely parametric, it does not indicate whether this is from base-pairing interactions, base stacking, other interactions, or all of the above. While some texts like to indicate which interaction is most important to DNA duplex stability, we will satisfy ourselves with the observed fact that the DNA double helix is not stable without water, counterions, base pairing, base stacking, and a pH between about 6 and 8.

5.6.3.2 RNA: Ribonucleic acids and folding

Most of the time, DNA under physiological conditions is in a double-helical conformation. There may be higher-level structure like supercoiling, triple helices, nucleosomes, and other important structures induced by interactions with proteins, but because of its primary role in information storage, the varieties of structure are limited. RNA, on the other hand, plays a number of different roles, though all are related to information. Because RNA is the intermediary between DNA information and protein activity, RNA needs to take on several roles. These roles require different structures. We will focus on three primary forms of RNA: messenger RNA (mRNA), transfer RNA (tRNA), and ribosomal RNA (rRNA). We will also mention a catalytic form of RNA, the hammerhead ribozyme. Other interesting forms exist in certain organisms: transfer-messenger RNA (tmRNA), small nuclear RNA (snRNA), double-stranded RNA (dsRNA), and many others. See http://en.wikipedia.org/wiki/List_of_RNAs for a list and references. These RNAs are not fundamentally different chemical entities but rather variants of RNA produced under different circumstances for different purposes.

Why can RNA take on many more conformations, even active, catalytic conformations? The differences between DNA and RNA are threefold: (1) RNA has an –OH group at the 2′ position of the ribose ring; (2) RNA replaces thymine, which has a base –CH₃ group, with uracil, which has a base –H group; (3) DNA and RNA are produced and degraded differently and in different places. RNA can, in fact, take on a double-helical form. DsRNA is found in nature in viruses, but is also routinely produced by several commercial suppliers of nucleic acid products (e.g., Harvard Medical School, http://flyrnai.org/DRSC-PRS.html.), so RNA is capable of forming structures similar to those of DNA. The small differences between DNA and RNA nucleosides (–OH versus –H on ribose and –H versus –CH₃) may not be so small, however. OH is a polar group, while H is not and will be more likely to be solvent exposed. H is smaller than CH₃ and will allow a wider variety of base orientations.

Let us look first at structures RNA will tend to take on. Figure 5.35a–b shows the seven-base heptamer sequence 5′-R(GpUpApUpApCpA)-3′. (The "p" between bases again

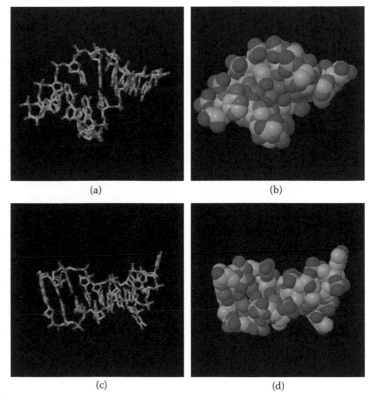

(a)

(b)

(c)

(d)

FIGURE 5.35

(a, b) X-ray crystal structure of a single strand of synthetic RNA, 5′-R(GUAUACA)-3′. Though a single strand, the formation of a compact structure through intrastrand base-pairing interactions, as well as base stacking can be seen. (PDB ID 1L3Z by Shi, K. et al., *Nucleic Acids Res.* 31, 1392–1397, 2013, displayed using Jmol software.) (From Jmol: An open-source Java viewer for chemical structures in 3D. http://www.jmol.org/; Protein Workshop: Moreland, J.L. et al., *BMC Bioinformatics* 6, 21, 2005; NGL Viewer: Rose, A.S. et al., *Bioinformatics*, 2018.) (c–d) X-ray crystal structure of a single strand of synthetic DNA, 5′-d(CGCGCGTTTTCGCGCG)-3′, which crystallizes as a Z-DNA hexamer, capped at one end by a T4 loop. (PDP ID 1D16, by Chattopadhyaya, R. et al., displayed using Jmol. Note the title of publication, "Structure of a T4 hairpin loop on a Z-DNA stem and comparison with A-RNA and B-DNA Loops," *J. Mol. Biol.* 211, 189–210, 1990.)

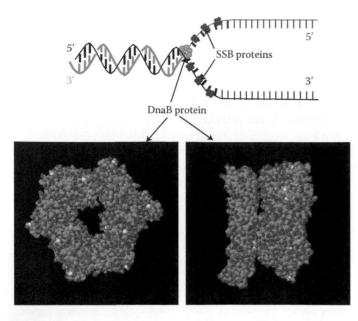

FIGURE 5.36

(*Upper*) The initial stage of DNA replication requires unwinding of dsDNA, done by helicases. The separated strands are prevented from immediately reforming dsDNA or forming intrastrand base pairs by single-strand binding proteins (SSBs). (*Lower*) Structure of the complex between the DnaB helicase and the DnaG primase from *Bacillus stearothermophilus*. (PDB ID 2R6E by Bailey, S. et al., *Science*, 318, 459–463, 2007, displayed using Jmol.) (From Jmol: An open-source Java viewer for chemical structures in 3D. http://www.jmol.org/; Protein Workshop: Moreland, J.L. et al., *BMC Bioinformatics* 6, 21, 2005; NGL Viewer: Rose, A.S. et al., *Bioinformatics*, 2018.)

stands for a phosphate group, not listed in the figure caption.) This is a short, single strand, but already the structural tendencies of RNA are apparent. The single strand does not have a random structure; neither is it even close to straight. On the contrary, the strand sharply bends, allowing *intra*strand H bonding between some of the bases. Base stacking is also clearly evident. The CPK space-filling model shows this RNA to be almost globular.

Before we conclude that RNA will take on quite different structures because of the two molecular group changes, let's see if single-stranded DNA may also take on "unusual" structures. Figure 5.35c–d shows PDB structure 1D16, which is a single strand of (5′-d(CpGpCpGpCpGpTpTpTpTpCpGp CpGpCpG)-3′). The title of the article in which the structure was first published is revealing: "Structure of a T4 hairpin loop on a Z-DNA stem and comparison with A-RNA and B-DNA loops." This short ssDNA sequence clearly has bent back on itself (the "loop") to form base pairs; base stacking is also apparent. Quoting the authors, "The T4 loop differs from that observed on a B-DNA stem in solution, or in longer loops in tRNA, in that it shows intra-loop and intermolecular interactions rather than base stacking on the final base-pair of the stem. Bases T7, T8 and T9 stack with one another and with the sugar of T7." Note that these structures are quite nonstandard for DNA as we have seen so far. Though they also differ in details from the RNA structures, we should disabuse ourselves of the notion that DNA is intrinsically incapable of forming RNA-like structures.

5.6.3.2.1 Single-strand binding proteins restrict structure formation by DNA

Important reasons why most DNA is B-DNA double stranded and RNA takes on interesting, odd structures is where and how the two are produced and whether there is a complementary strand or some other entity to prevent intrastrand structures. For example, as part of the process of DNA replication, the two strands in double-helical DNA are separated in preparation for duplication. The top panel in Figure 5.36 diagrams the initial process of unwinding of the dsDNA. Unwinding is done by helicase enzymes, one of which is shown in the bottom panel of Figure 5.36. This complex is made up of two hexameric protein complexes, DnaB and DnaG. After DNA strands are separated, they could immediately reform the double helix or form intrastrand base pairs. To prevent this, SSBs, single-strand binding proteins, bind to the strands. Farther away from the DNA "fork," the SSBs come off when other enzymes arrive to synthesize the new complementary strands. Without these single-strand binding proteins (SSBs), many interesting, but disruptive, DNA structures might be formed. Returning to the issue of DNA and RNA structural possibilities, we should note that the DNA 16-mer sequence was designed to produce such a loop structure. The two ends of the sequence are complementary and G–C forms strong base-pair H bonds, whereas the Ts in the middle do not strongly base pair (and have no A to pair with), so the strand will tend to wrap around and form an intrastrand double strand, as well as a loop.

Messenger RNA (*mRNA*) is single stranded and carries information about a protein sequence to the ribosomes, the protein synthesis factories in the cell. Every three nucleotides (a *codon*) correspond to one amino acid, as we noted earlier. In eukaryotic cells, a *precursor mRNA* (pre-mRNA) is first transcribed from DNA in the cell nucleus and is processed to mature mRNA by removal of *introns*—noncoding sections of the pre-mRNA. The mRNA is then transported from the nucleus to the cytoplasm, where it is bound to ribosomes, large complexes of RNA and protein (Chapter 6), and translated into its corresponding protein form with the help of tRNA. Prokaryotic

For helpful animations of the process of replication, see the PBS production called "DNA: The Secret of Life," part of which can be found at https://www.youtube.com/watch?v=d7ET4bbkTm0. See also https://en.wikipedia.org/wiki/DNA_replication.

More than 95% of the bases in DNA do not actually code for protein. Some act as control sequences, allowing or disallowing translation of nearby sequences into protein. Such control is directly related to cell response to the environmental challenges and to cancer induction, but the role of the noncoding DNA is not comprehensively understood.

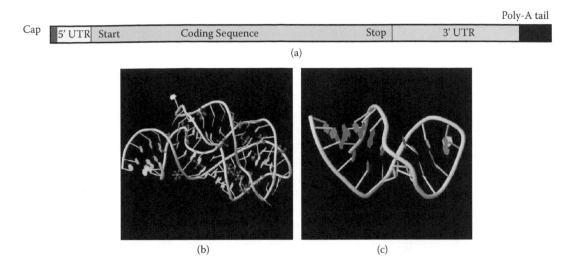

(a)

(b)   (c)

FIGURE 5.37
mRNA is not a straight, linear polymer: (a) "Structural" diagram of the of mRNA common in humans (UTR, untranslated region). See text. The spatial structure of single-stranded mRNA is complex, including *riboswitches* and *hairpins* that occur in the UTRs. (b) Guanine riboswitch (PDB ID 3G4M, by Gilbert et al., displayed by Jmol). (c) Selenocysteine insertion sequence (SECIS) hairpin structure. (PDB ID 1MFK, by Fourmy, D. et al., *J. Mol. Biol.*, 324, 137–150, 2002, displayed by Jmol.) (From Jmol: An open-source Java viewer for chemical structures in 3D. http://www.jmol.org/; Protein Workshop: Moreland, J.L. et al., *BMC Bioinformatics* 6, 21, 2005; NGL Viewer: Rose, A.S. et al., *Bioinformatics*, 2018.)

cells have no nucleus and cytoplasm compartments and mRNA will bind to the ribosome while it is being transcribed from DNA.

The structure of single-stranded mRNA is illustrated in Figure 5.37. The "structure" in Figure 5.37a shows a block diagram of the contents of the sequence of a typical mRNA in humans. The *5′ cap* is a modified guanine nucleotide added to the front (5′) end. This modification allows proper attachment of the mRNA to the ribosome. *Coding regions* (the codons) are decoded and translated into a protein by the ribosome. Coding regions begin with the "start codon" and end with a "stop codon." The start codon is usually an AUG triplet and the stop codon is UAA, UAG, or UGA. The coding regions tend to be stabilized by internal base pairs. *Untranslated regions (UTRs)* are sections of the mRNA that are not translated. These regions are *exonic* in the mature mRNA. Untranslated regions seem to enhance mRNA stability, localization, and translational efficiency. Some of the elements contained in untranslated regions form a characteristic secondary structure when transcribed into RNA (Figure 5.37b–c). These structural mRNA elements are involved in regulating the mRNA. Some, such as the SECIS element, are targets for proteins to bind. One class of mRNA element, the *riboswitches* (Figure 5.37b), directly bind small molecules, modifying transcription or translation. The "poly(A) tail" at the 3′ end of mRNA is a long stretch of adenine nucleotides often numbering in the hundreds. This tail promotes transport out of the nucleus and *translation*. It also helps prevent the mRNA from degrading too soon. The proper picture of mRNA, far from being a long, straight line of nucleotides, is rather a large, three-dimensional macromolecule with active structures designed to interact with proteins in the cell. The tendency of bases to stack and base pair, far from being greatly diminished by the H → OH and T → U changes, is exploited to produce a polymer that is structurally complex and active.

Transfer RNA (tRNA) is a small RNA molecule (74–95 nucleotides) that transfers a specific amino acid to a growing polypeptide chain at ribosome where a new protein is being assembled. Figure 5.38 shows a phenylalanine tRNA from yeast. It has a 3′ terminal site for amino acid attachment. The amino acid is covalently attached by an enzyme, aminoacyl tRNA synthetase. A three-base region called the *anticodon* that base pairs to the corresponding three-base *codon* on mRNA is marked at the bottom of the figure. The anticodon for Phe is AAG. Each type of tRNA molecule can be attached to only one type of amino acid, but because several different *codons* can specify some amino acids, tRNA molecules with different anticodons may carry the same amino acid. tRNA is L-shaped and narrow in the direction normal to the view of Figure 5.38. The narrowness is needed because two

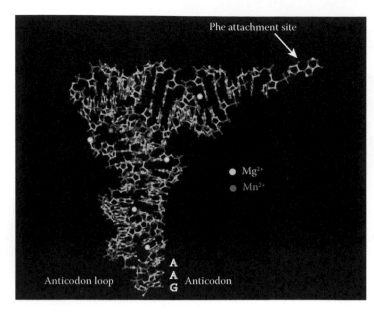

FIGURE 5.38

tRNA^{Phe} from yeast. Each amino acid will have a corresponding tRNA. Much smaller than mRNA, tRNA is the size equivalent of one localized mRNA structural element. The L-shaped structure is evident. The amino acid Phe will be bound to the ribose 3'OH of the adenine in the 3'-end CCA sequence. (PDB ID 1EHZ by Shi, H. and Moore, P. B., *RNA*, 6, 1091–1105, 2000, displayed using Protein Workshop software.)

tRNA molecules must bind near the same site at the ribosome. tRNA contains regions that are base paired and stacked and loops that have partially stacked, but nonpaired bases. These regions enable recognition by enzymes that bind to specific tRNAs.

There are many complex and intricate processes involving the RNAs and protein synthesis at the ribosome. As we are concentrating in this chapter on general structural features and not biochemical reactions, we cannot hope to describe even a small fraction of the processes here. However, we will note one unexpectedly accurate process, which reminds us to keep in close touch with our biochemical colleagues. Amino acids are added to a fully formed tRNA. Addition of a molecular group, like any physical–chemical process, is governed by free energies. For example, the addition of Ile to its tRNA does not occur with tRNA and Ile in isolation. There will be many other molecules and amino acids around which can compete with Ile. The selectivity for Ile should be governed by relative free energies: Ile should be preferentially added to tRNA^{Ile} because to add a different molecular group would be more costly in free energy. The creation of the amino acid charged tRNA is unexpectedly failure resistant. Linus Pauling pointed out that Val is structurally very similar to Ile, yet Val is incorrectly attached to the isoleucine tRNA, tRNA^{Ile}, only about one time in 10^4. Valine is identical to Ile, except for the absence of a methyl ($-CH_3$) group. If only the binding of the CH_3 group were involved, the estimated effect on the free-energy increase of Val with respect to Ile would be about +12 kJ/mol. Problem 5.17 will ask you to calculate the effect, but this free energy difference predicts an error rate of about 1%, not 0.01%. The solution to this conundrum was discovered by a biophysicist Paul Berg, who found that the enzyme that adds the amino acid to tRNA has another site where the incorrect Val can be bound and deactivated. The slightly larger Ile will not fit into that second site. The message here is that enzymes involved in genetic processes frequently have a *proofreading* as well as synthesizing ability.

Ribozymes, or ribonucleic acid enzymes, are RNA molecules that catalyze chemical reactions. A diagram on the cover of the third volume of a book series called The RNA World[26] gives a hint at the significance of ribozymes when compared to the *Central Dogma of Molecular Biology*. The Central Dogma (Figure 5.31) reflects life as it now exists: a complicated system of DNA, RNA, and enzymes, all of which are needed for inheritance to work properly. When life first began, the improbability of two or three different molecules appearing at the same site suggests that protolife consisted of a single system capable of both reproduction and energy conversion/processing. RNA has been proposed as the simplest such system, as in Figure 5.39. Present-day natural ribozymes catalyze the hydrolysis of their own phosphodiester bonds, or that of another RNA. Ribozymes can also catalyze the transfer of amino acids at the ribosome. Some laboratory-produced ribozymes are capable of catalyzing their own synthesis.[27]

Before the discovery of ribozymes, proteins were the only known biological catalysts. In 1967, Carl Woese, Francis Crick, and Leslie Orgel suggested that RNA could act as a catalyst, after the discovery of the complex, globular shapes that RNA can form. The first ribozymes were discovered in the 1980s by Thomas R. Cech and Sidney Altman. Cech found his ribozyme in the intron of an RNA transcript, which *removed itself* from the transcript. Altman found catalytic activity in the RNA component of the RNase P complex. They won the 1989 Nobel Prize in chemistry for their "discovery of catalytic properties of RNA."

Many ribozymes have either a hairpin- or hammerhead-shaped active center and a unique secondary structure that allows them to cleave other RNA molecules at specific sequences. It is now possible to make ribozymes that will specifically cleave any RNA molecule. These RNA catalysts may have pharmaceutical applications. For example, a ribozyme has been designed to

FIGURE 5.39

Hypothetical development from the "RNA World" to present-day inheritance. At the dawn of life, RNA could both specify and copy itself, in addition to other catalytic activity. Later, more sophisticated systems used DNA as the information-storage unit, protein as the catalytic unit, and RNA as an information and catalytic intermediate.

cleave the RNA of human immunodeficiency virus (HIV). If such a ribozyme was made by a cell, all incoming virus particles would have their RNA genome cleaved by the ribozyme, which would prevent infection. Some known ribozymes include peptidyl transferase 23S rRNA, RNase P, Group I and Group II introns, GIR1 branching ribozyme, leadzyme, hairpin ribozyme, hammerhead ribozyme, HDV ribozyme, mammalian CPEB3 ribozyme, VS ribozyme, glmS ribozyme, and CoTC ribozyme. The hammerhead ribozyme is probably the most well known.

5.6.3.2.2 DNA enzymes

Though DNA tends to be found in a double-stranded structure, we have seen that DNA can also make complex, single-stranded arrangements. Not surprisingly, a few DNA enzymes, deoxyribozymes, have been found or created, though not in cells.[28,29] The ribonucleic enzymes are more versatile, natural, and widespread and thus seem better candidates for primitive replication systems, but "predicting" the origin of life is a doubtful business.

5.6.3.2.3 Thymine in DNA, uracil in RNA

Though not directly related to RNA–DNA structural differences, we would be remiss not to mention an important reason why U cannot be used in DNA. If U occurred in DNA, mutations would disrupt DNA replication, cells could not properly divide, and organisms would quickly die. We will discuss DNA damage and repair further in Chapter 11, but for now we note that C is frequently converted to U by cellular processes, by spontaneous reactions, or induction by nitrites.[30] In human cells, this conversion happens more than 100 times per day. If U, rather than T, had been used in DNA, this C → U conversion would produce a dilemma: the resulting G·U mismatched base pair could not be identified by any repair mechanism. Did the mismatch G·U originate from A·U (the G is incorrect) or from G·C (the U is incorrect)? Half the time any repair would revert the DNA to an incorrect sequence, resulting in a mutation. In contrast, RNA is a transient species (compared to DNA) and 100 C → U conversions per day do not matter. The lesson is that we cannot focus only on the structure of a base but also on the structures of other bases and the many processes that modify, convert, and reprocess nucleic acid bases. If we then think of RNA as the more ancient form of the genetic material, we see one reason for a later change to use of DNA, with T replacing U, as a more permanent, durable genetic record in complex organisms.

5.6.3.3 What is a "random coil"?

An overall globular shape to a linear polymer would be a natural outcome of a chain whose direction changes randomly, where Brownian forces dominate over any long-range, ordering interactions. As we will see in Chapter 13, the radius of such a "random coil" is governed by the length of the polymer and its natural bending properties (persistence length), but it is important to realize that a globular shape is a natural tendency of polymer systems dominated by random collisions from solvent molecules. In this case, a significant fraction of the interior volume of such a random coil is occupied by solvent. A simple freely jointed chain model of a random polymer coil gives a radius of

$$r_N^{rms} = \sqrt{N} L \tag{5.11}$$

where $r_N^{rms} = \sqrt{\langle r_N^2 \rangle}$, L is the length of a unit, and N is the number of units. This gives an extended occupied volume of

$$V_{ext} = \frac{4\pi}{3} L^3 N^{3/2} \tag{5.12}$$

On the other hand, the actual volume occupied by the atoms of the polymer chain itself is

$$V_{chain} = vN \tag{5.13}$$

where v is the volume of one polymer unit. For any N large enough, the former volume is larger. Under most circumstances $V_{ext} \gg V_{chain}$. Most globular proteins in their active states are not this sort of random, extended coil. Nucleic acids in a double-stranded conformation do not have such strong residual interactions between exterior side chains that happen to approach each other, as proteins do. They are better candidates for "random coils" in terms of the path traced by the helix axis. On the other hand, DNA bases rarely if ever take on truly random orientations relative to their neighboring bases, even at high temperature.

The term *random coil* is often used loosely to indicate the structure of a biopolymer that is non-native. Frequently, the evidence for the randomness of a particular structure is absent. The randomness can refer to any of the following:

1. The native biological activity is greatly reduced, as measured by a reaction rate reduction.
2. The *sedimentation coefficient* has passed through an S-shaped change in magnitude as a function of increasing temperature or increasing concentration of a *denaturing agent*.
3. The molecule will no longer crystallize, or even becomes insoluble.
4. Some ordered structure has disappeared (e.g., the α-helical content has diminished), as measured by an optical measurement (Chapter 9).
5. The spatial fluctuations in the structure have become comparable to the dimensions of the structure, as measured by fluctuation spectroscopy.

Behaviors 1–3, though important chemical and physical characterizations, have nothing directly to do with randomness of structure and should be taken with a grain of salt. It is quite difficult to prove that a polymer has a truly random structure, but proof of randomness is usually not useful in biology. A demonstration of a significant deviation from native structure and native activity is what truly matters. Such changes can most easily be measured using behaviors 1 and 4 above.

5.6.4 Carbohydrates and Saccharides

Sugars, saccharides, and polysaccharides are the neglected stepchildren of biophysicists,[31,32] though they are essential to life and are the most abundant type of biomolecule. The year 1997 witnessed a multifaceted debate in the biological physics community centering on the areas of biology to which physicists could most usefully contribute to. This debate centered partly on sugars and sugar polymers, polysaccharides. On one of the (several) sides was an assertion that physicists could more effectively contribute to the understanding of saccharides than to proteins and nucleic acids because saccharides were so poorly understood and little studied. The fact that comparatively little of exceeding interest was known reflected insufficient study and experimentation. On another side was the assertion that young biological physicists should delve into the great mysteries of protein structure and activity and the genetic code; that we should work where the big problems are, not where the competition is low. Twenty years later, the situation has begun to change, partly due to the realization that saccharide complexes have an important role in cell–cell recognition, especially in some cancer cells' ability to evade detection by the immune systems.[33] Some assert that the "glycome"—an organism's collection of glycans (saccharides)—constitutes one of three fundamental "fingerprints" of the organism, along with the familiar genome.[34] (The "lipidome" is the third.) Large and stable funding for cancer research, especially in the U.S., has also played an important role in the growth of saccharide research. See report.nih.gov/categorical_spending.aspx and funded research.cancer.gov/nciportfolio/stats.jsp. Some work has been done on pristine polysaccharides, but much of the interest has to do with combinations (complexes) of saccharides with proteins and/or membranes: the glycocalyx, glycoproteins, cell walls, recognition, and nanotechnology. See also Section 6.3.3, Saccharide Aggregates.

Why, generally, should a glycoprotein interest a physicist? The membranes surrounding cells are often said to be made up of lipid molecules. However, a large fraction of the membrane mass is made up of proteins. Some of these proteins have segments extending to the outside of the cell. The *extracellular* segments of these proteins are often glycosylated: they have an oligosaccharide (sugar polymer with less than about 100 subunits) attached. Such protein-saccharide complexes vary in saccharide content from <1% to >90%. Glycoproteins play a central role in cell–cell interactions and cell recognition. Cell recognition is indeed one of the great mysteries of biology and is a flourishing area of research owing to its connections with cancer, cell *apoptosis* (naturally programmed cell death), and the immune response.

A serious study of glycoproteins and cell recognition will take us far beyond the boundaries of an introductory biological physics course. Are there, however, areas of saccharide bioscience that may be both manageable and of interest to a budding biophysicist? Molecular imaging through, e.g., atomic force microscopy (AFM) and electron microscopy is becoming practicable.[32] Computer modeling of the saccharide-mediated interaction of cells may also be within reach,[35] but a reader with near-future research career plans should be aware of the multidisciplinary requirements and obstacles to research.[36] We will treat AFM in some detail in Chapter 19, but leave details of polysaccharide chemistry to another course, after examining the basic structure of these polymers in Chapter 6. We should also note that, unlike proteins and nucleic acids, saccharides do not usually have a prescribed size or specific function, and commonly form extended, macroscopic structures. Nevertheless, the reader will find some structural information databases of simple saccharides, as well as some polysaccharides, categorized by organism. PolySac3DB© at polysac3db.cermav.cnrs.fr (last updated April 24, 2012, at the time of this writing) assembles 3D data from computer molecular modelling and experimental techniques like X-ray, electron, and neutron diffraction. The Carbohydrate Structure Database (CSDB) at csdb.glycoscience.ru/ collects sequences and branching information of a large number of organismal polysaccharides. Finally, the persistent reader will discover that some saccharide structures can be found in the PDB Protein Database.

5.6.5 Membranes

Membranes, made from constituent lipid molecules, separate an organism from its environment. Membranes separate one cell from another and thus constitute the basic framework for organelles and organs. The cell membrane is probably the single most distinguishing characteristic of life as we find it on earth. The membrane is also the most obvious example of a self-organizing system we can find in biosystems. However, unlike amino acids, nucleotides, and carbohydrates, lipids do not form covalent, linear polymers of size ranging from two to hundreds or thousands of units. The sizes of polypeptides and polynucleotides are generally determined by the number and type of units required for the activity performed or information stored, as prescribed by the genome (DNA). Lipids, on the other hand, form large, noncovalent, two-dimensional aggregates of size that varies according to what the membrane must enclose at a given time. It is an aggregate of a fundamentally different type than the polymers we have dealt with so far. We will leave these aggregates to Chapter 6.

5.7 MACROMOLECULES: WHEN DOES A MOLECULE BECOME A MACROSCOPIC OBJECT?

5.7.1 Visible to the Naked Eye?

Macroscopic normally carries the connotation of "visible to the eye," so we might propose that the criterion for a macromolecule be that it is visible to the naked eye. As the eye can distinguish objects of about 10^{-4} m or larger, a macroscopic object, if made of carbon atoms, would contain about

$$\frac{4}{3}\pi(0.5 \times 10^{-4} \text{ m})^3 = N_{\text{vis}} \frac{4}{3}\pi(1.4 \times 10^{-10} \text{ m})^3 ==> N_{\text{vis}} \approx 10^{16} \text{ atoms} \tag{5.14}$$

This definition relies on the particular characteristics of the human eye, however, and is not related to any general physical principle. Though it may be intuitive, it does not help us understand or quantify any biological, biochemical, or biophysical process.

5.7.2 Optical Resolution

Our next try might be to relate "macroscopic" to classical optical resolution, about half a wavelength. Ignoring the issue of what type of radiation we are talking about and choosing $\lambda\lambda = 500$ nm (green light), we get

$$\frac{4}{3}\pi(2.5 \times 10^{-7} \text{ m})^3 = N_{opt}\frac{4}{3}\pi(1.4 \times 10^{-10} \text{ m})^3 ==> N_{opt} \approx 10^{10} \text{ atoms} \tag{5.15}$$

or 6 orders of magnitude smaller than visible to the eye requires. We might press this further after noting that recent optical imaging developments have allowed resolution of $\lambda/10$. Choosing a near-UV (but still "visible") wavelength of 300 nm, we reduce the number of carbon atoms in a minimally macroscopic particle by a factor of almost 1000:

$$\frac{4}{3}\pi(3.0 \times 10^{-8} \text{ m})^3 = N_{opt}\frac{4}{3}\pi(1.4 \times 10^{-10} \text{ m})^3 ==> N_{opt} \approx 10^7 \text{ atoms} \tag{5.16}$$

Still, we have the issue of what business a classical or high-tech optical device has in deciding what macroscopic should mean.

The problem we are having here centers on allowing the properties of a measuring device to assign characteristics to a particle we are interested in. Rather than focusing on the device, we should focus on the particle and its behavior. The particles we are interested in are proteins, nucleic acids, saccharides, ribosomes, and complexes that are "pretty big," by some measure. We might even suggest that what qualifies as macroscopic may depend on the type of particles you are studying. A semiconductor nanoparticle, which has a rigid crystalline structure, might have a quite different limit on macroscopic size than a flexible polymer. In fact, we might even propose that the meaning of macroscopic can change, depending on what particle behavior is being observed. We are not backtracking in our thoughts, substituting "particle behavior" for "measuring instrument resolution." We are, in fact, coming closer to the truth: the measure of macroscopic depends on the particle behavior in which you are interested.

5.7.3 Quantum versus Classical Behavior

All objects, even those we intuitively understand as macroscopic and classical, have properties that can only be described by quantum mechanics. The elastic, electric, and optical properties of a 1-gram block of silicon, for instance, ultimately depend on the quantum interactions between the atoms in the solid. Likewise, the absorption of light by green leaves of a plant requires a quantum description of the electrons in chlorophyll and other pigments in order to explain photosynthesis. This does not imply that we must use the Schrödinger equation to describe the motions of the block of salt and a green leaf. On the other hand, we learn in elementary quantum theory that an efficient way to determine whether an object's quantum wave properties are important in any description of its motions is to determine if the De Broglie wavelength is comparable to the object's apparent size:

$$\lambda_{db} = \frac{h}{p} = \frac{h}{mv} \approx \left(\frac{4m}{3\pi\rho}\right)^{1/3} \tag{5.17}$$

where h is Planck's constant and p is the particle's momentum. The right side of the equation is the radius of a sphere of mass m and density ρ. Setting $m = Nm_C$, where m_C is the mass of a carbon atom, we can solve for the number of atoms in this "quantum" object:

$$N = N_{db} \approx \left(\frac{3\pi\rho h^3}{m_c^4 v^3}\right)^{1/4} \approx 8 \times 10^6 \, v^{-3/4} \tag{5.18}$$

where v is the (nonrelativistic) velocity and SI units are used. (Check the units in this equation!) You may be wondering what to put in for the velocity, perhaps thinking of statistical mechanics and the equipartition theorem. For now, however, we will just substitute a "reasonable" value of the velocity for a small biological object. A "small" object inside a cell might, at most, move 1/10th of a cell diameter in a "characteristic" biochemical time of 1 ms: $v \sim 10^{-3}$ m/s. Equation 5.17 gives

$$N_{db} \approx (8 \times 10^6)(10^{-3})^{-3/4} \approx 10^9 \text{ atoms} \tag{5.19}$$

If we had chosen $v = 100$ m/s, we would have estimated $N_{db} \sim 10^5$ atoms.

We are coming down to values of 10^5–10^{10} for the minimum number of atoms in a macroscopic object, as judged by one quantum (De Broglie) criterion. Ignoring how reasonable these estimation methods may be, let's plow on to other estimates.

5.7.4 Surface Atoms/Bulk Atoms

If a spherical chunk of matter has nearly as many atoms on the surface as in the bulk, some properties of the material will be significantly different because one side of the surface atoms will not be surrounded by like atoms. As the sphere radius gets smaller, the surface-volume ratio gets larger. Let's use this as an estimate of when the sphere makes the microscopic → macroscopic transition. Assuming a single layer of closely packed carbon atoms cover the surface of the sphere (Figure 5.40), the cross-sectional area of an atom, multiplied by the number of atoms on the surface, will approximately equal the surface area of the sphere:

$$4\pi R^2 = n_{surface}(\pi r_C^2) \tag{5.20}$$

where $n_{surface}$ is the number of atoms on the surface, R is the radius of the sphere, and r_C the radius of a carbon atom. The total number of atoms in the sphere is given approximately by

$$\frac{4}{3}\pi R^3 = n_{total}\frac{4}{3}\pi r_C^3 s \tag{5.21}$$

(We are ignoring empty spaces in the packing of the atoms.) A rough criterion defining the transition from microscopic to macroscopic in terms of surface atoms can then be set as follows:

$$f - \frac{n_{surface}}{n_{total}} = \frac{4r_C}{R}, \tag{5.22}$$

where f is a number, like 0.01 or 0.5, that we choose to define our surface- to bulk-dominated transition point. If $f = 0.01$, then $R \sim 400r_C$, and there are a total of about 10^8 atoms; if $f = 0.1$, there are 10^5 atoms.

5.7.5 Energies Compared to $k_B T$

To this point, we have discussed macroscopic in terms of geometrical features of a particle compared to various microscopic parameters—light or De Broglie wavelength, the size of a carbon atom. When we learned thermodynamics, the emphasis was on a different aspect of the "size" of a system: the number of particles making up the system and whether there was any hope of tracking each particle's properties, energy, velocity, and momentum, individually. The systems we first learned were uniform, ideal gases, then liquids and solids. We may not have specified how many identical particles would have to be in the system, but the assumption was that after 10 or so collisions of each particle, even 10 or 100 particles would be virtually impossible to track exactly. Instead of trying to track each particle, we spoke of the "average" particle and its average deviation from the average,

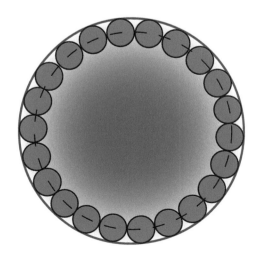

FIGURE 5.40
Spherical atoms packed at the surface in a spherical particle.

e.g., $\langle v_i^2 \rangle$ and $\langle v_i^2 - \langle v_i^2 \rangle \rangle$. We also learned that in a system of particles in thermodynamic equilibrium at temperature T, each degree of freedom (mode in which energy can exist as a function of a coordinate squared, like $\frac{1}{2}\,kx^2$ or $\frac{1}{2}\,mv^2$) would contain, on average, an energy of $\frac{1}{2}\,k_BT$. How can we apply this idea here? We would probably say that a water or O_2 molecule in a filled 1-L flask would qualify as a microscopic particle and would contain, on average, $\frac{1}{2}\,k_BT$ of energy each in mode $\frac{1}{2}\,mv_x^2$, $\frac{1}{2}\,mv_y^2$, $\frac{1}{2}\,I\omega_1^2$, etc. Would we say the same for a goldfish in the water? How about the smallest bacteria we observed in Chapter 1? How about a ribosome, or a protein of molecular mass 10^6 Da? Or 10^3 Da? Or a pollen grain undergoing Brownian motion?

We can get a handle on this thermodynamic view of the boundary between micro- and macroscopic by comparing characteristic energies to the value of k_BT at the temperature we are interested in, usually $k_BT \sim 4.1 \times 10^{-21}$ J.

5.7.6 Gravitational Energy

We assert that a particle becomes gravitationally microscopic—that the gravitational energy can change significantly by exchanging energy with the thermal environment—when the mass decreases to the point that

$$mg\Delta y \cong k_BT \qquad (5.23)$$

(We are ignoring density-buoyancy effects for now.) However, we seem to have an issue: the mass limit depends on the height variable Δy. This is actually not a problem, but rather an important truth: whether a mass m can be treated as in gravitational thermodynamic equilibrium depends on the height scale you are interested in. We can solve Equation 5.23 for m and obtain a definition of when an object becomes "gravitationally microscopic." Figure 5.41 shows a graph of this limit. For height ranges of $\Delta y \sim 10$ km (the Earth's troposphere), a particle made of a few carbon atoms ($m \sim 10^{-25}$ kg) is thermodynamically microscopic. This makes sense because 10 km is about the height of the Earth's atmosphere. For height changes corresponding to a cell diameter (~ 10 μm), a $\sim 10^{10}$ carbon-atom particle ($m \sim 10^{-16}$ kg or 10^{11} Da) will have its height randomized by thermal motion.

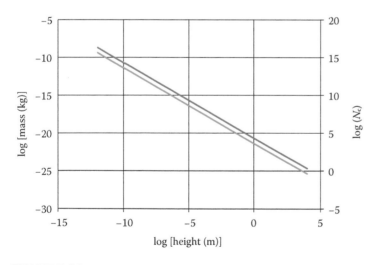

FIGURE 5.41
Graph of mass vs. y, the solution to mgy = kBT. Masses near or below the line may be considered "microscopic," in that their gravitational energy will be randomized by thermal fluctuations over a height y. mgy = kBT. (Red, number, blue, mass.)

5.7.7 Statistical Measures of "Large" Numbers of Particles

We are mixing up two somewhat different aspects of the "microscopic/macroscopic" border when we discuss how many particles we need for a system to be considered statistically large enough to treat thermodynamically. We must do this because biology forces us to. Biological systems force us to deal with molecules that (1) border on the macroscopic and (2) numbers of such particles that can range from 1 to 10^{10} per cell compartment. Both aspects force us to be careful with our statistical-thermodynamics treatments. If a particle is too large, it will not be able, as a single particle, to exchange k_BT packets of energy with the environment and be in thermal equilibrium. (On the other hand, if a macromolecule has 10^9 atoms, the individual atoms making up the particle may be in thermal equilibrium with the environment; so the molecule itself is borderline macroscopic.) Aspect 2 tells us that even if a particle is microscopic, there may not be enough of them for "thermal average" to mean much. We saw this latter effect in Chapter 1; Chapters 13 and 14 treat statistical averaging in more depth.

What is a handy estimate of the number of particles needed for a system to be considered "statistical," in the sense that statistical averages and average deviations make good predictions of particle behavior? Very roughly, if we want the average of a set of N measurements to differ from the average of a second set of N measurements by much less than the average deviation within a set, i.e., the average is accurate in some sense, we need

$$N^{1/2} \ll N \tag{5.24}$$

How much smaller must smaller than be? If $N = 10$, then $N^{1/2} \sim 3$ and our average is "accurate" to about 30%. This may be adequate. If we require the average to more closely reflect the true average, with $\sim 0.01\%$ accuracy, we must make 10^8 measurements. These statements apply equally well to measurements made at one instant on N particles or measurements made on one particle at N different times (appropriately spread over a sufficiently long time scale) under the *ergodic hypothesis*.

5.8 POINTS TO REMEMBER

- Biology on a scale smaller than the size of a cell often presents us with the following: a molecule may be made up of so many atoms that it may be considered a statistical assembly of atoms in itself, and those molecules may be too few in number for thermal averages to mean much. Know how to assess the microscopic-macroscopic boundary quantitatively.
- Amino acids and how they are bonded to form polypeptides.
- Protein α helices, β sheets, and the forces that govern their structures.
- The major differences between globular and structural proteins.
- The four standard DNA and RNA bases. Know something about the T versus U base.
- The major forces that govern nucleic acid structures.
- The major roles of DNA and the several forms of RNA.
- Use of the PDB and NDB.

5.9 PROBLEM-SOLVING

PROBLEM 5.1   VIEWING PROTEIN PROSTHETIC GROUPS

Go to the PDB Web site and find a protein with the following prosthetic groups. Print two versions of the structure: (1) a structure that allows a clear view of the prosthetic group and (2) a space-filling structure. Rotate the structure to select a good view. Describe where the protein is found, how many amino acids it has, and its molecular weight. If the protein exists as an aggregate of several component proteins, describe the aggregate.

1. Heme b
2. Bacteriochlorophyll b

PROBLEM 5.2   METAL IONS IN NUCLEIC ACIDS

Go to the PDB or NDB and find a nucleic acid (may also contain a protein) structure with the following bound metal ions. Print a structure that shows the DNA and the metal, state which part of the DNA the metal is associated with and describe the role of the metal ion, if known.

1. Magnesium
2. Cobalt
3. Zinc

PROBLEM 5.3   MOLECULAR FORMULAS AND MASSES

Find the molecular formula (e.g., $C_{10}H_{14}N5O_8P$) for each the following and calculate the molecular mass. Compare to values from a reliable source.

1. Tyrosine
2. Sodium monohydrogen phosphate
3. Guanosine diphosphate
4. Guanosine triphosphate

PROBLEM 5.4   NUCLEIC ACID BASE STACKING

Hand draw diagrams of the following pairs of nucleic acid bases and show the various ways the pairs can be positioned to form a hydrogen bond or bonds. Do not restrict yourself to the canonical DNA "base-pair" arrangement found in text books.

1. Adenine and thymine
2. Guanine and cytosine
3. Adenine and cytosine
4. Guanine and thymine
5. Adenine and guanine
6. Thymine and cytosine

PROBLEM 5.5   DRAWING NUCLEOSIDES

Take one of the purine and one of the pyrimidine mononucleosides and hand draw the molecular structure, including all C and H atoms.

PROBLEM 5.6 POLYPEPTIDE STRUCTURES AND ENERGETICS

1. Draw the "chemical formula" structure (à la Figure 5.14) of the pentapeptide Leu-Val-Asp-Phe-Trp. Include all C and H atoms that are in place at pH 7.0.
2. Assume this pentapeptide forms part of an α helix. Show any H bonds that form by dotted lines in the figure you just drew.
3. Calculate the electrical potential energy between the ends of this α helix. Take some care to justify the distance and its uncertainty. Assume a dielectric constant ε of bulk water. Compare this energy to $k_B T$. Is this electric force significant?
4. Repeat 3 for $\varepsilon = 4.0$.

PROBLEM 5.7 HELIX LENGTHS

Derive the relationship between the length L of a helix along the axis and the total path length PL traced along the helical strand from end to end. Use a clear diagram or diagrams to justify the relation. Hint: Find the cardboard core of an empty toilet-tissue roll and split along the seam.

PROBLEM 5.8 DIPOLES AND HELIX STABILITY

Assume a straight α helix of 8 turns has all backbone H bonds (N-H····O = C) intact, that the N-O distance is 0.28 nm and that the H bond effectively moves a charge of 0.3e from N to O.

1. Calculate the dipole moment in SI units (coul-m) and in debye.
2. Calculate the energy of interaction between all these dipoles. Refer to your first-year physics text for needed dipole interaction formulas or simply use Coulomb's law for the individual charges. Explain whether you need consider only nearest-neighbor dipole interactions or all possible dipole–dipole interactions. Use $\varepsilon = 6.0$.
3. Compare the energy in 2 to the total H-bond energy for the 8 turns. (Use H-bond energy estimates from text or from a reliable Web source.)
4. Grad problem: Find a published value for the energy of a backbone dipole–dipole interaction energy and compare to your results. Explain the differences.

PROBLEM 5.9 ENZYME ACTIVITY, SEQUENCE AND STRUCTURE

With clarity and illumination, explain the differences between the statements of Section 5.6.1

(i) "Enzymes with the same fold, catalyzing the same reaction, and sharing the same ancestor typically share less than 20 percent of their amino acids" (Andreas Wagner, *Arrival of the Fittest*), and

(ii) "One can randomly change 80% of the amino acids in an enzyme and still retain activity." Include issues such as local sequence and structure, roles of helices and sheets, and active sites.

MINIPROJECT 5.10 PREDICTING GLOBULAR AND TRANSMEMBRANE PROTEINS

This miniproject involves applying the Kyte–Doolittle hydropathy score distribution to protein amino acid sequences to predict which are globular—have an outer shell of hydrophilic amino acids and an inner core of hydrophobic amino acids—and which may be transmembrane—have hydrophilic ends and hydrophobic midsections. Go to the Web site http://gcat.davidson.edu/rakarnik/kyte-doolittle-background.htm#ref and do the sample problem or go to https://www.ncbi.nlm.nih.gov/protein and search for a particular protein, click on "FASTA Sequence," copy the displayed sequence information, and paste it into the Kyte–Doolittle online calculator at the davidson.edu address above. (Proteins have NCBI FASTA codes like "AAA88054.") (If this Web site set up by Soren Johnson, Rachel McCord, and Lisa Robinson has moved, follow any forwarding links or search for "Kyte–Doolittle hydropathy plot.")

MINIPROJECT 5.11 PROTEIN STRUCTURE, DISULFIDE BONDS, AND VENOM: A PROBLEM IN MEDICINE

1. Download a diagram of the structure of the venom toxin shown in Figure 5.25. Do the same for two or three other venom toxins that you find with several disulfide bonds.

2. Discuss reasons for the resistance of these toxins to degradation by *proteases*, enzymes that *cleave* (cut, chop up) the polypeptide chain of other proteins. Look up some properties of proteases, such as molecular mass (size) and modes of action.

3. Search your library database for articles on antivenoms for one or more of the above toxins. Choose a toxin or class of toxins where there seems to be a good number of articles available. (Be prepared to deal with articles in exotic journals such as that by C. Devaux and H. Rochat, Theoretical and experimental bases for treatment of scorpion envenomations, *Bulletin de la Société de pathologie exotique* 95 (2002):197–199 and J.-P. Chippaux and M. Goyffon, Epidemiology of scorpionism: A global appraisal, *Acta Tropica* 107 (2008):71–79.) Describe the primary antivenoms and how they work. Do they, for instance, degrade the venom toxin to a safe form? Is detoxification the primary treatment for bite victims or is symptomatic treatment?

PROBLEM 5.12   COLLAGEN TRANSPARENCY

Investigate and explain how corneal collagen can be transparent while that in tendon, skin, and cartilage is not. If you have never read Feynman's explanation of light transmission and reflection and why some things are clear, it is worth the effort. (R. P. Feynman, *QED: The Strange Theory of Light and Matter*. Princeton University Press, Princeton, NJ, 2014.)

PROBLEM 5.13   ENZYMES AND ATOMIC MOBILITY

Find an enzyme molecule in the PDB that has an active site, catalyzes a chemical reaction, and whose atoms can be displayed by temperature factor. (For example, use JSmol software and set Style/Scheme/CPK and Color/Atoms/Scheme/Temperature [relative].) Do not use myoglobin.

1. State the reaction this enzyme catalyzes and some important enzyme properties like molecular mass and number of amino acids.
2. Save the structure such that the active site can be seen. Mark this active site and insert the graphic file into your homework document.
3. Describe the relationship, if any, between the atomic mobility and the position with respect to the active site and with respect to protein interior/exterior.

PROBLEM 5.14   THE DNA HELIX AXIS

Go to the PDB, NDB, or other source of nucleic acid atomic coordinates. (NDB has an efficient search engine for nucleic acid structures, but for actual structure display and measurement, use the NDB link to PDB. You might also stay online with this structure for Problem 5.15.)

1. Find structures of A, B, and Z DNA and display/print a ball-and-stick model view down the helix axis for each helix.
2. Repeat but with a space-filling model.
3. For A DNA, estimate the diameter of the hole down the center of the helix, using tools in the display software. Explain how you define this diameter. Could a water molecule fit through this hole?

PROBLEM 5.15   H BOND DISTANCES IN DNA

1. Find a NDB/PDB structure for Z DNA. You may use that from Problem 5.14. Find a G–C base pair and determine the N–N, O–N, and N–O distances between the atoms that share a hydrogen atom in the three H bonds. Jmol software is efficient for distance measurements. Print the structure, and write down the structure ID number.
2. Repeat 1 for B DNA.
3. Discuss the variations in H bonding distances.

PROBLEM 5.16   DNA AND ANTITUMOR DRUG

One class of drugs that have antitumor properties is a relatively small molecule that binds to DNA. Go the PDB and find the structure 1CX3.

1. Describe how this structure was determined (20–30 words).
2. Briefly (75 words) describe the drug and its antitumor activities. Don't get too bogged down in topoisomerase and other complexities. Try to word the activity in laymen's terms.
3. Copy/paste a structural view from the side, clearly showing the orientation of the drug.
4. Determine the vertical distances (along helix axis) between adjacent base pairs in this DNA. By how much does the "intercalated" drug change the "standard" distance?

PROBLEM 5.17   ERROR RATES

Addition of Val to $tRNA^{Ile}$. In the section on tRNA, it was noted that the $-CH_3$ group of Ile adds about -12 kJ/mol of free energy of binding of Ile to the enzyme, isoleucine-tRNA synthetase, compared to Val, which has no $-CH_3$ group. Calculate the expected ratio of N_{Val}/N_{Ile}, the number of tRNA molecules with Val attached divided by the number with Ile attached.

PROBLEM 5.18   VIBRATIONAL ENERGY

The last section of this chapter considered gravitational thermodynamic equilibrium by comparing $mg\Delta y$ to $k_B T$. Make a similar graphical analysis of elastic vibrational energy as follows. The force required to compress a cube of side L_0 by ΔL is given by $F = YL_0\Delta L$, where Y is the Young's modulus. Take $Y = 2 \times 10^9$ N/m^2, near the value for a virus capsid. The elastic potential energy stored in such compression or extension is $\frac{1}{2}YL_0(\Delta L)^2$. (What quantity should be on the horizontal axis?) As in Figure 5.41, use two dependent axes, mass and number of carbon atoms. Look up reasonable values for density or any other quantities you may need. Briefly discuss when the macroscopic Young's modulus model begins to break down.

PROBLEM 5.19   RAMACHANDRAN DISTRIBUTION STATISTICS

In the following, show a calculation or justify your answer, as appropriate.

1. Calculate the number of all possible normal amino-acid *pairs*: ala-gly, tyr-cys, etc. Do not assume, e.g., that ala-gly is the same as gly-ala.
2. Calculate the number of all possible normal amino-acid *triplets*: ala-gly-trp, tyr-cys-pro, etc.
3. Calculate the number of possible Ala residues with a Val to the left and any normal amino acid to the right.
4. If a structural dataset includes 100,000 standard amino acids, how many of a specific amino-acid triplet should be found, on average. Assume the 20 amino acids are equally likely and ignore effects of finite lengths of polyamino acids.
5. Grad problem. Assume individual amino-acid probabilities are as in Table 5.2. Choose a high-percentage pair and a low-percentage pair and calculate the number of occurrences in the dataset of 100,000.

PROBLEM 5.20   PRION STRUCTURE

1. Go to the PDB and search for "human prion protein." How many structure entries do you find? How often were solution NMR and X-ray crystallography used to determine the structures?
2. Locate the early entries, with "1Q" at the beginning of the PDP ID. What method was used to determine these structures?
3. Find 1QLX and load the 3D display using JSmol or Jmol. When was this structure deposited? (If this viewer has been replaced on the PDP site, consult your instructor.) Download the default image that appears: File/Export/Export PNG Image. Change the display Style to three other styles and repeat. Paste these images into the file you turn in for your homework.
4. From the image display, estimate the fractions of α helix and β sheet in the molecule. What display style best facilitates your estimate?
5. Locate PDB ID 4KML. When was this structure deposited?
6. Repeat problem parts 3 and 4 for this structure.

MINIPROJECT 5.21   PREDICTING GLOBULAR AND TRANSMEMBRANE PROTEINS II

This followup to Miniproject 5.10 uses "TopGraph" to predict protein/membrane configuration. Go to the Web site https://www.ncbi.nlm.nih.gov/protein and search for a particular protein, preferably the one you used in Miniproject 5.10. Click on "FASTA" sequence. Open a second browser tab or window and connect to http://topgraph. weizmann.ac.il/bin/steps. Familiarize yourself with use of TopGraph. Enter your email address and follow directions. When asked for the protein sequence, copy the sequence from the other browser window and paste it into the appropriate TopGraph box. When initially asked for run type, choose "plain." After you have gone through the multiple "submit" windows, you will see "The request steps for the TransMembrane Prediction are now complete. TopGraph processing should start soon. Results, when available, will be sent by email to youremail@xxx.edu. You may check the status of this process *here*." (This process will take considerably longer than the Kyle–Doolittle process.) Paste the results of your run into your homework submission file, explaining the results. If possible, compare to results from Miniproject 5.10.

PROBLEM 5.22   POLYSACCHARIDE STRUCTURE

Go to the Web site polysac3db.cermav.cnrs.fr and locate the structure of M41 capsular polysaccharide from the bacterium *Escherichia coli*.

1. Describe the experimental method used to determine this 3D structure.
2. Write down the original reference in bibliographic form. An "Expert" search is suggested.
3. From information located in the "Expert" search, describe why the authors did this structural work.
4. Locate an image of the structure; copy and paste it into your solution.

REFERENCES

1. Kendrew, J. C., R. E. Dickerson, B. E. Strandberg, R. G. Hart, D. R. Davies, D. C. Phillips, and V. C. Shore. 1960. Structure of myoglobin: A three-dimensional Fourier synthesis at 2 Å resolution. *Nature* 185:422–427.
2. Perutz, M. F., M. G. Rossman, A. F. Cullis, H. Muirhead, G. Will, and A. C. T. North. 1960. Structure of hæmoglobin: A three-dimensional Fourier synthesis at 5.5-Å resolution, obtained by X-ray analysis. *Nature* 185:416–422.
3. Richardson, J. S. 1981. The anatomy and taxonomy of protein structure. In *Advances in Protein Chemistry*, eds. C. B. Anfinsen, J. T. Edsall, and F. M. Richards, 34:168–364. New York Academic Press.
4. Reilly, C. 2004. *The Nutritional Trace Metals*. Oxford, UK: Wiley-Blackwell.
5. (a) Schinn, S.-M., A. Broadbent, W. T. Bradley, and B. C. Bundy. 2016. Protein synthesis directly from PCR: Progress and applications of cell-free protein synthesis with linear DNA. *New Biotechnol* 33(4): 480–487; (b) Kumar, A., and S. Singh. 2013. Directed evolution: Tailoring biocatalysts for industrial applications. *Crit Rev Biotechol* 33(4):365–378; (c) Leemhuis, H., V. Stein, A. D. Griffiths, and F. Hollfelder. 2005. New genotype-phenotype linkages for directed evolution of functional proteins. *Curr Opin Struct Biol* 15 (4):472–478; (d) Acevedo-Rocha, C.G., S. Hoebenreich, and M. T. Reetz. 2014. Iterative saturation mutagenesis: A powerful approach to engineer proteins by systematically simulating Darwinian evolution. *Methods Molec Biol* 1179:103–28.
6. Richardson, J. S., and D. C. Richardson. Principles and Patterns of Protein Conformation. 1989. In *Prediction of Protein Structure and the Principles of Protein Conformation*, ed. G. D. Fasman. New York: Plenum Press.
7. Rost, B., G. Yachdav, and J. Liu. 2004. Predictprotein: The Predictprotein server. *Nucleic Acids Res* 32:W321–W326.
8. Voet, D., and J. G. Voet. 2004. *Biochemistry*. New York: John Wiley & Sons, 224.
9. Pace, C., and J. M. Scholtz. 1998. A helix propensity scale based on experimental studies of peptides and proteins. *Biophys J* 75:422–427.
10. Arora, N., and B. Jayaram. 1997. Strength of hydrogen bonds in a helices. *J Comput Chem* 18:1245–1252.
11. Engelman, D. M., T. A. Steitz, and A. Goldman. 1986. Identifying nonpolar trans-bilayer helices in amino acid sequences of membrane proteins. *Annu Rev Biophys Biophys Chem* 15:321–353.
12. Kyte, J., and R. Doolittle. 1982. A simple method for displaying the hydropathic character of a protein. *J Mol Biol* 157:105–132.
13. Elazar, A., J. J. Weinstein, J. Prilusky, and S. J. Fleishman. 2016. Interplay between hydrophobicity and the positive-inside rule in determining membrane-protein topology. *Proc Natl Acad Sci USA* 113(37):10340–10345.
14. Chaplin, M. 2009. *Water Structure and Science*. London, UK: Creative Commons, www.lsbu.ac.uk/water/.
15. Eckert, R. L. 1988. Sequence of the human 40-kDa keratin reveals an unusual structure with very high sequence identity to the corresponding bovine keratin. *Proc Natl Acad Sci USA* 85:1114–1118.
16. Barnard, K., N. D. Light, T. J. Sims, and A. J. Bailey. 1987. Chemistry of the collagen cross-links. Origin and partial characterization of a putative mature cross-link of collagen. *Biochem J* 244:303–309.

17. Robinson, C. R., Y. Liu, J. A. Thomson, J. M. Sturtevant, and S. G. Sligar. 1997. Energetics of heme binding to native and denatured states of cytochrome b562. *Biochemistry* 36:16141–16146.

18. Lin, L., R. J. Pinker, K. Forde, G. D. Rose, and N. R. Kallenbach. 1994. Molten globular characteristics of the native state of apomyoglobin. *Nat Struct Biol* 1:447–452.

19. Kitahara, R., H. Yamada, K. Akasaka, and P. E. Wright. 2002. High pressure NMR reveals that apomyoglobin is an equilibrium mixture from the native to the unfolded. *J Mol Biol* 320:311–319.

20. Frauenfelder, H., F. Parak, and R. D. Young. 1988. Conformational substates in proteins. *Annu Rev Biophys Biophys Chem* 17:451–479.

21. Frauenfelder, H., G. A. Petsko, and D. Tsernoglou. 1979. Temperature-dependent X-ray diffraction as a probe of protein structural dynamics. *Nature* 425, 21 (September 4, 2003). doi:10.1038/425021a 280:558–563.

22. Marraffini, L. A. 2015. CRISPR-Cas immunity in prokaryotes. *Nature* 526: 55–61. doi:10.1038/nature15386. See also Mestel, R. 2017. The original CRISPR, *Science News* 191(7): 22–26.

23. Davis, R. C., and I. J. Tinoco. 1968. Temperature-dependent properties of dinucleoside phosphates. *Biopolymers* 6:223–242.

24. Ke, C., M. Humeniuk, H. S-Gracz, and P. E. Marszalek. 2007. Direct measurements of base stacking interactions in DNA by single-molecule atomic-force spectroscopy. *Phys Rev Lett* 99:018302.

25. Voet, D., and J. G. Voet. 2004. *Biochemistry.* New York: John Wiley & Sons, 1122.

26. Gesteland, R. F., T. R. Cech, and J. F. Atkins, eds. 2005. *The RNA World.* 3rd ed. New York: Cold Spring Harbor Laboratory Press.

27. Johnston, W. K., P. J. Unrau, M. S. Lawrence, M. E. Glasner, and D. P. Bartel. 2001. RNA-catalyzed RNA polymerization: Accurate and general RNA-templated primer extension. *Science* 292:1319–1325.

28. Breaker, R. R., and G. F. Joyce. 1994. A DNA enzyme that cleaves RNA. *Chem Biol* 9(4):403–415, 1:223–229.

29. Schlosser, K., and Y. Li. 2009. Biologically inspired synthetic enzymes made from DNA. *Chem Biol* 16:311–322.

30. Voet, op cit., 1177.

31. Huebner, J. S., E. Jakobsson, H. G. Dam, V. A. Parsegian, and R. H. Austin. 1997. Harness the hubris: Useful things physicists could do in biology. *Phys Today* 50:11–14.

32. Brush, M. 1999. Sugars and splice. Glycobiology: The next frontier. *Scientist* 13:22–23.

33. Landhuis, E. 2017. Cancer cells cast a sweet spell on the immune system. *Science News* 191:24–27. See also Paszek, M. J., et. al. 2014. The cancer glycocalyx mechanically primes integrin-mediated growth and survival. *Nature* 511(7509):319–325; Kramer, J. R., B. Onoa, C. Bustamante, and C. R. Bertozzi. 2015. Chemically tunable mucin chimeras assembled on living cells. *Proc Natl Acad Sci USA* 112(41):12574–12579; Xiao, H., E. C. Woods, P. Vukojicic, and C. R. Bertozzi. 2016. Precision glycocalyx editing as a strategy for cancer immunotherapy. *Proc Natl Acad Sci USA* 113(37):10304–10309.

34. NIH, N. I. G. M. S. 2016. There's an 'ome' for that. *Sci Daily* www.sciencedaily.com/releases/2016/12/161216114449.htm (accessed April 26, 2017).

35. Paszek, M. J., et al. 2014. The cancer glycocalyx mechanically primes integrin-mediated growth and survival. *Nature* 511(7509):319–325.

36. Agre, P., et. al. 2016. Training the next generation of biomedical investigators in glycosciences. *J Clin Invest* 126(2):405–408.

37. Vojtechovsky, J. et al. See http://www.wwpdb.org or http://www.pdb.org.

38. Ting D., Wang G., Shapovalov M., Mitra R., Jordan M. I., Dunbrack R. L. Jr. 2010. *PLoS Comput Biol* 6(4):e1000763. https://doi.org/10.1371/journal.pcbi.1000763

39. Dickerson, R. E., and Geis, I., *The Structure and Action of Proteins*, W. A. Benjamin, Menlo Park CA, 1969.

40. Chattopadhyaya, R. et al. 1990. Displayed using Jmol. Note the title of publication, Structure of a T4 hairpin loop on a Z-DNA stem and comparison with A-RNA and B-DNA Loops, *J Mol Biol.*, 211: 189–210.

41. Ramachandran, G. N. and Sasisekharan, V. 1968. Conformation of polypeptides and proteins. *Adv Protein Chem* 23, 283–437.

42. Kauzmann, W. 1959. Some factors in the interpretation of protein denaturation. *Adv Protein Chem* 14, 1–63, 1959.

43. Friedman, R. A. and B. Honig. 1995. A free energy analysis of nucleic acid base stacking in aqueous solution. *Biophys J.*, 69:1528–1535.

44. RCSB PDB software references:

    Jmol: An open-source Java viewer for chemical structures in 3D. http://www.jmol.org/.

    Protein Workshop: Moreland, J. L., A. Gramada, O. V. Buzko, Q. Zhang, and P. E. Bourne. 2005. The Molecular Biology Toolkit (MBT): a modular platform for developing molecular visualization applications. *BMC Bioinformatics* 6:21.

    NGL Viewer: Rose, A. S., A. R. Bradley, Y. Valasatava, J. D. Duarte, A. Prlić, and P. W. Rose. 2018. NGL viewer: Web-based molecular graphics for large complexes. *Bioinformatics*. doi:10.1093/bioinformatics/bty419.

45. Montserret, R., M. J. McLeish, A. Bockmann, C. Geourjon, and F. Penin. 2000. Involvement of electrostatic interactions in the mechanism of peptide folding induced by sodium dodecyl sulfate binding. *Biochemistry* 39:8362–8373.

46. Guijarro, J. I., S. M'Barek, F. Gomez-Lagunas, D. Garnier, H. Rochat, J.M. Sabatier, L. D. Possani, and M. Delepierre. 2003. Solution structure of Pi4, a short four-disulfide-bridged scorpion toxin specific of potassium channels. *Protein Sci* 12: 1844–1854.

47. Lolis, E. and G. A. Petsko. 1990. Crystallographic analysis of the complex between triosephosphate isomerase and 2-phosphoglycolate at 2.5-. ANG. resolution: Implications for catalysis. *Biochemistry* 29: 6619–6625.

48. Okuyama, K., C. Hongo, R. Fukushima, G. Wu, H. Narita, K. Noguchi, Y. Tanaka, N. Nishino. 2004. Crystal structures of collagen model peptides with Pro-Hyp-Gly repeating sequence at 1.26 A resolution: implications for proline ring puckering. *Biopolymers* 76: 367–377.

49. Phillips, S. E. and B. P. Schoenborn. 1982. X-ray structure of sperm whale deoxy-moglobin refined at 1.4A resolution. doi:10.2210/pdb1MBD/pdb.

50. Kuriyan, J., S. Wilz, M. Karplus, and G. A. Petsko. 1986. X-ray structure and refinement of carbon-monoxy (Fe II)-myoglobin at 1.5 A resolution. *J Mol Biol* 192: 133–154.

51. Shi, K., B. Pan, and M. Sundaralingam. 2003. The crystal structure of an alternating RNA heptamer r(GUAUACA) forming a six base-paired duplex with 3′-end adenine overhangs. *Nucleic Acids Res* 31: 1392–1397.

52. Bailey, S., W. K. Eliason, and T. A. Steitz. 2007. Structure of hexameric DnaB helicase and its complex with a domain of DnaG primase. *Science* 318: 459–463.

53. Fourmy, D., E. Guittet, and S. Yoshizawa. 2002. Structure of Prokaryotic SECIS mRNA Hairpin and its Interaction with Elongation Factor SELB. *J Mol Biol* 324: 137–150.

54. Shi, H. and P. B. Moore. 2000. The crystal structure of yeast phenylalanine tRNA at 1.93 Å resolution: A classic structure revisited. *RNA* 6: 1091–1105.

CHAPTER 6

CONTENTS

First Pass at Supramolecular Structures

Assemblies of Biomolecules

6

The purpose of this chapter is threefold: (1) to introduce the types of large assemblies we observe in nature, and the methods to determine their characteristics; (2) to motivate the basic rules for assembly of such structures; and (3) to indicate the importance of such assembly to life, and to a lesser extent, to treatment of some diseases. Though this textbook is not designed to prepare students for a biomedical career, we would be remiss not to acknowledge the directions we can go to find an important, well-paying, fascinating career. We will get to these goals by first describing and quantifying what is in the cell (this chapter) and then applying the simplest physical laws to these systems and seeing what structures come out without complicated cellular assembly plants (Chapter 17). We cannot be comprehensive in these examinations and analyses; we cannot even be completely correct. There is much to be gained from some slightly faulty, but simple, treatments, and unaffordable additional effort for an only slightly improved treatment.

This chapter contains the most wide-ranging treatment of two-dimensional aggregates, membranes, in the text. Treatments of topics are not detailed, but a number of avenues for investigation of membrane behavior are introduced, ranging from thermodynamic, to molecular transport, to materials analysis. We will briefly return to membrane elastic properties in Chapter 12, but the interested student is directed toward other textbooks for more detailed treatments of biochemical[1] and materials properties.[2]

Material on aggregates of proteins, nucleic acids, lipids, and saccharides is often placed in individual textbook chapters on proteins, nucleic acids, lipids, and saccharides. That topic organization is effective for a student who bends her avocation toward one of the molecular types, say, nucleic acids, does not particularly care about lipids, and wants to use the textbook for nucleic acid reference material. Such material would be located in a few chapters, unmixed with lipids and proteins, would be easy to find, and would be a springboard to focused texts on nucleic acids. This textbook is organized differently. The organization is primarily by physical and chemical law and phenomena. Such organization may frustrate you when you have forgotten the formula for the hydrodynamic radius of DNA. Just use the index and be persistent.

6.1 MEASURING PROPERTIES OF THREE-DIMENSIONAL AGGREGATES

Before embarking on a tour of some biological aggregates, we must introduce one of the primary quantities used to describe aggregates—the sedimentation coefficient—as well as two of the standard methods for determining aggregate properties and separating subunits of noncovalent aggregates. We describe more of the theory in detail in Chapter 12, but we must be aware of the meaning of terms such as 70S that will be encountered in most descriptions of aggregates. We concentrate on classical, but still much used, techniques, with comments on how single molecule detection methods will gradually supplant many of these procedures in the research lab and later, in biotechnology. Single-molecule behavior is addressed throughout the text, especially in Chapters 13–16.

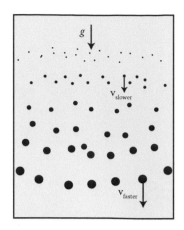

FIGURE 6.1
Diagram of "gravitational" separation of aggregates based on size. Particles will reach a terminal velocity that scales with mass (actually, mass less the mass of the displaced water) in water (unlike in air)—assuming the particles are the same geometrical size and shape and are denser than water. If they differ in geometry and density, the relation between terminal velocity and size depends on mass and these other quantities. These "size" descriptors are grouped together into a single quantity, the sedimentation coefficient, s: $v = s \cdot g$, where g is the effective gravitational acceleration. When the sample tube is placed in a spinning centrifuge, a centrifugal acceleration is applied, which is usually much greater than g, and g is replaced by its centrifugal equivalent: $v = s \cdot (\omega^2 r)$.

6.1.1 Sedimentation Coefficient

Centrifugation is a primary means for characterizing the size of proteins, RNAs, polymers, and subcellular components, as well as for separating particles of different sizes. The idea is simple. When particles are dispersed in water, large (heavy) particles will sink to the bottom. Larger particles sink faster than smaller particles. Very small particles, e.g., dissolved small molecules, do not sink at all. In Chapter 5 we discussed the issue of particle size and the macroscopic–microscopic boundary. Biological aggregates easily span this boundary line. One of the measures of this boundary involved comparing mgh with $k_B T$. Centrifugation is based on the dependence of drift velocity through water on gravitational force. The only added feature is that the rotating centrifuge increases the value of $g = 9.8$ m/s² by a large factor. See Figure 6.1. If we effectively increase the acceleration of gravity by a factor of 1000, particles that were "dissolved" before now may sink to the bottom of the tube. Figure 6.2 shows a typical type of rotor used in a laboratory centrifuge. This rotor allows the tube containing the sample to hang vertically when at rest and swing out horizontally when up to speed. The centrifugal force will, of course, depend on where the particle of interest is located in the tube. Laboratory centrifuges range in rotational speed from a few hundred to about 60,000 revolutions per minute (rpm).

The "size" of particles can be quantified by the speed at which the particle drifts through water, under the influence of a constant force. As noted in earlier chapters, the approximate equation relating velocity and force in a viscous liquid is $F = bv$, where b is a frictional coefficient. In centrifugation measurements, v is referred to as the sedimentation velocity, and the sedimentation coefficient s is defined by

$$v = s \cdot \left(\frac{F_{applied}}{m} \right) = s \cdot \omega^2 r \qquad (6.1)$$

where r is the effective radius of circular motion in the centrifuge and ω is the angular velocity of rotation. The sedimentation coefficient s has units of seconds, but convention in biology is to use units of "svedbergs," where 1 svedberg = 10^{-13} s. Figure 6.3 shows some values of typical biological aggregates, which range from about 1 to more than 10^5 svedbergs.

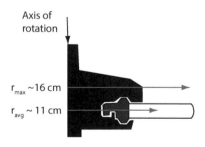

Axis of rotation

r_{max} ~16 cm

r_{avg} ~ 11 cm

FIGURE 6.2
"Swinging bucket" centrifuge. The tube, initially upright, swings horizontally when the rotor comes up to speed. This type of rotor generally has a maximum rotational speed of 40,000 rpm. (From Beckman SW rotor, image by permission of Scientific Calibration, https://scical.com.)

6.1.2 Separating Subunits of Small Aggregates: Chromatography

While the overall size of aggregates can be quantified by centrifugation, no information on individual subunits is obtained: the subunits normally maintain their aggregation state. The subunits can often be induced to loosen their grip on each other by other means, however. One of the most important of these separation methods is called *chromatography*. Two major types of chromatography are *ion-exchange* and *affinity* chromatography. The physical basis of these separation methods is the differing affinity, or attractive force, between subunits 1 and 2 and a specific group attached to a granular matrix of otherwise inert material, either an ionic group or a molecular group designed specifically to bind to the protein, nucleic acid, or subunit of interest. A classic type of chromatography ion-exchange column is prepared by packing a cylindrical tube, which ranges in size from a 10-cm-diameter by 2-m-long behemoth to a millimeter-sized chamber, with a packed ion-exchange resin in aqueous buffer at the proper pH. The buffer fills the small spaces between resin particles.

Ion-exchange resins are typically made of an insoluble base material like cellulose, polystyrene, or cross-linked dextran or agarose gels (both polysaccharides), which have ionic groups like carboxymethyl ($-CH_2COO^-$), diethylaminoethyl ($-CH_2CH_2N(CH_2CH_3)_2{}^+$), or sulfonate ($-SO_3{}^-$) attached to the surface. Note the common use of biomaterials such as cellulose and agarose to characterize other biological materials. The protein or other aggregate is then carefully layered on the top of the gel and buffer slowly flowed through. The buffer and protein aggregate flow in the spaces between resin particles, but as the resin's ionic groups are chosen to attract the protein subunits, subunits are slowed by factors that depend on the strength of their binding to the groups. The overall strength of the interactions can be adjusted by varying the free ion concentration in the buffer. (Buffer counterions will shield the charge on the resin's ion group.) Often the pH can also be adjusted to destabilize the forces holding the protein subunits together. This facilitates the separation of the subunits. A diagram of this method is shown in Figure 6.4.

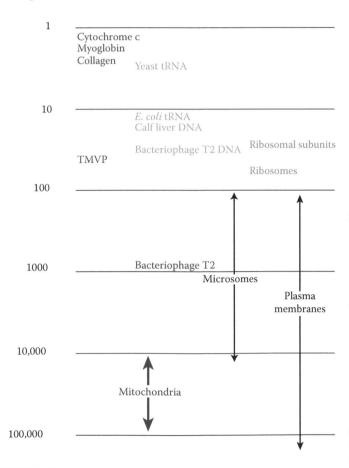

FIGURE 6.3

Sedimentation coefficients measure the sizes of common biological aggregates. Svedbergs have units of 10^{-13} s. Values for biological aggregates range from 1 to 2 for proteins with one subunit to more than 100,000 for structures such as intact plasma membranes. Lettering color code: proteins (red), DNA (green), membrane entities (black), other (brown).

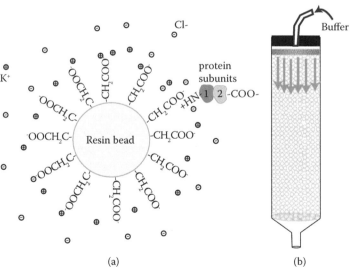

FIGURE 6.4

Ion-exchange chromatography. (a) Subunit 1 will be more strongly bound to the resin $-CH_2COO^-$ than subunit 2 and will travel more slowly past resin particles. If the subunit–subunit binding is weak enough, subunit 2 will, over time, detach and move ahead of subunit 2. The concentration of buffer ions will modulate the strength of the subunit–resin ionic interaction. (b) A cylindrical tube or column is packed with the buffer-surrounded resin, the sample (in buffer) layered on top, and more buffer flowed through.

6.1.2.1 Affinity chromatography

Ion-exchange chromatography is commonly used when a sample is relatively pure, consisting of several soluble components, all of which may be of interest. Because most proteins will have some amino acid side chains that interact with the resin charge group, the various proteins can be separated and collected as they pass through the chromatography column at different rates. If only one component in a crude sample is of interest, *affinity chromatography*, based on an interaction of that component with a more specific resin group, is used. Figure 6.5 diagrams the idea. A mixture of biomolecules passes down the column of resin. Those components that "fit" into the specially chosen affinity probe (ligand) will be delayed or even stopped. Those components that don't fit will pass undelayed. The "fit" is, of course, physical–chemical, with a specific chemical group attached to the end of the spacer, which must also be long enough to allow the target to get close enough to the resin bead. The probe/ligand can be as specific as an *antibody* to one specific molecular part of one specific virus or a more general probe that binds to a type of molecular group. A few of these probe groups are listed in Table 6.1. The dissociation constant of the ligand–target system should generally be between 10^{-4} and 10^{-14} M during the binding process:

$$K_D = \frac{[L][T]}{[LT]} \cong 10^{-4}\,\text{M to } 10^{-14}\,\text{M} \tag{6.2}$$

If the total concentration of ligand is 10^{-2} to 10^{-4} M, then K_D should be 10^{-4} to 10^{-6} M (or smaller) to bind a sufficient fraction of the target to the column. Note this dissociation constant has units of concentration. Because chemists make most frequent use of dissociation constants, we normally use units of molar, rather than molecules/m^3. $[L]$ is the concentration of free ligand (without target bound), $[T]$ the concentration of free target, and $[LT]$ the concentration of bound ligand–target. Think of this equation pictorially in two ways. First, at steady state, there is a constant, small ratio, $[T]/[LT]$: only a few target molecules are freely diffusing in the solution around the column matrix resin at any given time. Second, this steady state is set up by a constant on/off process of the molecules. If t_{on} and t_{off} are the times the target stays on and off the ligand, then

$$\frac{[T]}{[LT]} = \frac{t_{\text{off}}}{t_{\text{on}}} \tag{6.3}$$

Clearly, t_{off} is normally much smaller than t_{on}.

The ligand is the molecular group attached to the column matrix, which is an almost-solid, wet mass. Nevertheless, the concentration $[L]$ is a legitimate parameter as long as the ligand is accessible by random motions of the target molecule. If some fraction of the ligand is less accessible or inaccessible because of matrix obstructions, the treatment above will need correction. Clearly, during addition of the crude sample to the chromatography column, K_D must be small so that very little $[T]$ remains free. This equation can be approximated as[3]

We ignore a few technical complexities here, such as what happens if $[T] > [L]$, i.e., when there are more target molecules than ligand sites per unit volume. The extra target molecules just move farther down the column to where there are empty sites. We must not allow more total target molecules than ligands on the entire column, of course, but this happens only in a poorly planned experiment. Under most circumstances, the total number of target molecules $N_{\text{target}} \ll N_{\text{ligand}}$.

$$\frac{\text{Bound target}}{\text{Total target}} \cong \frac{L_0}{L_0 + K_D} \tag{6.4}$$

so if $L_0 = 10^{-2}$ M and $K_D = 10^{-4}$ M,

$$\frac{\text{Bound target}}{\text{Total target}} = \frac{0.01\,\text{M}}{0.01\,\text{M} + 0.0001\,\text{M}} = 0.99$$

The preceding conditions ensure that virtually all the target T in the crude sample will be immobilized on the column. Now, how do you get the target off? We must somehow reduce the value of K_D, or the target will stay there forever. A rule of thumb given by affinity-media manufacturers is that

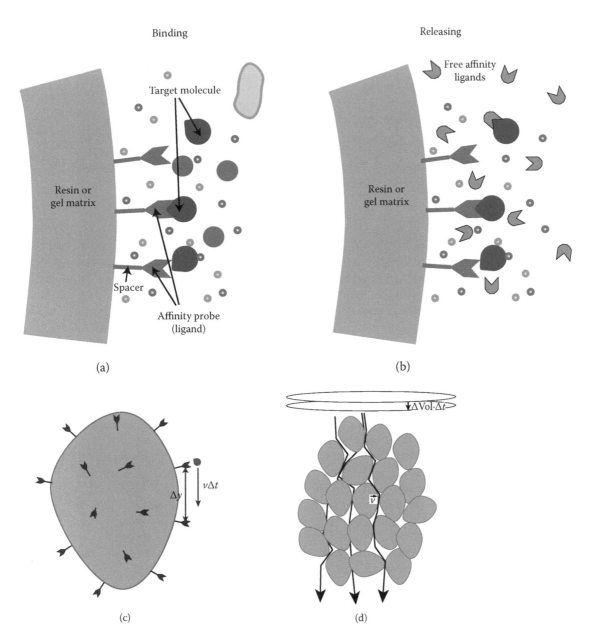

FIGURE 6.5
Affinity chromatography. (a) The crude sample, containing the *target* molecule among many others, is applied to a chromatography gel containing a *ligand* attached to the gel matrix by a molecular spacer. The ligand strongly binds to the target, immobilizing the target on the column. (b) To release the target from the column, a competing, free version of the ligand may be added to the column to bind those target molecules that have temporarily detached from their sites. Alternatively, ions or other agents that reduce the strength of the ligand–target binding can be added. As long as there is a constant flow of buffer through the column media, the target will eventually get washed out. (c) On a microscopic scale, an unbound target must travel a distance ~Δy to get to the next ligand site. (d) Fluid, as well as molecules, travel through a labyrinth of reduced volume down the column.

K_D should be increased to 10^{-1} to 10^{-2} M or more. Often a change in the solution ion concentration (e.g., [NaCl]), temperature, or pH will reduce the value of K_D, especially if Coulomb attraction is a major component of the binding.

    If we manage a set of recommended conditions, $K_D = 0.1$ M and $L_0 = 10^{-2}$ M, then $\frac{\text{Bound target}}{\text{Total target}} = \frac{0.01\text{M}}{0.01\text{M} + 0.1\text{M}} = 0.091$, or about 91% of the target is, on average, free. It would seem there would be no trouble with washing out most of the target. What if we managed only $K_D = 10^{-2}$ M

TABLE 6.1 Affinity Chromatography Probe Ligands

Probe Ligand	Target Molecule
Lectin proteins (e.g., concanavalin A)	Specific carbohydrate (sugar) molecules
Sugar	Lectin proteins
Metal ions (cobalt, nickel, copper, zinc, iron)	Proteins with exterior His amino acids
Amino acids	Any molecule (protein, RNA) that binds the amino acid
Avidin (a protein)	Any molecule with biotin (a protein) attached
Nucleic acid (ss)	Complementary strand, DNA-binding proteins
Any enzyme	The enzyme's substrate, cofactor molecules
Heparin (anticoagulant)	Coagulation factors

and $L_0 = 10^{-2}$ M. Then $\frac{\text{Bound target}}{\text{Total target}} = \frac{0.01\text{M}}{0.01\text{M} + 0.01\text{M}} = 0.50$, or 50% of the target is now free. Must we now be satisfied with collecting only half the target? To answer this, consider Equation 6.3 again: $\frac{\text{Free target}}{\text{Bound target}} = 0.50 = \frac{t_{\text{off}}}{t_{\text{on}}}$. Half of the time the target is free of the ligand. What does it do when it gets free? If buffer is flowing through the column, it flows down the column a bit, then gets bound again; then flows, gets bound, flows…. Eventually, every target flows out the bottom of the column! It's just a matter of how patient you are and how much volume of column effluent (the solution coming out the bottom) you are willing to deal with. Five days and 4000 L are usually unacceptable, but 5 h and 5 L may be manageable.

Unfortunately, there are two more complications, ones that would have been obvious had we started this computation from a first-year physics point of view. Suppose we start with one target bound to one gel-bound ligand. The idea of equilibrium between bound and free target implies that a bound target will periodically come loose, then will bind again later. What happens when the target is loose? As we said, the target would flow down the column, but how far? If the "off" time is only 10^{-9} s, the target cannot move far: if the buffer is flowing by the gel-resin particles at a brisk 10 cm/s, the target can move, at most $\Delta x = (0.1 \text{ m/s}) (10^{-9} \text{ s}) = 0.1$ nm. In all likelihood, the target would bind to the same ligand site and not move at all, on average. See Figure 6.5c. The second complication we have swept under the rug by dealing with "concentrations," as if this system were a beaker of water. The system is, rather, a tube mostly full of either macroscopic, spherical resin particles coated with ligands, or a cross-linked gel matrix, with ligands attached at various sites. Surrounding the gel particles is the buffer solution, whose volume is often comparable to or less than the (solid) matrix volume (Figure 6.5d). The first point is that any "free" molecule can only diffuse freely where the buffer is, not where gel particle is. The second part is the issue of how far it is between ligand sites on the gel. If $L_0 = 10^{-2}$ M $= 6 \times 10^{18}$ molecules/m³, the *average* distance between sites is

> Ignoring Brownian motion is a bad thing to do, but we must wait until later chapters to do this calculation correctly.

$$\langle \Delta y \rangle = \left(\frac{1}{L_0} \right)^{1/3} \tag{6.5}$$

Under our conditions this distance is about 550 nm. See Figure 6.5c again. What should we then do? (1) Measure the buffer volume flow rate through the column. (2) Convert this to an average linear speed of the buffer, taking into account the fractional volume the buffer occupies in the column. (3) Calculate how far a target molecule will move during its "off" time. (4) Compare distance 3 to the average distance to the nearest different ligand binding site; or maybe just to the distance the target needs to move to get out of the range of the force of its initial binding-site ligand. What, however, do equilibrium expressions like $\frac{[T]}{[LT]} = \frac{t_{\text{off}}}{t_{\text{on}}}$ mean in terms of microscopic distances moved and fluid flow?

How we answer the last question depends on how important the process is. If the issue is of fundamental importance to the mechanisms of life itself, we learners of biophysics should try to perform the theoretical analysis correctly. The question is not fundamentally important here. We should likewise solve the problem as correctly as possible if we are employed by a biotech firm trying to develop an improved, cheaper chromatography. In the end, for laboratory use of standard chromatography, guidelines from the manufacture for concentrations, flow rates, ion concentrations, pH, etc. are more reliable than most intricate computations. So just go to the lab and do it. When we come to describing movement of molecules and particles central to the cellular processes, intricate computations may become necessary. However, the geometry will not be that of a chromatography column, so let's breathe a sigh of relief and postpone the issue.

Directed evolution of proteins and high-throughput micro-chromatography. Demands on biological sample preparation in the last ten years have centered on high-throughput—preparation of hundreds to thousands of samples in a day or less—and small sample size—sample volumes of microliters. Though this does not alter the basic principles of separation chromatography, the devices and their control change considerably. Before the 1990s, the chromatography column shown in Figure 6.4b would typically have been 1 to 5 cm in diameter and 10 to 50 cm in length, would have purified one sample of volume 1 to 10 mL, and would have required 10 minutes to a few hours. Many recent approaches to the understanding protein structure and function require the preparation and (sometimes) purification of large numbers of variants of a protein, differing in one or many amino acids at known sites in the polypeptide chain.

Protein engineering before the advent of directed evolution and related approaches centered on the isolation and purification of the native protein, which was subjected to painstaking and expensive structural analyses based on x-ray crystallography, nuclear magnetic resonance, molecular dynamics computer simulations, etc. The analysis usually resulted in the identification of important parts of the structure, such as active or substrate-binding sites. Researchers then proposed what specific amino acids were primarily responsible for the structure and activity at those sites and replaced, one by one, those identified amino acids using genetic engineering techniques. In many cases, a few extra hours made little difference to the research effort, but even in the 1990s, such modified proteins could number in the hundreds, as techniques like CRISPR-cas9 were developed. Running 200 columns like in Figure 6.4b, each requiring a setup time of perhaps an hour and run time of 2 hours would be difficult in normal research labs.

A better approach to the production, isolation and purification of a large number of specific proteins is to run hundreds or thousands of Figure 6.4b columns in parallel, operated by a robotic system. Figure 6.6 shows a diagram of such a multi-column system, based on a 96-well rack that has been a standard for many years. (Racks with 384 wells are also a possibility.) Controlled by a robotic arm, a rack of 96 syringes obtains 96 samples from another rack, in which starting solutions have been prepared. The syringes are positioned over the chromatography cartridges, sample solutions applied, and after 0.5 to 5 minutes, the filtered sample appears in the well. Flow rates must be controlled and timed carefully to get desired sample purity. Systems on the market in 2017 can run trays of 9–12 racks, allowing 1000 samples to be run in parallel. As the reader may guess, issues of programming, fluidics, accounting and labeling, and cleanliness are major challenges to such large-scale chromatography.

6.1.2.2 Competitive binding

Another technique to remove target from the column is not to reduce K_D at all but rather to outcompete it. Suppose the ligand is the His amino acid and the target a His-binding protein. If a large amount of free His is added to the column, the protein molecules that happen to be free of the column at any given time will bind to one of the large number of added free His molecules. The Hisbinding protein is now bound to a free His and will no longer bind to the His ligand on the column matrix—the free His/protein complex will wash off the column. However, if K_D is too small, it is difficult to increase its value by large factors without drastic changes in conditions. The avidin–biotin system, with $K_D \sim 10^{-15}$ M, is one

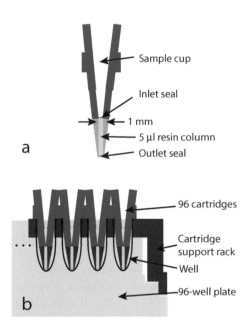

a

b

FIGURE 6.6
(a) High-throughput chromatography systems may contain as little as 5 µl of resin in each column, which has dimensions of millimeters. (b) Standard racks of 96 cartridges are robotically mounted in a 96-well plate. Liquid samples are applied to cartridges by a 96-syringe dispenser, with filtered materials collected in 96 wells. Samples run through cartridges in 0.5 to 5 minutes. Current systems can manage over 6,000 samples. See, for example, www.agilent.com/en/products, search high-throughput chromatography.

of the most commonly employed affinity probes, but the bound complex is difficult to pry apart. However, the binding strength, the well-developed chemistry, and the availability of the products makes this system attractive for answering "yes/no" questions: Is the target molecule present in a sample or not? An auxiliary reason for the common use of streptavidin/biotin is the ability to attach fluorescent molecules to biotin or streptavidin for easy optical detection of a target molecule.

Competitive binding is a process of fundamental importance to life. Competition for binding sites constantly occurs when a variety of amino acids are potential ligands for a binding site on a tRNA. The more negative binding energy of the correct amino acid allows production of correct polypeptide sequences to proceed. Carbon monoxide competes very effectively with oxygen for binding sites on hemoglobin: its free energy of binding is more negative than that of O_2. A small percentage (about 0.1% or more) of CO in the air breathed by a human will result in death in about 2 h. Only the replacement of the bad air by pure O_2, at 1000 times the CO percentage, can gradually replace the CO bound to Hb and other critical heme proteins. The majority of cancer therapy drugs employ competitive binding. (Skim through the National Cancer Institute's Drug Dictionary at https://www.cancer.gov/publications/dictionaries/cancer-drug?expand=A.) We encounter competitive binding again in Chapter 15.

A final issue of column chromatography, indicated in the second-previous paragraph, is how to determine if any target molecule is bound to the column. An extremely versatile and sensitive technique is fluorescence. We discuss this more in Chapter 9, but for now we note that even small amounts of protein or DNA can be detected if an efficient fluorescent label like rhodamine, Alexa Red, semiconductor nanoparticles, or another of the hundreds of commercially available fluorescent labels are also attached to the target. There are many suppliers of such fluorescent labels, but the Invitrogen Web site provides a comprehensive description at www.thermofisher.com/us/en/home/references/molecular-probes-the-handbook.html.[4] Some of these dye labels are so highly colored that they can be seen by the unaided eye on a chromatography column. As our goal is usually to quantify, visibility by the unaided eye is a mere convenience. The real significance of these strongly fluorescing dyes is the ability it has recently given us to *detect single molecules* of proteins or nucleic acids. Once single-molecule detection is perfected and made a common laboratory procedure, the need for chromatography or any other purification technique will be greatly diminished. The primary reasons to purify a protein or DNA molecule are (1) to increase their number to the level of detectability—not needed if one molecule can be detected—and (2) to increase the signal-to-noise ratio of the desired molecule to other molecules—again unnecessary if the other molecules are not fluorescently labeled. The promise is routine detection of single-molecule events as they occur in living cells… perhaps 5–10 years from now.

> The avidin–biotin system or streptavidin–biotin system.

6.2 SMALL AGGREGATES

6.2.1 Protein Aggregates

Many of the proteins and enzymes first discovered and isolated were found in *monomeric* form—a single polypeptide chain in a structure with some cellular activity. One reason for this is that the methods used to isolate components from cells were rather drastic and many natural aggregates of proteins were destroyed. Hemoglobin, one of the most abundant proteins in the human body, was an early exception. Isolation methods also sometimes created aggregates where none may have been. Alcohol added to cells, for example, can precipitate either protein or DNA or both. Such aggregates were usually regarded as unnatural, but reversible aggregate precipitation could be used effectively to separate protein from DNA. On the other hand, many large aggregates (e.g., chloroplasts) were visible to optical microscopes even in the late nineteenth century, so the idea of aggregates naturally occurring inside cells was not foreign. As isolation methods became more gentle and sophisticated, and especially, as electron microscopy came into use in the 1950s for biological imaging, aggregates took their place as important, intrinsic components of cells.

EXERCISE 6.1

Sizing chromatography is a type of chromatography that separates molecules according to molecular size or molecular mass. Assume that a chromatography material consists of 200-μm-diameter beads of a resin permeated with small (submicron diameter) tunnels. Sephadex G-50 is one such material, suitable for the separation of molecules of mass >30 kDa from molecules <1.5 kDa. Estimate the mean diameter of the tunnels in this medium.

Answer

A 30-kDa molecule must be too large to pass into the tunnels and be delayed. The diameter of such a molecule must then be approximately the tunnel diameter. $m = \rho(\frac{4}{3}\pi r^3)$, or $r = (\frac{3m}{4\pi\rho})^{1/3}$, where m is mass and ρ is density. Putting a density of 1.25 g/mL, we get $r = \left(\frac{3(30,000)\cdot(1.67\times10^{-27}\,\text{kg})}{4\pi(1250\text{kg}/\text{m}^3)}\right)^{1/3} = 2.1$ nm or a diameter of about 4.2 nm. Molecules between 1.5 and 30 kDa get delayed but apparently not much.

Aggregates of globular proteins are usually characterized by

1. The number of subunits (polypeptide chains) in the aggregate, n.
2. The number of *different* subunits in the aggregate.
3. The arrangement and symmetry of subunits in the aggregate.
4. Interactions between the subunits, both physical (forces) and chemical (reaction rate feedback).

The number of units in a globular protein aggregate is usually small, perhaps 2–16, and usually come in even numbers. Aggregates are often symmetric. The units may all be identical or there may be several distinct units.

We introduce only two small protein aggregates in this chapter. Neither can be regarded as a prototype for multipolypeptide proteins in biosystems; there are simply too many enzymes, too many different reactions to carry out, and too many different environments these enzymes must operate in. The first example, hemoglobin, is likely the most studied protein aggregate in history. We briefly introduce this protein and analyze its reactivity in Chapter 15.

6.2.2 Hemoglobin and Light Absorption

Hemoglobin (Hb) is a rather simple aggregate of four polypeptide chains. Figure 6.7 shows hemoglobin with O_2 molecules bound to the fours heme sites. The four hemes are contained in two types of subunits, α and β each of molecular mass about 17 kDa. Each subunit is similar to the myoglobin (Mb) molecule (Figure 5.1). One of the primary purposes of the two sets of subunits is to permit *cooperative* binding of four oxygen molecules: successive O_2 molecules bind more tightly than the previous. This cooperativity of activity is made possible by structural changes that occur after each O_2 is bound. An animation of these changes can be viewed at https://en.wikipedia.org/wiki/File:Hemoglobin_t-r_state_ani.gif, but the connection of such cartoons to biological oxygen binding demands a quite rigorous examination, which we sketch in Chapter 15.

The monomers of Hb can be separated from each other by treating with a chemical such as paramercurobenzoate, which loosens the bonds between the four subunits enough

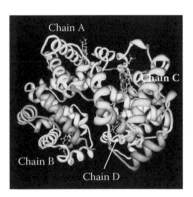

FIGURE 6.7
Crystal structure of human hemoglobin with oxygen bound at all four hemes. The cooperative binding of oxygen by hemoglobin results from interactions between the four chains, A, B, C, and D. Chains A and C are identical alpha subunits, B and D, beta subunits. Four heme groups can be identified. Chain C is oriented almost the same way as myoglobin (Mb) in Figure 5.1. (PDB 1GZX, by Paoli, M. et al., *J. Mol. Biol.*, 256, 775, 1996, displayed using Protein Workshop software.) (From Jmol: An open-source Java viewer for chemical structures in 3D. http://www.jmol.org/; Protein Workshop: Moreland, J.L. et al., *BMC Bioinformatics* 6, 21, 2005; NGL Viewer: Rose, A.S. et al., *Bioinformatics*, 2018.)

so that they will separate on an ion-exchange chromatography column. Each subunit can then be studied separately to determine detailed differences between the isolated subunit and the subunit as part of the Hb tetramer.[5]

Hemoglobin is the most recognizable, as well as the most well-studied, protein in history. Hb is responsible for the red color of blood, which occurs in countless novels and films (e.g., *The Red Badge of Courage*, by Stephen Crane, *Harry Potter and the Goblet of Fire*, by J. K. Rowling; see the blog by Jim Emerson, The color of blood: a study in scarlet[6]) and which is part of every human being's weekly, or at least monthly, experience. The bright color of hemoglobin, as well as the more subtle color changes—"blue" venous blood, "red" arterial blood—is indicative of an important quantum property, the efficient absorption of certain colors of light by the heme group, and the modification of heme's quantum properties through the binding of oxygen. Chapters 8 and 9 will deal with the formal issues of quantum theory and light absorption/emission in biophysical systems, but because of the dominance of optical methods in the biosciences, we make a first pass here, focusing on hemoglobin.

Two quantum properties have been critical to the scientific study of hemoglobin. The first is the high efficiency of light absorption, and the second is the change of absorption wavelength upon oxygen binding. The efficiency of absorption, the more difficult property to calculate from quantum theory, is usually quantified by the absorption or extinction coefficient ε and the absorbance A using the expressions:

> Values of ε for Phe, Tyr, and Trp amino acids are 200–5000 M^{-1} cm^{-1}

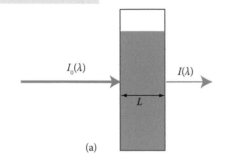

(a)

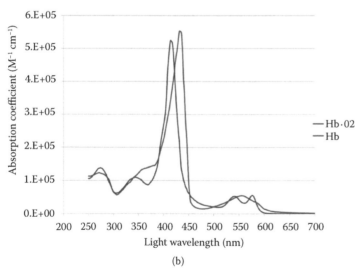

(b)

$$I = I_0 10^{-A}$$
$$A = \varepsilon c L \tag{6.6}$$

where L is the (optical) path length of the light beam in the sample and c is the concentration (Figure 6.8a). In the biosciences the power of 10 exponential, concentration units of molar, and path length in centimeters are traditional, which implies that the absorption coefficient ε has units of M^{-1} cm^{-1}. A brief manipulation of units shows that ε has SI units of m^2—it is a cross-sectional area—but those units are reserved for a base-e exponential and the symbol is changed from ε to σ. A graph of the absorption coefficient as a function of light wavelength (Figure 6.8b) shows the change in the shape of this graph, the absorption spectrum, when O$_2$ binds to Hb. Note three important features of this graph: (1) the values of ε are large compared to those of other ringlike organic molecules, especially in the 400-nm region; (2) the change of spectrum upon O$_2$ binding is significant; and (3) the wavelength regions of absorption are distinct. The 400-nm region is referred to as the "Soret," and the 500–600 nm region the "α–β." We will not dwell on this change here, but a variety of wavelengths can be used to measure quantitatively the ratio of oxygenated to non-oxygenated Hb.

FIGURE 6.8

Measuring the absorption spectrum of hemoglobin. (a) A collimated light beam of intensity I_0 and wavelength λ is directed into a sample in a rectangular cuvet of path length L. After emerging, the beam has a smaller intensity, I. A is absorbance; ε is the absorption coefficient. (b) The absorption spectra of oxy- and deoxy-Hb are distinctly different. Absorption graph based on values compiled by Scott Prahl using data from W. B. Gratzer, Med. Res. Council Labs, London, and N. Kollias, Harvard Medical School, Boston. See omlc.org/spectra/hemoglobin/.

6.2.3 Aggregated Membrane Channel Protein

A *voltage-gated potassium (K+) ion channel* in membranes is an example of a protein aggregate with three types of subunits. Figure 6.9 shows the dodecameric (12-subunit) membrane potassium channel KcsA-Fab protein aggregate.

Three distinct polypeptides each occur 4 times in this aggregate. The α-helical, intramembrane part of the aggregate can be identified at the top in Figure 6.9b. Potassium ions diffuse across the cell membrane in a single file through the narrow selectivity filter of potassium channels. The structure of the channel shows the chemical structure of the selectivity filter is due to four K^+ binding sites. There are 40 known human voltage-gated potassium channel alpha (membrane spanning) proteins.

While many types of globular protein aggregates exist, they generally form aggregates of a relatively small number of subunits, with no macroscopic structure like that found in fibrous proteins. Some major exceptions involve aggregates of protein and RNA. We introduce two of them now and examine RNA structure afterward.

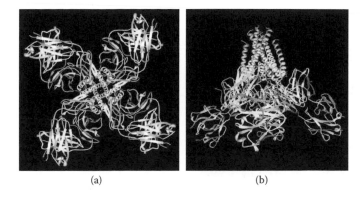

(a)   (b)

FIGURE 6.9

Dodecameric (12-subunit) membrane potassium channel KcsA-Fab protein aggregate with bound Tl^+: (a) along axis of ion flow (perpendicular to membrane); (b) side view. Three distinct polypeptides each occur four times in the aggregate. The α-helical, intramembrane part of the aggregate can be identified at the top in *b*. Potassium ions diffuse across the cell membrane in a single file through the narrow selectivity filter of potassium channels. The structure of the channel showed the chemical structure of the selectivity filter was due to four K^+ binding sites. (PDB 1R3J, From Zhou, Y. and MacKinnon, R., *J. Mol. Biol.*, 333, 965–975, 2003, displayed using Protein Workshop software.) (From Jmol: An open-source Java viewer for chemical structures in 3D. http://www.jmol.org/; Protein Workshop: Moreland, J.L. et al., *BMC Bioinformatics* 6, 21, 2005; NGL Viewer: Rose, A.S. et al., *Bioinformatics*, 2018.)

EXERCISE 6.2

What gives hemoglobin its strong red color? Examine Figure 6.8b.

1. To a human (not color-blind), what are the colors of light corresponding to the wavelengths of the peaks of the oxy-H spectrum?
2. What color of light is absorbed most strongly by oxy-Hb?
3. What determines the apparent color of hemoglobin? Assume you are looking through a vial of a solution of oxy-Hb held up to a white light.
4. What determines the apparent color of blood coming from a cut in your finger?

Answers

1. 420 nm: blue, bordering on violet; 535 nm: green; 580 nm: orange.
2. The blue/violet light is most strongly absorbed.
3. First, apparent color is a complex physical/psychological issue that can vary even among persons who are not color-blind. This is probably the reason for the wide variety of colored materials used to simulate "blood-red" in movies. However, moving to the physical issue (wavelengths and color), blue light will be strongly absorbed from the white light, with a bit of green and orange. The removal of the blue and some green will leave the transmitted light appearing red. The absorption of the 580-nm light, will remove a bit of orangish color, making the overall color more clearly red. (Note that 600–700 nm light is strongly red.) The transmission of significant amounts of 460–520 nm light (blue-green) means, however, that the color is not "pure" red.
4. Blood is not a clear solution, but hemoglobin is still the most colored component. The light you see has been scattered from the blood. Nevertheless, this scattered light comes from light that has penetrated partway through the red blood cells and then scattered outward. That scattered light is the light that has *not* been absorbed, similar to the case in part 3. So the color will again be red because of absorption by Hb.

FIGURE 6.10
A protein "vault," a *ribonucleoprotein* particle (1/3 of the structure). The function of this protein aggregate is unclear but may involve immunity. The highly symmetric, cage-like structure consists of a dimer of half vaults, with each half vault comprising 39 identical major vault protein chains—78 polypeptide chains. Each monomer folds into 12 domains: 9 structural repeat domains, a shoulder domain, a cap-helix domain, and a cap-ring domain. (NGL viewer, PDB ID 4V60 by Tanaka H. et al., *Science*, 323, 384–388, 2009. The structure of rat liver vault at 3.5 angstrom resolution. Science 323: 384–388.)

6.3 LARGE AGGREGATES

Many different large aggregates can be found in cells, as was the case with small aggregates. We survey only a few, as we return to the process of aggregation later in the text. We leave one of the most interesting and complex large aggregates, the cell nucleus and nucleosomes, to Chapter 17. As the last chapter of many one-semester textbooks gets postponed for "later," you might mark that chapter for further study if the mechanics of genetic information is of interest.

6.3.1 Large Protein Aggregates: Vaults[7]

Cells are filled with compartments that perform specific functions. Compartments such as *mitochondria* are large and enclose many different molecular machines. Other intracellular compartments are smaller, such as the membrane-enclosed transport *vesicles* that shuttle proteins and hydrophobic molecules inside the cell. Most of these compartments are surrounded by membranes. However, certain compartments are enclosed by a protein shell. In eukaryotic cells, *vaults* are an example of protein-enclosed compartments.

Vaults are composed of many copies of the major vault protein, which assemble to form a hollow football-shaped shell. Figure 6.10 shows a vault from rat liver cells. This vault contains 78 copies of the protein. The aggregate these proteins form is highly symmetric, which probably reflects their assembly process. In place in cells, the vaults have been observed to enclose several small RNA molecules, a protein that binds to RNA, and an enzyme that adds nucleotides to proteins. Human cells contain up to 10^5 vaults.

Vaults are a relatively new discovery and are not well understood. We do not yet know their function(s). They are found in many types of cells, but some organisms, such as fruit flies and yeast cells, don't have them at all. They are often found in the *cytoplasm*, where they are transported rapidly from site to site, but occasionally they are found in the nucleus. Because of this apparent mobility, vaults may be used for transport.

6.3.2 Ribosome

Aggregates bordering on macroscopic do exist and are critical to life. The *ribosome* is an aggregate of protein and RNA, with many other involved molecules. The ribosome is the central manufacturing complex for proteins. This manufacturing plant must consist of

1. A basic framework
2. A structure for handling for the codes—messenger RNA (mRNA)—for proteins to be synthesized
3. Input channels for raw materials (nucleotide and amino acid units, energy sources)
4. A structure for handling the output protein
5. Control units to start, stop, check, and repair

We can afford to deal with little of the ribosome's activity in the book, but right now we'll take a moment to peer at its amazing structure to whet our appetite for a brief study of molecular motors related to ribosome activity in Chapter 16. Figure 6.11 shows several views of parts of the 70S ribosome of the *Thermos thermophilus* microorganism.[8]

What does 70S mean? Any ribosome-related structure downloaded from the Protein Databank (PDB) will have such a number associated with it. The term 70S refers to a structure that moves with a sedimentation velocity of 70 svedbergs. Ribosomal fragments typically are named with this sedimentation parameter, because initially little was known about its detailed composition. Most researchers produced ribosome samples using preparative ultracentrifugation, which naturally resulted in a sedimentation velocity measurement, so the 70S notation was natural.

Protein components of the ribosomal fragments of *thermos thermophilus* are shown in a ribbon display, RNA in ball and stick. Figure 6.11a shows the 30S subunit of molecule 1, a 25-mer of molecular mass 0.882 MDa; (b) the 50S subunit of molecule 1, a 31-mer of molecular mass 1.36 MDa; (c) the 30S subunit of molecule 2, a 24-mer of molecular mass 0.882 MDa; (d) the 50S subunit of molecule 2, a 31-mer of molecular mass 1.36 MDa. The 30S, 50S, and 70S again refer to the size of the particle measured by sedimentation coefficient.

These ribosomal particles are mixed protein-RNA aggregates, with molecular mass typically more than 1,000,000 D and with several dozens of subunits. The fact that these structures were determined by X-ray crystallography means that the apparent glob of material is, in fact, a structure, most of whose individual atoms are in well-defined places, at least to a resolution of 0.1 nm or so. The structure is not amorphous or a randomly configured polymer aggregate. Students would be well advised to go to the PDB Web site to examine the structures on a finer scale, as well as to more clearly see the individual strands of protein and nucleic acid.

6.3.3 Saccharide Aggregates

6.3.3.1 Oligosaccharides

Oligosaccharides are commonly found attached to proteins following protein synthesis or to lipids in membranes. They determine factors critical to modern medicine, such as blood group classifications and tissue incompatibility/rejection in which cases they are typically bound in a cell membrane. They often appear as thick, fuzzy coats on the outside of cells and are called the glycocalyx. The human red blood cell (erythrocyte) has a glycocalyx up to about 140 nm thick. Erythrocytes, as well as other tissue cell types, have about 100 blood group factors. Two of these factors, the *ABO blood group* factor and the *rhesus (Rh) factor*, are of great clinical importance. Blood types are referred to as A, B, AB, or O. The ABO factors are glycoproteins that range in molecular mass up to 10^6 and are 85% oligosaccharide. The ABO and Rh glycoproteins trigger immune responses in individuals that genetically have one of the other cell types. The glycocalyx in Figure 6.12 is one of the thickest, about 500 nm, and is shown lining the inside of small blood vessels (endothelial cells). The link noted in the Figure 6.12 caption leads to a Web site of a physicist studying the mechanical properties of this thick coat using laser tweezers methods.

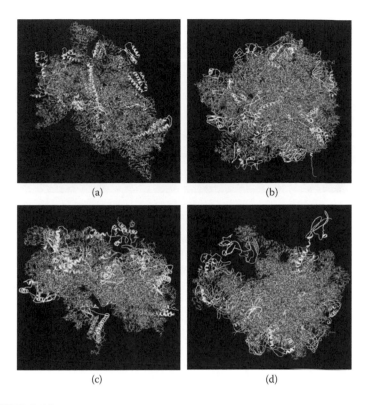

(a) (b)

(c) (d)

FIGURE 6.11

A ribosome—parts of the 70S ribosome of the *T. thermophilus* microorganism. Protein components are shown in a ribbon display, RNA in ball and stick. (a) PDB structure 2V46: 30S subunit of molecule 1, a 25-mer of molecular mass 0.882 MDa. (b) 2V47: 50S subunit of molecule 1, a 31-mer of molecular mass 1.36 MDa. (c) 2V48: 30S subunit of molecule 2, a 24-mer of molecular mass 0.882 MDa. (d) 2V49: 50S subunit of molecule 2, a 31-mer of molecular mass 1.36 MDa. The 30S, 50S, and 70S refer to the size of the particle measured by sedimentation means. (PDB IDs 2V46, 2V47, 2V48, 2V49 by Weixlbaumer, A. et al. *Nat. Struct. Mol. Biol.*, 14, 733–737, 2007, displayed using Protein Workshop software. See problem 6.12.) (From Jmol: An open-source Java viewer for chemical structures in 3D. http://www.jmol.org/; Protein Workshop: Moreland, J.L. et al., *BMC Bioinformatics* 6, 21, 2005; NGL Viewer: Rose, A.S. et al., *Bioinformatics*, 2018.)

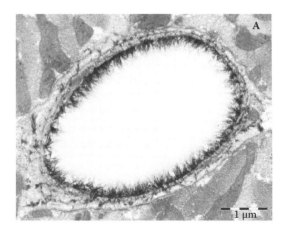

FIGURE 6.12

Glycocalyx structure formed by endothelial cells lining blood vessels is one of the thickest cell coating layers, approximately 500 nm, and is made up of proteins, lipids, and saccharides. (From Van den Berg, Vink, Spaan, Circ. Res. 2003, 92: 592–594; by permission.)

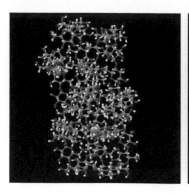 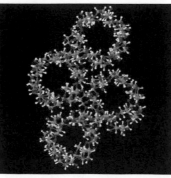

FIGURE 6.13

V-amylose is a linear, single-helical polysaccharide starch from plants. A- and B-amylose are double-helical polysaccharides. The structure shown is cyclo-amylose, containing 26 glucose residues; (left) side view; (right) end view. The cyclo-amylose is folded into two short left-handed V-amylose helices in antiparallel arrangement. Though a space-filling view shows the middle of the helices filled, there is room for water and other molecules. This cavity is where iodine molecules bind, producing a blue color in the grade-school iodine/starch demonstration. (PDP ID 1C58, by Gessler, K. et al. *Proc. Natl. Acad. Sci. USA*, 96, 4246–4251, 1999, displayed using Jmol.) (From Jmol: An open-source Java viewer for chemical structures in 3D. http://www.jmol.org/; Protein Workshop: Moreland, J.L. et al., *BMC Bioinformatics* 6, 21, 2005; NGL Viewer: Rose, A.S. et al., *Bioinformatics*, 2018.)

6.3.3.2 Polysaccharides

Polysaccharides usually function as either structural or (energy) storage polymers. Starch, a polymer of glucose, is used for storage in plants, being found in the form of the linear polymer, amylose, and the branched amylopectin. The amylose fraction of starch occurs in double-helical A- and B-amyloses and the single-helical V-amylose. The structure of V-amylose is shown in Figure 6.13. The structure shown is cyclo-amylose, containing 26 glucose residues. The cyclo-amylose is folded into two short left-handed V-amylose helices in antiparallel arrangement. Though a space-filling view shows the middle of the helices filled, there is room for water and other molecules. This cavity is where iodine molecules bind, producing a blue color in the grade-school iodine/starch demonstration.

In animals, the glucose polymer corresponding to plant starch is the more densely branched glycogen. Glycogen is made primarily by the liver and the muscles and is found in the form of granules in the *cytosol*. Glycogen forms an energy reserve that can be quickly mobilized to meet sudden energy needs for glucose, but glycogen's energy content per gram is less than that of triglycerides (fat). In muscle tissue, glycogen is found in lower concentration (1%–2% of the muscle mass), but the total amount exceeds that in the liver. The amount of glycogen stored in the body mostly depends on physical training, basal metabolic rate, and eating habits. Small amounts of glycogen are found in the kidneys, and even smaller amounts are found in certain glial cells in the brain and white blood cells. The uterus also stores glycogen during pregnancy to nourish the embryo. Only the glycogen stored in the liver can be made accessible to other organs. Glycogen, or rather the glycogen synthase kinase 3 (an enzyme), has been implicated in the onset of Alzheimer's disease.[9]

6.3.3.3 Structural polysaccharides

Cellulose and chitin are examples of structural polysaccharides based on glucose subunits. Figure 6.14 compares cellulose bonding structure to that of starch and glycogen, glucose polymers readily processed by human enzymes. The type of bonding in cellulose makes this material indigestible by humans because of lack of proper enzymes to cleave bonds. Cellulose is used in the cell walls of plants and other organisms and is probably the most abundant organic molecule on Earth. Paper, cardboard and cotton, linen, and other textiles are made from cellulose. Cellophane and rayon are identical to cellulose in chemical structure but differ in assembly and cross-linking. Table 6.2 summarizes some of the important industrial applications and materials based on cellulose.

Chitin has a structure similar to that of cellulose but has nitrogen containing side branches, increasing its strength. It is found in arthropod exoskeletons and in the cell walls of some fungi. It also has multiple uses, including surgical threads.

6.3.4 Composite Materials

Virtually all structural materials in biological systems are composites, i.e., are made of two or more materials blended or bonded together to achieve properties not allowed by one material alone. Figure 6.15 shows a manmade, bio-mimicking example of a microstructural arrangement of a cellulosic material designed for both structure and molecular chromatography. Superdex™ is a composite medium based on highly cross-linked porous agarose (polymer of galactose, a naturally occurring sugar) particles to which dextran (glucose polymer) has been covalently bonded.

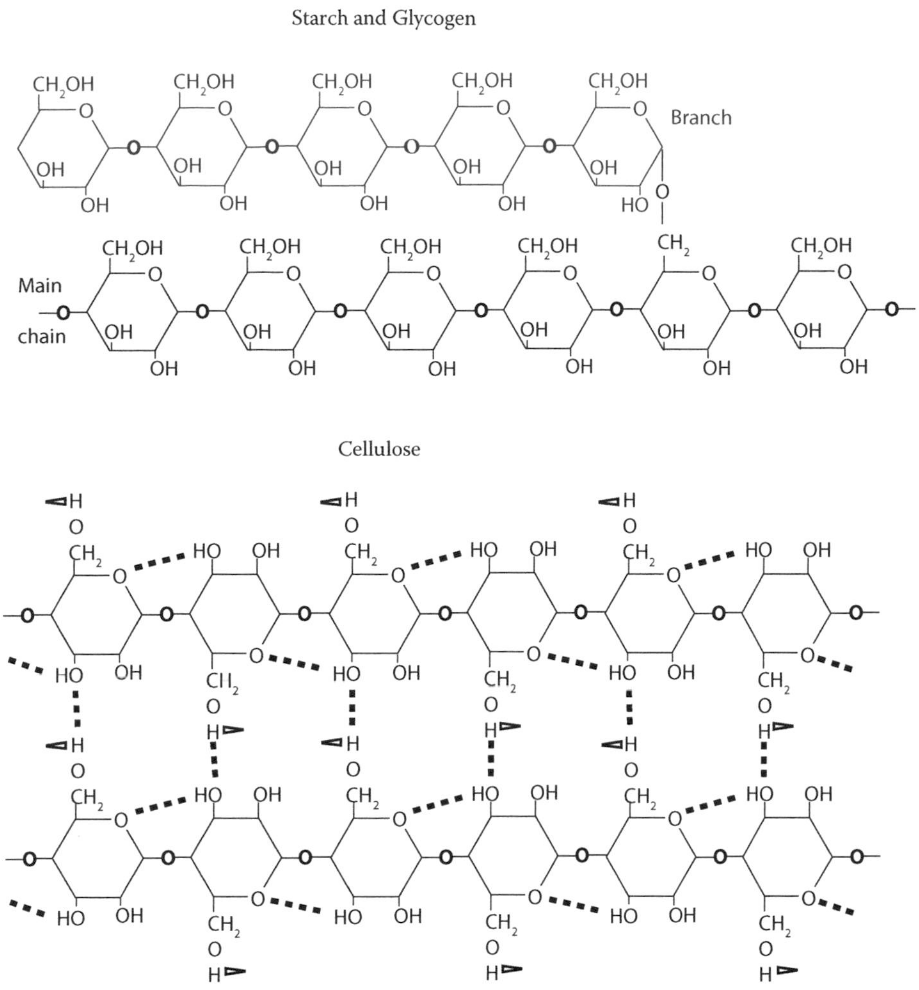

Starch and Glycogen

Cellulose

FIGURE 6.14
Diagram of structures of starches, glycogen, and cellulose, all made of glucose subunits. Starch has two forms, linear amylose (see Figure 6.13) and branched amylopectin. Branch points in the latter are about every 28 subunits. Glycogen has more frequent branch points, about every 10 subunits. Every other glucose subunit in cellulose in flipped, resulting in extensive hydrogen-bonded between strands. The small triangles next to H atoms indicates H bonding to layers above and below the plane, also. In plant cells, cellulose forms sheets of parallel fibers. Additional sheets are layered above and below at angles, "glued" to the other sheets by other saccharide polymers.

TABLE 6.2   Cellulose in the Human Economy

Material	Application
Lumber	Construction of houses, furniture
Paper, cardboard	Books, communication, packaging
Cotton, linen, rayon	Fabrics, clothing
Cellophane	Packaging, displays, protective layers, tape
Nitrocellulose	Celluloid, the first thermoplastic polymer (Hyatt Manufacturing Co., 1870) Celluloid film (movie, photographic film until ~1940) Smokeless gunpowder
Methyl cellulose, carboxymethyl cellulose	Water-soluble adhesives, chromatography materials
Microcrystalline cellulose	Fillers in drug tablets, processed food thickeners, stabilizers, noncaloric (indigestible) food components
Cellulose (proprietary)	Laboratory, biotech industrial chromatograph Sponges, building insulation

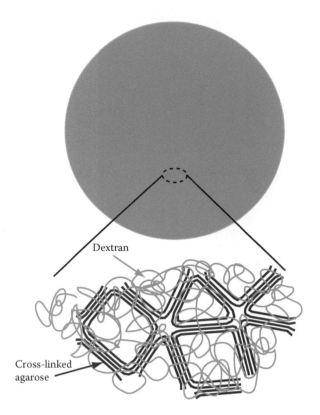

FIGURE 6.15
Superdex (GE Healthcare) composite polysaccharide material
for chromatography. This synthetic material uses two materials,
one highly cross-linked, for both strength and filtering properties. Dextran primarily determines the ability of the material
to pass macromolecules of different sizes and properties at
different rates. Cross-linked agarose adds mechanical and
chemical stability to the composite.

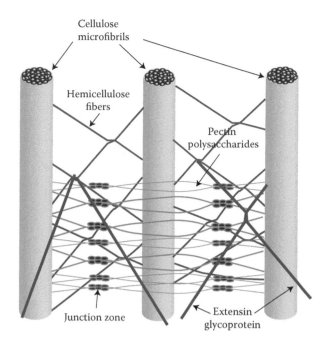

Cellulose
microfibrils

Hemicellulose
fibers

Pectin
polysaccharides

Junction zone

Extensin
glycoprotein

FIGURE 6.16
Diagram of a cell wall from a flowering plant. Strength and
elasticity are gained by use of several cross-linked polymers
of different elastic properties, with an adhesive cementing
structures together. Detailed orientations of components are
still not known. (Based on a more comprehensive diagram and
description from Carpita, N.C. and Gibeaut, D.M., *Plant J.*, 3,
1–30, 1993.)

The result is material with high physical and chemical stability,
owing mainly to the highly cross-linked agarose matrix, and gel
filtration properties determined mainly by the dextran chains. The
mechanical rigidity of Superdex allows high pressures and flow
rates, which are important features in industrial applications.

Figure 6.16 shows a diagram of a biological composite, a polysaccharide/protein, from the
cell wall of a flowering plant.[10] Strength and elasticity are gained by use of several cross-linked
polymers of different elastic properties, with an adhesive cementing structures together. Extensins
are highly abundant and usually have two repetitive peptide motifs, one hydrophilic and the other
hydrophobic. Extensins appear to be self-assembling and essential for cell-wall assembly and cell
expansion. Detailed orientations of components are not known, but extensive use has been made
of modern microspectroscopy, immunocarbohydrate microarray profiling, and other techniques to
measure cell wall mechanical dynamics.[11] Both structural strength/rigidity and elasticity are critical
to plant cells. Having no skeleton, plants must have rigid strength built into their cell walls. (Lipid
membranes provide little rigidity.) Plant cells must also be able to expand and contract rapidly in
response to changes in external conditions like rain after drought, so the elastic polymers are desirable in addition to the strengthening polymers.

Like plants, algae have cell walls that contain cellulose and a variety of glycoproteins. The
inclusion of additional polysaccharides in algal cells walls is used as a feature for classification
of algae. One of these, agar, is a polymer made up of subunits of the sugar galactose. Agar polysaccharides serve as the primary structural support for the algae's cell walls. An unusual polymer
called algaenan is especially chemically stable and facilitates preservation of algae in the fossil
record.[12]

Diatoms are a major group of *eukaryotic algae* and are a common type of phytoplankton.[13]
Most diatoms are unicellular, although they can exist as filamentous, ribbon- or fanlike colonies. Some of these forms are shown in Figure 6.17. Diatom cells are encased by a unique cell

wall, called a frustule, made of an inorganic-organic hybrid material, silica (hydrated silicon dioxide, orthosilicic acid, $SiO_2 \cdot 2H_2O$ or $Si(OH)_4$), to which various specific organic macromolecules (proteins, polysaccharides, long-chain polyamines) are attached. Rather than polymerizing organic sugars and amino acids, diatoms polymerize silicic acid monomers inside their cells and extrude the polymer to the exterior, where it coats the cell surface. These frustules show a wide diversity in form, some quite beautiful and ornate, but usually consist of two asymmetrical sides with a split between them, hence the group name. Archeologists believe they originated around the early Jurassic period. For those interested in nanotechnology, diatoms have been proposed as a system for the manufacture of micro- and nanodevices.[14,15]

The diatom cell wall seems to be an entirely different sort of biological structure, as it has an inorganic atom, silicon, as a major component. In a way the structure is different, and the interested reader should consult the review by Kröger and Poulson.[15] However, even humans use metal atoms for structures such as bone and teeth (calcium). Magnetotactic bacteria form Fe_3O_4 nanocrystals. At this point we must decide whether we want to pursue this issue, which will inexorably lead us to the areas of biomaterials and nanomaterials engineering. These are both extremely active research areas that some of you will want to pursue. However, in the context of an introduction to biosystems, this will be too much of an undertaking.

FIGURE 6.17
Assorted diatoms as seen through a microscope. These specimens were living between crystals of annual sea ice in McMurdo Sound, Antarctica. Image digitized from original 35-mm Ektachrome slide. These tiny phytoplankton are encased within a silicate cell wall. A copy of this National Oceanic and Atmospheric Administration image may be viewed at www.photolib. noaa.gov/htmls/corp2365.htm. (From corp2365, NOAA Corps Collection, 1983, photographer, Gordon T. Taylor, Stony Brook University, NSF Polar Programs.)

6.4 TWO-DIMENSIONAL AGGREGATES: MEMBRANES

The simplest two-dimensional aggregate of importance in cells is obviously the cell membrane, which is typically 5–7 nm thick. As Figures 1.7 and 6.12 suggest, the cell membrane can get rather complex because it usually consists of many types of lipid molecules, lipopolysaccharides, proteins/glycoproteins, and other miscellaneous molecules, sometimes with added layers of saccharides and/or a cell wall. Two-dimensional aggregates like membranes can be described by two approaches: (1) a molecular approach, where behaviors of the membrane are described in terms of molecular interactions and properties (with some thermodynamics added), and (2) a materials approach, where a sheet of membrane is assigned bulk elastic and permeability properties. Approach 1 better deals with particular structures that form in the membrane, like protein aggregates that form voltage-gated channels through which specific ions pass (controlled current flow) and structures responsible for cell signaling and recognition. Approach 2 can deal more readily with cell shape changes in response to applied forces, bulk diffusion through the membrane, and cell movement. Because elastic properties of real cell membranes depend on many structures besides the cell membrane (e.g., the filaments and microtubules forming the cytoskeletal network), it is challenging to connect an observed deformation of a real cell with properties of any particular part of the cell membrane and support structure. Artificial models for some intracellular membrane structures, liposomes and lipid vesicles, can be pure membrane. Deformations and energy costs can be more easily assigned to molecular characteristics. We will focus in this text, more on approach 1, with a brief foray into lipid energetics and motions. Unlike proteins, which can take on a wide variety of quite different structures that are difficult to predict *ab initio*, lipids that make up membranes have relatively few options. Because the number of structural possibilities is small, we are able to perform several crude computations related to structure and structural dynamics of membranes in general. Rather than postponing these general computations to a later chapter, we carry out a few of them in the following sections.

6.4.1 Two-Dimensional Membranes Result from One-Dimensional Properties of Lipid Molecules

The lipid molecules described in Chapter 5 have a hydrophilic (water-loving, polar) end and a hydrophobic (water-hating, nonpolar) end. These molecules will spontaneously form structures in aqueous media that maximize exposure of hydrophilic parts and minimize exposure of hydrophobic parts to water. One cannot find many types of structures that will accomplish this goal: polar head group surrounded by water, hydrophobic tail shielded from water. Two of the simplest structures are spherical structures called *micelles* and *vesicles*, shown in Figure 6.18.

Micelles, shown in Figure 6.18a, are normally formed by single-tailed lipids and consist of a single layer of the polar-nonpolar molecule. Such lipids are often referred to as detergents because our primary experience is with this application. Micelles can be a simple sphere of pure lipid, with tails oriented toward the center of the sphere. The single tail allows the close packing needed for this geometry. The center of the micelle can also be occupied by hydrophobic substances (Figure 6.18b), with almost no increase in free energy of the structure. (Of course, the free energy of the system would be *much* higher if the hydrophobic substance were exposed to the water.) The shape of such a micelle does not need to be spherical. Exactly how the oily substance gets from outside to inside the micelle is still an issue, however, as the micelle must at least temporarily disrupt its structure to allow molecules to transfer from an oil drop in the water to the micelle interior. From our experience with dishwashing, we suspect mechanical agitation is sufficient to transport oil droplets to the interior of micelles, at least when encouraged by a parent.

6.4.2 Lipid Vesicles, Liposomes, and Flat Bilayers

Lipid *vesicles* or *liposomes* (Figure 6.18c) are the more common structures found inside cells. Because of the bulk of their double tails, such lipids cannot normally form single-layer, highly curved micelles. Double-tail lipids will spontaneously form double-layered structures, with lipid tails contacting lipid tails for lowest free energy. The thickness of a pure lipid bilayer is about 5 nm; with protein and saccharide add-ons, 1 or 2 nm thicker. Vesicles are water-filled structures. The lowest free-energy form of a pure-lipid vesicle will be spherical, a shape that minimizes the curvature of the membrane. Flattening the sphere will increase the curvature and free energy. The flat *lipid bilayer* shown in Figure 6.18d would seem to be disfavored because of the high curvature and unfavorable packing, with resultant free-energy increase. However, if the flat part of the bilayer is extensive, the additional energy cost is relatively small. Such a structure would be at least metastable, especially if the edges of the bilayer could not contact an edge on the opposite side and fuse, forming a vesicle. In fact, we might expect that the energy required for the overall distortion of an existing flat bilayer into a spherical vesicle would be greater than the energy cost of the unfavorable edge packing. The answer can be found in observations of real cells, where extended, flat bilayers form permanent structures inside of cells.

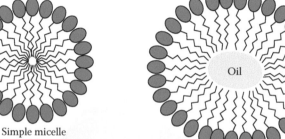

Simple micelle
(a)

Filled micelle
(b)

Aqueous solution

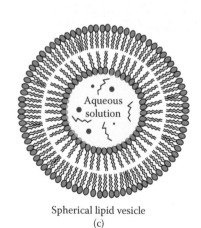

Spherical lipid vesicle
(c)

Aqueous solution

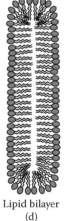

Lipid bilayer
(d)

FIGURE 6.18

Micelles and lipid structures. Micelles are normally formed by one-tailed lipids, usually referred to as detergents because they can efficiently surround small oil and dirt particles. Double-tailed lipids form vesicles and bilayers. (a) Empty spherical micelle; (b) ellipsoidal micelle surrounding oil droplet; (c) spherical lipid vesicle; (d) disk-like lipid bilayer (structure extends into and out of the page). Structural strain will occur at the edges.

Can we crudely model the relative free energies of a spherical vesicle and a flat bilayer with the same number of lipid molecules? What might we expect the primary energy differences between the structures in Figure 6.18c and d to be? Curvature is the obvious difference: the spherical vesicle has constant curvature everywhere, whereas the flat bilayer has zero curvature except at the ends, where it is high. Let us propose that the most stable conformation of a two-tailed lipid assemble is the flat double layer and that (1) crowding of the tails induced by membrane curvature and (2) a change of membrane area per lipid from some optimum induces a free-energy-raising strain.

6.4.2.1 Energy and membrane change of area

In a pure lipid bilayer there will be an optimum surface area per lipid molecule somewhat larger than πr^2, the average cross-sectional area of a lipid whose governing (lateral) radius is r. If we think of lipids connected to nearest neighbors by springs, the expansion (or contraction) of the membrane will entail an energy penalty. If the spring constants in the x and y directions are the same, the energy increase caused by a membrane area change, increase or decrease from the optimum (defined by the optimal area occupied by a lipid molecule), will be

$$\Delta G_{\text{area}} = \frac{\kappa_a}{2} \left(\frac{a - a_0}{a_0} \right)^2 \text{(membrane area)} = \frac{N\kappa_a}{2} \frac{(a - a_0)^2}{a_0} = \frac{\kappa_a}{2} \frac{(A - A_0)^2}{A_0} \tag{6.7}$$

where a_0 is the unstretched (uncompressed) area of a lipid molecule, N is the number of lipids in the membrane, and A and A_0 are total areas of the membrane.[16] ΔG_{area} is the energy change for the entire membrane. The area-stretch spring constant (modulus) κ_a has units of force/distance or energy/area. Values are typically in the range of 55–70 $k_B T/(\text{nm})^2$, or 0.2–0.3 N/m for bilayers. If we take a given membrane at equilibrium and increase its area by $\Delta A = A - A_0$, we can compute the energy change by the above expression.

Before computing values, we should think about the application of such an area stretch model to the two membrane structures we are considering. We are not simply distorting one membrane slightly and increasing or decreasing its area; the two structures are quite different. The spherical vesicle has a different outer and inner surface area; both cannot be the optimum, unless there are fewer lipid molecules on the inner layer than on the outer. In a purely theoretical sense, we should compute the free-energy change when we convert one structure into the lowest-energy form of the other. This means we could, for example, place fewer lipids on the inner layer than the outer layer in the vesicle, reducing the area-stretch energy penalty to zero. If we imagine bending the edges of the flat bilayer around onto itself and (somehow) fusing edges together, there is little we could do to prevent different numbers of lipid molecules from going onto inner and outer surfaces. So, we could calculate some area-stretch energy difference between vesicle (liposome) and flat bilayer, but its relevance would be questionable. We have hit a common barrier in simple physical calculations that purport to model mechanical distortions of materials: the expressions are valid only for relatively small changes (perturbations) from the lowest-energy structure.

6.4.2.2 Energy and membrane curvature

Can we escape the "perturbations-only" limitation on membrane area stretching for a membrane *curvature* model? Can we treat the two structures, vesicle and flat bilayer, as having small changes from a lowest-energy structure? Let us assume there is some optimal curvature of a given membrane and that this structure is flat. The vesicle has two sides that are oppositely curved from the optimum (flat). If the vesicle is large, the curvature magnitudes on the inside and outside are approximately equal. A complicating factor in the computation is that there may be differing numbers of lipids in the two layers. We will ignore this, group the two contributions to curvature energy together, and assume they are equal if the need arises.

When a membrane is bent or bowed, the energy increases. The free energy increase is given by

$$G_{\text{bend}} = \frac{K_b}{2} \int \left(\frac{1}{R_1(x, y)} + \frac{1}{R_2(x, y)} \right)^2 dx dy \tag{6.8}$$

where $R_1(x, y)$ and $R_2(x, y)$ are the radii of curvature in two directions at a point on the membrane and K_b is the bending rigidity. Note that K_b has units of energy, with values in the range 10–$20\ k_B T$ for lipid bilayers.[17] A spherical vesicle has constant curvature, and the integral results in

$$G_{\text{vesicle}} = 2K_b \left(\frac{1}{R}\right)^2 (4\pi R^2) = 8\pi K_b \tag{6.9}$$

This expression asserts that the spherical bending energy does not depend on the sphere radius, and values for energy are in the range 250–$500\ k_B T$. Independent from radius can make sense because a smaller sphere has a sharper bending but less curved area. Does this result give us any suggestions for the energy of the flat bilayer, with sharply curved, but single-layer edges? If we try to apply Equations 6.8 and 6.9 to a flat bilayer disk of radius R_{disk}, we could ignore the flat part of the bilayer, treat the edges as a half tube of radius about equal to the length of a lipid molecule, with one infinite radius of curvature and one equal to L_{lipid}, and divide K_b by two because the tube has only one layer of lipid, not two. Integration over the area of the tube would result in

$$G_{\text{bilayer disk}} = \frac{K_b}{4} \int \left(\frac{1}{R_1(x,y)}\right)^2 dx\, dy = \frac{K_b}{4}\left(\frac{1}{L_{\text{lipid}}}\right)^2 (2\pi L_{\text{lipid}})\pi R_{\text{disk}} = \frac{\pi^2}{2}\frac{R_{\text{disk}}}{L_{\text{lipid}}}K_b \tag{6.10}$$

(The length of the tube is half the circumference of the bilayer disk.) The length of a lipid molecule is about 2–3 nm. For a bilayer disk of radius 100 nm, the free energy would be about $40\ K_b \sim 400$–$800\ k_B T$. These values are somewhat higher than the spherical vesicle energy, but given the assumptions and approximations made, we should consider the energies comparable. The take-home messages are that we should look for other factors that may stabilize the high-strain edge of the disk (e.g., fewer lipids/nm^2, edge attachments to other structures, other molecules that reduce strain in the edge) and do a better calculation of its bending energy.

6.4.3 Multilamellar Membrane Structures

Figure 1.6e shows *multilamellar* lipid bilayer structures associated with chloroplasts in plant cells. This layered structure shows little if any space between bilayers. We started our discussion with the assertion that a membrane's primary purpose was to separate inside from outside to allow an interior space where processes could be shielded from the variables of the outside environment. In these stacked, layered structures, there is little or no interior aqueous volume for any processes to take place, so we must have stumbled upon a new application of membranes. We might logically suppose that this new application either utilizes a two-dimensional geometry or that molecules, signals, or energy can pass from one to adjacent layers. This is a topic for Chapter 10, but we are right on both accounts, though more so in the two-dimensional supposition. If we know that the multilamellar structure of Figure 1.6e is associated with photosynthesis, can we understand the layered geometry, at least crudely?

Photosynthesis must begin with the absorption of sunlight by plants. Green leaves, colored by chlorophyll molecules, absorb the light. What geometry will maximize light absorption of sunlight? We would naturally suppose that an extended flat surface is the best solution and that a flat membrane bilayer could be used. We would be right. However, why multiple layers of membrane? Why not just one bilayer? The answer lies in numbers and sizes and a bit of quantum mechanics. We saw that the chlorophyll molecule is comparable in diameter to a small protein. We also saw that chlorophyll is a prosthetic group (active site) of a protein. Proteins are generally several nanometers thick and only one or two molecules would fit across the thickness of a membrane. If we want multiple "backup" molecular absorbers for the first layer of chlorophyll exposed to the sun, we have to have multiple bilayers. Do we need backup absorbers, or can a single molecular layer of chlorophyll absorb most of the light? Chapter 9 will show us that one layer of virtually any molecule can absorb only a small fraction of the light incident upon it, so multiple layers of absorbers is a good design idea to "harvest" the maximum energy from sunlight. Thus multilamellar chloroplast structures can be expected from this particular application.

6.4.3.1 Energetics of membrane stacking

Is there a significant energy cost to stacking lipid bilayers rather than surrounding both sides with water? (Figure 6.19) Lipid head groups are polar or even charged. If all charges on the membrane surface were the same, then two identical membranes would repel each other upon close approach. On the other hand, a negatively charged membrane surface would attract a positively charged membrane surface, and the free energy of these two surfaces apposed would likely be lower than if each were surrounded by polar, but uncharged water molecules. (The charge–charge interaction is stronger than the charge–dipole interaction.) If both positive and negative charges were present in membranes, the interaction would depend on whether + and − charges on the two surfaces could come near each other. This would seem unlikely, unless the charges could move laterally in the membrane, which brings us to the subject of the next section: membrane molecules are in constant motion. If charges in the membrane can move, then they will move so as to reduce the free energy.

Before we get to the lipid mobility issue, however, we need to think briefly about the *entropic* part of the stacked membrane free energy. The Coulomb interactions discussed in the previous paragraph were the enthalpic, or potential energy, part of the free energy. Entropy reflects ordering of the system. Stacking two bilayers would appear to change little of the ordering in the membranes themselves, so we should look to the surrounding aqueous solution and ions for changes in their ordering. There are two types of ordering related to surfaces surrounded by a water, ions, and dissolved macromolecules. The first is the ordering of the water and ionic species in the solution. Figure 6.20 shows a molecular dynamics computer simulation of the structure of a membrane surrounded by water and ions. The still picture cannot show motion, but the simulation reveals a boundary layer of water next to the membrane. This layer is relatively immobile and hence is (likely) more ordered (lower entropy) than if that water were in the bulk solution. The entropic free energy of this effect is difficult to calculate, but we could reasonably expect there could be an entropy increase upon membrane stacking. The second part of the entropic free energy associated with the surfaces, the *osmotic* contribution, is relatively simple to calculate. We will learn how to apply such calculations in Chapters 13, 14, and 17 (especially), but a quick overview may be helpful here.

See "Membrane Builder" in the CHARMM-GUI Web site (http://www.charmm-gui.org/?doc=input/membrane).

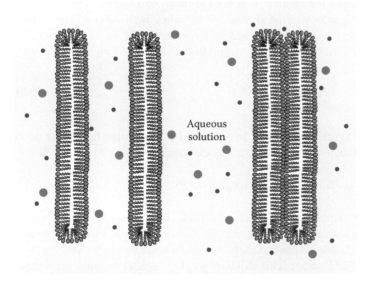

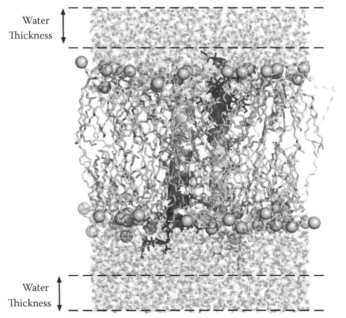

Water Thickness

Water Thickness

FIGURE 6.19
Stacks of bilayered membranes increase the enthalpy of the structure little, if any: polar head groups are exposed to (left) polar water molecules or (right) other polar head groups. Details depend on the structure and properties of the head groups. Depending on the sizes and concentrations of dissolved molecules (small spheres), the entropy of stacking the bilayers often reduces the entropic free energy, by reducing the volume of the exclusion layers around the membranes. See text.

Aqueous solution

FIGURE 6.20
Molecular dynamics simulation of the structure of a membrane with incorporated protein, surrounded by water and ions. (open-access diagram from Jo, S. et al., *PLoS One*, 2, e880, 2007.)

6.4.3.2 Exclusion zones and entropy

Consider the two bilayers in Figure 6.19. Suppose the larger, green molecules in the aqueous part of the solution have radii R. These molecules are free to occupy any position in the aqueous solution except for a layer of thickness R around the bilayers. (They cannot get closer than a distance R from the membrane surface.) Let's neglect the surface area at the ends of the bilayers (large bilayers) and assume a bilayer surface has area A on each side. When two bilayers are separated by a distance greater than $2R$, the green molecules are excluded from four bilayer surfaces, a total volume of

$$V_{\text{exc}}^{\text{unstacked}} = 4AR \qquad (6.11)$$

When the two membranes are stacked, large molecules are excluded from only two surfaces, a volume of

$$V_{\text{exc}}^{\text{stacked}} = 2AR \qquad (6.12)$$

When the bilayers stack, the excluded volume decreases: $\Delta V = -2AR$. When we see a volume change in the context of free energies (thermodynamics), what quantity, involving ΔV, should we think of that has units of energy? (Think a moment before reading on.) Such a term would be $P\Delta V$. (Remember the work done to compress a gas.) What pressure is associated with this exclusion volume change? What pressure are we talking about here? The "pressure" is that of the green molecules trying to force their way in between the two bilayers; just like the moving molecules in a gas provide the pressure against the side of their container. Let's go to what we know better: the ideal gas. The most important law we learned about ideal gases is the ideal gas law:

$$PV = Nk_BT \qquad (6.13)$$

where N is the number of molecules in the volume. It is not too much of a stretch to propose that an ideal solution, with dissolved molecules moving around and colliding with water and other molecules, as well as with the membranes, might be governed by a similar law. If so, the "pressure" of these dissolved molecules, called the *osmotic pressure*, would be

$$P_{\text{osm}} = \frac{N}{V}k_BT = ck_BT \qquad (6.14)$$

where c is the concentration of the (green) molecules, in molecules/m³. Equation 6.14 turns out to be exactly right. The free energy change when the bilayers stack is then

$$\Delta G = T\Delta S = P_{\text{osm}}\Delta V = (ck_BT)(-2AR) \qquad (6.15)$$

Is this significant? "Significant compared to what" is the question. In any free-energy, thermodynamic or statistical-mechanical calculation we *always* compare an energy to k_BT. This is convenient because k_BT is part of Equation 6.15. We conclude that the exclusion (osmotic) free energy will be significant in membrane stacking if $2ARc \gg 1$. This still isn't good enough. How much bigger than 1 does $2ARc$ need to be? If we were considering a small molecule, $2ARc$ about equal to 1 would be "big enough." Here, where we are dealing with an extended bilayer with many lipid molecules, we should propose an exclusion energy of k_BT per lipid molecule should be significant. How many lipid molecules are on a surface of area A? If the average radius of the lipid head group is R_{lip}, then each lipid occupies an area of about πR_{lip}^2, so the number of lipid molecules in the area A is $N_{\text{lip}} = \frac{A}{\pi R_{\text{lip}}^2}$. Our next proposal is that the exclusion free energy will be significant in bilayer stacking if

$$2AR_{\text{mol}}c_{\text{mol}} \approx \frac{A}{\pi R_{\text{lip}}^2} \text{ or } c_{\text{mol}} \approx \frac{1}{2\pi R_{\text{mol}}R_{\text{lip}}^2} \qquad (6.16)$$

where we have added the subscript on c_{mol} and R_{mol} to remind us they refer to the excluded molecule. We are done, for now, except to note that the units in Equation 6.16 appear correct and to acknowledge that enthalpic interactions may be even more significant.

6.4.4 Bilayers Are Dynamic

Though lipid molecules will form these bilayer structures on the average, "average" implies there will be deviations. Many molecular dynamics simulations have been done, showing the motions of atoms in lipid bilayers. A Web search on "membrane movie" will likely turn up a variety of research group Web sites with posted video clips of such simulations.[18] Though computer simulations do not prove that the simulated motions in fact take place on the time scale predicted, simulations rarely suggest motion where there is none. X-ray crystallography, as we have seen, can experimentally measure the extent of atomic motions in molecules. This has been done many times for protein molecules. Comparable crystallography measurements on extended membranes have not yet been done, but motions are indisputable because they are necessary for fundamental processes carried out by membranes. However, because membranes are intrinsically two dimensional (though 5 nm thick), we expect the motional properties of membranes also to be two dimensional, or at least much different in the $x-y$ plane than in the z direction normal to the membrane plane.

> Photobleaching is a photochemical reaction that renders the dye molecule to be non- or less fluorescent. The fluorescence will recover its initial intensity after other dye molecules diffuse laterally in the membrane from outside the illuminated spot.

6.4.4.1 Lateral motion in membranes

Other experimental methods besides crystallography directly demonstrate motions in membranes. One of these, amenable to Chapter 13 analysis, is called Fluorescence Recovery After Photobleaching (FRAP).[19] In this method, diagrammed in Figure 6.21, a spot on a membrane is illuminated with a low-intensity continuous-wave laser through a microscope. The membrane contains components (lipid, protein, and other molecules) that are either naturally fluorescent or (more commonly) are labeled with an attached fluorescent dye molecule. An intense laser pulse is then flashed on the membrane, photo-bleaching the dye molecules in the spot, greatly reducing the fluorescence intensity. The fluorescence will recover its initial intensity when components from outside the membrane spot diffuse laterally in the membrane into the spot. Such lateral diffusion in a membrane can be rapid and proves that lipids and other membrane-bound molecules move.

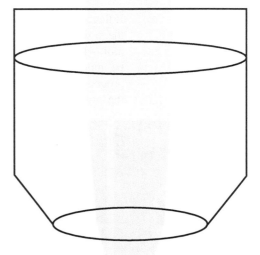

6.4.4.2 Vertical motion of lipids

Transverse motion of lipid molecules entails little energy penalty. Other molecules must move out of the way, but a different lateral position in the membrane entails no energy penalty in a uniform membrane. Vertical motion of the lipid, on the other hand, suffers an energy penalty (free energy increase) as soon as the hydrophobic tail moves into the water or the polar head moves into the membrane interior. The magnitude of this energy penalty is critical in determining what types of movements and particle transport membranes are capable of. For instance, if lipid molecules are added to or taken away from a membrane structure, certain types of lipid transport are not likely if a high-energy state is required, e.g., lipid hydrophobic tail in water. One of the secrets to biological activity is the reduction in energy of

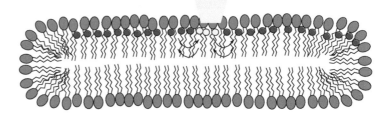

FIGURE 6.21
Fluorescence Recovery After Photobleaching (FRAP). Lateral motion in membranes is experimentally observed when unbleached dye-labeled lipids or membrane proteins (red dots) diffuse into a spot that has been bleached of active dye molecules by a laser pulse.

unstable or metastable states that are intermediate between initial and final state. Enzymatic catalysis, for example, is based on this energy reduction. A membrane protein or other molecule that must be added to an existing membrane presumably also has a hydrobic part that would form a high-energy state if it were simply pushed through water. A similar energy barrier would be faced by a lipid or membrane protein that needed, for some reason, to flip 180° in the membrane—a polar part of the molecule would have to pass through the hydrophobic core of the membrane.

Let's look at the magnitudes of these energies. Lipid molecules can, in fact, exist surrounded entirely by water. The molecular dynamics simulation that goes with Figure 6.20 implies that the lipid molecules move. We likewise calculated in Chapter 5 energies associated with hydrophobic amino acids moving from lipid core to water. The energy change was not large enough to completely rule out the presence of hydrophobic molecules in water. Molecular dynamics simulations and experimental measurements both agree that lipids do, in fact, flip across the membrane.[20] Simulations in the cited reference suggest that such a flip demands an extra energy of 15 kJ/mol, about $6k_BT$. The issue is now one of probability. $P_{flip} = \exp-(6k_BT/k_BT) \cong 2.5 \times 10^{-3}$ means that 1 in every 400 lipids will be in the process of flipping. (We take up the issue of how many times per second such flipping occurs shortly.) This extra energy does not include all of the rearrangement energy that would be required for a *stable*, partially flipped lipid state to occur: there is no such stable state. Nevertheless, the moderate energy barrier suggests that lipids can indeed flip under the right circumstances, especially if assisted by non-lipid structures in the membrane. We will shortly see such structures.

Moving a lipid from membrane to bulk water requires more energy. If the free energy difference between lipid molecule in-membrane and lipid molecule in-water is +80 kJ/mol (~32.4k_BT),[21] the ratio of the number in water to number in membrane would be

> We speak here of in the membrane versus in the water (cytosol). Motion of lipid molecules within the membrane is another issue.

$$\frac{N_{water}}{N_{membrane}} = \exp-\left(\frac{32.4k_BT}{k_BT}\right) = 9 \times 10^{-15} \qquad (6.17)$$

If there were 10^9 lipid molecules in a cell membrane, then at any given time about 10^{-5}, i.e., none of them, would be out in the water. On the other hand, if the free energy difference were, say, +30 kJ/mol (~12k_BT), then at any given time about 5000 lipids would be out in the water. Why might this be important? One of the enduring lessons biology teaches is that structures cannot be rigid, if life processes are to proceed efficiently. If lipid molecules were rigidly fixed in place in a membrane (i.e., the larger ΔG in Equation 6.17), lipid vesicles would be difficult to build. We will find that in real life, there are no pure lipid membranes and that "impurities"—proteins and other molecules—greatly change the resulting probabilities. Cells do not rely on bare lipids spontaneously diffusing through water to get to the site of vesicle construction. The lipids are normally moved through water while bound to proteins: lipoprotein complexes. Similarly, when a lipid molecule flips in a natural membrane, it does so at an impurity site, where the energy increase in the polar head moving through the hydrophobic membrane core is not so high. Nevertheless, in both these cases the lipids must at least temporarily expose hydrophobic parts to water or polar parts to hydrophobic membrane core.

Figure 6.22a shows one standard way that lipids, as well as hydrophobic molecules in general, are moved in humans. Human serum albumin (HSA) has been called the garbage collector of life because many different molecules can adsorb to HSA and be transported through the blood and excreted. HSA is a water-soluble protein but has its own hydrophobic regions to which "oily" molecules will spontaneously bind. The molecules can then move through blood without sticking to the inner surfaces of blood vessel. The structure in Figure 6.20a shows a total of seven binding sites that can be occupied by medium-chain

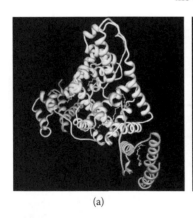
(a)

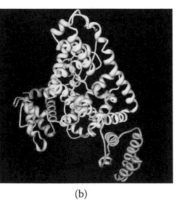
(b)

FIGURE 6.22

Lipid components are often transported in containers. Human serum albumin with (a) seven bound fatty acid molecules, palmitic acids; (b) two bound ibuprofen molecules. NGL viewer. (From Jmol: An open-source Java viewer for chemical structures in 3D. http://www.jmol.org/; Protein Workshop: Moreland, J.L. et al., *BMC Bioinformatics* 6, 21, 2005; NGL Viewer: Rose, A.S. et al., *Bioinformatics*, 2018.) (a) PDB ID 1E7H, Bhattacharya, A.A. et al., *J. Mol. Biol.*, 303, 721, 2000; (b) PDB ID 2BXG, Ghuman, J. et al., *J. Mol. Biol.*, 353, 38, 2005. The bound molecules are shown in ball-and-stick format. HSA binds and transports a variety of hydrophobic molecules.

and long-chain fatty acids. In addition, medium-chain fatty acids bind at four additional sites on the protein, yielding a total of 11 distinct binding locations.[22] Figure 6.22b shows two of another common molecule, ibuprofen, bound to HSA. This structure relates to the mechanisms of distribution and clearance of drugs like ibuprofen from the human body; we will not pursue this further.

6.4.4.3 Lipid flip-flop rates in membranes

Earlier, we alluded to the fact that the energy differences (e.g., ΔG) we have been using give information about equilibrium probabilities or relative populations of states, and do not directly shed light on the rates at which transitions between states take place. We develop this idea fully in Chapter 15, but because we do not address lipid-flip issues there, a brief comment is in order. As we might expect from the large free energy increase required to move a lipid from membrane to bulk water, spontaneous lipid flip-flops in pure (artificial) membrane structures is slow, with rates of one flip every several hours, at most. In contrast, rates measured in living cells or reconstituted membranes (membranes rebuilt using most of their natural components) can be as high as one every few seconds, a factor of 10^3 times faster.[23] The cellular mechanisms facilitating lipid flip-flops are poorly understood. Unlike the transport proteins identified earlier, no "lipid flip-flop assisting protein" has been firmly identified. In the case of the membrane of the *endoplasmic reticulum*, one of whose main jobs is manipulating and inserting membrane proteins into or through membranes, it has been proposed that the presence of almost any sort of transmembrane segment, especially an aggregated protein, could increase flip-flop rates of almost any membrane component. The fact that the protein aggregates in the membrane implies that the aggregate contact faces are of opposite hydrophobic/polar character to the surrounding lipids, providing a natural shielding for the hydrophobic and/or polar parts of a lipid as it attempts to flip. Physics students should think here of impurities increasing the conductivity of pure semiconductors. Biochemistry students should think of the energy-barrier models for enzymatic catalysis, which we take up in Chapter 15.

> We will not discuss the details that traditionally define the variety of "phase" and "melting" transitions and their order in condensed matter physics. Our use of such terms is looser.

6.4.5 Membrane Fluidity

Lipids generally come in large numbers of (mostly) identical units that are not covalently bound to each other. This makes a typical lipid system (membrane) act much more like a bulk material than nucleic acids or proteins. One of the characteristics of normal solids and liquids is the solid $\rightarrow$ liquid transition, termed *melting*. Many materials, such as iron and water, have a sharp temperature at which they change from solid to liquid: the materials properties change abruptly, within a small temperature range. Other materials, such as glasses, can be assigned a melting temperature, but materials properties, such as viscosity, change over a much wider temperature range. Lipid melting transitions behave more like the glasses.

Pure lipid membranes have a characteristic temperature above which the membrane fluidity, measured by the transverse diffusion rate of lipid molecules increases significantly, as diagrammed in Figure 6.23. The value on the vertical axis, the two-dimensional diffusion coefficient, measures the rate at which the average distance squared $\langle r^2 \rangle$ increases with time and will be discussed further in Chapter 13. The temperature at which the diffusion constant changes most rapidly is assigned the symbol T_m. Membranes made of mixed lipids, or containing proteins, glyco- and lipoproteins, cholesterol, and other molecules also have such a transition, but a single characteristic temperature often does not apply.

Like melting of a solid, the melting temperature of lipid membranes is determined by interactions between the molecules in the lipid bilayer. The most important intermolecular interactions are between the long, hydrophobic tails of the lipids. We would expect this transition

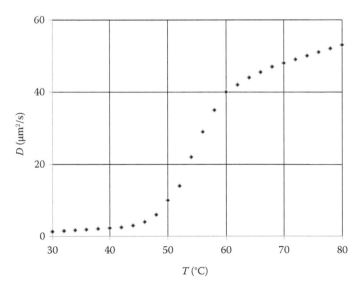

FIGURE 6.23
Simulated data for the increase in lipid mobility (two-dimensional diffusion coefficient) with increase in temperature.

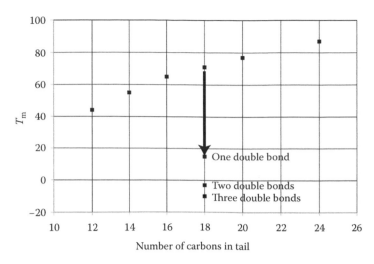

FIGURE 6.24

Dependence of fatty acid melting temperature on fatty-acid chain length (number of carbons) and on the number of double bonds in the chain. (Data points taken approximately from Figure 21.7 of Bergethon, The Physical Basis of Biochemistry, describing fatty acids, not lipids that have polar or charged head groups. The fatty acids with 2–3 double bonds are liquids even below 0°C.)

temperature to depend on the length of the fatty acid chains and on their structure. For example, if the chains are long and straight, their van der Waals interactions can be stronger (more negative energy) and polar molecules will be better excluded. *Saturated* fatty acids, those with as many H atoms bound to C atoms as possible, favor the rigid state because their straight hydrocarbon chains interact favorably. The resulting T_m value is high. On the other hand, a C=C double bond produces a bend in the hydrocarbon chain, which interferes with a ordered packing of the hydrocarbon chains, and T_m is lower. Longer fatty acid chains also increase T_m, as indicated: each additional $-CH_2$ group in the chain contributes about -2.1 kJ/mol $(-0.81 k_B T)$ to the free energy of interaction of two adjacent hydrocarbon chains. The dependence of T_m on the length of the fatty acid chain and the number of C = C double bonds is shown in Figure 6.24. The points below the arrow in the figure are for samples with double bonds present in the 18-carbon chain. Note the strong dependence on the latter quantity (bends in the hydrocarbon chain) make it significantly more difficult for the tails to pack together.

6.4.6 Membranes Must Have Controllable Holes in Them

When all is said and done, the primary role of membranes is still to separate inside from outside. The second most important role is to allow, but control, passage of materials (ions and molecules) through the membrane. Transport through the membrane implies that a hole of some sort must exist, at least for a time, in the membrane. Once he is told the ions or molecules that must be transported, the bright student will think of several ways in which such "holes" may operate. Many of these mechanisms have actually been identified in biological membranes. Structures involved range from (semi)permanent holes in the membrane, lined with charged or polar groups, to proteins that diffuse in the membrane until a transmembrane potential (voltage) is applied, whereupon aggregation forms a channel, to lipoprotein complexes that bind a molecule at the membrane surface and then rotate through the membrane. Membrane transport is a topic for a multivolume set of books, even when restricted to certain types of organisms.

We will look at only two structures involved in such transport. To be frank, these structures have been chosen because of their visual impact and apparent clean, elegant design. Keep in mind, however, that biology is also very practical: elegant design that comes at a high-energy price may lose out in the survival game to a cheap, "strings-and-tape" model.

Figure 6.25 shows a trimeric protein called Omp32 (outer membrane porin protein 32), which is an anion-selective *porin*, a protein that forms pores in membranes. The selectivity (anion/cation ratio) is about 20. This porin consists of a 16-stranded *beta barrel* with eight and seven loops or turns on either side of the membrane. A cluster of positively charged arginines (Arg 38, Arg 75, and Arg 133) determines the electrostatic field close to the constriction zone. The energy of the charged arginines is stabilized by a buried Glu amino acid (Glu 58), which strongly interacts with Arg 75 and Arg 38. A sulfate ion is bound to Arg 38 in the channel constriction zone in the X-ray–determined crystal structure of Figure 6.25. A small polypeptide of

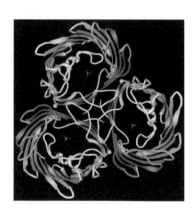

FIGURE 6.25

Omp32, the anion-selective porin from *Comamonas acidov-orans*, a Gram-negative bacterium. This porin occurs in the outer membrane and consists of a trimer of a 16-stranded beta barrel with eight external loops and seven periplasmic turns. Positively charged Arg residues create a charge filter in the constriction zone and a positive surface potential at the external and periplasmic faces. One sulfate ion was bound to Arg 38 in the channel constriction zone. Selectivity for relatively small negative ions is conferred by the positive potential and narrow constriction zone. You are encouraged to examine the electronic PDB structure, rotating 90° to see a view transverse to the membrane. (NGL viewer, PDB ID 1E54, by Zeth, K., K. Diederichs, W. Welte, and H. Engelhardt. 2000. Crystal Structure of Omp32, the Anion-Selective Porin from Comamonas Acidovorans, in Complex with a Periplasmic Peptideat 2.1 A Resolution. Structure 8: 981.)

mass 5.8 kDa appears bound to Omp32 on one side, close to the axis of the trimer. Omp32 selectivity for relatively small anions is conferred by the Arg positive potential, which is not attenuated by negative charges inside the channel, and by a narrow constriction zone. This porin protein membrane "hole" is typical of OMP proteins found in various organisms.

A membrane hole of an entirely different and nasty sort is shown in Figure 6.26—the ClyA porin. Pore-forming toxins (PFTs) are a class of virulence factors (proteins that cause disease symptoms and aid the infectivity and survival of a bacterium or virus) that convert from a soluble form to a membrane-integrated pore. PFTs supplement their toxic effects by destruction of the membrane permeability barrier, permitting delivery of toxic components through the pores. Diphtheria and anthrax toxins are two of the most lethal PFTs; others include PFTs from *Escherichia coli*, *Salmonella enterica*, and *Staphylococcus aureus*. "Good guy" PFTs also exist, e.g., the perforin and the membrane-attack complex, proteins of eukaryotic immune system. All PFTs whose molecular structures have been determined employ α helices or β sheets in the form of a barrel to span membranes. The conversion of the ClyA porin from a water-soluble monomer to a membrane spanning pore is accompanied by large structural changes. The conversion involves more than half of all residues. It results in large rearrangements, up to 14 nm, of parts of the monomer, reorganization of the hydrophobic

FIGURE 6.26
Cytolysin A (ClyA), a pore-forming protein toxin (PFT). PFTs are potent infective factors that convert from a water-soluble monomeric form to a membrane-compatible trimer pore, bypassing the otherwise large positive ΔG for moving from water to membrane. PFTs supplement their toxic effect by eliminating the membrane selectivity barrier and by allowing toxic components through the pores. ClyA (or HlyE is found in several *E. coli* and *S. enterica* bacteria. The structure of ClyA subunits in the pore is quite different from that in the soluble monomer, involving at least half the amino acids. Conformational changes are characterized by large rearrangements, up to 14 nm, of parts of the monomer, reorganization of the hydrophobic core, and conversion of β sheets and loop regions to α helices. (GL viewer, PDB ID 2WCD, by Mueller, M. et al., *Nature*, 459, 726, 2009.)

core, and transitions of β sheets and loop regions to α helices. What do we learn from this membrane protein example? In addition to providing an important vein to mine for future investigations of infectious disease and immune response, we find a good example of a pore protein that can both allow transport of molecules through a membrane and insert itself into a membrane through a monomer-to-aggregate process that changes the nature of the protein from water soluble to membrane content.

> We have now encountered another physical mechanism for efficient transport of a membrane-binding molecule: the molecule is *converted* from water loving to bilayer loving.

6.4.7 Charge Movement Across Membranes

We have been discussing the energetics and dynamics of the mechanical structure of the membrane itself. Such energetics ultimately depends upon Coulomb and dispersion interactions between fundamental charges in the membrane itself, but we have not yet addressed charge-charge interactions that govern another sort of dynamics central to life: motion of ions across membranes. We cannot perturb overall charge balance in any significant way, but many know from personal experience in hot climates that the balance of ions in their bodies directly affects health. Military personnel working in desert climates require about 4 gallons of drinking water per day to keep electrolyte imbalances from causing significant physical. (See www.armystudyguide.com/content/army_board_study_guide_topics/desert_operations/water-usage-in-desert-ope.shtml.) Illustrating the extreme of electrolyte imbalance, biochemical scientists employ a sudden lowering of external ion concentration to osmotically shock cultured cells into fragments, releasing cell contents. In this case, cell membranes are shattered when the osmotic surge of water into cells overwhelms both the forces holding the membrane together, as well as

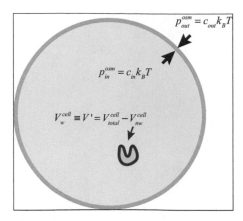

FIGURE 6.27
Ideal, spherical cell surrounded by a membrane, permeable only to water. The volume inside the cell consists of an aqueous volume V', in which the solutes are dissolved, and a volume, subscript nw, which does not contain water or solutes, indicated by the small brownish structure. Only volume V' is involved in osmotic movement. At equilibrium, the osmotic pressure inside and outside the cell must match, if the membrane is freely expansible.

the ability of the cell to increase the volume enclosed by the cell membrane by inserting additional lipid molecules into the bilayer.

How much should a cell expand if it suddenly immersed in pure water or a different external medium of solutes? In the case of real cells, especially plant cells that have a relatively rigid cell wall in addition to a membrane, the answer can often not be predicted theoretically. If we, however, assume a spherical cell, surrounded only by an easily-expansible membrane that is only permeable to water molecules, application of the osmotic pressure equation 6.14 provides guidance, even if the model is not very good for real cells. If the cell is initially at equilibrium, the osmotic pressure inside the cell must match that outside the cell: $p_{in}^{osm} = p_{out}^{osm}$. This implies

$$c_{in,i} = \frac{N_{in}}{V_i'} = c_{out,i} \tag{6.18}$$

Subscript i indicates initial quantities and we assume solutes inside the cell are only located in that part of the cell, volume V', that contains water that can move throughout the cell and through the outer membrane. See Figure 6.27. If the concentration of solutes outside is suddenly decreased, the aqueous volume V' inside the cell must increase as water passes into the cell to reduce the interior concentration. In this idealized model, the result of a sudden immersion of the cell in pure water would increase $V' \to \infty$, which certainly explodes the cell, releasing its contents. A reduction of exterior concentration to one-half the initial value would double the interior aqueous volume. This may also destroy our ideal cell, if forces holding the lipid bilayer cannot withstand the osmotic forces. While sudden immersion in distilled water would disrupt the membranes of most cells, a less severe change of solute bath may expand the cell by a lesser amount than predicted by Equation 6.18, since the overall (free) energetics of the system must also include the elastic energy increase of the membrane.

EXERCISE 6.3

Cell expansion energies, osmotic pressure, volume, surface area and energy per molecule.

Compute the elastic area free energy increase of a spherical bilayer vesicle, if osmotic forces cause a doubling of cell volume. The vesicle has an initial outer radius of 100 nm, the initial total solute concentrations are 0.30 M on both sides of the membrane. The exterior concentration is then suddenly cut in half. The membrane area elasticity is has elasticity constant $\kappa_a = 0.20$ J/m². You may ignore bilayer thickness and any non-aqueous interior contents of the vesicle.

The astute reader will find that assumptions can be misleading in this exercise. We deal with the issue at the end of the exercise.

Answer

The elastic membrane area change (Section 6.4.2.1) gives a free energy change $\Delta G_{area} = \frac{\kappa_a}{2} \frac{(A - A_0)^2}{A_0}$. The initial volume and surface area of the sphere are $V_i = \frac{4}{3}\pi (100\,nm)^3 = 4.2 \times 10^6\,nm^3$ and $A_i = A_0 = 4\pi(100nm)^2 = 1.26 \times 10^5\,nm^2$. Since the volume doubles, the final surface area will increase by a factor $2^{2/3} = 1.59$, to $2.00 \times 10^5\,nm^2$. The free energy change is then $\Delta G_{area} = \frac{(0.20 \times 10^{-6}\ pJ/nm^2)}{2} \frac{(0.74 \times 10^5\,nm^2)^2}{1.26 \times 10^5\,nm^2} = 0.0043$ pJ.

Two questions about this answer:

1. Is this energy large or small, and with what do we compare it?
2. Why have we not needed to consider the osmotic pressure, which allegedly caused the volume increase?

1. The size of the energy change. To better judge the size of the elastic energy increase, we convert the energy to multiples of $k_B T_R$, so we can compare the energy to the average molecular $0.0043\,\text{pJ} = (4.3 \times 10^{-15}\,\text{J})\,\frac{k_B T_R}{4.045 \times 10^{-18}\,\text{J}} \cong 1{,}100\,k_B T_R$ thermal energy.

This increase would be an extremely large molecular energy change, but a membrane is not a molecule. The energy is shared between all lipid molecules making up the membrane, so we should at least compute the number of lipid molecules in the membrane. The initial area of the membrane divided by the average area occupied by a lipid molecule, multiplied by about two (a bilayer), will give us an estimate of the number of lipid molecules. A quick search on the Web or use of the Protein Database to find the actual structure of a lipid molecule in a membrane will result in areas per lipid of 0.4–$0.7\,\text{nm}^2$. (See also Reference 15, p. 246.) As we found, the outer surface area is $A_{\text{outer}} \cong 1.3 \times 10^5\,\text{nm}^2$. Since the bilayer is about 5 nm thick, we can easily ignore the "ignore the thickness" advice and find the inner surface area as $A_{\text{inner}} = 4\pi R^2 = 4\pi(95\,\text{nm})^2 \cong 1.1 \times 10^5\,\text{nm}^2$, for a total surface area of $2.4 \times 10^5\,\text{nm}^2$. The number of lipid molecules is then $N_{\text{lipids}} \cong \frac{2.4 \times 10^5\,\text{nm}^2}{0.5\,\text{nm}^2} \cong 5 \times 10^5$. The surface elastic energy per lipid is then $\frac{1{,}100\,k_B T_R}{5 \times 10^5\,\text{lipids}} \cong 0.0022\,k_B T$ per lipid. This is a small energy compared to random changes in molecular energy due to thermal fluctuations, but we have ignored the fact that all 500,000 of the lipid molecules have simultaneously increased their elastic energy in the same way. The probability that 500,000 similar events occur simultaneously through exchanges of molecular energies during random collisions from water molecules is extremely low, so there must be some other sorts of free energy of the entropic sort involved. We will address one sort of entropic free energy in Section 6.4.3.2.

2. Osmotic pressure? Perhaps we should have equated osmotic to elastic energy change. We will find that Section 6.4.3.2 reasoning points us toward a second aspect of the energetics of vesicle volume and surface expansion—$\Delta G_{\text{osm}} = P_{\text{osm}}\Delta V = (c k_B T)\Delta V$. The term $P\Delta V$ should be familiar from introductory physics and chemistry: random collisions of molecules with a surface surrounding a volume accompanied by a change in that volume. But what molecules are colliding with what surface? Initially, the surface has both water and solute molecules on either side, in presumably identical concentrations. Water molecules can pass through the membrane, but solute molecules cannot. Nevertheless, both water and solutes collide with the membrane on both sides. When the outside solute concentration is suddenly cut in half, water passes into the cell, in order to equalize the solute concentrations. Does this mean that water molecules are doing the work and that "c" should be the concentration of water? Since the concentration of water is the same on both sides and does not change (at least if we assume that solute molecules intercalate between water molecules without causing a volume increase from that of pure water), the water should do no net work. The inside and outside osmotic pressures of the solutes are initially $1.8 \times 10^{26}\,\text{m}^{-3}$. After the outside solute concentration drops, the inside concentration decreases from $1.8 \times 10^{26}\,\text{m}^{-3}$ to $0.9 \times 10^{26}\,\text{m}^{-3}$. If we assume

that the given volume doubling is correct, we can try to calculate an osmotic energy change, remembering that the osmotic pressure decreases from its initial value to half of that (average pressure ¾ the initial):

$$\Delta G_{osm} = \left(\frac{3c_0 k_B T_R}{4}\right)\Delta V = \left(\frac{3(1.8 \times 10^{26}\,\text{m}^{-3})k_B T_R}{4}\right)\left(\frac{4.2 \times 10^6\,\text{nm}^3}{10^{27}\,\text{nm}^3/\text{m}^3}\right) = 5.7 \times 10^5\,k_B T_R$$

Dividing by 5×10^5 lipid molecules, we get $1.1\,k_B T_R$ energy per lipid. This work done by the osmotic pressure is 500 times larger than the surface elastic energy increase we found above. Is this just the way it is, or did we overlook something? The answers are that (i) we have overlooked a variety of issues that would be addressed in a more comprehensive treatment, and (ii) our two answers should be different. The osmotic pressure does not push until the elastic forces of the membrane rise to counter it. Equation 6.18 assumes, unlike a rubber balloon, that elastic forces in the membrane are negligible compared to osmotic. Our two energy calculations are consistent with this assumption. See Reference 24 for a more complete treatment.

If the solutes were ionic in nature, the exercise just completed would be an example of modification of interior charge by physicochemical forces beyond the cell's control. As implied by the term *osmotic shock*, a rapid and large change of exterior concentrations can force the cell to expand so rapidly that the cell membrane can rupture, causing cell death. In the exercise, the surface area per lipid molecule increased by a factor of 1.59 to $0.5\,\text{nm}^2 \times 1.59 \cong 0.8\,\text{nm}^2$, larger than the observed range 0.4–$0.7\,\text{nm}^2$. The lipids would have their hydrophobic tails so exposed to water that the bilayer would likely be unstable and rupture. If the exterior concentrations were changed more slowly, a living cell may have time to adjust by synthesizing and inserting more lipid molecules into the membrane. This synthesis mostly takes place in the endoplasmic reticulum or Golgi apparatus of mammalian cells, which we did not include in our ideal, spherical cell. See Chapter 7. The systems that signal these organelles to produce more lipids, transport lipids and insert them into a membrane under stress is complex and beyond the scope of this text. Whether these systems could insert new lipids into the membrane rapidly enough to prevent membrane rupture is a question posed in homework.

Living cells actively and selectively control their ionic contents. Two general mechanisms for electrolyte control exist: the flow of water through the membrane and flow of specific ionic species through specialized membrane sites, ion gates, pores or pumps, introduced in Section 6.2.3. The latter mechanism is what distinguishes living cells from nonliving lipid structures. Some of the ion pumps in photosynthetic cells are driven by the most fundamental energy source available on earth: sunlight. We deal with this aspect of ion transport in Chapter 10. See Figure 10.8 and note the movement of electrons and protons. The ATP synthase ion pump is part of a somewhat complex electrical energy system: energy from photons drives electron transfer, which is coupled to proton production, which raises the electrical potential difference across the membrane in which the synthase is embedded, which drives the generation of ATP from ADP. (Following the general rules of biochemical reversibility, one ATP molecule hydrolyzed to ADP can also cause the transfer of one proton across the membrane.) The action of the synthase itself is, in principal simple: the electrical potential energy of a proton on the higher-voltage side of a membrane is converted to ATP chemical energy when the proton flows across the membrane through the synthase's ion pore. Properly understanding of such bioelectric activity is not, however, a topic that can be dealt with quickly, as attested by Thomas Weiss's 500-page book, *Cellular Biophysics, Volume 2: Electrical Properties* (1996, MIT Press). An abbreviated, but illuminating treatment can also be found in Chapters 11 and 12 of Philip Nelson's *Biological Physics* (2008, W.H. Freeman), as well as several other textbooks.

6.5 POINTS TO REMEMBER

- Separation of multiple biomolecules in a sample is usually done by centrifugation (sedimentation) and chromatography: understand $v = s\omega^2 r$ and dissociation constants.
- Equation 6.6; an absorption spectrum, plotted as a function of wavelength, reflects the probability that light of wavelength between λ and $\lambda + d\lambda$ will be absorbed by a molecule. The spectrum represents a simple, nonintrusive way to characterize and quantify biomolecules.
- Exclusion forces and free energies contribute to the structural organization inside a cell. Understand, pictorially and quantitatively, the size and concentration dependence of these free energies.
- Pictorial structures and functions of (small) multisubunit proteins, membrane channel proteins, the ribosome, and saccharide aggregates.
- Membranes form the fundamental, dynamic two-dimensional walls between compartments of a cell. Know the nomenclature: lipids, vesicles, liposomes, etc. They must have controlled holes in them. Know how to compute the energy cost of membrane distortion, as well as the osmotic work that may cause distortion or cell destruction. Understand that Equation 6.9 does not imply that one can "blow up" a spherical vesicle with a constant number of lipid molecules with no energy cost.

6.6 PROBLEM-SOLVING

PROBLEM 6.1   CENTRIFUGATION

In the following, assume that a spherical particle of mass m and density 20% greater than the density of water is in a water-filled centrifuge tube being rotated horizontally (see Figure 6.2).

1. Calculate the centrifugal acceleration, as a multiple of g (9.80 m/s²), of the particle located 12.0 cm from the axis of rotation of a centrifuge rotating at 40,000 rpm.
2. Calculate the mass of a particle for which the centrifugal potential energy change for movement a distance of the particle's diameter is equal to $k_B T$. (Recall the $mg\Delta y \sim k_B T$ discussion in the text for the gravitational case.)
3. An object in water also has a buoyant force on it. In a nonmoving container of water, the buoyant force is upward (opposite to gravity), with magnitude equal to the weight of the displaced fluid. In the present case of centrifugation, do we need to account for the buoyant force? Explain.

PROBLEM 6.2   MOLECULAR SIZING CHROMATOGRAPHY

If beads of a chromatography resin are permeated with pores, molecules traveling down the column (see Figure 6.4b) will be delayed if they are smaller than the pore diameter. (They spend extra time inside the beads.) A cylindrical column of inside diameter 1.50 cm and length 35.0 cm is filled with molecular sizing resin. The "void volume," the buffer-filled volume between the resin beads in the column, is 30.0% of the total column volume. A small volume of a sample with high- and low-molecular mass proteins is added to the top of the column, and buffer is then passed through the column at a rate of 0.35 mL/min. The large protein cannot enter the pores in the resin and passes directly around the beads. The small protein will be delayed relative to the large but comes out roughly when an amount of buffer equal to the total column volume passes through.

1. Calculate the total column volume and column void volume.
2. Calculate the times at which the two proteins come out the bottom of the column.
3. Calculate the average linear velocity of the buffer flowing in the direction of the column axis.

PROBLEM 6.3 AFFINITY CHROMATOGRAPHY AND SPHERICAL MICROBEADS

The affinity ligands are attached to the surface of the microbeads. Two different affinity chromatography resins have the following characteristics:

Resin	Average Affinity Ligand Density[a] (µmoles/mL)	Average Bead Diameter (µm)
A	15	35
B	25	95

[a] Per unit (total) volume of resin packed in a column; assume 30% void volume (Problem 6.2).

In the following, draw diagrams.

1. Calculate the number of ligand molecules per square micron of bead surface for the two resins.
2. Calculate the number of ligands on a single sphere of each resin.
3. Calculate the average distance between nearest-neighbor ligands on the surfaces of each type of sphere.
4. How big (molecular mass) would a protein have to be such that protein binding at one ligand site would interfere with binding at an adjacent site?

PROBLEM 6.4 LIPID AGGREGATES

A diagram of a section of membrane can be found in Figure 1.7. (See also Exercise 1.1.) Membranes can form approximately spherical, double-layer structures. The smallest of these spheres has a diameter of about 50 nm. Find the dimensions of a typical lipid molecule and determine how many lipid molecules are in (1) this smallest spherical membrane and (2) in a large membrane of diameter 70 µm. Assume 60% of the membrane area is lipid.

PROBLEM 6.5 RIBOSOME STRUCTURE

Go to the PDB and load the structures in Figure 6.11.

1. Display each in a format that traces only protein and nucleic acid backbones. Print these out.
2. Display the structure in Figure 6.11a on a much finer scale. Examine a region where a protein strand contacts an RNA strand. Insert this display in your assignment, and describe the protein and nucleic acid at a point of closest approach. What part of an amino acid subunit is apparently interacting with what part of a nucleic acid subunit?

PROBLEM 6.6 MEMBRANE STACKING

Use Equation 6.16 to determine the approximate concentrations, in both molecules/m^3 and molar of (1) a 20-kDa, (2) a 200-kDa, and (3) a 2000-kDa protein needed to induce entropic (osmotic) stacking of two bilayers of a lipid membrane. Assume the proteins are spheres and their density is 25% greater than that of water. With each answer, be sure to show your calculated values of all parameters in Equation 6.16. If you need to find a lipid structure to determine size, search the PDB for structures (e.g., PDB ID 1BE2 shows a lipid bound to a protein).

PROBLEM 6.7 LIPID FLIPS IN MEMBRANES

Section 6.4.4 cited a reference that stated that the "energy barrier" to a lipid flip in a membrane is about 15 kJ/mol. This means that a lipid in the middle of a flip has a free energy 15 kJ/mol higher than when it is in its normal configuration.

1. Calculate the number of "flipping lipids" in a membrane made of 5×10^5 lipids for energy barriers of 5, 15, 30, and 60 kJ/mol.
2. Do some more investigating and find out many times per second lipid flipping occurs. Do not confuse the "time for a flip," which may be as small as 0.3 ps, with the "time between flips," which is the inverse of the flipping rate. Because the rate of flipping is directly proportional to the number of lipids in the membrane, normalize your rate for a 10^9 lipid system.

MINIPROJECT 6.8 MEMBRANE FLUIDITY

Below is a table of the data used in Figure 6.23.

T (°C)	D (µm²/s)
20	0.8
22	0.9
24	0.95
26	1
28	1.1
30	1.3
32	1.5
34	1.7
36	1.9
38	2.1
40	2.3
42	2.5
44	3
46	4
48	6
50	10
52	14
54	22
56	29
58	35
60	40
62	42

(Continued)

T (°C)	D (μm²/s)
64	44
66	45.5
68	47
70	48
72	49
74	50
76	51
78	52

1. Use a graphing program to plot both D and dD/dT (derivative with respect to T) versus temperature. Put both curves on the same graph. Describe how you calculate values for the derivative. (You may use the trapezoidal or similar rules for digital derivatives. If your graphing program has a built-in differentiator, state what method it uses.)
2. Describe a good way to graphically define T_m and determine T_m for this graph.
3. Graphically define and determine the half-width (half-width at half-maximum), ΔT, of the melting transition from the derivative plot. Describe how you deal with the "offset" of the derivative curve—the fact that dD/dT never reaches zero. Briefly discuss the physical meaning of this width. For example, what sorts of membranes will give narrow and wide widths. (Remember that these data are "made-up" data that only roughly correspond to real membranes.)

PROBLEM 6.9   LIGHT ABSORPTION BY HEMOGLOBLN

A light beam of power 3.0×10^{-6} W and variable wavelength is used as a source light (I_0) in an absorption spectrophotometer, which has an available cuvet with optical path length 1.00 cm. (You can use light power units for I and I_0.) See Figure 6.8. A 3.0-mM solution of hemoglobin is placed in the cuvet in the spectrophotometer.

1. Determine from Figure 6.8 the wavelengths and the values of the absorption coefficients at the peaks of the spectra for both oxy- and deoxy-Hb. Place these values in an appropriate table, leaving room for more columns of data below.
2. For light at each of the wavelengths in 1, calculate the absorbances ($A = \varepsilon cL$) of the samples. Enter these values in the table in 1.
3. For light at each of the wavelengths in 1, calculate the intensity of the light (I, in Watts) that will exit each sample. Calculate I/I_0 for each case and enter in the table in 1.
4. If you actually tried to make the measurements in 2, you would find that the absorption spectrophotometer would give you an error because the light detector could not accurately measure light intensities below 1 nW, which corresponds to an absorbance of well over 3.0. Determine the oxy-Hb concentration that would give an absorbance of 3.0 at the wavelength of the blue spectral peak. Compare this to the approximate concentration of Hb in blood.

PROBLEM 6.10   A POLYSACCHARIDE STRUCTURE

Find the cyclo-amylose structure of Figure 6.13 in the Protein Database.

1. Produce space-filling equivalents to the two views of the structure.
2. Discuss how water molecules might pass through the center of the helices. Refer to distances between atoms in the helices, and to the known size of the water molecule (Chapter 4).
3. Is there an evident need for mobility of amylose atoms to allow water passage? Explain.

PROBLEM 6.11   LIPID INSERTION TO PREVENT MEMBRANE RUPTURE

Exercise 6.3 dealt with the osmotic expansion of an idealized spherical cell that would likely have ruptured the cell, unless the exterior concentration change were gradual and the cell could add additional lipids to the membrane. You are to estimate the minimum time needed for a living cell to synthesize and insert enough lipids to prevent rupture. Investigate the following:

1. Search the library, Web, or other bio-information sources to find an estimate of the overall rate at which a particular cell can synthesize and insert new lipids into a membrane, in response to a challenge, such as reduced ionic strength or a need to grow a new membrane structure, such as a vesicle to export and transport proteins. You may have to calculate (estimate) the overall rate from several quantities or from a recorded video of a cellular process.
2. Estimate the minimum time over which the "osmotic shock" of the dilution in Exercise 6.3 would have to take place, in order that the cell would *not* rupture.

PROBLEM 6.12   RIBOSOME STRUCTURE

Structures of the ribosome shown in Figure 6.11 were "obsoleted" near the end of the year 2014, according to the Protein Database.

1. Search for the four PDB IDs listed in the figure at the RCSB.org Web site and find the ID which supersedes them.
2. Use the Protein Workshop app to display the structure in a way that highlights the interface between the 50S and 30S subunits. Orient the structure in four different ways, each rotated from the previous by 90° around an axis normal to the interface. You may want to use the "Tools/Surfaces/Transparent" to facilitate your identification of the subunit interface.) Use "Options/Save Image" to download each structural display to your computer. Use file names that do not overwrite the previous. (Look for files on your computer's desktop.) Be prepared for download times of 5 minutes or more.
3. Write down four important structural features that the authors of the structure point out.

REFERENCES

1. Voet, D., and J. G. Voet. 2004. *Biochemistry* (Chapter 12). New York: John Wiley & Sons.
2. Phillips, R., J. Kondev, and J. Theriot. 2009. *Physical Biology of the Cell* (Chapter 11). New York: Garland Science.
3. GE Healthcare. 2007. *Affinity Chromatography: Principles and Methods*. Uppsala, Sweden: GE Healthcare Bio-sciences. Download at www.gelifesciences.com/protein-purification.
4. Invitrogen, Staff. *The Molecular Probes Handbook*. www.thermofisher.com/us/en/home/references/molecular-probes-the-handbook.html. Accessed September 1, 2017.
5. Alberding, N., S. S. Chan, L. Eisenstein, H. Frauenfelder, D. Good, I. C. Gunsalus, T. M. Nordlund, M. F. Perutz, A. H. Reynolds, and L. B. Sorensen. 1978. Binding of carbon monoxide to isolated hemoglobin chains. *Biochemistry* 17:43–51.
6. Emerson, J. 2008. The Color of Blood: A Study in Scarlet. Web blog, http://www.rogerebert.com/scanners/the-color-of-blood-a-study-in-scarlet. Accessed Sepetember 1, 2017.
7. Tanaka, H., K. Kato, E. Yamashita, T. Sumizawa, Y. Zhou, M. Yao, K. Iwasaki, M. Yoshimura, and T. Tsukihara. 2009. The structure of rat liver vault at 3.5 angstrom resolution. *Science* 323:384–388.
8. Weixlbaumer, A., S. Petry, C. M. Dunham, M. Selmer, A. C. Kelley, and V. Ramakrishnan. 2007. Crystal structure of the ribosome recycling factor bound to the ribosome. *Nat Struct Mol Biol* 14:733–737.
9. Hernández, F., and J. Avila. 2008. The role of glycogen synthase kinase 3 in the early stages of Alzheimers' disease. *FEBS Lett* 582:3848–3854.
10. Carpita, N. C., and D. M. Gibeaut 1993. Structural models of primary cell walls in flowering plants: Consistency of molecular structure with the physical properties of the walls during growth. *Plant J* 3:1–30.
11. Mravec, J., X. Guo, A. R. Hansen, J. Schückel, S. K. Kracun, M. D. Mikkelsen, G. Mouille, I. E. Johansen, P. Ulvskov, D. S. Domozych, and W. G. T. Willats. 2017. *Pea border cell maturation and release involve complex cell wall structural dynamics*. *Plant Physiol* 174:1051–1066.
12. Briggs, D. E. G. 1999. Molecular taphonomy of animal and plant cuticles: Selective preservation and diagenesis. *Philos Trans R Soc B Biol Sci* 354:7–17.
13. Round, F. E., and R. M. Crawford. 2007. *The Diatoms. Biology and Morphology of the Genera*. Cambridge, UK: Cambridge University Press.
14. Drum, R. W., and R. Gordon. 2003. Star Trek replicators and diatom nanotechnology. *Trends Biotechnol* 21:325–328.
15. Kröger, N., and N. Poulsen. 2008. Diatoms—from cell wall biogenesis to nanotechnology. *Annu Rev Genet* 42:83–107.
16. Boal, D. 2012. *Mechanics of the Cell*. Cambridge, UK: Cambridge University Press, pp. 257–260.
17. Phillips, R., J. Kondev, and J. Theriot., op. cit., p. 424.
18. Karttunen, M. 2009. Membrane Molecular Dynamics. www.apmaths.uwo.ca/~mkarttu/pope-5ns.mpg. Accessed July 11, 2009.
19. Lakowicz, J. R. 1999. *Principles of Fluorescence Spectroscopy*. New York: Kluwer Academic/Plenum Publishers.
20. Martí, J., and F. S. Csajka. 2003. Flip-flop dynamics in a model lipid bilayer membrane. *Europhys Lett* 61:409–414.

21. Tieleman, D. P., and S.-J. Marrink. 2006. Lipids out of equilibrium: Energetics of desorption and pore mediated flip-flop. *J Am Chem Soc* 128:12462–12467.

22. Bhattacharya, A. A., T. Grune, and S. Curry. 2000. Crystallographic analysis reveals common modes of binding of medium and long-chain fatty acids to human serum albumin. *J Mol Biol* 303:721–732.

23. Sapaya, N., W. F. D. Bennetta, and D. P Tielemana. 2009. Lipid flip-flop: Influence of the bilayer composition and the presence of peptides. *Biophys J* 96:349a.

24. Blatt, M. R., Editor. 2004. *Membrane Transport in Plants Annual Plant Reviews*. London, UK: Blackwell.

25. Weiss, T.F. 1996. *Cellular Biophysics*, Vol 1: Transport. Cambridge, UK: The MIT Press. Chapter 4, Section 7.

26. Jo, S., T. Kim, W. Im. 2007. Automated builder and database of protein/membrane complexes for molecular dynamics simulations. *PLoS One* 2, e880.

27. RCSB PDB software references:

    Jmol: An open-source Java viewer for chemical structures in 3D. http://www.jmol.org/.

    Protein Workshop: Moreland, J. L., A. Gramada, O. V. Buzko, Q. Zhang, and P. E. Bourne. 2005. The Molecular Biology Toolkit (MBT): a modular platform for developing molecular visualization applications. *BMC Bioinformatics* 6:21. NGL Viewer: Rose, A. S., A. R. Bradley, Y. Valasatava, J. D. Duarte, A. Prlić, and P. W. Rose. 2018.

    NGL viewer: Web-based molecular graphics for large complexes. *Bioinformatics*. doi:10.1093/bioinformatics/bty419.

28. Beckman SW rotor, image by permission of Scientific Calibration, https://scical.com.

29. www.agilent.com/en/products, search high-throughput chromatography.

30. Paoli, M., R. Liddington, J. Tame, A. Wilkinson, and G. Dodson. 1996. Crystal structure of T state haemoglobin with oxygen bound at all four haems. *J Mol Biol* 256:775.

31. Zhou, Y. and R. MacKinnon. 2003. The occupancy of ions in the K+ selectivity filter: Charge balance and coupling of ion binding to a protein conformational change underlie high conduction rates. *J Mol Biol* 333:965–975.

32. Gessler, V., I. Usón, T. Takaha, N. Krauss, S. M. Smith, S. Okada, G. M. Sheldrick, and W. Saenger. 1999. V-Amylose at atomic resolution: X-ray structure of a cycloamylose with 26 glucose residues (cyclomaltohexaicosaose). *Proc. Natl. Acad. Sci. USA*, 96:4246–4251.

33. Ghuman, J., P. A. Zunszain, I. Petitpas, A. A. Bhattacharya, M. Otagiri, and S. Curry. 2005. Structural basis of the drug-binding specificity of human serum albumin. *J Mol Biol* 353:38.

34. Zeth, K., K. Diederichs, W. Welte, and H. Engelhardt. 2000. Crystal structure of Omp32, the anion-selective porin from comamonas acidovorans, in complex with a periplasmic peptideat 2.1 Å resolution. *Structure* 8:981

35. Mueller, M., U. Grauschopf, T. Maier, R. Glockshuber, and N. Ban. 2009. The structure of a cytolytic alpha-helical toxin pore reveals its assembly mechanism. *Nature* 459: 726.

CHAPTER 7

CONTENTS

Putting a Cell Together
Physical Sketch

Life has been around for a considerable length of time. Most organisms are quite complicated, as suggested by Figure 1.5. Life is often identified, quite legitimately, by the nonscientist because of its complexity. Even the latest Acura TSX built by Honda Corporation is dwarfed in its complexity, capabilities, and versatility by a fruit fly or goldfish. In this context it is quite hopeless to attempt to gain an overview of "life," a living organism or even a living cell in a one-semester biophysics course. This has been attempted in biochemistry and cell and theoretical biology books. The more successful of the first two sorts of books tend to be large (1,000–1,500 pages) and useful as reference texts. There are several that I admire immensely. However, it is quite hopeless for most humans to try to grasp the contents of those 1,500 pages, much less the totality of life's processes, structures, and organizing principles in less than a lifetime. We will not attempt to be comprehensive in this short overview of a cell. We will not even lay claim to scratching the surface of what constitutes a living entity. We can hope, however, to sketch enough of the cellular processes to create a new avocation for those who earn their rent through materials physics, complex systems analysis, or chemistry.

Chapter 1 introduced us to some living organisms and the hierarchical nature of their structure. Chapter 2 sketched two different approaches, the biologist's and the physicist's, to investigating living systems. Chapters 3, 5, and 6 got us into the building blocks of living biosystems—an exhausting task in itself. Now we attempt to see a few of the design principles of a cell: how the molecular structures are used and organized to allow a cell to exchange energy with the environment, keep itself organized and in good repair, and most importantly, to reproduce itself. Indeed, the single most clear-cut difference between man-made machines and a living cell is the inability of the machine to reproduce itself, at least as of 2017. Won't it be marvelous when the Ford Fusion you just bought presents you with a brand-new baby Fusion when it's 10 years old?

7.1 MINIMAL, PROKARYOTIC, AND EUKARYOTIC CELLS

There are three major categories of cells, or maybe there are two, or maybe one, or maybe it depends on your point of view. Two cell types, prokaryotic and eukaryotic, correspond to what we actually see in nature and the third, the "minimal" cell, corresponds to (some) scientists' drive to understand what the first living entity on Earth might have been. This third cell type, an imaginary cell, draws the interest of evolutionary biologists and theorists but should also demand the attention of anyone who wants to understand how the complicated multienzyme, membrane-structured, self-organizing systems in biology manage to coordinate their activity to allow life to proceed in even the most rudimentary way. The minimal cell has a long history; we will do a rough draft of one effort to organize ideas about this minimal cell and rapidly proceed through a description of real cells. While we are at it, we will *very* briefly address the idea that though existing eukaryotic cells are much larger and more complex than prokaryotic, eukaryotic cells came from prokaryotic.

In our pursuit of the cell, rather than attempting a complete understanding, we will become minimally knowledgeable about cells as they are, prokaryotic and eukaryotic, focusing on minimal, basic processes and structures.

7.1.1 Minimal Cell

A minimal cell contains the biomolecules, reaction pathways, and compartments necessary and sufficient for cell replication from small nutrient molecules. To keep ourselves close to what nature has actually provided us, we will assume the biomolecules are nucleic acid and polypeptide polymers encapsulated within a lipid membrane, which may be as simple as a vesicle. The outline of proposed minimal cell, in terms of words and rough organization, is shown in Figure 7.1, adapted from a 2006 proposal by Forster and Church.[1] Small molecules diffuse across the membrane. The biomolecules are ordered according to the pathways in which they are synthesized and act. The color-coding is as follows: blue, DNA synthesis; red, RNA synthesis and cleavage; green, RNA modification; purple, ribosome assembly; orange, posttranslational modification; and black, protein synthesis. The system could be started up with DNA, RNA polymerase, ribosome, translation factors, tRNAs, methionyl-tRNA formyltransferase (MTF), synthetases, chaperones, and small molecules (i.e., not all the molecules in the cell need be there at the start). This proposed minimal cell is described in some detail in the Forster and Church reference.

We have seen many of the structures in Figure 7.1 in Chapters 5 and 6. We could therefore start drawing what these structures might be like and how they could be packed together (or separated). Before we proceed, let's consider that each of the pathways (arrows) in Figure 7.1 involves a reaction that can most simply be described by a differential equation. Even the simplest of the processes in the proposed diagram, the diffusion of small molecules across the membrane, should be described by the diffusion equation (Equation 3.25, reproduced here):

$$\frac{\partial C\left(x,t\right)}{\partial t} = -D\frac{\partial^2 C\left(x,t\right)}{\partial x^2} \tag{7.1}$$

(Refer to Chapter 3 if necessary.) This diffusion is complicated. We must expand to three dimensions, but perhaps we could use spherical symmetry and deal only with a radial variable. Let's assume we can do this and wait until Chapter 15 to do so. Outside the membrane the diffusion could perhaps be described by such a diffusion equation. Equation 7.1 will need at least two spatial boundary conditions because of the second derivative. We also, of course, need the diffusion coefficients of the molecules. These are manageable if we know what they are and that the fluid is aqueous. Let's be hopeful and assert that Equation 7.1 will also apply inside the minimal cell. There are two more boundary conditions there. (Inside the membrane the small molecules may well find restricted geometries, greatly complicating the math.) Diffusion across the membrane is also described by a differential equation—not unmanageable, but we have to know the size of the holes in the membrane, or the energetics and kinematics of transfer of the small molecules from water to lipid bilayer to water again in order to calculate rates of passage through the membrane. Now, the rates of movement from outside to membrane to inside require that we match rates from the three differential equations. This is indeed getting complicated.

The coup de grace to our mathematical attempt is the fact that even if the boundary conditions that set the supply of small molecules outside the cell are well behaved and known, the processes inside the cell are usually unmanageable. The small molecules are presumably undergoing chemical reactions inside the cell. The difficulty is that all of these reactions inside the cell have at least one, if not multiple, feedbacks from other reactions and this means we must give up our assumption of simple diffusion inside the cell. Into the diffusion equation must be placed new "source" and "sink" terms for each process that uses or produces the small molecules. (All of the components and processes in Figure 7.1 use the small molecules, so the small molecule arrow inside the cell should point to each reaction.) What's even worse is that each of these other processes in the cell is affected by several of the other processes.

Let's propose to just throw every molecule and process into a computer code that numerically computes the solution to all the differential equations implied by Figure 7.1. This is indeed the avenue taken by many bright researchers at universities, including a colleague down the hallway from me. Their task is not easy, however. Most of you have not tried to solve a system of N-interconnected differential equations, where N is larger than 2. To get a feel for the problem, recall when you learned to solve systems of N linear algebraic equations. (Systems of differential equations can sometimes be converted into sets of linear algebraic equations.) When $N = 2$, life was easy: solve one equation for one unknown, substitute into the other equation, and you are done, after some brief algebra. $N = 3$ was much harder. The usual approach is to learn formal matrix algebra. This algebra leads to inversion of an $N \times N$ matrix. If $N = 3$, the inversion is manageable by hand. If $N = 4$ or more, most of us leave the table and watch our favorite DVD or go shopping. The rest of us think Mathematica, MathCAD, or some such computer program. These programs can indeed invert matrices bigger than 3×3, but even they give up for values of N not much larger than 3. The reality is that the differential problem is much more difficult, because not only does the feedback of one process on another involve numerical values, but these numerical values are time dependent. Even worse, when a realistic simulation of a single cell is attempted, many of these quantities are best described by random variables. This is why we have to learn about random walks and other processes in Chapters 13 and 14. The behavior of a single (or a few) DNA polymerase enzyme cannot be described by uniformity, with a molecule of ATP available and added to a DNA chain each and every 1.0-ms time interval. Many time intervals may go by when no ATP small molecule is around, because another process has used it up or the polymerase has drifted away from a spot where ATP is passing through the membrane. So, let's evaluate the minimal cell, but not with the illusion that we have a simple system to solve. The 20 or so molecules and processes in Figure 7.1 are not computable.

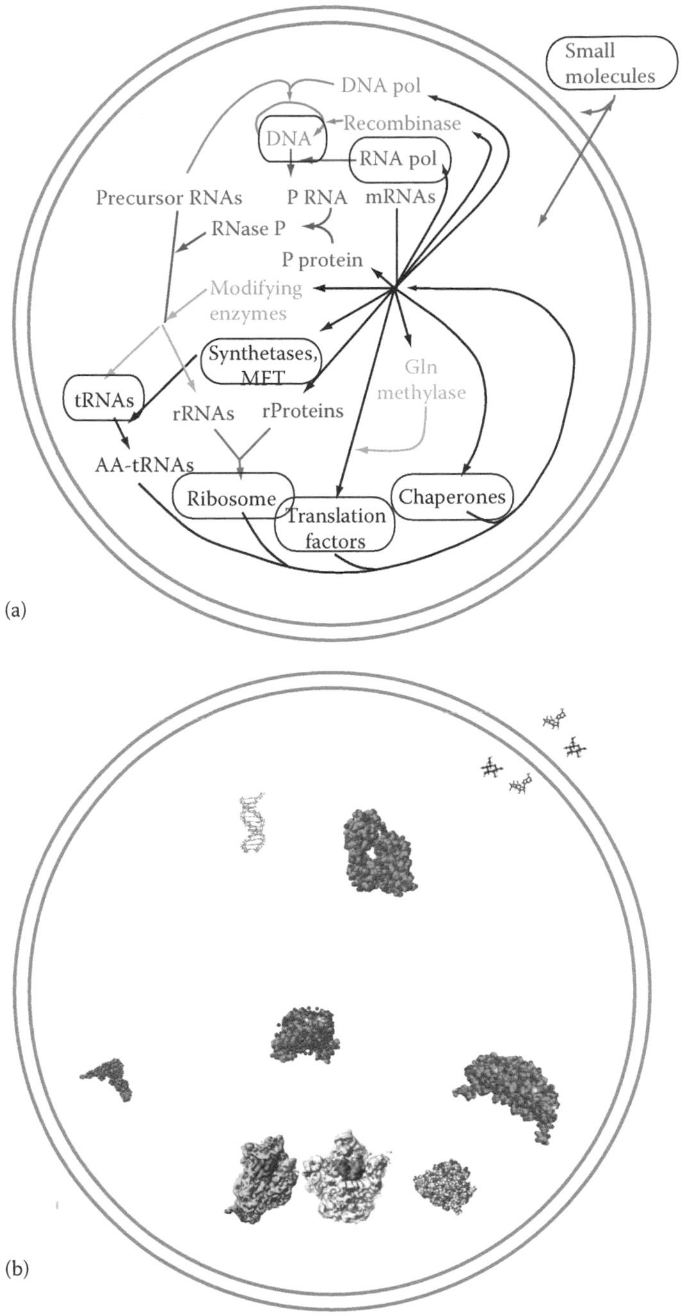

The words in Figure 7.1 are helpful in tracing what structures and processes must be part of this minimal cell, but let's be more specific. Most of the abbreviations we should be able to identify from Chapters 5 and 6. The "-ases" are enzymes; RNA polymerase is needed for construction of RNA chains equivalent to the DNA sequence; the "Gln methylase" must be an enzyme that adds a methyl group to something; RNase P is a ribozyme that cleaves RNA; MFT is methionyl-tRNA^{fMet} formyl-

(a)

(b)

FIGURE 7.1

Diagram of a minimal cell, focusing on active physical agents. (a) The minimal components needed to bootstrap this cell into "life" are claimed to be the circled items, but the protective environment of the cell membrane or its equivalent should also be required. (After Forster, A.C., and Church, G.M., *Mol. Syst. Biol.*, 2, 45, 2006.) (b) Display of the structures of the circled items in panel (a) illustrates their complexity. For a list of possible protein components, see Gil, R. et al., *Microbiol. Mol. Biol. Rev.*, 68, 518–537, 2004.

transferase; "-RNAs" are various forms of RNA; AA-tRNA are amino-acyl tRNAs (needed to transfer amino acids to the proper ribosome assembly point); the ribosome is the manufacturing plant for proteins, which we will see shortly in more detail; chaperones are protein "rooms" or utilities to aid the formation of other proteins' structures. It is claimed that the system of Figure 7.1 could be "bootstrapped" into life by only the initial presence of DNA, RNA polymerase, the

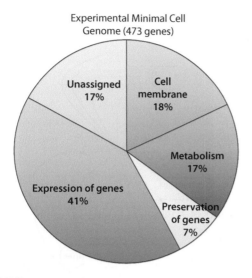

Experimental Minimal Cell Genome (473 genes)

FIGURE 7.2
Genes needed for a minimal cell. Experimental minimal cell JCVI-syn3.0, containing 531 kilobase pairs comprising 473 genes, has a genome smaller than that of any replicating cell found in nature. Hutchison, C.A. et al., *Science*, 351, aad6253-6251–aad6253-6211, 2016, synthesized the 1079–kilobase pair genome of the native *Mycoplasma mycoides* JCVI-syn1.0, then removed as many genes as possible, without destroying cell replication. Well-studied bacteria such as *Bacillus subtilis* and *Escherichia coli* carry 4,000–5,000 genes.

ribosome, translation factors, tRNAs, MTF, synthetases, chaperones, and the necessary small molecules. Why is this? Small molecules provide needed parts and energy sources; the DNA contains instructions for how to build all the protein molecules; the ribosomes construct the proper polypeptide sequence from the DNA instructions, with the intermediary aid of tRNAs (one for each amino acid), translation factors, RNA polymerase, and MFT and synthetases (for AA-tRNA synthesis); and finally, the chaperones assist in proper protein structure formation.

What about the viruses? Viruses attack not only plants and animals but also bacteria, in which they are believed to play a key role in the rapid adaptation of bacteria to antibiotics and other chemical agents, as well as to physical changes in the environment. Viruses will come up often during our introduction to biosystems because of their simplicity, the convenient size of their DNA and the interesting physical structure of the protein capsid, in which the DNA is stored. However, viruses are generally considered nonliving because of their inability to reproduce outside a host organism and therefore cannot be considered any type of minimal form of life.

In the end, theories about the minimal requirements for cellular life must be lead to experimental testing. Attempts to combine the components in Figure 7.1—proteins, nucleic acids and lipids—have not been the primary approach toward minimal cell investigation. Even before 2000, it was supposed that a minimal cell could be simpler than any naturally observed cell and that a profitable investigation could center on the genome for the cell.[2] The mycoplasmas, which usually grow in the nutrient-rich environment of animal hosts, have the smallest known genomes of any autonomously replicating cells. A comparison of the first two available genome sequences, *Haemophilus influenzae* [1815 genes (3)] and *M. genitalium* [the smallest known mycoplasma genome; 525 genes (2)], revealed a common core of only 256 genes, much smaller than either genome. This was proposed to be the minimal gene set for life (4). Experimental tests of this proposal soon showed that while many of the natural genes were not necessary for bacterial growth and survival in the laboratory more than 256 seemed needed. Hutchison et al.[2] noted that while some genes might not be absolutely necessary for a cell's existence under optimal conditions, its growth in a bacterial culture at a reasonable rate might require more components. Some of these genes seemed redundant, but more disconcertingly, the functions of 17% of the genes could not be assigned. See Figure 7.2. One striking feature of this experimental minimal genome is that nearly 50% of the genes function in the expression or preservation of genes. This implies that a large fraction of the mass of this functioning cell is also devoted to information and its coding—a sharp contrast to the machines that humans build.

7.1.2 Prokaryotes and Eukaryotes: Differences

The classes, phyla, and domains of Earth's present inhabitants, as well as how they are supposed to have developed over time, are shown in Figure 7.3. The figure, based on works by Gribaldo and Brochier-Armanet[3] and Cavalier-Smith,[4] reveal that there are many more types of prokaryotic than eukaryotic organisms. Most people think of prokaryotes as one-celled bacteria, which is close to the truth; but single-celled eukaryotic organisms, protists, also exist. Several characteristics distinguish single-celled eukaryotes from prokaryotes:

1. Eukaryotes tend to be more mobile.
2. Eukaryotes have two membranes (more below).
3. Eukaryotes have separated transcription (DNA→RNA) from translation (RNA→protein).
4. Eukaryotes have a genome distributed over several linear chromosomes (a cell nucleus), while most prokaryotes have a single, circular genome.

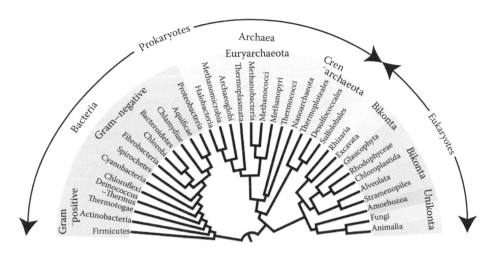

FIGURE 7.3
Classes and phyla of cells. Prokaryotic cell types, including the bacteria and archaea, outnumber the more complex eukaryotic types—a natural state if eukaryotic cells developed from a consolidation of several (or many) prokaryotic cell types in a colony.

5. Eukaryotes use introns—sections of DNA that are not eventually transcribed into protein—within transcribed genes, while prokaryotes do not.

Norris and Root-Bernstein describe these differences as a process of simplification, in going from a colony of different prokaryotic cells to a single eukaryotic cell:

> By integrating a community of prokaryotes into a single cell, eukaryotes integrated many of the best features of that community, thereby localizing their functions. By taking prokaryote functions that were previously distributed and duplicated within the community, integration into single cells would have produced more energetically efficient, better buffered, and more robust systems. Integration of partial duplications, however, would have resulted in incomplete overlapping of some features, resulting in some functions being distributed within eukaryotic cells, including features such as multiple internal membranes; multiple (partial) genomes; both nuclear and cytoplasmic (mitochondrial) DNA; and genes only partially integrated and therefore characterized by having exons punctuated by introns.[5]

In this view, eukaryotic organisms did not develop completely new structures or capabilities unseen in prokaryotes but rather organized them more effectively. Interpreting biological evolution in this way is not necessarily at odds with the traditional belief that eukaryotic organisms separated from the prokaryotic long ago and pursued a completely different strategy for survival.[6]

A different view, called a *Circos* diagram (circos.ca),[7] of the detailed genetic relationships between organisms is diagrammed in Figure 7.4. The complete set of chromosomes of two hypothetical organisms, A and B, form the upper and lower halves of the circle, respectively. Organism A has 26 chromosomes and A, 23. In the arc formed by each numbered chromosome is the complete DNA sequence of that chromosome. Identical or similar sequences are indicated by lines connecting the chromosomes of A and B. If only a short sequence runs are identical, the width of the connecting line is small. If long runs are identical or identical but reversed, the line is wide. The chromosome numbering in this diagram is such that if the genes on one chromosome are in reverse order of those on the other chromosome, the connecting band would be uniform in width. If the sequence order is not reversed, the connecting band must show a constriction point midway, so as to properly connect to the other chromosome. (Diagrams are often constructed so that reverse identity shows the constriction.) If there are

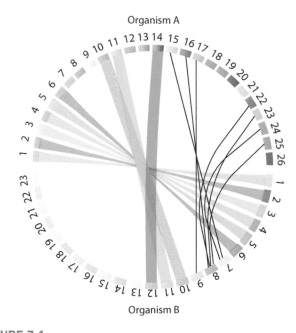

Organism A

Organism B

FIGURE 7.4
A diagram designed to compare the genetic sequences of two organisms, A and B. Numbered chromosomes are indicated in the outer part of the circle. A has 26 chromosomes, while B has 23. Bands with constrictions connecting chromosomes 1–5 of both organisms indicate these chromosomes are identical (or similar, depending on the illustrator's choice). The three wide bands with no constrictions indicate reverse identity of the chromosomes. See text. Narrow connecting lines indicate short sequences are identical. Based on Circos™ diagrams described by Krzywinski et al. (Ref. 7), but not drawn with Circos software. Circos software is free, licensed under GPL, www.gnu.org/copyleft/gpl.html.

no similar sequences on the organisms, there would be no connections in the diagram. In Figure 7.4, chromosomes 1 to 6 of the two organisms are apparently identical. Chromosomes 10 and 11 are reverse-identical. Chromosome 14 of organism A is reverse-identical to chromosome 12 of B. Small regions of chromosomes 15–17 of A connect to chromosomes 7–9 of B, although B chromosomes 7 and 8 also have connections to A chromosomes 22–25. The lack of connections between much of the rest of the chromosomes of A and B in this imagined example would suggest that while part of the phenotypes are the same, much is quite different. For comparisons between real organisms, see reference 7.

Such genetic connections between organisms can be used to inspect, in quantitative detail, the genetic similarities between any organisms of interest. If a Circos diagram of 6 organisms, believed to be evolutionarily related, were prepared, genetic similarities can then be quantified and communicated in a simple, visual way. Such diagrams are not uniquely specified by the genomes of the organisms, however, since the length of genetic sequences to be compared must be chosen. For example, genomes, consisting of hundreds of millions or billions of DNA bases, could be divided into blocks of 30-kilobase, 300-kilobase, 3-megabase, or whole genome sequences for comparisons to be made. Since the likelihood that a 30-kilobase sequence will have a similar sequence on another organism is intrinsically higher than that of a 3-megabase sequence, more connections will be found.

7.1.3 Prokaryotes: Parts

Except for the biologists among you, the classification scheme names in Figure 7.3 will be rather meaningless, and it is not the place of this book to imprint such names in your memories. Examples can help to familiarize us with at least some of the classes, phyla, etc. Among the bacterial prokaryotes are Gram-positive and Gram-negative bacteria. The "Gram" term is mostly historical, related to an experimental microscopic imaging technique that conveniently visualizes the differences, though does not to help explain them. To reveal details of microbe structure, most microscope techniques employ stains—organic dyes that bind selectively to certain structures. Modern-day stains, as discussed earlier, are quite sophisticated and can selectively bind to one particular protein that may be in a cell. The Gram protocol employs a less specific dye, crystal violet, a second dye, safranin, and a washing technique that distinguishes the colors of Gram-negative and Gram-positive bacteria. Gram-negative bacteria do not retain crystal *violet* color in the Gram staining protocol, and when the counterstain safranin or fuchsin is added after the crystal violet, Gram-negative bacteria are colored red or pink. Gram-positive bacteria retain the crystal violet dye. The structural differences causing this different staining behavior have to do with the bacterial cell wall. As we saw in Chapter 6, bacteria have a peptidoglycan layer made of a polymer of sugars and amino acids that forms a mesh-like layer outside the plasma membrane. Gram-negative bacteria have a much thinner peptidoglycan layer, have two cell membranes (versus one for Gram positive), and have a variety of lipoproteins and lipopolysaccharides in the outer membrane in addition to porin proteins. See Table 7.1 for some possibly familiar names of Gram-negative bacteria. Most of these bacteria can cause severe problems for us humans, but some (e.g., certain strains of *Escherichia coli*) exist naturally in the intestines, cause no trouble themselves, and probably benefit us by effectively blocking nastier bacteria from the environmental niche of the intestine.

The terms *cocci* (spherical) and *bacilli* (rodlike) refer to structural shapes of bacteria.

7.1.3.1 Prokaryotic cell structure

There are indeed a wide variety of prokaryotic cellular shapes and structures, which one can easily find through Web searches. Structures range from submicron diameter spherical *cocci* to 2- to 10-μm long rods, both straight and spiral, to macroscopic colonies of filaments, sheets, and balls, but the *E. coli* provides a wealth of understandable physical properties and behaviors for an introduction to biosystems text.

E. coli is the "hydrogen atom" of prokaryotes (arguably of all biological organisms). The reason for this designation is not primarily its small size or simple structure but rather its long research history since being discovered in 1885 by Theodor Escherich. We will follow the tradition of paying perhaps excessive attention to this cell, with the consequent danger of being misled as to the generality of its structure and behavior. *E. coli* is a much studied, rod-shaped bacterium common to the human internal and external environment. Figure 7.5a shows the structures responsible for the major activities of the bacterium. *Flagella* of length 2–4 μm drive the bacterium when an aqueous environment is available. (This will be a subject of Chapter 16.) *Pili* are responsible for interaction with external objects. In particular, pili have adhesin proteins that make sticky tips on the pili that bind to mucous membranes and cells. Some properties of this organism are listed in Table 7.2. As we have seen, ribosomes are responsible for protein synthesis in the cell. The genetic codes for more than 4,000 proteins are coded in the DNA, aggregated in the form of a nucleoid. The DNA is in the form of a single chromosome, but duplicates may be present in an active cell. Only about 2,600 of these proteins are actually in the cell at a given time, which tells us that even this simple organism must respond to environmental signals that trigger specific protein synthesis. The mesosome, an infolding of the plasma membrane, has a rather controversial history. The structure can be seen in electron micrographs, but several studies have shown that these were artifacts of the sample preparation. Similar structures have been observed after chemical or antibiotic treatment of the cells. The mesosome is indeed a real structure, but a reading of the literature suggests that at least some are formed in response to rather severe external perturbations and are not a normal part of *E. coli*'s suburban lifestyle. Similar structures in other bacteria have been observed with milder treatments of the cell, but the mesosome has largely disappeared

TABLE 7.1 Examples of Bacteria, Classified by the "Gram" Scheme

Gram Negative	Gram Positive
Escherichia coli	*Streptococcus*
Salmonella	*Staphylococcus*
Shigella	*Listeria*
Pseudomonas	*Enterococcus*
Heliobacteria[a]	Heliobacteria[a]
Legionella	*Mycoplasmas*
Cyanobacteria	*Spiroplasmas*
Green sulfur bacteria	*Lactobacilli*
Green nonsulfur bacteria	Eubacteria
Neisseria gonorrhoreae	
Neisseria meningitidus	
Hemophilus influenzae	
Klebsiella pneumoniae	
Legionella pneumophia	
Pseudomonas aeruginosa	
Heliobacter pylori	

[a] Photosynthetic heliobacteria stain Gram negative but are classified as Firmicutes, the largest group of Gram-positive bacteria, by RNA tests.

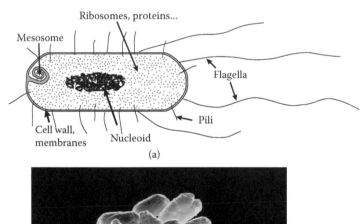

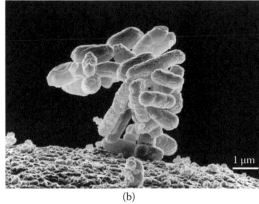

FIGURE 7.5

E. Coli bacterium. (a) schematic diagram showing the major components. *E. Coli* uses flagella for propulsion and pili for detection of and interaction with nearby objects or cells. DNA is stored as a nucleoid, rather than in a cell nucleus, as in eukaryotes. There are thousands of ~25-nm-diameter ribosomes in the cell, where proteins are synthesized. (b) electron micrograph of *E. Coli* colony. Sample preparation has likely removed pili and flagella. (Public domain image by Erbe, E. and Pooley, C., U.S. Department of agriculture.)

TABLE 7.2 *E. Coli* Physical–Chemical Properties

Characteristic	Property		Notes
Rod length	2 μm		
Rod diameter	0.5–1 μm		
Rod volume	0.6–0.7 μm^3		
Mass	2×10^{-15} kg		
Molecular composition	H_2O	70%	
	Protein	15%	~2600 different proteins in cell
	Nucleic acids		
	DNA	1%	$M_{DNA} \sim 2.5 \times 10^9$ D
	RNA	6%	
	Saccharides	3%	
	Lipids	2%	
	Other organics	1%	
	Inorganic ions	1%	

from the literature during the past 30 years.[8] Nevertheless, the structure is an attractive scheme for increasing the effective surface area of the membrane. The homework will probe the issue of cell surface/volume ratio and show that for small cells the ratio is only a borderline physiological problem for bacteria. For larger cells, the ratio implies invaginations of a slightly different sort may be indispensable.

7.1.3.2 Cyanobacteria (blue-green algae) and other photosynthetic bacteria

The intent of this discussion of prokaryotic and eukaryotic cell structures is not to comprehensively survey the many organisms and structures living on Earth, but to see representative structures and cell types, especially as they relate to physical analyses we do in this text. There are indeed many other fascinating and even simple prokaryotic organisms, but most bring no major new structure or process to the table—except for the cyanobacteria. The cyanobacteria, formerly called blue-green algae, are unique among the prokaryotes in their extraction of energy from the environment through photosynthesis. Table 7.3 lists a few of these photosynthetic prokaryotes.

Figure 7.6a shows an optical micrograph of an example cyanobacterium, *Anabaena scheremetievi*, from the Cyanobacterial Image Gallery http://www-cyanosite.bio.purdue.edu/images/images.html. Notice that the "bacteria" form filamentous structures of many individual cells. Many such filaments are not simply linear arrays of cells, but of cells inside a protective, tubular sheath made of polysaccharides and other polymers. In most cyanobacteria the photosynthetic apparatus, the set of protein complexes that absorb light, transfer electrons, produce O_2 and synthesize glucose, is embedded in disklike folds of the cell membrane, called *thylakoids* (Figure 7.7). Thylakoids usually form interconnected stacks called *grana*, but can also line the inner side of the cell membrane in certain cyanobacteria (Figure 7.7a). Photosynthesis in cyanobacteria generally uses electrons from H_2O or H_2S (hydrogen sulfide) and produces O_2 as a by-product. Carbon dioxide is reduced to form carbohydrates via the Calvin cycle. The oxygen in the atmosphere is believed to have been originally generated by the activities of ancient cyanobacteria. Owing to their ability to fix nitrogen in aerobic conditions, they are often found as symbionts with a number of other groups of organisms such as fungi, lichens, and corals.

Cyanobacteria are the only group of organisms able to reduce nitrogen and carbon in aerobic conditions. The water-oxidizing photosynthesis is accomplished by coupling the activity of photosystem (PS) II and I, generally believed to be organized in a "Z-scheme" (see Chapter 10). Without O_2 they are also able to use only PS I with electron donors other than water—H_2S, thio-sulfate ($S_2O_3^{2-}$), or H_2, like the purple photosynthetic bacteria (see, e.g., last row in Table 7.3). The plasma membrane of these photosynthetic bacteria contains only components of the respiratory chain, while the thylakoid membrane hosts both respiratory and photosynthetic electron transport.

Phycobilisomes act as light-harvesting antennae for the photosynthetic bacteria. Pigments bound to the phycobiliproteins are responsible for the blue-green color of most cyanobacteria. Variations are mainly due to carotenoids and phycoerythrins, which give the cells the red-brownish color. The color of the ambient light influences the composition of phycobilisomes in some cyanobacteria: in green light, cells accumulate more phycoerythrin; in red, more phycocyanin. The result is that the bacteria appear green in red light and red in green light, but the importance is that it maximizes use of available light for photosynthesis.

TABLE 7.3   Photosynthetic Bacteria

Name	Notes
Prochlorococcus	Accounts for over half the photosynthesis in the oceans; has the smallest cyanobacterium genome, 1.7 Mb (Mb = 106 base pairs)
Synechocystis sp. PCC6803	The third prokaryote, and first photosynthetic organism to have its genome sequenced
[Hormogonia]	Motile filaments that travel away from the main colony to start anew [A colonial structure, not a genus or species.]
Nostoc punctiforme	Has the largest known cyanobacterium genome, 9 Mb
Aphanizomenon flos-aquae, Arthrospira platensis (Spirulina)	Sold as food supplements; form mobile spiral colonies
Anabaena	Found as plankton; known for its nitrogen-fixing abilities, and they form symbiotic relationships with certain plants; produce neurotoxins; genome sequenced: 7.2 Mb
Chloroflexus	Isolated from hot springs; a green nonsulfur bacteria
Lyngbya	Causes "swimmer's itch"; some species damage aquatic ecosystems by forming dense floating mats; invasive species
Microcystis aeruginosa	Causes algal blooms in lakes; turns water an opaque green in the summer, resulting in little recreational use of the lake; standards for dissolved O_2 routinely violated; neurotoxins may be produced
Nostochopsis	Freshwater; grow attached to mosses and submersed wood in clean streams; tropical
Oscillatoria	Filamentous colonies; can slide back and forth against each to reorient to light; found in watering-troughs; *Membrancea* species can cause Black Band disease on reefs
Phormidium	Filaments are long, cylindrical, curved; The filaments move by gliding, creeping, rotating, or oscillating both inside and outside of the sheaths; capable of producing bioactive compounds
Prosthecochloris aestuarii	A green sulfur bacterium; anaerobic, found in sulfide-rich environments (e.g., hot springs); their unique light-harvesting antennae allow growth at light intensities under which no other phototrophs can survive; commonly used for photosynthesis research
Rhodospirillum rubrum, R. molischianum	Purple photosynthetic bacteria; can live either without O_2 and without light, or with O_2 and with light; used in photosynthesis research

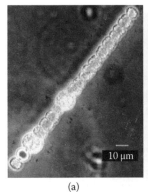 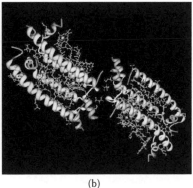

(a)   (b)

FIGURE 7.6

Cyanobacteria and light-harvesting protein complexes. (a) *A. scheremetievi* cyanobacterium 400× phase contrast optical microscope image. The bacteria form filamentous colonies. Distinct cell sizes can be seen; some of the smaller cells appear to be doubled or may have just divided. Depending on the species, cells can be surrounded by protective tubular sheaths, and observed moving through the sheaths or moving the sheath with them. (Courtesy of Cyanobacterial Image Gallery, http://www-cyanosite. bio.purdue.edu/images/lgimages/; credit: Roger Burks, University of California, Riverside, CA; Mark Schneegurt, Wichita State University, Wichita, KS.) (b) The crystal structure of the light-harvesting complex II (B800–850) from *R. molischianum*. The heart of photosynthesis in cyano- and other photosynthetic bacteria is bacteriochlorophyll and other pigment molecules attached to proteins and oriented at very particular angles and distances apart. Such light-harvesting complexes are usually located in thylakoid membranes (Figure 7.7). (PDB ID 1LGH, Koepke, J. et al., *Structure*, 4, 581–597, displayed using Protein Workshop, 1996.)

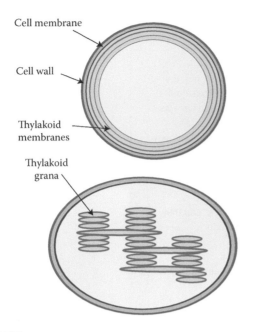

Cell membrane

Cell wall

Thylakoid membranes

Thylakoid grana

FIGURE 7.7
Diagram of thylakoid membranes. (*Upper*) Thylakoid membranes line the inner surface of the cell membrane in certain cyanobacteria. (*Lower*) thylakoid grana form stacks in the interior of cells. In plant cells, such grana and more extended chlorophyll-containing membranes fill most of the interior of chloroplasts, football-shaped subcellular structures.

Many cyanobacteria form "algal blooms" under certain conditions, as most pond owners know. Most years such an algal bloom occurs in a river, lake, bay, or gulf, some of them producing toxins that poison aquatic life and humans who may eat contaminated seafood or drink water that has had such a bloom. The cyanobacteria are proving to be a source of many novel organic compounds with biological activity. Information about such toxins, as well as cyanobacteria research in general, can be found at a Web site hosted by Purdue University, http://www-cyanosite.bio.purdue.edu/.

7.1.3.3 Halobacteria

A small but versatile class of bacteria, halobacteria, have been important to research for a number of reasons. First, they are a type of archaea found in environments where large amounts of salt, moisture, and organic material are available. Second, halobacteria can grow aerobically, anaerobically, or by means of photosynthesis. Parts of the membranes of halobacteria are purplish in color from the pigment bacteriorhodopsin (BR), related to the retinal pigment rhodopsin. This relationship to the visual pigment of humans plus the relative ease of handling the bacterial version (the visual pigments are extremely sensitive to light and O_2), has made the "purple membrane" patches of halobacteria a favorite research project. Purple membrane patches consist of bacteriorhodopsin plus lipids—five phosphatidyl glycerophosphates (per BR), two glycolipid sulfates, 0.5 phosphatidyl glycerols, and 0.5 phosphatidyl glycerosulfates, all containing dihydrophytyl chains, as well as one squalene molecule per bacteriorhodopsin monomer. This pigment absorbs light and makes a series of very rapid structural changes, which ends up providing energy to transport a protein and produce ATP. Figure 7.8 shows these structural changes. These X-ray crystallography structures represent a major experimental step forward, as prior to the late 1990s, crystals appropriate for crystallography could not be grown with both protein and lipid in a form adequate for high-resolution (0.3 nm or less) structures to be determined. Halobacteria also possess a second pigment, halorhodopsin, which pumps chloride ions in the cell in response to photons, creating a voltage gradient and assisting in the production of energy from light.

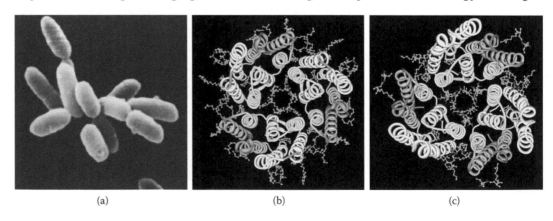

(a)   (b)   (c)

FIGURE 7.8
Halobacteria and the bacteriorhodopsin pigment-protein-lipid complex responsible for color light-induced structural changes that drive ATP generation. (a) Electron micrograph of a small colony of bacteria. (Public domain image from NASA, Washington, DC.) (b) Bacteriorhodopsin protein-lipid complex. (Luecke, H. et al., *J. Mol. Biol.*, 291, 899–911, 1999., PDB ID 1C3W, displayed by Protein Workshop software. 21) (c) Bacteriorhodopsin protein-lipid complex illuminated by strong light. The light energy drives the structure to a different state, which can be compared to the no-light structure in *a*. The conformational change ultimately results in transport of H+ and ATP generation. (Takeda, K. et al., *J. Mol. Biol.* 341, 1023, 2004., PDB ID 1DZE, displayed with Protein Workshop software.)

The process is unrelated to other forms of photosynthesis involving electron transport, however, and halobacteria are incapable of fixing carbon from carbon dioxide. The purple membrane is a prime example of the conversion of light energy to electric energy (H^+ transferred) to chemical energy (ATP) through the primary drive of a light-induced isomerization.

7.1.4 Eukaryotes: Parts

We usually think of eukaryotic organisms as big animals and plants. However, the fundamental way biologists presently classify organisms is through cellular architecture—the parts—not size. Figure 7.9 shows the smallest eukaryote presently known, *Ostreococcus*. *Ostreococcus* is a genus of unicellular coccoid (spherical) green alga, discovered in 1994 in the Thau lagoon (France) by Courties et al.[9] *Ostreococcus* has subsequently been found in many oceanic regions. It is the smallest known free-living eukaryote with an average size of 0.8 μm. Its genome sequence (12.56 Mb nuclear genome) was published in 2006. *Ostreococcus tauri* has a nucleus (with 14 linear chromosomes), one chloroplast, and several mitochondria. All three of these structures (nucleus, chloroplast, and mitochondria) indicate that *Ostreococcus* is eukaryotic.

In the following sections, we will use the structures inside the *Ostreococcus* cell to learn about the parts of eukaryotic cells in general. We will, of course, miss many details, but our purpose here is not to be complete but rather to be able to connect some of the analyses that we will do in later chapters (e.g., photosynthesis, molecular motors, energy production) to structures in the cell. We borrow heavily from a new type of publication that has appeared in recent years—the open-access article. The specific article is "3-D Ultrastructure of *O. tauri*: Electron cryotomography of an entire eukaryotic cell" (Henderson et al.,[10] which can be found at http://www.plosone.org/article/info:doi/10.1371/journal.pone.0000749). The three-dimensional structures we will examine in the following sections come not as sketched structures but from measured images, determined from electron microscopy tomography. A wonderful animation of the entire series of images can also be downloaded at the cited Web site and can also be found in the DVD associated with this textbook.

An overview of the subcellular structures found inside an *Ostreococcus* cell by electron tomography is contained in Figure 7.10. Figure 7.10a and b shows nondividing

> "Green algae" (as opposed to "bluegreen algae," renamed cyanobacteria) are photosynthetic eukaryotes, like green plants.

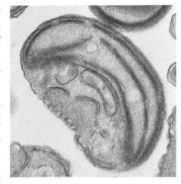

FIGURE 7.9
The picoplankton green alga *Ostreococcus* (Prasinophyceae, Chlorophyta) strain RCC 143 observed by transmission electron microscopy (TEM). *Ostreococcus* is the smallest known free-living eukaryote with an average size of 0.8 μm. (Courtesy of Eikrem, E., and Throndsen, J., University of Oslo, Oslo, Norway.)

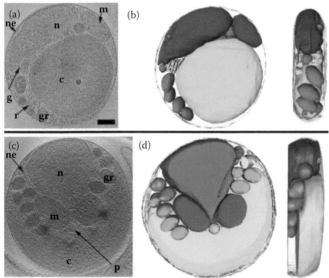

FIGURE 7.10
Overview of a small eukaryotic cell: cross sections and three-dimensional segmentations of *O. tauri* cells. Each row shows a single slice (a, 21.6 nm; b, 36 nm; c, 9.6 nm) through the middle (left) and a manually segmented model (two perpendicular views, right) of one particular cell: (a, b) a nondividing cell harvested at the dark-to-light transition; (c, d) a dividing cell harvested at the light-to-dark transition. Letters and colors identify nuclei (n, red), nuclear envelope (ne), chloroplasts (c, green), mitochondria (m, dark purple), Golgi bodies (g, yellow), peroxisomes (p, orange), granules (gr, dark blue), inner membranes including ER (er, light blue), microtubules (light purple), and ribosome-like particles (r). Scale bar 250 nm. (From Henderson, G.P. et al., *PLoS ONE*, 2, e749, 2007.) An open-access article distributed under the terms of the Creative Commons Attribution License, which permits unrestricted use, distribution, and reproduction in any medium, provided the original author and source are credited. A very helpful video clip of these images can be found at the journal Web site, http://www.plosone.org/article/info%3Adoi%2F10.1371%2Fjournal.pone.0000749. These images, as well as others throughout the book, were determined by cryo-electron microscopy, whose developers won the 2017 Nobel Prize in Chemistry.

Note you can do an Internet search on the "doi" (digital object identifier) number, 10.1371/journal. pone.0000749, to find the article. As online references sometimes disappear, be prepared to use Internet search engines, Wikipedia (look for article reference lists), and importantly, your library search facilities.

cell harvested at the dark-to-light transition; Figure 7.10c and d shows a dividing cell harvested at the light-to-dark transition. Letters and colors identify nuclei, nuclear envelope, chloroplasts, mitochondria, Golgi bodies, peroxisomes, granules, inner membranes including the endoplasmic reticulum, microtubules, and ribosome-like particles. The simple enumeration of the structures reveals some useful understanding about this eukaryotic cell:

1. Cells divide (reproduce) in response to some stimulus. In the case of this photosynthetic cell, a natural source of this stimulus is light, the source of the cell's energy.
2. The nucleus is enclosed by a nuclear envelope. In contrast to the nucleoid of prokaryotic cells, in which the DNA is exposed to the solution inside the cell, eukaryotic nuclei have a surrounding membrane. Access to the DNA is thus restricted, and cell division will require the envelope to somehow rupture.
3. The chloroplast takes up a major fraction of the intracellular volume. A relatively large power plant makes sense for this small, photosynthetic eukaryotic cell. That the chloroplast is disk-shaped brings up the question of whether the cell might preferentially orient its flat side toward the source of light, possibly maximizing light absorption. Problem 7.8 explores this issue.

7.2 PHYSIOLOGY: SELECTIVE OVERVIEW

Many of the "facts" stated in this section are gleaned from common resources, such as Wikipedia (no direct quotes, though) and several biochemistry textbooks, supplemented by recent referenced articles. However, the overview is intended to build a physiology framework into which later Chapter topics may be fit. It is neither a review of the latest research nor a comprehensive overview of the latest research nor a comprehensive overview of physiology.

Physiology, the study of the mechanical and biochemical functions of living organisms, is a major focus of the rest of this book. We will, however, go rather light on "biochemical" functions, maintaining some blissful ignorance but moving more efficiently toward the goal of understanding some major physical principles that apply to biological systems: reactions, motions, organization, management of quantum effects, management of a chaotic microworld. In the following sections we survey some of the major subcellular structures found in the *Ostreococcus* cell, with comparisons to larger eukaryotic cells.

7.2.1 Energy Transduction: Mitochondrion

The mitochondrion is a membrane-enclosed organelle found in most eukaryotic cells. The primary role of the mitochondrion is production of ATP, as reflected by the large number of proteins in the inner membrane for this task. ATP regeneration from ADP is done by oxidizing glucose, pyruvate, and NADH produced in the cytosol. Mitochondria generate most of the cell's ATP, used as an almost universal source of molecular chemical energy in living systems. Production of ATP is remarkably large scale. In humans, the ATP turned over (generated, used, and regenerated) each day is the equivalent to the weight of the entire body![11] This turnover is so startling that the authors of that article close with the statement:

> Thus, although a camel cannot pass through the eye of a needle, it is extraordinary to imagine that every day its body weight's equivalent in metabolites tunnel backward and forward through an integral membrane aperture ≈6 orders of magnitude smaller in diameter.

In 2016, Anna Karnkowska and Vladimir Hampl, discovered the first eukaryotic microbe, *Monocercomonoides*, which lacks a mitochondria.

We explore this number further in homework.

Mitochondria contain both an inner and outer membrane. ATP synthesis in the mitochondrion is coupled to a transmembrane voltage by the enzyme, ATP-synthase. ATP regenerated from ADP is actively (requiring energy) exported across the inner mitochondrial membrane by a transmembrane protein called the ADP/ATP "antiporter." In contrast, the transport of ATP, ADP, and other molecules across the outer mitochondrial membrane is passive, with the major pathway being through a voltage-dependent anion channel (VDAC), which is a porin protein. Mitochondria also manage other processes, such as signaling, differentiation/ specialization of the cell, cell death, control of the cell cycle, and cell growth.

Figure 7.11 shows the single mitochondrion from *Ostreococcus*. The number of mitochondria per cell varies by organism and cell type, ranging from one, like in *Ostreococcus*, to thousands in large cells. The placement of the mitochondrion, next to the cell nucleus, can be seen in Figure 7.10. The major components of the mitochondrion are the outer membrane, inner membrane, cristae (or internal compartments), and dense granules within the mitochondrial matrix. The spatial slices used in these tomographic images, as well as the scale indicates that major parts of the mitochondrion range from 5 to 50 nm. The overall dimensions of mitochondria in large eukaryotic cells are larger, 1–2 μm, larger than the entire *Ostreococcus* cell and comparable to many bacterial cells.

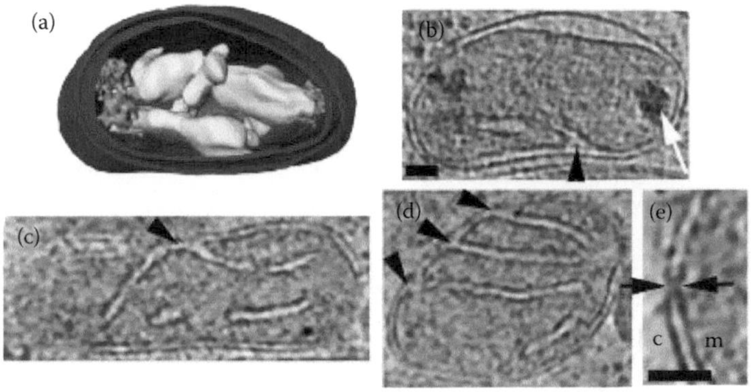

FIGURE 7.11

Mitochondria. (a) Three-dimensional segmentation of the mitochondrion from the cell shown in Figure 7.10. Membranes are colored purple (outer membrane), red (inner membrane), and yellow (cristae). Dense granules within the mitochondrial matrix are shown in green. (b–d) The 15-, 55-, and 29-nm-thick slices though three different mitochondria, showing crista junctions (arrowheads) and a dense granule (white arrow). (b) A slice through the cell shown in panel (a). Scale bar 50 nm (for b–d). (e) A 4.8-nm-thick slice through a junction or channel (black arrows) connecting the outer and inner membranes. c, cytoplasm; m, mitochondrion. Scale bar 50 nm. (*Source*: same as in Figure 7.10. doi:10.1371/journal.pone.0000749.g009.)

The compartments in mitochondria carry out specialized functions, most of which involve protein enzymes. In humans, more than 60 types of proteins have been identified from heart muscle mitochondria. The complement of enzymes in the mitochondria is not static but varies depending on what activities are being carried out. This variation of protein population implies that genes for protein production are being actively controlled. Most of a cell's DNA is in the cell nucleus, but the mitochondrion has its own DNA, allowing more rapid control of protein construction. The mitochondrial genome shows similarity to bacterial genomes, supporting assertions that eukaryotes consolidated functions of various different prokaryotic cells.

7.2.2 Energy Generation: Eukaryotic Photosynthetic Cells and the Chloroplast

Chloroplasts capture light energy to convert ADP to its higher-energy form, ATP, and NADP to its higher-energy and more electron-donating (H-donating) form, NADPH, in the process of photosynthesis. As we will spend an entire Chapter discussing photosynthesis, this section will be brief and focus on structural issues in chloroplasts.

Chloroplasts are observed as flat discs usually 2–10 μm in diameter and 1–2 μm thick. Refer back to Figure 1.6e. The chloroplast is housed in a double-membrane layer, with an intervening intermembrane space. The *Ostreococcus* cell has one chloroplast; higher plants have 10–200. Figure 7.12 shows the *Ostreococcus* cell with its single chloroplast.

Some biological definitions useful for further reading in the literature:

- Stroma: aqueous fluid inside the chloroplast
- Thylakoid: disklike membrane containing chlorophyll and other pigments and proteins
- Thylakoid lumen: inside of the thylakoid
- Grana: stacks of thylakoids

The chloroplast stroma contains one or more molecules of small, circular DNA, at least in higher plants, and a few ribosomes. Most chloroplast proteins are encoded by genes in the cell nucleus, produced outside and then transported to the chloroplast. Photosynthesis takes place in the thylakoid membrane, which appears in electron microscope images as about 10 nm thick. Antenna, or light-harvesting complexes, consisting of the light-absorbing pigments (chlorophyll, carotenoids,

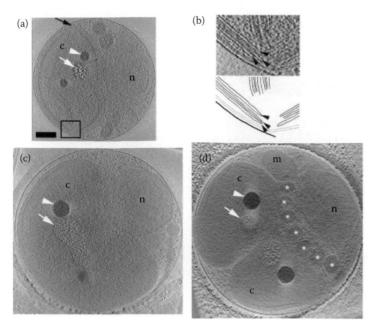

FIGURE 7.12

The chloroplast. (a) A 7.2-nm-thick slice through a nondividing cell. The starch granule (white arrow) has suffered damage from the electron beam. Besides it are two dark granules (white arrowhead). (b, Top) Enlarged view of the boxed area in panel (a). The three thylakoid membranes (black arrowhead) can be seen forming a stack. (Bottom) Schematic of above. Cell membrane (black), chloroplast membranes (green), outer thylakoid membrane (red), inner thylakoid membranes (blue). (c) A 24-nm-thick slice through an early predivisional cell, where the chloroplast is kidney-shaped rather than oval and the starch granule is elongated. (d) A 36-nm-thick slice through a late predivisional cell, where the chloroplast is deeply constricted and one dark and one starch granule is found in each side. Here the cytoplasmic granules (asterisk) are arranged in a V-shape, pointing to the division plane. Scale bar 250 nm. (*Source*: same as in Figure 7.10. doi:10.1371/journal. pone.0000749.g006.)

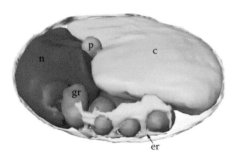

FIGURE 7.13

The endoplasmic reticulum (ER). Three-dimensional segmentation of a cell harvested at the dark to light transition. ER (light blue) forms a sheet near the edge of the cell, perforated by four granules (dark blue).

and some occasional other pigments) bound to proteins, are imbedded in the thylakoid membrane. Refer to Figure 1.6f. The energy of the absorbed photons is absorbed by the pigments and transferred to reaction center complexes through resonance energy transfer.

7.2.3 Endoplasmic Reticula and the Golgi Apparatus

Figure 7.13 shows the endoplasmic reticulum (ER) for the *Ostreococcus* cell. The ER is a membrane-based eukaryotic organelle that forms an interconnected network of tubules, vesicles, and sacs inside cells. ER structure is supported by the cytoskeleton. The ER is responsible for protein translation, folding and transport of proteins to be used in the cell membrane, storage and release of calcium, production and storage of glycogen, steroids, and some other macromolecules. The ER is part of a protein-sorting pathway—in essence, the protein transportation system of the cell. The basic structure and composition of the ER membrane is similar to the plasma membrane. Most larger eukaryotic cells contain three types of ER: "rough," "smooth," and sarcoplasmic.

The surface of the rough endoplasmic reticulum (RER) is studded with protein-manufacturing ribosomes, giving it a rough appearance. Early in a cell's life, however, the RER appears smooth, owing to a lack of bound ribosomes. Binding to the ER occurs once the ribosomes begin to synthesize a protein. The free ribosome synthesizes a polypeptide sequence until a cell signaling particle recognizes an initial sequence of 5–15 hydrophobic amino acids, preceded by a positively charged amino acid. The complex then loops the sequence through the RER membrane, and the short amino acid sequence is cleaved off within the interior of the RER.

The membrane of the RER is continuous with the outer layer of the nuclear envelope. Although there is no continuous membrane between the RER and the Golgi apparatus, where proteins are further processed, vesicles shuttle proteins between these two compartments. Vesicles are surrounded by coating proteins, which target vesicles to the Golgi. The coating proteins mark the vesicles to be brought back to the RER. The RER and the Golgi apparatus collaborate to target newly synthesized proteins to their proper destinations. (Sounds a bit like the Walmart transportation system.) Lipids and certain other small molecules can move out of the RER through membrane contact sites, where the membranes of the ER and other organelles come into close contact.

The smooth endoplasmic reticulum (SER) is involved in synthesis of lipids and steroids, drug detoxification, metabolism of carbohydrates, regulation of Ca^{2+} concentration (a universal signaling ion), and attachment of receptor groups to cell membrane proteins (cell recognition). The SER consists of tubules and vesicles that form a branching network, allowing increased surface area for the action or storage of important enzymes.

The sarcoplasmic reticulum (SR) is a special type of smooth ER found in muscle. The only structural difference between this organelle and the SER is the set of proteins. This protein difference reflects their respective functions: the SER synthesizes molecules, while the SR stores and pumps Ca^{2+} ions. The SR stores large amounts of calcium, which it releases when the muscle cell is stimulated. The SR's release of calcium upon electrical stimulation of the cell is necessary for contraction of muscle fibers: Ca^{2+} controls these molecular motors.

7.3 REPRODUCTION, DNA, AND THE CELL NUCLEUS

Students who want to learn about reproduction probably have, in one way or another. A few of you may have learned about the cellular processes in a biology or physiology course. We will not attempt to review or overview much of this. We will concentrate again on structures as they relate to physical processes of interest to us later in this book. Some of these processes involve untwisting DNA, wrapping DNA around 8-nm nanospheres (histone complexes), controlling random tangling of DNA, and stuffing DNA into a round bag without tangling it (viral DNA).

In eukaryotic cells, DNA is stored in the cell nucleus (Figure 7.14). The interior of the nucleus is not uniform. Several subnuclear structures exist, made up of proteins, RNA, and even particular parts of the chromosomes. The main structures in the nucleus are the following:

- Nuclear envelope: double membrane that encloses the entire organelle and separates its contents from the cellular cytoplasm
- Nuclear lamina: meshwork within the nucleus that adds mechanical support, much like the cytoskeleton supports the cell as a whole
- Chromosomes: contain DNA and proteins that aid in the organization of the DNA
- Nucleolus: manufactures ribosomes

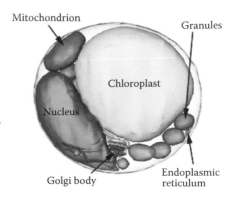

FIGURE 7.14
The nucleus takes up a major portion of the *Ostreococcus* cell. There is rather little room outside of the organelles: the cytoplasm is quite restricted. Compare this to the "minimal cell" of Figure 7.1.

7.3.1 Nuclear Envelope and the Nuclear Pore Complex

The nuclear envelope must have pores that extend through both membranes to allow movement of molecules and ions. Ions and small molecules passively diffuse through the pores, but transport of larger molecules (e.g., proteins) is controlled by other membrane proteins and requires energy input. Movement of enzymes and other molecules through the pores is required for both gene expression—the triggering of transcription of certain, specific regions of the DNA—as well as putting the chromosomal DNA back into its normal state after it has been partially unraveled to express genes or repaired by enzymes.

Some of the nuclear pores (nucleoporins) are similar to the porin proteins we reviewed before, though larger, to span the double membrane. There are about 2,000 nuclear pore complexes in the nuclear envelope of a vertebrate cell. Half of the nucleoporins contain either a membrane-spanning alpha solenoid or a twisted β sheet structures we saw in porins previously. The other nucleoporins appear to be unstructured, flexible proteins that contain many amino acid sequence repeats of Phe-Gly.

The entire nuclear pore complex (NPC) is large for a porin. The overall diameter is about 120 nm; the hole has an effective diameter of ~9 nm and depth of ~200 nm; the molecular mass is about 120 MD in mammals, with 30 types of proteins in the complex.[12] A protein complex of this large size will not likely be found in the Protein Data Bank (PDB) database of X-ray and nuclear magnetic resonance structures. However, electron microscopy, with its resolution of 10 nm or so, is an ideal tool to determine at least a coarse structure. This has been done, and one image is shown in Figure 7.15. The small end of the pore is shown in the diagram. Note that the authors believe the opposite, wider end of the pore is occupied by a particle in transport through the pore. Some other helpful electron microscope–based diagrams and animations of this complex can be seen at http://sspatel.googlepages.com/nuclearporecomplex2. The pore may be able to expand to 25 nm or so to allow passage of larger molecules. Small particles (<30 kDa) are able to pass through the NPC by passive diffusion. Passage of larger molecules through the pore requires several other proteins. Note that at least in the *Ostreococcus* cell (Figures 7.13 and 7.14), molecules that pass in and out of the nucleus are quite likely to come against or come from another organelle inside the cell; there is not a large free volume of cytosol outside the nucleus.

Finally, we may ask, how did such a large and complex structure like the nuclear pore complex get assembled in the first place? The "minimal cell" does not have any hint of such a

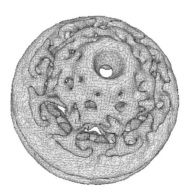

FIGURE 7.15
Nuclear pore complex (NPC). The nuclear membrane must have pores to control and allow transport of molecules and ions in and out. The large eukaryotic NPC shown here is from *Dictyostelium discoideum*, a species of soil-living amoeba. (Beck, M. et al., *Science*, 306, 1387–1390, 2004, EMDB ID 1097, displayed using EmViewer software.)

structure, though it could, of course, be manufactured later and put in place. The assembly of the NPCs is not completely clear, but it appears our old friend the vault protein has now been implicated in the process.[13]

7.3.2 Nucleolus

The nucleolus is an active body that is involved in the assembly of ribosomes. After being produced, these ribosomes are exported out of the nucleus. The nucleolus has also been studied extensively by electron microscopy. Many images can be viewed at http://www.uni-mainz. de/FB/Medizin/Anatomie/workshop/EM/EMNucleolus.html, though it helps if you can read German. Since nucleoli synthesize subunits that form ribosomes, the cell's protein-producing factories, the size and organization of the nucleolus depends on the ribosomal requirements of the type of cell in which it is found. In cells that produce large amounts of protein, significant numbers of ribosomes are needed, and the nucleolus is large, as much as 25% of the nuclear volume.

7.3.3 Cell Division: Some Major Structural Changes Are Needed

In higher eukaryotes, cell division (mitosis) requires partial or complete breakdown of the nuclear envelope, disassembly of nuclear pore complexes, and formation of mitotic "spindles." This last process, spindle formation, requires involvement of microtubules and molecular motors to form the structures. The nuclear envelope reforms and then encloses the chromosomes later. The traditional rule in lower eukaryotes, like yeast, has been that the envelope remains intact. The cited studies on *Ostreococcus*, however, show that its nuclear envelopes are mostly open, so the "rules" are not followed in every organism: gaps hundreds of nanometers in size were observed in most stages of cell growth. In some rare cases, the nuclear membrane did not come close to enclosing the DNA in the nucleus. We will not investigate the complicated process of cell division any further, but it is clear that major forces are involved—forces that require a sufficient magnitude, but more importantly, the organization and synchronization of those forces. A serious study of these processes is well beyond the scope of this text. A review of this sort of coordination of forces can be found in the article by J. Howard,[14] to which we will return in Chapter 16.

7.4 SENSORS AND RECOGNITION: RESPONDING TO THE OUTSIDE WORLD WITHOUT EYES

Individual cells have no "eyes" to see the environment around them, yet they can (and must) certainly sense and respond to their surroundings. Microorganisms could not survive very well if they could not respond, in some way, to the presence of nearby food. "Sensors" and "recognition" devices are very common to human-engineered systems. We recognize directions and rules of the road by seeing signs as we drive or hike on a trail. We put temperature sensors, thermometers, in our homes and ovens to trigger heating systems to turn on or off. The medical profession relies heavily on a variety of sensors to diagnose health and disease. At present, a database search, or a search through the index of a biochemistry or biophysics book, for biological sensors or recognition turns up disorganized, but numerous bits of information related to cellular recognition or sensing. Some of the specific terms related to cell recognition or sensing include cell adhesion molecules, membrane glycoproteins, fibronectin, laminin, membrane protein, membrane receptor, oligosaccharide, sphingolipid, B cells, protein function in cell membranes, cell recognition proteins, immune system, glycoconjugate, moesin, pattern recognition receptor, NK cells, immunoglobulin superfamily, O-glycans, antigens, immune system, T cells, CD48, TPBG, CD8, CD180, RP105/MD1, PCDH7, HLA-DPB1—on and on for hundreds of entries in the Wikipedia search engine. Clearly, eukaryotic and even prokaryotic cells have sophisticated systems to recognize specific particles around them. How might these work, in general?

7.4.1 Sensing through Diffusion

The simplest and slowest way to sense the environment is through passive diffusion of useful molecules through the outer cell membrane. If ATP or Ca^{2+} appeared in a cell's surroundings, the fact that many cellular processes depends on these molecules would influence the cell's behavior. In this case, there seems to be no sensor at all: the enzyme that may use ATP to pump charge across a membrane simply responds to the presence of more fuel with increased activity. Before we rule out random diffusion as an important part of sensing and recognition, however, consider our hydrogen atom of prokaryotic cells, the *E. coli*. We will discuss this issue more in the molecular motor chapter, but if a molecular motor rotating the flagellum in an *E. coli* cell runs more rapidly, or at least differently, in response to some dissolved ion or molecule in the environment, the *E. coli* will move differently than if the molecule was not there. This constitutes, or at least can constitute, a primitive sensing system. Of course, the simplest design for such a sensing system—a motor enzyme that speeds up and moves the cell in response to "food" in the surroundings—is exactly wrong. The bacterium would keep moving when food was present and stop when food was absent: not a good strategy for finding food, but a response system that does exactly the wrong thing every time is not far from the optimum. (More experienced code writers or spreadsheet aficionados may want to take on the homework miniproject related to such foraging.)

7.4.2 Purposeful and Random Movement: Preliminary Comments

One of the primary reasons for sensors is to do one or more of the following: (1) determine the better of several responses, (2) determine the direction of a response (increase/decrease, up/down, forward/backward), (3) determine the magnitude of a response, or (4) trigger another sensor to perform additional measurements. (With the hundreds or thousands of known biological sensors, there are certainly many more classes of actions that can be taken, but the few above will get us started in our thinking for Chapter 13.) These actions resulting from any sensing system in a cell have an intrinsic spatial scale of the cell size or less, if the system is a subcellular system: distances of 1 μm or less. As we saw from the Brownian motion images (video clips) in Chapter 1, random motions of objects the size of small cells occur over distances of micrometers on a time scale of seconds or less. How should any cellular sense/response system operate in the presence of constant random motion of comparable magnitude to its own capabilities? Suppose in our walking down a hallway we would, for every intentional forward step, randomly take a step backward, forward, left, right, or any angle in between? We would likely collide frequently with other pedestrians, but if we survived the collisions we would indeed eventually make it down the hallway. The higher likelihood of the forward step would eventually bring us to our destination, at least in terms of longitudinal distance. A biological molecular motor that relies on diffusion of ATP from its surrounding has a similar problem. If the motor takes a "step" once every millisecond, it will be able to step only if an ATP molecule diffuses to the motor faster than once per millisecond. Even if the diffusion time is comparable to the step rate^{-1}, there will statistically be many millisecond time intervals in which the motor will have no fuel. Can a microscopic biological system deal with such an environment? Obviously, cells do, but our question, still incompletely answered at present, is whether the "noise" inherent in the microscopic biosystem is simply tolerated by the cellular systems, like the random walker making her way down the hallway, or is the "noise" somehow filtered or rectified by the system? One can even imagine a rectifying system that allows steps in the forward direction, but not in the reverse... perhaps the system doesn't even need an energy input... perhaps a microscopic perpetual motion machine? We address these issues in Chapter 13.

7.4.3 Cell–Cell Recognition: Example

We cannot afford the time to review the many ways in which cells recognize the outside world. Cells have the following types of sensors:

- Photon receptors characteristic of photosynthesis, phototropism and phototaxis, all enabled by light-absorbing pigments whose electronic excitation triggers a multilayered response
- Vibration (sound) sensors
- Aroma (smell) sensors

The latter two are employed by multicellular organisms. Cells have no other direct way to detect outside entities at a distance.[15] Even smell involves direct contact with molecules originating from the exterior body. So, let's look at one example of a sensing or recognition process that occurs between two cells. The most common organization of this sort of sensing involves a cell surface receptor that interacts with a cell-surface structure on a foreign cell or, in the case below, of a fragment of an invader that is being carried around by an immune cell. The structures that are recognized by the immune system are usually proteins, glycoproteins, and/or saccharides.

T cells are white blood cells (lymphocytes) that play a central role in immunity. ("T" stands for thymus, where the T cells grow.) They are characterized by a special receptor on their cell surface called T cell receptors (TCRs) that interact with the *major histocompatibility complex* (MHC, a protein) on an *antigen presenting cell* (APC) that has managed to "bite off" a fragment of a foreign cell's (or virus's) protein or glycoprotein called an antigen (Figure 7.16). "Professional" APCs are cells that specialize in obtaining an antigen from an invading cell or virus and then displaying a fragment of the antigen, bound to an MHC molecule, on their membrane. The T cell recognizes and interacts with the MHC molecule complex on the membrane of the APC. (Note so far the T cell is not dealing directly with the invading cell.) When the T cell binds to the APC, the latter gives an additional signal that activates the T cell. The T cell then secretes a chemical (cytokine) that causes growth of more T cells, some of which become capable of killing the invading cell or cell that has been infected with virus. Huge amounts of biochemical and genetic information are known about the many components of this type of immune response, but we will focus on what takes place structurally when the T cell binds the MHC. For a more complete description of the immune recognition response, see K. Singleton et al.[16] Read the *Journal of Immunology* for continuing, up-to-date coverage of discoveries about the immune response.

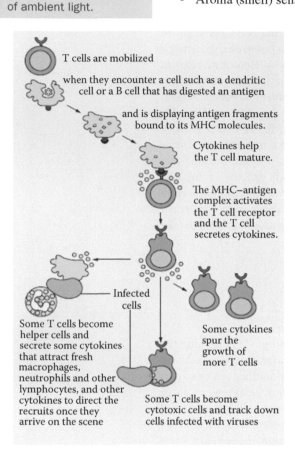

T cells are mobilized when they encounter a cell such as a dendritic cell or a B cell that has digested an antigen and is displaying antigen fragments bound to its MHC molecules.

Cytokines help the T cell mature.

The MHC–antigen complex activates the T cell receptor and the T cell secretes cytokines.

Infected cells

Some cytokines spur the growth of more T cells

Some T cells become helper cells and secrete some cytokines that attract fresh macrophages, neutrophils and other lymphocytes, and other cytokines to direct the recruits once they arrive on the scene

Some T cells become cytotoxic cells and track down cells infected with viruses

FIGURE 7.16

Activation of a T cell. The T lymphocyte is activated when a T cell encounters an antigen (a protein fragment from a foreign cell or virus) bound to a major histocompatibility complex (protein) on the surface of an antigen-presenting cell. T cells both direct immune responses and directly attack infected or cancerous cells. (Public domain image from NIH Publication 03-5423, September 2003 (modifications: September 4, 2006).)

The main question we have regarding T cell recognition of a foreign protein on the surface of another cell is what takes place structurally in the two cells during the interaction. We will focus on the spatial scale of 0.1–1 μm or so. Figure 7.17 shows a fluorescence micrograph of the structural changes in a T cell and an APC cell, which has an antigen from an invader on its surface, bound to the MHC. As we noted early in the book, the Green Fluorescence Protein (GFP) is an important technological development in microscopy. GFP has been employed here to visualize the distortion of the APC cell. In the experiment the APC can be imaged by selecting the green fluorescence and the T cell seen by scattered-light imaging ("bright field"). (This is done by rapidly switching optical filters in the microscope and does not require separate experiments.) The image cannot tell us what the initial interaction between the two cells is, but it clearly indicates that once the interaction starts, the T cell appears to suck the antigen-containing MHC on the APC into its interior. The lower panel of the figure shows the APC by itself (green fluorescence), in a false-color image that reflects not the real color of the fluorescence, but the intensity of the green fluorescence. It is clear that the extrusion into the T cell (out of the APC) contains a lot of the MHC with bound antigen. What might we learn from this? The authors of the article made many inferences but did not analyze what a physics student might: the force required, and the free energy expended, to pull part of the APC membrane into the T cell. We saw how this might be done from the elastic properties of a membrane in the previous chapter. A rough calculation could, in fact, be done, but because the local composition of the membranes obviously must change dramatically during the extrusion process and must contain a high fraction of proteins recruited to the site, the computation would be difficult to do accurately. We must let this example of recognition in this somewhat tantalizing but unsatisfying state. Note, however, that the word *recruitment* is a useful, appropriate, and descriptive word for the biological process of cell–cell recognition. When two cells recognize each other, a complicated process of recruitment of proteins to the site of interaction takes place. This recruitment process involves driven diffusion in the planes of the membranes of both cells: the interaction creates a driving force for specific proteins and glycoproteins (and lipids?) to move to the site of the action.

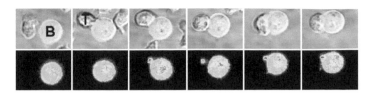

FIGURE 7.17
APC extensions and T cell invaginations. An interaction of a 5C.C7 T cell (T) with a CH27 APC (B) that has labeled with green fluorescence protein (GFP). (Top) Images are displayed (top) that have been overlaid with the GFP fluorescence in green. (Bottom) Matching maximum projections of the three-dimensional GFP fluorescence intensities in a false color scale (increasing intensity from purple to red reflects more GFP) are shown. A GFP extension from the APC can be seen, pulled deep into the T cell. (Figure 7.1, from Singleton, K. et al., *J. Immunol.*, 177, 4402–4413, 2006. Copyright 2006, The American Association of Immunologists, Inc. With permission. Video versions of diagrams can be downloaded from the journal Web site http://www.jimmunol.org/cgi/content/full/177/7/4402/DC1.)

7.4.4 The Bacterial "Brain" Has No Neurons

A single cell cannot have a brain, but a bacterium does have to detect and respond to a changing environment that can be complex. We discuss the propulsion systems of bacterial more quantitatively in Chapter 16, but insights into the physical/chemical control mechanism of *E. coli* movement have emerged in the last several years.[17] Such control is critical for the individual bacterium, even for the sorts of cooperative interactions between bacteria that have been noted.[18] A swimming bacterium needs to sense when food, toxins, light, and bacteria and other agents are encountered and then needs to decide whether to continue forward or change direction—chemotaxis. (The decision is usually binary, although stopping is also an option.) The task seems simple, but a single bacterium must coordinate its thousands of surface receptors such that a binary decision can be made. In addition to the organization and processing of these multiple signals, the bacterium must also "remember" the signals long enough to take action that aids their survival.

The molecular structures that control this brain-like activity involve large, highly organized chemosensory arrays, which establish a network of cooperative interactions that enable proper signal transmission and decision-making. Cassidy et al.[17] in the group of Klaus Schulten have combined cryo-electron microscopy, crystallography and other experimental data with molecular dynamic flexible fitting (MDFF) supercomputing to produce detailed atomic structures of these large assemblies of receptors, kinases and their motions. Figure 7.18 shows a small part of the assembly, with two long receptor "trimers of dimers" (TODs) and two types of kinases shown at the bottom. (Other structures and movies of molecular motions may be found in reference 17.) In a real bacterium, the membrane would support the upper part of a large number of coordinated TODs. Though many details are still not clear, a key part of signaling combines physical structural flexing of proteins with enzymatic generation of active chemical species and transfer to the next enzyme in the chain.

As is often the case with advances in biomolecular imaging and computing, a key factor was the purification and assembly of the array proteins by Peijun Zhang.

Klaus Schulten's research efforts over more than three decades have produced huge advances in many areas of large-scale biomolecular structure and function, using innovative imaging technology and large-scale computing. He brought those structures to life for established researchers and new students alike. During the composition of this book's second edition, Prof. Schulten died, to the great loss of the biophysics community.

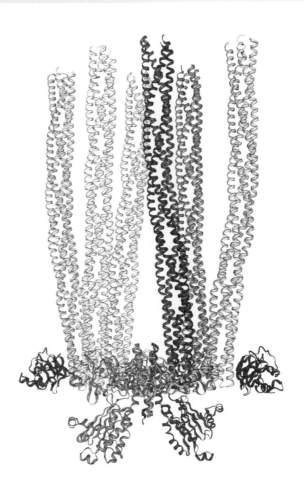

FIGURE 7.18
The "bacterial brain" of *E. coli*, displayed using the NGL 3D viewer at the Protein Database. A chemo-sensory receptor array is used by bacteria to detect food, toxins, light and other chemicals as they swim, then transmit the information to a deeper layer of protein kinases. The kinases make a binary decision: "Keep going" of "Change direction." If the latter decision is made, a phosphate is trans-ferred to a kinase protein, which then detaches, finds it way to the flagella that drive the bacterium, and causes the flagella to reverse direction. See Chapter 16. The red and orange proteins near the bottom are kinases and change conformation as part of the signaling process. The long receptor helices are organized as trimers of dimers. (Two are shown). PDB ID: 3JA6, Cassidy, C.K. et al., Cryo-electron tomography and all-atom molecular dynamics simulations reveal a novel kinase conforma-tional switch in bacterial chemotaxis signaling, DOI: 10.2210/pdb3ja6/pdb, NGL viewer.21.

7.5 POINTS TO REMEMBER

- Cells: minimal, prokaryotic, eukaryotic and how they are defined; cell sizes.
- Important cell parts: mitochondria, chloroplasts, endoplasmic reticula, Golgi apparatus, the nucleus and what function they each provide.
- Concentrations in cells and membranes
- Sensors, recognition and the bacterial "brain."

7.6 PROBLEM-SOLVING

PROBLEM 7.1 SURFACE/VOLUME RATIOS

1. Calculate the surface/volume ratio of a *Heliobacter pylori* cell. State clearly what dimensions you use and where you found the information.
2. Calculate the surface/volume ratio of one of the larger cyanobacterial cells in Figure 7.6a. State clearly what shape you assume and dimensions you use.
3. Calculate the surface/volume ratio of a eukaryotic cell $25 \times 15 \times 12$ μm.
4. Discuss the diffusion of small molecules from the cell perimeter to the cell center. If the mean-square diffusion $\langle (\text{distance})^2 \rangle$ scales linearly with time, calculate the times for a small molecule to diffuse from the edge to the center of the three cells above, normalizing t (*E. coli*) to 1.

PROBLEM 7.2 BACTERIAL CELL DIVISION

E. coli cells can divide every 20 min, under favorable conditions. Cells will divide until the cell density reaches a maximum concentration of about 10^{13} cells/L.

1. Assuming no cells die and the division time remains fixed until the maximum cell concentration is reached, calculate how long it will take one cell to fill a 5.0-L flask to saturation. Assume the *E. coli* is 2.0 μm long and 1.0 μm in diameter. Remember, one cell → two → four → eight....
2. Repeat 1 for a volume of 5 km^3.
3. Grad problem. Make a more reasonable mathematical model for cell proliferation to saturation. Part 1 assumes a constant division time up until saturation is reached, when division abruptly stops. Either look up a more correct mathematical model in the literature or make your own model for an increasing cell-division time as the population nears saturation. You might propose, for example, that the cell-division time increases as a function of cell concentration, 20 min at 0 cells/L and rising with density: $t_{\text{divide}}(c) = [20 \text{ min}]$ $[1 + c/c_0]$ or $t_{\text{divide}}(c) = [20 \text{min}]e^{c/c_0}$, where $c = c(t)$ is the time-dependent cell concentration and $c_0 = 10^{13}$ cells/m^3. (You can probably do better than these.) Recalculate the time in 2 using this model. If you have trouble with the analytic math, use an Microsoft Excel spreadsheet.

PROBLEM 7.3 CONCENTRATIONS IN CELLS

1. Calculate the number of H^+ ions inside an *E. coli* cell at inside pH's of 6.5 and 8.5.
2. Calculate the concentration of a protein if there are 50 protein molecules inside the *E. coli* cell.
3. An *E. coli* cell has 500 molecules of protein A, an enzyme, but these are all bound in the cell membrane. Determine and calculate the most physically/chemically meaningful protein A "concentration," keeping in mind where protein A is located.

PROBLEM 7.4   HOW MANY CELLS?

Estimate the number of cells in the entire human body. Show an approximate calculation (don't just look it up online).

PROBLEM 7.5   MOLECULES IN THE MITOCHONDRIAL OUTER MEMBRANE

The outer membrane on the mitochondrion in a typical eukaryotic cell is about 52% protein, 48% lipid (mass percentage).[6]

1. Estimate the volume and surface area of the *Ostreococcus* mitochondrial outer membrane.
2. Estimate the number of lipid molecules and number of protein molecules in this membrane. State and justify what average molecular masses and/or dimensions you use for protein and lipid molecules.

MINIPROJECT 7.6   CONCENTRATIONS AND ATP REGENERATION IN A MITOCHONDRION

The discussion in Section 7.2 referred to an assertion that in humans, "On any given day you turn over your body weight equivalent in ATP...."[11]

1. Calculate how many ATP molecules this is equivalent to.
2. Determine approximately how many ATP molecules per cell per day must be regenerated.
3. Determine approximately how many ATP molecules per mitochondrion per day must be regenerated. (Look up, justify, and cite the source for information on numbers of human mitochondria per cell.)
4. Find the average size (dimensions, volume, and surface area) of a human cell mitochondrion.
5. A protein called the voltage-dependent anion channel (VDAC) passively (no energy input required) transports ATP across the outer membrane of the mitochondrion. VDAC, a 294 amino-acid polypeptide, is the most abundant protein in the mitochondrion. Download and print the structure of this molecule (2JK4) from the PDB. Note the molecular mass.
6. Assume, as in the previous problem, that protein constitutes 52% of the membrane mass and that VDAC makes up 40% of the protein. Estimate how many VDAC molecules are in the outer mitochondrial membrane.
7. Calculate the average number of ATP molecules per second that pass through a VDAC channel in a human mitochondrion.

PROBLEM 7.7 SUBCELLULAR STRUCTURE DIMENSIONS IN *OSTREOCOCCUS*

1. Refer to the three-dimensional electron microscopy images cited in the text and figures and download original files from Henderson et al. *PLoS One* 2(8):e749, 2007, doi:10.1371/journal.pone.0000749). Print the overall cell structure so quick reference to the parts can be done.
2. Make a table of the average dimensions—dimensions, shape, volume, surface area, surface/volume ratio—of the interior structures in the *Ostreococcus* cell. Include the microtubule.
3. Discuss the issue of transport in the cell, why there are so few microtubules, and what the visible microtubule is believed to do.

PROBLEM 7.8 LIGHT ABSORPTION BY A CHLOROPLAST IN *OSTREOCOCCUS*

We know that plant leaves orient their leaves to maximize the surface area exposed to sunlight. Should individual cells of disklike chloroplasts similarly reorient themselves? Use Figures 7.10 and 7.12 for the following.

1. Sketch a diagram of the *Ostreococcus* chloroplast showing its major dimensions.
2. Find the average concentration of chlorophyll inside this chloroplast. You may look up this value on the Web or in a reference article or book. (Data on the chloroplast of any eukaryotic organisms may be used, if you cannot find such data on *Ostreococcus*.)
3. Find an optical absorption spectrum of chlorophyll in the visible region and write down the wavelengths (in nanometers) and values of the two highest absorption (extinction) coefficients (ε, in M^{-1} cm^{-1}).
4. Using the light absorption equations from the previous chapter, $I = I_0 10^{-A}$, $A = \varepsilon c L$, determine the approximate absorbance A and fraction of light absorbed, $(I_0 - I)/I_0$, for two orthogonal orientations of the chloroplast. Ignore any chloroplast curvature, possible edge effects, or light refraction or scattering.
5. Does the *Ostreococcus* chloroplast absorb more light in one of the orientations?
6. Discuss the chloroplast orientation effect on a typical plant chloroplast, which is about 5 μm in diameter and 2.5 μm thick.

This number is not terribly reliable. If you find a better number, use it.

PROBLEM 7.9 DNA AND STORAGE ISSUES: PROKARYOTIC CELL

1. Find the number of base pairs and length of *Saccharomyces cerevisiae* yeast DNA. Cite sources.
2. Calculate the molecular volume of the "bare" DNA: double-helix length times cross-sectional area.
3. Treat the DNA double helix as a single freely jointed chain (FJC) of N units of length L, where $L \sim 100$ nm (Chapter 5, Section 6.3.3). Calculate N (not the number of base pairs!), the radius and the volume of this random polymer.

4. Make a table like the one below and fill in the numbers from above calculations.

Property	Cell	"Bare" DNA	FRJ DNA
Length			—
Average radius		—	
Average volume			

5. Calculate the ratios of length, radius, and volume (as appropriate) of bare DNA and FJC DNA to those quantities for the cell.

PROBLEM 7.10   DNA AND STORAGE ISSUES: EUKARYOTIC CELL

Repeat Problem 7.9 for a human cell. Assume the cell is spherical, with radius 20 μm and that the DNA is in the form of 46 individual strands. Modify the computations and table to appropriately reflect the 46 DNA strands.

PROJECT 7.11   MOTION AND SENSORS

(This could be a several-week- or even semester-long project, so it must be managed and monitored carefully by both student and professor. Also, if you can think of a better, but still simple, way to do the following, please do so.) Design a spreadsheet or other type of computer code to simulate the motion of a rectangular bacterium that has no more than four sensors and two directional drive motors: forward or right, forward or backward, forward or rotate, left or right, etc. The bacterium moves on a two-dimensional grid; at any time, the bacterium occupies 2, 4, or 6 grid points. (Choose one or try all three bacterial sizes.) The bacterium can take N steps for every unit of food it consumes. (Try various N values: 1, 2, 3,….) Sensors that can be positioned on various points of the cell surface and that can signal one or both motors respond to the presence of "food" on an *adjacent* grid point. (No remote sensing allowed.) The bacterium consumes one unit of food on any grid point it covers. The food cannot be replenished. (The more sophisticated student may allow random diffusion of the food on the grid.) When a unit of food is consumed, the bacterium can move an additional N steps in one direction. Food on any grid point can be at a level of 0, 1, 2, or 3. Place an initial (variable) food distribution on the grid, and start the bacterium at some random point with some initial "store" of food-enabled steps. Design sensor placement on the bacterium and signals to onboard motors to maximize the bacterium's search efficiency and lifetime. You can measure the efficiency/ lifetime in various ways, including number of steps taken before "death," fraction of food consumed. Have fun, but don't spend all your time on this.

PROBLEM 7.12   NUCLEAR PORE COMPLEX: ONLINE VERSUS LOCAL COMPUTING RESOURCES

Search the PDB and locate the structure of the human nuclear pore complex (NPC).

1. View the structure using the NGL viewer. Produce side and top views of "Bioassembly 1 (Model 1)" and paste into your assignment. Use both "surface" and "cartoon" views of the structure—four pictures altogether. The NGL viewer uses on-site software to display structures.

2. View the structure using Protein Workshop. Produce two views, side and top, and save to your assignment. Protein Workshop downloads Java script to your computer, whose computing power is used to display structures. Compare the display and structure rotation time to that in part 1.

3. 3D printed structure. Optional: suggested for group projects, if your institution has ready access to (affordable) 3D printing resources. In the RCSB PDB Web site, click the "Analyze" tab and select "Third party tools," then locate "3D Printing Resources." Use the most appropriate tool to produce a 3D model file for the Bioassembly 1 above, surface view. Discuss with your instructor whether this (two-piece) file is too large to manage; if so, find a simpler and smaller PDB structure. Confirm that the file to be produced will work in your available 3D printer and prepare and submit the structural file for printing.

PROBLEM 7.13   THE BACTERIAL BRAIN

1. Locate and save to your assignment the 3JA6 structure in Figure 7.18, using the JSmol viewer, Cartoon style, Color by hydrophobicity. (Wait long enough for the Javascript and structure to load. Right-click or control-shift-click image to save image.) Orient the structure similar to that in the figure. Discuss the two main secondary structures present and what subunits have those structures.

2. Determine the approximate length of the long helices. (In the Scripting Options box below the structure, type "set picking distance"; Submit. Locate your first residue with the cursor; click; move cursor to second residue, noting the dotted line and distance label that appear. "Set picking off" to undo the distance picking.) Compare this length to the size of an *E. coli* cell and to the thickness of the cell wall.

3. If we presume that this sensor system extends from outside to inside the membrane in order to relay signals to the interior machinery, how might this system best be located and oriented with respect to the cell wall? No right/wrong, but give good arguments. Find published information on this issue.

PROBLEM 7.14   ENERGY WITHOUT A MITOCHONDRION

1. The diarrhea-causing microbe *Giardia intestinalis* was initially believed to be mitochondria-free, but this assertion was later revised. Describe, in 100–150 words, the evidence for the initial belief that *Giardia* was mitochondrion-free and for the later corrections.

2. Locate the publications evidencing the lack of a mitochondrion in *Monocercomonoides*. Is this evidence more convincing than that for *Giardia*? What are the weaknesses in this evidence? Explain, in 70–100 words.

3. Describe the most recent ideas for how *Monocercomonoides* survives without a mitochondrion.

REFERENCES

1. Forster, A. C., and G. M. Church. 2006. Towards synthesis of a minimal cell. *Mol Syst Biol* 2:45.
2. Hutchison, C. A., 3rd, R. Y. Chuang, V. N. Noskov, N. Assad-Garcia, T. J. Deerinck et al. 2016. Design and synthesis of a minimal bacterial genome. *Science* 351(6280):aad6253-1–aad6253-11. doi:10.1126/science.aad6253.
3. Gribaldo, S., and C. Brochier-Armanet. 2006. The origin and evolution of Archaea: A state of the art. *Philos Trans R Soc Lond B Biol Sci* 361:1007–1022.
4. Cavalier-Smith, T. 2004. Only six kingdoms of life. *Proc R Soc* 271:1251–1262.
5. Norris, V., and R. Root-Bernstein. 2009. The eukaryotic cell originated in the integration and redistribution of hyperstructures from communities of prokaryotic cells based on molecular complementarity. *Int J Mol Sci* 10:2611–2632. See also reference 11.
6. Voet, D., and J. G. Voet. 2004. *Biochemistry*, 7. New York: John Wiley & Sons.
7. Krzywinski, M., J. Schein, I. Birol, J. Connors, R. Gascoyne, D. Horsman, S. J. Jones, and M. A. Marra. 2009. Circos: An information aesthetic for comparative genomics. *Genome Res* 19:1639–1645. doi:10.1101/gr.092759.109.
8. Rucinsky, T. E., and E. H. Cota-Robles. 1974. Mesosome structure in Chromobacterium violaceum. *J Bacteriol* 118:717–724.
9. Courties, C., A. Vaquer, M. Troussellier, J. Lautier, M. J. Chrétiennot-Dinet, J. Neveux, C. Machado, and H. Claustre. 1994. Smallest eukaryotic organism. *Nature* 370:255.
10. Henderson, G. P., L. Gan, and G. J. Jensen. 2007. 3-D ultrastructure of *O. tauri*: Electron cryotomography of an entire eukaryotic cell. *PLoS One* 2(8):e749. doi:10.1371/journal.pone.0000749.
11. Törnroth-Horsefield, S., and R. Neutze. 2008. Opening and closing the metabolite gate. *Proc Natl Acad Sci USA* 105:19565–19566.
12. Beck, M., F. Förster, M. Ecke, J. M. Plitzko, F. Melchior, G. Gerisch, W. Baumeister, and O. Medalia. 2004. Nuclear pore complex structure and dynamics revealed by cryoelectron tomography. *Science* 306:1387–1390.
13. Vollmar, F., C. Hacker, R. P. Zahedi, A. Sickmann, A. Ewald, U. Scheer, and M. C. Dabauvalle. 2009. Assembly of nuclear pore complexes mediated by major vault protein. *J Cell Sci* 5:780–786.
14. Howard, J. 2009. Mechanical signaling in networks of motor and cytoskeletal proteins. *Annu Rev Biophys Biomol Struct* 38:217–234.
15. Krueger, A. P., B. B. Brown, and E. J. Scribner. 1941. The effect of sonic vibrations on phage, phage precursor and the bacterial substrate. *J Gen Physiol* 24:691–698. A 10-year-old hypothesis of subcellular "hyperstructures" that can emit and respond to specific vibrations exists in the literature: Norris, V., T. den Blaauwen, R. H. Doi, R. M. Harshey, L. Janniere et al. 2007. Toward a hyperstructure taxonomy. *Annu Rev Microbiol* 61:309–329. In this author's opinion, the latter hypothesis lacks in experimental verifications and checks.
16. Singleton, K., N. Parvaze, K. R. Dama, K. S. Chen, P. Jennings, B. Purtic, M. D. Sjaastad, C. Gilpin, M. M. Davis, and C. Wülfing. 2006. A large T cell invagination with CD2 enrichment resets receptor engagement in the immunological synapse. *J Immunol* 177:4402–4413.
17. Cassidy, C. K., B. A. Himes, F. J. Alvarez, J. Ma, G. Zhao, J. R. Perilla, K. Schulten, and P. Zhang. 2015. CryoEM and computer simulations reveal a novel kinase conformational switch in bacterial chemotaxis signaling. *eLife* 4. doi:10.7554/eLife.08419.

18. Brown, S. P., and R. A. Johnstone. 2001. Cooperation in the dark: Signalling and collective action in quorum-sensing bacteria. *Proc Roy Soc Lond Ser B Biol Sci* 268(1470):961–965.

19. Gil, R., F. J. Silva, J. Peretó, and A. Moya. 2004. Determination of the core of a minimal bacterial gene set. *Microbiol Mol Biol Rev* 68(3):518–537.

20. Cassidy, C. K., B. A. Himes, F. J. Alvarez, J. Ma, G. Zhao, J. R. Perilla, K. Schulten, and P. Zhang. Cryo-electron tomography and all-atom molecular dynamics simulations reveal a novel kinase conformational switch in bacterial chemotaxis signalling. doi:10.2210/pdb3ja6/pdb.

21. Koepke, J., X. Hu, C. Muenke, K. Schulten, H. Michel. 1996. The crystal structure of the light-harvesting complex II (B800-850) from Rhodospirillum molischianum. *Structure* 4:581–597.

22. Luecke, H., B. Schobert, B. Schobert, H. T. Richter, J. P. Cartailler, J. K. Lanyi. 1999. Structure of bacteriorhodopsin at 1.55 Å resolution. *J Mol Biol* 291:899–911.

23. Takeda, K., Y. Matsui, Y. Matsui, N. Kamiya, S. I. Adachi, H. Okumura, and T. Kouyama. 2004. Crystal structure of the M intermediate of bacteriorhodopsin: Allosteric structural changes mediated by sliding movement of a transmembrane helix. *J Mol Biol* 341:1023.

PART III

BIOLOGICAL ACTIVITY
Quantum Microworld

CHAPTER 8

CONTENTS

Quantum Primer

Introducing quantum mechanics immediately after considering the issue of cell division and cell–cell recognition, with their associated complex structures and structural changes, is rather shocking. After seven chapters, physics students may have just grasped the spirit of biological analyses; suddenly you are back into old territory. Biology students, after seeing a different view of your normal territory, now see yourselves entering a modern physics course. The fact is, "biophysics" is not a compact, coherent field. Introductory physics and biology courses at various universities are basically the same in terms of materials covered and tools learned. There will likely never be such standardization in biophysics. The huge variety of phenomena encountered, the huge number of principles, approaches, and methods required to probe and understand these phenomena, the rapid escalation in difficulty encountered when quantitative methods are applied, the general disagreement over what biophysics is, and a focus of professors on their research interests makes any standardization of a one-semester or 1-year biophysics experience unlikely. Even 4-year biophysics degree programs have trouble agreeing on what should be studied in 4 years of biophysics.

Why do we need quantum mechanics in biophysics? We have alluded to absorption of light and pointed toward chapters dealing with absorption spectra, photosynthesis, and ultraviolet (UV) effects. The place of quantum theory in these important is, in principle, clear: when a photon interacts with matter, it interacts with the quantum states of matter, usually molecular states, in the case of biosystems. While quantum effects are described in some biochemistry books covering photosynthesis and UV absorption by DNA, these texts are primarily relaying bottom line information to be remembered, not the quantitative methods that can be applied when a somewhat different phenomenon turns up. Nature is constantly surprising us with phenomena that may not be so surprising when put in the context of what general quantum theory encompasses. A UV photon hitting skin may be absorbed by applied sunscreen 1 and nothing further happens. When sunscreen 2 is used, an allergic response may result. Is this just a fact to note—that sunscreen 2 should have an allergic warning on it or even be taken off the market—or is this an example of one of the standard quantum effects of photons in biological systems?

Quantum mechanics is not just a set of principles that must be applied when photons are involved. Whenever a biochemical reaction takes place, one of those thousands of reactions that takes place in your body every second, quantum mechanics is involved. Whenever $ATP \rightleftharpoons ADP + P_i$, a covalent bond is broken or formed and quantum theory governs the changes in electronic structure. Of course, statistical mechanics also sticks its foot in the door here, as the amount of energy an enzyme can extract from ATP hydrolysis depends on local concentrations and structures that may restrict diffusion. That is what is so unique and interesting about biophysics: the same theory that governs the behavior of the Higgs boson applies to biosystems; statistical principles that apply to chaotic systems apply to biosystems. At the same time, traditional statistical thermodynamics used by biochemists for a century must not be forgotten.

Those who are not shocked when they first come across quantum mechanics cannot possibly have understood it.

Neils Bohr

It is an advantage to encounter the strangeness of quantum mechanics before studying biophysics. First, the shock of reading about quantum issues will not be new to you. Second, you will probably be shocked again by how nonshocking quantum theory in biosystems appears. This does not mean quantum calculations in biosystems are simple: they are, in fact, some of the most difficult calculations there are, in terms of complexity and required computer code. What we will encounter will look rather pedestrian, e.g., sketched waves, Pauli Exclusion Principle, and counted energy levels. Yet these simple tools allow an understanding of why hemoglobin is red and chlorophyll is green; why UV light may produce cancer; how plants can route the sun's energy to microscopic fuel-synthesis centers; and why molecular motors step.

The glossary in the back of the book has been used primarily to catalog and define various biological and physiological terms that may be new to physics students. Very few quantum terms will be found in that glossary. We define a few terms here and briefly sketch their role in quantum mechanics applied to biosystems.

8.1 QUANTUM GLOSSARY

8.1.1 Quantum, Quantum Behavior, Quantum Mechanics, and Wave Mechanics

"Quantum" refers to the rules by which the behavior of an object can be predicted: discrete values of physical quantities like energy, momentum, angular momentum, or other, depending on the environment. "Quantum behavior" usually refers to observed behavior of an object consistent with quantum mechanics, usually several allowed, fixed energy levels, emission or absorption of photons, sudden jumps in position, or outcomes of experiments which take on discrete values with definite probabilities. "Wave mechanics" refers to the prediction or description of an objects behavior in terms of its wave properties (e.g., evidence of a delocalized location).

In a quantum context, *classical*, or classical behavior, usually means that an object can be adequately described by classical laws of physics (usually Newton's laws of motion) in some defined context. The context is important because the path of a 3-mm-diameter copper sphere shot from a pellet gun can be accurately described by Newton's laws of motion—classical laws—but the sphere's electrical conductivity requires quantum theory. An object is not innately a "quantum" or a "classical" object. Usually, however, objects require quantum theory for a proper description of many properties if the object in small enough, below about 10 nm or so.

"Continuum" normally refers to the continuous allowed values of energy of a quantum particle once its energy becomes positive. Electrons bound to atoms have negative total energies, indicating they are in bound states. These negative-energy atomic states are quantized: only certain negative energy states are observed. If electrons are excited to zero or greater energy, the energy is allowed to take on any value, depending on the energy given to the electron by outside sources (e.g., a photon). Thus an electron excited by a photon into the continuum starts to behave according to Newton's laws of motion, though measurements can still show that the electron has wavelike properties. Is an electron excited to the continuum a quantum or a classical particle? It depends on what you need to calculate or what experiment you need to do.

The term *single particle* has nothing necessarily to do with quantum or classical behavior. Many single-molecule experiments are being done in biophysical research today, but few of them involve any quantum behavior. The single molecules being observed generally can be accurately described by Newton's laws, statistics and diffusion theory, Maxwell's electrodynamic equations, or Coulomb's law, with no quantum mechanics or wavelike behavior in evidence. Sudden jumps in position or between two distinct values of energy might well be observed, but these usually reflect classical, statistical behavior.

The following sections are designed to review quantum mechanics in terms of practical tools that can be used in introductory biophysics. If you do not "understand" quantum mechanics, you will probably not understand it from this chapter either, but you may polish up some of the instruments you learned in previous physics and chemistry courses. We spend only a little time on atomic electronic states, except when they exemplify certain aspects of quantum theory (e.g., discrete energy states). Atomic states are important to biomolecules that employ heavy atoms (Fe, Ni, Cu, etc.).

8.2 SCHRÖDINGER EQUATION AND OTHER TOOLS OF QUANTUM MECHANICS

The standard and most comprehensive tool for describing the quantum, wavelike behavior of objects is the time-dependent Schrödinger equation, introduced in Chapter 3:

$$\frac{-\hbar^2}{2m}\frac{\partial^2 \Psi(x,t)}{\partial x^2} + V(x)\Psi(x,t) = i\hbar\frac{\partial \Psi(x,t)}{\partial t} \tag{8.1}$$

where $V(x)$ is the potential energy (assumed time independent here), $\hbar$ is Planck's constant divided by 2π, x is position, and t is time. The Schrödinger equation explicitly encompasses the wavelike behavior of a particle in $\Psi(x,t)$, the particle wave function. The standard interpretation of the wave function is that

$$|\Psi(x,t)|^2 \equiv \text{probability for observing the particle at a}$$
$$\text{position between } x \text{ and } x + dx \text{ at time } t \tag{8.2}$$

8.2.1 Time-Dependence: Plane Waves

If the potential energy is zero, the Schrödinger equation becomes

$$\frac{\partial^2 \Psi(x,t)}{\partial x^2} = \frac{2m}{i\hbar}\frac{\partial \Psi(x,t)}{\partial t} \tag{8.3}$$

This equation is not like the wave equation that describes guitar string vibrations or Maxwell's wave equation for light:

$$\frac{\partial^2 \Psi(x,t)}{\partial x^2} = \frac{1}{v^2}\frac{\partial^2 \Psi(x,t)}{\partial^2 t} \tag{8.4}$$

The crucial difference here is the presence of a second time derivative in the classical wave equations. Equations 8.3 and 8.4 both describe traveling waves. In first-year physics we generally learned solutions to traveling plane wave equations in terms of sine and cosine functions:

$$\Psi(x,t) = \sin(kx - \omega t), \cos(kx - \omega t) \tag{8.5}$$

or any linear combinations thereof. See Figure 8.1. One of the particularly useful linear combinations is

$$\Psi(x,t) = A[\cos(kx - \omega t) + i\sin(kx - \omega t)] = Ae^{i(kx - \omega t)} \tag{8.6}$$

where A is a constant (usually real) and $i = \sqrt{-1}$. This exponential is still a traveling wave solution, because of the $(kx - \omega t)$ expression. Its big advantage is that derivatives are particularly easy to write down because $\frac{d}{dx}(e^{ax}) = ae^{ax}$, and the derivative is just a constant times the function. Let's convince ourselves this is a useful solution to one or both of the wave equations we have.

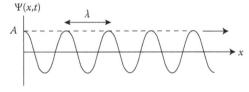

FIGURE 8.1
Sinusoidal wave traveling to the right along the x axis: $\Psi(x,t) = A\cos(kx - \omega t)$ or $\Psi(x,t) = Ae^{i(kx - \omega t)}$. For an electromagnetic wave, A represents the maximum electric (or magnetic) field strength, in volts per meter; $c\varepsilon_0|\Psi(x,t)|^2$, the intensity at (x,t), in Watts/m^2; and $\frac{c\varepsilon_0}{2}|A|^2 = \frac{c\varepsilon_0}{2}|E_0|^2$ the average intensity. The "per unit area" (m^{-2}) for the light wave refers to the area in the yz plane covered by the plane wave; the electric and magnetic field vectors also lie in the yz plane. For a matter wave, $|\Psi(x,t)|^2 = |A|^2\cos^2(kx - \omega t)$ is the probability, per meter, of finding the particle at a position between x and $x + dx$. The units of Ψ and A are therefore probability/m$^{\frac{1}{2}}$, or m$^{-\frac{1}{2}}$.

EXERCISE 8.1

Show whether $\Psi(x,t) = A \sin(kx - \omega t)$ and $\Psi(x,t) = B \cos(kx - \omega t)$ satisfy the Maxwell and Schrödinger wave equations. Interpret the meanings of k and ω.

Solution

Maxwell equation:

$$\frac{\partial \Psi}{\partial t} = \frac{\partial}{\partial t}\left(A \sin(kx - \omega t)\right) = -\omega\left(A \cos(kx - \omega t)\right)$$

$$\frac{\partial^2 \Psi}{\partial t^2} = \frac{\partial}{\partial t} - \omega\left(A \cos(kx - \omega t)\right) = -\omega^2\left(A \sin(kx - \omega t)\right) = -\omega^2 \Psi(k, x)$$

$$\frac{\partial \Psi}{\partial x} = \frac{\partial}{\partial x}\left(A \sin(kx - \omega t)\right) = k\left(A \cos(kx - \omega t)\right)$$

$$\frac{\partial^2 \Psi}{\partial x^2} = \frac{\partial}{\partial x} k\left(A \cos(kx - \omega t)\right) = -k^2\left(A \sin(kx - \omega t)\right) = -k^2 \Psi(x,t)$$

$$\frac{\partial^2 \Psi(x,t)}{\partial x^2} = -k^2 \Psi(x,t) \overset{?}{=} \frac{-\omega^2}{v^2} \Psi(x,t)$$

This equation is true if $\omega = kv$. If $\omega = 2\pi f$ is the (angular) frequency of the wave and k is 2π over the wavelength, $k = 2\pi/\lambda$, then $\omega = kv \rightarrow v = \lambda f$, which is the proper relation between wavelength, frequency, and velocity for a wave.

Schrödinger equation:

$$\frac{\partial \Psi}{\partial t} = \frac{\partial}{\partial t}\left(A \sin(kx - \omega t)\right) = -\omega\left(A \cos(kx - \omega t)\right)$$

$$\frac{\partial \Psi}{\partial x} = \frac{\partial}{\partial x}\left(A \sin(kx - \omega t)\right) = k\left(A \cos(kx - \omega t)\right)$$

$$\frac{\partial^2 \Psi}{\partial x^2} = \frac{\partial}{\partial x} k\left(A \cos(kx - \omega t)\right) = -k^2\left(A \sin(kx - \omega t)\right)$$

$$\frac{\partial^2 \Psi}{\partial x^2} = -k^2\left(A \sin(kx - \omega t)\right) \overset{?}{=} \frac{2m\omega i}{\hbar}\left(A \cos(kx - \omega t)\right)$$

The major problem here is that the functions on either side differ, so the equation cannot hold for all values of x and t. The imaginary $i = \sqrt{-1} = \frac{-1}{i}$ on the right side could possibly be dealt with by an imaginative definition of k^2.

The solution for $\Psi(x,t) = B \cos(kx - \omega t)$ is similar and is left for you to do.

EXERCISE 8.2

Show whether $\Psi(x,t) = Ae^{i(kx - \omega t)}$ and $\Psi(x,t) = Ae^{-i(kx - \omega t)}$ satisfy the Schrödinger and Maxwell wave equations. Interpret the meanings of Ψ, k, and ω.

Solution

Schrödinger equation:

$$\frac{\partial \Psi}{\partial t} = -i\omega Ae^{i(kx - \omega t)} = -i\omega \Psi(x,t)$$

(Note that derivative does not care about the imaginary constant in the exponent.)

$$\frac{\partial \Psi}{\partial x} = ikAe^{i(kx-\omega t)} = ik\Psi(x,t), \quad \frac{\partial^2 \Psi}{\partial x^2} = ikA\frac{\partial}{\partial x}e^{i(kx-\omega t)} = (ik)^2\Psi(x,t) = -k^2\Psi(x,t)$$

$$\frac{\partial^2\Psi(x,t)}{\partial x^2} = -k^2\Psi(x,t) \overset{?}{=} \frac{2m}{i\hbar}(-i\omega)\Psi(x,t) = -\frac{2m\omega}{\hbar}\Psi(x,t)$$

This equation is true if $k^2 = \frac{2m\omega}{\hbar}$ or $\frac{\hbar^2 k^2}{2m} = \hbar\omega = \frac{h}{2\pi}2\pi f = hf$. The expression on the right we recognize as the energy, at least for the case of a photon. In the Schrödinger equation, hf should be interpreted as the total energy of the object. In the case of an object that has zero potential energy, the total energy should then be just the kinetic energy (no free energy or entropy for a single particle!), and $\frac{1}{2}mv^2 = \frac{p^2}{2m} = \frac{(\hbar k)^2}{2m}$ We then know that the linear momentum of the particle is $\hbar k$. The wave function, Ψ, and $|\Psi|^2 = \Psi^* \cdot \Psi$ need some comment. The real part of Ψ is $A\cos(kx - \omega t)$. (Assume A is real.) The imaginary part of Ψ is $A\sin(kx - \omega t)$. $|\Psi|^2$, by definition real, equals $|A|^2(\cos^2(kx - \omega t) + \sin^2(kx - \omega t))$. We cannot think of $e^{i(kx-\omega t)}$ as "the same thing" as $\cos(kx - \omega t)$.

Maxwell equation:

$$\frac{\partial \Psi}{\partial t} = -i\omega Ae^{i(kx-\omega t)} = -i\omega\Psi(x,t), \quad \frac{\partial^2 \Psi}{\partial t^2} = -i\omega\frac{\partial}{\partial t}Ae^{i(kx-\omega t)} = (-i\omega)^2\Psi(x,t) = -\omega^2\Psi(x,t)$$

and again $\frac{\partial^2\Psi}{\partial x^2} = -k^2\Psi(x,t)$, so $\frac{\partial^2\Psi(x,t)}{\partial x^2} = -k^2\Psi(x,t) \overset{?}{=} \frac{1}{v^2}\frac{\partial^2\Psi(x,t)}{\partial t^2} = \frac{-\omega^2}{v^2}\Psi(x,t)$

This last equation is true if $kv = \omega$. We showed in the previous exercise that this relation is correct for a light wave.

We have thus found a traveling wave solution to the Schrödinger equation and two solutions to Maxwell's electromagnetic wave equation. It seems that the Schrödinger equation is a bit pickier in terms of traveling wave solutions.

We understand the meaning of the wave function for the Maxwell wave equation. The quantities k and ω are related to wavelength and frequency, as noted above, and the constant A is the amplitude of the field. For example, the electric field in the wave is described by

Electromagnetic wave: $\quad E(x,t) = E_0\sin(kx - \omega t)$ or $E_0 e^{i(kx - \omega t)}$ \hfill (8.7)

E_0 is maximum electric field strength (volts/meter).

We do have a slight problem of interpretation, however: the exponential function has a real and imaginary part:

$$E(x,t) = E_0 e^{i(kx-\omega t)} = E_0(\cos(kx - \omega t) + i\sin(kx - \omega t)) \tag{8.8}$$

We cannot have an imaginary number for an electric field strength in a free traveling wave. The usual approach is to state that we are interested in the *real part* of the electric wave function, so just ignore the imaginary part. What this really means is that the cosine part of the wave function was what we were really interested in, and writing the function in exponential form just simplified some derivatives for us. This principle of looking for the real part of a wave function that may have both real and imaginary parts ($\Psi = x + iy$) occurs frequently in quantum waves, though we shall mostly avoid them in this text.

Actually, we can have an imaginary part, but that is more advanced electromagnetic field theory.

How do we physically interpret the Schrödinger wave function? (Equation 8.9 summarizes the relations you frequently need in quantum theory.)

$$E_{tot} = \hbar\omega$$
$$\hbar k = p = \text{linear momentum}$$
$$\lambda = \frac{2\pi}{k} = \frac{h}{p} = \text{wavelength (de Broglie)} \tag{8.9}$$
$$A = \text{probability normalization constant}$$

Except for the *normalization constant*, these relations look much like those for electromagnetic waves. (Note that the name "de Broglie" is associated with the particle wavelength to distinguish it from photon wavelength.) There is, however, another major difference hidden in these equations. The wavelength-momentum relation at first looks like that for a photon:

$$\lambda = \frac{2\pi}{k} = \frac{h}{p} = \frac{v}{f} \text{(photon)} \tag{8.10}$$

This relation shows both the usual wavelength-frequency-wave-velocity relation as well as the expression for the momentum of a photon, $p_{photon} = h/\lambda$. For a particle with mass the expression with momentum p must guide our interpretation:

$$\lambda = \frac{2\pi}{k} = \frac{h}{p} \tag{8.11}$$

If we followed what we did with the light wave, we might write $p = mv$ for the particle (matter) wave to remind ourselves that particles have mass. In quantum mechanics, however, momentum is usually a good quantity to describe a particle's properties, but velocity is not. The fundamental reason is that while photons all travel at the same speed, $v = c = 3.0 \times 10^8$ m/s, independent of the photon's momentum, particles do not. A light wave pulse made up of various wavelengths will be made up of photons of various momenta ($p_{photon} = h/\lambda$), but they will all travel at the same speed, and we can legitimately speak of the wave-pulse speed. This is not so for a matter wave. Each (de Broglie) wavelength travels with its own speed, $\lambda = h/p = h/(mv)$, so a matter-wave pulse made up of different de Broglie wavelengths will spread out because the faster components travel ahead and slower ones lag behind (Figure 8.2). This is not a complicated quantum phenomenon: $\lambda = h/p = h/(mv)$ says that shorter-wavelength, higher-momentum components of a wave travel faster. If the wave represents probability, the probability of finding the particle somewhere in the wave pulse cannot change, so the area of the matter wave pulse must remain fixed. In fact, if the pulse represents one particle, the area should remain fixed at exactly 1.

So, matter waves represent *probability waves*: the probability for observing the matter particle at position x at time t. If the wave represents one particle, this probability is just 1, as long as the particle does not disappear or cause other particles to somehow join it. The total probability should be related to the absolute square of the wave function as in Equation 8.2. If we apply this to the wave function from Exercise 8.2, we get

$$\left|\Psi(x,t)\right|^2 = \left|Ae^{i(kx-\omega t)}\right|^2 = \left(A^*e^{-i(kx-\omega t)}\right)\left(Ae^{i(kx-\omega t)}\right) = A^*A = \left|A\right|^2 \tag{8.12}$$

For those unfamiliar with complex number notation, note that "absolute square" means to replace every imaginary number with its opposite, leaving real numbers unchanged. $|A|^2$ will always be a real, positive number.

Very short (fs) light wave pulses traveling in a nonlinear optical medium (e.g., glass, water) will also spread out or compress owing to differing travel speeds, but the physical causes differ from those governing matter waves.

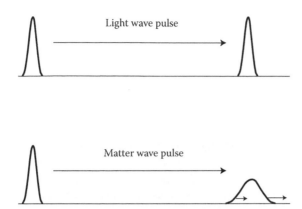

FIGURE 8.2
Traveling wave pulses are made up of various wavelengths (momenta). In an electromagnetic wave, photons of differing wavelengths (momenta) travel at the same speed, 3.0×10^8 m/s in a vacuum. In a matter wave, components of larger momenta travel at higher speeds, leading to spreading and flattening of the wave pulse. The area of the matter wave pulse, which represents the probability of particle location, remains constant.

Traveling matter wave pulses expand and reduce in amplitude as they move. Traveling planewave pulses are not common, so what do we learn in general from them? The lesson that applies to all matter wave functions is that confined matter waves "want" to expand. It takes force to keep a particle confined. This tendency can be expressed in several ways, but one of the simplest is the uncertainty principle.

8.2.2 Uncertainty Principle and Sketched Wave Functions

$$\Delta p \cdot \Delta x \geq \frac{\hbar}{2}, \ \Delta E \cdot \Delta t \geq \frac{\hbar}{2} \tag{8.13}$$

8.2.2.1 Confined particles want to expand

The expression of the uncertainty principle in terms of momentum uncertainty Δp and position uncertainty Δx says that if we confine a particle to a small region Δx, the matter wave will have a large range of momentum components: Δp is large. If the confinement of the particle is somehow released, an initially more confined particle will expand more rapidly because of its larger range of momentum components.

A second version of the uncertainty principle relates energy and time uncertainty. The usual first application of this uncertainty principle is to assert that states that are confined to more precisely defined values of total energy (small ΔE) have long lifetimes, while states with energy somehow distributed over a wide energy range are very short-lived. Atomic states, with energies precisely defined by the Bohr energy expression, can be long-lived.

8.2.2.2 Sketched wave functions and de Broglie wavelength

We can even think of the tendency of a confined particle to expand as a force. Consider the case of a confined particle with zero potential energy. In Figure 8.3a a particle is confined to a region of length L along the x axis. Because the walls of the well are infinitely high, the particle cannot

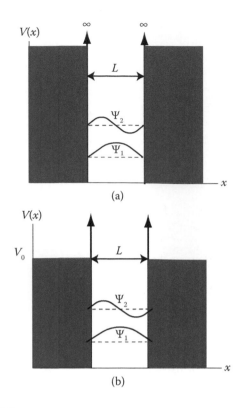

(a)

(b)

FIGURE 8.3

(a) A one-dimensional particle confined to a space L by infinitely high potential energy walls; referred to as a one-dimensional particle in a box. The potential energy in the allowed region is zero. Two sketched wave functions are shown in the lower part of the well. The lowest-energy state is Ψ_0, the one with the longest de Broglie wavelength, $2L$. The wavelength of Ψ_1 is L. Other wave functions will also fit in the well, if an integral multiple of half wavelengths fit: $n\lambda/2 = L$. (b) If the walls are not infinitely high, the wave functions will extend beyond the sides of the well.

exist at the wall boundaries, so Ψ must be zero at the walls. Two possible wave functions that satisfy this condition are shown. The longest de Broglie wavelength function corresponds to the lowest momentum and lowest energy state:

$$E_1 = \frac{p^2}{2m} = \frac{h^2}{2m\lambda_0^2} = \frac{h^2}{2m(2L)^2} = \frac{h^2}{8mL^2} \qquad (8.14)$$

If the walls (potential energy) are not infinitely high, as in Figure 8.3b, it is much more difficult to do a precise wave function calculation, but we can approximately say that the wave function can "leak" into the walls a bit because of the non-infinite potential. The wave function and energy levels will nevertheless not be far from those of the infinite well unless the well height V_0 becomes comparable to an energy level you are interested it.

We have argued in the past chapters that force is equal to the derivative of energy with respect to a coordinate. So, let's do a "hand-waving" calculation using this sketched wave function:

$$F_1 = -\frac{dE_1}{dL} = -\frac{d}{dL}\left(\frac{h^2}{8mL^2}\right) = \frac{h^2}{4mL^3} \qquad (8.15)$$

The subscript again means we are talking about the lowest-energy state that can exist in the well. Higher-energy states would have higher confining forces. Check units: $h = 6.63 \times 10^{-34}$ J·s, so the units are $[F]: \frac{(J^2 \cdot s^2)}{kg \cdot m^3} = \frac{(N \cdot m)^2 \cdot s^2}{kg \cdot m^3} = \frac{(N)^2}{kg \cdot m/s^2} = N$ (Newtons): correct. How big is the force confining an electron to a 1-Å box?

$$F_1 = \frac{h^2}{4mL^3} = \frac{\left(6.63 \times 10^{-34} \text{J} \cdot \text{s}\right)^2}{4\left(9.11 \times 10^{-31} \text{kg}\right)\left(10^{-10} \text{m}\right)^3} = 1.2 \times 10^{-7} \text{N} = 120 \text{ nN}.$$

This force applied to a free electron would impart an acceleration of $a = F/m = 1.3 \times 10^{23}$ m/s^2—a huge acceleration.

With a number this large, we better check that we have not gone out into left field and made a huge mistake. The force we calculated should not be that far from the force confining an electron in a hydrogen atom. The proton-electron distance in the H atom is a Bohr radius, 5.29×10^{-11} m. The proton–electron Coulomb force is

$$F_{\text{Coul}} = \frac{e^2}{4\pi\varepsilon_0 r^2} = \frac{\left(2.31 \times 10^{-28} \text{J} \cdot \text{m}\right)}{\left(0.529 \times 10^{-10} \text{m}\right)^2} = 8.3 \times 10^{-8} \text{N} = 83 \text{ nN}.$$

not at all far from our hand-waved value.

Just for the fun of it, let's pretend a 100-kDa biological molecular motor on a microtubule track were a quantum object, isolated from the rest of the world, except for confinement to a region 1 nm long. Equation 8.15 would give a confining force

$$F_1 = \frac{h^2}{4mL^3} = \frac{\left(6.63 \times 10^{-34} \text{J} \cdot \text{s}\right)^2}{4\left(10^5 \cdot 1.67 \times 10^{-27} \text{kg}\right)\left(10^{-9} \text{m}\right)^3} = 6.6 \times 10^{-19} \text{N},$$

which still corresponds to a large accelerating force on a 100-kDa particle in a vacuum. However, biomolecular motors are not in a vacuum, but in aqueous solution, constantly bombarded by solvent and other molecules. We will find that forces from these other sources are of order piconewton, so quantum forces are negligible. Where can we expect quantum energies and forces to matter in biological systems? We should look for general quantum effects governing electron and proton movement; possibly even small atoms and diatomic molecules, under restricted circumstances. When do electrons and protons move? Any charge transfer or oxidation–reduction reaction is a candidate. John Hopfield developed a theory in 1974, for example, for quantum-mechanical tunneling electron transfer in biomolecules that applies to a wide variety of cases.[1]

8.2.2.3 Quantum mechanics limits exactness

The uncertainty principle gives us information about how accurately we can simultaneously determine values of uncertainty-associated variables, (p, x), (E, t), (L, θ), etc. The idea is that precisely measuring x reduces the value of Δx, and by $\Delta p \Delta x \geq \hbar/2$ therefore increases the uncertainty of momentum and velocity: if you measure precisely where a particle is at one instant, the momentum gets large and unpredictable and you therefore quickly lose track of the particle. While this is as true in biosystems as well as in pristine quantum model systems, the application of this idea is rarely of any great consequence. An exception is the case of photosynthetic transfer of electrons between two sites. The usual way to determine where the electron is located during a transfer process is to send in a photon that is absorbed by the molecule where the electron is located. You may, then, not know where the electron is an instant later, but that often does not matter in a carefully designed experiment where you have large numbers of identical molecules.

L is angular momentum

8.2.3 Stationary States: Time-Independent Wave Functions

A time-independent quantum state is termed a stationary state. If we factor the wave function as a product $\Psi(x,t) = \psi(x)f(t)$ that satisfies the time-dependent Schrödinger equation, the function $\psi(x)$ is termed a stationary state and satisfies the time-independent Schrödinger equation:

$$-\frac{\hbar^2}{2m}\frac{d^2\psi}{dx^2} + V(x)\psi(x) = E\psi(x) \tag{8.16}$$

The wave functions sketched in Figure 8.3 are, in fact, stationary states of that system. In quantum mechanics, the time-independent Schrödinger equation is an expression of the addition of kinetic and potential energies to get the total energy. The first term, $-\frac{\hbar^2}{2m}\frac{d^2\psi}{dx^2}$, represents the kinetic energy, the second, $V(x)\psi(x)$, is the potential energy and the third, $E\psi(x)$ is the total energy. In "average values" below, we will see how to actually calculate values of these energies. We expect that large values of $V(x)$ correspond to large potential energies, but how do we recognize large kinetic energies? Large kinetic energy must correspond to large values of $\frac{d^2\psi}{dx^2}$, the second derivative. We will explore this derivative in the homework.

Stationary states of relevance in biophysics usually occur in cases of electrons or protons bound to atoms or molecules. We will examine some interesting and apparently complicated cases in photosynthesis and heme proteins in later chapters. These examples are, indeed, somewhat complex, but they take on surprising simplicity because of *symmetry*. Symmetry is an important concept in both quantum mechanics and in biosystems. You should have noted that many biological structures we have examined have high symmetry. If, for example, an electron can occupy a site on any of the monomers of a 12-mer aggregate and can move among the monomers, the wave functions of bound electrons become simple when the aggregate is a symmetric, 12-membered ring. Likewise, the case of electrons confined to the ring system of the heme group in hemoglobin or cytochrome enzymes becomes (approximately) very simple when the ring is symmetric. We will leave these interesting systems for later and simply write down the wave functions and energies for several simple potential energy cases. If you need to review these potentials further, refer to any modern physics or physical chemistry book. A summary of results for wave functions, wave numbers, and energies is shown in Figure 8.4 and is summarized here.

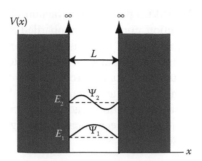

1D Particle in a box

$$\psi_n(x) = \sqrt{\frac{2}{L}} \sin\left(\frac{n\pi x}{L}\right) \quad (n = 1, 2, 3, \ldots)$$

$$k_n = \frac{2\pi}{\lambda_n} = \frac{n\pi}{L}$$

$$E_n = n^2 \frac{\pi^2 \hbar^2}{2ML^2}$$

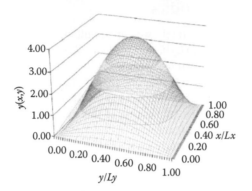

2D Particle in a box

$$\psi_{nm}(x) = \sqrt{\frac{2}{L_x}} \sqrt{\frac{2}{L_y}} \sin\left(\frac{n\pi x}{L_x}\right) \sin\left(\frac{m\pi y}{L_y}\right) \quad (n, m = 1, 2, 3, \ldots)$$

$$k_{xn} = \frac{2\pi}{\lambda_{1n}} = \frac{n\pi}{L_x}, \quad k_{ym} = \frac{2\pi}{\lambda_{ym}} = \frac{m\pi}{L_y}$$

$$E_{nm} = \frac{\pi^2 \hbar^2}{2M}\left(\frac{n^2}{L_x^2} + \frac{m^2}{L_y^2}\right)$$

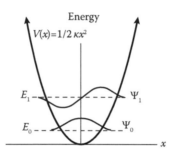

1D Harmonic oscillator

$$\psi_n(x) = H_n(x)e^{-\alpha x^2/2} \quad (n = 0, 1, 2\ldots) \quad \alpha = \sqrt{\frac{M\kappa}{\hbar^2}}$$

$$H_0(x) = \left(\frac{\alpha}{\pi}\right)^{1/4}, \quad H_1(x) = \left(\frac{\alpha}{\pi}\right)^{1/4}\sqrt{2\alpha}x, \ldots$$

$$E_n = (n + 1/2)\hbar\omega, \quad \omega = \sqrt{\frac{\kappa}{M}}$$

FIGURE 8.4
Wave functions and energies for several simple potentials.

Particle in a one-dimensional box:

$$\psi_n(x) = \sqrt{\frac{2}{L}} \sin\left(\frac{n\pi x}{L}\right)(n = 1, 2, 3, \ldots)$$

$$k_n = \frac{2\pi}{\lambda_n} = \frac{n\pi}{L}$$

$$E_n = n^2 \frac{\pi^2 \hbar^2}{2mL^2}$$

Particle in a two-dimensional box:

$$\psi_{nn'}(x, y) = \sqrt{\frac{2}{L_x}} \sqrt{\frac{2}{L_y}} \sin\left(\frac{n\pi x}{L_x}\right) \sin\left(\frac{n'\pi y}{L_y}\right)(n, n' = 1, 2, 3 \ldots)$$

$$k_{xn} = \frac{2\pi}{\lambda_{xn}} = \frac{n\pi}{L_x}, \quad k_{yn'} = \frac{2\pi}{\lambda_{yn'}} = \frac{n'\pi}{L_y}$$

$$E_{nn'} = \frac{\pi^2 \hbar^2}{2m}\left(\frac{n^2}{L_x^2} + \frac{n'^2}{L_y^2}\right)$$

One-dimensional harmonic oscillator:

$$\psi_n(x) = \frac{1}{\sqrt{2^n n!}} \left(\frac{\alpha}{\pi}\right)^{\frac{1}{4}} H_n\left(\sqrt{\alpha}\,x\right) e^{-\alpha x^2/2} \quad (n = 0,1,2\ldots)\ \alpha = \frac{m\omega}{\hbar}$$

$$H_0(x) = 1,\ H_1(x) = 2\sqrt{\alpha}\,x, \ldots$$

$$E_n = (n+1/2)\hbar\omega, \quad \omega = \sqrt{\frac{\kappa}{m}}$$

Unlike the box potentials above, the harmonic oscillator has potential energy $V(x) = \frac{1}{2}\kappa x^2$, where x is the displacement of the particle from the unstretched length of the spring, κ is the spring constant and ω is the angular frequency of oscillation. Note in Figure 8.4 the harmonic oscillator wave function does not equal zero at the position where the particle's total energy equals the potential energy. Classically, this is where the particle would turn back. (Wave functions must go to zero if the potential energy is infinite, however.)

An important quantum property has crept into our mathematics: some constants, n, m, etc., have appeared in the wave functions. These are referred to as *quantum numbers* and describe the *state* of the particle or object. The quantum numbers always appear in the wave function and usually appear in the expression for the discrete values of energy and other physical quantities. The quantum numbers above have appeared because of the confinement of the particle to some local region; other quantum numbers, like spin, will appear owing to intrinsic properties of the particle.

8.2.4 Energy-Level Diagrams

Stationary states, solutions to the time-independent Schrödinger equation, have quantized (discrete) energies. This contrasts with the traveling waves, which are allowed any value of total energy. The cause of quantized energies is a potential that confines a particle's wave function to a specific region: H atoms have the familiar $-E_0/n^2$ energies whenever the total energy of the electron is negative—when the negative Coulomb potential energy is greater than the kinetic energy.

8.2.5 Normalization Constants

If $|\psi(x)|^2 dx$ represents the probability for finding a quantum particle between x and $x + dx$, the integral over all x must give 1: the particle is definitely located somewhere between $-\infty$ and $+\infty$. Mathematically, this means

$$\int_{-\infty}^{\infty} |\Psi(x,t)|^2\, dx = 1 \tag{8.17}$$

When we are dealing with stationary states, the choice for the time-dependent part of the wave function is made $f(t) = e^{i\omega t}$, so that $|f(t)|^2 = f^*(t) \cdot f(t) = e^{-i\omega t} \cdot e^{i\omega t} = 1$, and we can write the normalization condition as

$$\int_{-\infty}^{\infty} |\psi(x)|^2\, dx = 1 \tag{8.18}$$

Let's apply this to our first wave function, $\Psi(x,t) = Ae^{i(kx - \omega t)}$. Remember that A is some constant, which we can now interpret as the normalization constant.

$$\int_{-\infty}^{\infty} |\psi(x)|^2\, dx = \int_{-\infty}^{\infty} A^* e^{-i(kx-\omega t)} Ae^{i(kx-\omega t)}dx = \int_{-\infty}^{\infty} A^* A\, dx = \int_{-\infty}^{\infty} |A|^2\, dx = 1.$$

Unfortunately, there is no value of A, except perhaps $A \rightarrow 0$, that will make this integral equal to 1. It is infinite. What we have found actually makes physical sense. If one free particle is traveling to the right with momentum exactly equal to $\hbar k$, it is equally likely to be found at any position along the axis, and the probability for being located at any particular position between x and $x + dx$ is vanishingly small. This is a peculiarity of this particular wave function. We can, in fact, produce a state of an electron that is at least approximately free: such states have existed for years inside television tubes that fired electrons at phosphor screens to produce pictures. However, in this case we know that the electron is somewhere between the tube's electron gun and the phosphor screen, a distance of about 40 cm. The limits of the integral are then $0 \rightarrow 0.40$ m, and the integral can be done:

$$\int_0^{0.40m} |A|^2 \, dx = |A|^2 (0.40m) = 1, \text{ or } A = 1.58m^{-1/2}$$

Notice again that this constant is not just an abstract mathematical parameter; it is directly related to the probability for finding the real particle.

Let's normalize a wave function that does not cause trouble: the two-dimensional particle in a box. (Actually, we will confirm that the wave function given above *is* properly normalized.)

$$\int_0^{L_x} dx \int_0^{L_y} dy \left(\sqrt{\frac{2}{L_x}} \sqrt{\frac{2}{L_y}} \sin\left(\frac{n\pi x}{L_x}\right) \sin\left(\frac{m\pi y}{L_y}\right) \right) \left(\sqrt{\frac{2}{L_x}} \sqrt{\frac{2}{L_y}} \sin\left(\frac{n\pi x}{L_x}\right) \sin\left(\frac{m\pi y}{L_y}\right) \right)$$

$$= \left(\frac{2}{L_x} \frac{2}{L_y} \right) \int_0^{L_x} \sin^2\left(\frac{n\pi x}{L_x}\right) dx \cdot \int_0^{L_y} \sin^2\left(\frac{m\pi y}{L_y}\right) dy$$

Making the variable substitutions $u = n\pi x/L_x$, $v = m\pi y/L_y$, $du = n\pi/L_x dx$, $dv = m\pi/L_y dy$ results in

$$\left(\frac{2}{L_x} \frac{2}{L_y} \right) \int_0^{n\pi} \sin^2 u \, du \frac{L_x}{n\pi} \cdot \int_0^{m\pi} \sin^2 v \, dv \frac{L_y}{m\pi} = \left(\frac{2}{L_x} \frac{2}{L_y} \right) \left(\frac{L_x}{n\pi} \frac{L_y}{m\pi} \right) \int_0^{n\pi} \sin^2 u \, du \cdot \int_0^{m\pi} \sin^2 v \, dv$$

$$= \left(\frac{2}{L_x} \frac{2}{L_y} \right) \left(\frac{L_x}{n\pi} \frac{L_y}{m\pi} \right) \left(\frac{1}{2} \cdot n\pi \right) \left(\frac{1}{2} \cdot m\pi \right) = 1$$

and the wave function is properly normalized. This wave function shows a choice that is generally made: if a wave function consists of products of functions of different variables, each function is separately normalized.

8.2.6 Average Values

Average values naturally occur in statistical phenomena, where processes are repeated more than once or when several or many identical (or almost identical) objects perform the same process and the outcome of the process can produce various values. In formal statistical applications, a distribution function is generally used to describe the probability that each outcome results. We will develop some of these statistical distribution functions in Chapter 14. A quantum wave function is similar to a statistical distribution function. If $|\psi(x)|^2$ is the probability per unit distance that the particle is located at position x, then for normalized wave functions

$$|\psi(x)|^2 dx = \psi^*(x)\psi(x)dx = \text{probability the particle is located between } x \text{ and } x + dx \quad (8.19)$$

and

$$\int_{x_1}^{x_2} \psi^*(x)\psi(x)dx = \text{probability the particle is located between } x_1 \text{ and } x_2 \quad (8.20)$$

If these are probabilities, then if we wanted to calculate the *average* position $\langle x \rangle$ of the particle we should do the following computation: $\langle x \rangle$ = probability for being at any position, x, summed over all positions. This is precisely the following computation:

$$\langle x \rangle = \int_{-\infty}^{\infty} \psi^*(x) x \psi(x) dx \tag{8.21}$$

Notice here that x is a variable that gets integrated over. It follows that if we want to calculate the average value of any other physical quantity that may (or may not) be a function of x, we should do the same sort of calculation:

$$\langle f(x) \rangle = \int_{-\infty}^{\infty} \psi^*(x) f(x) \psi(x) dx \tag{8.22}$$

In fact, if we want to find the average value of an "operator," like a derivative operator, on the wave function, for instance, we can do so:

$$\left\langle \frac{d}{dx} \right\rangle = \int_{-\infty}^{\infty} \psi^*(x) \frac{d}{dx} \psi(x) dx \tag{8.23}$$

Note here that it is critical that we wrote $\psi^*\psi$ rather than $|\psi|^2$, because the derivative operates only on $\psi(x)$ and must be placed between ψ^* and ψ.

Exercise 8.3 presents one case in which an average value of d/dx is zero. Does this have any physical meaning, especially for biophysics, or is this just a quantum/mathematical exercise that physics professors love to give? The answer is that this derivative has a very definite physical meaning, as does the second derivative, d^2/dx^2. We just observed that the second derivative, if multiplied by $-\frac{\hbar^2}{2m}$, gives the kinetic energy, when integrated over the entire wave function ($x = -\infty$ to $+\infty$). We won't take time to prove this, but if you multiply the first derivative by $-i\hbar$ (i again is the imaginary number $\sqrt{-1}$) you get the *momentum operator*:

> We place x between ψ^* and ψ because such placement is necessary when taking average values of some other quantities.

$$p = i\hbar \frac{d}{dx} \tag{8.24}$$

and

$$\langle p \rangle = -i\hbar \left\langle \frac{d}{dx} \right\rangle = -i\hbar \int_{-\infty}^{\infty} \psi^*(x) \frac{d}{dx} \psi(x) dx \tag{8.25}$$

Why do energy and momentum appear to be differential operators? We actually get differential operators for p and K because we chose to write our particle wave functions in terms of the coordinate x. We could have written the wave function as a function of p, in which case p would just be p, but x would be converted to $-i\hbar \, d/dp$. Most students agree d/dp derivatives are even more confusing than d/dx.

Is there any biophysics here? At this point we must admit to a bit of mathematical practice designed to prepare you for the chapter on statistics. However, if we need to find the average position or momentum of an electron, proton, or even an atom bound in some potential, we may need to do these integrals. For instance, we will find that a simple theory for chemical reactions—say the binding of O_2 to myoglobin's Fe^{2+} atom—assumes the O_2 is initially bound in an approximate parabolic well (harmonic oscillator), with quantized energy levels (Figure 8.5). If we want to estimate the speed at which this particle approaches one side of the well, we can use the methodology above. We can and will also use the equipartition theorem for this case, but if this oscillator is at low temperature or is fairly isolated, the equipartition theorem may not be valid (equilibrium not maintained). As stated before, quantum effects are most often observed with small particles, electrons, and protons, so we should look mostly look for such effects there.

> The harmonic oscillator is an amazingly useful model for many physical systems. Learn it.

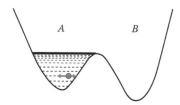

FIGURE 8.5
A reaction A ⇌ B can usefully be described by passage of a particle from well *A* to well *B*. In the bottom half of the well, the potential is approximately parabolic and energy levels are nearly equally spaced. Near the top of the intervening barrier, the energy levels become closely spaced because the effective spring constant of the well is much smaller. At lower temperature, lower energy levels will predominate.

> Protons and neutrons are also fermions, but bound states of two fermions, like a hydrogen atom [= proton + electron] or a deuteron [= proton + neutron], taken as a whole, are *bosons*. Bosons, with integral spin, do not obey an exclusion principle.

8.3 PAULI EXCLUSION PRINCIPLE

The wave functions we have written so far describe one electron. What happens if we place two or more electrons in the same potential energy well—around an atomic nucleus or in a one-dimensional box? Wolfgang Pauli proposed the exclusion principle in 1925: No two electrons in an atom may have all the same quantum numbers.

The purpose behind this proposal was to make spectroscopic observations on atoms agree with the Bohr theory of the atom. The emphasis in this statement should be on *electron*, not *in an atom*. This exclusion principle reflects the intrinsic nature of electrons as half-integral spin *fermions*, so electrons confined by forces other than Coulomb attraction of the nucleus should also obey the exclusion principle.

For example, if we place two electrons (fermions) in a one-dimensional box of length L, we would say they cannot have the same value of n. This is what we would say if we did not know about the intrinsic quantum numbers of electrons related to their spin. Electrons have spin ½, which reflects the spin quantum number s and relates to the magnetic moment of the electron: the spin will orient in a magnetic field. However, this spin can orient in two distinct orientations, up or down. These two distinct orientations reflect a second quantum number associated with electron spin, the z component of spin, s_z. An electron thus intrinsically has two possible spin states it can be in, independent of any particular confining potential well:

$$\text{Electron spin: } s = ½; \; s_z = ±½ \tag{8.26}$$

For now, we will trust you have studied the meaning of the spin and z component thereof.

EXERCISE 8.3

An electron is in the *ground state* (lowest-energy state) of a one-dimensional box of length 0.10 nm. Compute the average value of the operator d/dx for this electron.
Questions and thoughts that may go through your mind:

1. What does the average value of a differential operator mean? We never learned that in calculus.
2. Does calculate the average value mean calculate a numerical value?

Solution

1. A differential operator operating on a function just gives the function's slope at each point. It makes sense to say there is some average value of the slope.
2. If values of all parameters are provided, a numerical value can be found. Let's see.

$$\left\langle \frac{d}{dx} \right\rangle = \int_{-\infty}^{\infty} \psi^*(x) \frac{d}{dx} \psi(x) dx = \int_0^L \sqrt{\frac{2}{L}} \sin\left(\frac{\pi x}{L}\right) \frac{d}{dx} \sqrt{\frac{2}{L}} \sin\left(\frac{\pi x}{L}\right) dx$$

$$= \frac{2}{L} \int_0^L \sin\left(\frac{\pi x}{L}\right) \frac{d}{dx} \sin\left(\frac{\pi x}{L}\right) dx$$

$$\left\langle \frac{d}{dx} \right\rangle = \frac{2}{L}\int_0^L \sin\left(\frac{\pi x}{L}\right)\frac{\pi}{L}\cos\left(\frac{\pi x}{L}\right)dx = \frac{2\pi}{L^2}\int_0^L \sin\left(\frac{\pi x}{L}\right)\cos\left(\frac{\pi x}{L}\right)dx.$$ Now, when you do integrals, you should always look to see if the value might be zero (by symmetry or some other reason) before performing a lot of math. The best way to see if the integral might be zero is to sketch the integrand and see if it is negative as often as it is positive. The integral limits are from 0 to L, so $\pi x/L$ goes from 0 to π. The sine function goes from 0 to 1 back to 0: always positive. The cosine function goes from 1 to 0 to -1. If you multiply sine and cosine you get a function that is half positive and half negative—exact antisymmetry or an odd function—so the integral is 0. Again, we are talking here about a physical quantity, so it may have units. What are they? d/dx has units of m^{-1}, in SI units, so its average value is 0 m^{-1}.

What if we are asked for the average value $\langle d^2/dx^2 \rangle$? Hopefully, you can see that this is not zero, but homework asks you to confirm this.

Those of you who have studied the Bohr atom may protest that the $n = 2$ level can hold more electrons. You would be wrong. This is *not* a hydrogen atom; there is no orbital angular momentum about the central force of the nucleus here; there is no possibility of orbital motion at all.

Let's return to two electrons placed in the one-dimensional box. We know two things for sure: (1) they cannot have the same set of quantum numbers and (2) given the chance, particles try to find the lowest-energy state they are allowed. These two conditions can simultaneously be met if both electrons occupy the $n = 1$ state, but have opposite z components of spin. Figure 8.6a shows the lowest-energy configuration of two electrons in the box: both are in the $n = 1$ level, with associated wave function $\psi^{elec1}(x_1) = \sqrt{\frac{2}{L}}\sin(\frac{\pi x_1}{L})$, $\psi^{elec2}(x_2) = \sqrt{\frac{2}{L}}\sin(\frac{\pi x_2}{L})$, but the spin arrows are pointed in opposite directions, in conformity with the exclusion principle. If we place five electrons in the box, they will distribute among the energy levels as shown in Figure 8.6b.

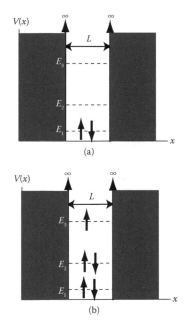

(a)

(b)

FIGURE 8.6
(a) Two electrons placed in a one-dimensional box will normally occupy the lowest energy level, with spins pointed in opposite directions. Arrows are placed in the $n = 1$ energy level to indicate the states of the two electrons. If energy is added to the system (e.g., photons), excitation of one or both electrons to higher energy levels can occur. (b) The lowest energy allowed state of five electrons placed in the box. There are no quantum numbers besides spin and n in this system.

8.4 FROM ATOMS TO MOLECULES

Quantum numbers of atomic electrons:

n: principle quantum number; $n = 1, 2, 3,\ldots$

ℓ, m_ℓ: orbital angular momentum quantum numbers;

$\ell = 0, 1, 2,\ldots (n - 1)$; $m_\ell = -\ell, -\ell + 1,\ldots, 0, \ldots \ell - 1, \ell$

s, m_s: spin angular momentum quantum numbers

$s = \frac{1}{2}$; $m_s = \pm\frac{1}{2}$

The hydrogen atom, as well as the basics of other atoms, should be familiar to all of you. Electrons in single atoms have quantum numbers as follows. The numbers ℓ, m_ℓ refer explicitly to motion in the *central force* field of the nucleus. These quantum numbers will *only* appear in a three-dimensional central force field and not, for instance, in a one- or three-dimensional box potential. As you will find in the homework, a quantum number that refers to clockwise and counterclockwise motion on a circular ring, corresponding to orbital angular momentum in two dimensions, will appear, but it is also different than ℓ and m_ℓ, which can take on n and $2n - 1$ different values, respectively.

What happens to quantum states when we move on to molecules? The answer: complicated stuff that takes an entire book to comprehensively explain. We have to summarize the issues and results succinctly.

1. Molecules consist of atoms covalently bound together by the sharing of *valence electrons*. These valence electrons do not behave like any of the pure atomic electronic states, though they are commonly referred to by the atomic states they came from in the isolated atom.

2. In molecules involving atoms of the mass of carbon or larger, the inner electrons, those in lower-energy states than the outermost shell of electrons, are approximately in their atomic-electron states and rarely affect interactions important for biology.

3. Certain electrons not directly involved in molecular bonding, may be left in "exposed" states that can interact readily with atoms of other molecules. "Lone-pair" electrons generally appear on one side of an atom whose other electrons are directly involved in bonding. The oxygen in H_2O, for example, has two lone pairs of electrons; N in NH_3 has one lone pair. We have already seen that the lone pairs in H_2O are intimately involved in H bonding in water and water's unusual properties.

4. The first and most important type of quantum behavior important to biology is the chemical reaction: the transfer of electrons, ions, or atoms from one molecule to another. Transfer of electrons, ions, or atoms constitutes the arena of chemical reactivity. While electron and proton transfer can be both measured and theoretically modeled much like energy transfer (below), the transfer of ions and atoms with attached electrons in biological reactions is generally limited to experimental measurements and simple reaction rate modeling. The first exception was a detailed quantum simulation of phosphoryl transfer by a protein kinase.[2]

5. The second type of molecular quantum behavior of relevance to biosystems involves vibrations and rotations, characterized by spring constants determined by the bonds and atomic masses, and moments of inertia, determined by interatomic distances and atomic masses. The spring constants can be approximately calculated but are most often determined through infrared vibrational frequency measurements. Moments of inertia can either be computed from atomic coordinates determined by X-ray crystallography or experimentally determined by rotational frequency measurements.

6. The third quantum effect of importance to biology is energy transfer. The obvious case is photosynthetic energy transfer, which we discuss later. Energy transfer occurs between two sites, separated by some distance, which interact through weak electromagnetic forces, often an electric dipole–dipole interaction. The weakness of the interaction allows a quantum "perturbation" treatment, which simplifies calculations significantly.

7. Finally, the harmonic oscillator model should always be kept near at hand. The model has a simple and exact solution and approximates many processes taking place in biomolecules: vibrations (recall the X-ray crystallography temperature factor), chemical reaction rates and oscillator reaction modeling, and modeling of forces between molecules, to name a few.

8.5 COLLISIONS OF ATOMS AND MOLECULES

Collisions of atoms constantly occur in gases, liquids, and solids. Particle collisions are the bread and butter of particle physics. For particles too small to see with a microscope, particle-particle collsion experiments and theoretical analysis offers scientists one of few ways to determine particle properties. The discovery of the "Rutherford atom" was made by collision experiments, Rutherford scattering, and first showed that an atom had a small, heavy nucleus surrounded by light electrons. Electrons were found to be light through collisions with heavier particles. Today, particle physicists are looking for the Higgs boson, the particle thought to be responsible for the fact that matter has mass, as a constitutive part of more common subnuclear particles. The understanding we have of the microscopic world stems largely from collision experiments done in the twentieth century. The quantitative description of these collisions has been largely based on quantum mechanics.

Are collisions of atoms and molecules important factors in the behavior of biological systems? The obvious answer is "Yes" because we have found that biomolecules, as well as the water and dissolved molecules in the water, are in constant motion at physiological temperature. Any chemical reaction taking place between two molecules requires that they first collide with each other. When we study reaction theory, the issue of the kinetic energies of the colliding molecules will be a primary determinant of the reaction rate. A second factor will be the relative velocities and orientations of the colliding particles, as in Figure 8.7. In the end, the actual bond-forming reaction depends on details of the quantum properties of the two colliding atoms or molecules. However, most of the (nonreacting) collisions of importance to biology can, thankfully, be treated classically. Why is this? The vast majority of the collisions taking place between molecules in biosystems do not result in the break-up of either molecule. If the molecules do not break apart, with electrons or atoms detaching, the analysis of the collision can proceed with classical computations. Though such collisions do not produce chemical reactions *per se*, they are responsible for osmotic pressure and other ordering forces in biosystems. The Equipartition theorem can be used if equilibrium is at least approximately maintained. This theorem says that the average translational kinetic energy of molecules is $3/2k_BT$. Furthermore, the Maxwell–Boltzmann distribution tells us the probabilities for any other particle kinetic energies. Even the assumed randomness of velocity directions gives some simplification because we can average over all angles.

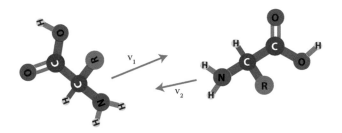

FIGURE 8.7

Two amino acids must come together (collide) in order for a peptide bond to form. To be successful, the two molecules must be oriented correctly and the relative velocities must be correct. Any successful molecular reaction must have these two qualities. In a cell, the amino acids are properly oriented by other structures to which they are bound; they do not react through random collisions.

8.6 CLASSICAL VS. QUANTUM: IS A 1-MM-LONG MOLECULE OF DNA A QUANTUM OBJECT?

Quantum objects are generally considered to be "small" objects. The primary way to judge "smallness" in a quantum mechanical sense is to determine the de Broglie wavelength and compare that to other measurable "sizes" of the object. A baseball moving at 100 miles per hour is not a quantum object (does not require quantum theory to accurately describe its behavior) because the quantity h/mv is vanishingly small even compared to the radius of a proton, while the collision of a baseball with a bat involves sizes of ~10 cm. We conclude the electron bound in a hydrogen atom *is* a quantum object because its de Broglie wavelength, h/mv, is the same size as the electron's orbital radius. (See your introductory physics or chemistry books.) We generally think of molecules as quantum objects, though we treat them with classical theories whenever we can.

The preceding example of molecular collisions is a good place to apply the de Broglie wavelength principle. Suppose a bulb is filled with H_2 molecules equilibrated at room temperature. The average translational kinetic energy is

$$K = \frac{p^2}{2m} = \frac{3}{2}k_B T \tag{8.27}$$

so the momentum is $p = \sqrt{3mk_B T} = \sqrt{3(2 \cdot 1.67 \times 10^{-27} \text{kg})(4.1 \times 10^{-21} \text{J})} = 6.4 \times 10^{-24}$ kg m/s and the average de Broglie wavelength is $\lambda_{dB} = \frac{h}{p} = \frac{6.63 \times 10^{-34} \text{J} \cdot \text{s}}{6.4 \times 10^{-24} \text{kgm/s}} = 1.0 \times 10^{-10}$ m. This length is comparable to the size of the electron clouds in the H atoms, so the wavelike properties of H_2 can be important to collisions. If we applied the same analysis to the myoglobin molecule (mass, 18 kDa), the resulting de Broglie wavelength would be $\sqrt{18,000} = 134$ time smaller, or 0.7 pm, clearly much smaller than a myoglobin diameter, so wave properties of Mb as a whole are not important. A strand of DNA 1 mm long would have an even smaller de Broglie wavelength, so by the measure of $\lambda_{dB} \ll$ particle size, DNA is a nonquantum object.

DNA *is* a quantum object when other phenomena are considered. DNA, or individual (or a few) DNA bases, interacts strongly with UV photons—a cause of DNA damage and in rare cases, mutations, and, rarer still, cancer (Chapter 11). In fact, the interaction of light with any molecule is a quantum interaction if the photon disappears, if an additional photon appears, or if a change in photon energy occurs. It is legitimate to say that any time a photon is absorbed a quantum event has occurred. But does it matter to biology? A damaged DNA base can certainly have important biological outcomes. Does photon absorption by myoglobin or hemoglobin have any biological consequences of importance? That depends. If hemoglobin has oxygen bound to it, there is a 5% chance that absorption of a photon will dissociate the O_2 from hemoglobin. The 5% number is referred to as the quantum yield of photo dissociation. If CO is bound, there is a 70% chance the photon will knock the CO off the Hb.[3] Suppose someone is unconscious as a result of breathing carbon monoxide gas. Can you think of any medical treatment protocol that might exploit this intrinsically quantum interaction of light with hemoglobin?

> Remember that any chemical reaction also involves quantum mechanics.

8.7 POINTS TO REMEMBER

- The de Broglie wavelength, $\lambda_{dB} = \frac{h}{p}$, its relation to energy (kinetic energy), $\text{KE} = \frac{p^2}{2m}$, and when these can be used to model the states of electrons
- The Pauli Exclusion Principle
- Light absorption and emission: $E_{\text{photon}} = hf = |E_n - E_m|$
- The sketching of wave functions

8.8 PROBLEM-SOLVING

PROBLEM 8.1 WAVE FUNCTIONS

Determine whether the function $Ae^{-\alpha x}$ satisfies the one-dimensional, *time-independent* Schrödinger equation if α is a real number for $V(x) = V_0 = $ const. Consider positive, negative, and zero values of V_0. Comment on other physical issues related to this proposed wave function.

PROBLEM 8.2 WAVE FUNCTIONS

Determine whether the function $Ae^{(\alpha x - i\omega t)}$ satisfies the one-dimensional, *time-dependent* Schrödinger equation if α is a real number for $V(x) = V_0 = $ const. Consider positive, negative, and zero values of V_0. Comment on other physical issues related to this proposed wave function.

PROBLEM 8.3 DE BROGLIE WAVELENGTHS

Calculate the de Broglie wavelengths for the following objects and state whether the object's quantum wave properties are likely to be observed.

1. An electron moving at 5.00×10^7 m/s
2. A proton moving at 5.00×10^7 m/s
3. A "thermal" neutron, with kinetic energy $= 3/2\, k_B T$ at room temperature
4. "Thermal" hemoglobin
5. An *Escherichia coli* cell moving at 25 μm/s

PROBLEM 8.4 ONE-DIMENSIONAL PARTICLE IN A BOX

1. Modify Equation 8.14 for the $n = 2$, $n = 3$, and $n = 4$ states. Sketch the wave functions (as in Figure 8.3) and calculate the expression for the energies of the $n = 2$, $n = 3$, and $n = 4$ states.
2. Modify Equation 8.15 to calculate the confining force expressions for $n = 2$, $n = 3$, and $n = 4$.
3. Calculate numerical values of energies and forces, as in 1 and 2 for $n = 1, 2, 3,$ and 4 for a proton in an $L = 0.15$ nm box.

PROBLEM 8.5 ELECTRON CONFINED TO A RING

Find the radius of a benzene ring.

1. Sketch the four lowest-energy (longest de Broglie wavelength) wave functions of this system by fitting an integral number of wavelengths around the ring. (The wave must smoothly connect to itself.) Determine these four wavelengths. Write both the algebraic expression for the wavelengths and the numerical values.

2. Calculate the numerical values of the lowest four energies of an electron confined to such a ring using the de Broglie wavelengths and the relation between kinetic energy and momentum. Write both the algebraic expressions and the numerical values. Ignore the presence of C atoms on the ring and assume the potential energy is zero along the ring.

3. Describe how you would place 1, 2, 4, and 6 electrons into the states of this system, remembering the exclusion principle. Consider both electron spin and orbital angular momentum—the direction of motion around the ring.

PROBLEM 8.6   TWO-DIMENSIONAL ELECTRON IN A BOX: ENERGIES

Calculate the numerical values of the lowest three energy levels of an electron confined to the area of a square that has the area of a benzene ring. Express these energies in joules, kJ/mol, electron volts (eV), and in multiples of $k_B T$.

PROBLEM 8.7   KINETIC ENERGY

Write down the wave functions for the three lowest-energy states of the one-dimensional particle in a box.

1. Compute the second derivative of each wave function.
2. From the results of 1 write down the ratio $K_3 : K_2 : K_1$, where K_3 is the kinetic energy of state $n = 3$.

PROBLEM 8.8   TWO-DIMENSIONAL PARTICLE IN A BOX: WAVE FUNCTIONS

Graduate problem. Figure 8.4 shows the lowest-energy ($n = m = 1$) wave function for a particle in a box.

1. Use Microsoft Excel or a graphing program to duplicate this graph.
2. Repeat 1 for $n = 2$, $m = 2$.

PROBLEM 8.9   EXPECTATION VALUES

1. Write down the lowest-energy wave function of the harmonic oscillator (see Figure 8.4).
2. Compute the average value $\langle K \rangle$ of the kinetic energy term of the time-independent Schrödinger equation.
3. Compute the average value $\langle V \rangle$ of the potential energy.
4. Compute the sum $\langle K \rangle + \langle V \rangle$ and compare to E for that state.

PROBLEM 8.10 EXCITED-STATE EXPECTATION VALUES

Graduate problem. Repeat the previous problem's computations using the next higher-energy wave function of the harmonic oscillator.

PROBLEM 8.11 H ATOM SPECTRA

The well-known formula for the energy levels of the hydrogen atom is $E_n = \frac{-13.6\text{eV}}{n^2}$. Calculate the photon energies (in electrovolts, in Joules, in kJ/mol, and in multiples of $k_B T$), frequencies, and wavelengths (in nanometers) of photons that will excite the hydrogen atoms from $n = 2 \rightarrow 3$, $3 \rightarrow 4$, $5 \rightarrow 6$.

PROBLEM 8.12 HEMOGLOBIN SPECTRA

1. Find the wavelengths of the Soret and α and β absorption band peaks of oxy- and of deoxy-hemoglobin.
2. From the wavelengths, calculate the photon energies (in electronvolts, in joules, in kJ/mol, and in multiples of $k_B T$), and the frequencies of these bands.

PROBLEM 8.13 HARMONIC OSCILLATOR SPRING CONSTANTS AND ENERGY SPACINGS

1. What is the approximate spring constant of the bond between N and N in N_2? (Do some research.)
2. What is the spacing between energy levels in this harmonic oscillator?
3. Suppose absorption of a photon can cause a transition from one level to higher levels. Calculate the energy (in eV, in Joules, in kJ/mol, and in units of $k_B T$) and wavelength of the photons that will promote an N_2 vibrational state 1, 2, and 10 energy levels higher. What sort of light do these photons correspond to (UV, visible, IR, X-rays, etc.)?

REFERENCES

1. Hopfield, J. J. 1974. Electron transfer between biological molecules by thermally activated tunneling. *Proc Natl Acad Sci U S A* 71:3640–3644.
2. Valiev, M., R. Kawai, J. A. Adams, and J. H. Weare. 2003. The role of the putative catalytic base in the phosphoryl transfer reaction in a protein kinase: First principles calculations. *J Am Chem Soc* 125:9926–9927.
3. Olson, J. S., E. W. Foley, D. H. Maillett, and E. V. Paster. 2003. Measurement of rate constants for reactions of O_2, CO, and NO with hemoglobin. In *Hemoglobin Disorders. Molecular Methods and Protocols*, ed. R. L. Nagel. Totowa, NJ: Humana Press, p. 74.

CHAPTER 9

CONTENTS

Light, Life, and Measurement

9

Life on Earth is sustained by constant energy input from the sun. Few of us need to be reminded of this, but reviewing some of the numbers and basic principles by which this energy is converted to biologically useful forms is a vital part of any introduction to biological physics. The energy that comes from the sun to Earth is in the form of photons ranging from ultraviolet to far infrared (Figure 9.1). Though the source of the energy in the sun is fusion, which produces gamma rays, multiple conversion processes result in a light emerging from the surface of the sun with a peak and shape characteristic of a ~5500 K blackbody radiator, a type of radiator similar to the glowing coil of a tungsten light bulb. Several characteristic wavelength bands are absorbed by gases in the atmosphere. Ozone, O_3, is responsible for much far-ultraviolet (UV) absorption and is important for protection of organisms from damage. Water absorption in the infrared is technologically important, as it restricts wavelengths that can be used in infrared (IR) astronomy and communications. Figure 9.1b shows an expanded view of the UV to near-IR region of the spectrum. The breadth of the spectral distribution, with almost a constant intensity from 450 to 600 nm, is responsible for the whiteness of sunlight.

Pay particular attention to the units of these irradiance spectra: power per unit area per unit wavelength, $W\ m^{-2}\ nm^{-1}$. These units technically are not SI units, which would be $W\ m^{-3}$, but they make the nature of "irradiance" more understandable. $W\ m^{-3}$ would appear to be a power density, Watts per cubic meter, which is misleading. Irradiance refers to light power hitting a square meter of surface in a spectral region between λ and $\lambda + d\lambda$. The total power hitting the 1 m^2 surface can then be calculated by integrating the irradiance spectrum over the wavelength region of interest. Integration over the entire UV–IR region results in about 1,400 W striking a 1 m^2 area normally (e.g., at the Earth's equator). This quantity is reduced by clouds and other atmospheric effects. Integration over the Earth's entire surface area would produce the total power hitting the Earth. Integration over the area of a spherical shell centered on the sun and passing through the Earth results in the total power radiated by the sun, about $3.9 \times 10^{76}\ W$. Several types of spectra will be used in this chapter. Pay special attention to the units of the spectra, as they indicate the nature of the spectrum and how conversions can sometimes be made from one to another.

> Note angular factors when citing sources of "energy flux" (W/m_2). For practical reasons, a square meter of surface on the Earth is often referred to. A square meter of surface at the north pole intercepts far less solar radiation than an equatorial square meter.

9.1 LIGHT: OUR ENERGY SOURCE

Sunlight is our energy source from several perspectives. From an astronomical point of view, sunlight hits the Earth, warming it to about 290 K, and the Earth then reradiates just as much power, but in a wavelength region characteristic of that lower-temperature blackbody. A steady state is created: power in equals power out. If this were not so, the Earth would either heat up or cool off indefinitely. This steady state reflects the ideal placement of the Earth from the sun, the sun's ideal size and resulting power output, and the biology-compatible temperature that results from this steady state. On a more local scale, 1000 W or so of light power striking each square meter of surface gives us a good understanding of several issues of importance to humans and other organisms:

> Distinguish "steady state" from "equilibrium," which would apply if the Earth were an isolated system (and also with no life).

1. The challenge of cooling an automobile or house during a tropical summer
2. The amount of solar energy striking a typical roof on a sunny day
3. The total optical power striking a typical plant leaf, and the maximum rate of fuel synthesis

Details of vital importance to living systems, however, require exploration of the details of the sun's spectrum. For instance, are all spectral regions of comparable significance to life? Examination of the power alone would imply that this is the case—misleading, indeed.

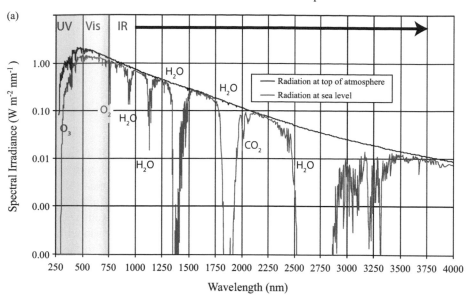

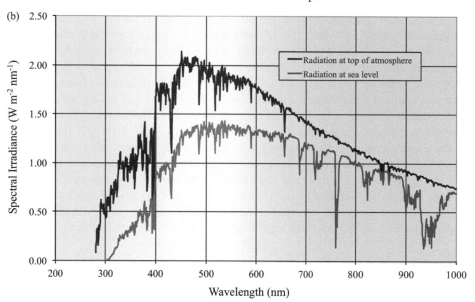

FIGURE 9.1

Solar radiation spectrum for direct light at both the top of the Earth's atmosphere and at sea level. The sun produces light similar to that from a 5525 K blackbody—approximately the sun's surface temperature. Atmospheric water vapor, ozone, and other gases absorb certain wavelength regions of the spectrum. Light is also scattered by Raleigh scattering, resulting in the sky's blue color. (a) Plot of log intensity versus wavelength from UV to far IR. (b) Linear graph of UV/Vis region of spectrum. Curves are terrestrial reference spectra for sunlight in North America. Graphs created from American Society for Testing and Materials (ASTM) original data posted at http://rredc.nrel.gov/solar/spectra/am1.5/by the U.S. Department of Energy, National Renewable Energy Laboratory.

9.2 CRUCIAL DIFFERENCES BETWEEN ONE 5-eV AND TWO 2.5-eV PHOTONS

The ultraviolet region of the solar spectrum is almost obscured in the graph in Figure 9.1a. The fact that the blackbody spectrum of the sun, as well as the absorption by ozone and other atmospheric gases, reduces the spectral spectrum sharply below 250 nm might lull us into ignoring UV, that is, if it weren't for sunburn. Two 2.5-eV ($\lambda = 500$ nm) photons impinging on a block of amorphous carbon may produce the same results as one 5-eV ($\lambda = 250$ nm) photon—heating of the carbon governed by the equation $E_{photon} = mc\Delta T$—but the same photons striking a biological system can produce radically different results. Fundamentally, the different responses have to do with (1) the energy-level structure of molecules making up the system and (2) the fact that absorption of light by biological systems is dominated by molecular (not atomic) energy-level properties.

Organic molecules, especially those with ring systems, allow multiple, delocalized electrons to occupy the quantum "boxes" and "rings" we studied in Chapter 8. Because the rings are spread over several (or many) atoms, the de Broglie wavelength of the electrons is longer, resulting in a smaller momentum ($p = h/\lambda$) and smaller kinetic energy ($K = p^2/2m$), causing the energy level differences among the higher occupied levels to correspond to the visible region (~2.5 eV) of the spectrum. On the other hand, a 5-eV photon from the sun will still be strongly absorbed, but the excitation may leave the electron with a positive energy and the molecule ionized. Many organic molecular excitations that result in ionization are not particularly favorable for the organism. An ionized molecule with a large ionization energy can transfer that energy to another molecule, leaving that second molecule ionized. If that molecule happens to be DNA, the result may be unfortunate. While in the majority of cases such solar *photochemical* damage is repaired by enzymes with a 99+% repair record, the small fraction of unrepaired cases usually produces a useless molecule, which is not a major problem. A few of those unrepaired instances result in cell death. A few cell deaths is, however, not a big deal: bacterial cultures continue with 10^7 cells, less a few; humans lose skin cells that are constantly being replaced. A potential advantage of DNA molecular damage in a single-celled organism is the minute chance that the modified DNA may code for a mutated cell-wall glycoprotein that does its basic job for the bacterium but cannot be recognized by the immune system of higher organisms. When a culture consists of 10^9 cells, a chance in a million of a favorable mutation is significant. A potential problem for the human organism is that in a tiny, tiny fraction of the cases where the 5-eV photon damages DNA, the damage is not repaired, and the result is not death of the cell but damage to the cell growth control systems. Twenty years later, the result may be cancer.

If the above assertions are correct, UV radiation from the sun should not be much of a problem for life. This has apparently been true for the past 10^7 years or so on Earth. Recent worries over the depletion of the ozone (O_3; see Figure 9.1a) layer have suggested that the safety of organisms from solar UV damage is borderline. In fact, UV light is a useful and clean tool for suppressing microbial life in certain circumstances. Those who care for ponds and aquariums know the problem of algae blooms and unwelcome bacterial infestations that sometimes arise. They also know that low-power UV lamps can successfully reduce populations of these unwanted organisms. You will briefly explore a theoretical analysis of this technique in the homework.

To summarize, solar photons have a wide range of energies. Higher-energy UV photons can do severe damage to biological systems: cells must be prepared to repair UV-induced damage. Lower-energy IR photons have relatively few biologically relevant interactions with individual cells and organisms. In the middle are photons that are "just right": photosynthetic organisms can harvest their energy for fuel production and higher organisms can employ them for remote detection (vision).

> Delocalized electrons are distributed over several atoms.

9.3 MEASURING THE PROPERTIES OF PHOTONS

The short introduction above introduces energies, wavelengths and spectra associated with biological life illuminated by light. These quantities should convince even the casual reader that measurement of light and the properties of light sources and light absorbers will be critical to our understanding of the roles of light in enabling life on earth. The importance of light measurement in biophysics has two aspects. First, light directly affects or enables most biological life. We must understand the physical and biological mechanisms of this interaction. Second, since the beginning of the twentieth century, light has been the premier tool for probing the structures and states of organisms and biological molecules. Such experimental methods can often be noninvasive, allowing measurement of an organism's or biomolecule's properties without damage or perturbation. Any routine trip to a medical clinic illustrates the ubiquity of light as a biomedical tool. Following measurement of a patient's weight and height, a small sleeve is clamped over the end of a patient's finger. This sleeve, a *pulse oximeter*, employs light to measure the oxygenation level of the patient's blood. Light can also be used to purposely cause a change. (See end-of-chapter problems.) To understand such changes and their photophysical or photochemical mechanisms, a clear understanding of the nature and measurement of photons is needed.

Since photons were conceived by physicists, it is not surprising that equations quickly appear in any serious analysis.

$$v = c = \lambda f \tag{9.1}$$

$$E = hf \tag{9.2}$$

Equation 9.1 is the photon-related equation that must never be forgotten. Whether you memorize it or reconstruct it from dimensional analysis $\frac{m}{s} = (m)(s^{-1})$, this equation must be a tool at hand for computing energies, replotting spectra, designing detection systems, and other routine computations involving light. Equation 9.2 is a close second. Its use is primarily for cases in which energy conversion efficiencies are of interest or in which determination must be made of the number of photons needed for a chemical process.

9.3.1 Waves versus Particles: How Many Photons Are in a Wave?

Figure 9.1 measures spectral irradiance in units of $W\ m^{-2}\ nm^{-1}$. $W = J/s$ measures how rapidly light energy is falling on a surface. Because we know light is made up of photons, we also know that the energy(ies) must correspond to a certain number of photons striking the surface per second. However, as Equation 9.2 points out, some photons carry more energy than others. Let's first see if we can convert spectral irradiance units to a quantity that refers to the number of photons in the beam arriving per second. As we learned in Chapter 3, when a quantity refers to a continuous distribution, it is helpful to deal with the "per unit" differential quantity on the horizontal axis explicitly. Although Chapter 3 referred to an "intensity" as a function of frequency as $I(f)$ and as a function of wavelength as $F(\lambda)$, let's refer to a quantity reflecting the "intensity" by using the letter I (because this is what occurs in the literature), making sure to include the dependent variable and explicitly attaching the differential(s) of immediate interest. The spectral irradiance of Figure 9.1 could then be notated $I(\lambda,A)d\lambda dA$, with units Watts per unit wavelength per unit area. Most commonly, conversions of this quantity to distributions as a function of frequency or energy are done, in which cases Chapter 3 conversion methods must be used. The "per m^2" unit, however, usually involves simply determining the area of interest and multiplying I by that area. To simplify notation, the explicit dependence on area is then left out, so we have $I(\lambda)d\lambda$, $I(f)df$, or $I(E)dE$. We are doing this for a practical reason: most measured data you encounter and will analyze use the letter I whether referring to quantities per unit wavelength, per unit frequency, with or without the area included. Rather than insisting that everyone use the same notation, we insist on understanding the quantity we are working with, whatever the notation.

"Spectral irradiance" ($W\ m^{-2}\ nm^{-1}$, irradiance per unit wavelength interval) is sometimes differentiated from "irradiance" ($W\ m^{-2}$, integrated over all wavelengths). Many texts, including this one, refer to the quantity with units $W\ m^{-2}$ as "intensity." This text will also use the term *intensity* in units of photons $s^{-1}\ m^{-2}$. Our goal is not to memorize photometric terms, but to understand what we are talking about.

Suppose we have a 1.0-mW "white-light" light-emitting diode (LED) flashlight beam with a cross-section beam area of 1.0 cm^2 at a distance of 10 cm from the end of the flashlight. The beam power is 1.0 mW. The beam irradiance (intensity) of this white light is assumed independent of wavelength between 400 and 700 nm, as in Figure 9.2. Can we determine the quantity on

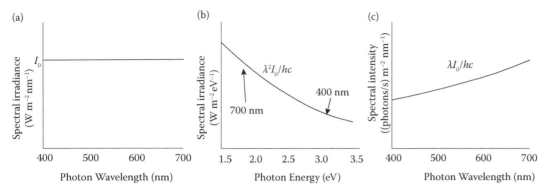

FIGURE 9.2
Spectral irradiance of an arbitrary "white light." The "standard" spectral irradiance is assumed to be constant as a function of wavelength between 400 and 700 nm. (a) Expressed as a function of photon wavelength, the spectral irradiance is constant, I_0 W m^{-2} area per unit photon wavelength interval (nm). (b) When expressed as a function of photon energy (eV) per unit photon energy interval, I_0 must be multiplied by λ^2/hc. (c) When expressed as a function of wavelength (nm), but as the number of emitted photons/s per unit wavelength interval, I_0 must be multiplied by λ/hc. Note that it is often convenient to use $hc = 1240$ eV nm.

the vertical axis—the power per unit area per unit wavelength? To get the total power of 1.0 mW we must integrate over both area and wavelength:

> Be prepared to also deal with "intensities" per unit solid angle (steradian).

$$\text{Power} \equiv P = \int dA \int d\lambda I(A, \lambda) \tag{9.3}$$

While a "beam" of light often varies in intensity over its beam area, usually via a radial dependence, we assume the intensity is constant over the beam area. The dA integral then simplifies to a multiplication by the beam area, $A = 1.0$ cm$^2 = 1.0 \times 10^{-4}$ m^2:

$$P = A \cdot \int I(\lambda) d\lambda \tag{9.4}$$

Because the intensity (irradiance) is wavelength-independent,

$$P = A \cdot \int_{\lambda_1}^{\lambda_2} I_0 d\lambda = A I_0 (\lambda_2 - \lambda_1) \tag{9.5}$$

For our case, this can be solved for I_0:

$$I_0 = \frac{P}{A(\lambda_2 - \lambda_1)} = \frac{1.0 \times 10^{-3} \,\text{W}}{(1.0 \times 10^{-4} \,\text{m}^2)(300 \,\text{nm})} = 3.3 \times 10^{-2} \,\text{W m}^{-2}\,\text{nm}^{-1}$$

How many photons are in this beam? Or, rather, how many photons per second come from the lamp? A photon carries an energy of $E = hf = hc/\lambda$. The 400-nm photons carry more energy than the 700-nm, so we cannot multiply or divide by a constant. Another version of this question: do photons of different wavelengths come out at the same rate (number of photons/s)? Apparently not: if there were the same number of blue photons/s as red photons, the Watts number at 400 nm would be higher than at 700 nm. (Keep in mind that this constant I_0 is not characteristic of real "white" LEDs.) What we must do first is convert $I_0(\lambda)$ to a distribution as a function of photon energy:

$$I_0(\lambda)d\lambda = I_0[\lambda(E)]dE \frac{d\lambda}{dE} = I_0[\lambda(E)]\left(\frac{\lambda^2}{hc}\right)(-dE) = \frac{\lambda^2 I_0}{1240 \,\text{eV} \cdot \text{nm}} dE,$$

where we have suppressed the negative sign, which just means that a large wavelength corresponds to a small energy. Note the important conversion factor λ^2/hc, where $hc = 1240$ eV nm is often useful. If we wanted to plot the number of photon/s emitted per unit wavelength interval, as a function of wavelength, a factor of λ/hc would appear that is multiplied by I_0. This factor comes about, not from change of the independent variable, but from

$$\text{No. of photons} = \frac{\text{Energy}}{hf} = \frac{\text{Energy} \cdot \lambda}{hc} \tag{9.6}$$

So, if we have the white light source emitting 1.0 mW, how many photons/s are coming out? Let's sidestep this and ask a simpler question, one that comes up more frequently in optical studies of biomolecules. Very few modern experiments direct white light at any sort of sample. Lasers or light filtered by narrow wavelength band pass filters ($\Delta\lambda <<$ wavelength of interest) direct light of nearly a single wavelength onto a sample. How many photons/s are emitted by a 1.0-mW helium-neon (HeNe) laser? The wavelength of the laser is 633 nm and the beam cross section is typically 1–2 mm^2. This example has no *spectral* irradiance associated with it: all power comes out at a single wavelength. The computation is simple:

$$\frac{\text{Photons}}{s} = \frac{\text{Energy}}{s} \frac{1}{\text{Energy/photon}} = \frac{P}{hf} = \frac{P\lambda}{hc} \tag{9.7}$$

In the present case,

$$\frac{\text{Photons}}{s} = \frac{P\lambda}{hc} = \frac{(10^{-3}\,\text{J/s})(633\times10^{-9}\,\text{m})}{1.99\times10^{-25}\,\text{J}\cdot\text{m}} = 3.2\times10^{15}\,\text{photons/s} = 3200\,\text{photons/ps}.$$

This last number gives us a handle on the question, "How many photons are in a wave?" In the case of this low-power HeNe laser, 3200 photons are emitted every picosecond. For those of you with laser experience, this number is small compared to the photon/s output of a pulsed laser but high compared to the number of sun photons/s striking a 1-mm^2 area. Modern laboratory pulsed lasers commonly emit single picosecond pulses with pulse *energies* of 1 mJ. As a quick exercise, calculate the number of photons and the number of photons/ps in such a laser pulse if $\lambda = 530$ nm.

9.3.2 Photon Momentum

The momentum of a photon can be calculated just as the de Broglie wavelength is calculated from a particle's momentum:

$$p_{\text{photon}} = \frac{h}{\lambda} = \frac{hf}{c} = \frac{E_{\text{photon}}}{c} \tag{9.8}$$

Photon momentum plays no direct role in biosystems in terms of the force it can exert: $F\Delta t = \Delta p$. Momentum is normally not an important conserved quantity in biological systems because the photon absorber is always embedded in water or another closely packed medium whose atoms can take up any excess momentum. Photon momentum has been employed to create a new nanoscale, noninvasive tool, however: the optical trap or laser tweezers (see Section 9.4.3).

9.3.3 Photon Angular Momentum

Photons carry a quantum of angular momentum, 1 $\hbar$. Unlike the momentum, the angular momentum of the photon plays an important role. Conservation of angular momentum requires that when an electron absorbs a photon, the photon disappears, and the electron must take up both the energy and the angular momentum of the photon. (Surrounding atoms cannot take up angular momentum from the excited electron because there is no efficient mechanism to do so.) The electron must therefore move between two energy levels that differ in angular momentum by $\hbar$. The two most important rules, *absorption and emission selection*, can now be stated:

$$E_{\text{photon}} = \left|E_f - E_i\right|$$

$$\Delta L_{\text{elec}} = \pm\hbar$$

(9.9)

The energy conservation requirement uses the absolute value to accommodate photon absorption and emission. Recall also that the magnitude of any angular momentum vector of $\left|\bar{L}\right| = \hbar\sqrt{L(L+1)}$. As this book is not focused on spectroscopy, we won't dwell on the selection rules, though they have played an important role in characterizing biomolecules by more sophisticated spectroscopic techniques.

9.4 SCATTERING AND REFRACTION

UV and visible light are capable of exciting electrons bound in molecules and atoms from one electronic energy level to another. This is the basis of many spectroscopies that seek to characterize directly those energy levels either because the associated excitations are biologically relevant or because the characteristic spectra of the process are coincidentally sensitive to biomolecular structure or structural changes of interest. The latter is more common. The classic example is the use of the hemoglobin absorption spectrum to monitor the binding of oxygen and other small molecules. The longest wavelengths known to excite electronic excitations in biological systems are between 800 and 900 nm and involve photosynthesis in cyanobacterial systems. We will deal with one of these in Chapter 10.

A single IR photon of wavelength 1000-nm or longer, in contrast, does not have enough energy to excite an electronic excitation. IR excitations can and do excite molecular vibrations, however, and there are numerous vibrational modes in biomolecules. In contrast to UV and visible absorption bands, vibrational bands are narrow, generally do not overlap, and are sensitive to the conformation of the vibrating atomic nuclei. IR and Raman vibrational spectroscopy has been a primary tool in identifying local structures of organic and biomolecules (e.g., DNA base orientations and base pairing), but we do not have time to explore this prolific area. Some recent reviews can be found in the literature.[1–5]

9.4.1 Rayleigh and Mie Scattering

The sky is blue because of Rayleigh scattering—because blue light is scattered more strongly by molecules in the atmosphere than is red light. The formula that describes this dependence, for the case where $R \ll \lambda$, where R is the radius of the particle (molecule) and λ is the wavelength of light, is

$$I_{\text{RS}}(\theta) = I_0 \frac{8\pi^4}{\lambda^4 r^2}\left(\frac{n^2-1}{n^2+2}\right)^2 R^6(1+\cos^2\theta)$$

$$= I_0 \frac{8\pi^4 \alpha^2}{\lambda^4 r^2}\left(1+\cos^2\theta\right)$$

(9.10)

RS means Rayleigh scattering, r is the distance of the observer from the particle, θ is the angle between the incident light and scattered light, and n is the particle index of refraction. A particle index of refraction is not a terribly good concept when dealing with individual molecules, so the latter expression, employing the molecular electric polarizability, is used. The sky's blueness comes from the λ^{-4} dependence.

A primary quantity used to describe any type of scattering is the *cross section*. The scattering cross section defines the apparent cross-sectional area the particle presents to the incident

> Electric dipole moment = $\alpha\cdot$electric field.

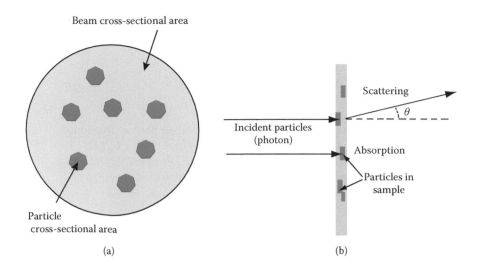

FIGURE 9.3

The scattering cross section. (a) The number of scattering particles present in the area illuminated by the beam times the particle cross section and divided by the beam area equals the fraction of beam photons (or beam particles) that are scattered: $\frac{N_s \sigma_s}{\text{beam area}}$ = fraction of photons scattered. (b) Photons (or other beam particles) may disappear from the transmitted beam because of scattering or absorption.

light beam (Figure 9.3). Unlike the shadows cast by centimeter-sized objects, however, optical cross sections are determined primarily by quantum properties, so that the scattering cross section can differ significantly from the physical size of the particle, usually defined by the van der Waals radius of the particle. If Equation 9.10 is integrated over all angles, a total scattering cross section results:

Gustav Mie was a German physicist born in 1869.

Search on "Mie calculator" if these links become inactive.

$$\sigma_{RS} = \frac{128\pi^5}{3} \frac{R^6}{\lambda^4} \left(\frac{n^2-1}{n^2+2}\right)^2 \tag{9.11}$$

This expression is only valid if $R \ll \lambda$. If $n = 1.5$, the Rayleigh cross section is approximately

$$\sigma_{RS} \cong 1100 \frac{R^6}{\lambda^4} \tag{9.12}$$

If $R = 1.0$ nm and $\lambda = 400$ nm, $\sigma_{RS} \sim 4 \times 10^{-8}$ nm^2 = 4×10^{-22} cm^2. The measured Rayleigh scattering cross section for N_2 is about 5.1×10^{-27} cm^2, though the alternate formula using polarizability must be used.

Though $R \ll \lambda$ has been cited above, a slightly better way to judge particle size for scattering purposes is the Rayleigh size parameter $x = (2\pi R)/\lambda$. The Rayleigh regime then corresponds to $x \ll 1$. Scattering from larger spherical particles is described by Mie theory for an arbitrary-size parameter x. Mie theory reduces to the Rayleigh when $x \ll 1$. Mie theory cannot be embodied by a single algebraic expression, but several online Mie calculators exist: omlc.ogi.edu/calc/mie_calc.html by Scott Prahl and www. lightscattering.de/MieCalc/by Bernhard Michel.

Figure 9.4 shows a plot of $\lambda = 633$ nm scattering efficiency as a function of particle diameter, based on the Michel Web site calculator. The scattering efficiency starts from low Rayleigh values (below about 0.1-μm), increases to a maximum, and proceeds to oscillate in magnitude as a function of particle size. As the diameter increases above about 10 times the wavelength, the scattering efficiency begins to reach an asymptotic value. For very large diameters, this efficiency should correspond to a scattering cross section

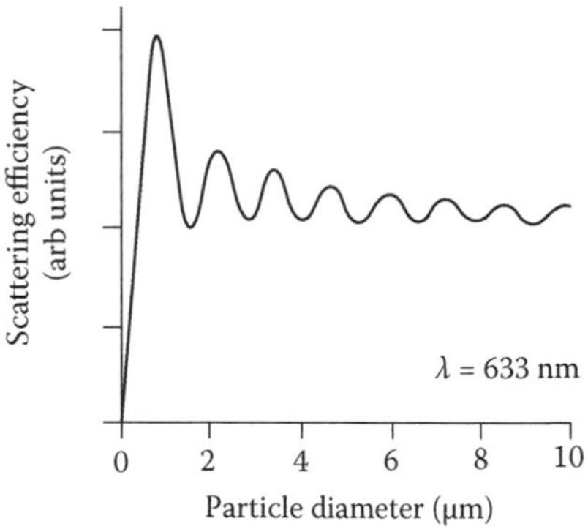

FIGURE 9.4

Mie scattering efficiency (arbitrary units) as a function of particle diameter. Parameters as follows: $\lambda = 633$ nm, particle index of refraction 1.50, medium index 1.00, no absorption. The Rayleigh approximation is valid for d less than about 0.1 μm. The graph uses results from www.lightscattering.de/MieCalc/by Bernhard Michel. Oscillations indicate the presence of interference effects, which can also be seen in the form of "whispering-gallery" effects in microparticle absorption, fluorescence, IR, and other optical spectroscopies.

equal to the particle cross-sectional area, πR^2, sometimes called the "shadowing area." Note that the maximum efficiency is nearly twice the asymptotic value: interference effects have increased the effective area of the particle. The interference effects noticeable in Mie scattering span the particle size regime from $R \sim \lambda$ to $R > \lambda$. Because the effects are based on refractive indices, this scattering can be described by ray optics.

9.4.2 Refraction: Interfaces and Whispering-Gallery Modes

A light ray impinging on a surface between two materials of different refractive indexes is reflected and refracted. The most familiar example of this in a biosystem is sunlight shining on water. Because the speed of light in water is less than in air, the transmitted beam has its wavelengths increased by a multiplicative factor $n_{water}/n_{air} \sim 1.3$, which makes little difference to organisms. The transmitted light intensity is less than the incident light intensity. This ~5% reduction in transmitted solar irradiance has some consequences for the energy balance of the overall biosystem but few consequences for individual organisms.

This should remind us again that a photon's frequency or energy, not speed or wavelength, is the constant during passage from one medium to another. Virtually all optical spectroscopies on biosystems are done in water, so the photon wavelength impinging on the biomolecule is not the spectrophotometer wavelength, but about 1.3 times that wavelength. We still refer to hemoglobin's Soret absorption band as the spectrophotometer-indicated 420 nm, however, and not 1.3 × 420 nm = 550 nm.

Cells also present boundaries to light: exterior water, to membrane, to interior cytosol, at a minimum. These boundaries differ from the ocean–air interface in that structures range in size from a few nanometers (the thickness of the membrane) to 1–100 μm (the diameter of the cell involved). Several optical effects can occur: refractive focusing, thin-film interference, total internal reflection, and scattering (Figure 9.5). All of these effects are obvious during imaging and spectroscopy of cells and microorganisms. The effect on spectra may be seen as oscillations around the average but usually appear as noise when many irregular structures are involved. Such effects should also occur in the everyday life of cells, but biological effects, if any, should be limited to photosynthetic cells.

The total internal reflection indicated in Figure 9.5 is the basis for a new technology in biosensing and microlaser design.[6] Because the ray (wave) traveling around the periphery of the sphere passes multiple times, the light can encounter and excite a molecule on the surface several times. If that molecule absorbs and emits light, it can be excited multiple times, or at least have its excitation probability greatly increased. These modes are generally discrete, as only certain wavelengths will properly fit around the periphery of the sphere. Such modes are called "whispering-gallery" modes, after the acoustical effects observed in domed cathedrals.[7] Arrays of dielectric microspheres have been produced that show the transfer of excitation in the array (Figure 9.6).[8] We will encounter a photosynthetic excitation transfer that appears similar to this excitation "percolation" in the antenna arrays of photosynthetic organisms; however, the mechanism is different. A single virus particle has also been detected and its mass measured (5.1×10^{-19} kg) from the shift in whispering-gallery resonance frequency when the virus binds to the surface of the microsphere.[9] We have seen that living cells rely on sensitive nonoptical methods for detecting foreign particles approaching the cell surface, but it is fascinating to think that we can perform a similar, though less sophisticated, feat with a spectrometer.

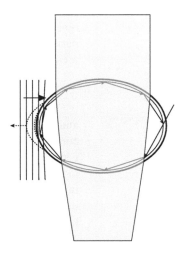

FIGURE 9.5
Light rays hitting a small dielectric sphere can produce interference, total internal reflection, scattering, and beam focusing (vertical beam). All of these effects occur during imaging and spectroscopy of cells and microorganisms. The effect on spectra may be seen as oscillations around the average but usually appear as noise when many irregular structures are involved. Such effects should also occur in the everyday life of cells, but biological effects, if any, should be limited to photosynthetic cells.

9.4.3 Optical Forces: Laser Tweezers

The change in photon direction upon hitting an interface at nonnormal incidence is the basis of another micro- and nanotechnology of great importance to modern biophysics: the laser tweezers or optical trap (Figure 9.7). The laser tweezers were invented in the 1970s

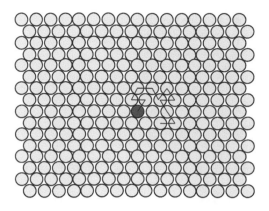

FIGURE 9.6

Movement of an optical excitation in arrays of microspheres. Microspheres containing fluorescent dyes can maintain an excitation for a short time in a whispering-gallery mode, then pass it to a neighboring sphere. This general scenario is reproduced in photosynthetic light antenna arrays (Chapter 10), though the mechanism is quite different.

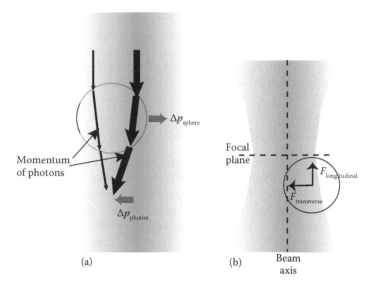

(a) (b)

FIGURE 9.7

(a) Ray-optics model for laser-tweezers force generated by the gradient of laser beam intensity hitting a dielectric microsphere. (b) A focused laser beam produces forces in both the transverse and longitudinal directions. The longitudinal force is often smaller and less reliable. This model correctly predicts optical forces for particles about 1 µm in diameter and larger, and laser wavelengths less than about 700 nm. An induced dipole model must be used for smaller particles.

by Arthur Ashkin and have been a key element in Ashkin's Keithley, Townes, Rank, and Ives awards in physics and engineering, as well as a Nobel Prize in physics for Steve Chu, the former Secretary of Energy of the United States. The most advantageous feature of the laser tweezers is its noninvasive nature: a force can be exerted on an object without physically touching it. Other features of importance in biological systems are the abilities (1) to simultaneously image in a microscope; (2) to sever covalent bonds by use of a second, shorter-wavelength laser; and (3) to impose a force in a crowded system with microscopic precision. Figure 9.7a shows that the basis for the optical force is the change of photon direction (momentum) upon entering and exiting a dielectric sphere, cell, or other small object. The force involves a difference between momentum changes of rays going through various parts of the microsphere and requires that the laser beam vary in intensity across its cross-sectional area. Good-quality laser beams generally have a Gaussian radial beam intensity:

$$I(r) = I_0 e^{-r^2/2\sigma^2} \tag{9.13}$$

and optical forces can be predicted with some accuracy. Figure 9.7 implies that momentum change correlates with the change in direction of a light ray, an effect of classical ray optics. Forces are also generated on smaller objects, even metallic nanoparticles, though the applicable theory involves an induced dipole moment ($\vec{p} = \alpha\vec{E}$) in the particle that then gets attracted to regions of higher light intensity. Figure 9.7b illustrates that since particles move toward regions of higher light intensity, the Gaussian transverse intensity variation produces a force toward the beam axis. Focusing the laser beam will produce a longitudinal intensity variation, so that the microsphere will be moved approximately to the focal spot. The magnitude and symmetry of the force field depend strongly on the quality (symmetry) of the focal spot, as well as on the beam intensity gradient. The transverse intensity gradient of a Gaussian beam can be calculated as

$$\frac{dI}{dr} = -\frac{r}{\sigma^2} I_0 e^{-r^2/2\sigma^2} \tag{9.14}$$

This function increases until r approaches σ; beyond $r \sim \sigma$, it decreases. A larger transverse force results from a higher-power beam with a narrower beam waist. Typically, microscope objectives of high numerical aperture and 60–100× magnification are used. Given a ~100-mW laser, a good microscope, some mirrors and mounts, and laser blocking filters, you can construct an optical trap in about 2 h. However, for a practical device, precise mechanical control (<100 nm, costs about $10,000) is needed. What objects can be trapped? Polystyrene microspheres (about $300 for a year's supply for a class) are ideal, but gold nanoparticles (several hundred dollars) can also be used. The surface of polystyrene offers many opportunities for controlling surface charge, adding DNA- or protein-binding agents, or creating a hydrophilic or hydrophobic environment. Gold has the advantage of selective chemical inertness: Au does not corrode in salt water, but it does allow the chemist useful sulfur chemical alternatives for attaching surface molecular groups. Because cell–cell interactions rely on the recognition of surface groups, this surface chemical versatility is valuable. We will soon see that the secret to a biophysically useful optical trap usually lies in surface chemistry.

The classical ray-based theory underlying the force on a dielectric sphere is not difficult, but we won't go through it. The theory for force generation on small, polarizable particles is straightforward.

$$\mathbf{p} = \alpha \mathbf{E}$$

$$U = -\mathbf{p}_{avg} \cdot \mathbf{E} = -\frac{\alpha}{2}E^2 \tag{9.15}$$

$$F_r = -\frac{\partial U}{\partial r} = \frac{\alpha}{2}\frac{\partial}{\partial r}(E^2) = \frac{\alpha}{2c\varepsilon}\frac{\partial}{\partial r}(I)$$

where α is the polarizability, ε the dielectric constant of the medium (usually water), c the speed of light in the medium, and I the intensity (power/m^2). We use subscript r on the force to indicate the radial force transverse to the beam direction. A simple end formula for the force generation of the laser tweezers trap has been developed. If you write a linear-spring model for the potential energy of the trap, with x, the displacement from the trap center (in any direction for a completely symmetric trap) and κ, the spring constant (Figure 9.8) can be approximated:

$$U(x) = \frac{1}{2}\kappa x^2, \; F_{opt} = \kappa x$$

$$\kappa = Q\frac{n^2 P}{\lambda c} \tag{9.16}$$

> The factor of ½ in the potential energy in (9.15) occurs because of the induced nature of the dipole: as you integrate from $0 \to E$ for the work, p goes from $0 \to \alpha E$, with average ½αE.

Q is the "quality factor" or trapping efficiency, which depends strongly on the beam and lens qualities, n is the index of refraction, λ is the laser wavelength, $c = 3.00 \times 10^8$ m/s, and P is the laser power. As Figure 9.8 suggests, the harmonic (quadratic) potential energy shape only applies near the center of the trap; the optical force does not keep increasing as x gets larger. When the trapped object is moved past a certain distance, x_{esc}, the trapping force has decreased significantly. This distance can be calculated if the shape of the potential is known. Practically, the escape distance can be found by moving the microsphere progressively to the perimeter of the trap. At some point, the random Brownian collisions of solvent molecules will knock the sphere out of the trap. As with most processes involving random collisions, the value of x_{esc} measured this way varies considerably, depending on how long the sphere must remain trapped.

Experimentally, laser traps can be calibrated by use of fluid frictional force. The sphere is dragged through the fluid at constant speed v, and the displacement of the sphere center from the trap center is measured (Figure 9.9). Because there is no acceleration, the frictional and optical forces are equal:

$$F_{friction} = bv = 6\pi \eta R = F_{opt}(x) \tag{9.17}$$

where b is the frictional coefficient, η is the fluid viscosity, and R the sphere radius. The magnitude of the force, and its dependence on x, and thus the potential energy of the trap, can be plotted.

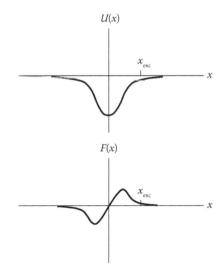

Typical values:
$P = 0.1$ W
$\lambda = 850$ nm
Sphere diameter ~1 μm
$Q \sim 0.1$
$\kappa = 0.01$–0.1 pN/nm
$x_{esc} \sim 0.2$–0.5 μm

FIGURE 9.8
Potential energy and force diagram and typical parameters for a good-quality optical trap. The force beyond x_{esc} is not large enough to overcome most Brownian forces.

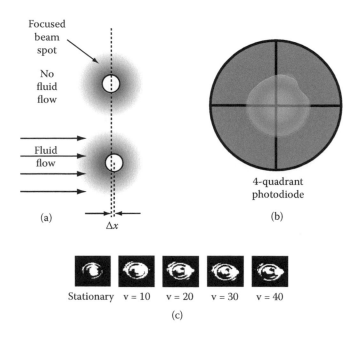

FIGURE 9.9

(a) Moving a microsphere through a fluid at fixed velocity is a good way to calibrate a laser trap. The fluid resistive force is quite accurately given by the Stokes–Einstein expression, $F = bv = 6\pi\eta Rv$. Large distances (0.2 to several μm) can be directly measured with the microscope camera pixel calibration. Smaller distances can be measured with (b) multisector photodiodes or (c) interference techniques. The four quadrants of the photodiode are precisely matched and difference currents measured by a difference amplifier. Interference fringes play a major role in any images formed by a particle in a laser trap.

A final characteristic of the optical trap compared to other force methods, such as atomic force microscopy (AFM), is the response time of the laser tweezers. If force is applied to an object in a fluid, it will not respond instantly. The response time is governed by the trap spring constant and the fluid friction:

$$F_{opt} = F_{friction}$$

$$\kappa x = bv = b\frac{x}{\tau} \tag{9.18}$$

$$\tau = \frac{b}{\kappa} = \frac{6\pi nR}{\kappa}$$

We have used the Stokes–Einstein expression for the frictional coefficient b. A typical value for τ in water is 200 μs. This response time is almost an order of magnitude faster than the AFM's.

One laser-tweezers design pitfall that must be avoided is absorption of laser light by the trapped particle. Even a tiny fraction of light absorbed by a 1-μm particle can cause irreparable heating. Because absorption by common materials for synthetic spheres rises toward the UV, near-IR lasers are mostly used for trapping. The second limitation is absorbance by water IR bands (see Figure 9.1). Cells pose a more severe problem, as certain molecular groups can absorb almost anywhere in the UV/Vis/near-IR spectrum. Still, the best bet for a nonabsorbed wavelength is in the 800–950-nm region.

> The 800–950-nm spectral region poses the problem of visualization: these wavelengths cannot be seen by the unaided eye (if you *do* see something, contact your eye doctor quickly!). Lasers of powers 1–100 mW cannot be treated carelessly.

9.4.3.1 Trapping of cells

Trapping microspheres is interesting, but where is the biophysics? Many cells are also of shape and size that can be trapped. Bacteria can be trapped, as can sperm cells, which can then be moved by the laser tweezers next to an ovum, a hole drilled in the ovum's cell wall, and the sperm implanted inside.[10] Other important interactions between cells have also been measured over the past 20 years. Recognition of erythroleukemia cells by natural killer cells has been explored by placing the latter next to the former using a laser tweezers. Complex biological recognition responses have then been observed: "After contact and a lag phase of a few tens of seconds, the target cell starts to change its morphology and membrane blebbing [blistering] occurs. The kinetics of the attack of the NK cell on K562 cells is not straightforward but governed by temporal oscillations in the shape of the target cell (zeosis)."[11] The laser tweezers/microscope system is an ideal tool to trigger and quantify such cell–cell interactions. Time-dependent forces can be measured, membrane structural changes observed, protein production quantified, and cell death determined.

9.4.3.2 Measurement of single-molecule interactions

Laser traps can indeed trap polymers like DNA. The optical force tends to collapse the DNA into a compact coil, which can be of some interest in terms of entropic forces (Chapters 6, 14, and 15). However, the greatest advances have been made by chemically tethering the ends of biopolymers, mostly DNA and RNA, to microspheres, exerting a trapping force on the sphere and attaching the other end of the DNA to a surface or allowing a surface-attached enzyme to bind to the free end of the DNA. The trapping force can then oppose the force exerted by the enzyme, *directly measuring a single-molecule force during a chemical reaction*, or the DNA can simply be stretched, probing its elastic properties. Many such experiments have been done during the past two decades.[12] One involves pulling out "knots" in RNA and is diagrammed in Figure 9.10, based on experiments done by Green, Kim, Bustamante, and Tinoco.[13] One can now find dozens of such experiments published every year. We will wait to analyze this experiment in more detail, but note the sophisticated surface and tethering chemistry employed by this study. RNA "constructs" that had the hybrid RNA/DNA "handles," with biotin and digoxigenin at the two respective ends were first made, as shown in the figure. The 2-μm diameter streptavidin-coated beads were added to a microfluidic chamber where a laser trap caught and transferred it to a micropipette attached to a piezoelectric positioner. The RNA "constructs" were attached to the other type of sphere by adding to a suspension of 3-μm antidigoxigenin-coated spheres and then to the chamber. One of these RNA-attached spheres was caught in the laser trap and the two spheres were moved close together repeatedly, until a biotin–streptavidin connection to the unattached sphere was made. The RNA construct now connects the two microspheres. Force was exerted on the molecule by moving the micropipette relative to the laser trap. The change in extension of the RNA was measured from the change in position of the spheres.

9.5 ABSORPTION SPECTRA

Optical spectroscopy is probably the most widely used noninvasive technique for general characterization of biomolecular structures. Virtually every undergraduate student in biology, chemistry, physics, and materials engineering learns this experimental method because she will likely use it in any technical, scientific career. Absorption spectroscopy is the simplest of these spectroscopies.

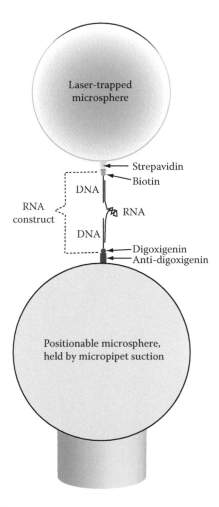

FIGURE 9.10

RNA-extension experiment designed by Green et al. (Green, L. et al., *J. Mol. Biol.*, 375, 511–528, 2008.) See text for details.

9.5.1 Energy-Level Diagram

Photons striking most materials cause electronic excitations. If the material consists of atoms that interact very little with each other (e.g., a gas), the excitations are atomic in nature and involve Bohr atomic energy levels. Figure 9.11 shows an example of the display of electronic energy levels in the form of an energy-level diagram. This sort of diagram is the most useful kind for understanding and describing absorption, fluorescence, and other spectroscopies of biomolecules.

9.5.2 Biomolecular Absorption Spectra Are Broad

Atomic absorption spectra usually occur in sharp lines, i.e., at well-defined wavelengths ($\Delta\lambda < 0.1\ \text{Å}$), corresponding to transitions among the atomic energy levels of the atom, as in Figure 9.11a. Many of the atomic absorption lines from the lower-energy, occupied states to the lowest-available excited state of heavier atoms occur in the X-ray region; transitions from higher occupied atomic states to the lowest-available excited state may be in the visible or UV. Molecules usually have absorption spectra greatly broadened by virtually continuous vibrational and rotational bands. The spectral line widths of biomolecules can be up to 50-nm.

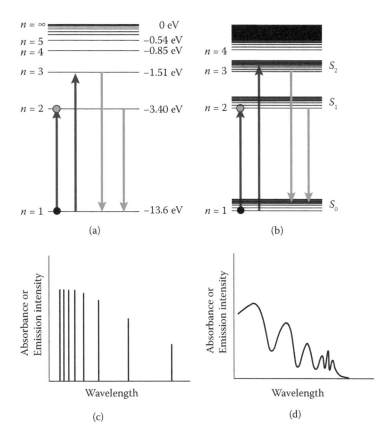

FIGURE 9.11

Energy-level diagrams. (a) An atomic system shows discrete levels, except at high energy, where the levels become very close together. (Numbers here are appropriate for hydrogen.) (b) Molecules show additional levels due to vibrational displacements of the lines. Sometimes these vibrational levels can be distinguished, as suggested in the lower-energy levels, but often they overlap expensively, creating (apparently) continuous bands. Panels (c) and (d) show types of absorption bands that could be observed.

9.5.3 Spectrum as a Digital Distribution Function

An optical spectrum is a graph of an excitation probability or optical intensity as a function of light wavelength, frequency, or energy. This is a distribution function. We have been dealing with manipulation of this sort of distribution function since Chapter 3, mostly in the form of continuous functions. We won't repeat our analyses here, but we should note that with the change from analog to digital over the past 40 years, spectra have also changed from analog (continuous) to digital. When an optical spectrometer measures a quantity $I(\lambda)$, the wavelength is in the form of a digital number, as is the intensity. The data array that is a spectrum may list discrete wavelength values $\lambda = 400.0, 401.0, 402.0$-nm, etc. Does this mean that for a biomolecule characterized by a broad spectrum, the intensity associated with each wavelength is the light intensity that came out *exactly* at 400.0-nm, with infinite precision? And that the measured light was all at 400.000000000000-nm and none was at 400.000000000001-nm? Absolutely not. The broad nature of the spectrum indicates that light is emitted or absorbed continuously as a function of wavelength. The spectrometer manufacturer decided, however, to group wavelengths together. For example, if data are collected every 1.0-nm, the intensity listed at 400.0-nm is typically the total intensity measured over the interval from 399.5–400.5-nm. The intensity emitted *precisely* at 400.0 nm would be *precisely* zero. A small wavelength interval, usually corresponding to width of slits transmitting light through a diffraction grating, must be included to get the intensity recorded. There is nothing physically complicated about this, but digitized data should not be mistaken for line spectra, as in atoms.

9.5.4 Absorbance and Absorption (Extinction) Coefficient

An absorption spectrophotometer (Figure 9.12) sends a light beam of measured incident intensity ($I_0(\lambda)$) and wavelength into a sample and then measures the transmitted light intensity $I(\lambda)$. Via

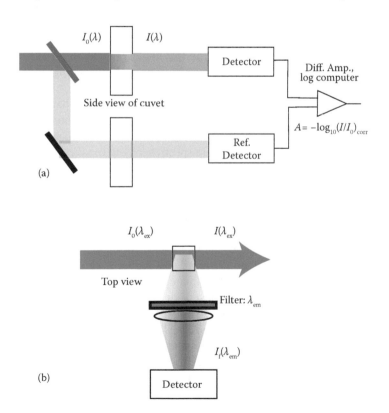

FIGURE 9.12
Measurement systems for (a) absorption and (b) fluorescence. The wavelength for absorption is selected and scanned by a diffraction grating. The excitation and emission wavelengths for fluorescence are both selected and scanned by diffraction gratings (not shown) and filters. There are also several correction systems built into the fluorescence apparatus (not shown).

software or electronic circuitry, the spectrophotometer then calculates the sample's transmission (percent) and absorbance A at that wavelength. The absorbance is defined by

$$I(\lambda) = I_0(\lambda)10^{-A} \tag{9.19}$$

The absorbance is related to molecular/atomic parameters by Beer's law:

$$A(\lambda) = \varepsilon(\lambda)cL \tag{9.20}$$

where ε is the *absorption (extinction) coefficient*, c is concentration of molecular species (moles/liter), and L is light path length in sample. The molecular *absorption cross section* σ is defined by

$$I(\lambda) = I_0(\lambda)e^{-\alpha} \tag{9.21}$$

$$\alpha = \sigma nL \tag{9.22}$$

where σ is absorption cross section (cm^2), n is molecular density (molecules/cm^3), and L is light path length (cm). The cross section represents, in a quantum-mechanical sense, the effective cross-sectional area the molecule presents to the light beam. In all of these equations, A, ε, α, and σ are a function of the light wavelength. Wavelength-scanning spectrophotometers have been built to automatically measure the absorbance as a function of wavelength.

The tradition in the biophysical sciences, as well as among the manufacturers of spectrophotometers, is to use the absorption coefficient, though it employs a strange (to a physicist) set of units. The units have the advantage of connection to quantities encountered in typical experiments: molar concentrations and centimeter-long sample cuvets. Common UV/Vis spectrophotometers measure from about 200–900-nm.

The absorbance you normally measure is that of ground- to excited-state transitions, usually singlet $S_0 \rightarrow S_1$, if a singlet is the lowest-energy state (Figure 9.11), because the ground state predominates at equilibrium near room temperature. Singlet and S_n refer to even, multielectron states that have total electronic spin equal to zero, with n signifying the order of the energy level. Most molecules have both singlet and triplet states, but the singlet is generally lower in energy. O_2 is an exception. One or more of the excited states may be observed, depending on the sample and the capabilities of the spectrophotometer.

Absorption spectroscopy of polymers can be an accounting challenge. What does the "concentration" in the above quantities refer to: the concentration of the entire protein or nucleic acid polymer molecule, or the concentration of polymer *units*, ignoring the fact that they are bound together? Nucleic acid absorption spectroscopy and protein spectroscopy have taken opposite approaches, primarily because the individual nucleotides are optically similar: their absorption coefficients are about the same, near $10^4 M^{-1} cm^{-1}$, and the absorption peaks are all in the 255–275-nm region. See Table 9.1.

Amino acids, on the other hand, vary widely in optical properties. The ones that absorb, do so in the 275–285-nm region, but, to a good approximation, only Trp ($\varepsilon_{280\,nm} \sim 5700\ M^{-1} cm^{-1}$), Tyr ($\varepsilon_{276\,nm} \sim 1300\ M^{-1} cm^{-1}$), Cys, and Phe ($\varepsilon_{260\,nm} \sim 200\ M^{-1} cm^{-1}$) absorb. If measurements at 280-nm are made, typically Phe is ignored and the data in Table 9.2 or Figure 9.13a are used.

This makes the overall absorbance of proteins dependent on the amino acid composition, though an average amino acid absorbance is sometimes used for proteins whose sequences are not known. This is based on the assumption that the protein is not far from the "average," with Trp 1.2%, Tyr 3.2%, Cys 1.6%, and Phe 4.1% of the amino acids. If the protein amino acid sequence and molecular mass is known, online calculators, such as http://www.biomol.net/en/tools/proteinextinction.htm, can be used.

Biochemists working with uncharacterized proteins often measure concentrations in terms of mass per unit volume. For example, if we are given a cell whose content is 15% protein, it would be straightforward for us to estimate the protein concentration in terms of mass/volume. Sometimes, units of µg/mL, mg/mL, or mg/L are also used. If an unknown protein concentration were 5.0-mg/mL, what would be its expected absorbance at

TABLE 9.1 Absorption Coefficients for Nucleotides

Nucleoside	λ_{peak} (nm)	ε ($M^{-1}cm^{-1}$)
dAMP	259	1.51×10^4
dGMP	252	1.42×10^4
dCMP	271	0.886×10^4
dTMP	267	0.949×10^4

Source: Cavaluzzi, N.J., and Borer, P.N., *Nucleic Acids Res.*, 32, e13, 2004.

TABLE 9.2 UV Absorption Data for Proteins at 280 nm

Amino Acid	Amino Acid Coefficient $\varepsilon_{280\,nm}$ (M^{-1} cm^{-1})	Number of Units per Protein	Whole Protein $\varepsilon_{280\,nm}$ (M^{-1} cm^{-1})
Trp	5690	N_{trp}	$5690N_{trp} + 1280N_{tyr} + 120N_{Cys}$
Tyr	1280	N_{tyr}	
Cys	120	N_{Cys}	

Source: Gill, S.C., and Von Hippel, P.H., *Anal. Biochem.*, 182, 319–326, 1989.

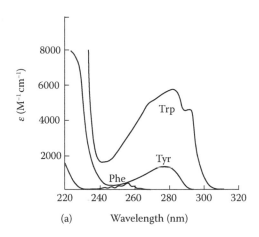

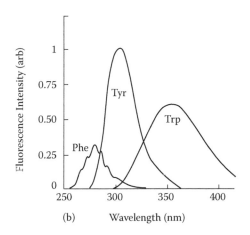

(a) Wavelength (nm)   (b) Wavelength (nm)

FIGURE 9.13
(a) Absorption and (b) fluorescence emission spectra of isolated amino acids, Trp (tryptophan), Tyr (tyrosine), and Phe (phenylalanine). Absorption is plotted as wavelength-dependent absorption coefficient. Fluorescence intensity is in arbitrary units. In proteins fluorescence emission from Tyr is usually less than from Trp. (After Lakowicz, J. R., *Principles of Fluorescence Spectroscopy*, 65, Springer, New York, 2006.)

280-nm? The expression for absorbance can be modified to $A = \varepsilon_{\%}c_{\%}L$ or $A = \varepsilon_m c_m L$, where the subscript % (m) refers to quantities with respect to mass percent (mg/mL concentration). National Institute of Standards and Technology (NIST) has developed a percentage standard, calibrated by absorbance at 280-nm by a BSA Fraction V standard solution, with $\varepsilon_{\%} = 6.67\%^{-1}$ cm^{-1}.

> Absorbance depends on whether disulfide bonds are in place.

9.5.5 Proteins That Unexpectedly Absorb Light

Many proteins have prosthetic groups whose absorbance is comparable to that of the amino acids: hemoglobin with heme, chlorophyll proteins, flavin-containing proteins. These cannot be described as absorbing light anomalously; they simply have additional absorbers besides amino acids. Collagen and elastin, however, are examples of proteins that have no amino acids in their primary sequences that absorb UV/Vis light and have no prosthetic groups, yet these proteins both absorb and emit. The absorption is highly variable and is now understood to stem from cross-linking between collagen and/or elastin strands.[14] Cross-linking appears to involve Lys and other modified amino acids and is related to mechanical and stability properties of skin and other tissues. Much remains to be done on collagen and elastin before their mechanical properties can be understood, as well as the relation and sensitivity of these properties to UV absorption.

> Readers who want to learn more about practical biophysical fluorescence measurements and analysis than what's covered in this brief section should consult the fine text: Lakowicz, J. R., *Principles of Fluorescence Spectroscopy*, 3rd ed. Springer, New York, 2006.

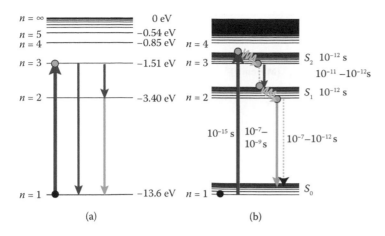

FIGURE 9.14

De-excitation of an excited state in (a) an atomic state and (b) a molecular state. Downward pointing vertical arrows signify *radiative decay* and are accompanied by photon emission; slanting, squiggly arrows, *vibrational relaxation*, are accompanied by phonon (vibrational quanta) emission. Downward pointing, dotted lines signify nonradiative de-excitation, sometimes labeled "internal conversion" if vibrational levels of the two electronic states overlap. In nonradiative decay, energy is conserved through nonradiative means, commonly by transfer of energy to vibrational modes, to another molecule through collisions or other nonradiative energy transfer. In (b), typical time scales for the processes in organic molecules are noted.

9.6 EMISSION SPECTRA

An excited state in an atom must eventually transit back to the ground state. When this occurs, a photon may be emitted. In the case of atoms, any decay of an excited state that is not "forbidden" to do so is accompanied by photon emission. Because of the line structure of the energy levels, the photon emission spectra are also in the form of lines, corresponding to $3 \to 2$, $2 \to 1$, $3 \to 1$, etc. transitions in Figure 9.11a. Also, because of this line structure, the energy and wavelength of the emitted photon is equal to that of the equivalent absorbed photon causing the excitation. However, in measurements near room temperature, excitations only from the ground state to excited states occurs, e.g., $1 \to 2$, $1 \to 3$, because there are usually no electrons already in state 2 or 3 to excite to yet higher states. Thus an absorption may cause excitation from $1 \to 3$, but the reverse process could be $3 \to 1$ or $3 \to 2$, $2 \to 1$, with two emitted photons in the latter case. Energy is still conserved: $E_{31} = E_{32} + E_{21}$, where the energies refer both to energy-level differences and the energy of the corresponding photon. See Figure 9.14a. Note that the *fluorescence excitation spectrum*, measured by the intensity of emission at some fixed wavelength as a function of the wavelength of the exciting light, should be identical in shape to the absorption spectrum. This is approximately true for biomolecules like proteins, so the absorption spectra in Figure 9.13a can be considered as excitation spectra corresponding to the emission spectra in Figure 9.13b.

> We will mostly ignore the issue of allowed and forbidden transitions, though important examples can be found in biosystems.

The case of molecular de-excitation is more complex (Figure 9.14b). If the second excited state (the third-lowest energy level) is occupied by an electron, the most common de-excitation pathway is not direct transition to the ground state, but $3 \to 2 \to 1$ ($3 \to 1$ occasionally occurs, however). A second difference from the atomic case is that the initial excited state cannot usually be ascribed to a specific state in the $n = 3$ state broadened by vibrations. To understand this, we think of the normal characteristics of experimental measurements. Most commonly in molecular electronic spectroscopy, a beam of light whose wavelength is spread over $\lambda \pm \Delta\lambda$ (photon energy spread over $E_p \pm \Delta E_p$) is directed at the molecule. If any part of the beam energy band overlaps the energy difference between the vibrational sublevel of the ground state and a vibrational sublevel of the excited state, that excited-state sublevel, or rather those sublevels that overlap, will be excited. The particular sublevels that are excited depends on the *degeneracy* of the sublevels—how many quantum states have that particular sublevel energy. The particular vibrational sublevel excited is usually not the lowest-energy sublevel of the electronic band. As these sublevels are predominantly vibrational, vibrations ensue. Because the various vibrational modes interact with each other (atoms hit other atoms), energy is transferred from mode to mode. Eventually, atoms exposed to protein or biomolecular backbone atoms or water transfer their vibrational energy to water molecules. The excitation then ends up, temporarily, in the lowest vibrational level of the $n = 2$ state, which we now relabel the traditional S_1, indicating the first excited state of the (assumed) singlet ("S") manifold of states. The ground state is S_0. If we found that we were dealing with triplets, we would label T_0 and T_1. If the spin were half integral, as in transition metal atoms, we would label states with $S = 3/2$, $1/2$, etc.

> Recall water's high thermal conductivity. High thermal conductivity means vibrations and kinetic energy rapidly and efficiently move in the fluid.

The fluorescence emission spectra of three amino acids are shown in Figure 9.13b. These spectra were measured by the fluorescence intensity as function of the emission wavelength, when some excitation wavelength was chosen near the peak of the corresponding absorption spectrum shown in Figure 9.13a.

9.6.1 Stokes Shift

The vibrational relaxation that precedes fluorescence emission from molecules implies that the energy of the emitted photon is less than the energy of the absorbed photon. The difference, often referred to as a wavelength difference, is called the *Stokes shift*, after G. G. Stokes, who first quantified the shift. The Stokes wavelength shift allows us to distinguish scattered light from fluorescent light: the color of the light is different. Because scattered and transmitted light are generally much more intense than fluorescent light, multiple methods of suppressing scattered and transmitted light are employed in measurements:

1. Fluorescence is observed at a 90° angle from the excitation direction.
2. Observed light is wavelength filtered to pass λ_{fluor} and suppress λ_{exc}.
3. A delayed time window allows any excitation pulse to disappear, leaving the longer-lived fluorescence.
4. Polarizing filters may distinguish between excitation and emitted photons.

Stokes shifts in typical fluorescent organic molecules typically range from 10 to 70-nm; that for tyrosine, shown in Figure 13a and b is about 305–275 nm = 30 nm.

> The story is that Stokes discovered this shift when he peered through a glass of wine and observed fluorescence from a quinine solution illuminated by sunlight. The colored wine filtered out scattered sunlight that otherwise overpowered the much weaker fluorescence.

9.6.2 Time Scales and Excited-State Kinetics

As you learned from previous chemistry and physics courses, the frequency of typical molecular vibrations is 10^{12}–10^{13} s^{-1}, which implies that the time scale for relaxation of this vibrational energy—transfer to other modes and to water—occurs over a few vibrational periods, roughly 10^{11}–10^{12} s. The excitation time, listed as 10^{-15} s, corresponds to the period of a UV/Vis photon. While it is true that a single molecule is excited by a single photon in 10^{-15} s, the average excitation time for a population (e.g., a 10^{-9} M solution) of molecules excited by a light pulse of duration 10^{-9} s is 10^{-9} s—the excitation time is limited by the measurement apparatus. This is the case in time-resolved spectroscopies of most types.

The kinetics of the excited state, diagrammed in Figure 9.15, are often described as follows, assuming that initially, the excited state is instantly excited:

$$\frac{dN}{dt} = -(k_{\text{rad}} + k_t + k_{\text{nr}} + k_q + k_{\text{isc}})N$$

$$N(0) = N_0$$

(9.23)

where k values reflect the associated rate of the process and $N(t)$ is the (singlet) excited-state concentration, population, or probability. For an experiment in which N_{ph} photons in a short pulse of light illuminate a sample, the number of initially excited molecules can be calculated:

$$N_0 = N_{\text{ph}}(1 - 10^{-\text{A}})$$

(9.24)

(See Equations 9.19 and 9.20.) These excited molecules will be distributed over the beam path in the sample according to beam intensity. The processes include radiative decay, fluorescence resonance energy transfer (FRET), nonradiative decay, fluorescence quenching by other molecules (non-FRET), and intersystem crossing to the triplet spin state. Crossing to the triplet state is often negligible. When intersystem crossing (isc) does occur, the triplet state is usually long lived and the decay of triplet to ground singlet state is slow, accompanied

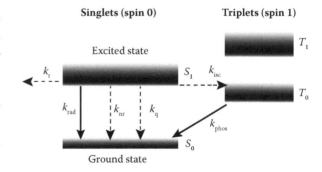

FIGURE 9.15
Analyzing the excited-state kinetics of a molecule. Only the solid arrows are (normally) accompanied by photon emission. Transfer and intersystem crossing from the S_1 state may be accompanied by phonons (vibrational excitations) to conserve energy.

by phosphorescence light emission at a longer wavelength. This differential equation has the familiar exponential decay solution,

$$N(t) = N_0 e^{-t/\tau}$$
$$\tau = (k_{\text{rad}} + k_t + k_{\text{nr}} + k_q + k_{\text{isc}})^{-1}$$

(9.25)

The excited-state lifetime is τ, and any measurement of the fluorescence from this system will have a characteristic exponential time dependence given by τ: $I_f(t) = I_0 e^{-t/\tau}$ This lifetime is shorter than the characteristic time scale of any of the processes, e.g., $\tau < k_{\text{nr}}^{-1}$, k_{rad}^{-1}, etc.

9.6.3 Non- or Multiexponential Decay and Solvent Relaxation

The majority of measurements of fluorescence from biological systems does not show single-exponential behavior predicted by this simple theory. Reasons for the nonexponential time dependence vary widely but often stem from variations in the environment of the population of molecules excited or multiple excited states.

Unlike atoms in a gaseous sample, biomolecules that may be excited by photons are surrounded by water and other molecules. The energy levels reflect the energy of the molecule as it interacts with water and other molecules. If water or an ion is moved closer to the ring system of a Trp or Tyr amino acid, for example, the energy levels will shift. After photon absorption by a molecule is surrounded by water, a process called *solvent relaxation* always occurs. When a molecule changes from ground to excited state, the distribution of electrons in the molecule changes: the promotion of the electron in the energy-level diagram is accompanied by a change in physical location of that electron. For common cases of photon absorption, the most important change resulting from this charge movement is in the electric dipole moment of the molecule. As we learned in Chapter 4, water also has a large dipole moment and responds to this change in biomolecule dipole moment by reorienting itself to minimize the free energy. This energy minimization lowers the energy of the excited state and shifts the spectrum of emitted light to longer wavelengths. The process of solvent relaxation in water normally takes 10^{-10} s or less, and can be observed by judicious choice of the emission wavelength in a time-resolved measurement. In steady-state measurements (excitation and emission intensity constant), time-averaging occurs: the spectrum of the emitted light is a weighted sum of the spectra before and after solvent relaxation.

$$I_{\text{em}}(\lambda) \cong \frac{\tau I_{\text{em}}^{\text{relaxed}}(\lambda) + \tau_{\text{sr}} I_{\text{em}}^{\text{unrelaxed}}(\lambda)}{\tau + \tau_{\text{sr}}}$$

(9.26)

where τ is the excited-state lifetime and τ_{sr} is the solvent relation time. For usual cases, $\tau \sim 10^{-9}$ s and $\tau_{\text{sr}} \sim 10^{-10}$ s, and the spectral change is small. Consult the Lakowicz text for details.[15]

9.6.4 Fluorescence Quantum Yield

The quantum yield of fluorescence, usually notated Q, is defined as the number of emitted photons divided by the number of absorbed photons, or equivalently as the probability of photon emission by an initially excited molecule (Table 9.3):

$$Q = \frac{\text{Number photons emitted}}{\text{Number photons absorbed}}$$

(9.27)

In the case of the simple exponential decay described by Equation 9.23, Q can also be calculated as

$$Q = \frac{k_{\text{rad}}}{k_{\text{rad}} + k_t + k_{\text{nr}} + k_q + k_{\text{isc}} + \dots} = \frac{\tau}{\tau_{\text{rad}}}$$

(9.28)

TABLE 9.3	Fluorescence Quantum Yields of Some Chromophores Used in Biophysical Research			
Chromophore	**Solvent**	**λ_{ex} (nm)**	**T (°C)**	**Q**
Phenylalanine	Water	260	23	0.024
Tryptophan	Water	280		0.13
Tyrosine	Water	275	23	0.14
Quinine sulphate	0.1 M H_2SO_4	350	22	0.58
Rhodamine 6G	Ethanol	488		0.94

Source: Values are from Lakowicz, J.R., *Principles of Fluorescence Spectroscopy*, 3rd ed., Springer, New York, 2006, Table 2.4.

where $\tau_{rad} = k_{rad}^{-1}$ is the *radiative lifetime*. The intensity of any fluorescence measurement is related to this quantum yield: systems with low Q are generally difficult to measure. Equation 9.27 implies a simple, experimental measurement of photons in and photons out but offers little understanding of the molecular reasons for a small or large value. Use of Equation 9.28 demands a more sophisticated measurement of the time dependence of fluorescence to get the value of τ, and (usually) a theoretical calculation that involves the absorption spectrum, to determine the value of τ_{rad}.

9.6.5 Nanoparticles: New Fluorescence Labels for Biological Applications

One of the most visually stimulating advances in biological imaging during the past decade has been in fluorescence imaging on the 0.1–100-μm scale. Even a cursory Google search on the Internet will turn up hundreds of impressive fluorescence micrographs and video micrographs that could not have been made 20 years earlier. We should not downplay the technical improvements in optical microscopy,[16] but fluorescence labeling has made great advances. These advances have been in two areas: (1) the versatility and durability of the fluorescent molecules themselves and (2) the chemical specificity of attachment of these fluorescence labels to particular proteins or other parts of the cell. To be frank, more of the progress so far has been due to progress in clever labeling of specific sites in cells with existing organic fluorescent dyes, but we shall briefly study the physics of a new type of fluorescence probe: the metallic or semiconducting nanoparticle. There is a recent review of imaging aspects of organic dyes and nanoparticles for those who desire a more detailed treatment.[17] The treatment is from a simple quantum, particle-in-a-box approach, so does not provide the whole truth of the matter. Other aspects of modern fluorescence labeling and imaging may be easily found by electronic database searches and should include the following: *The Molecular Probes Handbook*,[18] the Lakowicz text,[15] single-particle tracking applications,[19] membrane probes,[20] and development of new probes.[21]

A metallic particle is similar to a three-dimensional quantum box for electrons. Electrons in the box find a zero potential until they encounter the wall of the particle, where the potential becomes very high. The quantum wave functions resulting from this potential are therefore like those shown in Figure 8.4, except for the third (z) dimension. The wave functions cannot be plotted as in Figure 8.4 for the three-dimensional case, but Figure 9.16 shows two of the dimensions for the case $n = m = 10$. Imagine the z dependence. The energy levels are virtually continuous for high-energy levels because of the large number of (n, m, l) triplets that can give

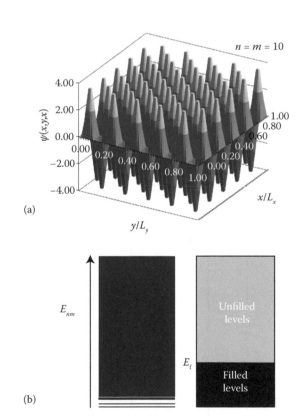

FIGURE 9.16

Quantum-box model for a metallic nanoparticle. (a) The x and y dependence of the wave function for $n = m = 10$. (b) Depending on the size of the box (the number of atoms in the metal particle) there may be many electrons occupying the box. For macroscopic three-dimensional particles, levels are filled up to large values of (n, m, l), where energies appear almost continuously due to the large number of (n, m, l) triplets per unit energy.

This model is developed in many modern physics books as the theory of electrical conductivity. E_f refers to the Fermi energy in that theory.

rise to any energy interval $E \pm \Delta E$. Metal atoms have a certain number of loosely bound electrons that are effectively free when many atoms are put together in a lattice (a "metal"). The number typically ranges from 30–250-nm^{-3}. For silver and gold (typical metals used for nanoparticles) the numbers are 58.6-nm^{-3} (Ag) and 59.0 nm^{-3} (Au).[22]

Consider a 1-nm^3 silver or gold nanoparticle. There will be about 59 free electrons placed in the box. The lowest-energy wave function will be $(n, m, l) = (1,1,1)$. Because of electron spin, two electrons can occupy this state. The next lowest-energy wave functions will correspond to (1,1,2), (1,2,1), and (2,1,1). There will be 6 electrons in these levels. The exact energies of these states will depend on the dimensions of the box. Some (n, m, l) states may even have the same energy (degenerate states), but we will ignore this. Proceeding further: (2,1,2), (2,2,1), (1,2,2), 6 states; (3,1,1), (1,3,1), (1,1,3), 6 states; (2,2,2), (3,2,1), (3,1,2), (2,3,1), (2,1,3), (1,3,2), (1,2,3), 14 states; (3,3,1), (3,1,3), (1,3,3), 6 states; (4,1,1), (1,4,1), (1,1,4), 6 states; (3,3,2), (3,2,3), (2,3,3), 6 states; (4,2,1), (4,1,2), (2,4,1), (2,1,4), (1,4,2), (1,2,4), 12 states, etc. Though we may not have the energy ordering correct, we have reached 58 electrons in our 1-nm^3 box and are still in the range where energy levels should be distinguishable, rather than continuous, as in the macroscopic particle case.

Will the electrons in this 1-nm^3 box absorb light? The answer is that they will—after all, they are "free electrons" that can oscillate in the electromagnetic field—unless a transition happens to be forbidden or unless the electric field is attenuated by the metal. Introductory physics reveals that electric fields could not exist inside conductors, however. Intermediate physics teaches that there is a "skin depth," below which the electric field is zero: as with all real physical quantities, the field does not instantaneously go to zero as the electromagnetic wave crosses the surface. For optical wave frequencies, the skin depth in silver is several nanometers—slightly larger than the size of our 1-nm^3 particle. Figure 9.17 shows what this radiation might look like when it strikes a nanoparticle. The particle shown is larger to illustrate the attenuation of the wave in the metal sphere. (The separation between wave fronts of Vis/IR light is 500–1000 nm, but to keep lines on the page, the separation shown is much smaller.) This would imply that for a $10 \times 10 \times 10$ nm particle, the radiation may not penetrate to the core of the particle. But does this matter? A skin depth less than the particle size implies that *all* radiation that "hits" the particle is absorbed.

The word "hits" is in quotes because, as we learned earlier in the chapter, radiation is absorbed by a particle according to its absorption coefficient, which corresponds to an absorption cross-sectional area.

We conclude that transitions that are not quantum forbidden will indeed absorb light. Which transitions might be quantum forbidden? The total spin of the system of free electrons should remain unchanged, so an electron that absorbs a photon should keep its spin in the higher-energy state. What about the angular momentum of the photon and the $\Delta L = \pm 1$ rule for atoms and molecules? Do the electrons in the box have angular momentum besides spin? We are in danger of getting in deeper than our quantum sophistication warrants, but let's try to go a step further. If the "box" were, in fact, spherical, the electronic states would have angular momentum just like electrons in atoms. Any spherically symmetric potential results in conservation of angular momentum and states characterized by L values. The metal sphere should then have some ΔL-disallowed transitions. Such forbidden transitions should be detectable in a 1-nm^3 spherical particle that does not have a large number of free electrons. What about a rectangular solid nanoparticle? First, let's note that there is no such thing as a 1-nm "cube" or rectangular solid: the corners will always be curved according to the limitations of the particle synthesis technology and the surface forces pulling back on protruding atoms. Such curved particles should then have some states characterized by angular momentum. The tentative conclusion is that selection rules should operate also in nanoparticles, though the details are beyond our level of treatment.

Using the particle-in-a-box model we can, in fact, count the number of free electrons, place them in the energy levels, and calculate transition energies. What photon energy would be required to promote an electron in the highest-occupied state to the next higher state? Let's assume the 1-nm^3 silver particle is a 1-nm cube. Remember the energies depend only on the (de Broglie) wavelength of the particle, which is not very different for a 1-nm^3 sphere. The energies for the cube are

$$E_{nml} = \frac{\pi^2 \hbar^2}{2ML^2}(n^2 + m^2 + l^2) \tag{9.29}$$

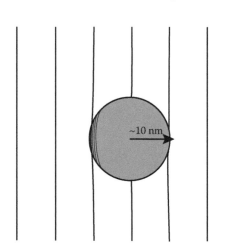

FIGURE 9.17
The field of a Vis/near-IR plane electromagnetic wave hitting a metal nanoparticle will extend several nanometers into the metal. Some of this energy can be converted into longer-wavelength fluorescence. Scattered and fluorescence waves are not shown.

(M is the electron mass). From our deliberations above we suspect the highest-energy levels correspond to (n, m, l) values in the 1–4 range. Consider the energies of (4,2,1) and (4,2,2) states. The energy difference is $\Delta E = E_{422} - E_{421} = \frac{3\pi^2\hbar^2}{2ML^2} = 1.8\times10^{-19}$J $\cong 44k_BT = 1.1$eV. This is in the near-IR region, $\lambda \sim 1000$ nm. Slightly different values of (n, m, l) result in visible wavelengths. We conclude that nanoparticles should probably absorb light in the UV to IR region.

The second major issue we have left unsolved is that of the absorption coefficient or absorption cross section: how strongly do these metal nanoparticles absorb light? A 1-mm-radius silver sphere will have an absorption cross section of $\pi r^2 \sim 3$ mm^2 = 3×10^{-2} cm^2 = 3×10^{-6} m^2—simply the cross-sectional area of the sphere. This is because both the wavelength and skin depth of UV/Vis/IR light is much smaller than the particle diameter. The large sphere simply "shadows" the light. At the other extreme, we have a single silver atom: what is its absorption cross section? Turning back to first-year physics and chemistry books, we find—nothing; much written about wavelengths and frequencies, but nothing about silver or even hydrogen atom absorption coefficients or cross sections. Even senior-level quantum-mechanics books and graduate-level texts include nothing (usually). The take-home message is that quantum absorption cross sections are not easy to calculate. That does not leave us with much hope for the silver nanoparticle case.

A further complication of nanoparticle absorption is that, as Figure 9.17 suggests, absorption is done by electrons near the surface. The surface atoms and their electrons are not quite the same as atoms and electrons in the "bulk." For nanoparticle diameters about 10 nm and smaller, all the atoms may, in fact, differ from bulk atoms. In any case, the importance of the nearby surface has led to the assignment of the term *surface plasmon* absorption to light absorption by nearly free electrons in a metal near the surface. "Plasma" is a group of charged particles (e.g., electrons) that are basically free particles but in an excited state. Because one cannot have a "plasma" made up of one electron, use of "plasmon" for metal surface absorption implies that the nanoparticle absorption involves the correlated, simultaneous excitation of many electrons. This is exactly right. Furthermore, the reason metallic (and semiconductor) nanoparticles have been developed into fluorescent labels for use in biological and other systems is that the absorption (and emission; see next section) is strong. When 50 electrons collaborate to absorb a photon, their absorption coefficient (cross section) should roughly be 50 times as large as that of one electron! Nanoparticles are very efficient at absorbing and emitting light near the visible band. Depending on their dimensions and shape, a suspension of nanoparticles can appear bright red, bright blue, any color in between, or colorless. Figure 9.18a shows a picture of several nanoparticle suspensions. Nanoparticle fluorescent labels have been used increasingly in the past 5 years to detect single molecules or single virus particles.[9] Figure 9.18b illustrates the application of semiconductor nanoparticles to do three-color

Semiconductor nanoparticle absorption is characterized by *excitons*, electron hole pairs that move about in a correlated fashion.

HeLa cells are "immortal" cells derived from cervical cancer cells taken from Henrietta Lacks, who died from her cancer in 1951.

You should be able to figure out what the 655, 585, etc. refer to.

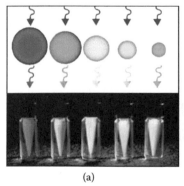
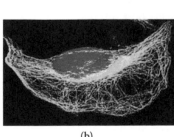
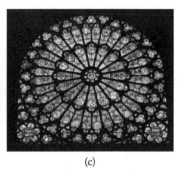

(a)   (b)   (c)

FIGURE 9.18

Fluorescent nanoparticles, or quantum dots. (a) The emission wavelength is determined by the particle size (primary), as well as material and shape. Nanoparticles are semiconductor (cadmium selenide or telluride). (b) Nanoparticle conjugates, nanoparticles with surface-attached molecular that bind to specific cellular proteins or other structures, have been used to create highly colored micrographs and video micrographs of specific structures in cells. (c) Metal nanoparticles have been used for hundreds of years to color glass: the north rose window in the cathedral of Notre Dame, Paris. Colors are as follows: red, gold nanoparticles; blue, cobalt nanoparticles; yellow, silver nanoparticles. ([a–b] Courtesy of Invitrogen Corporation/Thermofisher; [c] Courtesy of Krzysztof Mizera.)

staining of HeLa cells using fluorescent Qdot® nanocrystals, bound using immune antibody recognition bonding. Intracellular structures were visualized using a red-fluorescent Qdot 655, a yellow-fluorescent Qdot 585 for the Golgi apparatus, and a green-fluorescent Qdot 525 streptavidin conjugate for microtubules. The success of this sort of imaging derives primarily from the selective binding of the fluorescent nanoparticles, secondarily from the strong nanoparticle fluorescence and its resistance to photochemistry.

Persons interested in the historical development of metallic nanoparticles should note that highly colored nanoparticles are not a recent invention. They have colored the windows of cathedrals around the world, as shown in Figure 9.18c. These church windows have retained their color after hundreds of years of exposure to the sun and other elements of the weather, illustrating their resistance to photobleaching. In contrast, plastics and other materials colored with organic dyes lose their color after hours to months of sun exposure. As has been written: there is nothing new under the sun. Nanostructures have been around for a long while, as well as technologies for their manufacture, but only recently have we discovered how they can be manipulated and controlled on a microscopic scale.

Nanoparticle optical spectroscopy is primarily an experimental science, though the theories for absorption and emission are progressing. Nanoparticles can be synthesized in various sizes and shapes and by several different methods (sol–gel [wet colloid chemistry], laser ablation/metal vapor deposition, pyrolysis/plasma jets, lithography). We have focused on metal nanoparticles for simplicity, but a large effort has been put into semiconductor nanoparticles. These materials and synthesis methods are often the subject of a separate course on nanomaterials. We will not pursue them further, except as they appear in biological measurements.

Nanoparticles have been developed in large part to supplement organic dyes as labels for optical microscopic imaging of cells. The several advantages of nanoparticles over organic dyes are as follows:

- *Stability:* nanoparticles do not photobleach (lose their fluorescence via photoionization or other photochemical processes).
- *Fluorescence intensity:* a nanoparticle is equivalent in intensity to several 10's of dye molecules.
- *Fluorescence lifetime:* may be as long as 10^{-6} s, as compared to organic dye's 10^{-8} s.
- *Excitation wavelength:* nanoparticles of differing emission wavelengths can have the same excitation wavelength; one wavelength can thus excite several different probes.
- Nanoparticles can be manipulated with other systems, such as laser tweezers and magnetic fields (with suitable materials).

The primary disadvantage of nanoparticles is the relatively large size (10–100 nm, as opposed to 0.5 nm), along with less highly developed biochemistry. A 20-nm nanoparticle designed to bind to a mitochondrial protein cannot easily pass through the several membranes to get to its binding site. Special procedures designed to let the particle pass may disrupt the cell processes being studied. Issues also remain as to in vivo use of nanoparticles to study living tissues. Metal or semiconductor nanoparticles are not normal entities encountered by organisms and represent possible hazards for both scientist and organism being studied. Organic dyes may also be hazardous, but their metabolism in humans is better understood and handling procedures well established. (More below.)

A hybrid of nanoparticles and organic dyes has also been devised. *Cornell dots,* a play on the term *quantum dots* (nanoparticles) and Cornell University, were devised in 2005. Relatively inert spherical cores of silicon dioxide, about 2 nm in diameter, contain several dye molecules and are surrounded by a protective silica shell. Overall, the particle is about 25 nm in diameter. Like semiconductor quantum dot nanoparticles, Cornell dots are 20–30 times brighter than single dye molecules and resist photobleaching. In principle, these dye/nanoparticle hybrids can employ a wide variety of dyes, producing a large assortment of colors.

As suggested, almost no nanoparticle synthesized for use in biological systems (e.g., salt water) is made from a pure metal or semiconductor. As you might expect, the surface of silver can corrode, and since much of a nanoparticle is "surface," much of a silver nanoparticle could quickly change its properties. An additional issue is that the chemistry usually changes for nanometer-sized particles: they do not act like pure, macroscopic blocks of metal or semiconductor nor like dissolved molecules or ions. Chemists have had to learn a new synthetic art for the nanoscale. Nanoparticles also have issues of toxicity. Semiconductor nanoparticles are commonly made from cadmium selenide or telluride, known biohazards, but even gold and carbon nanoparticles have been identified as potential poisons.[23] Carbon nanotubes, a special type of nanoparticle not used for its fluorescence properties, but for its structural and "container" properties, have been identified as hazards,[24] for example, though coatings of various sorts hold promise of mitigating such hazards. Many institutions are presently studying the hazards, as well as biological benefits, of nanoparticles, but one of the largest efforts has been at the Smalley-Curl Institute, Rice University (https://sci.rice.edu/).

9.7 EINSTEIN RELATIONS BETWEEN ABSORPTION AND EMISSION OF ATOMS (GRADUATE SECTION)

The Einstein relations between the A and B coefficients are usually thought to apply to laser physics. Einstein published these relations in 1917, however, long before a working laser was a glimmer in mind of the (disputed) inventor(s) of the laser.[25] The Einstein A-B relations apply to any system that absorbs light. A summary of the Einstein A-B relations is: *Any molecule that absorbs light also emits light.* The only remaining issue is how strong the absorption and emission are. The answer is: *A molecule that absorbs strongly will emit strongly, unless another process intervenes.*

Those more experienced in quantum mechanics know that energy levels are commonly predicted by the Schrödinger equation or related approximations. Such theories are good at predicting wavelengths of radiation that are absorbed and emitted by the atoms or molecules under study. They do not, however, predict how strongly radiation may be absorbed or emitted other than to declare that certain types of transitions are "forbidden" and should have zero intensity. In this short section we introduce one small tool for predicting the intensity of emitted light if we know the strength of the corresponding absorption.

The symbol A represents the probability or rate of spontaneous photon emission from level $2 \rightarrow 1$ in Einstein's notation (Figure 9.19). A is related to the fluorescence emission we have been dealing with. B_{21} is the probability for (or rate of) stimulated photon emission $2 \rightarrow 1$. Stimulated emission only occurs when a photon whose energy equals the difference $E_2 - E_1$ passes near an atom excited to state 2. B_{12} is the probability for (stimulated) photon absorption $1 \rightarrow 2$. Both B terms are dependent on the intensity of radiation present. The two results Einstein derived were

$$B_{21} = B_{12} = B$$
$$\frac{A}{B} = \frac{8\pi h f^3}{c^3} \tag{9.30}$$

These results state that the ratio of the spontaneous emission rate A to absorption rate B is a constant that depends on the frequency (energy) of the radiation. As Figure 9.19a shows, there are no issues as to different absorption and emission wavelengths or frequencies. In the two-level atomic system, there is only one wavelength, and this wavelength produces both absorption and (stimulated) emission. Einstein's derivation does not apply to molecular systems, where broadened energy levels result in distributions of wavelengths. In 1962, Strickler and Berg made the modification of Equation 9.30 for a molecular system.[26] The results are considerably more complicated, but the spontaneous emission rate, which is k_{rad} in our notation, can be written in terms of τ_{rad}:

$$k_{rad} = \frac{1}{\tau_{rad}} = \frac{2303 \cdot 8\pi c n^2}{N_A} \left\langle \overline{v}_{em}^{-3} \right\rangle_{Avg}^{-1} \frac{g_l}{g_u} \int \varepsilon(\overline{v}) d(\ln \overline{v}) \tag{9.31}$$

where $\overline{v}$ is the "wave number" (λ^{-1}) in cm^{-1}, $c = 3.0 \times 10^{10}$ cm/s, g_l and g_u are the quantum degeneracies of the upper and lower levels, and 2303 stems from use of the logarithm term and molar units in ε. (The units here are unusual for historical reasons. You are invited

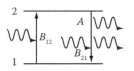

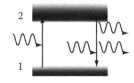

Atoms: line spectra   Molecules: band spectra

FIGURE 9.19
Einstein relations between absorption and emission of radiation. Einstein-derived relations for discrete, atomic energy levels in 1917; Strickler and Berg modified the relation for broadened molecular levels in 1962. The atomic case is simple because absorption and emission wavelengths are identical.

to convert this expression to a more familiar format.) The integrand is the wave number–dependent absorption coefficient, in units of $M^{-1}\,cm^{-1}$. Noting the correspondence of B with the integrated absorption coefficient and of $\langle \bar{\nu}_{em}^{-3} \rangle_{Avg}^{-1}$ with f^3 rationalizes the equivalence of Equations 9.30 and 9.31. Where does the degeneracy ratio come from? If there are two allowed degenerate (equal energy) levels in the lower state for every excited state, the decay rate will be twice as much. Equation 9.31 is often used in the calculation of fluorescence quantum yields via Equation 9.28, because it is rare that the decay rate of a molecule can be measured in the absence of all nonradiative decay processes. Most commonly, a graduate student is assigned the task of measuring Q via Equation 9.27 using a steady-state fluorometer, measuring τ with a time-resolved fluorometer, calculating Q via Equations 9.28 and 9.31, and concluding which nonradiative processes are important in the denominator of Equation 9.28.

9.8 INTERSYSTEM CROSSING: SINGLETS (S = 0) TO TRIPLETS (S = 1)

The rate k_{isc} in Figure 9.15 is the "intersystem crossing" rate. The ground state of organic molecules typically has total electronic spin equal to zero. The set of spin 0 electronic states is called the singlet system or singlet manifold of states. We noted before that since photons carry one unit of angular momentum, photon absorption normally requires a change of one unit of electronic orbital angular momentum during the excitation process. $\Delta L = \pm 1$ is the first selection rule for transitions involving photon absorption or emission. The "intersystem crossing" from singlet states to triplet ($S = 1$) states is to a triplet state of equal energy, if it is available. If not, crossing can still occur, but some thermal, vibrational energy will have to be found to simultaneously make up any energy difference, making the transition less likely (slower).

The second selection rule is $\Delta S = 0$: the total spin of the system does not change. Photon absorption and emission thus normally take place within either the singlet system or within the triplet system; crossing from one to the other is forbidden. Of course, as *Jurassic Park* (the movie) has taught us, "Nature will find a way." Perhaps more aptly, selection rules are not absolute because the simple assumptions behind them are not strictly true. If a strong magnetic perturbation accompanies the transition, e.g., if a heavy atom is nearby, a spin can flip and the transition takes place, allowing intersystem crossing or phosphorescence. Both k_{isc} and k_{phos} are rates of these selection rule-violating rates. They should be, and usually are, slow compared to the other rates in Figure 9.15. Typical phosphorescence decay times are $10^{-6} - 10^{-2}\,s$, indeed much longer than the times associated with the other rates. It is much more frequent that k_{isc} is fast, up to $10^8\,s^{-1}$ or so. It is usually the case that k_{isc} is much larger than k_{phos}. Why might this be?

9.9 ENERGY TRANSFER (FRET)

Figure 9.15 has a "k_t" at the left side. This represents the "resonance" energy transfer rate and describes a process whereby the excitation energy is transferred from one molecule (the "donor") to a nearby molecule (the "acceptor"), resulting in an electronic excitation on that second molecule. In the literature this process has acquired the acronym FRET, or fluorescence resonance energy transfer. The basic phenomenon is not unexpected: an electronic excited state in a molecule is the equivalent of an oscillating excited dipole, which can interact with a nearby dipole via a dipole–dipole interaction, as illustrated in the classical analog in Figure 9.20. The interaction force in this figure is due to a dipole–dipole force, which some of you have calculated in a previous physics course. Resonance refers to the increased transfer rate when the energy of the photon that could be emitted from the donor matches the energy of the photon needed to excite the acceptor. Classically, energy will more efficiently pass from one oscillator to the other when the oscillator frequencies, $\sqrt{k/m}$, are equal. No photon actually passes from donor to acceptor, though one quantum model for such transfer employs a "virtual" photon.

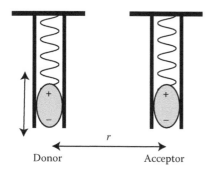

Donor   Acceptor

FIGURE 9.20
An oscillating dipole can transfer its energy to a nearby dipole via a dipole–dipole interaction. The interaction depends on the magnitudes of the dipole moments and the distance of separation.

The dipole model suggests that the efficiency of the energy transfer should increase with the strength of the dipoles and with decreasing distance. This is indeed the case. The energy-transfer rate via the so-called Förster transfer mechanism is

$$k_t(r) = \frac{1}{\tau_{\text{donor}}}\left(\frac{R_0}{r}\right)^6 \qquad (9.32)$$

τ_{donor} is the excited-state lifetime of the donor in the absence of the acceptor. (The lifetime of the donor will be shortened by the presence of the acceptor.) R_0 is the major new quantity and is called the Förster distance. R_0 can be calculated as follows:

$$R_0^6 = \frac{9000 Q_D \left(\ln(10)\right)\kappa^2 J}{128\pi^5 n^4 N_A}$$

$$J = \int_0^\infty F_D(\lambda)\varepsilon_A(\lambda)\lambda^4 d\lambda \qquad (9.33)$$

$$\int F_D(\lambda)d\lambda = 1$$

Q_D is the fluorescence quantum yield of the donor in the absence of the acceptor, κ is a factor related to the relative angles of the dipole moments of donor and acceptor, J is the overlap integral, n is the index of refraction of the medium, $N_A = 6.02 \times 10^{23}$, $F_D(\lambda)$ is the donor emission spectrum (its integrated area normalized to 1), and $\varepsilon_A(\lambda)$ is the acceptor absorption spectrum, measured in terms of the absorption coefficient. Out of lack of angle information, the value of κ^2 is often taken as 2/3, corresponding to dynamic averaging of both dipoles over all possible angular orientations. Figure 9.21a diagrams the process of energy transfer: transfer can create a new excited state on

> There are actually several "Förster" energy transfer mechanisms that depend differently on distance. The biochemical literature has unfortunately reduced them all to only one possible meaning.

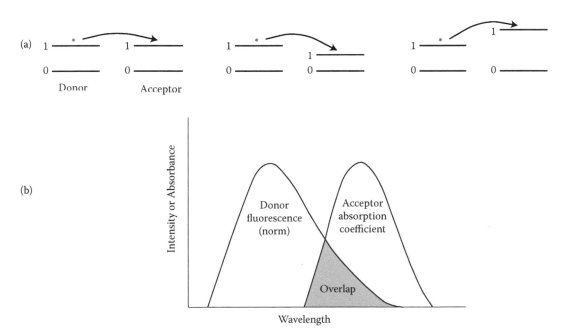

FIGURE 9.21
Resonance electronic energy transfer can be efficient if the acceptor of the energy has a lower- or equal-energy excited state as the donor. (a) The asterisk stands for an electronic excitation: the excitation, not the electron, is transferred. Transfer of energy to a molecule with higher excited-state energy would seem to violate energy conservation but does not if energy can be extracted from the thermal energy surrounding the molecules. Because organic molecules have spectral bands, not lines, the requirement translates to the overlap of the emission band of the donor with the absorption band of the acceptor. (b) The overlap integral J along with the distance between donor and acceptor dipoles and the angular orientation factor κ primarily determine the magnitude of the transfer rate k_t.

TABLE 9.4 Förster Distance for Donor–Acceptor Pairs Useful in Biophysics

Donor	Acceptor	R_0 (nm)
Trp	Dansyl	2.2
Trp	DPH	4.0
Dansyl	FITC	3.3–4.1
FITC	TMR	3.7–5.0
BODIPY	BODIPY	5.7
Terbium	Rhodamine	6.5
Europium	APC	9.0

Source: Lakowicz, J.R., *Principles of Fluorescence Spectroscopy,* Springer, New York, 2006. Tables 13.3 and 13.4. Those who must know the chemical abbreviations should refer to Lakowicz or search the NIST chemical name database at http://webbook.nist.gov/chemistry/name-ser/.

the acceptor molecule that has energy equal to, lower than, or higher than the original energy. Any excess or needed energy is dissipated to or extracted from the thermal environment: at a cost in probability, of course. (If thermal energy needs to be extracted from or added to the environment, the probability of the associated process will decrease.) The dependence of the transfer rate on the energies of donor and acceptor excited states is contained in the overlap integral, J, which is illustrated in Figure 9.21b. A word on units in Equation 9.33: if $\varepsilon_A(\lambda)$ is measured in our standard units, $M^{-1}\,cm^{-1}$, and λ in cm, J will have units of $cm^3\,M^{-1}$.

As we have seen, the radiative rate is directly proportional to the (integrated) absorption coefficient, which turns out to be proportional to the magnitude of the transition dipole moment squared. The distance dependence of the interaction, $1/r^6$, dies off quickly with distance, but typical values of the constant R_0 make the transfer rate significant in the 1–10 nm region. See Table 9.4. A bit of thought about distance will reveal the primary assumption behind the transfer theory of Equations 9.32 and 9.33: the distance between donor and acceptor dipole moments can only be defined when r is significantly larger than the diameters of the molecular groups responsible for the dipole moments. Such groups might be tryptophan, heme, dansyl, Alexa fluor, etc., indicating that the Förster approximation commonly breaks down below about 1 nm.

The three most important applications of FRET in biosystems have been

1. Measurement of distances and distance changes in proteins and nucleic acid systems (the molecular "ruler")
2. Energy transfer in photosynthetic systems, coupled to charge transfer
3. Imaging of connections and interactions between subcellular structures

The photosynthetic application of energy transfer is the only known use nature has conjured on its own. We will explore this more in Chapter 10. Application 3 has received wide attention in the past 5–10 years because of technological advances in donor–acceptor pairs, excitation sources, and microscopes. Striking two-color video micrographs of cellular structures approaching each other, reducing the donor-acceptor separation r, and causing acceptor fluorescence enhancement may be found by searching the Web and library literature resources. For more detailed information on FRET as it is currently used in biophysical research, consult the Lakowicz text (Chapters 13 through 15), Clegg,[27] and the many FRET research books published since 2000 (e.g., search "Fluorescence Resonance Energy Transfer" on www.amazon.com).

9.10 POINTS TO REMEMBER

- The fundamental wave relations $E_{photon} = hf$ and $c = \lambda f$
- Light intensity and power units and conversions between them
- Absorption and emission strengths measured by transition-state or oscillator dipole moments
- Beer's law for light absorption, $I = I_0 10^{-A}$, $A = \varepsilon c L$, and the nature of absorption spectra
- Ringlike molecular groups are responsible for most absorption in biological systems
- Laws describing excited-state decay
- The nature of Förster energy transfer

9.11 PROBLEM-SOLVING

MINIPROJECT 9.1   SOLAR POWER

Go to http://rredc.nrel.gov/solar/spectra/am1.5/ and download the Microsoft Excel-formatted data for the sun's irradiance.

1. Reproduce the two graphs shown in Figure 9.1 using a graphing program. (The coloring, gas, and region labels may be omitted.)
2. Using your software, integrate the surface irradiance spectrum from 200 to 300 nm to get the total UV power per m^2 hitting the Earth.
3. Repeat 2 for the visible region, from 300 to 850 nm.
4. Repeat 2 for the IR region, 850–4000 nm.
5. Repeat 2 for the radiation striking the Earth's upper atmosphere. Compute the fraction of the UV light transmitted by the atmosphere (cloudless day).
6. Repeat 3 for the radiation striking the Earth's upper atmosphere. Compute the fraction of the visible light transmitted by the atmosphere (cloudless day).

PROBLEM 9.2   SPECTRA: VARIABLE ON THE HORIZONTAL AXIS

1. Using the spreadsheet-formatted upper atmospheric data from http://rredc.nrel.gov/solar/spectra/am1.5/, convert the irradiance data to W m^{-2} per unit frequency interval; construct the corresponding graph. Review Chapter 3 if necessary.
2. Convert the data to W m^{-2} per unit photon energy (eV^{-1}).

The 800-nm light is visible to the human eye, but the region out to 900 nm is included with "visible" light because some bacterial photosynthesizers absorb radiation in this region.

PROBLEM 9.3   BLACKBODY SPECTRA (GRADUATE PROBLEM)

Using the results of Problem 9.1, generate another graph of (only) the upper atmospheric irradiance data from http://rredc.nrel.gov/solar/spectra/am1.5/. Using software, compute a blackbody spectrum that can be directly compared with the irradiance data. The blackbody temperature should be about 5500 K, but optimize this for the best fit to the sunlight data. Consult a modern physics text or reliable online sources for the blackbody formula. Use a normalization factor to make the peak of the blackbody spectrum equal to that of the sunlight data.

PROBLEM 9.4   LIGHT ABSORPTION AND HEATING OF MICRO-OBJECTS

1. Calculate how much heat it takes to raise the temperature of a 0.80-μm diameter *Escherichia coli* from 20°C to 99.9°C. Assume the cell has the thermal properties of water.
2. If a 100-mW Nd:YAG laser ($\lambda = 530$ nm) is used for optical trapping experiments on the above cell, how long will it take to heat the cell if (1) 5.0% of the laser power is absorbed; (2) 0.1% is absorbed. Neglect heat conduction away from the cell.

3. How long will it take the above system to melt a 2.0-μm diameter polystyrene sphere. Assume polystyrene has an absorbance of 0.00010 per cm at 530 nm. ($A = 10^{-4}$ for an optical path length of 1.00 cm.)

4. Graduate problem. Absorption of light by materials such as polystyrene or glass is normally done by defects. However, the absorption of a photon by a defect often produces a second defect (or a larger defect). Defects then build up during laser irradiation. In the end, the microsphere should not be used again because of defect buildup. Treat the sphere in part 3 as a 1.0-μm cube with an initial concentration of 5.0×10^{20} defects/m^3, and that laser irradiation is uniform throughout the cube. How long will it be before the defect concentration increases by a factor of 1,000? (Remember that the number of photons/s absorbed (= number of defects/s produced) is proportional to $I_0 - I(t)$.)

PROBLEM 9.5   LASER TWEEZERS BIOPHYSICS RESEARCH

Using your library's electronic journal search engine, search "RNA laser tweezers" from 2007 to the present.

1. Print out all the bibliography references, including titles. You do not need to print the abstracts.

2. Choose one of these references that has a force versus extension curve somewhere in it. Print the abstract and the force-extension curve. Describe what is physically happening during the extension and what magnitude forces are involved.

PROBLEM 9.6   EXCITATION TRANSFER IN AN ARRAY: QUALITATIVE FIRST TRY

Suppose the microspheres in a two-dimensional array like Figure 9.6 each contain many fluorescent dye molecules. A short pulse of light excites a microsphere near the center of this array. If the excited sphere were isolated, the excitations would decay much like nuclear radiative decay: $N(t) = N_0 e^{-t/\tau}$, where $N(t)$ is the number of excited dye molecules on the sphere at time t, N_0 is the number initially excited (assume only a small fraction of molecules are excited), and τ is the lifetime characteristic of the dye. N_0 is also the number of photons initially absorbed. The intensity of fluorescence from this microsphere would have time dependence $I(t) = I_0 e^{-t/\tau}$. Suppose that excitations in the first sphere can "hop" to an adjacent sphere in the array and excite a dye molecule there, but that it takes a time τ_h to hop over. Assume that any excited dye molecule can emit a fluorescence photon.

1. Qualitatively describe the fluorescence from the array for the two cases $\tau_h \sim 100\tau$ and $\tau_h \sim \tau/100$. Discuss where fluorescence will come from and how long it will last. What total number of photons will be emitted?

2. In your answers, did you assume that the "clock" started over each time a dye molecule received energy transferred to it from a neighboring sphere? (Does the receiving dye molecule start fresh with a remaining excited-state lifetime of τ or has part of its lifetime expired?)

PROBLEM 9.7   UV LIGHT (GRADUATE PROBLEM)

From our experience, solar UV radiation, though damaging, is not intense enough to destroy a major fraction of microbial life on Earth. However, UV lamps are used for killing cyanobacteria, as well as pseudomonas, *E. coli*, and other bacterial life in pools and ponds. Assume a 25-W UV lamp emits 0.80 W of power in the 250–300 nm spectral region. The lamp is cylindrical, and water from the pond flows around the lamp in a cylindrical geometry, as shown in Figure 9.22.

The UV-transmitting fused-quartz envelope of the lamp is 1.5 mm thick.

1. Compute the UV irradiance this lamp produces midway through the water. Compare this value to the solar-produced irradiance in the 250–300 nm region.
2. Compute the average number of photons/(s-m²) passing through a surface midway through the water. You may treat all the photons as average 275-nm photons.
3. Assume a bacterium is a 1.5-μm radius sphere, which consists of protein, DNA, and other materials in percentages as described in Chapter 7. (Note that H_2O, 70% fraction, has concentration ~55 M.) Determine the approximate concentrations of DNA nucleotides and protein Trp, Tyr, and Phe amino acids.
4. Assume also that only protein and DNA absorb UV light and that the NIST standard protein absorbance applies: $A = \varepsilon_\% c_\% L$, with $\varepsilon_\% = 6.67$. The absorption coefficient of DNA or RNA can be expressed in a mass concentration form as follows: at $\lambda = 260$ nm, the average extinction coefficient for double-stranded DNA is 0.020 $(\mu g/mL)^{-1} cm^{-1}$, for single-stranded DNA and RNA it is 0.027 $(\mu g/mL)^{-1} cm^{-1}$. Using Beer's law, calculate the average fraction of UV light absorbed by a cell's protein and by DNA/RNA.
5. Calculate the average number of photons/s absorbed by a cell's DNA and RNA as it passes through the UV device. Discuss the likelihood that the cell's ability to reproduce (divide) is hampered by the UV absorption.

FIGURE 9.22
Cylindrical UV system used to partially sterilize pond water. The core of the coaxial system is a UV-discharge lamp (see Problem 9.7).

PROBLEM 9.8   EXCITED-STATE LIFETIME

Calculate the first excited state lifetime for a molecule whose natural excited-state radiative rate is $1.00 \times 10^9\,s^{-1}$, and which collides with nearby molecules ("quenchers") at an average rate of $2.00 \times 10^{10}\,s^{-1}$, each collision having a 10% probability of transferring the excitation to the quencher.

PROBLEM 9.9   FLUORESCENCE EXPERIMENT

You are asked to design an experiment to measure fluorescence from a phenylalanine (Phe) amino acid in a protein that contains one each of Phe, Tyr, and Trp. Use Figure 9.13 and select the best excitation and emission (observation) wavelengths to use. For your chosen wavelengths, discuss the expected ratio of Phe:Tyr and Phe:Trp fluorescence intensity. Also, discuss the best bandwidths ($\pm\Delta\lambda$) to use for each wavelength.

PROBLEM 9.10   TRIPLETS AND RATES

Intersystem crossing takes place whenever the excited molecules undergo a spin change. Both k_{isc} and $k_{phos} = 1/\tau_{phos}$ involve spin changes, yet k_{isc} is often much larger than k_{phos}. Discuss why this might be. Examine Figure 9.15 in your deliberations.

PROBLEM 9.11   TRANSFER AND EXCITED-STATED LIFETIMES

1. Assume the radiative lifetime of tryptophan is 19.0 ns and various other de-excitation processes result in a net nonradiative de-excitation rate of $3.50 \times 10^9\,s^{-1}$ in a certain protein. Calculate the excited-state lifetime of this Trp.
2. A DPH molecule, an acceptor of excitation from Trp, is placed nearby. Using a graphing program, plot a graph of the resonance energy transfer rate of this tryptophan to DPH located at distances ranging from $r = 0.50$–15.0 nm away.
3. Calculate the Trp excited-state lifetime with DPH present at $r = 1.0$ nm and $r = 10.0$ nm.

PROBLEM 9.12   FLUORESCENCE MICROSCOPY

Figure 9.18b shows an image that includes well-resolved microtubules. Microtubules have tube diameters of about 25 nm. Microfibers like actin are even thinner, about 5 nm; DNA finer yet, at about 2 nm. These fibers and microtubules can be visualized very nicely using fluorescence microscopy, yet they are generally not seen in scattered-light ("bright-field") microscopy. The explanation for the latter is, as you learned in first-year physics, that the resolution of a classic optical system is limited to about half the wavelength of the illumination light, or about 200 nm. Explain why fluorescence microscopy can so easily detect these small structures. (Hint: This is not a complex question.)

PROBLEM 9.13 DIPOLE–DIPOLE ENERGY TRANSFER: CLASSICAL ANALOG

Though FRET is a quantum phenomenon, we can compute a classical analog in terms of the force between two dipoles. Assume two charges, $+Q$ and $-Q$, are separated by a fixed distance d to form a dipole. Two of these dipoles are then positioned a variable distance r apart, with $r \gg d$. In the following, you may want to make use of the approximations introduced in Chapter 3, e.g., an approximation for $1/(x + \delta)^n = (x + \delta)^{-n}$, when $\delta \ll x$.

1. Place one dipole at the origin and the other on the y axis at $y = r$, with both dipoles pointed upward ($+Q$ above $-Q$). Calculate the magnitude and direction of the force between the two dipoles.
2. Calculate the magnitude and direction of the force if the dipole at $x = r$ is rotated $90°$ counterclockwise.

PROBLEM 9.14 QUANTUM DOTS IN FLUORESCENCE RESEARCH AND HDTV

As discussed in the text, emissive nanoparticles, quantum dots, have been used in fluorescence research in biosystems since the early 2000s. Recently, quantum dots have also become a major player in the TV market. Investigate the Q dots used in commercial HD TVs and compare them to those discussed in Section 9.6.5. Focus on physical properties such as size and geometry, materials, light wavelengths, excitation processes, and why these properties are useful in quantum-dot TVs. Do not get distracted by characteristics of commercial TVs themselves.

MINIPROJECT 9.15 BLOOD OXYGENATION AND SPECTROSCOPY

Hemoglobin oxygenation in solution can be determined from data like that in Figure 6.7, which shows the absorption spectra of oxy- and deoxy-hemoglobin from wavelengths 250 to 700 nm. However, when the hemoglobin is located in red blood cells circulating in a human body, these wavelengths are impractical to use, because they are absorbed or strongly scattered by skin, blood vessel walls and other tissues. For this reason, pulse oximeters usually employ wavelengths between 600 and 1,000 nm. In the following, use of *Wikipedia* is a suggested starting point.

1. Describe, in about 100 words, how a fingertip pulse oximeter physically works.
2. Find a diagram of a pulse oximeter showing the optical generation and detection components of the device. Copy this diagram into your assignment. Describe how the light enters the peripheral blood vessels and how it is detected. Indicate the path of the light beam from the light generator to light detector.
3. Locate an absorption spectrum of hemoglobin, such as that at the Wikipedia site; copy and insert the spectrum in your assignment. The spectrum must clearly show the absorbance values between wavelengths 650 and 950 nm.

4. Write separate equations for the absorbance at 680 and 950 nm in terms of the extinction (absorption) coefficients at those wavelengths, the concentrations of deoxy-hemoglobin, [Hb], and oxy-hemoglobin [HbO_2], and an arbitrary path length, L. Solve these two simultaneous equations for the concentrations [Hb] and [HbO_2], in terms of absorbance at 680 nm, absorbance at 950 nm, and numerical parameters. Calculate the ratio, $100[HbO_2]/([Hb]+[HbO_2])$, which is the % oxygenated hemoglobin. (Note that pulse oximeters automatically and quickly calculate this %, but using adjusted transmitted light intensities rather than absorbances.)

5. Estimate the scattering attenuation of light in the fingertip as follows. Locate an online Mie scattering calculator and calculate the relative scattering efficiencies for wavelengths 400, 700, and 900 nm. Assume the tissue is adequately described by a suspension of 40-µm spherical particles in water.

REFERENCES

1. Barth, A., and C. Zscherp. 2002. What vibrations tell us about proteins. *Q Rev Biophys* 35:369–430.

2. Pelton, J. T., and L. R. McLean. 2000. Spectroscopic methods for analysis of protein secondary structure. *Anal Biochem* 277:167–176.

3. Matthäus, C., B. Bird, M. Miljković, T. Chernenko, M. Romeo, and M. Diem. 2008. Chapter 10: Infrared and Raman microscopy in cell biology. *Methods Cell Biol* 89:275–308.

4. Harz, M., P. Rösch, and J. Popp. 2009. Vibrational spectroscopy—A powerful tool for the rapid identification of microbial cells at the single-cell level. *Cytometry A* 75:104–113.

5. Zhuang, W., T. Hayashi, and S. Mukamel. 2009. Coherent multidimensional vibrational spectroscopy of biomolecules: Concepts, simulations, and challenges. *Angew Chem Int Ed Engl* 48:3750–3781.

6. Schwesyg, J. R., T. Beckmann, A. S. Zimmermann, K. Buse, and D. Haertle. 2009. Fabrication and characterization of whispering-gallery-mode resonators made of polymers. *Opt Express* 17:2573–2578.

7. Bate, A. E. 1938. Note on the whispering gallery of St. Paul's Cathedral, London. *Proc Phys Soc* 50:293–297.

8. Astratov, V. N., and S. P. Ashili. 2007. Percolation of light through whispering gallery modes in 3D lattices of coupled microspheres. *Opt Express* 15:17351–17361.

9. Vollmer, F., S. Arnold, and D. Keng. 2008. Single virus detection from the reactive shift of a whispering-gallery mode. *Proc Natl Acad Sci USA* 105:20701–20704.

10. Clement-Sengewald, A., K. Schütze, A. Ashkin, G. A. Palma, G. Kerlen, and G. Brem. 1996. Fertilization of bovine oocytes induced solely with combined laser microbeam and optical tweezers. *J Assist Reprod Gen* 13:259–265.

11. Seeger, S., S. Monajembashi, K. J. Hutter, G. Futterman, J. Wolfrum, and K. O. Greulich. 1991. Application of laser optical tweezers in immunology and molecular genetics. *Cytometry* 12:497–504.

12. Yin, H., M. D. Wang, K. Svoboda, R. Landick, S. M. Block, and J. Gelles. 1995. Transcription against an applied force. *Science* 270:1653–1657.

13. Green, L., C. H. Kim, C. Bustamante, and I. Tinoco Jr. 2008. Characterization of the mechanical unfolding of RNA pseudoknots. *J Mol Biol* 375:511–528.

14. Avery, N. C., T. J. Sims, and A. J. Bailey. 2009. Quantitative determination of collagen cross-links. *Methods Mol Biol* 522:103–121.

15. Lakowicz, J. R. 2006. *Principles of Fluorescence Spectroscopy*, 3rd ed. New York: Springer.

16. Huang, B., M. Bates, and X. Zhuang. 2009. Super-resolution fluorescence microscopy. *Annu. Rev. Biochem.* 78:993–1016.

17. Resch-Genger, U., M. Grabolle, S. Cavaliere-Jaricot, R. Nitschke, and T. Nann. 2008. Quantum dots versus organic dyes as fluorescent labels. *Nat Methods* 5:763–775.

18. Johnson, I. 2010. *The Molecular Probes Handbook*, 11th ed. Carlsbad, CA: Life Technologies. www.thermofisher.com/us/en/home/references/molecular-probes-the-handbook.html.

19. Levi, V., and E. Gratton. 2007. Exploring dynamics in living cells by tracking single particles. *Cell Biochem Biophys* 48:1–15.

20. Demchenko, A. P., Y. Mély, G. Duportail, and A. S. Klymchenko. 2009. Monitoring biophysical properties of lipid membranes by environment-sensitive fluorescent probes. *Biophys J* 96:3461–3470.

21. Lee, J. S., Y. K. Kim, M. Vendrell, and Y. T. Chang. 2009. Diversity-oriented fluorescence library approach for the discovery of sensors and probes. *Mol Biosyst* 5:411–421.

22. Thornton, S. T., and A. Rex. 2006. *Modern Physics for Scientists and Engineers*, p. 315. Belmont, CA: Thomson (Brooks/Cole).

23. Hoshino, A., N. Manabe, K. Fujioka, K. Suzuki, M. Yasuhara, and K. Yamamoto. 2007. Use of fluorescent quantum dot bioconjugates for cellular imaging of immune cells, cell organelle labeling, and nanomedicine: Surface modification regulates biological function, including cytotoxicity. *J Artif Organs* 10:149–157.

24. Helland, A., P. Wick, A. Koehler, K. Schmid, and C. Som. 2007. Reviewing the environmental and human health knowledge base of carbon nanotubes. *Environ Health Perspect* 115:1125–1131. See also www.osha.gov/dsg/nanotechnology/nanotech_healtheffects.html.

25. Einstein, A. 1917. *Zur Quantentheorle der Strahlung* (On the quantum mechanics of radiation). *Phys Z* 18:121–128.

26. Strickler, S. J., and R. A. Berg. 1962. Relationship between absorption intensity and fluorescence lifetime of molecules. *J Chem Phys* 37:814–822.

27. Clegg, R. M. 1996. Fluorescence resonance energy transfer. In *Fluorescence/Imaging Spectroscopy and Microscopy*, ed. X. F. Wang, and B. Herman, pp. 179–252. New York: John Wiley & Sons.

CHAPTER 10

CONTENTS

Photosynthesis

10

Photosynthesis is the most important biophysical process taking place on Earth.[1] Without photosynthesis, there is no life—period. On the other hand, the topic of photosynthesis occupies a rather small part of most modern textbooks in the biosciences. The page allocation of five well-known textbooks published during the past 25 years ranges from nominal ~0.6% to 5.9%, with an average (±standard deviation) of 2.06% ± 2.26%. Later publication dates correlate with less coverage, as shown in Figure 10.1. This graph is not a serious historical attempt to track coverage of the topic, but photosynthesis has indeed lost its patina since the energy crisis of the 1980s. In addition to reduced national energy policy focus on solar energy generation during the 1985–2008 period, a possible reason for neglect of photosynthesis in bioscience courses has been the rise in importance of quantum physics in the topic. Photosynthesis is fundamentally a quantum process: the use of photon energy to move an electron and synthesize high-energy chemical compounds. While any chemical reaction can be described correctly as a quantum process at its core, the primary processes in photosynthesis cannot be understood or even measured without reference to quantum efficiencies and spectroscopy. The attention we have paid in this text to quantum mechanics and spectroscopy will hopefully serve us well.

Photosynthesis is one of the oldest and best studied areas of the biological sciences. Because human survival has relied on the successful cultivation of plants since being booted out of the Garden of Eden, it is perhaps not surprising that scientific studies can lead a researcher well back into the archives. Historical reviews are fascinating and turn up many surprising facts, such as major advances made by church ministers and scientists in the middle of war. You can read about some of these at the following Web sites: www.life.illinois.edu/govindjee/photoweb/subjects.html#history and www.life.illinois.edu/govindjee/history/articles.htm. (If these sites are inactive when you read this chapter, try the search terms "photosynthesis and history.")

Because many individual and multivolume books have been written about photosynthesis, as well as thousands of research articles, we must narrow our focus by taking a particular point of view to understand at least some of the major physical principles governing photosynthesis. Our point of view or focus will be the harvesting and transport of energy and charge on a microscopic scale. Because of this focus on transfer, an intrinsically quantum process, a reader with only a modest background in quantum mechanics should be prepared to review Chapter 8 before attacking Section 10.4, Light-Harvesting (Antenna) Proteins: Arrays of Absorbers. This chapter will not be a good source of up-to-date "facts" about photosynthesis, which can quickly become obsolete. Instead, our point of view is this: if you forget everything else in this chapter, don't forget how light harvesting, energy transfer, and electron transfer occur. These events are often called the *primary processes of photosynthesis*.

10.1 GLOBAL NUMBERS

Photosynthesis provides about 4×10^{21} J of energy per year, which is now (only) about six times the energy use by humans.[2] About 50% of photosynthetic activity presently derives from phytoplankton. See Figure 6.16. Photosynthesis captures 10^{14} kg of carbon from CO_2 and converts it into the form of higher-energy carbohydrates (food) and produces an equivalent mass of O_2. Atmospheric oxygen that allows life as we know it has been produced by photosynthetic activity, though the date of the start of this activity is not known for certain.[3] Photosynthesis regulates the climate. The global distribution of plant and bacterial photosynthetic activity is shown in Figure 10.2 in terms of chlorophyll production. Close examination of this map shows the expected higher production in areas of heavier rain, but other concentrations of chlorophyll can be seen at mouths of large river systems and in certain lakes and in arctic regions. Figure 10.3 shows a global map of the average solar *insolation*, the solar energy actually received at ground level (reduced by clouds). There is little correlation of

The older, 1998 image is in the public domain.

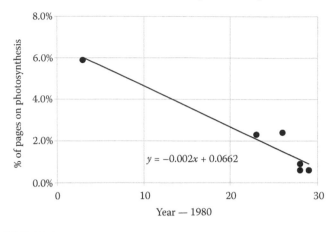

FIGURE 10.1
Coverage of photosynthesis in six well-known "biophysics" and "biochemistry" books as a percent of pages. The horizontal axis is the year since 1980.

solar insolation and photosynthetic activity on these maps. We will not pursue issues of rainfall- and pollution-related photosynthetic activity in this text, but the maps of Figures 10.2 and 10.3, as well as comparable maps of rainfall and other quantities, provides a basis for such investigations.

The global distribution of human solar energy use is presently an entirely different matter than photosynthetic activity. If human energy-requiring activity mirrored activity of organisms as a whole, solar power maps would replicate Figure 10.2 or perhaps Figure 10.3. Human energy use does not follow photosynthesis patterns. Maps of solar power generation activity are crude in comparison to maps of photosynthetic activity and solar insolation, though the International Energy Agency tracks installed photovoltaic generation by country (Table 10.1). The primary centers of solar cell manufacturing are now in China, Taiwan, Japan, Malaysia, Germany, and the United States. A collaboration between German and Italian designers installed a "solar tree" on a street in Vienna, Austria, an area of frequent cloud cover. The "tree" successfully provided lighting for the entire month of October 2008. This tree can be viewed at http://www.digitaljournal.com/article/248029.

10.2 OVERALL PROCESS

Photosynthesis in a nutshell can be summarized from several different points of view. From a biochemical reaction view,

$$CO_2 + H_2O \xrightarrow{\text{light}} (CH_2O) + O_2 \tag{10.1}$$

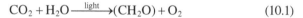

CO_2 is "fixed" and reduced, while H_2O is oxidized to yield carbohydrates and molecular oxygen. (Physics students note that there are fewer O atoms and more H atoms attached to the carbon on the right; it is thus reduced. Note also that "CH_2O" is not an independent molecule but rather an additional backbone unit in a sugar molecule.) Carbohydrates and O_2 provide two of the main necessities for animal life on Earth. Oxidative carbohydrate metabolism in essence is the reverse of this photosynthetic chemical reaction. The free energy of the molecular forms on the right is higher than that on the left,

FIGURE 10.2
Global distribution of photosynthetic activity in terms of the magnitude and distribution of global primary chlorophyll production, oceanic (mg/m³ chlorophyll a) and terrestrial (normalized difference land vegetation index). Provided by the SeaWiFS Project, NASA/Goddard Space Flight Center and ORBIMAGE; from en:Image:Seawifs global biosphere. jpg; SeaWiFS Global Biosphere September 1997–August 1998. Animations and updated (2007) images may be viewed at oceancolor.gsfc.nasa.gov/SeaWiFS/HTML/SeaWiFS. BiosphereAnimation.70W.html, www.nasa.gov/vision/earth/lookingatearth/seawifs_10th_feature.html, and oceancolor. gsfc.nasa.gov/SeaWiFS/Sites accessed November 5, 2017. The SeaWiFS project ended in 2011.)

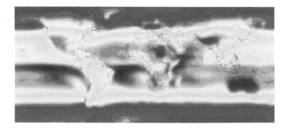

FIGURE 10.3
Global map of annual solar insolation, the energy received at ground level across the earth, averaged over 1993–2005. (Courtesy of NASA, https://asdc-arcgis.larc.nasa.gov/sse/. Individual maps for specific areas and periods of time can be customer-produced at this Web site, See also https://neo.sci. gsfc.nasa.gov/view.php?datasetId=CERES_NETFLUX_M.)

courtesy of the photon energy. Some important intermediate forms of energy and charge (H) storage not shown in Equation 10.1 are photogenerated ADP→ATP and NADP⁺ → NADPH.

Another legitimate, partial view of photosynthesis is the photon-assisted *splitting of a water* molecule to form H_2 and O_2:

$$2H_2O \xrightarrow{\text{light}} 2H_2 + O_2 \qquad (10.2)$$

Hydrogen gas is not, of course, one of the products of natural photosynthesis, though continuing research efforts are underway to modify photosynthesis toward this end. Equation 10.2 correctly implies that the oxygen atoms in the O_2 come from H_2O, and not from CO_2 in Equation 10.1. The source of photosynthetic O_2, H_2O, not CO_2 was demonstrated in 1941, after isotopic ^{18}O became available from nuclear research developments.

From another point of view, photosynthesis amounts to the conversion of photon energy into several other forms, as shown in Figure 10.4. The first conversion is to singlet electronic excitation energy, as discussed in Chapter 9. This singlet excitation energy is rapidly transferred to other nearby molecules and finally converted into the electric energy of charge separation, and ultimately to chemical energy. While the quantification of energy is reasonably clear in the cases of photon and molecular excitation energy, $hf = E_2 - E_1$, the identification and computation of the energy of charge separation and molecular *free* energy is complex. There are two main complicating issues for the chemical energy: (1) the concentration-dependent entropy is part of the free energy, and (2) the extractable energy depends on what states (chemical compounds) are accessible. Though it may be theoretically possible to convert compound X to compound Y, if that conversion is not possible under prevailing biological conditions, that energy is not extractable from the compound. A simple example is the chemical free energy of cellulose. While the cattle and termites may exist quite comfortably on the energy in cellulose, humans cannot, because systems are not in place to degrade the cellulose to a lower-energy form.

After a brief summary of the overall process of conversion of light energy to chemical (some "facts to remember"), we focus in this chapter on the physics of the early energy-conversion processes of photosynthesis: photon absorption, energy transfer, and charge separation. These are parts of the "light reactions" of photosynthesis, because they are directly driven by photon absorption. These light reactions are, surprisingly, better understood in bacterial photosynthetic systems, so we spend most of our time on them, neglecting the green tree leaves suggestively posed at the beginning of this book and chapter. Green leaves, after all, are more pleasing to the eye than brownish pond scum.

TABLE 10.1	Installed Photovoltaic Power Capacity, 2016	
	Country	**PV Capacity**
1	China	78.1
2	Japan	42.8
3	Germany	41.2
4	USA	40.3
5	Italy	19.3
6	UK	11.6
7	India	9
8	France	7.1
9	Australia	5.9
10	Spain	5.5

Source: Report IEA PVPS T1-31:2017, IEA International Energy Agency, www.iea-pvps.org/fileadmin/dam/public/report/statistics/IEA-PVPS_-_A_Snapshot_of_Global_PV_-_1992-2016__1_.pdf.

10.2.1 Finding Cellular and Subcellular Structures

Several links to useful photosynthesis Web sites will be cited in this chapter. Wikipedia, in which several photosynthesis articles have been cowritten by Govindjee (University of Illinois at Urbana–Champaign) and Larry Orr (Arizona State University) also has credible discussions and links. You already know about the Protein Data Bank (PDB) and Electron Microscopy Data Bank (EMDB) and can likely search the Web for several others. One you may want to use for general cellular structure searches is the Cell Centered Database (National Center for Microscopy and Imaging Research), at http://ccdb.ucsd.edu/sand/jsp/lite_search.jsp).

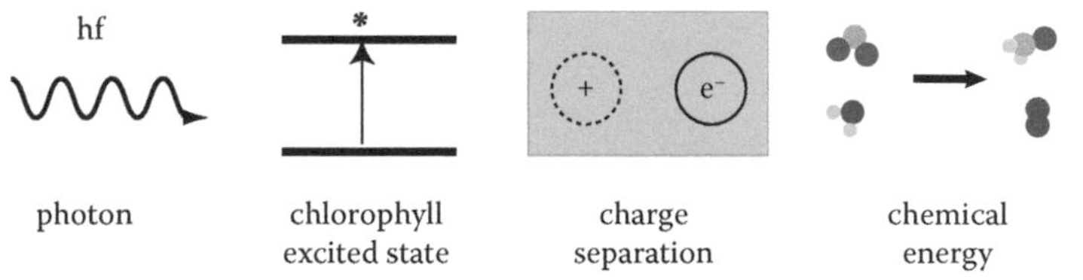

FIGURE 10.4

Photosynthesis as energy conversion, from photon energy to chemical energy. The symbolic "CH_2O" produced is part of the carbon backbone of a sugar. The chemical energy is in the form of atoms redistributed to higher free-energy forms, including charge separation.

10.2.2 Light and Dark Reactions: Products and Yields

The observation that CO_2 does not provide the oxygen given off as O_2 suggests, though does not prove, that there may be two parts to photosynthesis. There are indeed:

- The light reactions, which use photon energy to produce ATP (chemical energy) and NADPH (chemical energy in the form of reducing [electron-transferring] power)
- The dark reactions, which use ATP and NADPH to generate carbohydrates

The light reactions are the main focus of this chapter. The dark reactions can be summarized by Equation 10.3, which shows the involvement of two other major products of photosynthesis, ATP and NADPH, in the dark reactions:

$$6CO_2 + 12H_2O + 18ATP + 12NADPH \xrightarrow{\text{dark}}$$
$$C_6H_{12}O_6 + 18ADP + 18P_i + 12NADP^+ + 12H^+ + 6O_2 \tag{10.3}$$

where the subscript i stands for inorganic, not an integer index. As you may guess, the mechanism of this overall reaction is complex, though fairly well understood. The book by Bacon Ke describes the systems and reaction mechanisms behind Equation 10.3.[4]

The quantum model of light and matter accurately explains the interaction of electromagnetic waves with atoms and molecules. In the quantum model, atoms, electrons, protons, and photons occur as distinct entities and can be counted. It is natural, then, to ask how many photons are needed to produce one O_2 molecule or one transferred electron in the form of NADPH. It is natural to think that one chlorophyll excited state is caused by absorption of one photon; perhaps also that one transferred electron results from one chlorophyll excited state. We must be careful here, however, for two reasons. While it may be true that each transferred electron results from a chlorophyll excited state and that each excited state results from an absorbed photon, not every chlorophyll excited state may move an electron. As we saw in Chapter 9, excited states have several ways in which they can dispose of their energy. Emission of a fluorescent photon from the excited state will divert energy from any electron transfer process; so will some of the nonradiative de-excitation processes. As we will see, the chlorophyll that initially absorbs the photon has to transfer the energy to several (or many) other chlorophylls in order for eventual electron transfer to take place. We will also see, at the end of Section 10.3, that natural photoprotective systems sometimes intervene to divert and convert photon energy to heat. The scheme of energy relay by many molecules to the *reaction center*, where charge separation takes place, implies that probability rules. Out of 1,000 absorbed photons, an integral number results in charge transfer but that integer may not correspond exactly to a small, integral number of photons per electron transferred, or to exactly 8 photons absorbed per O_2 molecule generated. Second, we must be aware that the production of any form of chemical energy like ATP in a cell involves a *free* energy that depends on concentrations. ATP "fuel" is not like the gasoline we burn in our car engines, which produces the same combustion energy per liter whether we put 2 or 200 L in the tank, whether we have recently been burning fuel or not. If ATP happens to be generated when the prevailing ADP and ATP concentrations are near equilibrium, the free energy available from the new ATP will be less than if a higher ratio of ATP/ADP prevails. The same principle holds for separated charge, if that charge produces a change to a voltage differential across a membrane, where the energy may be used for a variety of purposes and the prevailing voltage may vary. Finally, available free energy depends on temperature, through the $T\Delta S$ term.[5] This complex dependence of energetics on existing conditions would make quantification of yields of product per unit energy input seem impossible, except for the fact that life ensures that prevailing concentrations are usually held within narrow ranges. Likewise, though ambient temperature variation causes free energies to change, photosynthesis mostly takes place when the temperature is in the range 10°C–40°C, or 283–310 K—not a wide range of $T\Delta S$. Given this approximately constant physiological environment and the fact that one NADPH carries the energy equivalent of about three ATPs, it has been estimated that *one absorbed photon results in about 1.25 ATP generated energy equivalents.*[6]

Normal intracellular concentrations of ATP and ADP are about 4 and 0.013 mM, respectively, a ATP/ADP ratio higher than at equilibrium. The standard free energy of the reaction ATP→ADP + P_i is −30.5 kJ/mol.

TABLE 10.2 Quantities Associated with Photosynthetic Production of One O_2 Molecule

Quantity	Number	Notes
Photons absorbed	~8–10	Not all photons are involved in every process
Chlorophylls required	~2400	
Chlorophylls per photon absorbed	~2400/8 = 300	Interpreted as the number of chlorophylls absorbing photons for one "photosynthetic unit"
H_2O split	2	
O_2 produced	1	
ADP→ATP	~4	
$NADP^+ \rightarrow$ NADPH	~2	
CO_2 consumed	1	Part of the "dark reaction"
(CH_2O) produced	1	Part of the "dark reaction"

If the reaction scheme in Equation 10.1 is correct, one O_2 molecule and one CH_2O sugar unit are produced for every CO_2 molecule consumed, but the number of photons needed is unknown. In 1932, Robert Emerson and William Arnold measured the photosynthetic output of O_2 in *Chlorella* green algae when flashed with red light that was absorbed by the algae's chlorophyll.[7] This was the beginning of accurate measurements of yields of products, as well as reaction mechanisms in green plant, green algae, and cyanobacterial photosynthesis. The results of these studies are summarized in Table 10.2. Note we are also not distinguishing between green plant and bacterial systems.

Emerson and Arnold proposed that about 300 chlorophylls were associated with one green-plant "photo-synthetic unit" (PSU) and that most of these chlorophylls had the job of absorbing photons and making their energy available to the photosynthetic reaction center, where catalysis occurred. We should note that convincing evidence for this arrangement of 300 chlorophylls per reaction center awaited the PhD thesis of Louis Duysens,[8] who demonstrated energy transfer between bacteriochlorophylls in purple bacteria using a differential spectrophotometer he built in the lab.[9] (Duysens estimated 200 bacteriochlorophylls associated with the single reaction center in purple bacteria.)

10.2.3 Time Sequence of Photosynthesis[10]

The initial absorption of light takes a time of about one electromagnetic wave cycle: 10^{-15} s. Growth of a green plant takes a much longer time, of course. Figure 10.5 shows the distribution of various processes related to photosynthesis on a time scale from a light cycle to days. Electronic excitation and transfer of both excitation energy and electrons are rapid because they involve motions of electrons within their potential wells: particles of small mass that do not need to move far. Governing physical mechanisms are primarily quantum. These processes are enclosed in the dashed box in the figure. For the most part, processes to the right of the dashed box demand diffusion and directed transport of atoms and molecules, as well as cell division and growth—classical (non-quantum) processes that take place over distances of microns to meters and require much more time. Photochemical and electron transport processes can span both regions, depending on whether molecular diffusive processes are included. For example, the primary charge transfer reaction in reaction centers takes place within a ps after photon absorption, though the replenishment of the "pool" of charge available for transfer requires μs to ms.

Figure 10.5 shows the rough time scale of photosynthetic processes in the horizontal direction, but the vertical direction indicates the energy, at least inside the dashed box. For example, Chl*, chlorophyll in the excited state after photon absorption, has more energy than Chl. Likewise, the two electrons from split H_2O and available for transfer, gain energy from the de-excitation of Chl* → Chl so that they can be added to $NADP^+$ to form NADPH.

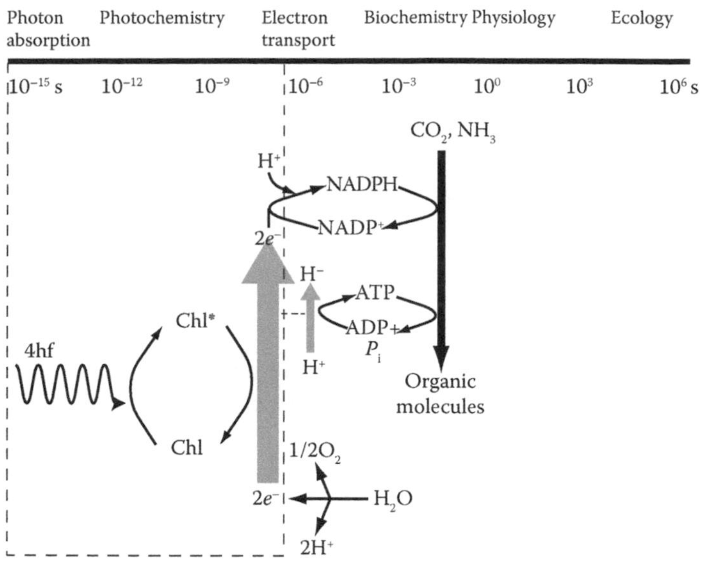

FIGURE 10.5

Time scales of major events of photosynthesis, from 10^{-15} to 10^6 s. The large arrow near the center shows the chlorophyll excitation energy transport of two electrons. We concentrate on the processes inside the dashed box. (Based diagrams in Mohr, H., Schopfer, P., and Ke, B., 36.)

10.2.4 "Z Scheme"

The Z scheme was proposed by Hill and Bendall in 1960 to explain the quantum yield of products of photosynthesis in green plants in terms of absorbed photons. As we learned in Chapter 9, not all photons are the same in terms of energy or absorption by a molecule of interest. During World War II, Robert Emerson and Charlton Lewis measured the wavelength-dependent ability of photons to cause O_2 generation in a (eukaryotic) green algae, *Chlorella pyrenoidosa*. They found that the yield of O_2 per photon absorbed remained almost constant from 660 to 685 nm, then dropped off sharply at longer wavelengths, similar to data shown in the lower curve in Figure 10.6a. This was called the *red drop* in O_2 yield. The red drop was not understood because chlorophyll *a* absorbs beyond 685 nm and the available energy from this excitation was thought to be the same as for 660-nm excitation. These excitations are within the first excited state, where rapid vibrational relaxation to the lowest vibrational level should make 660-nm excitation indistinguishable from 685 nm after 10^{-12} s, as in Figure 9.14.

In 1957, Emerson, Chalmers, and Cederstrand[11] made a small change in measurement procedure and found a surprising difference. When a supplementary light of fixed, 650-nm wavelength was added to the variable wavelength of excitation, the yield of O_2 was not only higher but remained high from 685 to 710 nm. This was some of the first data to suggest that there were two photosystems, now called *photosystems I and II*, responsible for photosynthetic generation of O_2. These two photosystems must be in series, as will be demonstrated in the homework. When a mechanistic model for the coupling of these two photosystems was made, the result was the *Z scheme*, shown in Figure 10.6b. These Z-scheme processes are called the *non-cyclic* light reaction. (We ignore the "cyclic" process, where photosystem I returns some electrons to the cytochrome b_6f enzyme.)

10.2.5 Absorption Spectra of Photosynthetic Pigments

The preceding discussion of red drop reminds us that we should examine the absorption spectra of the molecules that absorb light in photosynthetic systems. Though we will shortly focus on bacterial systems, where spectra are farther to the red and near infrared, we need to put the historic measurements on eukaryotic photosynthetic systems in the context of their photon-absorbing energy levels.

In Figure 10.6a, note that the dependence of the absorption coefficient versus wavelength has already been taken into account: the yield is per *absorbed* photon, not per incident photon.

See also an updated (2017) Z scheme at www.life.illinois.edu/govindjee/publication/Z-scheme_final%20(2017-09-09).pdf.

Figure 10.7 shows the absorption spectra of the major components of green plants. Chlorophylls *a* and *b* are the most recognizable and are responsible for the green color of plants and algae. The chlorophylls have two major absorption bands, 420–440 nm and 640–660 nm. Low absorption in the 470–620 nm region would leave a major fraction of solar power untapped were it not for the phycocyanins and phycoerythrins, accessory photosynthetic pigments. We will shortly see that such accessory pigments are closely integrated with the chlorophylls in larges complexes of chlorophyll proteins and that excitation energy of carotenoids, phycocyanins, and phycoerythrins can be (but is not always) transferred to chlorophylls. Note that the longest-wavelength bands of the chlorophylls are longer (lower energy) than those of the accessory pigments, so transfer of excitation energy can be efficient (see Section 9.9).

Where are the absorption bands of P680 and P700 in the spectra of Figure 10.7? If the Z scheme has any hope of being correct, shouldn't we be able to identify 680- and 700-nm peaks? The answer is "Yes," and the reason we cannot see such peaks reveals another important and, in retrospect, expected feature of the photosynthetic apparatus. P680 and P700 were identified as chlorophyll-containing "reaction centers" of a photosynthetic unit (PSU) containing 250–300 chlorophylls. If the reaction center chlorophylls comprised only one or two of the chlorophylls, their absorbance would be extremely small. The large majority of the chlorophylls, the ones that produce the green in leaves, are "light-harvesting" or "antenna" chlorophylls—pigments that absorb light and transfer energy to the reaction centers. (The other fact is that Figure 10.7 was assembled from spectra of isolated pigments; it is not the absorption spectrum of an intact leaf or algae suspension.) Such samples indeed show apparent absorption in the 680–700 nm region, but scattered light dominates the absorption. The only way to eliminate such scattering is to disintegrate the cells, which, unfortunately, also destroys reaction center complexes.

We now have the background spectral information to help us understand the red drop and the Z scheme. Let's briefly sketch the Z-scheme's operation before we concentrate on the light-harvesting process.

In plants and green algae, the light-dependent photosynthetic reactions occur in the thylakoid membranes of the chloroplasts. The primary products of these reactions are ATP and NADPH. ATP, of course, is the common energy currency of life: its stored chemical free energy can be used for almost any energy-requiring reaction in the cell. NADPH also represents stored energy but in a form more specific to electrical processes—the addition of an electron (usually with an accompanying proton to ensure charge neutrality) to another molecule or the replenishment of charge that may have been pumped across a membrane. The photons are absorbed by light-harvesting antenna protein complexes of photosystem II by chlorophyll and accessory pigments. When the reaction center chlorophyll (actually, a chlorophyll dimer) at the core of the photosystem II unit accepts this excitation energy, the center transfers an electron to an electron-accepting molecule, pheophytin. The electron is then shuttled through the Z scheme's first electron transport chain, which can either generate a

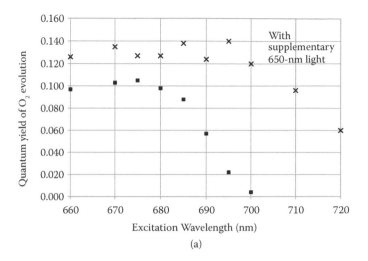

(a)

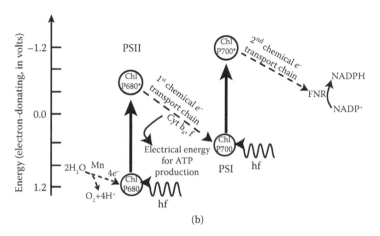

(b)

FIGURE 10.6

(a) Quantum yield for O_2 evolution from *Chlorella* algae as a function of excitation wavelength from 660 to 720 nm, with and without supplementary excitation light of wavelength 650 nm. The yield is higher and remains higher at excitation wavelengths beyond 680 nm when a second light of wavelength 650 nm is used. The hypotheses of two photosystems and the Z scheme are based on this data. (Graph created based on Emerson, R. et al., *Proc. Natl. Acad. Sci. U. S. A.*, 43, 137, 1957.) (b) The Z scheme. The vertical axis reflects energy in terms of electron-donating power. More negative indicates more donating power. The energy can be thought of in eV, but that energy reflects the free energy of an equilibrium process, not a fixed packet of energy, like a photon. PS-I, photosystem I; PS-II, photosystem II; Chl, chlorophyll; P680/P700, chlorophyll-containing reaction center, where photon energy separates charge; Cyt b_6, f: one of the primary recipients of the electron donated from PS-II. Many chemical intermediates along the diagonal lines are omitted, for clarity. For more details, see http://www.life.illinois.edu/govindjee/photoweb/subjects.html#Light-reactions.

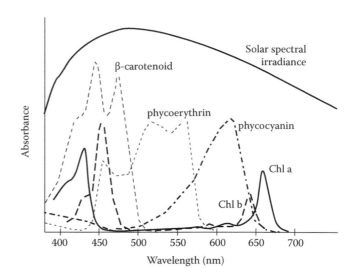

FIGURE 10.7

Absorption spectra of pigments of green plants and algae. Note that the chlorophylls have two main absorption bands, 420–440 and 640–660 nm. Relative amounts of pigments vary from plant to plant and season to season. Several types of carotenoids, phycoerythrins, and phycocyanins also exist. (Based on spectral drawings of Ke, B., Photosynthesis: An overview, in *Photosynthesis: Photochemistry and Photobiophysics*, series Ed., Govindjee, Kluwer Academic, New York, p. 13, 2001.)

chemiosmotic potential across the thylakoid membrane or transfer to the PS-I photosystem. Note that in this electron transport chain are no less than six electron acceptors, one of which is actually a pool of reduced molecules that can be used as needed. Besides shuttling the electrons to the correct places, these multiple acceptors act as a buffered pool for stored energy and reducing power. An ATP synthase enzyme uses the membrane potential energy obtained from this pool of stored energy to make ATP. Some electrons, however, enter the PS-I system. The electron gains energy from light absorbed by the P700 chlorophyll protein of PS-I. A second electron transport chain accepts the electron. The electron-donating energy carried by the electron acceptors is used to move H+ ions across the thylakoid membrane into the lumen, the aqueous phase inside the thylakoid (see Figures 7.5, 7.7, and 7.12), as well as to reduce NADP.

> The rumor is that this large number of electron acceptors was invoked by biochemists to form a pool that physicists dare not enter.

Numerous cartoons of the believed spatial structure of the components of the Z scheme have been published. Figure 10.8 shows one of these that applies to green plant photosynthesis. Note that the majority of the heavy machinery of photosynthesis resides in the thylakoid membrane. The two photosystems are identified, as well as several other protein complexes: cytochrome b6f, ATP synthase, and ferredoxin/ferredoxin-NADP reductase. Though the Z scheme of Figure 10.6 does not require this sort of structure, note that the downward pointing first electron transport chain of 10.6b corresponds to the passage of the electron from outside (chloroplast stroma) to inside (thylakoid lumen) the thylakoid membrane. Because membranes are low dielectric constant layers, this physical structure guarantees that membrane electrical potential differences play a major role in photosynthesis and that the enzyme machine, ATP synthase, a rotary molecular motor, can be powered electrically. The ATP synthase is, in fact, one of the classic examples of biomolecular motors. We will see it again.

The Z scheme is also not absolute. The first exception was published in 1995 by Elias Greenbaum, whose group showed that a mutant of *Chlamydomonas reinhardtii*, deficient in photosystem I, is still capable of photosynthesis.[12] Before this result, most believed that the conversion of light energy to chemical energy required both PS-I and PS-II reaction centers. The editors of the journal *Nature Biotechnology* stated, "The significance of this discovery is that, in principle, one should be able to double the maximum theoretical conversion rate of light to chemical energy from 10% to 20%. Since these mutant algae are capable of synthesizing hydrogen without the participation of CO2 fixation, Greenbaum's group has proposed that they might be used to create a commercially viable hydrogen-production method. At present, the coproduction of oxygen with the hydrogen obviously means that an additional scheme needs to be developed to separate these potentially combustive gases."[13] As many years have passed since this result and commercial H2 fuel is still a decade or more away, we should not underestimate the practical, technical, financial, and political issues restraining these discoveries. See also Section 10.6.

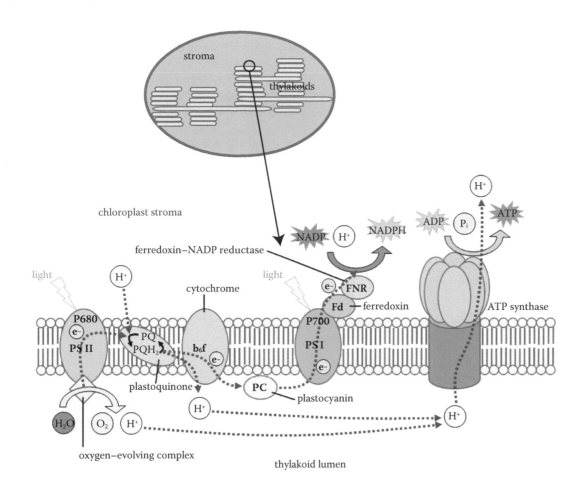

FIGURE 10.8
Physical model of the thylakoid membrane, containing the two photosystems, the ATP synthase protein complex, and charge-transferring enzymes. There is evident motion of charge from one side of the membrane to the other, implying involvement of electrical forces. The lumen is the inside of the thylakoid. Many of these structures are now known from electron microscopy and X-ray crystallography. (Modified from Somepics, upload.wikimedia.org/wikipedia/commons/4/49/Thylakoid_membrane_3.svg, license creativecommons.org/licenses/by-sa/4.0/deed.)

EXERCISE 10.1

How much difference does one elementary charge, pumped into the lumen of a thylakoid, make? Calculate the change in pH that would be produced when one H^+ is pumped into the lumen.

Answer

$pH = -\log_{10}([H_0^+] + [\delta H^+])$, where H_0^+ is the initial H^+ concentration (let's assume 10^{-7} M) and δH^+ is the concentration increase due to one generated H^+. If we want to convert one H^+ into a concentration, we need to know the volume. We could go on the Web and get an average green plant volume, but let's use what we have from Figure 7.10. The thylakoid there is revealed as a very thin structure. We should even expect that one H^+ ion, transported across the membrane at a particular point, would not be able to diffuse throughout the interior lumen for some time. The distance across Figure 7.10b is 250 nm, so we estimate the thickness of the lumen space to be about 35 nm. Let's assume that the length and width of the lumen over which the H^+ can disperse in a "reasonable" time is about 500 nm. The volume is then

$$V = 35 \times 500 \times 500 \text{ nm}^3 = 8.8 \times 10^{-21} \text{ m}^3 = 8.8 \times 10^{-18} \text{ L}$$

Because it is a common mistake, we will first do an *incorrect* calculation. We assert $\Delta pH = -\log[\Delta H^+]$. The proton concentration change is $\frac{1 \text{ mole}}{6.02 \times 10^{23} \cdot 8.8 \times 10^{-18} \text{ L} = 1.89 \times 10^{-7} \text{ M}}$, so $\Delta pH = -\log(1.89 \times 10^{-7} \text{ M}) = 6.72$. Seems a bit large. Perhaps we should subtract 7.0, so that $\Delta pH = -0.28$.

This type of thought suggests something is wrong, but what? Recall the definition of pH: $pH \equiv -\log[H^+]$. We cannot stick in the symbol Δ whenever we choose to. The correct computation is

$$pH_f = -\log\left(\left[H_0^+\right] + \left[\Delta H^+\right]\right) = -\log\left(\left[H_0^+\right]\left(1 + \frac{\left[\Delta H^+\right]}{\left[H_0^+\right]}\right)\right) = pH_0 - \log\left(1 + \frac{\left[\Delta H^+\right]}{\left[H_0^+\right]}\right), \text{ so}$$

$$\Delta pH = pH_f - pH_0 = -\log\left(1 + \frac{\left[\Delta H^+\right]}{\left[H_0^+\right]}\right)$$

$$= -\log\left(1 + \frac{1.89 \times 10^{-7} \text{M}}{10^{-7} \text{M}}\right) = -0.46$$

We have learned an important lesson: one transported charge can make a large difference in a small space. Even if we underestimated the volume by a factor of 10, movement of only a few protons is needed to change the pH by about 0.1 pH unit or more. Problem 10.3 examines the small-change regime, where ΔpH depends linearly on transferred charge.

Exercise 10.1 tells us that a single transported charge can make an appreciable difference in the small volume inside a thylakoid membrane. Problem 10.4 extends our understanding a bit further by estimating how many times per second light will produce such charge separation. Remember that our goal here is not to find the exact number and remember it. We could search the literature or the Web to find the "correct" numbers for many of these quantities, but having Google or Alexa tell us the answer does little for our understanding of how things work.

10.3 STRUCTURAL ORGANIZATION OF PHOTOSYNTHETIC UNITS

Most photosynthetic organisms that have access to oxygen have two photosystems, PS-I and PS-II, described earlier in this chapter, and connected by the Z scheme. Green plants and cyanobacteria are examples of such organisms. PS-II is (usually) where the overall photosynthetic process starts and is also the site of the water-splitting reaction. Both of these photosystems require input of photon energy, so both have their own antenna systems. We should expect similarities between the organizations of the antenna systems of both photosystems. A few photosynthetic microorganisms live where oxygen is absent or scarce, but H_2S is available. Some developed with a single reaction center. *Rhodopseudomonas acidophila*, *Rhodospirrilum molischianum*, and *Rhodobacter sphaeroides* are examples of these whose reaction centers and antenna systems have been studied successfully with X-ray crystallography. Schematic diagrams of three types of photo-systems, green plant, cyanobacteria, purple bacteria, and green bacteria are shown in Figure 10.9. Green plants seem to have the least orderly organization of the antenna systems. Antenna systems for both PS-I and PS-II are evident, but that for PS-II is apparently able to associate with PS-I under certain conditions. The cyanobacteria also have two photosystems, but their antennae are primarily associated with a *phycobilisome*, a complex of protein and phycoerythrin and/or phycocyanin, that is exposed to the solvent. These pigments are used rather than chlorophyll because the ecological niche of these organisms is located where chlorophyll-absorbed wavelengths have largely been removed from the sunlight. The purple bacteria have one reaction center (RC), but two types of light-harvesting complexes, LH-I and LH-II. LH-I associates with the RC, while LH-II forms more mobile arrays that transfer energy to LH-I and then to RC. We will consider this system in more detail. The green bacteria also have one RC but a more elaborate dual antenna system. A *chlorosome*, a liposome attached to the cell membrane through a baseplate that contains bacteriochlorophyll *a* (BChl *a*), contains rodlike arrays of BChl *c* that function as primary antenna. BChl *c* transfers energy to the baseplate and intramembrane BChl *a*, which then transfer to the RC.

These physical structures represent working systems in successful photosynthetic organisms found in various ecological niches in virtually every location around the Earth. These four are undoubtedly not the only PSU types that exist, but they exhibit variations on the theme of two arrays of light-harvesting, pigment-containing proteins surrounding a reaction center that contains only a few absorbing pigments. We will shortly see that the key component of these reaction centers is a (bacterio) chlorophyll dimer, with other nearby *accessory pigments*. A key property of these antenna arrays is precise positioning of the large, electronically active chlorophyll, carotenoid, phycocyanin, and other light-absorbing pigments for

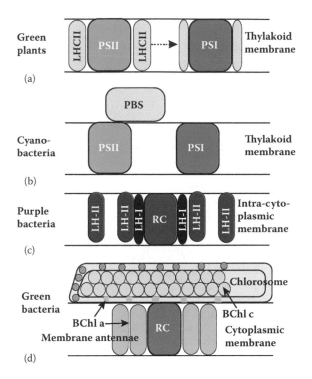

FIGURE 10.9

Four different organizations of photosynthetic units (PSUs). (a) Green plants have two photosystems (PSs) with two antenna systems, though one can apparently associate with either PS. PS-I and PS-II each contain a reaction center, where charge-separation takes place. (b) Cyanobacteria have two PSs, but the primary antenna is the phycobilisome (PBS), a protein with absorbers phycoerythrin and/or phycocyanin. (c) Purple bacteria have a single reaction, with two types of light-harvesting proteins bound in the same intracytoplasmic membrane (the inner cell membrane [see Chapter 7]). (d) Green bacteria have two types of light-harvesting complexes, with BChl *a* and BChl *c*, respectively. BChl *c* is contained in rodlike structures inside a membrane-enclosed chlorosome, which pass excitation energy to BChl *a* in the baseplate and cytoplasmic membrane-bound antennae. (Based on Figure 10.9 of Hu, X. et al., 37.)

energy transfer and charge separation. We have seen in Chapter 9 that energy transfer via the Förster mechanism depends sensitively on the distance between the transition dipoles. The primary tool nature has to position any elements precisely with respect to each other, minimizing random Brownian motion, is to bind the elements to much larger proteins and imbed them in membranes. The larger protein moves less because of its large mass and because attachment of the protein to a membrane associates the protein with the even larger mass of the membrane structure.

From a purely strategic point of view, a photosynthetic organism should be able to adjust to high and low levels of light. There may be times during which sunlight is too intense and would cause too many photon absorptions and too many energy transfers to reaction centers. Because the primary elements that absorb (harvest) the photons are the antenna arrays, it is essential that the organism have an ability to adjust the proximity of antenna pigments to the reaction center. All of the PSU designs in Figure 10.9 incorporate antenna adjustment. In green plants, LHC-II can move; in purple bacteria, LH-II can move: organisms that include two photosystems also retain the ability to control even the relative excitation of the two photosystems.

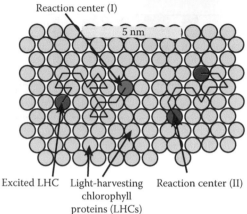

FIGURE 10.10

Structural extremes of the photosynthetic antenna system, from 5 cm to 5 nm. On a 5-cm scale, the antenna turns its surface toward the sun over a period of minutes to hours. On a 5-nm scale, the excited state resulting from an absorbed photon is transported to a reaction center on a time scale of picoseconds to nanoseconds. The red circle indicates an LHC initially excited by a photon; light-green circles other LHCs, and dark circles the reaction centers. Green plants and cyanobacteria generally have two reaction centers, with somewhat different sets of LHC molecules.

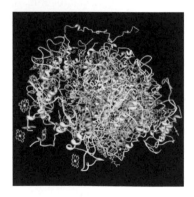

FIGURE 10.11

PS-I light-harvesting "supercomplex" from pea plants contains 17 protein subunits, 168 chlorophylls, 2 phylloquinones, 3 [4Fe-4S] clusters, and 5 carotenoid pigments. Several chlorophylls appear at the periphery of the complex, but the large majority permeate the structure. Fe-S clusters and quinones are also almost universally associated with electron transfer. The molecular mass of this complex is 504, 252 Da. (PDB 2001, by Amunts, A. et al., *Nature*, 447, 58–63, 2007; displayed using Protein Workshop software.) (From Jmol: An open-source Java viewer for chemical structures in 3D. http://www.jmol.org/; Protein Workshop: Moreland, J.L. et al., *BMC Bioinformatics* 6, 21, 2005; NGL Viewer: Rose, A.S. et al., *Bioinformatics*, 2018.)

Two other discoveries involving photosynthetic system adjustment to changing light levels should be mentioned. First, a primary role of carotenoids is believed to be photoprotection from high light levels, which can produce potentially damaging amounts of triplet excitations in the chlorophyll arrays. Yellow-colored carotenoid xanthophylls assist in photoprotection of green leaves. The plant and light intensity control enzymatic interconversion of two major xanthophyll forms, violaxanthin and zeaxanthin. Zeaxanthin's triplet quenching protects lipids and the light-harvesting antennae (next section), and triggers reorganization of PSII antenna pigment-proteins. At high light levels, zeaxanthin production from violaxanthin is stimulated, reducing levels of triplets and excitations that reach the reaction centers, thus protecting the plant. Kromdijk et al. found a potentially important agricultural application of this xanthophyll cycle. The green plant response to changing light levels does not quite match the rate at which variable cloud cover typically changes the light intensity reaching plants in farm areas.[14] See Figure 10.10. After production of higher levels of zeaxanthin and array reorganization to suppress high-light photodamage, plant leaves needed minutes to hours to readjust to reduced light caused by cloud cover. During this readjustment period, photosynthetic activity is corresponding reduced. Kromdijk bioengineered an accelerated interconversion and found that tobacco plant biomass production could be increased by about 15% under real field conditions. Though engineered plants would have to be produced in high quantity for this to be economically significant, the potential crop yield increase may justify the effort. The reader should notice the very biochemical/molecular-biological "flavor" of Figure 10.10: enzymes and genetic engineering are the focus; structures and mechanisms are not—a challenge for the next cadre of biophysicists. An issue deserving some thought relates to the genetic approach to this bioengineering of photosynthesis. If the new strain of crops survived, the modification would be a permanent addition to the (local) ecosystem, with its possible benefits and dangers. See miniproject 10.10. A final note on xanthophylls—readers with sensitive or aging eyes may have noticed zeaxanthin featured prominently on the labels of eye-vitamin supplements, where it purportedly supports eye health in humans.

The second recent discovery related to photoprotection involves energy dissipation in cyanobacteria (Section 10.5). Leverenz et al. determined the atomic-resolution crystal structure of the orange carotenoid protein (OCP), which senses light and effects photoprotective energy dissipation. OCP consists of protein and a noncovalently bound carotenoid pigment (Figure 10.11a). Leverenz found that OCP photoactivation, causing nonphotochemical quenching (NPQ) is accompanied by a 12Å movement of the pigment within the protein and a reconfiguration of carotenoid-protein interactions, which affects the interaction with light-harvesting array, the phycobilisome, Figure 10.11b.[15] A 12-Å movement should be considered quite large in the context of protein structure. Activation of the OCP occurs when its dark (orange) state, absorbs blue light and forms the quenching active (red) state. The red state binds to the phycobilisome and initiates NPQ protection. The key biophysical mechanism in this process is the large change in protein-bound carotenoid structure triggered by light absorption. As we noted in the previous chapter, light absorption causes electron redistribution in the absorber, the carotenoid, but this initially small, localized movement of charge triggers large, biologically essential changes that adapt the cyanobacterium to a changing environment.

10.4 LIGHT-HARVESTING (ANTENNA) PROTEINS: ARRAYS OF ABSORBERS

The large majority of chlorophylls and other pigments associated with photosynthetic chloroplasts, cyanobacteria, and other photosynthetic organisms have the job of capturing photons. There are 200–300 chlorophylls associated with a single photosynthetic unit (PS-I + PS-II), though numbers vary by organism. PS-I of green pea plants has 168 chlorophylls,[16] suggesting that in this organism chlorophylls are roughly equally distributed between the two photosystems. Figure 10.9 shows us how these light-harvesting and reaction center chlorophylls are organized in a general way; now we want a few more details. Rather than attempt to put together an accurate picture of all the photosystems in all the organisms—a challenging task that is still being investigated—we will examine a few better characterized systems.

Chlorophylls, bacteriochlorophylls, carotenoids, phycoerythrins, phycocyanins, and other light-absorbing pigments in photosynthetic systems are almost universally bound to proteins. We asserted that one reason for this is to ensure precision orientations of excitation donors and acceptors. A second reason for protein-pigment complexes is chemical stability. Chlorophyll and the other light-absorbing pigments are unstable and would quickly degrade, if they existed as isolated molecules in solution or in a membrane. Most of these pigments are not soluble in water, so we correctly expect them to be located in the membrane. (Phycocyanins are exceptions.) Even the membrane would not provide an environment sufficiently friendly to stabilize the pigment structure from solar photochemistry and oxidation, however. We thus find the photosynthetic pigments bound to specific sites in proteins. A protein-bound pigment has an added geometric advantage: when placed in a membrane, proteins cannot move very quickly like smaller dye molecules can. Precise positioning, maintaining precise positioning over time, but retaining the ability to adjust positioning when conditions dictate, are all critical requirements if an antenna pigment is to transfer excitation energy to another molecule efficiently.

Figure 10.12 models what generic light-harvesting, or antenna, arrays look like on a 5-cm scale (upper) and a 5-nm scale (lower). The upper figure is the entire leaf of a tree in spring. This leaf antenna can position itself for

> Pigments are essentially no different than "dyes." Pigment usually refers to a dye found in nature.

> We will not pursue this interesting photomechanical adjustment capability, called *phototropism*.

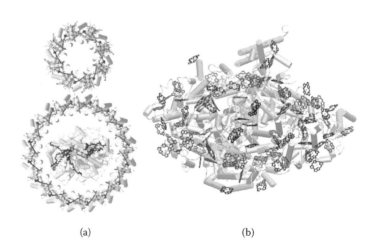

(a)   (b)

FIGURE 10.12

A comparison of light harvesting structures in the (a) absence and (b) presence of oxygen. (a) The single photosynthetic unit (PSU) of purple bacteria: 5–10 peripheral light harvesting complexes LH-II (small rings) transfer harvested light energy to an LH-I complex, which transfers its energy to the central reaction center. (b) Photosystem I from cyanobacteria combine the peripheral antenna array, the reaction center, and the electron transfer chain in to a single, large protein complex. (Reproduced from Schulten, K. 38 With permission.)

maximum reception of solar intensity. The lower figure is, of course, a hand-constructed diagram, but illustrates the basic role of *light-harvesting chlorophyll proteins* (LHCs): absorption of light and transfer of the excitation energy to a reaction center. The path of energy transfer in the diagram appears to be random, which is partly true, but near the reaction center chlorophylls are precisely positioned in both physical location and in excitation-energy level to draw the excitation into the reaction center. The remainder of this section and the next will address the structures responsible for this energy transfer and subsequent charge separation at the reaction center. We will start with green plants and end with antenna systems in purple photosynthetic bacteria, because they illustrate the primary physical principles of light harvesting and charge transfer, because detailed structures have been determined and the structural arrays are somewhat simpler.

10.4.1 Green Plant Light-Harvesting Complexes

Before we begin on the bacteria, however, we will briefly look at the PS-I "supercomplex" of green pea plants. Figure 10.13 shows this protein complex, whose structure was determined by X-ray crystallography by Amunts et al.[16] This supercomplex is indeed super because it contains 17 protein subunits, 168 chlorophylls, 2 phylloquinones, 3 [4Fe-4S] clusters, and 5 carotenoid pigments. Any molecule with *quinone* in its name is usually an electron acceptor in a photosynthetic system. Fe-S clusters are also almost universally associated with electron transfer—not surprising since one of the two primary purposes of the photosystems is electron transfer. The molecular mass of this complex is 504,252 Da. What does this complicated structure teach us about photosynthesis in green plants? The sheer number of chlorophylls shows that many light-harvesting chlorophylls occur together with the (few) reaction center chlorophylls in the same supercomplex: they are not independent entities. While it is likely that more PS-I light-harvesting chlorophylls exist than are found in this particular complex, more than 100 chlorophylls seem to be a part of one large structure. Several of the chlorophylls in this structure appear to be at the edge of the structure, not bound to anything in particular. The fact that such chlorophylls are resolved by the X-ray measurements indicates that they are relatively immobile in the crystal but that they may have become associated with the complex during the crystallization procedure. One take-home suggestion from this structure is that not all light-harvesting chlorophylls are equal: ones nearer to the reaction center may well differ in properties from those farther away. We will see in the cyanobacterial PS-I that chlorophylls (Chls) farther from the reaction center are not as well optimized for energy transfer as nearer Chls. In the single reaction center of the purple bacterium, there are indeed different light-harvesting pigments and the arrangement of the pigments is such that "downhill" energy transfer occurs: the excitation energies of the pigments decline as distance from the reaction center decreases.

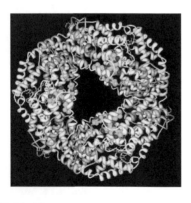

FIGURE 10.13
Three-dimensional structure of C-phycocyanin from PS-II of the thermophilic cyanobacterium *Synechococcus elongatus*. One of the three phycocyanins (green and red) per unit has been enlarged to emphasize the trimeric nature of the crystalline complex. Other phycocyanins can be discerned by their color. Though a trimer in the crystal does not prove a trimer exists in the living cyanobacterium, the tendency of the complex to form symmetric structures is clear. (PDB ID 1JBO Nield, J. et al., *J. Struct. Biol.*, 141, 149–155, 2003, displayed using Protein Workshop software.) (From Jmol: An open-source Java viewer for chemical structures in 3D. http://www.jmol.org/; Protein Workshop: Moreland, J.L. et al., *BMC Bioinformatics* 6, 21, 2005; NGL Viewer: Rose, A.S. et al., *Bioinformatics*, 2018.)

10.4.2 From Green Plant to Cyanobacterial to Purple-Bacterial Light-Harvesting Systems

The Schulten Research Group has published a comparison between green plant PS-I and that from cyanobacteria.[17] While differences can be found in the compartmentalization of antennae, similarities dominate. A very readable, online version of this comparison, *The Tale of Two Photosystems*, can be found at http://www.ks.uiuc.edu/Research/psres/plantps1.html.

A comparison between PS-I from two-center cyanobacteria and the single purple bacteria photosynthetic system can also be found online at http://www.ks.uiuc.edu/Research/psres/. This latter comparison, shown in Figure 10.14, is not only revealing but points toward a common issue in biosystems: life must deal with randomness and chaos on a microscopic scale. The simpler purple bacteria, which live in the absence of oxygen and have the oldest known photosynthetic system, have two highly symmetric, light-harvesting protein complexes (Figure 10.14a). The more peripheral antenna complex, LH-II, can move with respect to the RC. Depending on conditions, a variable number (~1–10) of these peripheral LH-II antenna can associate with the RC, which is located in the highly symmetric PH-I antenna. In contrast, the photosystem I pigment networks of oxygenic green plants and cyanobacteria form a single, rather complex array with a high chlorophyll:protein ratio.

The circular arrangement of pigments in purple bacteria may be a consequence of subunit symmetry and self-assembly processes. We will see that these primitive, symmetric systems allow elegant quantum treatments, but do the more closely packed and seemingly random chlorophyll networks found in cyanobacteria and plants, have an advantage? The profusion of green plants and cyanobacteria on Earth suggests that their photosystems are well adapted to many ecological niches, but does a more unified, complicated structure have a light-harvesting advantage?

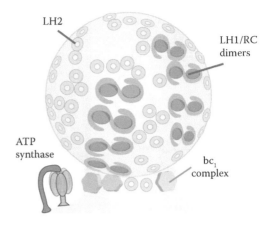

FIGURE 10.14
Schematic of the purple-bacterial membrane vesicle-like photosynthetic chromatophore from *Rba. sphaeroides*. The LH1-RC dimers and LH2 complexes, similar to those in *Rs. molischianum*, are closely packed in the bulb of the chromatophore, as seen in AFM images, and the bc1 complex (quinone electron acceptor and cytochrome enzyme) and ATP synthase, which are absent from AFM images, are tentatively placed near the neck of the chromatophore. (Redrawn from a picture by Svinarski, O. 39.)

10.4.3 Quantum Efficiency of Electron Transfer

It turns out that the quantum efficiency of primary electron transfer in PS-I of green plants and cyanobacteria is exceedingly high, above about 96%. This means that for every 100 photons absorbed by PS-I, 96 or more electrons are transferred to the primary acceptor. What may be even more important is that these systems (at least the cyanobacterial) are also very *robust* to system noise and environmental changes that may alter positions and orientations of chlorophylls. Sener et al. found that thermally accessible fluctuations of chlorophyll orientations, chlorophyll site energies, as well as even loss of individual chlorophylls from the network did not significantly reduce the electron transfer efficiency.[18] The largest reduction was found to be from 97.2% to about 96%. The "random bag" of chlorophylls in a cyanobacterium or green pea is highly efficient at converting photon energy to charge separation and highly adaptable as well. Sener et al. found that the placements and orientations of chlorophylls near the reaction center were optimized for transfer, while placement of the more peripheral chlorophylls was not. But such optimization of peripheral pigments seems not to matter much in the presence of so many closely packed chlorophylls.

10.4.4 Phycocyanin Antennae

Photosynthetic organisms that live in seawater, where the majority of photosynthesis occurs, have a problem: if they attempt to survive below about 1 m, the intensity of light in the wavelength region where chlorophyll absorbs, 400–460 nm and 620–670 nm, is quite small. The light has been absorbed by water and chlorophyll-containing organisms in the top 1 m of water. This would greatly restrict the available oceanic volume for photosynthetic activity. But as we have learned from our *Jurassic Park* lesson, "Nature will find a way." Photosynthetic cyanobacteria in water have developed an accessory pigment, phycocyanin, that acts as an antenna in the region of about 500–620 nm (see Figure 10.7). Because these pigments function as light-harvesting accessory pigments, they must transfer energy to the reaction center and must therefore be carefully positioned. Not unexpectedly, these pigments are contained in protein complexes. One of these, found in the thermophilic (heat-seeking) cyanobacterium *Synechococcus elongatus*, is C-phycocyanin, shown in Figure 10.15. This complex of light-harvesting pigments does not have a directly associated reaction center like the PS-I supercomplex in Figure 10.13. This C-phycocyanin structure should be viewed as one component of the PBS structure in Figure 10.9b.

10.4.5 Purple Bacterial Photosystem

In this critical section we will mix together light-harvesting and reaction center structures from similar bacteria, *Rhodopseudomonas (Rps.) sphaeroides*, *Rhodospirillum (Rs.) molischianum*, *Rhodopseudomonas (Rps.) acidophila*, and *Rhodopseudomonas viridis*. The structures are not

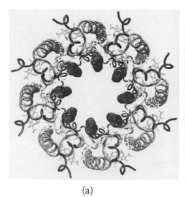

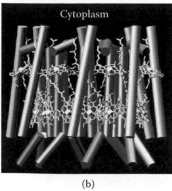

(a)　　　　　　　(b)

FIGURE 10.15

Light-harvesting LH-II complex of *Rs. Molischianum* purple bacteria. (a) View from above the membrane. (b) View from the side. BChl *a* molecules are shown in green. This complex has a ring of 16 BChl-a molecules parallel to the membrane (lower ring of BChls in b). The porphyrin rings of these BChl-a molecules are perpendicular to the membrane plane. These BChl's are called *B850* because they absorb at 850 nm. Eight B800 BChl-*a* molecules, whose porphyrin planes are nearly parallel to the membrane surface, form a second, larger ring parallel to the membrane. (Reproduced from Hu, X. et al., *Q. Rev. Biophys.*, 35, 1–62, 2002. With permission.)

> Rps. Sphaeroides: also known as *Rhodobacter sphaeroides*

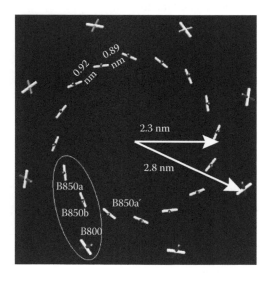

FIGURE 10.16

Absorption transition dipole moments of the BChl-*a* molecules in the LH-II complex. The orientation of a dipole moment determines the direction of an electromagnetic wave's electric field that is most efficiently absorbed. The relative orientations of two dipoles, as well as their separation, determine the efficiency of excitation energy transfer via the Förster transfer mechanism (Chapter 9). A complication: each BChl molecule has two, Q_y and Q_x, transition dipole moments. The Q_y transition primarily governs energy transfer. Three circled BChl-*a* molecules represent one unit of the octamer. The distances between the outer B800 and the nearest B850 (from an adjacent unit) is 1.91 nm. The B850s in the same unit are 2.02 and 2.55 nm away from B800. (Reproduced from Koepke, J. et al., *Structure*, 4, 581–597, 1996. With permission.)

identical in all three organisms, but close enough for our purposes. Most of the figures are based on those created by Klaus Schulten's molecular-modeling research group. Schulten's models are, in turn, based on X-ray crystallographic and atomic force microscope (AFM) structures determined by several groups. Two of the earliest such structures were determined for *sphaeroides* and *viridis* and formed the basis of the 1988 Nobel Prize in Chemistry for Johann Deisenhofer, Hartmut Michel, and Robert Huber.[19] Figure 10.16 shows a recent model for an entire *chromatophore*, the vesicle/protein/pigment structure for *Rps. sphaeroides*. This assembly contains reaction center, two types of antenna proteins (LH1 and LH2), accessory pigments, electron transfer entities, and a vesicle-like sac that contains them all. LH1 and RC occur as dimers; LH2 complexes are closely packed in the bulb of the chromatophore, and the bc1 complex (a quinone–cytochrome enzyme complex) and ATP synthase are placed near the neck of the chromatophore, though no firm image data yet exists for the latter. The conformation of the chromatophores varies among the purple bacteria, from stacks of flat lamellar folds in *Rps. acidophila* and *Rs. modischianum* to the spherical vesicles in *Rba. sphaeroides* and *Rb. capsulatus*. There are two antenna complexes, LH-I and LH-II (or LH1 and LH2), in all these organisms, LH-I closely associated with the reaction center and LH-II a "floating" antenna.

The LH-II antenna complex of *Rs. molischianum* appears as two concentric cylinders of 16 protein helices spanning the membrane, each with two symmetric, concentric rings of BChl-*a* molecules (Figure 10.17). Sixteen B850 BChl-a molecules,

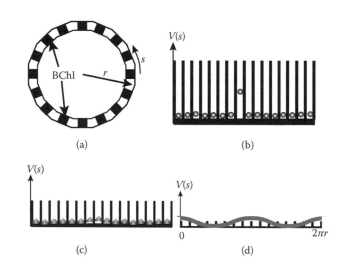

(a)　　　　　　　(b)

(c)　　　　　　　(d)

FIGURE 10.17

Idealized model of 16 quantum boxes on a circular ring. Electrons are bound by this potential array. (a) Dark areas indicate sites of zero potential. (b) No interactions between electrons at different sites. Excitations are local. (c) Weak interactions between neighboring sites causes correlations in excitations over a short range. (d) Strong interactions: site binding potentials are comparatively weak, electrons are delocalized and excitations are of the system. In purple bacteria, the 16 LH-II sites alternate in separation, 0.89 and 0.92 nm. The ring radius (inner ring in Figure 10.16) is about 2.3 nm.

perpendicular to the membrane plane, form one ring. The other eight B800 BChl-a molecules that are nearly parallel to the membrane plane form the other ring. Eight lycopenes (a carotenoid accessory pigment in this complex) stretch out between the B800 and B850. The structure of one quarter of this complex can be found in the Protein Data Bank: 1LGH.

In LH-II of *Rs. molischianum*, the Q_y transition dipole moments of neighboring B850 and B800 BChl-as are nearly parallel to each other—optimally aligned for Föster transfer B800→B850 (Figure 10.18). The transfer is also downhill, so that the fluorescence spectrum of B850 overlaps well with the absorption spectrum of B850. Another type of transfer involving a quantum electron exchange mechanism, Dexter energy transfer, between these chlorophylls is also possible. Accessory pigments such as lycopene seem to be involved in this transfer. Overall, the double ring structure of the B850 BChl-a molecules optimizes light absorption because it samples all possible spatial absorption directions. Spectral coverage is maximized because the accessory pigments absorb lower light wavelengths and facilitate transfer to

> B850: bacteriochlorophyll a whose absorption maximum is at 850 nm

> The transition dipole moment represents the relative displacement of electric charge when a light-absorbing molecule makes a transition from ground to excited state. It is computed quantum mechanically as $\vec{d}_{10} = \int \psi_1^*(\mathbf{r}) e\mathbf{r}\, \psi_0(\mathbf{r}) d^3\mathbf{r}$, where ψ_0 is the ground-state wave function and ψ_1^* is the complex conjugate of the excited-state wavefunction. It differs from the static dipole moment of the ground state of the molecule, $\vec{d}_{00} = \int \psi_0^*(\mathbf{r}) e\mathbf{r}\, \psi_0(\mathbf{r}) d^3\mathbf{r}$.

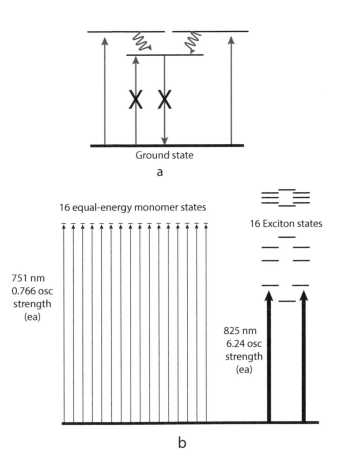

FIGURE 10.18
Excitonic states of the *Rs. molishianum* purple bacteria B850 bacteriochlorophyll-*a* 16-mer. Rather than 16 equal-energy, first-excited states associated with each BChl, there are 16 states of differing energies associated with excited states delocalized, to a limited extent, over the 16 molecules. (a) Of the three lowest-energy excited states, only the higher (equal-energy) states are accessible by photon absorption. Because the lowest excited state cannot absorb, it cannot emit, and excitations that arrive by thermal relaxation from the higher states will remain trapped for an extended period. The ring of LH-II can thus act as an excitation-energy storage ring. (b) The 16 equal-energy individual BChl states become 16 unequal exciton states when the BChls strongly interact. The two optically active exciton states absorb as strongly as the 16 individual states. The 825-nm wavelength is a theoretical result that neglects the presence of solvent; it corresponds to the observed 850 nm absorption.

Q_y and Q_x are transition dipole moments associated with first- (850 nm) and second-excited state (590 nm) absorption by B850 (similarly for B800). The dipole moments are almost perpendicular to each other, as suggested by the subscripts.

bacteriochlorophylls. Finally, the B850 chlorophyll ring is designed to effectively accept transferred energy: the emission characteristics of B800 match the high oscillator strength (dipole strength) of the low-energy, thermally accessible B850 *exciton* states.

10.4.6 Exciton States

The inner ring of 16 B850 bacteriochlorophyll-*a* molecules in LH-II has a radius of 2.3 nm, with B850s separated by alternating distances of 0.89 and 0.92 nm. The outer ring of 8 B800 BChls has an average separation of 2.2 nm between BChls, much larger than that of the inner ring. The B800 bacteriochlorophylls interact weakly with each other; B850s interact strongly with each other.[20] The strong BChl–BChl interactions result in excited states that are delocalized over the ring of BChls. Figure 10.19 crudely models this. Figure 10.19a shows the geometry of the ring and (Figure 10.19b–d) a particle-in-a-box models, with a potential wall of varying height between the zero potential energy sites of the BChls. High potential walls eliminate BChl–BChl interactions, while low walls allow delocalization. In Figure 10.19b electronic states of BChls are excited individually, with no effect on neighbors. In Figure 10.19c, weak interactions permit transfer of excitation energy to nearest neighbors, via Förster interaction[21] or other mechanisms. Interactions in Figure 10.19d are strong enough that excited electrons behave as if they were delocalized over the ring of 16 BChls. Under conditions of strong intersite interactions, the optical spectra are characteristic of delocalized *excitonic states*.[22]

The delocalization is technically of the excited electrons.

Since we started with 16 LH-II BChls, each with identical states, we must end up with 16 delocalized excitonic states, but Cory et al. showed that only 2 of these 16 states absorb light. These two light-accessible states are the second- and third-lowest excitonic energy states. What about the lowest-energy (first) excited state? You cannot excite from the ground state to this state by photon absorption, but you can get there by nonradiative relaxation from the next two higher states (Figure 10.20). Once excitation

Remember the case of the hydrogen molecule, H_2, where two equal-energy H states become two unequal-energy states when the H atoms approach each other.

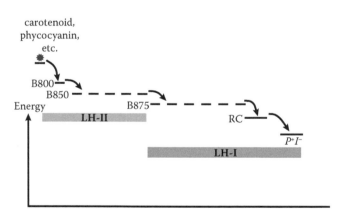

FIGURE 10.20
Downhill energy transfer in purple bacteria photosynthetic systems. The transfer process shown starts with absorption by an accessory pigment in the light-harvesting LH-II antenna system, followed by transfer to B800 and then to B850, also in the LH-II system. B850 transfers to B875 in the LH-I antenna, which then transfers to the reaction center. The special pair of the reaction center (P) is the final acceptor of excitation energy, except for a possible assist from an accessory BChl, before energy is converted to charge separation (P^+I^-), with the electron probably transferred first to a bacteriopheophytin (I^-). Strictly speaking, P^+I^- should not be on the same excitation-energy plot, as its energy is a free energy. However, this free energy is mostly enthalpic, so the comparison is at least partially legitimate.

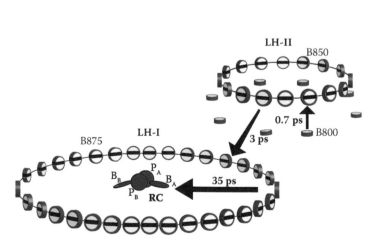

FIGURE 10.19
Hierarchy and timing of energy transfer in purple bacteria (*Rs. molischianum*). Excitation energy is transferred on a picosecond time scale, but most rapidly within the LH-II light-harvesting complex. (Based on Cory, M.G. et al., *J. Phys. Chem. B.*, 102, 7640–7650, 1998.)

arrives at this lower level, however, it will remain there for an extended period because fluorescence from this state is also forbidden.

Sixteen individual bacteriochlorophylls absorb 16 times as much light as one. What about these 16 excitonic states? We noted that at least one of the 16 states did not absorb light; all but the two we noted are also weakly absorbing. Have we lost light-absorbing capability by assembling our quantum energy storage ring? It turns out that an absorption *dipole-strength sum rule* operates; the

EXERCISE 10.2 (TOO) CRUDE QUANTUM MODEL OF A 16-BCHL RING

We know that quantum mechanics is simplest in cases of high symmetry. The particle in a box is the simplest case. We have electrons from BChl molecules distributed almost symmetrically on a ring. Assume each BChl contributes N electrons, which are free (zero potential energy) to roam the LH-II inner ring of 16. Calculate the energy of the ground state and first-excited state, and the photon wavelength that will excite from ground to first-excited state. Start with $N = 1$; then try $N = 9$.

Solution

Use Figure 10.18 as our guide for geometry. The ring radius is $r = 2.3$ nm, which makes the ring circumference $= 2\pi r = 14.5$ nm. Our electrons are confined to the circumference of the circle. We could do better, but why not start with the assumption that the potential energy on the ring is zero. We can then use de Broglie analysis, with the boundary condition that the de Broglie waves must smoothly connect: $\psi(\theta) = \psi(\theta + 2\pi)$ and likewise for the first derivative. This amounts to fitting an integral number of wavelengths onto the circumference. If we know the wavelengths, we find the momentum by $p = h/\lambda$ and kinetic energy by $K = E = p^2/2m$ (potential energy is zero). Note in the following the judicious choice of constants, e.g., $(hc)^2/mc^2$. The mass should be the mass of the electron, in our crude model.

$$n\lambda_n = 2\pi r = 14.5\,\text{nm}$$

$$E_n = \frac{h^2 n^2}{8\pi^2 m r^2} = \frac{(hc)^2 n^2}{8\pi^2 mc^2 r^2} = \frac{(1240\,\text{eV}\cdot\text{nm})^2 n^2}{2(511\times10^3\,\text{eV})(14.5\,\text{nm})^2}$$

$$= 0.00716 n^2 \ \text{eV} = 1.15\times10^{-21} n^2 \ \text{J} \approx 0.3 n^2 \, k_B T$$

Note that n reflects the quantum state. If we have 16 electrons ($N = 1$), we will fill up to the $n = 8$ level, with two electrons per level. Wait: we remember that for systems of spherical symmetry (circular in two dimensions) angular momentum is conserved and therefore quantized. We include two possibilities for each momentum: one clockwise and one counterclockwise. This corresponds to $\pm$ values for p (angular momentum $L = \pm rp = \pm n\hbar$) in the de Broglie wave treatment, and the wave functions for clockwise and counterclockwise rotation should differ. (Note there are only $\pm$ angular momentum z components in this two-dimensional system.) So, let's put four electrons in each n state. Sixteen electrons would then fill up to $n = 4$ and the first excited state would consist of one $n = 4$ electron promoted to $n = 5$. The energy difference would be

$$E_5 - E_4 = 0.00716(25-16) \ \text{eV} = 1.15\times10^{-21}(25-16) \ \text{J} \approx 0.28(25-16) K_B T$$

$$= 0.0644 \ \text{eV} = 10.4\times10^{-21} J \approx 2.5 k_B T$$

The corresponding photon wavelength would be $\lambda_{photon} = hc/E_5 - E_5 = 1240\,eV \cdot nm/$ 0.0644eV = 19,200 nm which we know is much longer than the observed 850 nm. Now since we know (or suspect) there is more than one electron associated with the porphyrin ring of the bacteriochlorophyll, let's put a more realistic number of electrons, $N = 9$ per BChl. There are then 144 electrons on the LH-II ring. This would fill all levels up to $n = 36$, and the energy we need is $E_{37}-E_{36}$:

$$E_{37} - E_{36} = 0.00716(1369 - 1296)eV = 0.523\ eV,$$

> If we feigned ignorance of the two angular momentum states per energy level, our answer here would have been 1190 nm—not far from correct.

with corresponding photon wavelength $\lambda_{photon} = hc/E_5 - E_4 = 1240\ eV \cdot nm/0.523\,eV =$ 2,370 nm. We are still a factor of 2.8 times too low in excitation energy. To get a photon wavelength of 850 nm, we would have to assume levels up to $n = 101$ were filled with 404 electrons (25 electrons per BChl). While Cory et al.[22] put more than 2,000 electrons into their computation (>125 per BChl), the vast majority would not be of the sort that could be considered free to roam around the LH-II ring.

rule says that the two absorbing states absorb as strongly as the original 16 molecular states: we have not lost the ability to absorb light.

The fundamental problem with our simplistic attempt is that the electrons can be considered to be "free" to wander around ring circumference only when the BChl electron clouds overlap each other. For this we would need to have about twice as many BChls on the LH-II ring. We can refer to excitations as "delocalized," but this does not mean the electrons are no longer associated with individual chlorophylls. If we are serious about the business of accurately modeling LH-II as it exists, we need to learn better computation techniques, like INDO/S-CI.

The 32-member 4.5-nm radius BChl LH-I ring is much like the LH-II ring, leaving the BChls about the same distance apart as those in LH-II (see Figure 10.14a). There are again three lowest-energy excited states, two of which are accessible by photon absorption. The LH-I BChl molecules of the ring of 32 are called *B875*, because they absorb at 875 nm. This is convenient because the energy of the B850 bacteriochlorophylls of the LH-II antenna is higher, and downhill energy transfer applies to the B850 → B875 transfer. LH-I has accessory pigments that can both transfer energy to the BChls and protect the complex from stray triplet excitations, but there is no secondary BChl ring corresponding to LH-II's B800.

10.4.7 Hierarchy and Timing of Energy Transfer

The overall organization of purple bacteria energy transfer is LH-II → LH-I → RC (Figure 10.21). The overall efficiency is high, about 95%. Within LH-II, transfer from B800 → B850 occurs, as well as transfer from accessory pigments. The time scale is very rapid—about 0.7 ps. What entity does energy transfer to, a single B850 molecule or to the B850 "quantum ring"? To go about answering this question, we must calculate the extent of delocalization of excitations on the B850 ring. Figure 10.19c shows effective delocalization over 2 or so adjacent BChls, and Figure 10.19d, over all 16 BChls. If Figure 10.19d applies, B800 → B850 transfer is to the 16-mer of BChls. If Figure 10.19b or c applies, B800 → B850 transfer is to an individual B850 molecule or two. In the end, if the B850 delocalization is small, but the coupling between adjacent B850s so strong that energy rapidly circulates around the ring, the difference between small and extensive delocalization does not really matter. The energy just needs to equilibrate on the LH-II ring in 3 ps or less—the transfer time from LH-II → LH-I (B875) when LH-II is associated with LH-I.

Do we have the analytic tools to compute transfer times in these ringlike photosynthetic systems, perhaps using (the inverse of) Equations 9.32 and 9.33? The answer is "Yes, sort of." We can try them and see what we get, but it is not easy and is highly questionable. The first problem is that the sort of Förster theory behind Equation 9.32 requires the donor–acceptor separation to be much larger than the size of donor and acceptor molecules. This may be okay for B800 → B850 transfer, but only if one B850 is the acceptor. Such theory is not okay for transfer among the B850s. What would we need, to use these FRET equations from Chapter 9? Equation 9.33 indicates we must know κ^2, the relative angular orientation of the donor–acceptor pair. We can get this information from X-ray crystal structures. J, the absorption-fluorescence overlap integral, must come from measurements. Q_D, the fluorescence quantum yield of the donor in the absence of the acceptor, but with otherwise identical surroundings, must be measured. The index of refraction term n^4 seems pretty straightforward, but it is not. That index refers to

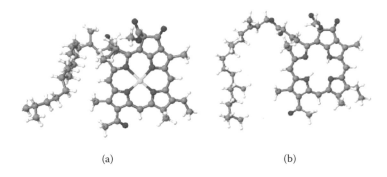

(a)   (b)

FIGURE 10.21

(a) Bacteriochlorophyll b (BChl b) and (b) bacteriopheophytin b (BPhe b). The reaction center special pair (P in Figure 10.20) consists of two BChls, plus two accessory BChls that speed up transfer from B875 to RC and one of which may act as a short-lived (~1 ps) first acceptor of an electron from the special pair. BPhe (I in Figure 10.20) is the next acceptor of a transferred electron, where charge remains localized for about 200 ps, after which the electron is transferred through a cascade of quinones, iron centers, etc. Like the accessory BChls of the RC, there are two, almost symmetrically placed BPhe molecules, but only one accepts the electron.

the refractive index in the nanometer (or less) region between the donor and acceptor in the photosynthetic unit.[23] The quantity $(1.2)^4 = 2.1$, while $(1.6)^4 = 6.6$, so the index matters. Finally, the excited-state lifetime of the donor molecule (Equation 9.32), in the absence of the acceptor, but with otherwise identical surroundings, must be measured. This process, when carried out by the physics or chemistry graduate student, generally requires a semester of work for a first result.

We have a similar question of extent of delocalization and/or transfer time around the ring of B875 bacteriochlorophylls. Again, we can say with certainty that delocalization is either extensive or transfer around the LH-I ring should occur in less than 35 ps, the transfer time to the reaction center (RC). These questions of delocalization have been addressed in the literature and involve difficult experimental measurements and difficult, and sometimes tricky, theoretical analysis.[22–25] The ambitious student is welcome to explore this literature.

10.5 REACTION CENTERS AND CHARGE SEPARATION: PURPLE BACTERIA AND CYANOBACTERIA

The "center" of the photosynthetic unit is the reaction center. We have seen, however, that there are several important reactions in photosynthesis: water cleavage, electron transfer to acceptor molecules in both PS-II and PS-I, reduction of NADP⁺, synthesis of carbohydrate. We have prepared ourselves for the final step of energy transfer, B875 → RC, followed by charge separation, in the purple bacteria. We will concentrate on this last step of energy transfer and first step of electron transfer in the purple bacterial system, with comments on the same process in PS-II of cyanobacteria.

Figure 10.21 indicates that transfer of energy from the final antenna bacteriochlorophylls of the purple bacteria, B875, to the reaction center takes place in about 35 ps. The entire process, from photon absorption to final transfer to the RC takes about 40 ps—quite fast indeed. What, exactly, is this reaction center? The diagram in Figure 10.21 shows four bacteriochlorophylls: the same molecules used to absorb light and transfer energy. These are only the primary optically and electronically active parts, however, not the only molecules in the reaction center. The frame that holds the BChls of the reaction center is, as expected, protein. The protein frame also holds other optically active molecules, which we will see later in the cyanobacterial RC. These RC BChls differ from their light-harvesting counterparts in their relative orientation to each other: the primary two BChls in the RC, called the *special pair*, P_A and P_B, are close together, with porphyrin-ring planes nearly parallel and separated by 0.32–0.34 nm, compared to the ~9-nm distance between B850 or B875 bacteriochlorophylls. Because that latter are already found to produce excitations delocalized over more than one BChl, it is certain that the special pair operates opto-electronically as one unit. Figure 10.20 and associated discussion already taught us that when two identical molecules are brought close together, the initial two localized, equal-energy states become two delocalized, unequal-energy states, one lower and one higher in energy than the original localized state. Without knowing any further details, we expect that the lower-energy state of the special pair will act as a "sink" for transferred energy: once energy is transferred, there it will have low probability to reverse transfer (Figure 10.22). The special pair is sometimes referred to by the Pnnn designation, reflecting the wavelength of absorption. For *Rs. molischianum* the designation is P862; *Rs. viridis*, P960; *Rb. sphaeroides*, P870.

How efficiently does the RC center convert photon energy into charge separation? By the time the excitation gets to the special pair, the excitation energy is lower—excited P862 has 99% as much energy as P800—not much energy has been lost, and even lost energy has been put to good use, to ensure that energy funnels toward the reaction center and not backward. In other words, some of the photon energy has been traded for order (entropy).

What is the purpose of the accessory BChls, B_A and B_B, in the RC? In X-ray crystal structures (PDB 1PCR) these two BChls are in van der Waals contact with the special pair, with porphyrin planes almost perpendicular to those of the pair. A "startling" finding by the Schulten group was that calculations of the energy transfer rate from the ring of B875s to the RC slowed by a factor of 10 when the B_A-B_B accessory molecules were omitted. The B_A-B_B accessory BChls have apparently acted as a quantum tunneling pathway between the P875 BChls of LH-I and the special pair, located about 4 nm distant.[24]

When does excitation energy become energy of separated charges? Energy conversion happens rapidly: it must, because excited states of organic molecules rarely last for more than 1 ns. After energy gets to the special pair, the B_A molecule apparently (also) acts as a first, short-lived acceptor of electronic charge. The electron moves to B_A within about 3 ps. About 1 ps later, the electron moves to a bacteriopheophytin (BPhe) molecule, where it stays for about 200 ps before transferring to various quinone, iron center, and other acceptor molecules in the electron transfer chain (see Figure 10.23). Because the separated electron leaves a hole in the BChl, the state generated by charge separation is notated P^+I^-, where P indicates the special pair and I the electron acceptor (see Figure 10.22).

Why should a BPhe act as the first major acceptor of electronic charge? First, the pair of BPhes are, like the accessory BChls in the RC, close to the BChls. Second, they are almost identical to BChl except that the Mg atom in BChl is replaced by two protons: the molecule is ready to accept an electron. But why, like the special *pair*, should there be two bacteriopheophytins? The obvious, but wrong, answer is that two electron acceptors can accept electrons twice as fast as one, and faster conversion of excitation energy to charge separation means higher conversion efficiency. Measurements have shown that electrons are transferred preferentially to one BPhe and not to the other,[25] so we conclude that nature has again employed the means of a slight asymmetry in a pair of excitable molecules to efficiently relay energy to the proper place.

> The fraction of energy lost varies among the purple bacteria.

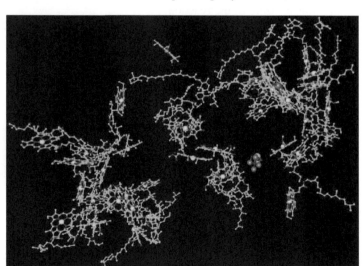

FIGURE 10.22
Optically and electronically active molecules in the PS-II reaction center from the cyanobacterium *Thermosynechococcus elongatus*. Protein structure has not been displayed. Two almost parallel chlorophylls, as well as the manganese O_2-generating site are evident. (PDB ID 1S5L by Ferreira, K.N. et al., *Science*, 303, 1831–1838, 2004, displayed with Protein Workshop software.) (From Jmol: An open-source Java viewer for chemical structures in 3D. http://www.jmol.org/; Protein Workshop: Moreland, J.L. et al., *BMC Bioinformatics* 6, 21, 2005; NGL Viewer: Rose, A.S. et al., *Bioinformatics*, 2018.)

How much energy is preserved in the first conversion of electronic excitation energy to charge separation energy? As discussed earlier, care must be taken in "energy" comparisons, because the separated charge eventually contributes to a pool of charge, separated across a membrane. As soon as the generated electron can interact with any pool of charge, the *free energy* of the P^+I^- state, including entropy, must be considered. Measurements in the purple bacterium *Rb. sphaeroides* suggest that the entropic component of the free energy change $P^* \rightarrow P^+I^-$ is small, and that $\Delta G \sim \Delta H = 0.263$ eV $= 25.4$ kJ/mol $\sim 10.3\ k_BT$.[26] This represents an 18% drop in energy from the P862 of *sphaeroides'* special pair. Figure 10.22 indicates a roughly similar percentage drop for *Rs. molischianum.*

> Photosynthetic systems cannot use one 400-nm photon to generate two pigment excited states in a purple bacterial or other photosynthetic system.

We have reached the end of our consideration of the primary energetic processes of purple bacterial photosynthesis. We have left the electron poised, after the first charge separation, at a bacteriopheophytin site for further transfer. The quantum efficiency of conversion of photons to electrons is above 95%. The (free) energy conversion efficiency from first excitation to first separated charge is roughly 70%–75% for absorption of a 700–800 nm photon. The next task for the reaction center is to move separated charge to a site where its energy can generate needed chemical energy or oxygen. Purple bacteria do not generate O_2, but the PS-II PSU we have introduced does. The PS-II complex is found in cyanobacteria and green plants and includes, in addition to an O_2-generating site, many of the general features of the features of the purple bacteria: light-harvesting chlorophylls (though not in ringlike assemblies), a reaction center, special pair of chlorophylls, and accessory pigments. These pigments are displayed in Figure 10.24 for the cyanobacterium *Thermosynechococcus elongatus.* We leave for homework the exploration of this structure in terms of its special pair and manganese oxygen center.[27]

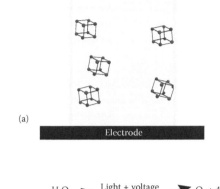

(a)

Electrode

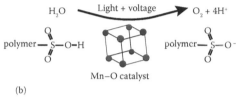

(b)

FIGURE 10.23
Manganese-based catalyst for splitting of water. (a) Core cubic Mn/O catalyst for water-splitting reaction. (b) Photon energy and an electric field are needed for splitting of water. See original publication for chemical balancing and other details. (After Scheme 1 of Brimblecombe, R. et al., *Inorg. Chem.*, 48, 7269–7279, 2009.)

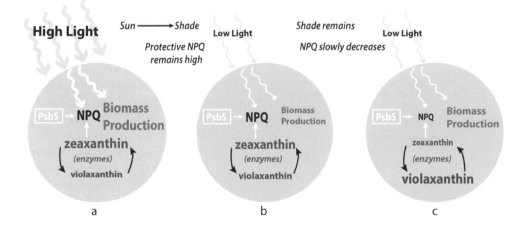

FIGURE 10.24
Strategy for increasing crop yields by speeding the response time to changes in sunlight intensity. NPQ = Nonphotochemical quenching of chlorophyll fluorescence; turns on to dissipate excess energy as heat. PsbS = PSII subunit S; over-expression of PsbS increases rate of induction and relaxation of energy-dependent quenching, which is also aided by conversion of violaxanthin to zeaxanthin. (a) In bright sunlight, biomass production is high and zeaxanthin and NPQ are high to protect from excess intensity. Two enzymes adjust the interconversion of violaxanthin and zeaxanthin. (b) Clouds obscure the sun, light intensity decreases. Biomass production goes down rapidly, because zeaxanthin does not rapidly convert to non-quenching violaxanthin. (c) If clouds remain, zeaxanthin converts to violaxanthin and NPQ quenching decreases over a 10–15-minute time; biomass production under the low-light conditions increases. A strategy to more rapidly adjust to low light by bioengineering faster enzymatic interconversion of zea- and violaxanthin, increasing tobacco crop yields by 15%, was found by to be superior to engineering only the PsbS levels. (After Kromdijk, J. et al., *Science*, 354, 857–861, 2016.)

10.6 ARTIFICIAL MODELS AND NONPOLLUTING ENERGY PRODUCTION

At the beginning of this chapter we noted that photosynthesis presently produces about 6 times as much (chemical) energy as humans use in any time interval. On the other hand, the solar energy hitting the Earth in 1 h is about equal to that consumed by mankind in a year, a factor of nearly 10^4. Though this number exaggerates the practical energy production possibilities by a factor of 10^2 or more (for a variety of reasons), a potential source of 10–100 times our present energy usage, with less environmental damage, seems attractive. Mimicking the process of photosynthesis but diverting the process toward energy in the forms humans predominantly use, electricity and burnable fuel (e.g., H_2), would seem natural. In view of political, economic, and environmental developments in the first decade of the twenty-first century, such efforts are well-advanced. A quick search on *artificial photosynthesis*, *solar hydrogen production*, and related terms will result in hundreds of articles and books since 2010. A March, 2017, Faraday Discussion in Kyoto, Japan, focusing on artificial photosynthesis has recently appeared.[28]

In 1845, Julius Robert von Mayer proposed that the sun is the ultimate source of energy for life on Earth and proposed that photosynthesis is a conversion of light energy into chemical energy. See Photosynthesis Timelines (photobiology.info/History_Timelines/Hist-Photosyn.html).

Photosynthesis has been an energy resource model for commercial power production for many decades, arguably for at least 165 years. Efforts to mimic natural photosynthesis but divert (e^-, H^+) to H_2, rather than to NADP+ and eventually to sugar (CH_2O), have not been successful on a commercial scale. Nature's production line employs carefully designed enzymes that trap excitation energy, transfer electrons, store energy and charge in chemical form, and inhibit reverse processes. The primary processes of photosynthesis are, as we noted, quite efficient, but the energy we can extract from generated plant material amounts to a fraction of a percent of the solar energy that fell on that plant. While nature doesn't care about this low efficiency, we, with our machines that demand large energy outputs in small spaces and short times, do. Redesign of the photosynthetic network is complex.[29] However, simpler, nonenzymatic systems that employ more traditional schemes of metal catalytic conversion may prove successful in the next decade. An NSF-funded collaboration called *Powering the Planet*, directed by Harry Gray, recently studied technologies needed to produce an integrated device for H_2, current, and O_2 production. A low-cost, integrated system of an efficient photovoltaic membrane,[30] a catalyst for hydrogen production, and an oxygen catalyst is the goal.[31] In this design, H_2 and O_2 are produced using light during daytime, where sunlight drives a standard photovoltaic cell, whose current splits H_2O using the Co catalyst. A fuel cell then uses the generated gases to produce clean energy after dark. In this approach, high efficiency is sacrificed for low cost. Currently, this research effort continues in CCI (*Center for Chemical Innovation*) *Solar—an NSF Center for Innovation in Solar Fuels*.

Other researchers are attempting to follow nature's model of photosynthesis more closely by eliminating the photovoltaic cell and using sunlight to directly split water on the surface of a catalyst.[32] The catalyst is based on four manganese (Mn) atoms that are also at the heart of green plant photosynthesis: the oxygen-evolving complex of the *photosystem II* protein (see Figures 10.24 and 10.25).

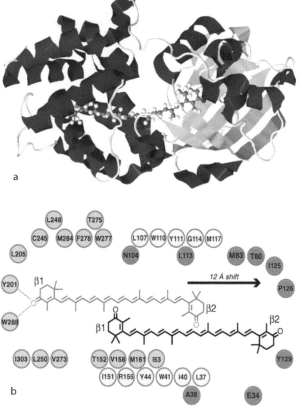

FIGURE 10.25

The cyanobacterial orange carotenoid protein (OCP), which acts as a light sensor and agent of photoprotective energy dissipation. (a) Crystal structure of OCP, showing carotenoid bound to the N-terminal domain of protein, which is responsible for interaction with the antenna and induction of excitation energy quenching, while the C-terminal domain is the regulatory domain that senses light and induces photoactivation. (PDB ID 5HGR, displayed with JSmol software; Lopez-Igual, R. et al., *Plant Physiol.*, 171, 1852–1866, 2016.) (b) Diagram of 12Å movement of carotenoid, which engages the photoprotective state (NPQ, as in Figure 10.24) of the protein. (Reproduced from Leverenz, R. L. et al., *Science*, 348, 1463–1466, 2015. With permission.)

The Mn cluster is doped into proton-conducting films, which are then coated onto a conducting electrode. A claim was made in 2008 that one of these technologies will be commercially viable by fall 2013.[33] This goal seems largely to have been fulfilled in the "Bionic leaf 2.0," though one may quibble about the date and commercial availability.[34] Daniel Nocera (Harvard University) and Pamela Silver (Harvard Medical School) created a system that uses sunlight to split water molecules plus hydrogen-eating soil bacteria, genetically engineered *Ralstonia eutropha*, to produce liquid fuels such as isopropanol. The system can convert solar energy to biomass at 10% efficiency, nearly an order of magnitude higher than the efficiency of natural photosynthesis. Of course, nature creates the tree, while the Bionic leaf produces only the liquid fuel as "biomass."

Photosynthetic models are by no means the sole basis for design of solar energy generation. The great majority of commercialized products are traditional semiconductor solar cells in the form of inexpensive arrays or sheets that directly generate electric power. The watchwords in commercial solar power are still *cheap to manufacture* and *efficient*. A record solar cell efficiency was set in 2013 by Sharp Corporation, 44.4% for a "concentrator" three-layer—InGaP, GaAs, InGaAs—semiconductor cell. A year later, a French-German collaboration raised the record to 46%.[35] Hazardous materials are still an issue, however, with manganese from the photosynthesis-based system more benign than catalytic cobalt and gallium and arsenic in semiconductor models (see http://www.scorecard.org/index.tcl).

10.7 POINTS TO REMEMBER

- About 1 kW/m^2 of solar energy strikes the Earth, with an approximate blackbody spectrum.
- "Light reactions": $CO_2 + H_2O \xrightarrow{light} (CH_2O) + O_2$; $2H_2O \xrightarrow{light} 4H + O_2$.
- "Dark reactions": $6CO_2 + 12H_2O + 18ATP + 12NADPH \xrightarrow{dark}$
  $C_6H_{12}O_6 + 18ADP + 18P_i + 12NADP^+ + 12H^+ + 6O_2$.
- One absorbed photon produces a bit more than 1 ATP chemical energy equivalent.
- Three hundred chlorophylls collaborate to absorb 1 photon.
- Nature has two approaches to antenna arrays: symmetric rings and spheres or complex, probably redundant assemblies.
- The primary processes of photosynthesis are absorption, energy transfer, and electron transfer (charge separation).
- Energy transfer involves donor–acceptor separations a few nanometers or less; some too close for Förster transfer theory; time scale approximately picoseconds.
- The *reaction center* consists of a chlorophyll dimer and performs charge separation.

10.8 PROBLEM-SOLVING

PROBLEM 10.1 SOLAR AND PHOTOSYNTHETIC ENERGY

1. Use the Microsoft Excel–formatted solar irradiance data at http://rredc.nrel.gov/solar/spectra/am1.5/ to determine the total solar energy hitting the Earth (the ground) in one year in the spectral region 400–900 nm.
2. From data cited in this chapter, determine the fraction of this energy converted to chemical energy by photosynthetic organisms.

PROBLEM 10.2 PHOTOSYSTEMS IN PARALLEL AND IN SERIES

Suppose two photosystems exist, photosystem A and photosystem B. Photosystem A absorbs light only between 670 and 690 nm, while system B absorbs only between 690 and 710 nm. A variable-wavelength light source is shined on this combined system. These two photosystems are involved in the production of a product called P_2, a dimer of two P molecules, and that absorption of one photon by either system produces one P. (The quantum yield of one P is 1.00 for either system.) Any P produced will stay around until another P is produced, when it will instantly form P_2. Assume the two photosystems act in parallel and independently.

1. Plot the quantum yield of P production versus wavelength from 650 to 750 nm.
2. Plot the quantum yield of P_2 production versus wavelength. (Consider: If absorption of one photon produces 1 molecule of P, does absorption of one photon produce 0.5 P_2 molecules?)
3. Assume a second light beam of wavelength 670 nm, with intensity equal to that of the variable-wavelength light, simultaneously illuminates the system. Repeat 1 and 2. Define the yield as the yield per photon of the variable-wavelength beam.

Now assume the two photosystems are arranged in series and that the only way P_2 can be produced is when photosystem A produces a product P and passes it to photosystem B, where another P is added.

1. Repeat part 1 for this series arrangement.
2. Repeat part 2 for this series arrangement.

PROBLEM 10.3 MEMBRANES: pH AND VOLTAGE

Exercise 10.1 estimated the local pH change inside a thylakoid structure due to transport of one H^+ ions across the membrane. For the volume given, $\Delta pH = 0.46$.

1. There is a regime where the ΔpH depends linearly on the amount of transported charge. Examine the Exercise 10.1 computation,

$$\Delta pH = -\log\left(1 + \frac{\left[\Delta H^+\right]}{\left[H_0^+\right]}\right) = -\log\left(1 + \frac{1 \text{ mole}}{6.02 \times 10^{23} \cdot V \cdot \left[10^{-7} m\right]}\right)$$

and determine the volume V above which the logarithmic dependence is approximately linear. Express the volume in liters and in μm^3. Note

$$\ln(1 + x) = x - \frac{x^2}{2} + \frac{x^3}{3} - \cdots$$

2. Estimate the electrical energy required to move a proton across a membrane. Assume the proton goes from one side of a flat, 9.0-nm-thick membrane slab to the other side and that the initial potential difference across the membrane is 35 mV. (Positive work must be done to transport the proton.) If you believe you need the dielectric constant, explain and use a value of $\varepsilon = 6.0\varepsilon_0$.

MINIPROJECT 10.4   RATE OF PHOTOSYNTHETIC CHARGE SEPARATION

Assume that a green plant chloroplast is irradiated by the sunlight.

1. Determine the approximate number of chlorophyll molecules in a typical thylakoid of a chloroplast and the approximate dimensions of a thylakoid. (You will have to search some of the Web sites noted in the text to find numbers needed for this. Start with Wikipedia.)
2. Assuming these chlorophyll molecules are distributed uniformly throughout the thylakoid volume (they are not, actually), determine the average chlorophyll concentration.
3. Chlorophyll absorption coefficient data appropriate for a Microsoft Excel or other spreadsheet/graphing program can be found at http://omlc.ogi.edu/spectra/ PhotochemCAD/abs_html/chlorophyll-a%28MeOH%29.html. (If this site is unavailable, search for another.) Load these data into a spreadsheet. Also download solar spectral irradiance data at the link noted in Figure 9.1.
4. If the broadside of the thylakoid is exposed to the sunlight (normal light incidence), use your spreadsheet to determine what fraction of the incident light is transmitted at each wavelength. Note that Beer's law, $I = I_0 10^{-A}$, implies that the fraction of light absorbed is 10^{-A}.
5. By appropriate multiplication and integration of your spreadsheet data, using the area of the thylakoid exposed to light, determine the number of photons absorbed per second by the thylakoid chlorophylls.
6. About how many separated charges will be produced per second in the thylakoid?

PROBLEM 10.5   MOBILE ANTENNAE

Green plants have the ability to move LHC-II light-harvesting chlorophyll proteins, normally associated with PS-II, over to PS-I. Describe a situation in which this would benefit the organism.

PROBLEM 10.6   ENERGY TRANSFER EFFICIENCY AND THERMAL FLUCTUATIONS

Two chlorophyll molecules in a light-harvesting complex are separated by 1.91 nm (transition dipole-to-dipole distance). Assume each chlorophyll is bound to its site with an effective spring constant of 100 pN/nm. (You may treat this problem in one dimension and you will need Chapter 9 energy transfer results.)

1. If the chlorophylls are displaced from their equilibrium positions (distance of separation) by thermal fluctuations equivalent to ½ $k_B T$ at room temperature, by what percentage can the Förster energy transfer rate maximally increase or decrease?

2. (Graduate problem.) Assume, for simplicity, that the two chlorophylls have Gaussian-shaped optical spectra, with full width at half maximum 40 nm and that the emission spectrum of one (the donor) is identical to the absorption spectrum of the other (the acceptor). See Figure 9.23. Suppose thermal fluctuations cause the excitation energy of the acceptor to increase by ½ $k_B T$. Compute the wavelength shift of the acceptor's absorption spectrum and calculate the percent decrease in the Förster energy transfer rate (decrease in the overlap integral).

PROBLEM 10.7 PURPLE-BACTERIAL (*RS. MOLISCHIANUM*) LH-II DISTANCES

1. Compare the diameter of the ring of a single bacteriochlorophyll-*a* molecule with the distances between adjacent BChls in the 16-member LH-II ring. Discuss the validity of Förster theory (rate ~r^{-6}) to describe energy transfer within this ring.
2. Compute the average distance between BChls in the eight-member LH-II ring. Discuss the validity of Förster theory (rate ~r^{-6}) to describe energy transfer within this ring.

PROBLEM 10.8 FÖRSTER THEORY ANGULAR DEPENDENCE

Look up the vector expression for the angular dependence of Förster energy transfer (κ^2).

1. Show that transfer is maximum if the two dipole moments are parallel.
2. Find an orientation such that the transfer rate is zero. Justify.
3. The value of κ^2 is often taken as 2/3, the value obtained as the average over all angles. What sorts of donor–acceptor environments would justify this average? Must both donor and acceptor be free to randomly reorient? How likely are the chlorophylls in the PS-I light-harvesting "supercomplex" from pea plants (Figure 10.13) to satisfy conditions for random averaging? (Examine the structure on the Protein Database.)
4. Consider a random solution of a donor molecule, with fluorescence lifetime 10 ns. An acceptor is added to the solution and the donor lifetime is observed to decrease to 8 ns. Both donor and acceptor rotate freely, with rotational correlation time of about 70 ps. Assuming the intrinsic donor fluorescence properties were unchanged, would the lifetime decrease be the same if the rotation times were 200 ns, rather than 70 ps?

PROBLEM 10.9   REACTION CENTER STRUCTURE

Access the structure of the cyanobacterial PS-II reaction center, oxygen-generating protein published by Ferreira et al. in 2004 at the PDB Web site. (Use a computer with relatively fast Internet access.)

1. Display, save, and print the structure with protein as ribbons and ligands (chlorophyll, carotenoids, iron, calcium, manganese, etc.) as atoms and bonds.
2. Display, save, and print the structure with only ligands displayed. Use the same magnification as in 1.
3. Locate the special pair and the Mn O_2-generating site. Zoom in so that both these structures are clear, with the pair of chlorophylls approximately sideways to the screen. Try to leave other ligands displayed, unless they obscure visibility. Identify other atoms that may be interacting with Mn atoms. You do not need to include amino acids, though they are critical to function.
4. As measured by special pair chlorophyll edge-to-edge distances, determine the closest and farthest distances between the two chlorophylls. Determine the corresponding angle between the special pair of chlorophylls.
5. Determine the approximate distance between the center of the special pair and the Mn center. How many Mn atoms are in this structure?

MINIPROJECT 10.10   PHOTOPROTECTION, CROP YIELDS AND GENETIC MANIPULATION

Section 10.3 described a genetic engineering approach to increasing leafy crop yields by accelerating the interconversion of zeaxanthin and violaxanthin, which then caused the light-harvesting system to more rapidly adjust to changing sunlight intensity. As with many applications of fundamental bioscience discoveries to society's (perceived) needs, the applied bioscientist must consider social and ecosystem consequences.

1. Study the 2016 Kromdijk et al. publication (reference 14) and summarize, in about 200 words, the genetic modification done and method for introducing modified crops to growing fields.
2. Search for subsequent publications by this research group and others. For each publication, describe in 50–100 words the additional results obtained.
3. A well studied and debated genetic crop modification is the Roundup™-resistant seed strains developed and marketed by Monsanto. Find two or three recent review articles that discuss the ecological and social aspects of this crop introduction. As this issue is highly polarizing, make sure your articles represent a balanced discussion. Summarize, in about 200 words, the primary benefits and concerns about this genetic modification of crop seeds.
4. Similarities and significant differences in ecological and social concerns exist between the Roundup™-resistant and modified xanthophyll cycle crops. Summarize, in about 200 words, these similarities and differences.
5. Is one of these crop modifications significantly more worrisome than the other? Develop and logically support your arguments and assertions.

REFERENCES

1. For useful Web-based references on photosynthesis, see Photosynthesis Web Resources: 2013, Orr, Larry and Govindjee, www.life.illinois.edu/govindjee/photoweb/index.html and https://bioenergy.asu.edu/education, accessed August 18, 2017.
2. Mead, I. 2017. International Energy Outlook 2017, Center for Strategic and International Studies, U.S. Energy Information Administration. www.eia.gov/pressroom/presentations/mead_91417.pdf, accessed November 5, 2017.
3. Buick, R. 2008. When did oxygenic photosynthesis evolve? *Philos Trans R Soc Lond B Biol Sci* 363:2731–2743.
4. Ke, B. 2001. *Photosynthesis: Photochemistry and Photobiophysics*. Dordrecht, the Netherlands: Kluwer Academic Publishing.
5. Parson, W. W. 1978. Thermodynamics of the primary reactions in photosynthesis. *Photochem Photobiol* 28:389–393.
6. Voet, D., and J. G. Voet. 2004. *Biochemistry*, 3rd ed. New York: John Wiley & Sons, p. 896.
7. Emerson, R., and W. Arnold. 1932. The photochemical reaction in photosynthesis. *J Gen Physiol* 16:191–205. For a description of the history of photosynthesis research, see Govindjee, and D. Krogmann. 2005. Discoveries in oxygenic photosynthesis (1727–2003): A perspective. In *Discoveries in Photosynthesis*, ed. Govindjee, J. T. Beatty, H. Gest, and J. F. Allen. New York: Springer, pp. 63–105.
8. Duysens, L. N. M. 1952. Transfer of excitation energy in photosynthesis. PhD thesis, State University at Utrecht, Utrecht, the Netherlands.
9. Parson, W. W. 2006. Electron donors and acceptors in the initial steps of photosynthesis in purple bacteria: A personal account. In *Advances in Photosynthesis and Respiration*, ed. Govindjee, J. T. Beatty, H. Gest, and J. F. Allen, pp. 213–224. Dordrecht, the Netherlands: Springer.
10. For a more complete introduction to timescales and other topics in photosynthesis, see Ke, B. 2001. Chapter 1: Photosynthesis: An overview. In *Photosynthesis: Photochemistry and Photobiophysics*. New York: Kluwer Academic Publishing, p. 13.
11. Emerson, R., R. Chalmers, and C. Cederstrand. 1957. Some factors influencing the long-wave limit of photosynthesis. *Proc Natl Acad Sci* 43:133–143.
12. Greenbaum, E., J. W. Lee, C. V. Trevault, S. L. Blankinship, and L. J. Mets. 1995. CO_2 fixation and photoevolution of H_2 and O_2 in a mutant of *Chlamydomonas* lacking photosystem I. *Nature* 376:438–441.
13. Editors, 1996. Biohydrogen production deserves serious funding. *Nat Biotechnol* 14:799.
14. Kromdijk, J., K. Głowacka, L. Leonelli, S. T. Gabilly, M. Iwai, K. K. Niyogi, and S. P. Long. 2016. Improving photosynthesis and crop productivity by accelerating recovery from photoprotection. *Science* 354(6314):857–861.
15. Lopez-Igual, R., A. Wilson, R. L. Leverenz, M. R. Melnicki, C. Bourcier de Carbon, M. Sutter, A. Turmo, F. Perreau, C. A. Kerfeld, and D. Kirilovsky. 2016. Different functions of the paralogs to the N-Terminal domain of the orange carotenoid protein in the cyanobacterium anabaena sp. PCC 7120. *Plant Physiol* 171:1852–1866;

Leverenz, R. L., M. Sutter, A. Wilson, S. Gupta, A. Thurotte et al., 2015. A 12 Å carotenoid translocation in a photoswitch associated with cyanobacterial photoprotection. *Science* 348(6242):1463–1466.

16. Amunts, A., O. Drory, and N. Nelson. 2007. The structure of a plant photosystem I supercomplex at 3.4 A resolution. *Nature* 447:58–63.

17. Sener, M. K., C. Jolley, A. Ben-Shem, P. Fromme, N. Nelson, R. Croce, and K. Schulten. 2005. Comparison of the light harvesting networks of plant and cyanobacterial photosystem I. *Biophys J* 89:1630–1642.

18. Sener, M. K., S. Park, D. Lu, A. Damjanovic, T. Ritz, P. Fromme, and K. Schulten. 2004. Excitation migration in trimeric cyanobacterial photosystem I. *J Chem Phys* 120:11183–11195.

19. Allen, J. P., G. Feher, T. O. Yeates, D. C. Rees, J. Deisenhofer, H. Michel, and R. Huber. 1986. Structural homology of reaction centers from *Rhodopseudomanas sphaeroides* and *Rhodopseudomonas viridis* as determined by x-ray diffraction. *Proc Natl Acad Sci* 83:8589–8593.

20. Hu, X., A. Damjanovic, T. Ritz, and K. Schulten. 1997. Architecture and function of the light harvesting apparatus or purple bacteria. In *Theoretical Biophysics Technical Reports*, pp. 1–28. Urbana, IL: University of Illinois at Urbana–Champaign.

21. Förster, T. 1948. Zwischenmoleckulare energiewanderung und fluoreszenz (Intermolecular energy transfer and fluorescence). *Ann. Physik* (ser. 6) 437:55–77. English translation by Robert Knox available in Mielczarek, E., E. Greenbaum, and R. S. Knox, Eds. 1993. *Biological Physics*, pp. 148–160. New York: American Institute of Physics. Note again there are several forms of the "Förster interaction." We refer to the weak and very weak sorts.

22. Cory, M. G., M. C. Zerner, X. Hu, and K. Schulten. 1998. Electronic excitations in aggregates of bacteriochlorophylls. *J Phys Chem B* 102:7640–7650.

23. Knox, R. S., and H. van Amerongen. 2002. Refractive index dependence of the Förster resonance excitation transfer rate. *J Phys Chem B* 106:5289–5293.

24. Hu, X., A. Damjanović, T. Ritz, and K. Schulten. 1998. Architecture and mechanism of the light-harvesting apparatus of purple bacteria. *Proc Natl Acad Sci* 95:5935–5941.

25. Borisov, A. Y. 2000. Why is electron transport in the reaction centers of purple bacteria unidirectional? *Biochemistry (Moscow)* 65:1429–1434.

26. Goldstein, R. A., L. Takiff, and S. G. Boxer. 1988. Energetics of initial charge separation in bacterial photosynthesis: The triplet decay rate in very high magnetic fields. *Biochim Biophys Acta Bioenerg* 934:253–263.

27. Ferreira, K. N., T. M. Iverson, K. Maghlaoui, J. Barber, and S. Iwata. 2004. Architecture of the photosynthetic oxygen-evolving center. *Science* 303:1831–1838.

28. Royal Society of Chemistry editors. 2017. *Artificial Photosynthesis: Faraday Discussion 198*. Faraday Discussions (Book 198). Cambridge, UK: Royal Society of Chemistry.

29. There have been small-scale successes in enzymatic production of H_2, some partially mimicking photosynthesis. See Woodward, J., M. Orr, K. Cordray, and E. Greenbaum. 2000. Biotechnology: Enzymatic production of biohydrogen. *Nature* 405:1014–1015; Zhang, Y. H. P. 2009. A sweet out-of-the-box solution to the

hydrogen economy: Is the sugar-powered car science fiction? *Energy Environ Sci* 2:272–282; Hambourger, M., A. Brune, D. Gust, A. L. Moore, and T. A. Moore. 2005. Enzyme-assisted reforming of glucose to hydrogen in a photoelectrochemical cell. *Photochem Photobiol* 81:1015–1020; Takeuchi, Y., and Y. Amao. 2004. Photo-operated glucose-O_2 biofuel cell based on the visible-light photosensitization of chlorophyll derivatives adsorbed on nanocrystalline TiO_2 film. *J Japan Petrol Inst* 47:355–358; see also www.ansercenter.org/ for a description of a new center at a national lab exploring a variety of photosynthesis-mimicking and non-mimicking methods for solar generation of electrical and chemical energy.

30. Chabi, S., K. M. Papadantonakis, N. S. Lewis, and M. S. Freund. 2017. Membranes for artificial photosynthesis. *Energy Environ Sci* 10(6):1320–1338. doi:10.1039/c7ee00294g. Huber, R. C., A. S. Ferreira, R. Thompson, D. Kilbride, N. S. Knutson et al. 2015. Long-lived photoinduced polaron formation in conjugated polyelectrolyte-fullerene assemblies. *Science* 348(624):1340–1343.

31. Kanan, M. W., and D. G. Nocera. 2008. In situ formation of an oxygen-evolving catalyst in neutral water containing phosphate and Co^{2+}. *Science* 321:1072–1075. Walter, M. G., E. L. Warren, J. R. McKone, S. W. Boettcher, Q. Mi, E. A. Santori and N. S. Lewis. 2010. Solar water splitting cells. *Chem Rev* 110:6446–6473; Mi, Q., A. Zhanaidarova, B. S. Brunschwig, H. B. Gray, and N. S. Lewis. 2012. A quantitative assessment of the competition between water and anion oxidation at WO3 photoanodes in acidic aqueous electrolytes. *Energy Environ Sci* 5(2):5694–5700.

32. Brimblecombe, R., G. F. Swiegers, G. C. Dismukes, and L. Spiccia. 2008. Sustained water oxidation photocatalysis by a bioinspired manganese cluster. *Angew Chem (Int Ed)* 47:7335–7338; Brimblecombe, R., D. R. J. Kolling, A. M. Bond, G. C. Dismukes, G. F. Swiegers, and L. Spiccia. 2009. Sustained water oxidation by $[Mn_4O_4]^{7+}$ core complexes inspired by oxygenic photosynthesis. *Inorg Chem* 48:7269–7279.

33. Nocera, D. 2008. *Fall Meeting of the American Chemical Society*, Philadelphia, PA, quoted by Katharine Sanderson in Chemistry for the climate, *Nature Reports* (online), October 2008, pp. 124–125. doi:10.1038/climate.2008.96.

34. Liu, C., B. C. Colón, M. Ziesack, P. A. Silver, and D. G. Nocera. 2016. Water splitting–biosynthetic system with CO2 reduction efficiencies exceeding photosynthesis. *Science* 352(6290):1210–1213. https://news.harvard.edu/gazette/story/2016/06/bionic-leaf-turns-sunlight-into-liquid-fuel/.

35. https://cleantechnica.com/2014/12/03/new-solar-cell-efficiency-record-set-46/

36. (a) Mohr, H., and P. Schopfer. 1995. *Plant Physiology.* New York: Springer, p. 152; (b) Ke, B. 2001. *Photosynthesis: Photochemistry and Photobiophysics.* Dordrecht, the Netherland: Kluwer Academic Publishing.

37. Hu, X., A. Domjanovic, T. Ritz, and K. Schulten. *Theoretical Biophysics Technical Report UIUC-TB-97-22.* Urbana, IL: Beckman Institute, University of Illinois at Urbana–Champaign.

38. Schulten, K. 1995. *Theoretical and Computational Biophysics Group.* Urbana, IL: Beckman Institute, University of Illinois at Urbana–Champaign. http://www.ks.uiuc.edu/. Accessed 2009.

39. Svinarski, O. *Theoretical and Computational Biophysics Group*. Urbana, IL: Beckman Institute, University of Illinois at Urbana–Champaign. www.ks.uiuc.edu/. Accessed 2009.

40. Koepke, J., X. Hu, C. Muenke, K. Schulten and H. Michel. 1996. The crystal structure of the light-harvesting complex II (B800–850) from Rhodospirillum molischianum. *Structure* 4(5):581–597.

41. Cory, M. G., M. C. Zerner, X. Hu, and K. Schulten. 1998. Electronic excitations in aggregates of bacteriochlorophylls. *J Phys Chem B* (102):7640–7650.

42. Hu, X., T. Ritz, A. Damjanović, F. Autenrieth, and K. Schulten. 2002. Photosynthetic apparatus of purple bacteria. *Quart Rev Biophy* 35(1):1–62.

43. RCSB PDB software references:

    Jmol: An open-source Java viewer for chemical structures in 3D. http://www.jmol.org/.

    Protein Workshop: Moreland, J. L., A. Gramada, O. V. Buzko, Q. Zhang, and P. E. Bourne. 2005. The Molecular Biology Toolkit (MBT): a modular platform for developing molecular visualization applications. *BMC Bioinformatics* 6:21.

    NGL Viewer: Rose, A. S., A. R. Bradley, Y. Valasatava, J. D. Duarte, A. Prlić, and P. W. Rose. 2018. NGL viewer: Web-based molecular graphics for large complexes. *Bioinformatics*. doi:10.1093/bioinformatics/bty419.

44. Somepics, upload.wikimedia.org/wikipedia/commons/4/49/Thylakoid_membrane _3.svg, license creativecommons.org/licenses/by-sa/4.0/deed.

45. Nield, J., P. J. Rizkallah, J. Barber, and N. E. Chayen. 2003. The 1.45 Å three-dimensional structure of C-phycocyanin from the thermophilic cyanobacterium Synechococcus elongatus. *J Struct Biol* 141:149–155.

CHAPTER 11

CONTENTS

Direct Ultraviolet Effects on Biological Systems

Photosynthetic organisms have complex, but elegant, systems for the harvesting and conversion of light energy into useful chemical forms, as long as the wavelength of light is in the visible to near-infrared region. Given the spectrum of solar light arriving at the surface of the Earth (Figure 9.1), whose half-maximum irradiance points are at about 400 and 900 nm, the targeted use of this wavelength ranger makes perfect sense. If we were photosynthetic designers, we might quibble with nature's choice of chlorophyll, with first excited state absorption in the 600–800 nm region. Why not place the first excited state in the 450–600 nm region, where the solar intensity is somewhat higher and where the entire energy of a 500-nm photon might be used to generate chemical energy? After all, a 450-nm photon carries 44% more energy than a 650-nm photon. As it stands, when a short wavelength photon is absorbed by chlorophyll's second excited state in the 420–460 nm region the excitation rapidly relaxes to the first excited state, dissipating the excess energy mostly to heat (see Figures 10.7 and 9.14). Only the first excited-state energy is relayed to the reaction center. Why not place the first excited-state absorption at 500 nm? Perhaps base photosynthesis on carotenoids and phycoerythrins, rather than chlorophylls.

There are several answers to such a proposed photosynthesis redesign. The first is aesthetic: humans might not be enamored of fields and forests of yellow-orange (Figure 11.1). The second category of answers relates to photochemistry: photons of sufficiently high energy can cause damaging chemical reactions. Figure 11.2 shows typical *molecular* effects of radiation of various frequencies. Starting at wavelengths about 600 nm and shorter, photon absorption can cause bond breakage or rearrangement. The purpose of photosynthesis *is* to cause bond breakage (e.g., $2H_2O \rightarrow 2H_2 + O_2$), but this sort of bond rupture is catalyzed by a complex system of enzymes, not by direct use of chlorophyll excited states (Figure 11.3). An enzyme's primary energy function is to lower the activation-barrier energy to bond breakage (Chapter 15): while the free energy difference between a structure with intact C–C bond and a structure with a broken bond may be 2.5 eV ($\sim 40 k_B T$), an energy 50% higher may temporarily be required to distort the entire structure before the bond can be broken.

We have stated the secret to *bio*chemistry: enzymes reduce the *activation energy* and force required to break bonds. Enzymes, however, perform this function under carefully controlled circumstances. The raw energy provided by a 5-eV photon is not so specific. If that energy can be transferred to the atoms forming a bond, the bond will break, whether biochemical equilibrium requires breakage or not.

The topic of this chapter is the effect of higher-energy photons on biosystems. To keep it rather compact, we have chosen to focus on effects of ultraviolet (UV) radiation, wavelengths from about 200 to 400 nm, on DNA and other components of human tissue such as skin. Another interesting direction, which we will not take, lies in the study of short-wavelength photon effects on photosynthetic systems, starting with absorption of visible light by carotenoids and other accessory pigments in green plants, and ending with UV effects on photosynthetic structures and function.[1]

A useful compendium of facts related to light, emitters, units, and light detection principles is the *RCA Electro-Optics Handbook*.[2]

NASA predicted "non-green" photosynthetic systems in an April 11, 2007 press release: "NASA Predicts Non-Green Plants on Other Planets. NASA scientists believe they have found a way to predict the color of plants on planets in other solar systems. Green, yellow or even red-dominant plants may live on extrasolar planets, according to scientists whose two scientific papers appear in the March issue of the journal *Astrobiology*. The scientists studied light absorbed and reflected by organisms on Earth, and determined that if astronomers were to look at the light given off by planets circling distant stars, they might predict that some planets have mostly non-green plants." (www.nasa.gov/centers/goddard/news/topstory/2007/spectrum_plants.html, accessed November 22, 2017.)

FIGURE 11.1
Moving chlorophyll's 620–650 nm absorption band to the 500–550 nm region would place more energy into the first excited state, potentially increasing the efficiency of chemical energy production, but the aesthetic effects might not be appealing. The shift would also increase unwanted photochemical side effects.

Wavelength (nm)	Photon Energy (eV)	Molecular Excitation	Bond Energies
1000	1.24	Vibrations	
900	1.38	Electronic excitations	
800	1.55		
700	1.77		
600	2.07		C–N (2.1 eV)
500	2.48	Molecular dissociation	C–C (2.5 eV)
400	3.10		
300	4.13		C=C (4.4 eV)
200	6.20	Ionization	C=O (6.3 eV)
100	12.4		C–H (11.1 eV)

FIGURE 11.2
Photon wavelengths shorter than about 600 nm have enough energy to disrupt or break covalent bonds. (Based on Smith, K.C. and Hanawalt, P.C., *Molecular Photobiology*, Academic Press, New York, 1969. Figure 1.1.)

11.1 TYPES AND SOURCES OF UV LIGHT

Ultraviolet light can include light of any wavelength between 4 and 400 nm.[3] The energy of such photons ranges from 310 eV ($12,000k_BT$) to 3.1 eV ($120k_BT$). Although the precise definition is not important, the physical effect such radiation has on biosystems is. UV is divided into various regions according to who is using it and for what purpose: health, environment, materials and engineering, or spectroscopy.

11.1.1 Extreme or Vacuum UV: 4–200 nm

Radiation in this range is strongly absorbed by air: measurements must normally be done in a vacuum. As Figure 9.1 suggests, little vacuum UV light from the sun penetrates to the surface of the Earth: the effective 5780°C blackbody irradiance of the sun at 200 nm is considerably higher than the measured solar irradiance at Earth s surface. Hotter stars have even greater vacuum UV emission, though their distance from Earth makes their current earthly biological consequences insignificant. Such stellar UV, however, may be responsible for complex hydrocarbons, even amino acids, that may have reached Earth long ago and still reach Earth, riding on meteorites and comets, so this region of the spectrum may once have been important.[4–8]

A major part of atmospheric vacuum UV absorption is due to O_2 and its cousin, O_3 (ozone): a pure N_2 atmosphere will transmit well, at least in the 150–200 nm region. Likewise, any radiation in this regime produced by a lamp or discharge, including low-pressure Hg-vapor lamps (the shortest, 185 nm line) and deuterium lamps (112–200 nm), is quickly absorbed by air and other materials, such as the material enclosing the lamp discharge (Figure 11.4). The energy of vacuum UV photons can be high enough to produce effects like Compton scattering,[9] with a recoil electron scattered out of a thin sample, or X-ray scattering or diffraction, but these effects are topics for another course of study. Of more direct interest are the molecular and atomic electronic excitations that produce photoproducts with biological consequences.

Your astronomy or exobiology colleagues will tell you that hot stars, e.g., *β Centauri, Sirius, Achernar, Rigal, Altair, Canopus*, produce a considerable amount of vacuum UV. *β Centauri's* blackbody emission peaks at about 130 nm (*RCA Electro-Optics Handbook*, 1974, p. 67). Such UV radiation appears to be heavily involved in interstellar generation of organic compounds.

The molecule with greatest biological consequences presently produced on Earth by vacuum UV light is ozone O_3. On the one hand, ozone is produced by irradiation of O_2 by wavelengths shorter than 200 nm:[10]

These lamps also produce longer wavelengths.

$$O_2 \xrightarrow{\text{UV}(<200 \text{ nm})} O + O, \quad O + O_2 \rightarrow O_3 \qquad (11.1)$$

On the other hand, ozone absorbs strongly in the 200–300 nm region, where O_2 does not. We cannot afford to investigate the atmospheric, engineering, environmental, and political details of ozone and ozone depletion at this point. Measuring amounts of ozone are complicated because ozone of any consequence is not confined to a 1-cm optical sample holder that can be placed into a standard spectrophotometer (Section 9.5). Specially designed Dobson and Brewer spectrophotometers are used for this purpose.[11] Nevertheless, if you manage to place ozone into a container of length L, its optical absorption can be described by a modified Beer's law,

$$I = I_0 e^{-\sigma c L} \qquad (11.2)$$

where σ is the absorption cross section (usually in cm^2), c is the concentration of ozone (molecules/cm^3), and L is the thickness of sample in centimeters. The UV absorption cross section of ozone as a function of wavelength is shown in Figure 11.5. There is a large decrease in the absorption cross section from 260–300 nm. (Note the vertical log scale in this diagram.) We will see later that 260 nm is the direct absorption (and UV damage) peak of DNA, so ozone would seem to be a savior of the DNA genetic code, protecting it from damage. However, UV has ways of damaging DNA other than by direct photon absorption. The actual ozone in the stratosphere is concentrated over an altitude range of 20–40 km at a (partial) pressure of about 10^{-2} Pa (N/m^2) and temperatures ranging from −50°C to −30°C. If the normal ozone above North America were brought to 0°C and 1 atm pressure (STP), it would be about 3 mm thick, which corresponds to 300 Dobson units (DU).

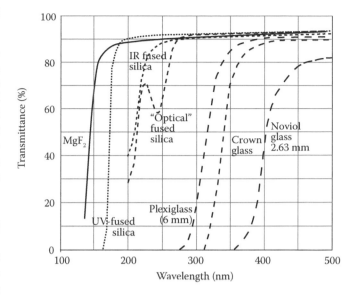

FIGURE 11.4

Ultraviolet transmission of magnesium fluoride (crystal), fused silica, glass, and Plexiglas. Because none of these materials transmits below about 120 nm, vacuum UV lamps typically do not emit below this wavelength. "Noviol O" is welder's UV-protective glass. Thicknesses are 10 mm, except as marked. (After graphs in *The Book of Photon Tools*, Oriel Instruments Catalog, 2001, Stratford, CT, 15–18; Smith, K. C. and Hanawalt, P. C., *Molecular Photobiology*, Academic Press, New York, 1969, Figure 2.1; Koller, L. R., *Ultraviolet Radiation*, Wiley, New York, 1952, Figure 21; and International Crystal Laboratories Web site, www.internationalcrystal.net/optics_11.htm, accessed December 2, 2017.)

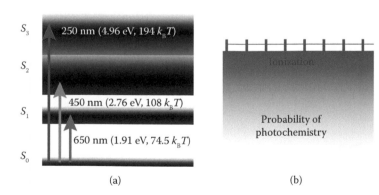

(a)

(b)

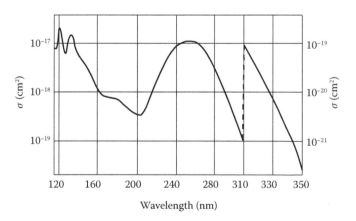

FIGURE 11.3

(a) Energy levels of chlorophyll (approximate). S_0 signifies the ground singlet state, S_1 the first excited, etc. (b) Only the third excited state carries enough energy to entail a significant probability for photochemistry (nonenzymatic bond modification). If the first excited state were moved to 450 nm, improper disposition of the excited-state energy of first and second excited states could result in damaging photochemistry.

FIGURE 11.5

The absorption cross section σ (cm^2) for ozone: $I = I_0 e^{-\sigma N L}$. The axis break at 310 nm follows the traditional division of the curve into "Hartley" and "Huggins" bands. (Based on Warneck, P., *Chemistry of the Natural Atmosphere*, Academic Press, Vol. 71 (First Edition), San Diego, CA, 1988.) The 310–350 nm data have been smoothed, removing vibrational spectral details. These hard-to-find data are said to be "freely available" in digital form.)

11.1.2 Far UV: 200–300 nm

Although not uniformly defined, the "far" UV is generally considered the wavelength range of 200–300 nm. Virtually all aromatic hydrocarbons, including DNA bases, amino acids, $NADP^+$ and NADPH, lipids, and many other molecules, absorb in this spectral region, especially at the lower wavelengths. There are two major biophysical implications of this absorption: (1) these absorbing biomolecules can be quantified and their structures studied through absorption spectroscopy, and (2) UV light in this wavelength region can cause chemical (structural) damage.

11.1.3 Near UV: 300–400 nm

The intensity of UV light from the sun and man-made light sources is much higher in the 300–400 nm region than in shorter-wavelength regions. This would be a cause of great photochemical damage to biosystems, except that the common absorbances of proteins and DNA are much smaller, less than about 1%, than at shorter wavelengths. Nevertheless, "smaller" does not mean zero. We shall see that though the absorbance of isolated DNA is small in this region (see also UV-A in the following section), the apparent ability of DNA to be damaged by such light is much higher.

Most of the biological molecules that absorb in the 300–400 nm region are prosthetic groups of proteins, which have primary absorption in the visible and second or third excited state bands in the near UV. Table 11.1 lists a few of these molecules, with their associated absorption peaks. Peaks in the 300–400 nm region are in boldface.

11.2 DIVISIONS OF THE UV FOR HEALTH PURPOSES: UV-A, UV-B, AND UV-C

Regions of the ultraviolet have been separated into spectral regions based on classifications of biological effects. Such classifications often seem arbitrary, and because they are based on national and international agreements or traditions, they are often hard to change, even when the biomedical evidence changes. Nevertheless, most regions of the world accept classifications as follows:

UV-A: 320–400 nm
UV-B: 280–320 nm (or 290–320 nm, or 270–320 nm)
UV-C: 100–280 nm (or 200–290 nm)

TABLE 11.1 Protein Prosthetic Groups and Absorption Maxima, Including Near-UV[a]

Molecule/ Prosthetic Group	Enzyme or System	$\lambda_{1,max}$ (nm)	$\varepsilon_{1,max} \times 10^{-3}$ $(M^{-1} cm^{-1})$	$\lambda_{2,max}$[b] (nm)	$\varepsilon_{2,max} \times 10^{-3}$[b] $(M^{-1} cm^{-1})$
Chlorophyll	Reaction centers, light-harvesting proteins	640–660	55–85	430–450 390–420	105–155 30–50
Flavins (FMN, FAD)[c]	Amino acid oxidase, flavodoxin, monoamine oxidase	455–460	9.1–13	355–438	7.9–15
Heme (Fe^{III})	Hemoglobin, myoglobin, cytochromes	540–580	13–15	410–440 315–350	120–130 12–20
Retinal	Rhodopsin (eye)	500	42	350	11
Fe^{IIII} sulfide	Ferredoxin	420	9.8	330	13

[a] Sources: Cantor, C. R. and Schimmel, P. R., *Biophysical Chemistry*, Part II, Table 7.2, Freeman & Co., San Francisco, CA, 1980; Urry, D.W., *J. Biol. Chem.*, 242, 4441–4448, 1967.
[b] Second and/or third excited state.
[c] FMN, flavin mononucleotide; FAD, flavin adenine dinucleotide.

Choice of the lower limit of UV-B, 270, 280, or 290 nm is somewhat important because absorption by protein Trp and Tyr is strong at 270 and 280 nm and much weaker at 290 nm. The lower limit of UV-C is often moved up to 200 nm because very few natural or artificial light sources produce wavelengths below 200 nm.

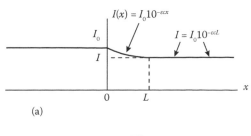

11.3 UV DAMAGE TO ORGANISMS: "ACTION SPECTRA"

Very little UV damage is done to any biosystem that is not preceded by photon absorption: absorption deposits the photon's energy in a molecule of the system. Therefore the most fundamental indicator of light's tendency to damage is the system's absorption spectrum. We recall that absorption is described by Beer's law, $I = I_0 10^{-A} = I_0 10^{-\varepsilon cL}$ or $I = I_0 e^{-a} = I_0 e^{-\sigma nL}$, where ε is the absorption coefficient (M^{-1} cm^{-1}), σ the absorption cross section (cm^2 or m^2), c (n) is the particle concentration in moles/liter (particles/cm^3 or particles/m^3), and L is the optical path length. While light hitting tissue or a microorganism does not precisely obey Beer's law—elastic scattering, reflections, and refractions often have a combined effects at least as large—Beer's law for absorption nevertheless gives us a reasonable starting point for describing the damaging effects of UV light.

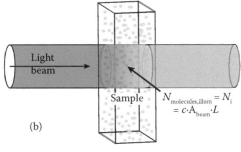

Where Beer's law applies, I and I_0 can be any measure of the quantity of light that scales linearly with the number of photons at a given wavelength: power (W), intensity (W/ m^2), photons, photons/s, photons/s·m^2, etc., as long as the light intensity is not high enough for nonlinear optical effects and as long as proper attention is paid to the area of the sample actually illuminated by the beam. Suppose, for example, that 10^{-9} moles of photons (10^{-9} Einstein) impinges on a sample that has an absorbance $A = \varepsilon cL = (10^4$ M^{-1} $cm^{-1})$ $(10^{-5}$ M$)(1.0$ cm$) = 0.10$. $I_0 - I = (6.0 \times 10^{23})(10^{-9})(1 - 10^{-0.10}) = 6.0 \times 10^{14} (0.206) = 1.2 \times 10^{14}$ photons, or 20% of the photons, are absorbed. Again assuming no nonlinear optical effects, 1.2×10^{14} molecules will have absorbed a photon. In the volume illuminated by the beam (Figure 11.6), there are

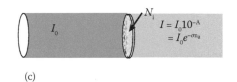

FIGURE 11.6
Light beam hitting a sample: Beer's law as sample optical path length decreases to a thin layer. (a) Beer's law, path length L. (b) The light beam often illuminates only part of the sample. (c) If the sample is thin, it is more convenient to write Beer's law in a form using the particle/ molecular absorption cross section and number of particles/molecules per unit area. σ (cm^2) $= 3.824 \times 10^{-21} \times \varepsilon$ (M^{-1} cm^{-1}).

$$N_{mol} = (10^{-5} \text{mol/L})(A_{beam} \cdot L)N_A$$

$$= (10^{-5} \text{mol/L})(\pi(0.5 \text{ cm})^2 \cdot 1.0 \text{ cm})\frac{1 \text{ L}}{1000 \text{ cm}^3}(6.0 \times 10^{23} \text{ molecules/mol})$$

$$= 4.7 \times 10^{15} \text{molecules}$$

so only 2.6% of the molecules will have absorbed a photon.

As Figure 11.6b and c shows, if the sample is thin, e.g., a monolayer layer of cells or thin layer of molecules on a microscope slide, it is more convenient to write Beer's law of absorption in terms of surface density of particles/molecules (particles/cm^2) and particle absorption cross section σ (cm^2). The relation between the cross section and absorption coefficient is

$$\sigma(cm^2) = 3.824 \times 10^{-21} \times \varepsilon(M^{-1}cm^{-1}). \tag{11.3}$$

Action spectra are defined by the extent of a specific photon-initiated effect as a function of the wavelength of excitation light. The O_2 production spectrum measured by Emerson and Arnold (Chapter 10) was such an action spectrum. In the present case the "action" we are considering is some sort of quantifiable damage or impairment to an organism or molecule caused by photon absorption. Because photon absorption is statistical in nature, specific chemical outcomes that may follow absorption in only a small fraction of the cases will be even more strongly governed by statistics. Finally, if the measured outcome is death of a cell, the statistical variation in cell "hardiness" will come strongly into play. Such spectra

Physicists entering into a research career in the biosciences had better polish up their statistics credentials.

EXERCISE 11.1: TYPICAL NUMBERS OF PHOTONS IN AN ABSORPTION SPECTROPHOTOMETER

If an absorption spectrophotometer uses a 100-W light bulb (assume 25% of the power comes out as light and the dispersive element, a diffraction grating, deflects 40% of the maximum possible light to the sample) that produces a constant power per nm from 350–650 nm (and none outside these wavelengths), scans at a rate of 5 nm/s, and has a bandwidth of 1 nm, calculate (1) the number of photons hitting the sample at 350 and 650 nm, and (2) the fraction of the photons absorbed, if the sample has absorbances of $A(350\text{ nm}) = 1.0$ and $A(650\text{ nm}) = 0.010$.

Answer

The spectrophotometer sends $(25\%)(40\%) = 10\%$ of its power toward the sample. In each 1-nm interval, the power sent to the sample is $\frac{Power}{nm} = (100\text{ W})(0.10)\frac{1\text{ nm}}{300\text{ nm}} = 0.0333\frac{W}{nm}$.

$$N(350\text{ nm}) = \left(0.0333\frac{J}{s\cdot nm}\right)(1.0\text{ nm})(0.20\text{ s})\frac{1}{E_{photon}} = \left(6.67\times10^{-3}\text{ J}\right)\frac{\lambda}{hc}$$

$$= \left(6.67\times10^{-3}\text{ J}\right)\frac{350\text{ nm}}{1.99\times10^{-16}\text{ J}\cdot nm} = 1.17\times10^{16}\text{ photons}$$

$$N(650\text{ nm}) = \left(0.0333\frac{J}{s\cdot nm}\right)(1.0\text{ nm})(0.20\text{ s})\frac{1}{E_{photon}} = \left(6.67\times10^{-3}\text{ J}\right)\frac{\lambda}{hc}$$

$$= \left(6.67\times10^{-3}\text{ J}\right)\frac{650\text{ nm}}{1.99\times10^{-16}\text{ J}\cdot nm} = 2.18\times10^{16}\text{ photons}$$

1. $N(350\text{ nm, absorbed}) = N(350\text{ nm})(1 - 10^{-1.0}) = (1.17 \times 10^{16})(0.90) = 1.1 \times 10^{16}$
2. $N(650\text{ nm, absorbed}) = N(650\text{ nm})(1 - 10^{-0.01}) = (2.18 \times 10^{16})(0.0228) = 1.05 \times 10^{16} = 0 \times 10^{14}$

An absorption coefficient of 10,000 M^{-1} cm^{-1} corresponds, for example, to a cross section of 3.8×10^{-17} $cm^2 = (0.062\text{ nm})^2$.

Action spectra are usually measured after a fixed amount of light energy is delivered to a sample. When wavelength is varied, the number of photons delivered is not the same. When wavelength is varied over a wide range, the "per photon" and the "per unit light energy" spectra will differ.

A similar, but more difficult, experiment would be to measure the fraction killed for a fixed dose.

can be measured on an absolute scale, such as the amount of energy needed to cause 50% of the maximal effects (e.g., death of 50% of the cells in a sample), or more commonly, on a relative scale.

An example of an absolute action spectrum is shown in Figure 11.7. The inverse of the amount of energy per square millimeter ($[J/mm^2]^{-1}$) needed at various wavelengths to kill 50% of the *Escherichia coli* bacteria reflects the sensitivity of the bacterium to various wavelengths of light. Such an experiment is typically done as follows. A wavelength band, $\lambda_1 \pm \Delta\lambda$, from the beam of a UV lamp is selected by a diffraction grating or UV-selective filter (e.g., an interference filter). The power at this wavelength is measured, and the amount of time needed for a certain number of Joules or photons from the beam to hit the sample is computed. The number of Joules or photons is chosen because some photochemical effect had previously been observed at this energy. The sample is then illuminated for this computed amount of time. The process is repeated at other desired wavelengths. As we noted, typical spectrophotometers optically excite only a small fraction of the molecules. In addition the quantum yield of photochemistry—the probability an excited molecule will undergo photochemistry—is usually small, so a bright UV source, with wider $\Delta\lambda$ band and long illumination times is usually needed. The selected $\Delta\lambda$ must be significantly smaller that the width of the action spectrum.

You will note from the caption of Figure 11.7 that the study was done in 1930, when efficient ways to sterilize medical-care facilities were being developed. The author of this study concluded his discussion with the statement,

… a characteristic curve of bactericidal effectiveness with a striking maximum between 260 and 270 [nm]. The reciprocal of this abiotic energy curve suggests its close relation to specific light absorption by some single essential substance in the cell.

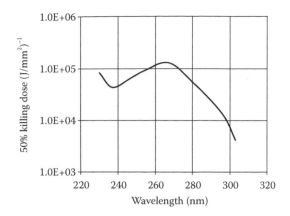

FIGURE 11.7
Action spectrum for bacterial killing: inverse of light energy per mm[2] needed to kill 50% of *E. coli* bacteria as a function of wavelength. (Based on Gates, F. L., *J. Gen. Physiol.*, 14, 31–42, 1930.)

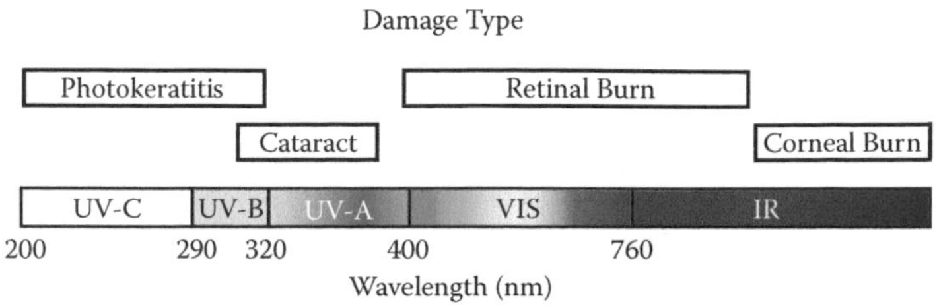

Damage Type

| Photokeratitis | | Retinal Burn | |
| | Cataract | | Corneal Burn |

| UV-C | UV-B | UV-A | VIS | IR |
| 200 | 290 320 | 400 | | 760 |

Wavelength (nm)

FIGURE 11.8
Definitions of UV-A, UV-B, UV-C, Vis, and IR for health purposes and types of eye damage that can result. (Based on Tuchinda, C. et al., *J. Am. Acad. Dermatol.*, 54, 845–854, 2006.)

We now know, of course, what this "essential substance in the cell" is: DNA (Section 11.5).

As some of us have experienced in the snow at high elevations, UV light can damage eyes. "Snow blindness," caused by a sunburn or "coldburn" of the cornea, will usually heal itself after a day or two. Depending on the wavelength of light, various types of damage can be done to the eye, some reparable, some not. Figure 11.8 summarizes the types of damage done at various wavelengths. Lasers are a cause of particularly severe damage. Occupational Safety and Health Administration (OSHA) and institutional safety regulations usually apply (www.osha.gov/dts/osta/otm/otm_iii/otm_iii_6.html).

11.4 WAVELENGTH-DEPENDENT PHOTOCHEMICAL YIELDS AND PROTEIN DAMAGE

In comparison to near UV, far UV has often been found experimentally to cause more photochemical damage per absorbed photon, even when the photon is not directly absorbed by the molecular group that will be damaged. An example of this is found in bacterial photosynthesis, where absorption of a 280-nm photon by the protein (amino acid) part of a light-harvesting complex is 10–100 times more likely to cause damage to bacteriochlorophyll *a* than absorption of a 365-nm photon directly by the bacteriochlorophyll. Quantum yields for 280-nm damage are of order 10^{-4}–10^{-3}, which means that one of every 1,000–10,000 photons absorbed causes bacteriochlorophyll destruction.[12] This is an illustration of three important facts: (1) higher photon energy correlates with more molecular damage, (2) light-harvesting molecules such as bacteriochlorophyll are efficient at the safe disposal of energy, and (3) life relies on the filtering of far UV (especially 260–280 nm) by atmospheric ozone. A second illustration of point 3 is that peptide bonds, of which there are hundreds in a typical protein, are efficiently broken by 180–190 nm light. Because excitation at 254 nm, where the peptide bond absorbs a hundred times less strongly, is still measurable (Table 11.2), polypeptides could not maintain their integrity under illumination by UV of wavelengths <200 nm.

TABLE 11.2 Photochemical Damage Liabilities of Protein Components, $\lambda = 254$ nm[a]

Component	Absorption Coefficient ε (M^{-1} cm^{-1})	Damage Quantum Yield Φ_c	Damage Coefficient ($\varepsilon \times \Phi_c$) (M^{-1} cm^{-1})
Disulfide bond (–S–S–)	270	0.13	35
Trp	2870	0.004	12
Phe	140	0.013	1.8
Tyr	320	0.002	0.6
Peptide bond[b]	0.2	0.05	0.01

[a] Adapted from Smith, K. C., and Hanawalt, P. C., *Molecular Photobiology*, Academic Press, New York, 1969, Table 5.1.

[b] Peptide bonds absorb maximally between 180 and, 190 nm where excitation efficiently breaks the bonds. The absorbance of a typical at 190 nm is about three times that at 280 nm.

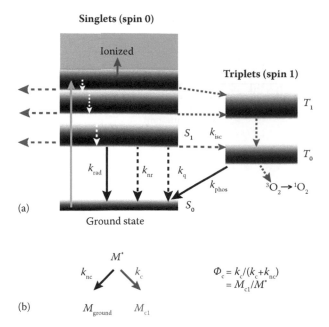

FIGURE 11.9
Excited-state kinetics when photochemistry occurs. (a) Many decay routes (colored in red) may lead to photochemical products. Routes available depend on the initial excited state. States closer to the ionized state, an electron detached from the molecule, are more likely to lead to photochemistry, though even an ionized molecule may recombine with its freed electron. In the presence of oxygen (ground state, triplet $^3O_2$) a triplet-state molecule and $^3O_2$ can exchange spin states, producing singlet oxygen, $^1O_2$, which can oxidize the molecule or another part of the molecule. Depending on the type of photochemistry considered, more than one molecule or molecular group may be involved. (b) The photochemical quantum yield of a certain photoproduct, M_{c1}, is defined as the ratio of the sum of all the photochemical rates leading to that photoproduct (k_c) divided by the sum of all decay rates from the initially excited state, or by the ratio of the number of product molecules divided by the number of excited molecules initially produced.

A radical is a chemical group that contains too few or too many electrons. OH$^+$, an OH$^-$ ion with a missing electron, is called an O–H radical, for instance. Radicals are defined with respect to the most stable form in the system at hand. OH$^-$ is a radical when surrounded by water because [OH$^-$ + H$^+$] is much lower energy than [OH$^-$ + H]. Although more often referred to as a reducing agent, the hydrogen atom can also be considered a radical because under most circumstances energy is lowered if the "extra" electron is at least partly shared with another nucleus, e.g., H_2 or H_2O. Needless to say, radicals often oxidize or reduce other molecules they encounter.

A photochemical quantum yield Φ_c (subscript stands for chemistry), appears in Table 11.2. How does this quantity fit in with other excited-state parameters we have used in Chapter 9 (e.g., Figure 9.15)? Figure 11.9 illustrates how complex an issue photochemistry can be. The primary complexity is due to multiplicity of complex molecules in a biosystem: biology simply refuses to abide by isolated-atom rules. The red arrows in the diagram indicate processes that can, but do not always, lead to photochemical damage. Leftward pointing, dotted arrows indicate energy transfer to some nearby molecular group. Of course, energy transfer by itself is not damaging, but if the entity receiving the energy is ionized or if a *radical* is formed, damage to that acceptor may well occur. Similarly, conversion of the absorbing molecule to a triplet spin state, usually because of interactions with nearby atoms, does not in itself constitute photochemical damage. The excitation energy may not perform its "intended" role, as in photosynthesis, but no damage necessarily results. As we noted in Chapter 10, photosynthetic systems have carotenoids pigments, one of whose roles is to "quench" triplets—to accept the triplet excitation energy and safely convert it to heat. Triplet excitations are more likely to result in photochemical damage, however, even if the excitation energy is not large. The reason? Oxygen. The energy difference between ground-state (triplet) and singlet oxygen is 94.3 kJ/mol ($\sim38k_BT$) and corresponds to a transition in the near-infrared at ~1270 nm. Although $38k_BT$ is not a large amount of energy compared to most chemical bond energies, the fact that this extra energy is contained in oxygens makes those atoms much more likely to oxidize another material it contacts. (While many materials oxidize in $^3O_2$, the probability is much higher in $^1O_2$.)

Generally, excitations to higher excited-state energies are more likely to cause damage to some molecular entity in the system. Outside of *photoisomerization*, which we address shortly, *photoionization* is the simplest form of photochemistry. Suppose a molecule absorbs a high-energy UV photon: high enough energy to take an electron from its current state to the *continuum*—no longer bound to the original molecule. When we study the isolated hydrogen atom, we learn that absorption of a 13.6-eV or higher photon will ionize the hydrogen atom: the electron goes off into free space. In "free space" the electron could return to its parent, emit a 13.6-eV photon and return to it bound, ground state only if the initial photon energy was exactly 13.6 eV, so that the ejected electron had zero kinetic energy. In biosystems any molecule has a dense surrounding of other molecules: there is no "free space" or vacuum state. Ionization therefore constitutes the transfer of the electron from the parent atom(s) to an adjacent atom. With the variety of molecules surrounding a biomolecule, the amount of excitation energy needed to transfer an electron from parent to some "electron-friendly" neighbor is much less than the free-space ionization energy. Photoionization is a biosystem, or any condensed system, for that matter, is better described as photoelectron transfer. Also keep in mind that in condensed systems, the photon energy need not supply the entire energy. The constant presence of the thermal bath of vibrational and kinetic energy can always supply a few extra k_BT of energy. This is referred to as a phonon or thermally assisted process. The issue is a matter of probability: if the photon has too little energy by an amount k_BT, one time out of a (finite) zillion, a random thermal vibration or two will provide just the right nuclear motion to reduce the needed ionization or transfer energy by k_BT.

Voila! We have photoionized a molecule in a condensed system even with too little photon energy (Figure 11.10). We will see this thermal assistance in quantitative form in Chapter 15.

So, absorption of a photon promotes an electron into an excited state, where it can decay via Chapter 9 processes to the ground state or where it can, with the collaboration of surrounding atoms and molecules, cause photochemical damage to itself or to a neighbor. In the case of proteins or DNA, the "neighbor" may be the same protein molecule, just a different amino acid, a prosthetic group, or even a different part of the absorbing molecule. It has been known for 40 years, for example, that excitation of Phe amino acids (at $\lambda_{exc} = 258$ nm) results in formation of Tyr, Asp, several other acids and a higher molecular weight, melanin-like polymer. The most likely damage was reported in 1969 to be release of CO_2 (i.e., oxidation) from $-COO^-$.[13] See Table 5.2. It is now known that such irradiation of Phe results in production of the OH radical, which is a very reactive oxidizing agent that can diffuse in solution. It is so unstable that it reacts with almost any biomolecule it encounters within about a nanosecond.

The topic of diffusing radicals brings us to a decision point: do we seriously pursue the important area of photoinitiated radical production, followed by diffusing, reaction rates, and concentration-dependent reactions—classical photochemistry—or do we focus on more direct effects of the photon (Figure 11.11)? We will be ignoring many types of damage actually done in biosystems, but we must choose the more restricted view of photoeffects. Diffusing radicals, scavengers in solution, multiple reaction products, reaction products that cause further reactions—these bring us to a set of possibilities that are too complicated to cover here. Some of the photochemistry we discuss will, in fact, involve primary and secondary molecules, but the role of diffusing chemical species will be minimal (short-distance diffusion).

11.5 UV DAMAGE TO DNA

DNA bases absorb UV light very efficiently in the 240–280 nm wavelength range. Table 9.1 shows peak absorption coefficients from 9,000 to 15,000 M^{-1} cm^{-1}, with those for the purines (A, G) 60% higher than for pyrimidines (C, T). These absorption coefficients are for the isolated bases, or more precisely, isolated deoxymononucleoside phosphates. In contrast, DNA bases in a strand are close to each other and absorb about equally well, independent of their identity, A, T, G, or C. The difference from isolated-base absorption suggests that adjacent DNA bases may interact, perhaps even to the extent that excited states are no longer associated with single bases. Our discussion of the bacteriochlorophyll ring in Chapter 10 leads to the issue of whether DNA bases act as isolated absorbers of UV photons, or whether the excited states, as well as the ground state, consist of electrons delocalized over adjacent bases. Nucleic acids could then present a variation on the energy transfer that occurs in photosynthetic systems: an excited base on a DNA strand passes its excitation energy along the strand, perhaps to an energy "sink" or drain located some distance away (Figure 11.12). This mechanism was proposed in the 1960s to explain why UV and other types of damage tend to localize at thymine dimer (consecutive T–T bases) sites. Such a transfer-mediated damage mechanism would have to be very rapid

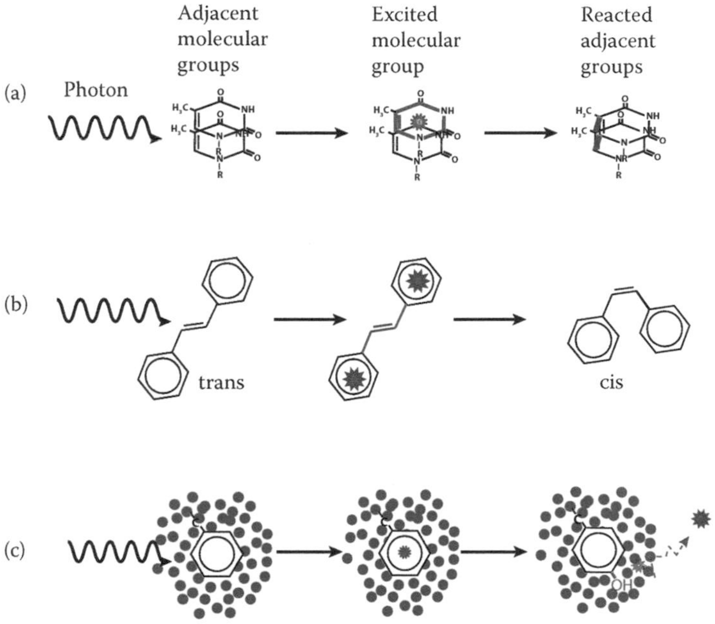

> Melanin: the major dark pigment in skin.

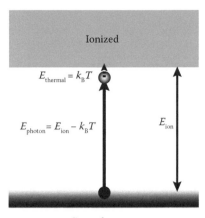

FIGURE 11.10
In biosystems and other condensed systems, photons can be "assisted" by the energy of thermal vibrations and collisions, so that less than the prescribed photon energy may cause ionization (photoelectron transfer). Electron transfer is assisted by the photon, which is assisted by thermal vibrations. The probability for thermally assisted process is less than for the direct, one-photon process, however. Chapter 15 has an example of a thermally assisted tunneling process.

FIGURE 11.11
Examples of three different types of photochemistry. (a) Excited electrons in the initially excited molecular group form bonds to a neighbor. (b) Photoisomerization: excitation of electrons cause nuclei of excited molecule to find new stable (or metastable) positions. (c) Excitation of a molecular group causes reaction of the group itself with some neighboring molecules plus creation of a radical that can diffuse and react with other molecules.

FIGURE 11.12

UV absorption by DNA. Absorption is almost uniform along the DNA strand because absorption coefficients of all four bases are comparable. The initial excited state can migrate (energy transfer) along one strand for a short distance, up to about five bases. Energy transfer is more efficient along regions of consecutive adenines (AAAAA). Radicals or $^1O_2$ that may be produced can also diffuse through the solvent shell along the helix and react with bases removed from the initial site of photon absorption. Damage along the DNA helix has been shown to localize at preferential sites of several adjacent thymine (T) bases along the DNA.

because singlet excited states decay in picoseconds. Alternatively, the excited state could transfer energy to a dissolved molecule such as O_2, which could then diffuse to other DNA sites, reacting at the most susceptible location. Although several different types of damage sites are observed, reaction is clearly not as equally distributed as absorption. Measurements show that (singlet) excitation energy transfer or excitation delocalization does occur in DNA strands, but the distance over which transfer occurs is at most about five bases.[14] Damage at localized sites separated by tens or hundreds of bases cannot be easily explained by such short-range energy transfer, suggesting involvement of longer-lived, diffusing agents.

We should remember our negative lesson from extended rings of bacteriochlorophyll about applying simplistic quantum models to biomolecules, but perhaps the small, benzene-like ring of a pyrimidine might succumb to our quantum ring model. In Exercise 11.2 we pursue such modeling more successfully. The interested student could push simple modeling even further and propose an interacting linear array of these rings on which energy can migrate.

EXERCISE 11.2: ELECTRONS ON RINGS

Apply the model of electrons confined to a ring of radius R to see if absorption spectral peaks (excited-state energy levels) of pyrimidine bases (T or C) can be approximated. This would suggest that some of the electrons in the ring are free to move from atom to atom. First, find the radius of a thymine base. Refer to Figure 5.11 for structures.

Solution

Try a PDB search on "thymine." A lot of apparently irrelevant references come up. This is primarily because thymine crystallography was done in the 1960s and will not be at the top of PDB's index. Persevere: find any structure with a thymine in it, such as 1CN0. Use Jsmol to display the structure; set Style/Scheme/Ball and Stick. Set Select/Display Selected Only/Nucleic/AT base pairs. Set Measurements/Show measurements. (Reset display to show only AT base pairs only, if necessary.) Find a thymine, zoom in and click on an atom on one side of the thymine ring. Point to an atom on the other side of the ring and double click. A dotted line with the length 0.266 nm appears (Figure 11.13a). This is the diameter of the ring. OK, so our ring has radius 0.133 nm. We can return to quantum ring example in Chapter 10 (Example 10.2), or we can quickly set $n\lambda_{dB}$ equal to the circumference again, use $p = h/\lambda_{dB}$ and $E = p^2/2m$ to get $E_n = \frac{(hc)^2 n^2}{8\pi^2 mc^2 r^2} = \frac{(1240 \text{ eV·nm})^2 n^2}{8\pi^2 (511\times10^3 \text{ eV})(0.133 \text{ nm})^2} = 2.15 n^2$ eV. If we again have four electrons per n value, then six electrons (a guess, for now), would fill $n = 1$ and place two electrons in $n = 2$, leaving room for two more. Candidates for the first excited state are then an $n = 1$ to an empty $n = 2$ spot and $n = 2 \rightarrow n = 3$. The corresponding photon energies would be

$$E_2 - E_1 = 2.15(4-1)\text{eV} = 6.45 \text{ eV}$$

$$E_3 - E_2 = 2.15(9-2)\text{eV} = 10.75 \text{ eV}$$

, with wavelenghts

$$\lambda_{21} = \frac{1240 \text{ eV·nm}}{6.45 \text{ eV}} = 192 \text{ nm}$$

$$\lambda_{32} = \frac{1240 \text{ eV·nm}}{10.75 \text{ eV}} = 115 \text{ nm}.$$

For such a crude model, these results are really not far from the measured dTMP peaks at 267 and 205 nm. If the ring radius were "adjusted" to 0.157 nm, the computed wavelengths would be 267 and 160 nm. We could argue that the two O atoms bound to ring carbons effectively enlarge the ring because O is more electronegative than C.

We're clearly fudging here, but let's give ourselves a break: there aren't many molecular orbital theorists who can calculate the first two thymine absorption bands on half of the back of an envelope. One reason this calculation works better than the attempt at the light-harvesting complex LH-II bacteriochlorophyll ring states is that the four C and two N atoms making up the thymine ring are separated by less than 0.1 nm.

For discussion: Should we consider an $n=0$ state?

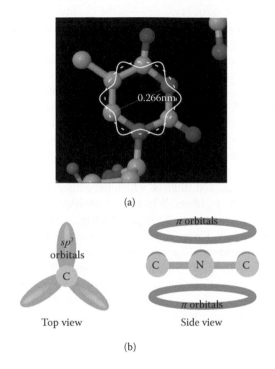

(a)

(b)

FIGURE 11.13
(a) Thymine ring diameter, found from PDB crystal structure, and quantum-ring model for electrons. Using a diameter of 0.30 nm in the ring model gives a first excited state energy close to that observed. (PDB ID 1CNO, displayed using Jmol software.) (b) Sketch of sp^2 hybrid orbitals centered at each N and C atom in the ring; delocalized π orbitals formed from the $2p_z$ orbitals of each ring atom.

11.5.1 Molecular Orbitals of DNA Bases

A good basic description of the electronic states of DNA bases can be found in the article "Excited States of Nucleic Acids," by Gueron, Eisinger, and Lamola.[15] They identify four types of electron molecular orbitals of the pyrimidine bases that govern the UV absorption and fluorescence:

1. $1s$ orbitals of N and C, similar to those in isolated atoms.
2. σ: The so-called σ framework of electron orbitals; made up of $1s$ H orbitals and $2s$ and $2p$ orbitals from C and N, hybridized to delocalized sp_2, in-plane orbitals.
3. π: $2p_z$ orbitals, one each from C and N—combined to form the delocalized π orbitals that extend vertically-upward from the ring plane. There are six of these orbitals, three of lower (π) and three of higher energy (π^*). In this model there are six electrons in the π and none in π^* orbitals.
4. n: Two nonbonding n orbitals, formed from sp_2, located at each N atom.

See Figure 11.13b. In order of increasing energy (higher is less tightly bound): atomic $1s$, σ, n, and π, with n and π comparable. The n and π orbitals give rise to the dominant UV absorption bands: $n \rightarrow \pi^*$ and $\pi \rightarrow \pi^*$ transitions. Our quantum-ring description of thymine corresponds most closely to the π orbitals and $\pi \rightarrow \pi^*$ transitions.

Like the quantum ring model, these orbitals are approximations to reality. Sophisticated computations and X-ray crystallography on systems as complex as thymine or cytosine rarely precisely confirm the integral number of electrons in simplistic orbital model.

11.5.2 UV Excitations and Spin

Our UV damage business with DNA does not demand that we go a great deal further into orbital theory, but because we have noted that singlet oxygen, $^1O_2$, is sometimes created when spins change, we should briefly consider spin in this system. In both our ring and the slightly more sophisticated orbital description, there is an even number of electrons. In the ground state, all of the electrons will be spin paired ($\uparrow\downarrow$), giving total spin zero, consistent with the usual description of photon excitation of singlet (spin 0) states. When an electron is excited—in the quantum-ring description $1 \rightarrow 2$ and in the orbital description $\pi \rightarrow \pi^*$—the excited electron has a choice of retaining its original spin, preserving the singlet state, or spin-flipping, creating a triplet state. Photon absorption will select for the singlet excited state, but it is possible that interactions between the excited electron and other spins in the surrounding could flip the electron spin, creating a triplet ($S = 1$) excited state. A triplet excited state is a multiple threat. On the one hand, the triplet will remain excited longer than a singlet, because of the forbidden nature of the $S = 1$ to $S = 0$ transition. This allows the excited DNA base a longer time to encounter another molecule and chemically react with it (i.e., form a damage product). On the other hand, the ubiquitous presence of oxygen, and its tendency to react with excited triplets, making the $^3O_2 \rightarrow {}^1O_2$ transition, produces singlet oxygen, which can diffuse further and cause damage.

So, triplet excited states seem to have a greater likelihood for causing damage but are less likely to occur than the predominant excited singlets, which have a shorter excited-state lifetime and less time to encounter and react with neighbors. Which is responsible for observed UV damage to DNA? In a sense, the answer does not matter: both singlet and triplet excited states can result in damage, but both with very low probability. Under conditions optimal for damage (pure solutions of nucleic acids) the probability for photochemical alteration of pyrimidines is about 10^{-3} and for purines 10^{-4}. Furthermore, in biological systems the vast majority of such damage is repaired. However, we all know UV-induced skin "lesions" and cancer occur and have serious consequences: these very unlikely events sometimes govern the long-term destiny of an organism. So, what is the answer? At least in the case where neighboring DNA bases are held relatively in place, e.g., in polynucleotides or in frozen solution, the excited singlets seem to be responsible for most damage.[16]

11.5.3 DNA Damage Products

The most common photoproduct of UV irradiation of DNA is the pyrimidine dimer. As our goals do not include a comprehensive inventory of the possibilities nature and chemistry affords, we will quit after a brief examination of thymine dimers. Most pyrimidine dimers are thymine T–T dimers, rather than C–C, C–T, or T–C. Such dimerization is common: when exposed to solar UV light, about 50–100 dimers/s form in each cell of your skin, according to pdb101.rcsb.org/motm/91. It would seem incumbent on biosystems to have ways to repair such frequent damage, and they do have such systems. We will first examine a nonbiological repair mechanism, and then see how biology improved upon it. There are many classic papers on DNA damage. A search on J. K. Setlow and R. B. Setlow will get you started.

11.5.4 Mutagenesis and Carcinogenesis (Brief!)

1. Absorption of UV light is not the only cause of DNA damage. As should be obvious, UV damage predominates in the skin, where UV light can penetrate (Section 11.6). DNA anywhere can be damaged by reactive chemicals in the environment, drink, or food. (We will not consider this.)
2. Damage to a DNA base is not the same thing as a "mutation." A mutation is only effected when DNA has a defect (e.g., a damaged base) that causes the organism to translate the DNA sequence into an unnatural protein. The mutation is preserved (passed to a daughter cell) only if the defective DNA is replicated, in which case the defect is converted into a normal DNA base or bases, but not the same one that was present before the initial damage.
3. Neither DNA damage nor a mutation necessarily results in cancer, which is an extremely complicated malady that we will not delve into.

11.5.5 Photon Damage and Photon-Assisted DNA Repair

One of the important lessons from physics and chemistry is that most things are in principle reversible. The only issue is probability. Of course, if the probability for a reverse process is so low that it will likely not occur even once during a lifetime or the duration of the researcher's measurements, it can be called irreversible. Is the process

$$(UV\ photon) + (T\text{–}T) \rightleftharpoons (T\text{=}T)$$

where T–T is a pair of neighboring thymine bases on DNA and T=T is a damaged thymine dimer, reversible? Not in our lifetime, but a related "reverse" process does occur:

$$(UV\ photon') + (T\text{=}T) \rightarrow (T\text{–}T)$$

The prime on the photon indicates that the UV-induced dimer can revert to the original by absorption of a photon of different wavelength. We will not go into details of thymine dimer structure. There are several variants, but they all are similar to the doubly bonded, neighboring thymines on the same DNA strand (not across the DNA strand), shown in Figure 11.11a and, in more detail, in Figure 11.14. In that figure, note the two interbase covalent carbon–carbon bonds. In addition, the two bonded thymines are 0.16 nm apart, rather than the normal 0.34 nm that characterizes B DNA. Besides the distortion of the base orientations, the dimer introduces a kink in the double helix that affects its mechanical behavior on a long-range scale. On the surface, this reversal is hard to understand: UV light causes damage and UV light cures damage? It almost sounds like the student and the dorm room. From an energetics and spectra point of view, however, reversal is expected: if the T=T dimer has an absorption spectrum, as it certainly should, there is no reason why energy absorbed by the dimer could not break the dimer bonds. Let's see how this can occur. Figure 11.15a shows the absorption spectrum of thymine and of the thymine dimer, formed by UV irradiation of poly(T). If poly(T) is irradiated at 280 nm, undamaged thymines will absorb and dimers will form with a certain probability. If the dimer quantum yield is 1.0, every absorbed photon results in a dimer former. Any dimers formed do not absorb at 280 nm, so no photoreversal occurs. The percent dimerization rises to the maximum possible, about 65%. See Figure 11.15b. (Why might the maximum not be 100%?)

Exercise 11.3 strongly suggests that the quantum yield for dimer formation is 10–100 times smaller than for breakage.

FIGURE 11.15
Thymine dimer formation and reversal by UV irradiation.
(a) Absorption spectra of thymine and thymine photodimer.
(Modified from Smith, K. C. and Hanawalt, P. C., *Molecular Photobiology*, Academic Press, New York, 1969, Figure 47, which is modified from Setlow, R., *Biochim. Biophys. Acta.*, 49, 237–238, 1961). (b) Formation of thymine dimers by 280-nm radiation (0–50 dose), followed by irradiation at 240 nm. After an extended dose at 280 nm, 65% of the thymines are dimerized; 240-nm irradiation of either the 280-irradiated sample or a fresh (lower curve) sample brings an equilibrium of 17% dimers. The horizontal axis is also proportional to time. Sample: polythymidylic acid. (Based on Deering, R. A., *Sci. Am.*, 207, 135, 1962.)

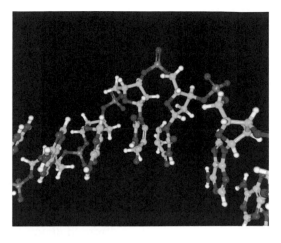

FIGURE 11.14
Solution structure of a thymine dimer. The dimer is in the center, but lines corresponding to the dimer bonds are not shown. The pair of carbons on the two bases could not be so close together without covalent bonds between them. Notice the distortion of the backbone and the much larger distance to the adenine on the right side. The DNA is double stranded, but only the dimer-containing strand is shown, for clarity. This structure was computed from nuclear magnetic resonance data. (PDB ID 1TTD by McAteer et al., 1998.)

(a)

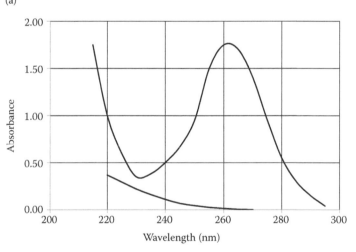

(b)

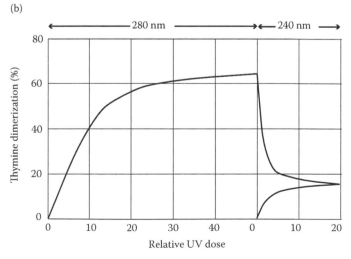

EXERCISE 11.3

At any given UV excitation wavelength an equilibrium will be set up between thymine and thymine dimers in a poly(T) sample. Using the absorbance values in Figure 11.15, determine this equilibrium at an excitation wavelength of 240 nm. (1) Assume quantum yields for formation (ϕ_f) and for breakage (ϕ_b) of a dimer are both 1.0. (2) Assume $\phi_b = 1.0$ and determine the value of ϕ_f that will explain the 240-nm data on the right side of Figure 11.15a.

Solution

$T + T \rightleftharpoons D$, where T is a thymine monomer and D is a dimer. It takes two monomers to make one dimer. We will assume a large number of thymine units and ignore neighbor exclusion or site-separation effects—if there are two thymines left anywhere, they can form a dimer. Also, assume the entire sample is uniformly irradiated throughout. (This means that the absorbance must be less than about 0.05 [Figure 11.6], so that the incident light intensity is not reduced much through the sample: the sample is about 1/10th as concentrated as in Figure 11.15.)

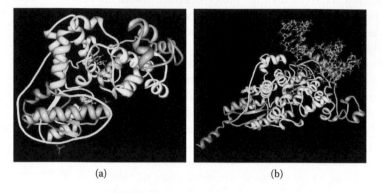

(a)   (b)

FIGURE 11.16

(a) Photolyase from *Thermus thermophilus*. A flavin mononucleotide (FMN) is visible in the lower center and a flavin adenine dinucleotide (FAD) is in the upper center. The blue light-absorbing FMN act as an antenna, and transfers absorbed excitation energy to FADH·, which transfers an electron to a nearby bound thymine dimer. (PDB ID 2J09, by Klar et al., 2006, displayed using Protein Workshop software.) (b) Photolyase from *Drosophila melanogaster* has a single blue light-absorbing, electron-donating 7,8-dimethyl-8-hydroxy-5deazaflavin (DHF) group. A bound strand of DNA, with damaged $T = T$ pulled in near the DHF can also be seen. (PDB ID 3CVV, by Glas, A.F. et al., 2009, displayed by Protein Workshop software.)

The rate of dimer formation should be proportional to the rate at which photons are absorbed by monomers. If I_0 is the number of photons/s hitting the sample, then Beer's law implies that $I_0(\lambda)(1 - 10^{-\varepsilon_T c_T L})$ photons/s are absorbed in the sample. Because we have assumed the absorbance is small, $I_0(\lambda)(1 - 10^{-\varepsilon_T c_T L}) \cong I_0(\lambda)(1 - (1 - 2.303 \cdot \varepsilon_T c_T L)) = I_0(\lambda)(2.303 \cdot \varepsilon_T c_T L)$, where c_T is the thymine concentration. (Remember $e^{-x} = 1 - \frac{x}{1!} + \frac{x^2}{2!} - \cdots$ and the 2.303 factor converts to base-10 exponentials.) So we can write the rate at which thymine absorbs photons and produces dimers $= I_0(\lambda)(2.303 \cdot \varepsilon_T c_T L) \cdot \phi_f$. Similarly, the rate at which dimers are broken is $I_0(\lambda)(2.303 \cdot \varepsilon_D c_D L) \cdot \phi_b$. We can then write

$$\frac{dc_D}{dt} = I_0 2.303 L \left(\varepsilon_T c_T \phi_f - \varepsilon_D c_D \phi_b \right)$$

In a steady state, $dc_D/dt = 0$, so $c_D = \frac{\varepsilon_T \phi_f}{\varepsilon_D \phi_b} c_T$. We have to be careful because a dimer has two T units. In Figure 11.15 the absorbances correspond to the same number of bases, whether they are dimerized or not. If we use conservation of bases, $c_T + 2c_D = c_0$, the initial concentration of thymines, and count one dimer as two thymines, then values of dimer absorbance in Figure 11.15a really correspond to $\varepsilon_D(c_D/2)L$. At 240 nm, the ratio $\frac{\varepsilon_T}{\varepsilon_D} \cong \frac{0.50}{2 \cdot 0.15} = 1.7$. So $c_D = \frac{\varepsilon_T}{\varepsilon_D}\frac{\phi_f}{\phi_b}(c_0 - 2c_D) \equiv \alpha(c_0 - 2c_D)$. Solving this for c_D, $c_D = \frac{\alpha c_0}{1+2\alpha}$.

1. If both quantum yields are 1.0, then $\alpha = 1.7$ and $c_D = \frac{\alpha c_0}{1+2\alpha} = 0.39c_0$, or $2 \cdot 39\% = 78\%$ is dimerized. (The value of c_D can be at most $0.50c_0$.)
2. The data in Figure 11.15a show that $c_D \approx 0.085c_0 = \frac{\alpha c_0}{1+2\alpha}$, which gives $\phi_f \approx 0.060$. The author of the article referred to in the figure caption states that $\phi_b \approx 1$ and $\phi_f \approx 10^{-2}$.

You should ponder this and hopefully come up with an answer that goes more or less as follows. As you will demonstrate in Problem 11.9, thymine dimers are unusually close together because of the interbase covalent bonds. Normally, there is only the stacking interaction between adjacent bases. When UV light strikes an undamaged strand of DNA containing two or more consecutive thymine bases, two bases must either, for some reason, be unusually close together when the photon arrives, so that the excited electrons can move across and be shared by the other base, or the bases must move together after excitation. Neither of these are highly probable events, so the probability (quantum yield) is low. On the other hand, a dimer has electrons in the interbase covalent bonds already tightly coupled to the ring electrons responsible for photon absorption, so 240-nm excitation can funnel energy to the dimer bonds very efficiently, making the dimer-breaking probability (quantum yield) high. Problem 11.10 asks you to fill in some more details on this issue.

The net result of the preceding discussion is that 260–300 nm light can cause T=T dimer formation, but wavelengths below about 240 nm break up dimers. What is the biological relevance? There is an appreciable amount of solar radiation in the 260–300 nm region, but little at 240 nm and below. We cannot therefore expect UV disruption of thymine dimer photodamage to be of much significance. However, the mere possibility that a photon could reverse the damage that another photon caused apparently intrigued nature.

11.5.6 Photolyases: Thymine Dimer Repair via Collaboration of an Enzyme and a Photon

In the previous paragraph it was shown that a photon of shorter wavelength, 240 nm (higher energy, 5.17 eV), could reverse the photodamage caused by a longer-wavelength, 280 nm (lower energy, 4.43 eV) UV photon. Is this because more energy is needed to break the dimer covalent bonds than is needed to create them? The answer is "No": a photon has to be absorbed before any energy can be directed toward bond-breaking. The absorption peak of the dimer is at a shorter wavelength than thymine's, so a higher-energy photon must be used to excite the dimer. Although the new dimer covalent bonds have moved the first excited-state energy higher, this excited-state energy is not the same as the covalent bond energy. The energy needed to break the two thymine dimer covalent bonds could be as small as 120 kJ/mol ($\sim 50 k_B T$),[17] corresponding to a photon wavelength above 900 nm, so the energy of a 240-nm photon is not needed. As we might expect, nature's better way of repairing photodimers involves enzymes. In the following paragraph we explore one of these enzymes that assists photoreversion of the dimer to two separated thymines. Several other important types of repair, involving excision and replacement of the dimer, are important, especially in humans. As you might expect, excision repair is complex.

DNA photolyase is a repair enzyme that reverts UV-induced thymine dimers in DNA by an electron transfer reaction between photoactivated FADH, an electron donor, and the dimer in the DNA-enzyme complex.[18] Why should an *electron transfer* reaction be involved in repair? The original, undamaged thymines each had an intraring C=C double bond. The dimer, in effect, has one of each of the double bonds displaced to the adjacent thymine. There seems to be no electron transfer involved in the damage formation. Why should electron transfer be needed for reversion of the dimer? The answer: nature has apparently found it easier to destabilize the dimer structure by adding an extra electron, at relatively low energy cost, rather than to forcibly redirect the covalent bonds. Adding an extra electron to the dimer structure allows the intraring C=C double bond in one of the thymines to reform, as in the original thymine. The interthymine C–C bonds are then destabilized, causing them to break. But, you say, there is an extra electron floating around somewhere. That's right—if nothing were done about the extra electron, this scheme would not work. One of the thymines has an extra electron that must go somewhere. Why not put the electron back where it came from? The electron-donating FADH only donated the electron because it was excited by a 300–500 nm photon (see below). After time has elapsed, the FAD excited state (it was the excited state that donated the electron) has decayed back to the ground state, so its more amenable to taking its electron back—which it does. Is that not incredibly clever?!

The crystal structures of the DNA/photolyase complexes from several organisms are known.[19] Figure 11.16 shows two such structures. The first is from the bacterium *Thermus thermophilus* and has FMN and FAD(H·) groups. Both groups are based on three-ring flavin but have distinct functions. FMN acts as a blue light-harvesting antenna, which rapidly transfers its energy to the FADH group, increasing its electron-donating tendency. If a T=T dimer is bound to this protein, an electron is then transferred, reverting the dimer to separated thymines. Figure 11.16b shows another version of a photolyase, this one from *Drosophila melanogaster*. This photolyase has a dual-purpose blue-light-absorbing/electron-donating 7,8-dimethyl-8-hydroxy-5deazaflavin (DHF) group. The excited flavin donates an electron to the damaged bases, which can be seen pulled in near to the flavin group. The issue of why a more advanced organism (*drosophila*) has a simpler photolyase system is an active research area. In 2005 it was discovered that humans have a photolyase *homolog*, but this human version does not, by itself, photoreverse damage. It may stimulate other, more sophisticated repair machinery and may be involved in circadian-rhythm activity. Search on *photolyase* and *cryptochrome* to find more information.

11.6 OPTICAL PROPERTIES OF THE SKIN

Skin is the largest organ in our bodies. Like the membrane and wall of a cell, the primary function of skin is to provide a barrier between a living organism (e.g., you) and its environment. Like any part of a biological system, a structural hierarchy underlies both the structure and function of skin: molecules form long polymers, which aggregate into bundles and sheets with useful structural characteristics. The DNA polymer constitutes the instructional manual for the replacement of skin cells constantly dying and sloughing off. As many young and old people know, the condition and operations of skin are complex and tricky to keep balanced. From our present point of view, however, skin is much simpler. It is a biological tissue made up of the normal components: water, protein, nucleic acids, lipids, polysaccharides, etc. In the context of effects of UV light, the primary properties we need clarify are interactions with photons: light scattering, absorption, fluorescence, photochemistry, energy transfer, and dissipation. Restricting consideration to this set of phenomena (a very small, specialized subset of what professional dermatologists study) will provide us with enough unanswered questions for the semester.

11.6.1 Structure of the Skin

Gray's Anatomy (the book, not the TV series) is the classic of biomedical education. Figure 11.17 pays its respects to this reference, with a few labeling enhancements. The three major layers of the skin—epidermis, dermis, and hypodermis (subcutis)—are home to a variety of cells and organs. Nerve and muscle cells run through the bottom two layers of the skin, as do sebaceous and sweat glands, blood arteries and veins, hair follicles, and other active elements of the skin organ. The absence of blood vessels from the top layer, *stratum corneum*, suggests the absence of living cells. While true in a general sense—the exterior surface of the *s. corneum* indeed is "dead" and lower levels are "on their way out"—the bottom layers are indeed active. We will not need to deal with these complex cellular structures in the lower levels of the skin because UV light primarily affects the outer layers: the s. corneum and s. germinativum.

11.6.2 Light Penetration into Skin

11.6.2.1 Wavelength dependence

Less than 1% of any light of wavelength below 300 nm penetrates more than a few tenths of a millimeter into the skin (Figure 11.18). After passing through a millimeter of skin or so, white light would appear distinctly red because of strong absorption in the UV to blue. The molecules absorbing this UV–blue light are not surprising: proteins (Trp, Tyr, Phe, Cys), DNA, flavins, NAD, and a few other common ring-containing molecules dominate. There are two other molecules that deserve further comment, however.

1. *Collagen.* Chapter 5 introduced the structure of collagen: a triple helix constructed primarily from polymers of the amino acid triplet Pro-Hyp-Gly (Hyp is hydroxyproline). The polymers collagen and elastin are primarily responsible for the strength and elasticity of skin. None of these three amino acids absorb UV light except near 200 nm, so we would expect (incorrectly) that any UV effects on skin would be due to Trp, Tyr, Phe, flavins, NAD, etc., contained in some other protein or polymeric material.

   As we learned in Chapter 9, any molecule that absorbs also reemits light as fluorescence. Absorption measurements on skin are difficult because of the dominating intensity of scattered light (see below) and the need to detect a transmitted beam at the same wavelength as the incident. Fluorescence measurements, on the other hand, rely on detection of an emitted beam at a different wavelength and in a different direction, so that scattered light can be minimized by optical filtering and judicious choice of observation angle. Molecular excitation (absorption) bands that are hidden in data such as that in Figure 11.18 are revealed. Measurement of the fluorescence from skin has revealed the presence of two characteristic bands at 295 and 335 nm,[20] neither of which can be discerned in the spectra of Figure 11.18. The first is at 295 nm and has been identified as primarily tryptophan (Trp). A second band at 335 nm is collagen. Because the three collagen amino acids we mentioned do not fluoresce, we must have missed something. Indeed, we have. Besides the primary Pro-Hyp-Gly bulk collagen material, cross-links between strands are found. The number of the cross-links apparently increases with age. Furthermore, the cross-links absorb at 335 nm. Many of the cross-links appear to be lysine derivatives but are not well characterized. Figure 11.19 diagrams these collagen cross-links.

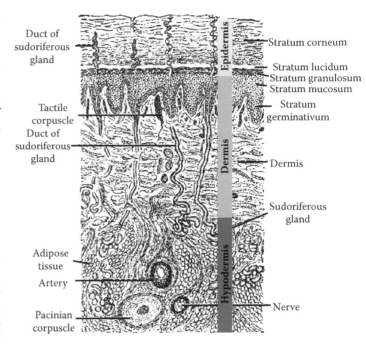

FIGURE 11.17

Diagram of human skin, based on a lithograph from *Gray's Anatomy of the Human Body* (20th U.S. ed., 1918), with a few labeling upgrades. The proportion of living cells increases with depth, but even parts of the *stratum corneum* contain metabolizing cells. Missing from this diagram are melanocytes, cells with dark pigments that absorb light. Modern micrographs, which generally convey less detail than this hand-drawing, can easily be found through Wikipedia and Web search engines.

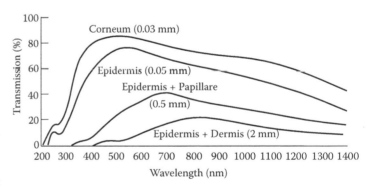

FIGURE 11.18

Transmission of light through human skin, based on historic measurements by Bachem and Reed. (From Bachem, A. and Reed, C. I., *Am. J. Physiol.*, 97, 86–91, 1931. With permission.) The ethnicity of skin samples was not mentioned. The transmission spectra are approximate, owing to limitations on spectral resolution in the UV and IR. Estimated corrections for "living" skin, with blood flow, have been made. Presence of dark melanins, with absorption coefficients ranging from 14,000 to 4,000 $M^{-1} cm^{-1}$ from 200 to 300 nm and decreasing at longer wavelengths, would significantly reduce transmission in the UV. (Steven Jacques, Oregon Medical Laser Center, http://omlc.ogi.edu/spectra/melanin/extcoeff.html, accessed November 30, 2017.)

> Scattering is still usually the "sun" to the "candle" of fluorescence.

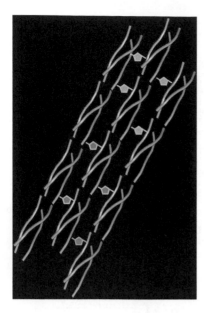

FIGURE 11.19
Collagen cross-links (diagrammatic). Cross-links between collagen fibers build up with age of the organism. Cross-link structure is indicative only. UV irradiation near 335 nm can break these cross-links, weakening the structure. Other, more complex cellular processes are also triggered, such as skin thickening and coarser wrinkles.

The presence of these collagen cross-links, absorbing at 335 nm, now presents us with a skin system that has important components absorbing over the UV-B and UV-A region: DNA at 250–270 nm, protein at 270–290 nm, and collagen at 330–350 nm. We have discussed the effect of UV light on DNA and, to a lesser extent, proteins, but what effect does UV have on this new structural component absorbing at 335 nm? The answer relates to skin "aging"—an amorphous idea usually associated with wrinkles and drooping cheeks. If collagen cross-links absorb light, we expect that the cross-linking bonds can be broken by the absorbed UV, which is true—except that UV can also *cause* cross-linking in some systems. In this case, UV radiation in the UV-A region, 300–400 nm, removes collagen cross-links, or at least reduces their fluorescence. Such irradiation has been correlated with skin thickening and coarser wrinkles.[21] (Skin aged with little UV exposure can still be wrinkled, but wrinkles are finer and skin is thinner.) From our experience with UV creation and UV reversion of thymine dimers, we can speculate that some UV wavelength may create cross-links, as well as break them. Effects on collagen are not minor issues to an organism like a human because collagen constitutes about 80% of the dry weight of the dermis and skin is our primary protection against unwanted bacteria and hazardous materials.

So much for wrinkles and old skin; on to tanning and young skin....

2. *Melanin* is another major pigment in the skin. The absorption peak is near 335 nm, but a long tail of absorption extends into the red spectral region, making melanin appear brown, red, or black, depending on spectral details. (Collagen does not have this tail.) The amount of melanin determines the color and darkness of the skin and hair of animals. The rumor is that melanin helps protect the skin from damaging effects of solar UV and that tanning, which increases the amount of melanin, will further this goal. The first part of the rumor is true—dark-skinned people incur less skin damage than light-skinned—but the second apparently is not. There is a difference between "constitutive" pigmentation and "induced" pigmentation. Miles Chekedel concluded a melanin conference as follows: "While it is clear that constitutive melanogenesis is photoprotective, the operative mechanism(s) are not clear. Whether or not induced pigmentation (tanning) is as protective as constitutive pigmentation is still an open question. No matter how high the level of epidermal melanin, it is not possible to exclude all UV light from reaching the dermis."[22] Twenty years later, the consensus is that induced pigmentation, either from a "base tan" or from chemical tanning products, does not reduce damage to skin. (www.scientificamerican.com/article/fact-or-fiction-a-base-tan-can-protect-against-sunburn/)

Eumelanins and pheomelanins are two classes of these polymers, the former found in brown/black hair and the retina; the latter in red hair and feathers. Melanin pigments are tyrosine-based polymers formed in melanocytes within specialized organelles called melanosomes. From these latter two statements and (some of) our experience with tanning, we may guess another instance of UV-induced cross-linking is at hand. We cannot seriously deal with all of the issues of melanin and skin pigmentation, but we should disavow ourselves of the notion that melanin is a simple material with merely cosmetic implications. One of the world's long-time melanin experts has posted his most important conclusions about a century or more of melanin studies (not all his):

*The chemical study of melanins have given poor results until now because of using of raw material in chemical and physical analyses and scarce knowledge of solid state by chemists and biologists. The discovery of the **particle** [my emphasis] may change our strategy in the study of structure and function of melanins. Melanin is necessary for brain, ear, eyes, skin to function. It would not be possible to see blue or green colours in animal or plant kingdom without melanin. No blue or green eyes to admire.... If you don't believe it imagine to remove the melanin from the biological tissue and observe what happens.*

Rodolfo Alessandro Nicolaus[23]

A professor of organic chemistry is teaching us that melanin does not just provide Caucasians with an attractive summer complexion, but enables sight, hearing, and brain function. Furthermore, melanin in its active form is a *nanoparticle*. What an invitation for physicists to join in! Collagen, melanin—biomaterials or biological physics thesis topics. We will not explore this area further, but the promise of many physics and chemistry dissertations lies before us: nanoparticles made up of organic, colored polymers.

11.6.2.2 Random walk of light in tissue

Despite the strong UV absorption of skin, the dominant process governing the interaction of light with tissue is scattering. Chapter 9.5 dealt with the theory of scattering as a function of particle size. For particles much smaller than the wavelength, Rayleigh scattering dominates and scattering efficiency increases with decreasing wavelength. Mie scattering applies when particles comparable to and larger than the wavelength scatter. In this case, scattering may decrease with decreasing wavelength. Scott Prahl has posted a simple-to-use online Mie calculator at omlc.ogi.edu/calc/mie_calc. html. Problem 11.11 and Miniproject 11.12 explore this wavelength dependence.

In contrast to the single-beam absorption of a clear sample, light passing through skin can scatter at large angles. Besides being more difficult to measure—the light detector must be either movable over wide angles or have a 2D recording surface—the scattered light is difficult to correlate with specific structures in the tissue. Furthermore, after passing through a millimeter of skin most photons have been scattered multiple times. This means it is difficult to associate a photon that emerges at some angle with the specific path it followed in the tissue. If a beam of light were directed into the skin to detect a small tumor, for instance, multiple scattering would result in some of the photons coming back out of the skin. Some of those photons may have even passed through the tumor, but how can you tell? And if you could tell that a given photon encountered a tumor, perhaps because a metallic nanoparticle or fluorescent dye were bound to the tumor, how could you trace the path of the photon to find where the tumor was? Some physical scientists, inspired by their astrophysics/cosmology colleagues, are not daunted by difficult tasks, however. We all know that the photons produced in the sun are (very) multiply scattered, absorbed, and reemitted before they enter the freedom of interstellar space, to arrive at Earth in the orderly package of a 5800 K blackbody emitter.

Contributions to this detection problem have been made by undergraduate researchers, some of whom have used the most common possession of a student, a smartphone, to record photon diffusion data. See ASEE 2014 Zone I Conference www.asee. org/documents/zones/ zone1/2014/Student/ PDFs/221.pdf

The field of *photon diffusion* has been created to deal with multiple scattering in skin and the tissue beneath, even in the skull. Photon diffusion theory has similarities to diffusion of molecules in liquids: multiple collisions take place and the mean free path is short. Figure 11.20 shows a simplified diagram of apparatus that might be used to attempt to detect and image tumors. Tumors and cysts are characterized by a different density and effective index of refraction, both real and imaginary (refraction and absorption, not make-believe). If a short pulse of light (0.1–10 ps) enters the material from the left, light will be scattered from the material and the scattered light detected in the various detectors will be delayed relative to the initial pulse according to the path length of the detected photon and the speed of light. Note that the speed of light in water is about 2.2×10^8 m/s. If we can resolve time delays of one laser pulse width (0.1 ps), we can resolve distances of 2.2×10^{-5} m. If the material is mostly clear, as in Figure 11.20a, a distinct backscattered pulse, reflected from the wall, will arrive in one of the detectors at a time given by twice the material thickness divided by the speed of light. Materials that are weakly scattering and absorbing, except for large, discreet objects will return a series of discreet scattered pulses, depending on the detector location. The peak locations and amplitudes will be governed by the locations of the objects and their optical properties. Highly scattering tissue containing a few discreet objects, like a tumor, will return a scattered signal, rising slowly and continuously from early times because of diffuse scattering from nearby tissue, with a pulse-like peak reflected from the tumor.

Time-resolved optical imaging, or its complement, frequency- and phase-resolved imaging, reconstructs images according to time delays. This works well for clear materials but is not needed there. In highly scattering media, where novel imaging methods are needed to supplement X-ray, magnetic resonance imaging (MRI), and ultrasound imaging, you do not know whether a scattered light signal arriving at some time delay comes from material close by, the light having traveled a circuitous route to get to the detector, or from farther away, via a straighter path (Figure 11.21).

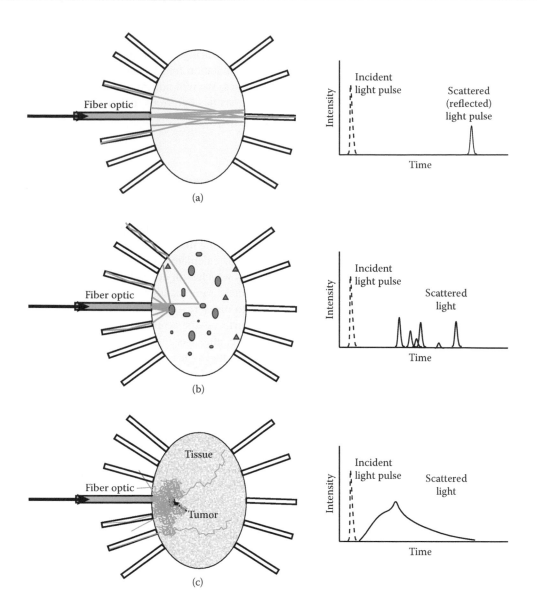

FIGURE 11.20

Time-resolved photon diffusion/imaging: a light pulse enters a material through an input optical fiber. Scattered and re-radiated light is detected through multiple fibers, whose locations are known. Often the light consists of short pulses (ps) or is modulated at high frequency, so that travel time information as well as exit location are known. Information about tumor location and size can be obtained. (a) Clear material with reflecting wall: a reflected pulse returns, with time delay given by path length and refractive index (light speed). (b) Almost clear material with large, discreet scattering particles: scattered pulse returns, with time delay given by path length and refractive index, amplitude reduced by absorption and diffuse scattering of intervening material. (c) Highly scattering medium (e.g., skin and tissue) with one large, discreet scattering object. Backscattered light begins to appear quickly from diffuse scattering. Scattering intensity from deeper tissue is delayed longer and reduced in intensity by intervening scattering and absorption. The tumor causes a pulse-like rise in scattered light. Terahertz imaging is based on these ideas.

However, it is not hopeless. The two photon paths are not equally likely. For example, it is very improbable that multiple scattering will occur along a straight line. Even without the aid of time resolution (pulsed excitation), the more likely paths can be determined. The mathematics underlying imaging/image processing via photon diffusion is rather daunting, is done via computer analysis, and is beyond the scope of this book. An example of an image analyzed with one algorithm is shown in Figure 11.22. A research center in this area can be explored at www.lrsm.upenn.edu/pmi/.

Now, we were discussing UV light illuminating skin before this diffusive interlude. UV light does not diffuse far into the skin because of the strong absorption, coupled with scattering. It would be silly to attempt to image a millimeter-sized pre-tumor located 1 cm below the skin using UV light that is absorbed in the top 0.5 mm of skin. Infrared and red light travel much deeper into tissue because such light is both absorbed and scattered less, and is more suited to imaging purposes. Nevertheless, the same physical processes, scattering and absorption, govern all these wavelength regions; only the scattering and absorption coefficients differ.

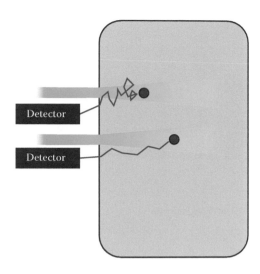

FIGURE 11.21
Optical signal from a nearby object may have traveled a distance d via a circuitous route, with many scattering events, or from a more distant object, with fewer scatterings. The signals would arrive at the detector at the same delay time. The probabilities of these two sets of scattering events are not the same, however.

11.7 SUNSCREENS

This last half of Chapter 11 doesn't discuss cosmetic appearance versus healthful living. It involves serious study of the physics of biosystems. Sometimes physics brings us to unexpected places; in this case, sunscreen is one of them. We again take a particular point of view in looking at sunscreens: conservation of energy and the disposal of absorbed UV solar energy. As we have seen, absorbed UV, especially in the UV-B region (280–320 nm), is especially liable to cause production of radicals and highly reactive molecular species in general. Sunscreens, whose primary job is to absorb solar UV before proteins, DNA, or other biologically active molecules do, must then perform their second most important job: disposing of the energy safely. The safest disposal mode of this energy is in the form of heat—vibrational energy that is rapidly dissipated throughout the tissue and especially into water, where it can do little harm other than to raise the local temperature a bit.

Location, location, location. Most sunscreens are rubbed into the skin. We have learned through our parents and health advisors that we should rub the sunscreen in and let it sit there for at least 15 minutes, so it won't wash off when we plunge into the pool or surf.[24] There is some science behind this advice, science related to materials properties of the skin and sunscreens.

The outer layer of skin is dead, mostly. This is of great advantage in terms of protection of lower levels and our freedom to place UV-absorbing agents in this dead layer with minimal fear that any resulting radicals might damage critical cellular systems. The goal is to develop a sunscreen that will penetrate throughout the stratum corneum (Figure 11.17) but stop before it gets into

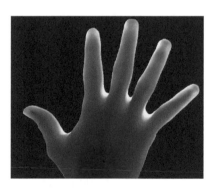

FIGURE 11.22
Image enhancement through photon diffusion processing. (Image presented at the Eurographics Symposium on Rendering (2007), *Rendering Translucent Materials Using Photon Diffusion*, Craig Donner and Henrik Wann Jensen, University of California, San Diego. Used by permission.)

more active tissue. The sunscreen cannot simply remain on the surface of the skin, because it can then easily be washed off. The design of sunscreen molecules must then have two major features: (1) an efficient UV-absorbing part, usually an organic, ring structure, and (2) a part that allows efficient passage into the stratum corneum, but not into the more water-dominated lower levels. The design solution for most sunscreens has been a ring attached to a hydrophobic hydrocarbon tail. For the most part, modern designs have succeeded, except that some of the sunscreen agents have been too successful at permeating the stratum corneum and have not stopped before entering the more "alive" layers. For example, 1%–2% of applied sunscreen molecules were recovered in urine.[25] Because of their efficiency at penetrating the skin, several active sunscreen ingredients have been proposed as transdermal delivery vehicles for hormones (birth control patches).[26]

Structures of several sunscreen molecules are shown in Figure 11.23. With the exception of para-aminobenzoic acid (PABA) and the particulate TiO_2, the molecules are hydrophobic (oily). PABA has largely disappeared from the market because it could not dispose of excitation energy in an acceptable way (see below).

Or particles—many modern sunscreens are based on nano- or microparticles.

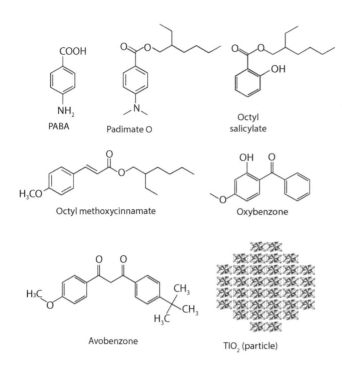

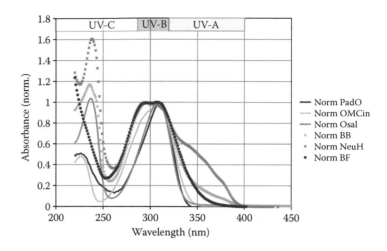

FIGURE 11.24

Normalized absorption spectra of three pure sunscreen ingredients—octyl methoxy cinnamate (OMCin), octyl salicylate (Osal), and padimate O (PadO)—along with those of three commercial sunscreen lotions—Banana Boat (BB), Neutrogena (NeuH), and Bull Frog (BF). The absorption coefficients of the pure chemicals are about 37,000 $M^{-1}cm^{-1}$ (OMCin), 5800 $M^{-1}cm^{-1}$ (Osal), and 38,000 $M^{-1}cm^{-1}$ (PadO), with some dependence on the medium/solvent.

FIGURE 11.23

Active sunscreen agents. Oily, organic molecules still dominate the ingredient list, but nano- and microstructured particles are coming into more widespread use.

11.7.1 Absorption Spectra

The UV-B region of the spectrum has been the primary target for development of effective sunscreens, until about the year 2000. This makes sense, because UV-C is presently of borderline concern and because the higher energy of UV-B, as well as its overlap with important protein and nucleic acid absorption bands in the 250–290 nm region, allows UV-B to directly damage protein and DNA molecules. UV-A has less energy per photon than UV-B and is absorbed little by amino and nucleic acids, so why concern ourselves about near-UV and blue light? You will remember two facts we have come across in our study of photophysics and photochemistry of biomolecules: (1) molecules besides amino and nucleic acids absorb light, and (2) the net energy needed to produce damaged DNA is only the equivalent of a red photon. We have seen that cross-linked collagen, of which there is a *large* amount in skin, flavins, NAD, and other biological molecules, absorb in the 320–400 nm region. If these molecules absorb and manage to produce reactive radicals, these radicals could still result in significant damage to important biological systems. Although our primary concern may be DNA damage, mutations, and cancer caused by direct DNA absorption, we cannot ignore the absorption of other species.

Figure 11.24 shows the normalized absorption spectra of three pure sunscreen ingredients, octyl methoxycinnamate, octyl salicylate, and padimate O, along with three commercial sunscreen preparations, Banana Boat Quick Dry Sport UVA & UVB Sunblock SPF 30, Neutrogena Sunblock Gel with Helioplex Broad Spectrum UVA/UVB SPF 45, and Bull Frog Sun Block Quik Gel UVA/UVB Protection SPF 36. As you note from the long commercial names, there has been some effort to extend the absorption spectrum into the UV-A region. The spectra show that some success has been achieved.

11.7.2 Sun Protection Factor

The Sun Protection Factor (SPF) of a sunscreen is determined two different ways: experimentally and theoretically. Experimentally, the SPF number is determined by exposing human subjects to a light spectrum meant to mimic noontime sun. Some subjects wear sunscreen and others do not.

> The commercial preparation spectra should be treated as approximate only, as the high scattering of the lotions makes classical absorption measurements difficult.

The amount of light that induces redness in sunscreen-protected skin divided by the amount of light that induces redness in unprotected skin is the SPF. Theoretically, it can be calculated as

$$SPF = \frac{\int A(\lambda)E(\lambda)d\lambda}{\int \left(A(\lambda)E(\lambda)/MPF(\lambda)\right)d\lambda} \qquad (11.4)$$

where $A(\lambda)$ is the "erythemal" (sunburn) action spectrum, $E(\lambda)$ is the solar irradiance spectrum, and $MPF(\lambda)$ is the "monochromatic protection factor" spectrum (approximately the inverse of the transmission). Both $A(\lambda)$ and $E(\lambda)$ are from experiment, so the calculation has some direct connection to reality. Figure 11.25 shows approximate spectra for A and E. The erythemal action spectrum, consisting of straight-line segments, is evidently a sunburn effectiveness judgment call—a model for the susceptibility of Caucasian skin to sunburn (erythema). It was proposed by McKinlay and Diffey in 1987 and adopted as a standard by the Commission Internationale de l'Éclairage (International Commission on Illumination). SPF values computed from Equation 11.4 have been observed to differ in some cases from the experimental (human subject measurements).[27] SPF values in commercial products ranging from 2 to 75 can be found.

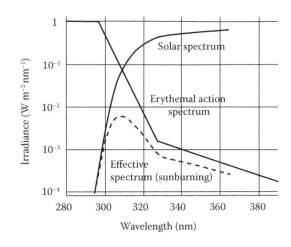

FIGURE 11.25
Spectra (excluding the sunscreen transmission spectrum) that go into "theoretical" (in vitro) determination of the Sun Protection factor (SPF). The (CIE) erythemal action spectrum is a model for the susceptibility of Caucasian skin to sunburn (erythema), proposed by McKinlay and Diffey[28] and adopted as a standard by the Commission Internationale de l'Éclairage (International Commission on Illumination).

11.7.3 Excited States of Sunscreens

What should a sunscreen do? The answer, or rather, the multiplicity of answers to this question, underlies the continuing debate about the use, safety, and effectiveness of sunscreens. One answer is, "Prevent sunburn." Another is "Prevent skin cancer," or "Prevent melanoma," or "Prevent basal cell carcinoma." Yet another view is encompassed by the more general statement, "Prevent skin damage." This last answer might include skin aging and wrinkling, and the "actinic keratosis" word that most Caucasians hear about from their doctors by age 50. Finally, the physical scientist might answer, "Absorb UV light, so it cannot do any damage." We will take this last view, with added emphasis on "cannot do any damage." In particular, we will consider what the sunscreen does with the UV energy following absorption. In some cases we find that the disposal mode of the energy is not favorable to living cells.

> We will not address the issue of the safety of the sunscreen chemicals themselves, when applied to skin.

11.7.3.1 Getting rid of absorbed UV energy

We have already explored most of the ways a molecule's excited state can decay: fluorescence, energy transfer, nonradiative decay, collisional quenching, ionization, triplet and radical generation (e.g., $^1O_2$). See Figures 11.3 and 11.9. Fluorescence, the re-radiation of a lower-energy photon, would seem a fairly benign method of disposal, as long as the emitted photon did not cause problems. We will see below one example where a re-radiated photon *is* at least a potential problem. Safe modes of energy disposal must involve conversion of 2/3 or more of the initial UV photon energy to *heat*. The energy left from 300-nm absorption would then be the equivalent of a 900-nm photon, which is not likely to cause significant photochemistry.

11.7.3.2 High absorption without a high radiative rate

We learned from Einstein and Strickler-Berg that a large absorption coefficient brings with it a large radiative rate. A ground and excited state pair cannot be designed so that the upward, photon-absorption process is highly probable while the downward, photon-emission process is improbable

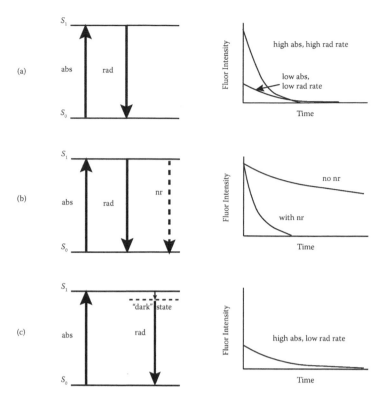

FIGURE 11.26

(a) For a simple ground- and excited-state pair, a high absorption coefficient requires a high radiative rate; low absorption, to low radiative rate. (b) If, in addition to the radiative decay route, a nonradiative (nr) decay exists, the excited state will decay faster. The fraction that decays via the nonradiative decay, usually converts excited-state energy to vibrations (heat). (c) Fast conversion of the excited state to a "dark" state introduces the radiative rate of the dark state, which may be significantly lower. Besides being lower compared to any nonradiative rate, a lower radiative rate also implies a lower-energy transfer probability.

(Figure 11.26a). The only way to reduce the probability of photon emission is to outcompete the radiative process with a faster, nonradiative process (Figure 11.26b), wherein the excited state decays more quickly. Why is a high radiative rate unfavorable? (1) If photons are emitted, they may do damage, and (2) a higher radiative rate implies a higher-energy transfer rate, if any acceptors are nearby. We certainly do not want the full quantum of excited-state energy to transfer to any other molecule; if it did, the sunscreen would have completely failed its appointed task. While some sunscreen molecules have a high nonradiative rate, corresponding to a short excited-state lifetime, others do not.[29] Have these more slowly decaying sunscreen excitations then failed in their task of converting most of the energy to heat? Apparently not. At least two of the sunscreens seem to have bypassed Einstein's high-absorption rate → high-emission rate rule by rapidly converting from the initial excited state to another, less highly emitting state (Figure 11.26c). Because this new state does not radiate as rapidly, it is less likely to transfer energy and nonradiative decay processes compete more easily.

11.7.3.3 $^1O_2$ and other radicals produced from sunscreen excited states

The sunscreen PABA has largely disappeared from the commercial sunscreen market. As the author understands it, the reason was related to allergic responses in certain users. It was later shown that UV excitation of PABA resulted in production of singlet oxygen, $^1O_2$, which is a likely culprit for producing biologically unfavorable responses in living tissue.[30] Singlet oxygen has also been observed to result from UV irradiation of other sunscreen agents.[31,32] We see that designing sunscreens is not such a simple matter. In fact, the latter study suggests even the use of sunscreens is a tricky business. We have seen that sunscreening agents move down into the skin, as they are designed to. As time goes on, more and more of the sunscreen will be in the deeper layers, where a more aqueous environment dominates and more active cells are located, and less sunscreen in the upper layers. Increasingly, deeper-lying molecules will absorb; if they produce radicals or other unpleasant products, more living cells will be affected. A possible bottom line of this matter: either reapply sunscreen frequently, replenishing the concentration in the upper layers of skin, or consider not using sunscreen at all and not staying in the sun very long. Protecting the skin from solar UV damage is not a trivial matter: designing agents that absorb efficiently and also dispose of the energy in a safe way is difficult.

11.7.3.4 Padimate O case

PABA is not a favorable sunscreen because it is water-soluble (washes off) and was found to produce singlet oxygen when irradiated. Modifications of PABA were made to exploit PABA's large absorption coefficient but reduce water solubility and undesirable photoproducts. The result was padimate O. However, a recent study revealed that when padimate O applied to skin was irradiated with UV light absorbed by the padimate, blue fluorescence was observed.[33] While fluorescence is not necessarily hazardous to living cells, this fluorescence was thought to be from skin collagen: energy initially absorbed by the sunscreen was somehow transferred to the cross-links in collagen. While this may not cause the more feared DNA damage, collagen cross-link absorption is not likely to have favorable effects in terms of graceful skin aging.

11.7.4 Beneficial Effects of UV

We close this chapter with one story of the *beneficial* effects of UV radiation on humans. While the psychological benefits of sun exposure are clear, it turns out that the active form of vitamin D is synthesized in the skin, with the requirement of UV-B radiation. Optimal synthesis occurs between 295 and 300 nm. This narrow UV band is sometimes referred to as D-UV. It further turns out that taking vitamin D supplements does not help much because little of the vitamin-pill vitamin D turns into the biologically active form of vitamin D.

Remember melanin and your summer tan? Melanin functions as a UV filter in the skin, and the concentration of melanin in the skin reduces the ability of UV-B light to penetrate the epidermal layer and reach the level where vitamin D is synthesized, below the stratum granulosum (Figure 11.17). Under normal circumstances, that sun tan does not eliminate your body's ability to create its vitamin D, but individuals with higher skin melanin content will need more time in the sun to produce the same amount of vitamin D.

> You can bet there are some benefits to UV when your mother told you, as a kid, to play outside and get some sun. Mothers do enunciate most of the important truths in life.

11.8 POINTS TO REMEMBER

- The most biologically relevant regions of solar UV
- The absorption spectra of proteins, DNA, and the few other biomolecules that absorb in the 250–400 nm region
- Excited-state processes: rates and quantum yields for various results
- The shorter the absorbed wavelength, the higher the excitation energy and the higher the likelihood for photochemistry
- How to calculate the number of absorbed photons from Beer's law

11.9 PROBLEM-SOLVING

PROBLEM 11.1 SOLAR BLACKBODY

The surface of the sun radiates as a Planck blackbody of effective temperature 5780°C:

$$I(\lambda, T)d\lambda = \frac{2\pi c^2 h}{\lambda^5} \frac{1}{e^{hc/\lambda k_B T} - 1} d\lambda$$

1. Show what the units of this expression are.
2. Find the wavelength corresponding to the peak of the distribution by differentiation.
3. Use an arbitrary multiplicative constant to force the peak height to equal the corresponding "top of atmosphere" value from Figure 9.1. Compute the solar blackbody irradiance (W m^{-2} nm^{-1}) at 250 nm and the ratio $I(500 \text{ nm})/I(250 \text{ nm})$.

PROBLEM 11.2 OZONE ABSORPTION AND DEPLETION

1. Using Equation 11.2 and Figure 11.5, calculate the UV transmission, I/I_0, at 265 nm and at 280 nm for 300 Dobson units (DU) of ozone, the "normal" amount of atmospheric ozone. (Use the ideal gas law, as necessary.)
2. Calculate the factor by which the solar UV intensity reaching the ground at 265 and 280 nm would increase if atmospheric ozone decreased to 220 DU. (220 DU is considered "depleted"—an ozone "hole.")

PROBLEM 11.3   UV ABSORPTION BY DNA

1. Compute the absorption cross section in m^2 from the absorption coefficient of a single thymine base at 265 and 285 nm. See Figure 11.15 and Table 9.1. Take this as the coefficient for an average DNA base. (Grad students: obtain access to the article Tataurov et al. 2008. *Biophys. Chem.* 133(1–3): 66–70, including the Excel spreadsheet appendix, and design a more average DNA base. Explain.) This is the effective cross-sectional area exposed to a beam of light.
2. Ignoring absorption "flattening" and other effects, calculate the effective absorption cross sections of a 1 MB (million basepair) strand of DNA.
3. Using solar irradiance data (see Figure 9.1 and rredc.nrel.gov/solar/spectra/am1.5/), calculate the number of solar photon absorptions per second by the 1-MB DNA (i.e., the number of photons striking the cross-sectional area in part 2 in the wavelength regions 260–270 and 280–290 nm. You will have to extrapolate the log(solar irradiance) data to the 260–270 nm region. You may treat the two wavelength regions as having one average energy per photon and an average spectral irradiance (W m^{-2} nm^{-1}).

PROBLEM 11.4   ACTION SPECTRA

Figure 11.7, as well as the original 1930 publication on which it is based, plots the log of the inverse of the energy/mm^2 needed to kill 50% of the bacteria. In your explanations of the following, recall that an absorption spectrum is the absorbance A plotted as a function of wavelength, based on $I = I_0 10^{-A(\lambda)}$.

1. Explain why the inverse of the energy/mm^2 is taken.
2. Explain why the log of the inverse in 1 is taken.
3. Redraw Figure 11.7 with a linear vertical scale.

PROBLEM 11.5   NUMBER OF UV PHOTONS NEEDED TO KILL BACTERIA

Figure 11.7 provides you with information about the photon energy deposited in a 1 mm^2 area. From earlier chapters, you know the dimensions of an *E. coli* bacterium.

1. Calculate the average number of photons at 260 nm that hit the surface of a single *E. coli* bacterium before it dies (with 50% probability). Assume the bacteria were in a monolayer and less than 30% of the incident UV light was absorbed.
2. Repeat at 303 nm.
3. (Graduate problem.) Find the amount of DNA (number of DNA bases) present in an *E. coli* cell. Find the average absorption coefficient of a DNA base. From these numbers, estimate the fraction of the DNA bases that actually absorbed a photon. Draw diagrams, use Beer's law, and recall that $I_0 – I$ gives the number of photons per second that are absorbed. Discuss the extent of possible damage.

PROBLEM 11.6 LOCAL HEATING AFTER PHOTON ABSORPTION

A 254-nm photon is absorbed by a Trp amino acid side chain in a protein of molecular mass 85 kDa. See Figure 9.13 and Table 11.2. (Henry et al.[34] studied this effect in heme protein absorption by the heme group.)

1. Calculate, approximately, the volumes of the Trp side chain and the entire protein.
2. The photon is absorbed within ~10^{-14} s, initially depositing its energy within the Trp ring. Calculate the energy/m^3 in the volume of the Trp.
3. Assuming a specific heat about half that of water, calculate the temperature rise of the Trp.
4. Assuming the photon energy is initially dispersed into 20 vibrational (or other) modes of the Trp, calculate the temperature rise of the Trp.
5. Henry et al. found that 50% of the deposited energy was dispersed to protein vibrational modes within about 5 ps. Assuming a specific heat about half that of water, calculate the temperature rise of the entire protein.
6. Assume that the heat from the protein is dispersed into the interior of an *E. coli* bacterium in about 1 ms. Calculate the temperature rise of the entire bacterium.

PROBLEM 11.7 THYMINE DIMER FORMATION

Figure 11.15b and 11.11a show formation of thymine dimers in poly(T), a long, single-stranded polymer of thymines.

1. Explain why 100% of the thymines do not form dimers from a statistical view. Draw a 10-mer of thymine and determine all possible sets of dimers on the strand when the 280-nm UV dose is asymptotically high (all possible dimers form). Calculate the ensemble average number of dimers and the percent of the thymines involved in a dimer. Note that you cannot have more than two T's bonded sequentially together.
2. Local distortion. Dimerization forces the dimerized thymines closer together, locally distorting the polymer chain. Assume that if a dimer forms, the thymines on either side of the dimer *cannot* dimerize with their next neighbors farther away from the initial dimer. Calculate the ensemble average number of dimers in the 10-mer.

MINIPROJECT 11.8 (PROGRAMMING) THYMINE DIMER FORMATION 2

Problem 11.7 has end effects because of the small number of bases. It also ignores the issue of the decreasing photon-absorption probability as the number of dimers increases.

1. Repeat Problem 11.7, part 2, for a 100-mer (or larger) using simulation/programming techniques. For example, write a routine or construct a spreadsheet that uses a random number generator (or other random technique) to generate damage at site n_i upon absorption of the photon number $m = 1$. This causes a $[n_i - n_{i+1}]$ dimer to form but also removes thymines n_{i-1} and n_{i+2} from the pool of potential dimers: one photon takes out four bases. Send in the photon $m = 2$. If the selected site is not among those already eliminated, another four bases are "removed." If the site has already been removed, the photon does nothing (it is not absorbed).
2. The probability that a photon is absorbed by an undamaged site decreases as more sites get damaged. Assuming the first photon ($m = 1$) absorption probability $P(m = 1)$ is 1.0, determine, from your simulation, the dependence of $P(m)$ on m.

PROBLEM 11.9 THYMINE DIMER INTERBASE DISTANCES

Access PDB ID 1SM5.

1. Locate the thymine dimer and print out a structure similar to Figure 11.14. Determine the carbon–carbon distances for the both pairs of bonds between the two thymines. (The interbase C–C bonds may not be displayed in the structure.)
2. Go to the pairs of atoms on the side of the bases opposite the dimer bonds and determine the interbase distances between those atoms.
3. Repeat for PDB ID 1RYS
4. Compare these distances to the normal interbase distances in B DNA.

PROBLEM 11.10 LOWER DIMER FORMATION QUANTUM YIELD AND STRUCTURAL DYNAMICS

Schreier et al.[35] argue that (1) the low quantum yield of UV-induced T–T dimer formation and (2) the observed rapid (<1 ps) formation of the dimer in poly(T) together imply that the dimer forms only at those rare times when the two T bases happen to be close together when the UV photon arrives. Fill in some details of this argument.

PROBLEM 11.11 MIE SCATTERING: WAVELENGTH AND SCATTERING-ANGLE DEPENDENCE

Go to the Web site http://omlc.ogi.edu/calc/mie_calc.html.

If this site, accessed November 29, 2017, disappears, search for another such site. There are others that can be found through the Wikipedia Web site.

1. Calculate Mie scattering for $\lambda = 700$ nm, sphere diameter 10.0 μm, sphere and medium refractive indices 1.50 and 1.33, respectively, 0 imaginary index (no absorption), sphere concentration 0.1 sphere/μm³. Record the total attenuation coefficient and the angle (full-width half-maximum) of the scattered radiation (magnitude vs. angle plot). Reduce the wavelength in steps of 50 nm and repeat the above until the program gives an error message or the wavelength reaches 200 nm. Create a spreadsheet for these results (see below).
2. Repeat 1 for 5.0-μm spheres.
3. Repeat 1 for 1.0-μm spheres.
4. Plot a graph of attenuation coefficient versus wavelength for the three spheres.
5. Plot a graph of scattering angle full width at half maximum as a function of wavelength for the three spheres.
6. Describe the scattering (attenuation coefficient and angle versus wavelength) from a sample consisting of a mixture of the above three spheres.

MINIPROJECT 11.12 SCATTERING AND ABSORPTION (GRADUATE PROBLEM)

Repeat Problem 11.11 but include the absorption of Trp (Chapter 9) at 280 nm. Assume the spheres are typical cells, with typical protein content and that Trp constitutes one out of every 80 amino acids. Determine the average Trp concentration in the sphere. The online Mie calculator allows for an imaginary index of refraction. Explore the Web site to find the relation between imaginary index and absorbance.

REFERENCES

1. Vass, I., A. Szilard, and C. Sicora. 2005. Adverse effects of UV-B light on the structure and function of the photosynthetic apparatus (Chapter 43). In *Handbook of Photosynthesis*, ed. M. Pessarakli, Baton Rouge, LA: CRC Press, pp. 827–842.
2. RCA Corporation. 1974. *Electro-Optics Handbook. Technical Series EOH-11.* Lancaster, PA: RCA Corp.
3. Koller, L. R. 1952. *Ultraviolet Radiation*, 3. New York: John Wiley & Sons.
4. Allamandola, L. J., S. A. Sandford, and B. Wopenka. 1987. Interstellar polycyclic aromatic hydrocarbons and carbon in interplanetary dust particles and meteorites. *Science* 237:56–59.
5. Shock, E. L., and M. D. Schulte. 1990. Amino-acid synthesis in carbonaceous meteorites by aqueous alteration of polycyclic aromatic hydrocarbons. *Nature* 343:728–731.
6. Greenberg, J. M., O. M. Shalabiea, C. X. Mendoza-Gomez, W. Schutte, and P. A. Gerakines. 1995. Origin of organic matter in the protosolar nebula and in comets. *Adv. Space Res.* 16:9–16.
7. Arnoult, K. M., T. J. Wdowiak, and L. W. Beegle. 2000. Laboratory investigation of the contribution of complex aromatic/aliphatic polycyclic hybrid molecular structures to interstellar ultraviolet extinction and infrared emission. *Astrophys. J.* 535:815–822.
8. Sandford, S. A., et al. 2006. Organics captured from comet 81P/Wild 2 by the Stardust spacecraft. *Science* 314:1720–1724.
9. Thornton, S. T., and A. Rex. 2006. *Modern Physics for Scientists and Engineers*, 114–118. Belmont, CA: Thomson (Brooks/Cole).
10. Koller, L. R. 1952. Ultraviolet Radiation, 223–227. New York: John Wiley & Sons. 1952.
11. Ozone Commission. 1991. Guidance for the use of new ozone absorption coefficients in processing Dobson and Brewer spectrophotometer total ozone data beginning 1 January 1992. International Ozone Commission Report, Geneva, Switzerland. www.esrl.noaa.gov/gmd/ozwv/dobson/papers/coeffs.html.
12. Solov'ev, A. A., Z. K. Makhneva, and Y. Erokhin. 2001. UV-Induced destruction of light-harvesting complexes from purple bacterium *Chromatium minutissimum*. *Membr. Cell Biol.* 14:463–474.
13. Smith, K. C., and P. C. Hanawalt. 1969. *Molecular Photobiology*. New York: Academic Press, pp. 87–88.
14. Xu, D., and T. M. Nordlund. 2000. Sequence dependence of energy transfer in DNA oligonucleotides. *Biophys. J.* 78:1042–1058.

15. Gueron, M., J. Eisinger, and A. A. Lamola. 1974. Excited states of nucleic acids. In *Basic Principles in Nucleic Acid Chemistry*, ed. P. O. P. T'so, 311–398. New York: Academic Press.

16. Gueron, M., J. Eisinger, and A. A. Lamola. 1974. Excited states of nucleic acids. In *Basic Principles in Nucleic Acid Chemistry*, ed. P. O. P. T'so, 50. New York: Academic Press.

17. Aida, M., F. Inoue, M. Kaneko, and M. Dupuis. 1997. An ab initio MO study on fragmentation reaction mechanism of thymine dimer radical cation. *J. Am. Chem. Soc.* 119:12274–12279.

18. Zheng, X., J. Garcia, and A. A. Stuchebrukhov. 2008. Theoretical study of excitation energy transfer in DNA photolyase. *J. Phys. Chem. B* 112:8724–8729.

19. Klar, T., G. Kaiser, U. Hennecke, T. Carell, A. Batschauer, and L.-O. Essen. 2006. Natural and non-natural antenna chromophores in the DNA photolyase from *Thermus thermophilus*. *Chembiochem* 7(11):1798–1806.

20. Tian, W. D., R. Gillies, L. Brancaleon, and N. Kollias. 2001. Aging and effects of ultraviolet A exposure may be quantified by fluorescence excitation spectroscopy in vivo. *J. Invest. Dermatol.* 116:840–845.

21. Uitto, J. 2008. The role of elastin and collagen in cutaneous aging: Intrinsic aging versus photoexposure. *J. Drugs Dermatol.* 7:s12–s16.

22. Zeise, L., M. R. Chedekel, and T. B. Fitzpatrick, Eds. 1995. *Melanin: Its Role in Human Photoprotection*, 299. Overland Park, KS: American Society for Photobiology/Valdenmar Publishing.

23. Rodolfo Alessandro Nicolaus. http://www.tightrope.it/nicolaus/index.htm (accessed September 15, 2009). See also Nicolaus, R.A. 1968. Melanins, Chemistry of Natural Products (Paris) no. 8. Paris: Hermann Publishing.

24. Diffey, B. L. 2001. When should sunscreen be reapplied? *J. Am. Acad. Dermatol.* 45:882–885.

25. Hayden, C. G., M. S. Roberts, and H. A. Benson. 1997. Systemic absorption of sunscreen after topical application. *Lancet* 350:863–864.

26. Morgan, T. M., B. L. Reed, and B. C. Finnin. 1998. Enhanced skin permeation of sex hormones with novel topical spray vehicles. *J. Pharm. Sci.* 87:1213–1218.

27. Diffey, B., and P. Farr. 1991. Sunscreen protection against UVB, UVA and blue light: An in vivo and in vitro comparison. *Br. J. Dermatol.* 124:258–263.

28. McKinlay, A. F., and B. L. Diffey. 1987. A reference action spectrum for ultraviolet induced erythema in human skin. In *Human Exposure to Ultraviolet Radiation: Risks and Regulations*. W. F. Passchier, and B. F. M. Bosnjakovich, editors. New York: Excerpta Medica Division, Elsevier Science Publisher. pp. 83–87.

29. Krishnan, R., and T. M. Nordlund. 2008. Fluorescence dynamics of three UV-B sunscreens. *J. Fluoresc.* 18:203–217.

30. Allen, J. M., C. J. Gossett, and S. F. Allen. 1996. Photochemical formation of singlet molecular oxygen ($^1O_2$) in illuminated aqueous solutions of p-aminobenzoic acid (PABA). *J. Photochem. Photobiol. B Biol.* 32:33–37.

31. Allen, J. M., C. J. Gossett, and S. K. Allen. 1996. Photochemical formation of singlet molecular oxygen in illuminated aqueous solutions of several commercially available sunscreen active ingredients. *Chem. Res. Toxicol.* 9:605–609.

32. Hanson, K. M., E. Gratton, and C. J. Bardeen. 2006. Sunscreen enhancement of UV-induced reactive oxygen species in the skin. *Free Radic. Biol. Med.* 41:1205–1212.

33. Krishnan, R., S. Pradhan, L. Timares, S. K. Katiyar, C. A. Elmets, and T. M. Nordlund. 2006. Fluorescence of sunscreens adsorbed to dielectric nanospheres: Parallels to optical behavior on HaCat cells and skin. *Photochem. Photobiol.* 82:1557–1565.

34. Henry, E. R., W. A. Eaton, and R. M. Hochstrasser. 1986. Molecular dynamics simulations of cooling in laser-excited heme proteins. *Proc. Natl. Acad. Sci. U. S. A.* 83:8982–8986.

35. Schreier, W. J., T. E. Schrader, F. O. Koller, P. Gilch, C. E. Crespo-Hernández, V. N. Swaminathan, T. Carell, W. Zinth, and B. Kohler. 2007. Thymine dimerization in DNA is an ultrafast photoreaction. *Science* 315:625–629.

36. Employees, Oriel Instruments. 2001. *The Book of Photon Tools.* Stratford, CT: Oriel Instruments Catalog, pp. 15–18.

37. Deering, R. A. 1962. Ultraviolet radiation and nucleic acid. *Sci. Am.*, 207:135–145.

38. Warneck, P. 1988. *Chemistry of the Natural Atmosphere.* San Diego, CA: Academic Press.

39. Glas, A. F., M. J. Maul. et al., 2009. The archaeal cofactor FO is a light-harvesting antenna chromophore in eukaryotes. Proc.Natl.Acad.Sci.USA 106: 11540–11545.

40. Bachem, A., and C. I. Reed. 1931. The penetration of light through human skin. *Am. J. Physiol.*, 97:86–91.

41. Donner, C., and H. W. Jensen. 2007. *Rendering Translucent Materials Using Photon Diffusion.* San Diego, CA: University of California.

42. Gates, F. L. 1930. A study of the bactericidal action of ultra violet light: III. The absorption of ultra violet light by bacteria. *J. Gen. Physiol.*, 14:31–42.

43. Tuchinda, C., S. Srivannaboon, and H. W. Lim. 2006. Photoprotection by window glass, automobile glass, and sunglasses. *J. Am. Acad. Dermatol.,* 54:845–854.

44. McAteer, Jing, Y. et al., 1998. Solution-state structure of a DNA dodecamer duplex containing a Cis-syn thymine cyclobutane dimer, the major UV photoproduct of DNA. *J.Mol.Biol.* 282: 1013–1032.

PART IV

BIOLOGICAL ACTIVITY
(Classical) Microworld

CHAPTER 12

CONTENTS

Classical Biodynamics and Biomechanics

12

The previous few chapters have emphasized the truth that quantum mechanics underlies all major biological events: every chemical reaction, photosynthesis (the source of energy for biological life), the structures of biomolecules and their balance of stability and instability, the properties of materials that make up tissues in biosystems, etc. What is there to life that does not rely on these fundamentals? The answer is that everything biological relies on the laws of quantum mechanics for its function, but it is not always useful to attempt a quantum description of everything biological, especially when the scale of moving objects is larger than 100 nm or so. Where then do we go for the quantitative tools to describe larger-scale motion? We need two types of laws: *conservation laws* and *statistical mechanics/thermodynamics*. The former deals with single objects of macroscopic size, and the latter deals with microscopic objects involved in motion over macroscopic distances. Movement of cells, subcellular organelles, and even molecular motors and biopolymers can largely be viewed as macroscopic. A chemical reaction involving many identical molecules over a spatial scale of microns or larger is an example of the second type of law. We again find that biology is distressingly on the borderline of micro- and macroscopic. The next few chapters deal with these laws and their implementation in a biological context.

Before we begin the laws, we could use a refresher on definitions for describing how things move (Table 12.1):

12.1 CONSERVATION LAWS, NEWTON'S LAWS, FORCES, AND TORQUES

Newton's laws of motion are often the first analytical tools we learn in our first course of physics. They are not the most fundamental, however, and can be viewed as arising from the conservation laws in nonrelativistic form:

CONSERVATION LAWS

$$\textbf{Conservation of energy: } E_i = E_f \qquad (12.1)$$

Condition: isolated system (no external input of energy)

$$\textbf{Conservation of Momentum: } \vec{p}_i = \vec{p}_f, \textbf{ with } \vec{p} = m\vec{v} \qquad (12.2)$$

Condition: absence of a net external force

$$\textbf{Conservation of Angular Momentum: } \vec{L}_i = \vec{L}_f, \textbf{ with } \vec{L} = I\vec{\omega} \qquad (12.3)$$

where I is the moment of inertia and $\vec{\omega}$ is the angular velocity (frequency), direction defined by the right-hand rule. Condition: central potentials, absence of a net external torque.

TABLE 12.1 Variables Describing Motion

Quantity	Notation(s)	Relation to Position	Average Value for the Discreet Case
Position	$\vec{r}$, (x,y,z)	—	—
Displacement	$\Delta\vec{r}$, $(\Delta x, \Delta y, \Delta z)$	$\Delta\vec{r} = \vec{r}_2 - \vec{r}_1$, $(x_2 - x_1, y_2 - y_1, z_2 - z_1)$	(same)
Velocity	$\vec{v}$, (v_x, v_y, v_z)	$\vec{v} \equiv \dfrac{d\vec{r}}{dt} = \left(\dfrac{dx}{dt}, \dfrac{dy}{dt}, \dfrac{dz}{dt}\right)$	$\vec{v} \equiv \dfrac{\Delta\vec{r}}{\Delta t} = \left(\dfrac{\Delta x}{\Delta t}, \dfrac{\Delta y}{\Delta t}, \dfrac{\Delta z}{\Delta t}\right)$
Acceleration	$\vec{a}$, (a_x, a_y, a_z)	$\vec{a} = \dfrac{d\vec{v}}{dt} = \dfrac{d^2\vec{r}}{dt^2} = \left(\dfrac{dv_x}{dt}, \dfrac{dv_y}{dt}, \dfrac{dv_z}{dt}\right)$	$\vec{a} = \dfrac{\Delta\vec{v}}{\Delta t} = \left(\dfrac{\Delta v_x}{\Delta t}, \dfrac{\Delta v_y}{\Delta t}, \dfrac{\Delta v_z}{\Delta t}\right)$
		$= \left(\dfrac{d^2 x}{dt^2}, \dfrac{d^2 y}{dt^2}, \dfrac{d^2 y}{dt^2}\right)$	$= \left(\dfrac{\Delta x}{(\Delta t)^2}, \dfrac{\Delta y}{(\Delta t)^2}, \dfrac{\Delta y}{(\Delta t)^2}\right)$

How do Newton's laws derive from these conservation laws? First, let us state them.

NEWTON'S THREE LAWS OF MOTION

First Law: In the absence of a net, external force, an object in uniform motion (constant velocity) remains in uniform motion and an object at rest remains at rest.

$$\text{Second Law: } \vec{F}_{net,ext} = m\vec{a} \tag{12.4}$$

Third Law: $\vec{F}_{21} = -\vec{F}_{12}$ If object 1 exerts a force $\vec{F}_{21}$ on object 2, then object 2 exerts an equal and opposite force on object 1.

All three of these laws can be seen as a consequence of conservation of momentum, using the definition of the force and momentum,

$$\vec{F}_{net,ext} \equiv \frac{d\vec{p}}{dt} \text{ and } \vec{p} = m\vec{v} \tag{12.5}$$

Conservation of angular momentum implies Newton-type laws for angular motion:

NEWTON'S THREE LAWS OF ANGULAR MOTION

First Law: In the absence of a net, external torque, an object in uniform motion (constant velocity) remains in uniform motion and an object at rest remains at rest.

$$\text{Second Law: } \vec{\tau}_{net,ext} = \vec{r} \times \vec{F} = \frac{d\vec{L}}{dt}$$

$$\tau_{net,ext,z} = I_z \frac{d\omega_z}{dt} = I_z \alpha_z \tag{12.6}$$

Third Law: $\vec{\tau}_{21} = -\vec{\tau}_{12}$. If object 1 exerts a force $\vec{\tau}_{21}$ on object 2, then object 2 exerts an equal and opposite force on object 1.

In addition to Newton's laws, conservation of energy gives the Work-Energy Theorem:

$$E = \text{const} = KE + PE$$

$$\Delta E = 0 \Rightarrow \Delta(KE) = -\Delta(PE) \tag{12.7}$$

12.2 FRICTION: FAMILIAR AND LESS FAMILIAR EXAMPLES OF MOTION

We often preface the solution to a first-year physics problem with, "Neglecting friction...." Though this approximation is often not very accurate even in those problems, it leads in the wrong direction for most situations in biosystems. Let's look at Figure 12.1 for situations we have encountered before. We preface our considerations with the admission that all treatments of frictions are approximate. The equations we write so confidently to describe frictional forces do not emerge from fundamental principles but from approximate, phenomenological treatments. The fundamental problem is that friction results from interactions and reorganization of atoms and molecules on a nanometer scale, but we usually make approximations that amount to the nanoscale structure being constant over many cm^2 of area.[1] When we get down to cellular and subcellular distances and areas, these uniformity assumptions are simply not accurate enough. Consideration of even the simplest realistic cases from a microscopic view generally leads to 5-year doctoral research projects. In the case of biosystems, fluid (water) generally surrounds moving objects: the uniformity assumptions deal with the contents of the local volume around the moving particle. This can be as variable as a surface on a nanometer scale. If we intend to make any progress on a first approximation to microscale motions and dynamics in a biosystem, we must start by ignoring many details, assuming uniformity, and comparing with experiments to judge how close we have come to reality.

12.2.1 Surface Friction

Surface friction *per se* is not relevant to biological systems. Surface *forces* and the work done by them are, of course, critical, as many types of cells rely on attachment to and motion across surfaces for their livelihood. There also is a frictional energy loss, but this sort of loss is a result of receptors and motors on the cell surface attaching to a structure on a surface (often another cell), de-attaching and moving in the cell membrane, and reattaching, with the entire membrane flowing. This cannot be described by the force diagram in Figure 12.1a. Some recent nanoscale analysis confirms that frictional force is proportional to the number of atoms that interact between two surfaces and that at the nanoscale materials in contact behave more like large rough objects rubbing against each other rather than as two smooth surfaces.[1] Nevertheless, we solve the standard surface friction problem to contrast the behavior with friction in fluids, which is more relevant to biology. Consult your first-year physics text for more information, if necessary. As usual, we assume uniform, flat surfaces characterized by some coefficient of friction μ. We ignore the fact that "uniform and flat" contradicts the existence of a non-zero coefficient of friction, but we understand that there are no holes or bumps comparable in size to the moving object.

If the applied force in Figure 12.1a is in the x direction and exceeds the static frictional force, the net force is to the right and Newton's second law gives

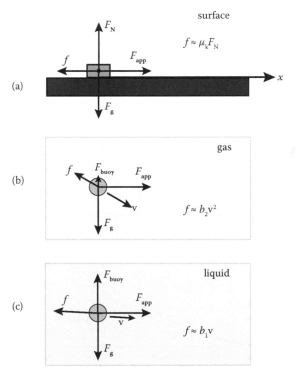

FIGURE 12.1
Force diagrams with friction (a) on a surface, (b) in a gas, and (c) in a liquid. Though force diagrams look deceptively similar, the behavior of the moving object is very different in the three cases.

$C_d \sim 0.9$, a typical bicycle plus cyclist; 0.4, rough sphere; 0.1, smooth sphere; 0.25, Toyota Prius; 0.295, bullet.

$$a = a_x = \frac{F_{net}}{m} = \frac{\left(F_{app} - f\right)}{m} = \frac{F_{app} - \mu_k F_N}{m} = \frac{F_{app} - \mu_k F_g}{m} = \frac{F_{app} - \mu_k mg}{m} \quad (12.8)$$

In typical cases F_{app} and F_g are of the same order of magnitude, μ_k is between 0 and 1 and the acceleration is some fraction of $g = 9.80$ m/s.^{2} This acceleration persists forever, resulting in a linearly increasing velocity. Figure 12.2a graphs the velocity and position of the object, as well as the power generated and work done by the force F_{app} as a function of time. The power generated by the constant applied force must increase linearly because the velocity is linearly increasing with time.

12.2.2 Air Friction (High Reynolds Number)

For a relatively large object moving through a gas at relatively high speed (high Reynolds number), the frictional force, sometimes called the *drag force*, is approximately proportional to the square of the speed, and is directed opposite to the direction of motion.

The force on a moving object due to fluid (gas) friction is

$$f_{air\ drag} = b_2 v^2 = \frac{1}{2} \rho_2 A C_d v^2 \quad (12.9)$$

where ρ_a is the gas density (about 1.20 kg/m³ for air), A is the "reference area" (approximately the cross-sectional area of the object perpendicular to the direction of motion), C_d is the (dimensionless) drag coefficient, and v is the speed of the object through the gas. The friction coefficient b_2 is given the subscript 2 because it is the (dominant) second term in an expansion of the friction coefficient as a function of speed: $f = \sum_i b_i v^i$. The power needed to counteract the aerodynamic drag is

$$P_{air\ drag} = f_{air\ drag} \cdot v = b_2 v^3 = \frac{1}{2} \rho_a A C_d v^3 \quad (12.10)$$

If a constant force is applied to an object, the object accelerates until the drag force rises to equal the applied force, when the net force is zero and the speed reaches the (constant) terminal speed. If only the gravitational and drag forces exist in Figure 12.1b,

$$F_{net} = ma = m\frac{dv}{dt} = F_g - f_{air\ drag} = F_g - b_2 v^2 \quad (12.11)$$

The terminal speed occurs when $dv/dt = 0$:

$$v_{term} = \sqrt{\frac{F_g}{b_2}} = \sqrt{\frac{2mg}{\rho_a A C_d}} = \sqrt{\frac{2\rho_{obj} V_{obj} g}{\rho_a A C_d}} \cong \sqrt{2g \langle R \rangle \frac{\rho_{obj}}{\rho_a}} \quad (12.12)$$

In the last expression V_{obj} is the volume and $\langle R \rangle$ the average radius of a spheroid-shaped object. Figure 12.2b shows graphs of velocity, position, power, and work as a function of time for one-dimensional motion under the influence of gravity and air friction.

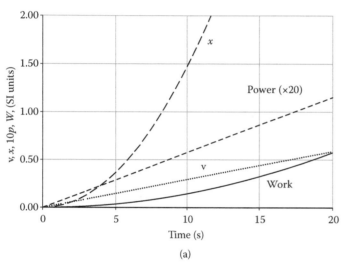

(a)

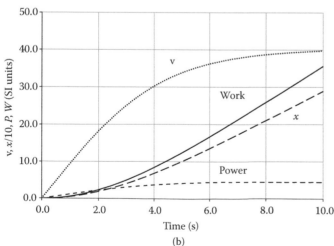

(b)

FIGURE 12.2

Dynamics of a particle under the influence of a constant applied force and (a) surface friction and (b) air friction. (Air friction corresponds to a low density fluid at high Reynolds number.)

12.2.3 Liquid Friction (Stokes, Low Reynolds Number)

The frictional force on a "small" object moving through a fluid (liquid) at "low" speed is given by

$$f_{\text{liq drag}} = b_1 v \tag{12.13}$$

If a force F_{app} is applied to the object, the net force is $F_{\text{app}} - f_{\text{liq drag}}$ and

$$F_{\text{net}} = ma = m\frac{dv}{dt} = F_{\text{app}} - f_{\text{liq drag}} = F_{\text{app}} - b_1 v \tag{12.14}$$

The terminal velocity is given by $0 = F_{\text{app}} - b_1 v$ or $v_{\text{term}} = \dfrac{F_{\text{app}}}{b_1}$. The friction coefficient b_1 is given by the Stokes equation for a sphere:

$$\text{Stokes frictional coefficient: } b_1 = b_{\text{stokes}} = 6\pi\eta R \tag{12.15}$$

where η is the liquid viscosity in $N{\cdot}s/m^2$ or $Pa{\cdot}s$ ($\eta_{H2O} = 1.0 \times 10^{-3}\,Pa{\cdot}s$).
   So

$$v_{\text{term}} = \frac{F_{\text{app}}}{6\pi\eta R} \tag{12.16}$$

The solution to the differential equation (Equation 12.14) can be guessed: $v(t) = A(1 - e^{-t/\tau})$, so $\dfrac{dv}{dt} = A\left(\dfrac{1}{\tau}e^{-t/\tau}\right) = \dfrac{A}{\tau} - \dfrac{1}{\tau}A\left(1 - e^{-t/\tau}\right) = \dfrac{A}{\tau} - \dfrac{1}{\tau}v = \dfrac{F_{\text{app}}}{m} - \dfrac{b_1}{m}v$, $A = F_{\text{app}}\tau/m$, and $\tau = m/b_1$.

$$v(t) = \frac{F_{\text{app}}}{b_1}\left(1 - e^{-b_1 t/m}\right) = \frac{F_{app}}{6\pi\eta R}\left(1 - e^{-6\pi\eta R t/m}\right) \tag{12.17}$$

Let's calculate some values of these terminal velocities and time constants.
   Table 12.2 should be shocking to those uninitiated in the implications of Newton's laws on a microscale in liquids: "inertia" is an irrelevant concept. Objects 1 μm or smaller coast less than about the diameter of one atom when an applied force is removed. Things start and stop moving almost instantly: it appears that Newton's first law in liquids could be stated, "Objects in uniform motion remain in uniform motion only if a constant force is applied; if the applied force is removed, motion stops instantly." Note that these numbers result primarily from the dimensions of the system: objects micron-sized or smaller do not coast. Figure 12.3 shows the response times of 1-mm and 1-nm objects to the sudden application of a 10-pN force in water. The two graphs look identical, but

TABLE 12.2   Terminal Velocity in Water and Time to Reach Terminal Velocity[a]

R	Volume (m³)	Mass (kg)	b_1 (N·s/m)	τ(s)	v_{term}(m/s)	Distance[b]
1 nm	4.2×10^{-27}	1.3×10^{-23}	1.9×10^{-11}	6.7×10^{-13}	5.3×10^{-1}	0.35 pm
10 nm	4.2×10^{-24}	1.3×10^{-20}	1.9×10^{-10}	6.7×10^{-11}	5.3×10^{-2}	3.5 pm
100 nm	4.2×10^{-21}	1.3×10^{-17}	1.9×10^{-9}	6.7×10^{-9}	5.3×10^{-3}	35 pm
1 μm	4.2×10^{-18}	1.3×10^{-14}	1.9×10^{-8}	6.7×10^{-7}	5.3×10^{-4}	0.35 nm
10 μm	4.2×10^{-15}	1.3×10^{-11}	1.9×10^{-7}	6.7×10^{-5}	5.3×10^{-5}	3.5 nm
100 μm	4.2×10^{-12}	1.3×10^{-8}	1.9×10^{-6}	6.7×10^{-3}	5.3×10^{-6}	35 nm
1 mm	4.2×10^{-9}	1.3×10^{-5}	1.9×10^{-5}	6.7×10^{-1}	5.3×10^{-7}	350 nm

[a] Sphere density 3,000 kg/m³, applied force, 10 pN; frictional force assumed to scale linearly with velocity
[b] Distance moved after applied force is removed.

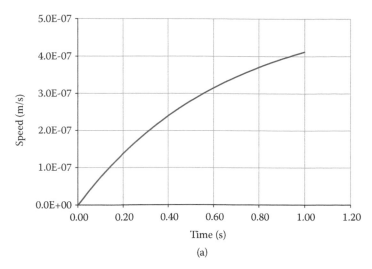

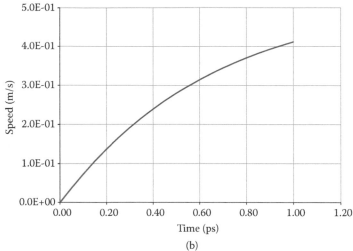

FIGURE 12.3

Dynamics of a particle moving in water with a constant, 10-pN applied force. (a) Particle radius 1 mm, density 3,000 kg/m³. (b Particle radius 10 nm, density 3,000 kg/m³. The two graphs look identical, but the time scales differ by 10^{12} and the velocity scales by 10^6. For particles 100 nm in size or less, travel distances to reach terminal speeds are the diameter of an atom or less.

the time scales differ by 10^{12} and the velocity scales by 10^6. To be sure, the velocity scales directly with the magnitude of the applied force, as in Equation 12.17, so applying a 10 μN force to the 1-mm sphere would make the velocities of the two spheres identical. The response time, however, does not depend on the magnitude of the applied force. We probe more of the dependence of these coasting times and distances in Problem 12.2.

A legitimate criticism of Table 12.2 is that it ignores nanoscale structure and the constant, random collisions with solvent molecules. We discover that the directional "drift" motion caused by the applied force can roughly be superimposed on the random Brownian motions caused by solvent collisions. Might our continuum approximation, $f_{\text{liq drag}} = b_l v$, be leading us astray, such that when we include the detailed molecular motions might we then find gross corrections and inertial coasting, as in our normal life in air? There are two answers. First, even if we're off by a factor of 10, we still conclude that nanometer-sized objects stop within an atom or two of distance. Second, the response times and distances are about the same as typical vibration times and distances in the closely packed solvent (water) system, so we have no serious reason to doubt our approximation.

12.3 GRAVITATIONAL FORCES

No one has yet figured out how to eliminate or shield systems from the gravitational force, so we need to understand how it affects typical biological systems. For our applications, we won't need to know how birds fly or ducks swim, but we do need to modify our first-year physics approach to gravitational and other forces when particles are small and immersed in water. Chapter 5 dealt with the issue of defining "small" in various contexts. The rule of thumb was to compare the magnitudes of the gravitational and thermal energies: if $mgh << k_B T$, where h is the height range in the system at hand (e.g., the height of a test tube or height of a cell), then the particle can be considered small. The implication of this type of smallness is that random thermal motion and collisions overwhelm the tendency of the system to minimize its gravitational energy by reducing the height of the mass m. The primary modification of these gravitational considerations in a biosystem relate to water, the density of water, the relatively strong interactions between water molecules and dissolved or suspended particles, and the significance of thermal energy. In water and other liquids, the gravitational force is opposed by a buoyant force. In contrast to molecular solutions, where mgh is usually much smaller than $k_B T$ and forces between dissolved molecules are smaller than forces between dissolved molecule and water, colloidal suspensions of particles of mass $mgh \sim k_B T$ can form because of significant interactions between colloid molecules. We do not consider colloids further, though they can be significant in biosystems involving particulate-rich water. We deal with a specific biophysical application of the gravitational/buoyant force principle in Section 12.6.1.

Figure 12.4 shows how the gravitational force causes heavier particles to sink to lower heights. This may be precise enough when the fluid surrounding the particles is gaseous or a vacuum, but when the fluid has a density comparable to the particles, the buoyant force must be included. As Archimedes taught us, the magnitude of the buoyant force is equal to the weight of the displaced volume of water:

$$F_{\text{buoy}} = m_w g = \rho_w V g \tag{12.18}$$

where ρ_w is the density of the fluid (water), V is the volume of the object, and g is the acceleration due to gravity. The direction of this buoyant force is opposite to the force of gravity. This means that the net force on an object submerged in water is

$$F_{net} = mg - m_w g = (\rho - \rho_w)Vg \qquad (12.19)$$

The density of many biomolecules is only about 25% greater than the density of water, so the buoyant force is significant: $(m - m_w) \sim 0.25m$. The primary relevance of the gravitational and buoyant forces in biology on a microscopic scale is in the biophysical technique of centrifugation (Section 12.6 and Chapter 6).

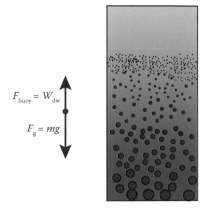

FIGURE 12.4
Buoyant force, usually negligible in gases, opposes the gravitational force in fluids. The magnitude of the buoyant force is equal to the weight of the displaced fluid (water): $F_{buoy} = \rho_w Vg$.

12.4 VOLUME CHANGES AND COMPRESSIBILITY

$$\beta \equiv -\frac{1}{V}\frac{\partial V}{\partial P} \text{ (compressibility)}, \quad B \equiv \frac{1}{\beta} \text{ (bulk modulus)} \qquad (12.20)$$

Physicists know how to change the volume of an object: change the pressure. The compressibility β of a bulk substance is defined as (–) the fractional change in volume per unit incremental pressure increase. Water is generally described as an incompressible fluid. While the majority of liquids are, in fact, incompressible to good accuracy, the incompressibility of water is directly related to the strong interactions and close packing between water molecules. At 0°C and low pressure the compressibility is $5.1 \times 10^{-10}\,Pa^{-1}$, so for most practical purposes water does not compress. The bulk modulus of water is 2.2 GPa, the inverse of the compressibility. The low compressibility of water means that even at an ocean depth of 4 km, where pressures are 40 MPa (~400 atm), there is only a 1.8% decrease in volume.

> 1 atm = 1.013×10^5 Pa = 0.1013 MPa.

Under what circumstances does the volume of a cell or a protein change in water? Volume changes are important to cells because their membranes may burst. If the volume of a protein changes during a chemical reaction, there may be a significant free energy change due to a $P\Delta V$ free energy term. Because a cell is filled with water, increasing the outside pressure by even large amounts does not appreciably change the cellular volume (Figure 12.5a): the interior water won't let the cell shrink. A second factor is that cells are not like steel drums sealed to the outside world. They are permeable to water, so if the pressure change were to occur slowly, water molecules could pass through the membrane. This permeability is the source of most cellular volume changes. If, for example, salt is added to the outside solution, osmotic forces tend to equalize concentrations inside and outside the cell: water leaves the cell interior to dilute the exterior salt. The cell then shrinks (Figure 12.5b). If a cell equilibrated in 0.15 M NaCl is suddenly transferred to distilled water, the cell expands as water moves inside to dilute the interior salt concentration. If the expansion is too rapid and too great, the cell membrane bursts. This is one of the common biochemical sample preparation methods to fragment cells and extract their interior materials.

> The *French press* (not for coffee) can also be used to fragment cells. It looks like Figure 12.5a, except that a large pressure is suddenly removed, causing a pressure wave to pass through the sample, breaking the cell membrane through shear stress.

Biomolecules and other structures in biosystems have compressibilities that are important in two situations: in deep water and in high-pressure experiments designed to measure whether conformational changes of proteins (or other biomolecules) result in volume changes. However, the compressibility itself is not the biophysical issue but rather the net volume change of the reaction. If a protein, for example, undergoes a reaction in which the volume changes, there is a free-energy change associated with the volume change, in addition to other free-energy changes.

$$A \rightleftarrows B, \ \Delta G = P\Delta V \text{ (+other)} \qquad (12.21)$$

The rule of thumb for underwater pressure is that pressure increases by 1 atm for every 10 m of depth below the water's surface.

Suppose a 5-nm-diameter protein undergoing a binding reaction shrinks in volume by 1%. (These numbers correspond to a large change. More typical values appear to be in the region of 0.1%.[2]) The volume change is $-(4\pi/300)(2.5\times10^{-9}\text{m})^3 = 6.5\times10^{-28}\text{m}^3 = 0.65 \text{ nm}^3 = 390 \text{ cm}^3/\text{mol}$. If the ambient pressure is 1 atm (1.0×10^5 Pa), then the energy change is $P\Delta V = (1.0 \times 10^5 \text{ N/m}^2)(6.5 \times 10^{-28} \text{ m}^3) = 6.5 \times 10^{-23} \text{ J} = 39 \text{ J/mol} \cong 0.016 \, k_B T$. Under most circumstances this energy is much smaller than other contributions to the free energy change. Four kilometers deep in the ocean, however, this $P\Delta V$ energy change rises to 15 kJ/mol or about 6 k_BT—enough to alter some biochemical reaction equilibria. This brings us to one of the author's favorite books he still has not managed to read: *An Introduction to Physical Oceanography*, by William S. von Arx.[3] (There are many newer books.) Biological physics would seem to change at a fundamental level in the deepest parts of the ocean because of pressure effects. Common biochemical reactions at the surface may not appreciably occur at high pressure; side reactions may become important; polymers may not form the same way. Some old and obvious observations about life in the deep, such as why fish seem to all have long, needlelike teeth,[4] remain to be answered. Answers may be related to subtle changes in biochemical reactions that could be the fodder for many MS and PhD dissertations.

There is a "however" to our developing enthusiasm about radically different life in the deep being related to high-pressure effects. The "however" has to do with experiments that are now in progress to test the enzymology of the deep. There are several recent X-ray crystal structures of deep-sea enzymes. One of them is dihydrofolate reductase, the deep-sea version of whose structure is in the Protein Data Bank (PDB) ID 3IA4, shown in Figure 12.6a. The unusual feature of this structure is that it is *not* terribly unusual: "A C$_\alpha$ [polypeptide alpha carbon] superposition of a representative monomer of the MpDHFR ternary complex and its EcDHFR equivalent (1DRE) reveals an rms deviation of 1.0 Å."[2] This means the shallow and deep-sea versions of the protein have almost identical structures. In contrast, the deep-sea flashlight fish in Figure 12.6b is clearly not the cousin of a rainbow trout. The lesson here is that while some physical factors seem to be important, they may not be the ones that govern biology. In the case of Figure 12.6a the "low-pressure" version of the enzyme seems perfectly capable of adapting to high-pressure life with no major structural changes.[2] On the other hand, the adaptation of the flashlight fish in Figure 12.6b to life under low-light conditions is much more dramatic. We should again ponder the importance of light to biosystems. Think of it: the creature in Figure 12.6b may be nature's adaption to the rules of quantum mechanics when photons are rare.

12.5 STRESS AND STRAIN

The application of one or more forces to an object of finite size creates a *stress* on that object. The object's response to that stress is called *strain*, although we usually reserve that term for solids, which can bend and break. Fluids respond to stress, of course, by flowing, but if the stress is too great even the fluid can break up into turbulent flow.

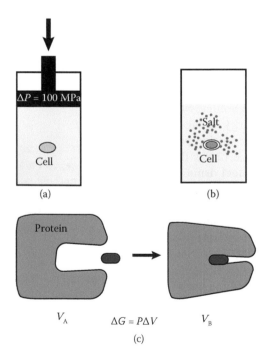

FIGURE 12.5
Volume changes in biosystems. (a) Increasing the pressure does not result in a significant change in volume of a cell. (b) Adding salt to the water around a cell causes the cell to shrink. (c) Volume changes in protein molecules can be caused by conformational changes. Volume changes in proteins correspond to a free energy change $P\Delta V$. The free energy change is usually only significant when the ambient pressure is several hundred atm or more.

12.5.1 Shear Stress

When a force is applied to a bulk object in one direction, molecules in the object tend to slide in the direction of the force but exert an equal and opposite force, if there is no acceleration. When a viscous fluid is placed between two plates and one plate is moved with velocity v_0 (Figure 12.7a), fluid next to a plate tends to stay with that plate. There is then a fluid velocity gradient in the transverse (vertical) direction, causing fluid molecules to rub against each other and resist that gradient. The proportionality constant η in the descriptive equation

$$\text{Viscous fluid shear:} \quad \frac{f}{A} = \frac{\eta v_0}{d} \qquad (12.22)$$

defines the viscosity. There is a constant dissipation of energy to heat. More on dissipation shortly.

If the same sort of shear force is applied to a solid, the solid does not flow at a constant velocity, but deforms a certain distance Δx in the direction of the force until the "springiness" of the material equals the applied force (Figure 12.7b):

$$\text{Shear of a solid:} \quad \frac{F}{A} = S \frac{\Delta x}{L_0} \qquad (12.23)$$

The constant S is called the *shear modulus*. This diving board is the classic example of shear stress, but the board would be oriented vertically in Figure 12.7b, with the area A the cross-sectional area of a cut through the board. One of our applications is the bending of rodlike cilia and flagella, where we find that function demands that the shear modulus differs, depending on the direction of the applied force: the biomaterial is an anisotropic solid.

When a solid object of length L_0 is pulled or pushed, it tends to stretch or compress a certain distance ΔL (Figure 12.7c). The constant relating the length change to the force and material characteristics is called *Young's modulus*:

$$\text{Stretching of a solid:} \quad \frac{F}{A} = Y \frac{\Delta L}{L_0} \qquad (12.24)$$

There can be a variety of microscopic events occurring during stretching. Most of us have observed the thinning of a rubber band in the transverse direction as it is stretched, but the amount of thinning varies by material.

Equations 12.22 through 12.24 do not represent fundamental laws of physics but rather descriptive parameterizations that have been applied to bulk, man-made materials. Are they useful in a biological context? We make constant use of the viscosity, primarily because it gives us a useful and quick estimate of frictional losses as particles move through water. The viscosity characterizes the water, or rather the aqueous solution surrounding the biomolecular structures we are interested in. We can argue that water in a cell might be close to bulk water. We have seen the limitations of this assertion in previous chapters. What about the solid bulk materials? Are there any bulk solids in cells? The examples we gave above were the flagella and cilia used in cell motility. What about proteins and DNA? Here we run into the issue of the spatial scale. On the one hand, a piece of tendon (in large part collagen, a protein) can be, and has been, studied in sample sizes of several cm³, clearly qualifying as "bulk." On a slightly smaller scale, skin and hair also have elastic properties that can be measured like those of a steel rod.

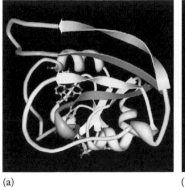

(a)   (b)

FIGURE 12.6
(a) Crystal structure of dihydrofolate reductase (DHFR) from the psychropiezophilic bacterium *Moritella profunda*, which was isolated from the deep ocean at 2°C and 280 atm. The structure is not much different from DHFR from 1-atm species. (PDB ID 3IA4, by Hay, et al., *Chembiochem.*, 6, 1419–1422, 2009.) (b) Flashlight fish *Photostomias* from NOAA archives (http://oceanexplorer. noaa.gov/explorations/04deepscope/background/deeplight/ media/fig3a_400.jpg) clearly differs from shallow-water fish, but the differences seem to center on adaptation to low light, not high pressure. (Image is now copyright by Edith Widder: used by permission.)

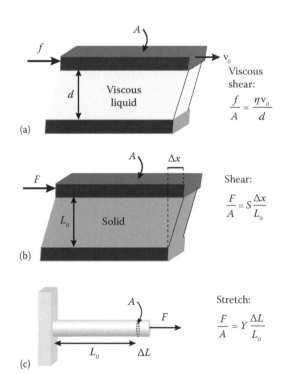

FIGURE 12.7
Materials under stress. (a) Shear stress applied to a viscous liquid. The upper plate is moved to the right at velocity v_0 by application of a force f, which is equal and opposite to the fluid frictional force. The *viscosity* is a proportionality constant in this equation. (b) Shear stress applied to a deformable solid. The *shear modulus* is the proportionality constant. Applied to the bending of a rod, the vertical distance L_0 should be interpreted as the rod length. Δx is the transverse distance the rod is bent. (c) Stretching of a rod is governed by Young's modulus.

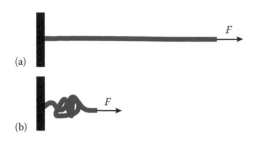

FIGURE 12.8
Polymer stretching. (a) Stretching of a single polymer molecule initially in a straight line can usefully be described by Young's modulus. (b) Except for arrays of molecules like muscle, tendon, bone, and, to a certain extent, microtubules and fibers, linear bio-polymers exist primarily in coiled form. Depending on the extent of stretching, the resistance is related to a combination of stretching and shearing forces.

On the other hand, nucleic acids, polysaccharides, and many of the active and structural elements in cells can often not usefully be described by Young's and shear moduli. Imagine, for example, the stretching of a DNA molecule or any long polymeric, single molecule. These molecules do not lie along a straight line in a cell or solution waiting to be stretched in a longitudinal or transverse direction. Entropy forces these polymers to coil into a diffuse sphere as the lowest free-energy state (Figure 12.8). When one manages to grab the two ends of a DNA molecule and pull, the coiled molecule must first uncoil before it can resist the force with its Young's modulus. Such a response could be described from a combination of stretching (Young's) and shearing, but these bulk parameters are not intended to describe single molecules. See also Figure 5.31.

12.6 FORCE OF FRICTION, DISSIPATION, INERTIA, AND DISORDER

Chapter 6 introduced the method of centrifugation, or sedimentation, to determine the molecular mass of proteins, aggregates, and other biomolecules. Sedimentation uses centrifugal forces to increase the settling rate of molecules by, in effect, increasing the acceleration of gravity g. We saw that the sedimentation rate is described by the sedimentation coefficient, in units of svedbergs (10^{-13} s), with values of 2–10 S for moderate-sized proteins like hemoglobin. Equation 6.1 defined the Svedberg sedimentation coefficient s: $v = s\, g_{eff} = s \cdot \omega^2 r$, where g_{eff} is the effective acceleration, ω is the angular frequency of rotation, and r is the average distance of the sample molecule from the axis of rotation (Figure 12.9). This equation is a differential equation that can be solved fairly easily:

$$v = \frac{dr}{dt} = s \cdot \omega^2 r \Rightarrow \frac{dr}{dt} - s \cdot \omega^2 r = 0 \tag{12.25}$$

Hopefully, your guessed solution would be $r(t) = r_0 e^{kt}$; $\frac{dr}{dt} = kr$, so $k = s\omega^2$ and

$$r(t) = r_0 e^{s\omega^2 t} \tag{12.26}$$

Density gradient centrifugation is a variant in which the density of the fluid medium increases toward the bottom of the tube, usually through a sucrose concentration gradient.

This solution becomes infinite as $t \to \infty$, but this limit is quite irrelevant because there is a bottom to the centrifuge tube.

To estimate s, we also saw that Archimedes instructs us to replace the molecular mass by the net mass, the difference between the molecular mass and the mass of the displaced solvent. For typical globular proteins this means $m_{protein} \to 0.25 m_{protein}$. We could also write the potential energy as a function of distance from the axis as $V(r) = -1/2\, m_{net} \omega^2 r^2$, but because equilibrium is never reached (all particles would eventually be plastered on the bottom of the centrifuge tube), we cannot make use of any Boltzmann factor.

At the beginning of a centrifuge run the sample molecules may be located at the top of the tube or distributed uniformly throughout. The molecules then move downward at (approximately) constant speed, governed by the sedimentation coefficient. Because we have an applied force, but movement at constant speed, we therefore know that energy is being dissipated. Thus fluid friction is our next concern.

FIGURE 12.9
Centrifugal acceleration depends on the distance from the axis of rotation.

12.6.1 Viscosity, Scaling Exponent, and Friction

If we return to the gravitational model for sedimentation, g represents the applied force/mass. For motion through a high-density fluid (at low Reynolds number) at a constant drift speed the frictional force f is related to the speed by $f = bv_{drift} = m_{net}g$, so

$$v_{drift} = \frac{m_{net}g}{b} \cong \frac{m_{net}g}{6\pi\eta R} \qquad (12.27)$$

for roughly spherical particles we can write $b = 6\eta\pi R$, where R is the particle radius and η the viscosity. The sedimentation coefficient, defined as $s = m_{net}/b$, is then

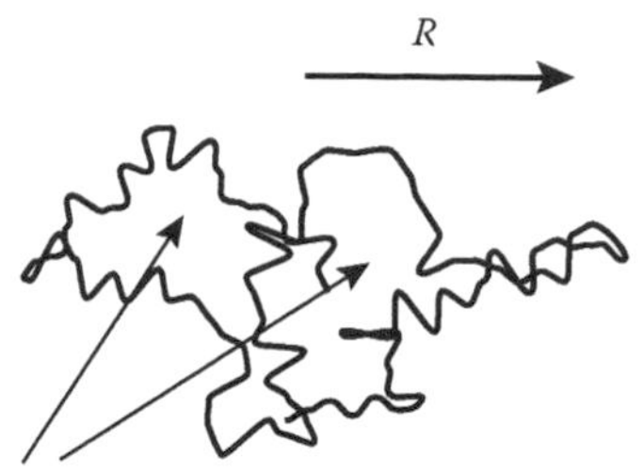

$$s = \frac{m_{net}}{b} \cong \frac{m_{net}}{6\pi\eta R} \qquad (12.28)$$

FIGURE 12.10
A compact, globular protein may behave like a solid, but a random-coil polymer has solvent dispersed throughout its volume.

12.6.2 Scaling Exponent

If the protein or other particle we are centrifuging is a solid of density ρ, we write $m_{net} = m - \rho_{solv}V = (\rho - \rho_{solv})V$ and we have gone about as far as we can go. Some compact, globular proteins behave like solids; other biopolymers behave like polymer chains that take up random conformations in solution. We first dealt with this case in Chapter 6. Solvent molecules are then interspersed in the interior of the polymer (Figure 12.10). For the freely jointed chain, the average radius squared of the coiled polymer is

$$\left\langle R_N^2 \right\rangle = 3NL^2 \propto m \qquad (12.29)$$

where the last proportionality reflects the fact that the mass scales with the number of units in the chain. This can be written as

$$R_{FJC} \propto m^{1/2} \qquad (12.30)$$

The radius/mass *scaling exponent* ρ is then ½ for the freely jointed chain.

EXERCISE 12.1

Determine the scaling exponent ρ for a solid: $R_{solid} \propto m^p$.

Answer

This one is easy. Mass = $\rho \cdot$volume. The volume of a sphere is $4\pi R^3/3$; that of a cube is a^3; in general, the volume scales with the cube of the average radius. Therefore

$$R_{solid} \propto m^{1/3} \quad \rho = 1/3.$$

Returning to the freely jointed chain,

$$s = \frac{m_{net}}{6\pi\eta R} = \frac{m - \rho_{solv}V}{6\pi\eta R} = (const)\frac{m - \rho_{solv}V}{6\pi\eta m^p} \qquad (12.31)$$

For a "loose" polymer, it is usually the case that the volume of the displaced solvent is directly proportional to the mass of the polymer—only the polymer chain itself precludes solvent from occupying a space—so $V \propto m$ and Equation 12.29 becomes

$$s \propto \frac{m - (const) \cdot m}{6\pi\eta m^p} \propto \frac{m}{6\pi\eta m^p} \propto \frac{m^{1-p}}{\eta} \qquad (12.32)$$

The sedimentation velocity (coefficient) therefore depends on a fractional power of the polymer mass divided by the solution viscosity. For a spherical solid, $1 - p = 2/3$; for a solid, rod, $1 - p = 0.15$; for a freely jointed chain in a good to ideal solvent, $1 - p = 0.4$–0.5.

12.6.3 Object Shape and Frictional Loss

The frictional coefficient for a sphere in a fluid is $b = 6\pi\eta R$. What if the particle is not spherical? The frictional coefficient for an arbitrary shape is a complicated matter, but for shapes not far from spherical—prolate and oblate spheroids—the results can be written as

$$b_{\text{spheroid, avg}} = b_{\text{sphere}} \cdot F(u) = 6\pi\eta R \cdot F(u) \tag{12.33}$$

where u is the axial ratio a/b and $F(u)$ is a function that typically ranges from 1 to 4. $F(u)$ is sometimes called the *Perrin function*. Table 12.3 lists Perrin functions for spheroids and rods shown in Figure 12.11. We have had to insert a subscript "avg" on b for the spheroids because the friction depends on the orientation of the spheroid. Unless the axial ratio is large, the orientation asymmetry is not too great, and because control over a microparticle's orientation is often not controllable (except by a microorganism), the average value is used. We return to the matter of orientation shortly.

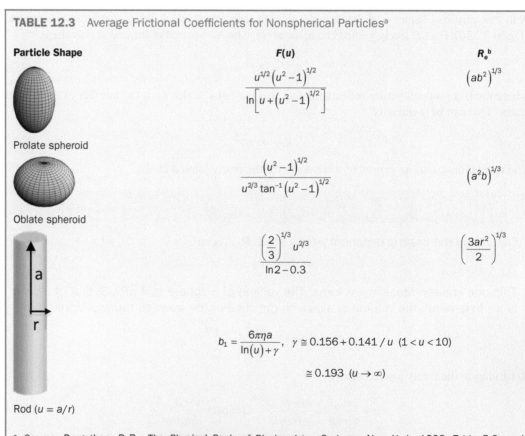

TABLE 12.3   Average Frictional Coefficients for Nonspherical Particles[a]

Particle Shape	$F(u)$	R_e[b]
Prolate spheroid	$\dfrac{u^{1/2}\left(u^2 - 1\right)^{1/2}}{\ln\left[u + \left(u^2 - 1\right)^{1/2}\right]}$	$\left(ab^2\right)^{1/3}$
Oblate spheroid	$\dfrac{\left(u^2 - 1\right)^{1/2}}{u^{2/3}\tan^{-1}\left(u^2 - 1\right)^{1/2}}$	$\left(a^2 b\right)^{1/3}$
Rod ($u = a/r$)	$\dfrac{\left(\dfrac{2}{3}\right)^{1/3} u^{2/3}}{\ln 2 - 0.3}$	$\left(\dfrac{3ar^2}{2}\right)^{1/3}$
	$b_1 = \dfrac{6\pi\eta a}{\ln(u) + \gamma}, \quad \gamma \cong 0.156 + 0.141/u \ \ (1 < u < 10)$	
	$\cong 0.193 \ (u \to \infty)$	

[a] *Source:* Bergethon, P. R., *The Physical Basis of Biochemistry*, Springer, New York, 1998, Table 5.2; and Serdyuk, I. N. et al., *Methods in Molecular Biophysics*, Cambridge University Press, New York, 274, 2007. Expressions for the rod are from Bergethon (upper, corrected R_e) and Serdyuk (modified, lower). Programs, e.g., SOLPRO, to calculate functions can be found at http://leonardo.inf.um.es/macromol/programs/programs.htm.
[b] Equivalent radius of equal-volume sphere.

12.6.4 Asymmetry of Frictional Force

In the case of rodlike structures such as cilia and flagella the direction dependence of the frictional coefficient is important. Zahn et al.[5] have measured the gravitational drift velocity of rods constructed of micron-sized magnetic spheres, kept in a straight line be a magnetic field (Figure 12.11, lower right, and Figure 12.12). They found good experimental agreement with the following theoretical expressions:

$$b_n^{\parallel} = \frac{2}{3} \frac{n}{\ln 2n - \gamma^{\parallel}} b_0 \qquad b_n^{\perp} = \frac{4}{3} \frac{n}{\ln 2n - \gamma \perp} b_0 \qquad (12.34)$$

$$\gamma^{\parallel} (\gamma^{\perp}) = 0.649 \;\; (-0.418) \;\text{ for chain of spheres}$$

(Because this experiment involved linear chains of spheres, n, the ratio of the rod length to diameter, also the number of spheres in the chain, was used rather than p.) In this expression we have used b_0 for the (Stokes) friction coefficient of a single sphere. Measurements showed good experimental agreement for values of n from 3 or 4 to 100. For a short chain (three spheres), Equation 12.34 gives a ratio $\frac{b^{\perp}}{b^{\parallel}} = \frac{2(\ln 2n - 0.649)}{\ln 2n + 0.418} \cong 1.03$; for $n = p = 100$, $\frac{b^{\perp}}{b^{\parallel}} \cong 1.63$. For those interested in more details of a variety of shapes, consult Serdyuk et al.[6]

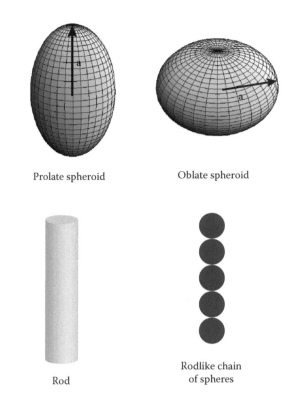

Prolate spheroid   Oblate spheroid

Rod   Rodlike chain of spheres

FIGURE 12.11
Spheroids, rods, and chains of spheres. Fluid frictional coefficients for these objects are tabulated in text. (Spheroids from Wikipedia.)

12.6.5 Note on Cilia and Flagella

Cilia and flagella, made from bundles of microtubules and molecular motors that travel on them, provide the motile force for some bacteria as well as for eukaryotic cells. Cilia occur in single-celled eukaryotic *ciliates*, a group of *protists* characterized by the presence of hairlike cilia, similar in structure to flagella but typically shorter and present in much larger numbers. The difference between cilia and flagella is usually understood to be the sort of motion: cilia wave back and forth while flagella rotate. Cilia are also sometimes considered to be shorter than flagella. Many of these short cilia undulate in synchronism to generate forces used in swimming, crawling, attachment, and feeding. Upon stimulation by an external force, cilia may also transmit signals to the interior of the ciliate. In larger organisms such as humans, cilia are generally attached to cells that are fixed in place, and have the job of moving fluid and particles by the cells. Prokaryotic bacteria, including *Escherichia coli* also employ cilia and flagella for movement, though usually in smaller numbers. The multiple flagella of *E. coli* and other bacteria sometimes form bundles that intertwine and rotate together. As they drive the bacterium forward, they also may induce rotation. Some examples of these cilia and flagella are shown in Figure 12.13. Many others can easily be found by a Web search, including the famous example of paramecium's large number of cilia. The terms *cilia* and *flagella* are occasionally used interchangeably (en.wikipedia.org/wiki/Flagella), so it is important to know the features of the cilium/flagellum of interest.

Directional asymmetry of the frictional and bending (shear) coefficients for rodlike structures underlies the ability of cilia and flagella to create forces in a particular direction. Cilia and flagella occur in both bacteria and eukaryotic cells. Figure 12.14 shows a crude model for the waving ciliary mechanism of a cell. (Two or more waving cilia positioned on either side of a bacterial cell can create a frictional force to drive the microorganism through the water, but rotating flagella seem to be more common.) It is clear that the forward (drive) stroke and the reverse (recovery) stroke cannot be symmetric, or recovery would undo what drive did. The cells get around this by holding the cilia relatively rigid (in this model) during the drive stroke and rather limp in the recovery stroke. The net effect is that during the drive stroke, the rodlike cilia are moving mostly transverse to their length and mostly along their length during recovery. Because the transverse frictional coefficient ($b^{\perp}$) is larger than the longitudinal ($b^{\parallel}$), there is, on average, a forward force.

An interesting connection between an object's friction in a fluid and its electrical capacitance was made in the 1990s. The electrical capacitance of a sphere is $C = 4\pi\varepsilon_0 R$. For a sphere we can then write $b = 6\pi\eta C/4\pi\varepsilon_0$ (SI units; in electrostatic units where $4\pi\varepsilon_0 = 1$, the similarity is even more striking). It was shown that this connection also holds for a variety of shapes, so better developed electrostatic computational techniques can be used in place of hydrodynamic calculations.

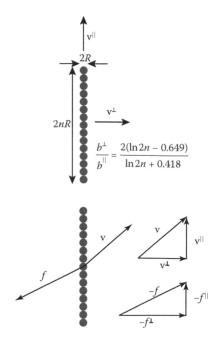

FIGURE 12.12
Orientation dependence of frictional coefficients of a rod constructed out of small spheres. The transverse friction coefficient is larger, so frictional force is not exactly opposite in direction to the velocity. The frictional coefficients do not differ much for a slender prolate spheroid of similar axial ratio.

Problem 12.7 explores this model. Both a shape-induced anisotropy in the frictional coefficient and a materials anisotropy in the bending stiffness of the cilia must come into play for this cilia drive mechanism to work. In reality, the cilia are not as crude as this model but consist of bundles of microtubules, along which molecular motors crawl and pull on adjacent tubules, creating ciliary bends.

The fluid frictional force dissipates work into fluid heating. The power expended in moving an object of frictional coefficient b through a liquid at velocity v is

$$\text{Power} = f \cdot v = bv^2 \tag{12.35}$$

In the case of centrifugation, the (negative) work done by the friction in moving particles through the fluid to a larger distance from the axis is converted to an amount of heating that is quite miniscule compared to the overall energy used by the machine. When cell cilia or flagella employ the frictional force to move cells or fluid, the work done is often not negligible, however. Most motive processes use ATP or a membrane electrical potential as an energy source for motion and the fluid friction may be an important part of the energy balance. We examine this energy balance more closely when we deal explicitly with molecular motors.

12.7 FLUIDS AND TURBULENCE

On a cellular and subcellular scale, pure water "looks" like a viscous fluid. We showed this by calculating that a micron-sized particle stops almost instantly when a driving force is removed. Particles appear to have no inertia; there is no turbulence. Because water has one of the smaller intrinsic viscosities, we should seek a quantity other than viscosity to describe whether inertial effects are significant. Because a submarine and fish in water clearly show inertial effects, we expect the quantity to reflect properties of both fluid and the object. We first write what this quantity is, then motivate its appropriateness.

$$\text{Reynold's number:} \quad \mathcal{R} = \frac{\rho v_0 R}{\eta} \tag{12.36}$$

$\mathcal{R} \gg 1$: inertial effects, turbulence are important
$\mathcal{R} < 1$: inertial effects, turbulence are negligible

where ρ is fluid density, v_0 is speed of object of radius R in fluid, and η is fluid viscosity.

The Reynolds number has the simplest (linear) functional dependence on the four quantities we expect should govern acceleration, turbulence, and inertia. When the fluid density is large, it is more likely to keep flowing after a force is removed. Likewise, when a larger object moves at higher speed, the fluid is more likely to swirl. Finally, a low-viscosity liquid has weaker forces between its molecules, forms turbulent flows more easily, and particles accelerate more easily through the liquid.

12.7.1 Critical Force

The Reynolds number contains two parameters describing the liquid and two describing the particle and its motion. If we think of applying a force to the pure liquid, we expect that there would be some critical force, above which groups of liquid molecules would begin to accelerate. This force should depend only on the liquid. Because p and η refer to the liquid, we could use dimensional analysis to predict this critical force:

$$\left[f_{\text{crit}}\right]=\left[\rho\right]^{x}\left[\eta\right]^{y}=\left[\frac{\text{kg}}{\text{m}^{3}}\right]^{x}\left[\frac{\text{N}\times\text{s}}{\text{m}^{2}}\right]^{y}=\left[\frac{\text{kg}}{\text{m}^{3}}\right]^{x}\left[\frac{\text{kg}}{\text{s}\times\text{m}}\right]^{y}=\left[\frac{\text{kg}^{x+y}}{\text{m}^{3x+y}\text{s}^{y}}\right]=\frac{\text{kg}\times\text{m}}{\text{s}^{2}},$$

so $y = 2$ and $x = -1$. The critical force is therefore, outside of a dimensionless constant,

$$f_{\text{crit}} = \frac{\eta^2}{\rho_{\text{fluid}}} \tag{12.37}$$

Any force that is much larger than this critical force creates turbulence; any much smaller does not. We could now do a detailed derivation of the fluid acceleration and streamlines to find when the force on the fluid approaches this critical force, but we already know what the fluid frictional force on a spherical particle is: $f = 6\pi\eta Rv_0$. The force on the fluid is the same, so if we take the ration of the frictional force to the critical force, we have a dimensionless number,

$$\frac{f}{f_{\text{crit}}} = \frac{6\pi\eta Rv_0}{\eta^2/\rho_{\text{fluid}}} = 6\pi\frac{\rho_{\text{fluid}}Rv_0}{\eta} = 6\pi\mathcal{R} \tag{12.38}$$

Outside of the factor of 6π, this is exactly the Reynolds number previously defined. Table 12.4 shows Reynolds numbers for several systems common to our experience.

12.7.2 Viscous Flow through Tubes

A classic, nonbiological application of the Reynolds number is to determine when flow of a fluid through a cylindrical pipe becomes turbulent. Experiment has answered this for smooth pipes: $R > 1,000$ for turbulent flow, so the Reynolds number should not be treated as a precise indicator of turbulence. In the case of no-slip, laminar, low-R flow, the velocity profile, $v(r)$, of a fluid of viscosity η throughout the diameter of a cylindrical pipe of radius R, length L with an applied pressure differential P can be found as

Flow through a cylindrical tube: $v(r) = \dfrac{(R^2 - r^2)P}{4L\eta}$ (12.39)

It is useful to derive formulas from dimensional analysis, but doing so does not necessarily give us intuitive and quantitative ways of visualizing and connecting to well-known physical ideas. For instance, we assume that fluid friction has something to do with "rubbing" between molecular surfaces or layers of materials and that turbulence is connected with velocity differences between volumes of fluid, where the volumes are more or less sharply defined. But in a fluid, molecules rub against molecules: how do we quantify if energy is dispersed randomly, in small packets to many individual water molecules (heat) or to a clump of water molecules that move together with some collective kinetic energy? Figure 12.15 shows fluid as it moves around a particle of radius R traveling to the left at speed v_0.

(a)

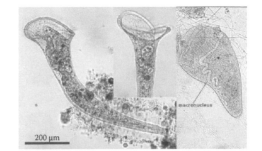

(b)

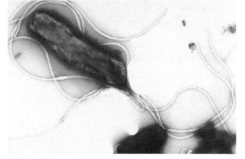

(c)

FIGURE 12.13

Cilia and flagella in biological systems. Cilia generally wave back and forth, while flagella rotate. (a) *Stentor roeseli*: cilia move food into the organism's mouth. (Public domain image by Protist Image Database, http://protist.i.hosei.ac.jp/Science_Internet/TI_2001E/text.html.) (b) Scanning electron microscope image of lung trachea epithelium. Cilia move foreign particulate matter out of the lungs. (Public domain image by Charles Daghlian, remf.dartmouth.edu/images/mammalianLungSEM/source/9.html.) (c) Flagella of *Heliobacter pylori* drive the bacterium even through cell membranes. (Public domain image by Yutaka Tsutsumi, Department of Pathology, Fujita Health University School of Medicine, en.wikipedia.org/wiki/File:EMpylori.jpg.)

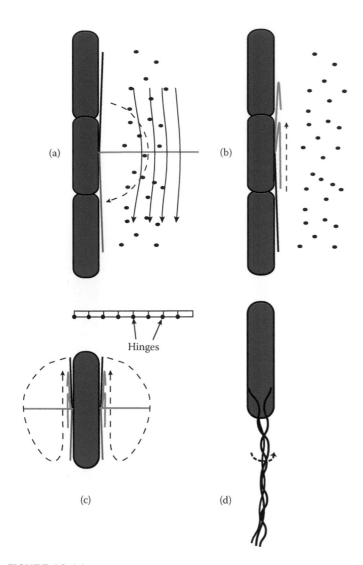

FIGURE 12.14
Simplistic model cilia beat or wave to drive fluid-borne particles along a surface. (a) During the drive stroke the cilia behave like rigid rods. (b) During the recovery stroke, the cilia bend freely and move almost longitudinally. Closer to reality, the cilia are made of bundles of microtubules, along which dynein (or other) molecular motors travel and force neighboring tubules to slide. By wavelike coordination of motors along the length of the cilia, bending of the cilia can be created. Cilia are typically 5–10 μm long and about 0.2 μm in diameter. (c) Crude model for cilia-driven movement of bacteria. (d) Bacteria with flagella normally rotate them to drive forward. Several flagella may form a bundle and rotate together.

The Stokes friction coefficient we use corresponds to "no-slip" boundary conditions, which is appropriate as long as the sphere is not too small. Proteins of molecular mass 5,000 Da or more usually obey no-slip boundary conditions. See Serdyuk et al.[6] (pp. 273–293), for a discussion of this and fluid flow lines.

Three types of physical effects should be considered. First, the sphere must push fluid forward and out of the way as it moves. Second, fluid molecules tend to stick together, as measured by the viscosity. Third, fluid molecules may stick to the sphere surface—"stick" or "no-slip" boundary conditions on the hydrodynamic flow. If fluid sticks to the sphere, it tends to be dragged along with the sphere. That boundary layer then drags fluid farther out along with it, depending on the viscosity. (Under less common "slip" conditions, the fluid is still pushed forward and transversely by the sphere, but fluid must also travel around the surface of the sphere in the time it takes the sphere to travel one diameter's distance.) In the reference frame of the sphere, a small volume ℓ^3 of fluid traveling on a curved path has a centripetal acceleration. The viscous force, on the other hand, distributes energy to random rotations, vibrations, and translations of fluid molecules as a surface of area A ($=\ell^2$ in this diagram) moves against another surface with different velocity. The velocity of the fluid changes by about v_0 over a distance R from the surface: $-dv/dx \approx v_0/R$, so in a distance ℓ, the velocity decreases by an amount $\frac{\ell}{R}v_0$. (This same ℓ^3 volume is moving with velocity $\frac{\ell}{R}v_0$ relative to its neighbor farther from the sphere.) The ratio of the centripetal to viscous force in $\frac{f_{\text{cent}}}{f_{\text{visc}}} = \frac{\rho\,\ell^3 v_0^2/R}{\eta\,\ell^3 v_0/R^2} = \frac{\rho v_0 R}{\eta} = \mathcal{R}$, the Reynolds number. When the centripetal force gets too large the volume can collectively shear away from its neighbors, creating turbulence.

12.7.3 Motor Design and the Reynolds Number

We humans know how to design motors to propel large objects through water at high Reynolds numbers. Rotating, curved propellers propel (accelerate) water along the axis of rotation, driving a boat or submarine forward (Figure 12.16). By looking at the rear of the boat, we can see the turbulent, accelerating water with our own eyes. There are also propulsion designs that attempt to minimize turbulence by using pump jets or magnetohydrodynamic impellers, but these systems still accelerate water to the rear of the vessel.

What happens when a bacterium flails a cilium or rotates a flagellum? The low Reynolds condition implies that the frictional force created by the flagellum is transferring energy to fluid, but the viscosity is high enough that the liquid molecules stick together on the spatial scale of the propeller, rather than shearing and moving as a turbulent mass (Figure 12.17) The low Reynolds number that applies to such a flagellum or cilium does not mean that no forces accelerate fluid but that the viscous forces do work on individual fluid molecules, imparting random rotations and translations to the fluid molecules around the flagellum.

12.7.4 Viscosity Variations: Temperature and Dissolved Materials

The viscosity of water depends on temperature (Table 12.5). We typically use the value 1.0×10^{-3} Pa·s at 20°C. Occasionally, we require the value at other temperatures, as shown in Table 12.5.

We have seen many times that the interior of a cell is not pure water. It is occupied by a variety of structures of various sizes and rigidity. It can be difficult to model the complex and ever-changing space inside a cell, including the effects of dissolved molecules. Solutions of polymers often cannot be described by a single value of viscosity. These are termed *non-Newtonian fluids*. Blood is a classic example of a non-Newtonian fluid. We initially defined the viscosity by a linear velocity gradient of a fluid between two plates, with one plate moving at velocity v_0. If this gradient is not linear, or if the force depends nonlinearly on the velocity v_0, the fluid is non-Newtonian.

12.7.5 Dissolved Solutes Change the Viscosity

For Newtonian fluids the viscosity change is quantified by

$$[\eta] = \lim_{\phi \to 0} \frac{\eta - \eta_0}{\eta_0 \phi} \tag{12.40}$$

where ϕ is the volume fraction of the solute, η_0 is the viscosity in the absence of the solute and $[\eta]$ is a dimensionless number. Einstein showed that the value of this number is 2.5 for rigid spheres:

$$\eta - \eta_0 = 2.5 \, \phi \eta_0 \tag{12.41}$$

A fluid with 20% volume fraction of dissolved solids has its viscosity increased by about 50%.

12.7.6 Bacterial Foraging: Escaping Low-$\mathcal{R}$ and the Peclet Number

One of the implications of the low-$\mathcal{R}$ rule for microscopic life in water is that bacteria that need to chase their prey have difficulty catching it without either temporarily exceeding the critical force ($f_{crit} = \eta^2/\rho \sim 10^{-9}$ N) or employing another trick. The physical corollary to the laminar flow rule is that a clump of viscous fluid associates with the moving particle (bacterium). Small molecules like oxygen can diffuse through this "bow-wave" of fluid, but another small microorganism pursued

TABLE 12.4 Reynolds Numbers

System	$\mathcal{R}$
Whale swimming in water at 10 m/s	3×10^8
Neon tetra (tropical fish) swimming at 1 cm/s	200
Bee flying in air at 14 cm/s	100
Protein molecule (6 nm diameter) diffusing in water at 8 m/s	5×10^{-2}
Bacterium (1 μm) swimming in water at 30 μm/s	3×10^{-5}

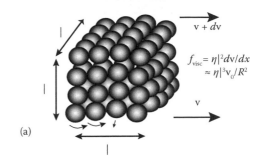

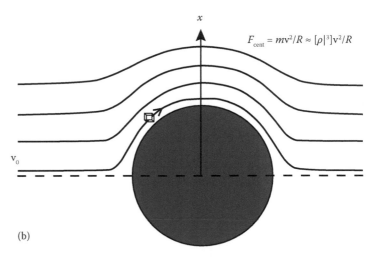

FIGURE 12.15
(a) Centripetal force on a small volume of fluid, a cube of side ℓ, as it accelerates around the equator of a spherical particle, radius R. The sphere is moving to the left at speed v_0 or equivalently, fluid is moving to the right. (b) The velocity of the fluid decreases from $\pi v_0/2$ at the sphere surface to about v_0 at a distance R from the surface: $-dv/dx = 0.6v_0/R \approx v_0/R$, so in a distance ℓ, the velocity decreases by $\frac{\ell}{R}v_0$. The ratio of the centripetal to viscous force in $\frac{f_{cent}}{f_{visc}} = \frac{\rho \ell^3 v_0^2/R}{\eta \ell^3 v_0/R^2} = \frac{\rho v_0 R}{\eta} = R$, the Reynolds number. The viscous force distributes energy to random rotations, vibrations, and translations of fluid molecules (small arrows).

In the field of aerodynamics, "laminar flow" often refers to high-$\mathcal{R}$ situations where shape design minimizes obvious turbulence, or even where turbulence exists, but is measured by a laminar flow (sheetlike flow) method.

FIGURE 12.16
Extreme turbulence caused when a boat propeller imparts a large acceleration to water. The shearing forces create *cavitation*, bubbles of vacuum that begin filling with dissolved gases and water vapor and then collapse. Continual cavitation can damage steel surfaces. (Cavitating propeller in a water tunnel experiment at the David Taylor Model Basin. Image, taken by an unnamed sailor or U.S. Naval employee, now at https://commons. wikimedia.org/wiki/File:Cavitating-prop.jpg.)

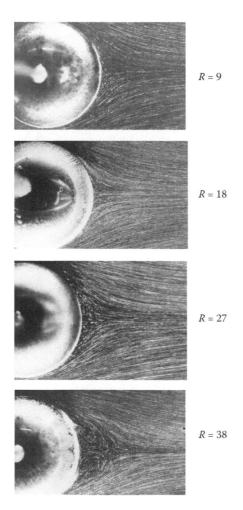

$R = 9$

$R = 18$

$R = 27$

$R = 38$

FIGURE 12.17
Fluid flow around a sphere for Reynolds numbers 9–37. The pattern does not change much for numbers lower than 9. (Images from Taneda, S., *J. Phys. Soc. Jpn.*, 11, 1104, 1956. With permission.)

by the bacterium is pushed away. In order to pass through this shell of solvent the bacterium must manage to accelerate through it or use another technique to reach through the fluid. Figure 1.2 suggests a few ways that larger microorganisms could employ to catch prey: using long cilia or catching prey associated with a larger object that cannot be pushed. Even small, rapidly diffusing molecules like sugar can present a challenge. Though we have postponed a detailed discussion of diffusion for later chapters, we introduced the diffusion coefficient D that describes the random movement of a small molecule. The average distance a molecule diffuses in time t is given by

$$\langle x^2 \rangle = 2Dt \text{ (one dimension)}$$

$$\langle r^2 \rangle = 6Dt \text{ (three dimensions)}$$

Suppose some sugar or other molecular food is localized some distance away from a microorganism. That food disperses according to the diffusion law cited. The diffusion coefficient for glucose is about 10^{-9} m²/s, for instance, so in 1 s glucose moves about $d = \sqrt{\langle r^2 \rangle} = \sqrt{6Dt} \cong 8 \, \mu$m. If a bacterium wants to locate and consume this glucose, it should do so in a time less than the diffusion time. Approximately, you want the time for the bacterium to move a distance d, $\Delta t_{\text{bact}} = d/v_{\text{bact}}$, to be less than the time, $\Delta t_{\text{diff}} \sim d^2/D$, for the glucose to diffuse the same distance. In other words, $\frac{d}{v} < \frac{d^2}{D}$ or $vd > D$. The ratio vd/D, called the *Peclet number*, should be large for successful bacterial foraging:

$$\text{Peclet number} = \frac{vd}{D} > 1 \tag{12.42}$$

The maximum speed of an *E. coli* is about 30 μm/s, so $\frac{(30 \mu\text{m/s})d}{10^{-9} \text{m}^2/\text{s}} > 1$ and $d > 30$ μm, about 20–30 *E. coli* lengths, is the distance the bacterium should travel in a given direction in each run, before stopping and changing direction.

12.7.7 Concluding Thoughts

In this chapter we have applied Newton's laws to dynamical behavior in water on a spatial scale of microns. The physical laws are exactly the same as we used in first year physics. In spite of the same governing equations, the mechanical and kinematic behaviors are completely different. If humans happened to be the size of bacteria and lived in water, it is doubtful Newton's Laws would have ever been discovered or accepted, even if a micron-sized brain could conceive of them. In our 1-m-sized existence, friction is a lousy irritant. In the life of a bacterium, friction dominates everything. In the next few chapters, we further explore the 10^{-9} to 10^{-5} m spatial scale and examine more closely the random, diffusive behavior of individual and large populations of molecules.

12.8 POINTS TO REMEMBER

- The Stokes frictional coefficient $b = 6\pi\eta R$. It plays a prominent role in diffusion and reaction theory.
- Sedimentation, the drift of particles through a liquid under the influence of gravitational or centrifugal forces, is governed by the sedimentation coefficient $s = \frac{m_{net}}{b} \cong \frac{m_{net}}{6\pi\eta R}$, where m_{net} is typically $m/4$ for globular proteins. Also, $s \propto \frac{m^{1-p}}{\eta}$, where p is the scaling coefficient. Values are $p = 1/3$ for a solid and $1/2$ for a freely jointed chain.
- A micron-sized or smaller object in uniform motion remains in uniform motion only if constant external force is applied. A micron-sized or smaller object stops in a time less than a nanosecond and at a distance less than the diameter of a single atom when a driving force is removed. Know how to prove this.
- Geometric asymmetry in both frictional and elastic properties is critical to the function of motile microstructures like cilia.
- Understand the transition from laminar (smooth) flow to turbulent flow, how the Reynolds number $\mathcal{R} = \frac{\rho v_0 R}{\eta}$ determines the transition and why the frictional force changes from $f = bv$ to $f = b_2 v^2$ (or other higher powers of v).
- Bacterial foraging requires that a bacterium be able to exceed the critical frictional force for a short time, accelerating in order to catch its prey; otherwise, prey is pushed away by the cohesive fluid clump in front of the bacterium.
- The computation of frictional dissipation (conversion to heat). Dissipation plays two key roles in later chapters: the necessity of a force for coordinated molecular drift and the contribution to entropy (disorder).

TABLE 12.5 Viscosity of Pure Water versus Temperature

Temperature (°C)	Viscosity (Pa·s)
10	1.30×10^{-3}
20	1.00×10^{-3}
30	0.798×10^{-3}
40	0.653×10^{-3}
60	0.467×10^{-3}
80	0.355×10^{-3}
100	0.282×10^{-3}

12.9 PROBLEM-SOLVING

PROBLEM 12.1 AIR DRAG

1. Solve the differential equation (Equation 12.11) using a guessed solution of $v(t) = A \tanh(t/\tau)$. Find values of A and τ.
2. Determine the position as a function of time and velocity.
3. Graph both of these functions.
4. Determine the terminal velocity if the applied force is the gravitational force for spheres of radii 0.50 m, 0.50 cm, 0.59 mm, and density 2.8 g/(cm)^3.
5. For the spheres in 4, determine the time τ it takes to reach the terminal velocity.

PROBLEM 12.2   LIQUID DRAG

1. Calculate the value of τ in Table 12.2.
2. Derive the expression for the velocity of an object as a function of time after an applied force has brought the particle to v_{term} and then has been removed. Integrate this expression over the appropriate time to obtain the distance in Table 12.2.

PROBLEM 12.3   PRESSURE EFFECTS ON AN ENZYMATIC REACTION

Find the reference publication to Figure 12.6a (PDB ID 3IA4).

1. In Table 1 of that paper (DOI: 10.1002/cbic.200900367), write down values of the "activation volumes" listed. These are (temporary) volume changes, often notated ΔV^*, that occur in the protein as it is in the process of reacting. (Chapter 15 explains this more fully.)
2. Find the molecular mass of the protein and estimate the volume, using a density of 1.2 times that of water. What fraction of the molecular volume is the "activation volume"?
3. Calculate the value of $P\Delta V^*$ as a multiple of $k_B T$ for 1 and 280 atm pressure.

PROBLEM 12.4   CENTRIFUGATION TIMES

Myoglobin (Mb) has a sedimentation coefficient of 1.71 svedbergs (10^{-13} s) in water. The viscosity of water is 1.0×10^{-3} Pa·s; that of glycerol is 1.4 Pa·s (at room temperature).

1. Calculate the sedimentation coefficient of Mb in glycerol.
2. Calculate the time for a Mb molecule to move from $r = 11.0$ cm to $r = 13.0$ cm in a centrifuge tube rotating at 10,000 revolutions per minute (rpm) in water and in glycerol.
3. Repeat 2 for a rotational rate of 75,000 rpm, corresponding approximately to the maximum rate of an *ultracentrifuge*.
4. Repeat 3 for cowpea mosaic virus ("empty shell"), a particle with s = 58 svedbergs.
5. Suppose you are assigned to perform the above procedures in the lab. Comment.

PROBLEM 12.5 FLUID FRICTION AND ASYMMETRIC SHAPE

An *E. coli* bacterium has an equivalent spherical radius of 0.55 μm.

1. Calculate the fluid frictional coefficient b ($f = bv$) for a spherical *E. coli* in pure water and in blood, with a viscosity about 4 times that of water.
2. If the bacterium can propel itself at a speed of 85 μm/s in the equivalent of pure water, what is its maximum speed in blood?
3. Recalculate the frictional coefficient in water assuming the bacterium is a prolate spheroid of axial ratio $a/b = 2.1$. Use the averaged coefficient (Table 12.1).
4. Repeat 3, but compute the frictional coefficients for transverse and longitudinal motion.

PROBLEM 12.6 FALLING ROD IN WATER

Consider a rod 4.0 μm long and 1.5 μm in diameter. The rod is solid, with density 1.3 times that of water.

1. Calculate the terminal velocity of the rod falling through water in a centrifuge rotating at 5,000 rpm, at a distance of 10.0 cm from the centrifuge axis of rotation, if the motion is transverse (rod is horizontal).
2. Repeat 1 if the rod orientation is vertical.
3. When the faster-moving rod orientation has traveled 1.0 cm, how far will the other rod orientation have moved? Neglect the change in centrifugal force with radial distance.

PROBLEM 12.7 WAVING CILIA DRIVING A BACTERIUM? (GRADUATE PROBLEM)

A crude model for a proposed cilia-driven movement of bacteria is shown in Figure 12.14. The body of the bacterium is rodlike, with a length of 3.5 μm and a diameter of 1.5 μm. In the model, cilia behave like rigid rods during the drive stroke and limp strings during the recovery stroke. (The motion of both strokes is driven by many molecular motors distributed throughout the cilia.) Assume two cilia exist, that they are both 7.5 μm long and 0.25 μm in diameter, and that their movement is coordinated. The strokes take 0.15 s each, with a constant rotational rate during the drive stroke. In recovery, the bend in a cilium starts at the cilium attachment site and travels toward the other end at constant speed. The fluid medium has a viscosity of 2.0×10^{-3} Pa·s.

1. Calculate the generated force during the drive stroke, when the cilia are perpendicular to the body of the rod-shaped bacterium. (Neglect any forward movement of the body of the bacterium.)
2. Repeat 1 when the cilia are 45° from a complete drive stroke. Calculate both x and y components of force of each cilium and show that one component cancels out.
3. Calculate the average reverse force during the recovery stroke. Assume the cilia remain parallel to the bacterium's body, except for the (traveling) sharp bend, and that, on average, half of a cilium length moves a distance of one cilium length in a time 0.10 s.
4. If the average drive force is 2/3 that at 90° minus the average reverse force, calculate the average speed of the bacterium through the fluid.
5. Calculate the average power generated by the bacterium in its movement.
6. (Physics graduate students may recognize the many approximations made above for even this crude model. They are welcome to solve the model exactly as preparation for a graduate mechanics qualifying exam.)

PROBLEM 12.8   VISCOSITY OF BLOOD

1. Use the Einstein formula (Equation 12.41) to estimate the viscosity of blood plasma. Blood plasma is about 10% dissolved solids. Compare to the measured value of plasma viscosity.
2. Whole blood contains a large volume fraction of red blood cells (RBCs, erythrocytes). A typical human erythrocyte has a diameter of 6–8 μm and a thickness of 2 μm. The RBC count, the number of RBCs per mm^3 ranges from 3.8–7.2 x 10^6. Use Equation 12.41 to estimate the viscosity increase due to the RBCs for the lower and upper counts. Compare to the measured value of blood viscosity and discuss the difference.

PROBLEM 12.9   BLOOD FLOW

The viscosity of normal blood is about 3 times that of water. (Your solution to the previous problem should have confirmed this, with perhaps better accuracy.) The heart pumps blood through blood vessels that are approximately cylindrical; the maximum total volumetric flow rate is about 500 cm^3/s, and the average diameter of the aorta, into which blood is first pumped, is about 2.5 cm in diameter. (The diameter expands and contracts with pressure.) The length of the aorta as it ascends from the heart to where it branches into the carotid arteries is about 5 cm.

1. Calculate the pressure drop through the ascending aorta.
2. Starting with Equation 12.39, compute the volume flow rate (m^3/s), Q, for a cylindrical tube.
3. Rewrite the results of 1 as $P = QZ$, where Z is the hydrodynamic resistance to flow.
4. If a blood vessel radius decreases (via plaque buildup) by 15%, by how much does hydrodynamic resistance increase?
5. Derive an equation for the power needed to pump the blood through the ascending aorta.

REFERENCES

1. Mo, Y., K. T. Turner, and I. Szlufarska. 2009. Friction laws at the nanoscale. *Nature* 425, 21 (September 4, 2003); doi:10.1038/425021a 457:1116–1119.

2. Hay, S., R. M. Evans, C. Levy, E. J. Loveridge, X. Wang, D. Leys, R. K. Allemann, and N. S. Scrutton. 2009. Are the catalytic properties of enzymes from piezophilic organisms pressure adapted? *Chembiochem* 6(8), 1419–1422.

3. von Arx, W. S. 1962. *An Introduction to Physical Oceanography.* Reading, MA: Addison-Wesley.

4. Keller, E. http://www.extremescience.com/zoom/index.php/life-in-the-deep-ocean/39-viperfish-fangtooth, accessed September 5, 2010.

5. Zahn, K., R. Lenke, and G. Maret. 1994. Friction coefficient of rod-like chains of spheres at very low Reynolds numbers: I. Experiment. *J. Phys. II* 4:555–560.

6. Serdyuk, I. N., N. R. Zaccai, and J. Zaccai. 2007. *Methods in Molecular Biophysics.* Cambridge, UK: Cambridge University Press, 272–290.

7. Taneda, S. 1956. Experimental investigation of the wake behind a sphere at low Reynolds numbers. *J. Phys. Soc. Jpn.* 11:1104–1108.

CHAPTER 13

CONTENTS

Random Walks, Diffusion, and Polymer Conformation

This chapter begins a more formal treatment of what may be called *identical-particle biophysics*, which implies that we are dealing with more than one particle and should investigate the extent of the identical natures of the particles. In the treatment of a monatomic ideal gas, each gas particle is assumed to be identical to all the other particles, but, of course, the "identicalness" did not apply to velocity, momentum, kinetic energy, etc. Such kinematic properties of the particle were described by probability distributions: each particle has a certain probability of having speed v or kinetic energy $\frac{1}{2} mv^2$. For diatomic identical particles the situation is almost the same, except now a molecule can rotate and vibrate. Nevertheless, because any of the diatomic molecules can have the same vibrations and rotations, they can be considered identical. When the number of identical particles is very large ($\sim 10^{22}$), the distribution description is adequate because the averaging process over such a large number is very accurate, at least when the exchange of energy and momentum between particles is fast compared to any observation time. (Roughly, the average deviation of most physical quantities measured for all particles is $\sqrt{N}/N \sim 10^{11}/10^{22} \sim 10^{-11}$.) Nevertheless, we discover that many processes involve a small minority of the molecules that have an unusual amount of energy or some other quantity.

13.1 REVIEW OF KINETIC THEORY OF GASES: IMPLICATIONS FOR BIOMOLECULAR AVERAGING

The Equipartition Theorem states that each allowed mode contains an average energy of $\frac{1}{2}k_{B}T$.

Maxwell speed distribution. For an ideal gas (does not need to be monatomic), the probability distribution for speed v is

$$F(v)dv = 4\pi \left(\frac{m}{2\pi k_{B}T} \right)^{3/2} \exp\left(\frac{mv^2}{2k_{B}T} \right) v^2 dv \tag{13.1}$$

$$\text{Mean velocity: } \langle v_x \rangle = \langle v_y \rangle = \langle v_z \rangle = 0 \tag{13.2}$$

$$\text{Mean speed : } \langle v \rangle = \frac{4}{\sqrt{2\pi}} \sqrt{\frac{k_{B}T}{m}} \tag{13.3}$$

$$\text{Most probable speed : } v_{mp} = \sqrt{\frac{2k_{B}T}{m}} \tag{13.4}$$

$$\text{Root-mean-square (RMS) speed: } v_{rms} \equiv \sqrt{\langle v^2 \rangle} = \sqrt{\frac{3k_{B}T}{m}} \tag{13.5}$$

See F. Reif, *Statistical Physics (In SI Units): Berkeley Physics Course Vol. 5* (New York: McGraw Hill Education (India) 2010). Most modern physics and introductory physics textbooks also contain the needed review information.

$$\text{Mean relative speed of two particles: } v_{rel} = \sqrt{2}\langle v \rangle \tag{13.6}$$

$$\text{Mean free path: } \ell_{mf} = \frac{1}{\sqrt{2}N\sigma} = \frac{k_B T}{\sqrt{2}\sigma P} \tag{13.7}$$

$$\text{Mean time between collisions: } \tau = \frac{\ell_{mf}}{\langle v \rangle} = \frac{1}{N\sigma v_{rel}} \tag{13.8}$$

where N is the number of gas particles per cubic meter and the scattering cross section of spherical gas particles is $\sigma = 4\pi R^2$.

$$\text{Monatomic ideal gas: } \langle E \rangle = \langle KE \rangle = \frac{3}{2}k_B T = \frac{1}{2}mv^2_{RMS} \tag{13.9}$$

$$F(E)dE = 2\pi \left(\frac{1}{\pi k_B T} \right)^{3/2} \exp\left(-\frac{E}{k_B T} \right)(E)^{1/2}dE \tag{13.10}$$

Many or most of these relationships should be familiar to you. We could go on, but the above set of equations suffices for now. Chapter 14 explores distributions in more detail.

Returning to the issue of averages, with biological systems we have three new issues to confront: (1) the number of particles is often not very large, (2) important biophysical quantities often depend on those particles that are far from the average, and (3) the "particles" are far from point-like. Issue 1 we addressed in Chapter 1. Issue 2 is common to most reaction chemistry—the molecules most likely to react are those with high energy, which we address again in Chapter 15. Issue 3 in its most extreme form characterizes biological systems, where single "particles" may be made up of hundreds to millions of bonded atoms. Gas molecules made up of only a few atoms do not have many modes in which energy and momentum can be contained, and any given particle changes rapidly between most possible energy and momentum states. In Problem 13.1 we see that even in a 1-atm ideal gas, the exchange of momentum and energy (collisions) takes place at a rate of more than 10^9 times per second for any given molecule. In molecules with more atoms, various modes can also share and exchange energy that may be acquired from a collision, speeding up the averaging process.

The initial statement of this Levinthal paradox addressed the question of how macromolecules could ever find the lowest free-energy state, if so many possible states existed. The answer appears to be related to an entropic "funnel," which continually steers the macromolecule into a much smaller set of "likely" conformational possibilities, among which is the active or native state of the macromolecule, and the fact that the "ground state" of a macromolecule is actually any among a set of low-energy states.

The Levinthal paradox. In contrast to gas molecules, a biomolecule made up of 150 amino acids (about 2,000 atoms), on the other hand, may have on the order of $3^{300} \sim 10^{143}$ possible conformational states.[1] (Each amino acid is assumed to have three orientational degrees of freedom with respect to each of its neighbors.) Even at an unphysical exchange rate of 10^{15} s^{-1}, an insignificant fraction of these states could be explored in the lifetime of the universe.

What does this mean in terms of averaging in biological systems? An even more basic question—"How can biosystems function at all, with such a huge number of variants and possibilities?" The answer is that nature deals with variability and randomness quite well. Small numbers of biomolecules, coupled with many possible different structural versions of the molecules, show that variations from the average (if an average even exists) should dominate in biological systems. In this and the following few chapters, it's important to keep in mind the big picture. When biology is looked at closely, variability and randomness are seen to abound, yet this unpredictability does not matter much in the end. If one molecular motor in a muscle fiber decides to go the wrong way, it doesn't matter, as long as 10 others go the right way. It similarly does not matter if the 10 others have differing intrinsic speeds. The muscle system is designed to tolerate these molecular "failures and disagreements." In some cases the molecular variability is actually an advantage when unusual circumstances develop. In Chapter 15 we show that one of the simplest reactions involving a protein, the binding of oxygen by myoglobin, is characterized by an ensemble of myoglobin conformations that differ significantly in intrinsic binding rate, yet the binding at physiological temperatures is nearly a single exponential process in time, usually taken to indicate a single binding process. We also find the involvement of quantum mechanics into this "simple" reaction.

13.2 ONE-DIMENSIONAL RANDOM WALK: PROBABILITIES AND DISTRIBUTIONS

We return now to the random walks with which we started in Chapter 1 (Figure 1.3). First, however, we should note some caveats. When we look at a sample of micron-sized spheres in a microscope (Figure 1.3b), we visibly see them move randomly. In fact, depending on the sphere size and fluid viscosity, it may appear that the particles are moving continuously and smoothly, rather than suffering sudden collisions. How can this be, if motion of the sphere is caused by collision with a fluid (water) molecule? The collision should produce a sudden change in momentum. Let's look at this momentum change more closely: water of molecular mass 18 D (3×10^{-26} kg), collides with a 1-μm-diameter sphere. The sphere mass should be about $m_s \cong \frac{1000 \text{ kg}}{m^3} \frac{4}{3} \pi (0.5 \times 10^{-6} \text{m})^3 \approx 5 \times 10^{-16} \text{kg} \approx 3 \times 10^{11} \text{D}$. What happens when 18 D meets 10^{11} D in a collision? The 18 D mass retains virtually all of its kinetic energy when it bounces off the sphere; the sphere acquires virtually no kinetic energy. So much for Brownian motion caused by solvent collisions! But because we see the sphere move and because Albert Einstein's most cited paper explained Brownian motion in terms of solvent-sphere collisions, we must be missing something. (In the next paragraph we do some hand waving. For the correct version, see Einstein.[2])

What we are missing is the fact that there are 3×10^{10} water molecules per cubic micron. At any given time, half of them travel toward the sphere. Consider the shell of water around the sphere (Figure 13.1). Water molecules that are within one water *mean-free-path length* can possibly collide with the sphere. Water molecules farther away collide with intervening water molecules. What is the mean free path of water in water? We could use the ideal gas theory to guesstimate the mean free path, but we actually know the answer already. Water is closely packed, so a water molecule can only move about one H_2O diameter before colliding with something. This means that only the closest layer or two of water collide with the sphere. In a layer 0.3 nm thick around the 1-μm sphere, there are about $4\pi R^2 (\Delta R)(30 \text{ } H_2O/nm^3) = 4\pi (1{,}000 \text{ nm})^2 (0.3 \text{ nm})(30 \text{ } H_2O/nm^3) \approx 10^8 H_2O$. Because water molecules must either move toward or away from the sphere, a significant fraction of the molecules collide with the sphere. Of course, on average, the collisions occur symmetrically on all sides, but average is not what we are interested in. Let's think: we have 10^8 18-D water molecules that are within first-collision distance with a sphere of mass 10^{11} D. If those water molecules behave like one-dimensional random walkers, what is the probability that an excess of m of them collide in one direction with the sphere? Before going farther, let's

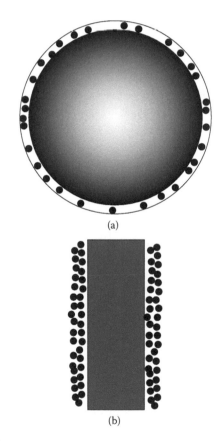

(a)

(b)

FIGURE 13.1
(a) Micron-sized sphere surrounded by water molecules. Water shown greatly enlarged for clarity. Water molecules move randomly in three dimensions. Molecules within about a molecular diameter or two of the sphere are within a mean-free-path length and can strike the sphere. There are about 10^8 such water molecules around a 1 μm-diameter sphere. (b) A simplified scheme for estimating the effects of molecular collisions with the sphere, now a rectangular solid.

simplify the geometry to that in Figure 13.1b: water molecules strike a rectangular solid from left and right. Letting

$$n = \text{number of right-steppers, } n' = \text{number of left-steppers,}$$

$$N = \text{total number of steppers,}$$

$$m = n - n' = n - (N - n) = 2n - N,$$

the answer to this question is given (approximately) by the binomial distribution:[3]

$$P'(m)\frac{N!}{\left(\frac{N+m}{2}\right)!\left(\frac{N-m}{2}\right)!}p^{\frac{N+m}{2}}q^{\frac{N-m}{2}} \tag{13.11}$$

where p is probability for a step to the right and q is probability for a step to the left; m is the net number that go to the right. The p-q factor becomes $(\frac{1}{2})^N$ for the case where $p = q = \frac{1}{2}$. A more usual form of the binomial probability can be expressed as $P(n)$, the probability that n steppers step right and the remaining $N - n$ step left is given by

$$P(n) = \frac{N!}{n!(N-n)!}p^n q^{N-n} \tag{13.12}$$

Notice the difference between m and n. $P(n)$ and $P'(m)$ are *not* the same function, though they refer to the same system. In particular, $P'(m) = P\left(n = \frac{N+m}{2}\right)$. Though we have not derived them, Equations 13.11 and 13.12 form a core of our ability to simulate random walks on paper. The binomial distribution can also be applied to a single stepper, where m is interpreted as the number of steps to the right and N the total number of steps.

There are other, more sophisticated ways to simulate random walks. For instance, are double steps allowed? What about a particle staying in place, rather than stepping to the right or left? Allowing only one of two choices is a bit restrictive.

When should we use the expression $P'(m)$ and when $P(n)$? $P(n)$ is the probability that n of N particles step to the right (or point up in an up/down orientation choice). $P'(m)$ is the probability that m more particles step to the right than to the left. We also asserted that $P'(m)$ describes the probability that one particle ends up m steps to the right after a total of N steps. If we are interested in single-particle probabilities, $P'(m)$ seems more useful. Likewise, if we are measuring some property of a material that depends on the difference between number to the right and number to the left, we want $P'(m)$. The classic example of this is the magnetization of a material in which atomic spins can point up or down: the net magnetization depends most directly on m, not n. Below we use both expressions for the case of Brownian motion.

Think of the meaning of the distribution of results for N identical objects, then think of the case where the N objects were actually one object observed N times: n times it jumped to the right, $N - n$ times it jumped to the left. After all N jumps, the (one) stepper ended up $m = n - n'$ steps to the right.

EXERCISE 13.1

Binomial distribution average values and widths. Find the averages and widths of the $P(n)$ and $P'(m)$ distributions for $p = 1/2$ and $N = 5$.

Equations 3.58 through 3.60 showed how to calculate average values and deviations for discrete distributions: $\langle n \rangle = \sum_{n=1}^{N} nP(n)$ and $\langle \Delta n \rangle = \sqrt{\langle n^2 \rangle - \langle n \rangle^2}$. The latter reminds us that we should always consider the average value of n^2, in addition to average n, especially when the value can be negative.

$$\langle n \rangle = \sum_{n=0}^{N} nP(n) = \sum_{n=0}^{N} n \frac{N!}{n!(N-n)!} p^n q^{N-n} = N! \left(\frac{1}{2}\right)^N \sum_{n=0}^{N} n \frac{1}{n!(N-n)!}$$

$$= 5! \left(\frac{1}{2}\right)^5 \sum_{n=0}^{5} \frac{n}{n!(5-n)!}$$

$$= 5 \cdot 4 \cdot 3 \cdot 2 \cdot \left(\frac{1}{32}\right) \left(\frac{1}{1 \cdot 4 \cdot 3 \cdot 2} + \frac{2}{2 \cdot 3 \cdot 2} + \frac{3}{3 \cdot 2 \cdot 2} + \frac{4}{4 \cdot 3 \cdot 2} + \frac{5}{5 \cdot 4 \cdot 3 \cdot 2 \cdot 1}\right)$$

$$= \frac{15}{4} \left(\frac{1}{24} + \frac{1}{6} + \frac{1}{4} + \frac{1}{6} + \frac{1}{24}\right) = \frac{5}{2} = 2.5$$

This is, of course, not surprising.

$$\langle n^2 \rangle = \sum_{n=0}^{N} n^2 P(n) = \sum_{n=0}^{N} n^2 \frac{N!}{n!(N-n)!} p^n q^{N-n} = N! \left(\frac{1}{2}\right)^N \sum_{n=0}^{N} n^2 \frac{1}{n!(N-n)!}$$

$$= 5! \left(\frac{1}{2}\right)^5 \sum_{n=0}^{5} \frac{n^2}{n!(5-n)!}$$

$$= 5 \cdot 4 \cdot 3 \cdot 2 \cdot \left(\frac{1}{32}\right) \left(\frac{1}{1 \cdot 4 \cdot 3 \cdot 2} + \frac{4}{2 \cdot 3 \cdot 2} + \frac{9}{3 \cdot 2 \cdot 2} + \frac{16}{4 \cdot 3 \cdot 2} + \frac{25}{5 \cdot 4 \cdot 3 \cdot 2 \cdot 1}\right)$$

$$= \frac{15}{4} \left(\frac{1}{24} + \frac{1}{3} + \frac{3}{4} + \frac{2}{3} + \frac{5}{24}\right) = \frac{15}{2} = 7.5$$

Now the variance of n and standard deviation:

$$\text{Variance } (n) = \langle n^2 \rangle - \langle n \rangle^2 = \frac{15}{2} - \left(\frac{5}{2}\right)^2 = \frac{5}{4} = 1.25 \text{ and}$$

$$\text{Standard deviation } (n) = \sqrt{\frac{5}{4}} = \frac{\sqrt{5}}{2} = 1.12$$

The corresponding quantities for $P'(m)$:

$$\langle m \rangle = \sum_{m=-N}^{N} mP'(m) = \sum_{m=-N}^{N} \frac{mN!}{\left(\frac{N+m}{2}\right)! \left(\frac{N-m}{2}\right)!} p^{\frac{N+m}{2}} q^{\frac{N-m}{2}}$$

$$= 5! \left(\frac{1}{2}\right)^5 \sum_{m=-5,\text{odd}}^{5} \frac{m}{\left(\frac{N+m}{2}\right)! \left(\frac{N-m}{2}\right)!}$$

$$= \left(\frac{15}{4}\right) \left(\frac{-5}{5!} + \frac{-3}{4!} + \frac{-1}{2!3!} + \frac{1}{3!2!} + \frac{3}{4!} + \frac{5}{5!}\right) = 0$$

(This should be zero because the average net number of steps is zero for $p = q = 1/2$. Note the restriction to odd numbers because an odd total number of steps cannot result in an even value of m. See Figures 13.4 and 13.5.)

$$\langle m^2 \rangle = \sum_{m=-N}^{N} m^2 P'(m) = \sum_{m=-N}^{N} \frac{m^2 N!}{\left(\dfrac{N+m}{2}\right)!\left(\dfrac{N-m}{2}\right)!} p^{\frac{N+m}{2}} q^{\frac{N-m}{2}}$$

$$= 5!\left(\frac{1}{2}\right)^5 \sum_{m=-5,odd}^{5} \frac{m^2}{\left(\dfrac{N+m}{2}\right)!\left(\dfrac{N-m}{2}\right)!}$$

$$= \left(\frac{15}{4}\right)\left(\frac{25}{5!} + \frac{9}{4!} + \frac{1}{2!3!} + \frac{1}{3!2!} + \frac{9}{4!} + \frac{25}{5!}\right) = \left(\frac{15}{4}\right)\left(\frac{4}{3}\right) = 5 = Variance\ (m)$$

The standard deviation is then $\sqrt{5} \cong 2.24$, exactly twice the standard deviation of $P(n)$. Why should this be? Think about the fact that every time n changes by 1, m must change by 2, and in the case $p = 1/2$, the $P'(m)$ distribution is symmetric about $m = 0$.

Let's apply Equation 13.11 to our case of 10^8 water molecules. Note that we are now assuming all water molecules are identical, have the same magnitude of momentum, etc. This isn't quite right, but let's plow ahead. What is the probability that an excess of 50% of the molecules travel to the right (3/4 of them travel to the right)?

$$P'(m) = \frac{N!}{\left(\dfrac{N+m}{2}\right)!\left(\dfrac{N-m}{2}\right)!}\left(\frac{1}{2}\right)^N = \frac{10^8!}{\left(\dfrac{3\cdot10^8}{4}\right)!\left(\dfrac{10^8}{4}\right)!}\left(\frac{1}{2}\right)^{10^8}$$

Don't bother taking out your calculator. Most cannot do factorials of large numbers. Instead, learn Stirling's formula and use high-school math.

$$\text{Stirling's formula: } \ln(n!) \cong n\ln(n) - n + \frac{1}{2}\ln(2\pi n) \tag{13.13}$$

Taking the natural log of $P'(m)$ and using $m = N/2$, we get

$$\ln\left(P'(m)\right) = \ln(N!) - \ln\left(\frac{N+m}{2}!\right) - \ln\left(\frac{N-m}{2}!\right) + N\ln\left(\frac{1}{2}\right)$$

$$= \ln(N!) - \ln\left(\frac{3N}{4}!\right) - \ln\left(\frac{N}{4}!\right) + N\ln\left(\frac{1}{2}\right) \tag{13.14}$$

We can now use Stirling's formula on the factorials. As usual, we try to use as few terms in the Stirling expansion as possible. Trying just the first Stirling term for the case $m = N/2$, we get $\ln(P'(m))_1 \cong -0.131N$. However, we note along the way that the 0.131 resulted after several larger numbers almost cancelled out. This suggests that we may need more terms from the Stirling formula. We quickly find that the second Stirling term contributes nothing to Equation 13.14. The third Stirling term gives some numbers not far from 1 and a $\ln(N)$, all of which are negligible compared to $-0.131N$. Now, however, we recognize that the probability that ¾ of the 10^8 molecules travel to the right is exceedingly small—$\ln(P'(N/2)) \cong -0.131N = -1.3 \times 10^7$ or $P'(N/2) = \exp-(1.3 \times 10^7)$ $\cong 10^{-5,600,000}$. Three quarters of the 100,000,000 water molecules are never going to conspire to hit the sphere toward the right.

Let's go to the other extreme. What about just one more to the right than to the left? A brief examination of either Equation 13.11 or 13.12 suggests that we again compute a very small probability.…

Something should be dawning on us all by now. Equations 13.11 and 13.12 are virtually unmanageable for large values of N. The probability for any particular result is exceedingly small. What we have (re)discovered is the truth that for a distribution function with a virtually continuous set of possible outcomes, the probability for any particular (exact) outcome is virtually zero. For $N = 5$, 20, or 100, as in Figure 13.2a, Equations 13.11 and 13.12 are very helpful, but not for $N = 10^8$. If we want to use the probability distribution, we must convert it to a functional form and ask questions like, "What is the probability that m lies in the interval $(0–0.1)N$ or $(0.1–0.3)N$?" This demands an integral (or at least multiplication of $P'(m)$ by a width Δm).

13.2.1 Gaussian Distribution

We want a functional form of Equation 13.12 valid for large values of N, so that allowed values of n are almost continuous. We do not carry out this derivation. It can be found in the cited Reif text and in other sources. The result is a Gaussian distribution:

$$P(n)\,dn = \frac{1}{\sqrt{2\pi Npq}}\exp-\left((n-Np)^2/2Npq\right)dn \qquad (13.15)$$

The limits on n are 0 to N, but if N is large and p isn't too close to 1.0, the upper limit can be extended to ∞. This approximation of the binomial distribution is valid for large values of N, so that the range of n is large.

In general, a Gaussian distribution is written

$$P(x)\,dx = \frac{1}{\sigma\sqrt{2\pi}}\exp-\left((x-x_0)^2/2\sigma^2\right)dx \qquad (13.16)$$

where x_0 is the average and most probable value of x and σ is the standard deviation:

$$\sigma^2 = \left\langle(x-x_0)^2\right\rangle = \int_{-\infty}^{+\infty}(x-x_0)^2\,P(x)\,dx \qquad (13.17)$$

(a)

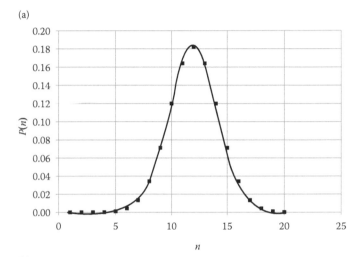

(b)

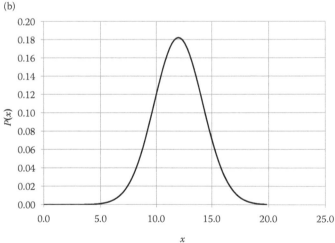

FIGURE 13.2
Probability distributions. (a) Discrete probability for n of N particles stepping to the right. The total number of particles is $N = 20$; $p = 0.60$ and $q = 0.40$ are the intrinsic probabilities to step right and left, respectively. Only integer values of n are allowed. The sum of the probabilities is 1: $\sum_{n=-\infty}^{\infty}P(n)=1$. (b) Continuous prob-ability distribution. Continuous values of x are allowed. The distribution function $P(x)$ must be multiplied by an interval dx to get a probability. The integral of the distribution is 1: $\int_{-\infty}^{\infty}P(x)dx=1$.

Figure 13.2a and b shows these two versions of the Gaussian distribution. Note that the two distributions look almost identical, but the choice of plotting a line graph versus a point scatter plot is not cosmetic preference. The discrete case $P(n)$ allows *only* integer values of n; $P(x)$ allows continuous variation of x.

What about the Gaussian limit for $P'(m)$? Using the principle that $P'(m) = P\left(\frac{N+m}{2}\right)$, we might write $P'(m) = \frac{1}{\sqrt{2\pi Npq}}\exp-$ $((m-2N(p-1/2))^2/4Npq)$, but we must take care to account for the new, continuous nature of $P'(m)$: $P(n)dn = P'(m)dm$. Because $m = n - (N - n) = 2n - N$, $dm = 2 \cdot dn$, and $P'(m) = P(n)/2$, so

$$P'(m)dm = \frac{1}{2\sqrt{2\pi Npq}} \exp\!-\!\left(\left(m - 2N\left(p - 1/2\right)\right)^2 \middle/ 4Npq\right)dm$$

$$= \frac{1}{\sqrt{2\pi \cdot 4 \cdot Npq}} \exp\!-\!\left(\left(m - 2N\left(p - 1/2\right)\right)^2 \middle/ 2 \cdot 4Npq\right)dm$$

(13.18)

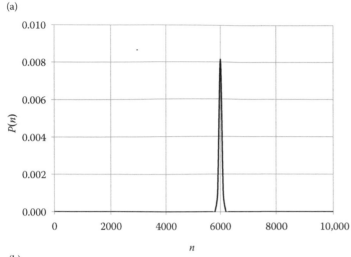

(a)

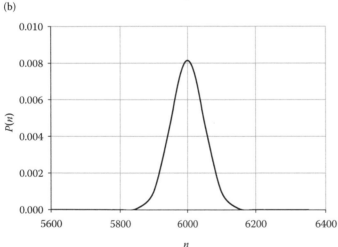

(b)

FIGURE 13.3
Gaussian probability function (Equation 13.15) for $p = 0.60$ and $N = 10,000$. (a) Plotted on a scale of 0–10,000, the probability at $n = 0$ is seen to be vanishingly small. (b) Plotted on a scale from $n = 5600$–6400, the plot looks much like those in Figure 13.2.

We see here what we noted explicitly in Exercise 13.1, that the standard deviation (width) of the binomial distribution as a function of m is twice as wide as the width of $P(n)$. This apparently also holds for the Gaussian limit of the binomial distribution.

We also need to remind ourselves that approximations like Equation 13.15 usually have regions of inapplicability. Those who have worked with Gaussian functions like Equation 13.16 know that negative values of x are allowed. When the center of the distribution is $x_0 = 0$, half of the probability lies to the left of $x = 0$. Is this allowed in Equation 13.15? Is n allowed to be negative? Physically, it is certainly not allowed: we cannot have a negative number of particles that move in any direction. The value of $m = 2n - N$ can be negative but n cannot be. If the value of Np is large, then we shouldn't worry about negative n values: the value of $\exp\!-\!\left(\frac{(n-Np)^2}{2Npq}\right)$ is vanishingly small even if we mathematically allow negative values of n (Figure 13.3).

So, back to the Brownian motion of the 1-μm sphere in water. We can now use Equation 13.15, treating n as quasi-continuous, and ask, "What is the probability that the value of n is between n_1 and n_2?" We find that Equation 13.15 is still awkward to use because N is so large. The normalization constant in front of the function is of order 10^{-4} and the Gaussian function has a large number ($10^8/2$) to be subtracted from n in the exponent: $\exp\!-\!((n-10^8/2)^2/10^8/2)$. Variable substitution helps some, but the essential problem appears when you try to sketch a graph of the function. Letting $\Delta n = n - Np = n - 10^8/2$, the exponential has value 1 when $\Delta n = 0$. It is also 1 when $\Delta n = 1$; and when $\Delta n = 10$, and 100, and 1,000. Only at about $\Delta n \approx \sigma = \sqrt{Npq} = 5000$ does the exponential decrease (to about 0.6). This means there is high probability that anywhere from 1 to about 5,000, more water molecules bump the sphere toward the right than toward the left. We already know that if one water molecule collides with the 10^{11}-D sphere, not much happens. If 5,000 molecules simultaneously collide, the equivalent of a 10^5-D particle hits the sphere. The effect is much larger, though still small.

We can alternatively use Equation 13.18, which would simplify the present case. The same computational issues appear in Equation 13.18 if $p \neq 1/2$, however.

$\Delta n = m/2$ if $p = 1/2$.

EXERCISE 13.2

Calculate the relative probability that (1) 10,000 and (2) 50,000 more water molecules than the average bump the previous sphere toward the right.

For $\Delta n = 10,000$, the value of the exponential is $\exp\!-\!((10^4)^2/10^8/2) \approx 0.14$, a significant relative probability. For $\Delta n = 50,000$, the value of the exponential is $\exp\!-\!((5 \times 10^4)^2/10^8/2) = e^{-50} \approx 10^{-22}$, a very small relative probability: the Gaussian function drops off very quickly. Table 13.1 shows values of the Gaussian function for the present case.

13.2.2 Attempt Frequencies and Probabilities

The preceding probability computations relate the relative probability that Δn more molecules bump to the right than average in any given trial. How often do such "trials" occur? In a liquid or solid system, the first guess at a collision attempt frequency should be the typical vibrational frequency of atoms, 10^{12} s^{-1}. This frequency is not always appropriate, but in general we can write the probability per unit time that some event occurs is

$$\text{Probability per second} = (\text{Attempt frequency}) \times (\text{Probability per trial}) \quad (13.19)$$

We find that this principle of multiplying a probability by an attempt frequency is the basis of chemical reaction theory. In the present microsphere Brownian motion context, Table 13.1 indicates that about half of the total probability lies between $\Delta n = \pm 5,000$, which means the other half of the probability consists of an excess of more than 5,000 molecules bumping the sphere toward the right at any given time, so 5,000 to 10,000 more molecules hitting to the right is not uncommon. Even n between 29,000 and 31,000 ($dn=2,000$) is not rare: $P(n)dn \approx [P(30,000)][2,000] = [8.0 \times 10^{-5} \cdot 1.5 \times 10^{-8}]$ $[2,000] = 2.4 \times 10^{-9}$ (Table 13.1), but when collisions occur at a rate of 10^{12} times per second, we have the equivalent of 5×10^5 D mass of water hitting the sphere every millisecond or so. Now, while our math has been a bit sloppy, we have uncovered a basic truth: we can describe motions of particles in water from a framework of probabilities for various outcomes, and that even unlikely outcomes occur if we observe long enough. A remaining issue is to resolve what "long enough" means in terms of the health of a biological organism. If a result takes 0.5 s to occur with a reasonable probability, it cannot be relevant to a cell that needs that result to occur every 10 ms.

TABLE 13.1 Relative Gaussian Probabilities That Δn More Water Molecules Hit a 1-Micron Particle from One Side Than from the Other

$\Delta n = n - 10^8/2$	$\exp-[(\Delta n)^2/(10^8/2)]$
0	1
±1	1
±10	1
±100	1
±1,000	0.98
±3,000	0.84
±5,000	0.61
±10,000	0.14
±30,000	1.5×10^{-8}
±50,000	0.00 (10^{-22})

13.3 SPREADSHEET MODEL FOR A ONE-DIMENSIONAL RANDOM WALK

We have just seen that a binomial calculation can be applied to random walks in one dimension and that the binomial calculation turns into a Gaussian probability calculation if there are either a large number N of walkers or, equivalently, if we observe one walker but repeat the walk a large number of times and take the average of the outcomes. Our first application, however, has only considered one step—walkers (water) stepped either to the right or to the left and we stopped. What if we take one walker, let it step more than once and ask where it ends up? We already have the math for this: $P'(m)$. For small numbers of steps we can use the binomial version of $P'(m)$; for large numbers of steps, the Gaussian expression is more efficient.

When (for what value of m) should we switch from the binomial to the Gaussian equations? The mathematically correct answer is that we switch when the Gaussian formula is close to the binomial. For most applications this occurs when the binomial $P(n)$ probability for small n (near $n = 1$) is small: the center of the Gaussian is far from the origin, as measured by its width. The Gaussian expression suggests that this is the case when the width σ is much smaller than the most probable value:

$$\sigma \ll x_{mp} \text{ or } \sqrt{Npq} \ll Np \text{ or } \sqrt{\frac{1-p}{Np}} \ll 1 \quad (13.20)$$

The support Web site contains this spreadsheet, though it is not difficult to reconstruct.

The $p \sim 1$ or 0 cases are not of much interest because the "random" steps would almost always to the right or always to the left. If p is not too far from ½, we would say that when N is 100 or perhaps more, the Gaussian should be OK to use. Another "practical" way to decide when to give up on the binomial is when your calculator and/or computer spreadsheet gives errors. As of 2010, this typically happens in about the same range as estimated above. A Microsoft Excel spreadsheet begins giving errors for the binomial $P'(m)$ at about $N = 170$, because 171! cannot be computed.

Let's consider a spreadsheet calculator for binomial stepping. Figure 13.4a and b shows output graphs of a spreadsheet computation of Equation 13.11 for $N = 0$ and $N = 20$ steps, with equal intrinsic right/left probabilities, $p = q = 1/2$. The shape of the distribution is symmetric. Note the allowed values of m. For this particular spreadsheet the computation of $P'(m)$ is successful up to $N = 170$. The factorial 171! cannot be done by the spreadsheet. For an everyday, 7-year-old program available to most undergraduate students, this is not bad performance at all! Figure 13.4c shows a portion of the spreadsheet used to calculate the graphs in a and b.

Figure 13.5 repeats the computation of $N = 20$ steps, but for $p = 0.60$, 0.70, 0.80, and 0.90. There are two characteristics that are very clear, and perhaps unexpected:

- The probability distribution $P'(m)$ becomes more and more asymmetric as p increases from 0.5 to 1.0 (or as p decreases to 0).
- The position, as well as the speed of the distribution, gets tricky to define no matter what the value of p.

These are two *very* important observations. A good mathematician would carefully define the average "position" and "position-squared" and try to calculate them:

$$\langle x_N \rangle = L_s \langle m_N \rangle = L_s \sum_{m=-N}^{N} m P_N(m) = L_s N (2p-1) = \frac{L_s(2p-1)}{t_s} t \tag{13.21}$$

$$\langle x_N^2 \rangle = L_s^2 \langle m_N^2 \rangle = L_s^2 \sum_{m=-N}^{N} m^2 P_N(m) = L_s^2 4N \left[pq + N \left(p - \frac{1}{2} \right)^2 \right] = \frac{L_s^2 4 \left[pq + N \left(p - \frac{1}{2} \right)^2 \right]}{t_s} t \tag{13.22}$$

where L_s is the length of a step, assuming steps are all of identical length, with the variance and standard deviation

$$\langle x_N^2 \rangle - \langle x_N \rangle^2 = L_s^2 4Npq = \frac{L_s^2 4pq}{t_s} t$$

$$\Delta x_N \equiv \sqrt{\langle x_N^2 \rangle - \langle x_N \rangle^2} = L_s \sqrt{4Npq} = L_s \sqrt{\frac{4pq}{t_s}} t^{1/2} \tag{13.23}$$

The "drift velocity" can likewise be defined as

$$\langle v_N \rangle = \frac{\langle x_N \rangle - \langle x_{N-1} \rangle}{t_s} = \frac{L_s}{t_s}(2p-1) \tag{13.24}$$

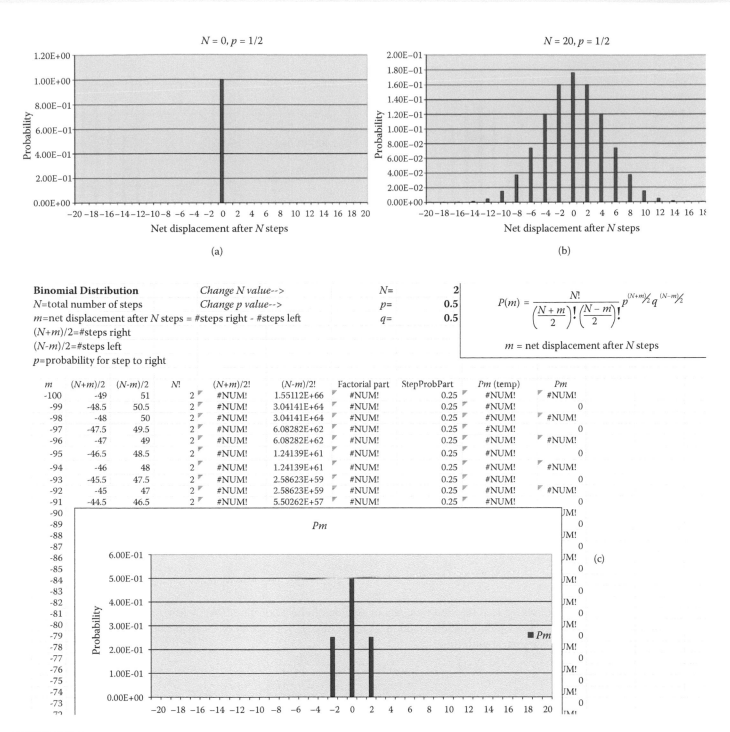

FIGURE 13.4
Spreadsheet binomial stepper. (a) Probability distribution $P'(m)$ before the first step. (b) Probability distribution $P'(m)$ after $N = 20$ steps: $p = q = 1/2$. (c) A portion of the spreadsheet. Note how the sheet is designed for easy variation of N and p. The errors ("#NUM!") for large, negative values of m do not cause the spreadsheet to malfunction for "good" values of m, which range from -2 to $+2$ for $N = 2$. (Microsoft Excel 2007 spreadsheet.)

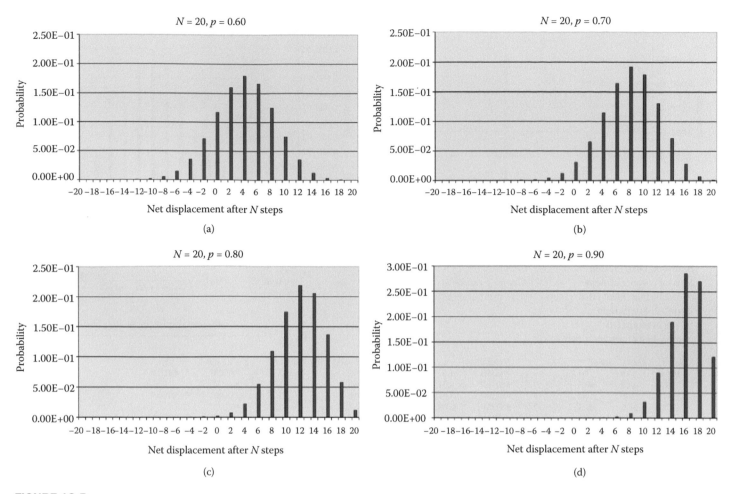

FIGURE 13.5
Spreadsheet binomial stepper. (a) $N = 20$, $p = 0.60$; (b) $N = 20$, $p = 0.70$; (c) $N = 20$, $p = 0.80$; (d) $N = 20$, $p = 0.90$. Note how the asymmetry of the distribution increases as the intrinsic right-step probability, p, increases. For 20 steps the maximum range is 20, of course. (Spreadsheet as in Figure 13.4.)

where t_s is the step time, again assumed to be a constant. (We note $v_N = $ const.) We could also ask how quickly the distribution average outpaces the spread of the distribution because both quantities increase with time:

$$\frac{x_N}{\Delta x_N} = \frac{L_s N (2p - 1)}{L_s \sqrt{4Npq}} = \frac{\sqrt{N}(2p - 1)}{\sqrt{4p(1 - p)}} \propto \sqrt{t}$$

$$\text{with } N = \frac{t}{t_s}$$

(13.25)

where t_s is the step time and t the total time for N steps. The skewness and kurtosis of the distribution could also be considered, though we will not. If the computations are already in a spreadsheet, these expressions can be used with other than the assumed binomial probabilities.

13.3.1 Stepper Motion with More Possible Steps

What happens if more than just left/right steps are possible? A real case of Brownian motion in a fluid involves steps of a variety of sizes. Though many one-dimensional molecular motors can be described by simple left/right stepping, some may not, especially if resistive forces are too large. We are, however, going to leave the analytical treatment of such cases to other texts.[4] The general approach is to start with a probability distribution for step size and direction: $K(k)$, where $k = $ step $=...-2, -1, 0, 1, 2,...$ and K is the probability of such a step. (For the binomial case, $K(-1) = q$ and $K(1) = p$, with all other step probabilities 0.) This distribution could be set up to emulate expected Brownian jumps in a liquid or to model a molecular motor that does not fit the simpler left/right p-q model. Below we show that a spreadsheet can also be set up to simulate variable step sizes.

We remind ourselves that our goal is biophysics: treatment of a random walker is a model for a biomolecular motor. If we have a molecular motor moving a biological payload along a track, $P_N(m)$ represents the *probability* of where the motor stands (m) after taking N steps. A given motor could, with reasonable probability, be at position $m = 10$ after 10 steps or after 50 steps for $p = 0.70$. (Try such a simulation with your own spreadsheet.) In the former case, the stepper happened to take 10 consecutive steps to the right—a 3% probability. In the latter case, 50 total steps, most back and forth, ending up at $m = 10$—a 4% probability. How many steps, and correspondingly, how much time, did it take for the stepper to get to $m = 10$? Did it move at speed $10L_s/10t_s = L_s/t_s$ or at $10L_s/50t_s = L_s/5t_s$? Or if the cell has 10 such motors and needs only one to deliver its cargo, how long does it take? In a given situation, the probability that 1 of the 10 binomial motors travels significantly faster than the global average speed can be high. Such questions can be answered from a statistical point of view, but a cell with a few motors may not live off of the average performance of its motors.

13.3.2 First-Transit Time

An active transport system in which a stepper moves from $m = 0$ to $m = 10$ is likely delivering a product that will be used at position $m = 10$. If this is the case, the speed and transport time gets to be even more interesting. If the product is grabbed by an enzyme that needs it as soon as it gets to $m = 10$, the "first-transit" time, the time at which the stepper first arrives at $m = 10$, is appropriate: when the stepper first gets to $m = 10$, the product is delivered. If the enzyme is not so quick—perhaps it's sleeping and takes a time Δt to wake up after the stepper first arrives at $m = 10$—the stepper may go beyond $m = 10$ and later (randomly) come back to $m = 10$, at which point the enzyme grabs the delivery. This would correspond to a later time and longer delivery. We can have great fun thinking about various possibilities and biological implications, the answers to which are active research topics.

13.3.3 Average Movement of N Steppers

We started with an expression for $P(n)$, the binomial probability that n of N objects in a system had one of two either/or properties: up/down, right/left, etc. The width of the corresponding distribution is characterized by a standard deviation $\sigma = \sqrt{Npq}$, where p is the intrinsic probability for up or right. The distribution describing the difference (#up-#down) or (#right-#left) $= n - n' = 2n - N \equiv m$, on the other hand, is $P'(m)$, which has a standard deviation $\sigma = 2\sqrt{Npq}$. We applied this latter distribution to the movement of a molecular motor (stepper) along a one-dimensional track, with m being the final position after N steps. The one-dimensional steppers we have considered have taken single steps to the right or left. If $p = q = 1/2$, the average position of the stepper remains at zero, but the probability distribution spreads with time. If $p \neq 1/2$, the average drifts to the right with average drift velocity $(2p - 1)L_s/t_s$. This drift can be extended, as shown above, to a variable step size and direction, with variable probabilities for each step. The net result is to replace pq in Equation 13.23 by the variance of the step-size probability distribution.

Figure 13.6 shows two examples of one-dimensional random steppers where $p = 1/2$, so stepping is equally likely to right and left. There is therefore zero net "drift" velocity. As we have seen, however, $\langle x \rangle = 0$ does not imply that $\langle x^2 \rangle = 0$. Equation 13.23 states that $\langle x_N^2 \rangle - \langle x_N \rangle^2 = \frac{L_s^2}{t_s}t$ or $\Delta x_N = L_s \left(\frac{t}{t_s}\right)^{1/2}$ for $p = q = 1/2$ and $N = t/t_s$: an initial distribution of particles spreads if they are allowed to freely step. In the upper three panels an initial number N_0 of particles is confined as closely

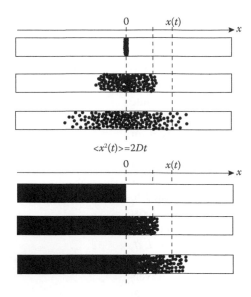

FIGURE 13.6

Random walk/diffusion in one dimension. In the *upper* scenario, a "pulse" of N_0 molecules is placed in a narrow tube at $x = 0$ at time zero. The high concentration disperses with time as the "front" spreads in both directions (ignoring tube boundary effects, that are often important). The average position of the front is described by $\langle x(t) \rangle = 2Dt$, where t is the time elapsed since initial placement of N_0 molecules and D is defined as the diffusion coefficient. In the *lower* scenario, the left side of the tube is filled with a large supply of molecules such that the concentration at $x = 0$ is maintained at c_0. The location of the front is again described by $\langle x(t) \rangle = 2Dt$, though the concentration distribution is not identical to that in the *upper*. See text for mathematical solutions.

> This is for spreading of $P'(m)$. For $P(n)$, replace t_s by $2t_s$, where t_s is the time for one step. $P(n)$ corresponds more closely to the diffusion of N particles in one dimension.

> As we will shortly see, Figure 1.3a is a three-dimensional simulation, with results projected onto two dimensions.

as possible to the region near $x = 0$ and then allowed to freely spread in both directions. This is the so-called pulse diffusion situation. The position of the front of the distribution, described by Δx, moves in both $+x$ and $-x$ directions symmetrically. The lower three panels picture the situation where the concentration is fixed to a value c_0 at $x = 0$ by some means. This situation can be approximated experimentally when a large supply of particles sits to the left of the origin. Those with more experience in differential equations also should recognize this as a simple, though somewhat artificial, boundary condition on a differential equation. We address the solutions to these problems shortly, but the physics to recognize now is that the speed at which particles (randomly) move down the tube does not depend on boundary conditions at $x = 0$ but rather on particle and fluid properties, in addition to temperature.

13.3.4 Casino of Real Life

The binomial $P'(m)$ distribution allowed us to model probabilities for a single random walker. The probability distributions all appeared smooth and well behaved. Where was the "randomness"? The intrinsic wildness of random processes is described by the probabilities we have derived, but probabilities are probabilities: they are not what you actually observe if you measure the process once or even 100 times. Often, the smooth probabilistic behavior does not appear until you have repeated the same process N times, where $\frac{N}{\sqrt{N}} \gg 1$, sometimes demanding $N \sim 10^8$ or more for sufficient accuracy. To see a more realistic simulation of life in the "small-number lane," we should go to the casino and bet our money rather than computing probabilities safely from the comfort of our offices. Actually, we won't bet money, but we will create a small casino on our computers. Figure 1.3a shows the result of the random walk, in two dimensions for clarity. When the 50-step simulation is run again, the path of the particle is completely different. We can run the simulation 100 times and never get two paths that appear almost identical.

13.4 THREE-DIMENSIONAL RANDOM WALK

How do we construct such a random walk simulator as used in Figure 1.3a? There are many ways, including the (extremely tedious) hand-thrown dice method to decide the direction of each throw. A simple spreadsheet again provides a quite useful solution to the task. The key spreadsheet function to learn is the random number generator. Figure 13.7a shows part of a spreadsheet that employs random number generators to decide on each step a random walker takes. Constructing such a spreadsheet is a good exercise because it forces us to consider exactly what we mean by a "random walk." There are many options. In Figure 13.7 it is assumed that steps are equally likely in $\pm x$, $\pm y$, or $\pm z$ directions. One option would be to restrict any given step to one of these six possible directions. The rather loose connection to the reality of a liquid, which has no preferred axes, would be compensated for by the many steps that would occur in a typical case of bulk chemical reactions. It may even approximate the diffusion of molecules within a cell, if there are enough molecules and enough free space. The simulator for Figure 13.7 allows for more than six possible directions. Each step can have relative (x, y, z) components of ± 1 or 0; the magnitude of the step size is then

t	x	y	z	R	Random del_x	Random del_y	Random del_z	Step_size _unnorm	Notes
0	0.0	0.0	0.0	0.0	−1	1	0	1.414	3D random walk simulator. Steps assumed to be of
1	−0.7	0.7	0.0	1.0	1	−1	−1	1.732	length −1, 0 or +1 in x, y and z directions. Total step
2	−0.1	0.1	−0.6	0.6	0	−1	1	1.414	length is then normalized to 1 (or left at 0).
3	−0.1	−0.6	0.1	0.6	−1	1	0	1.414	
4	−0.8	0.1	0.1	0.9	0	0	1	1.000	
5	−0.8	0.1	1.1	1.4	−1	1	0	1.414	
6	−1.5	0.8	1.1	2.1	1	−1	0	1.414	command+= to recalculate sheet
7	−0.8	0.1	1.1	1.4	1	1	1	1.732	
8	−0.3	0.7	1.7	1.9	1	1	0	1.414	Col F,G,H: Randbetween(−1, 1)
9	0.4	1.4	1.7	2.3	1	1	−1	1.732	B3: Iferror(B2+F2/I2,B2)
10	1.0	2.0	1.1	2.5	1	−1	−1	1.732	Iferror corrects for error if no
11	1.6	1.4	0.6	2.2	0	0	−1	1.000	step is taken in any direction.
12	1.6	1.4	−0.4	2.2	−1	0	−1	1.414	(Other B's, as well as column
13	0.9	1.4	−1.2	2.0	1	0	0	1.000	C & D are similar.)
14	1.9	1.4	−1.2	2.6	0	1	0	1.000	

(a)

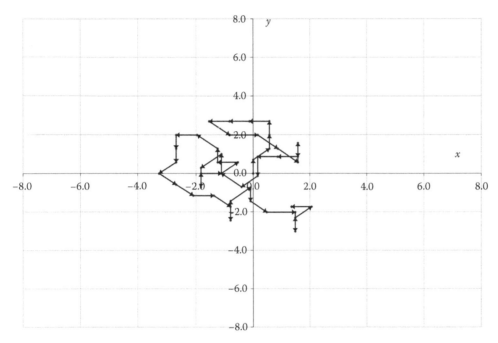

(b)

FIGURE 13.7
Random-walk simulation on a spreadsheet. (a) Part of the computational sheet. This sheet is set up for a three-dimensional walk. Step components of 0 or ±1 are allowed along x, y, and z axes, but the total length of the step is then normalized to 1, unless it has all-zero components. (See text for discussion of this normalization.) This allows steps in directions along ±x, ±y, ±z, and certain other allowed angles. The walker can also remain in place. Most of these options should be considered and adjusted for the application, but all are approximations to Brownian motion in a liquid. (b) Result of one 100-step simulation. The apparent step sizes are not identical because only x and y (not z) components are plotted. (Microsoft Excel spreadsheet.)

normalized to 1. For example, if the randomly chosen (x, y, z) components were $(0, 1, 1)$, the step would be normalized to $\left(0, \frac{1}{\sqrt{2}}, \frac{1}{\sqrt{2}}\right)$, shown in the inset to Figure 13.7b. This step direction would be in the yz plane, 45° up from the +y axis toward the +z axis. Directions along any of the ± axes, 45° between any of these axes, or in the middle of any octant (26 directions) are allowed. A more sophisticated spreadsheet might allow more directions or normalize the step size randomly to a size distribution function, perhaps similar to the Maxwell–Boltzmann (M–B) speed distribution.

   The correspondence between three-, two-, and one-dimensional simulations of random-walk motions should be considered carefully. Figure 13.7 plots the (x, y) components of the three-dimensional coordinates of the moving particle. Although the three-dimensional step sizes are all fixed at 0.00 or 1.00, the components in the xy plane are not. If the simulation had allowed only steps along axes, the trajectory in the xy plane would show only steps of length 0.00 or 1.00. The effect would be similar if only the x component were plotted. Under what conditions does a one-dimensional (two-dimensional) simulation give the same results as a three-dimensional simulation, with one (two) component(s) plotted? Suppose our three-dimensional simulation had allowed steps of size 1 only along $\pm x$, y, and z axis directions: no steps of 0 allowed. A one-dimensional plot of the x, y, or z coordinate of the particle would all show steps of size 1. In this case the motions along x, y, and z appear independent. Likewise, if 0 and 1 step sizes are allowed, but the total three-dimensional step size is not normalized, the walks along x, y, and z axes are independent. As soon as the total step size is normalized, as in Figure 13.7, we begin to see x step components that would not have appeared in a pure one-dimensional simulation. Which is closer to physical reality: normalized or un-normalized steps? We find that when we move to a continuum, diffusion description of random motion, the math assumes there is no such indirect effect of motion in the different dimensions. If we recall our treatment of the ideal gas, the x, y, and z motions of a gas particle were governed by the M–B speed distribution and the equipartition theorem, with $\frac{1}{2}m\langle v_x^2\rangle = \frac{1}{2}m\langle v_y^2\rangle = \frac{1}{2}m\langle v_z^2\rangle = \frac{1}{2}k_BT$ and $\frac{1}{2}m\langle v^2\rangle = \frac{3}{2}k_BT$ If v_x happened to be above average, that did not force v_y and v_z to be lower to compensate: x, y, and z are independent modes/degrees of freedom, each separately governed by the M–B statistical distribution. If we apply that independent-mode assumption to our random walker, our starting position should be *not* to normalize the three-dimensional step size. Did we just learn something from this "mistake" in Figure 13.7?

13.5 DIFFUSION IN THE BULK

Brownian motion of a spherical particle in bulk water can should be described by a random step of length and direction, both continuously variable, but governed by a Boltzmann-type distribution of step length. We saw in our 1-micron sphere-in-water case that during a given time instant one or 10^7 water molecules could collide from the left or any other direction. The step could have size zero or could be large. We again leave this task to others and move on to the continuum diffusion treatment. In the simulator used for Figure 13.7, step sizes of 0 or 1 are allowed in either $+$ or $-$ directions. We began with a single random-walker, described by the binomial distribution and preceded to the large-N limit, where the binomial became the Gaussian. The Gaussian distribution directly leads to the theory of diffusion, which underlies chemical reaction theory, where we are heading in the next two chapters.

13.5.1 Diffusion Equation

The equation governing bulk diffusion derives from two principles: the phenomenological Fick's law and the conservation of particles. *Fick's law states that particle flow is proportional to concentration gradient.* Particles in a bulk fluid diffuse, move randomly in x, y, and z directions. However, if a number of particles are initially located at $x = 0$ and none anywhere else, an observer at $x = +1$ μm initially detects particles traveling in the $+x$ direction. If there are also particles initially at $x = +2$ μm, the observer detects particles traveling in both directions. If more were at the origin, the right-moving particles outnumber the left-moving particles. This effect is quantified by

$$\text{Fick s law:} \quad j(x,t) = -D\frac{\partial c(x,t)}{\partial x} \quad (13.26)$$

where $j(x, t) \equiv$ particle flux $=$ number of particles crossing a 1-m^2 area oriented perpendicular to the x axis at position x, SI units (particles)/s·m²; $D \equiv$ diffusion constant $=$ proportionality constant, SI units m²/s; and $c(x, t) \equiv$ particle concentration, SI units (particles)/m³. Similar equations can be written for the y and z components of the flux and concentration when the particle concentration varies in x, y, and z. Figure 13.8 motivates this proportionality between particle flow and the gradient of the concentration in the x direction.

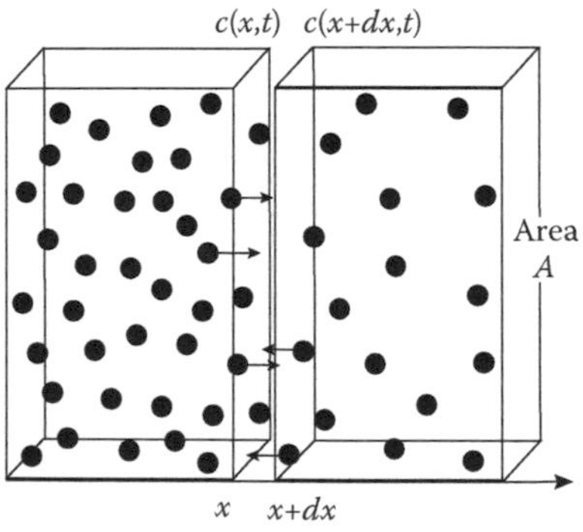

FIGURE 13.8

Fick's law and continuity. The net flux (j) of particles (particles per unit area A per unit time) flowing into the volume between x and $x + dx$ changes the number of particles in that volume: $A\Delta c\Delta x = A\Delta j\Delta t$ or $dc/dt = dj/dx$.

Continuity: particles don't disappear. In the flow of particles we assume that none disappear and none are created. This means that if we count the number of particles in the small volume $A dx$ in Figure 13.8, any change must come from particles flowing in and out. The change in the number of particles in a short time interval Δt (dt) must equal the volume, $A\Delta x$, times the concentration change Δc. The change in number must also equal the net number of particles flowing in per second times, $A\Delta j$, times the elapsed time Δt. We thus have $A\Delta x\Delta c = A\Delta j\Delta t$, or

$$\text{Continuity equation:} \quad \frac{dc}{dt} = -\frac{dj}{dx} \quad (13.27)$$

(Note the negative sign: if the net flux to the right is larger at $x + dx$ than at x, particles are draining out of the volume.) If we now take the derivative with respect to x of Fick's law, $\frac{\partial j}{\partial x} = -D\frac{\partial^2 c}{\partial x^2}$ and combine with the continuity condition we get the diffusion equation:

$$\text{Diffusion equation:} \quad \frac{\partial c(x,t)}{\partial t} = D\frac{\partial^2 c(x,t)}{\partial x^2}, \quad (13.28)$$

sometimes called *Fick's second law*.

The diffusion equation is, of course, very general and useful, but it must be dealt with carefully. Roughly, the equation states that for the concentration (in a small volume located) at x to change with time, there must be a nonzero second derivative with respect to x: $c(x)$ must have some curvature with respect to x at the point. Thus, for example, the initial concentration distribution in Figure 13.9a, which has zero first and second derivatives in the region $0 < x < x_1$, would be expected to result in a time-independent concentration distribution. If there were walls at $x = 0$ and $x = x_1$, we would be right: $c(x, t) = c_0$ between $x = 0$ and $x = x_1$ for all time.

What if there were only a wall at $x = 0$ in Figure 13.9a and that the initial concentration distribution were generated, for example, by photo-irradiating the region between $x = 0$ and $x = x_1$? Do we still expect $c(x, t)$ to be time independent? Intuitively, we would say "No; the molecules would diffuse to the right." But how does this come from the math, from the diffusion equation? The key is the derivatives right at $x = x_1$: $c(x,0) = c_0$ just to the left of x_1, but $c(x,0) = 0$ just to the right. The derivatives are, in fact, infinite (or undefined). We can proceed to determine how to mathematically deal with concentration discontinuities in the context of the second-order, partial differential diffusion EQUATION, but this would be rather useless because no such discontinuity could be generated, in reality. (Remember, *nature abhors sharp edges.*) We would be better off modeling the edge more realistically, giving it negative curvature just to the left of $x = x_1$, positive curvature to the right, and a small distance, δ, over which the concentration decreases to 0. The concentration distribution would then change with time: the region of negative curvature would decrease in concentration with time; the positive-curvature region would increase in concentration with time. As the concentration near $x = x_1 - \delta/2$ begins to decrease, the region of negative curvature extends to the left, propagating the concentration decrease further.

The initial concentration distribution, $c(x,0) = c_0(1 - x/x_1)$, in Figure 13.9b also appears to have zero curvature d^2c/dx^2 between $x = 0$ and $x = x_1$, though $dc/dx = -c_0/x_1$. Is this distribution therefore time independent? Suppose we say that walls exist at $x = 0$ and at $x = x_1$. Does $c(x, t) = c_0(1 - x/x_1)$ for all time?

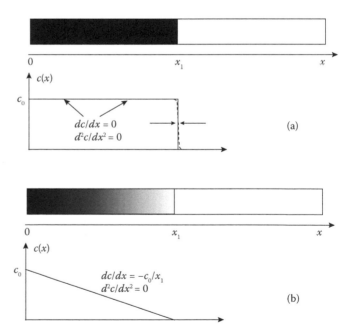

FIGURE 13.9
Two initial concentration distributions as a function of x. (a) Concentration is constant between $x = 0$ and $x = x_1$. Both dc/dx and d^2c/dx^2 are zero for $0 < x < x_1$; not really because a physical concentration cannot drop instantaneously to 0 (dotted line). (b) Concentration gradient between $x = 0$ and $x = x_1$; d^2c/dx^2 is still zero between $x = 0$ and $x = x_1$.

Intuitively, we know the answer is, "No, the distribution would relax to $c(x,\infty) = \frac{1}{2} c_0$ between the walls." How does this come from the math? The answer again comes from boundary effects. Both the first and second derivatives have discontinuities at $x = 0$ and $x = x_1$. The discontinuity at $x = x_1$, if made physically correct, would consist of a positive curvature, indicating that the concentration should increase with time. At $x = 0$, the opposite would be the case. The curvatures would then propagate toward the center of the distribution until $c(x,\infty) = \frac{1}{2} c_0$.

It is also useful to think of the flow of particles in terms of Fick's law. Fick's law states that particles move from regions of higher to lower concentration. If the concentration gradient in Figure 13.9b is to be maintained, particles must flow from left to right. For positions on the right side of the box, it's easy to see that flow can come from higher-concentration regions to the left. But what flow maintains the concentration c_0 at $x = 0$? Careless analysis using the diffusion equation may have assumed that the rising concentration in Figure 13.9b continues to the left of zero when it does not.

13.5.2 Diffusion from an Abrupt Front (Graduate Section)

Let's briefly look at the solution to the idealized case of Figure 13.9a, where the left side of a tube or box has uniform concentration c_0 up to a position $x = x_1$, where it drops abruptly to zero. To simplify a bit more, let's assume the concentration c_0 initially extends to $x = -\infty$. And as long as we're extending to $-\infty$, we might as well set $x_1 = 0$, so our initial boundary is at $x = 0$. As noted, we should always consider both Fick's equation and the diffusion equation when confronting a new situation. Qualitatively, we know that the abrupt boundary should smooth and broaden as time proceeds. The particle flux should be greatest at the edge ($x = x_1$) and should decrease away from the edges; flow should always be left to right. What flux function might describe such a case? It should also be smooth—no sharp edges. So far, we have seen one mathematical function come up in random-walk, diffusion problems: the Gaussian function. To describe the flux, remembering $x_1 = 0$ and the extension of the concentration to $-\infty$, let's try

$$j(x,t) = A\exp-\left(\frac{x^2}{4Dt}\right) \tag{13.29}$$

This function has a maximum at $x = 0$ and, because of the $4Dt$ denominator in the exponential, also has the spreading characteristic of diffusion—just what we expect. Fick's equation would then imply that $j(x,t) = -D\frac{\partial c}{\partial x} = A\exp-\left(\frac{x^2}{4Dt}\right)$ or $c(x,t) = B\int_{-\infty}^{x} \exp\left(\frac{(x')^2}{4Dt}\right)dx' + C$. C is an integration constant. Letting $y = \frac{x'}{\sqrt{4Dt}}, dy = \frac{dx'}{\sqrt{4Dt}}$,

$$
\begin{aligned}
c(x,t) &= \sqrt{4Dt}\,B\int_{-\infty}^{\frac{x}{\sqrt{4Dt}}} \exp-y^2 dy + C = \sqrt{4Dt}\,B\left[\int_{-\infty}^{0} \exp-y^2 dy + \int_{0}^{\frac{x}{\sqrt{4Dt}}} \exp-y^2 dy\right] + C \\
&= \sqrt{4Dt}\,B\left[\frac{\sqrt{\pi}}{2} + \int_{0}^{\frac{x}{\sqrt{4Dt}}} \exp-y^2 dy\right] + C = \sqrt{4Dt}\,B\frac{\sqrt{\pi}}{2}\left[1 + Erf\left(\frac{x}{\sqrt{4Dt}}\right)\right] + C
\end{aligned}
\tag{13.30}
$$

The integral of the familiar Gaussian becomes the less familiar "error function" or probability integral. It has rather simple behavior, however, and *should* become familiar to you: Figure 13.10a. Think of the error function as that function whose first derivative is the Gaussian: that is, in fact, exactly how we constructed it from $j(x)$. The error function approaches -1 as the argument goes below about -2, approaches $+1$ as the argument gets larger than about 2, and is zero for argument 0. The limit for $x \to -\infty$ seems to set the value of $C = c_0$, but we must be careful, since $t \to 0$ also sends the argument to infinite values.

Our lack of mathematical sophistication, combined with the rather unphysical initial conditions, are catching up with us. We should also note that we have only been using Fick's law and ignoring continuity and the time derivatives. As far as Fick's law is concerned, with its x derivative, t is a constant. The quantity B could be a function of time. So let's convert the equation to $c(x,t) = B'\left[1 + Erf\left(\frac{x}{\sqrt{4Dt}}\right)\right] + c_0$, with the proviso that B' could depend on time. For negative x and t slightly bigger than 0, the Erf argument is large and negative, so Erf $= -1$ and $c(x, t) \rightarrow c_0$, which is correct. For positive x and t slightly bigger than 0, Erf $= 1$ and we conclude $B' = -c_0/2$. For x exactly zero and t slightly positive, Erf $= 0$, and we get $c(0,0) = c_0/2$. This seems like a reasonable description of the concentration at $x = 0$, which is double valued, 0 and c_0, from the initial conditions. We can also trace back the value of the factor $A = \frac{c_0 D}{\sqrt{4\pi Dt}}$. So we have a mostly guessed solution of

$$c(x,t) = \frac{c_0}{2}\left[1 - Erf\left(\frac{x}{\sqrt{4Dt}}\right)\right] \qquad (13.31)$$

Is this a solution of the continuity equation, $\frac{dc}{dt} = -\frac{dj}{dx}$, and the diffusion equation? This task is left to the homework. Equation 13.31 is plotted in Figure 13.10b for three values of $4Dt$. For short times, the concentration distribution looks like the original step function. The increasing-concentration front then proceeds to the right, while a decreasing-concentration front proceeds to the left. For your consideration—What would change if a boundary condition $c(0,t) = c_0$ were imposed? This might correspond to a narrow tube connected to a large reservoir of particles at concentration c_0, with care taken to eliminate flow and convection.

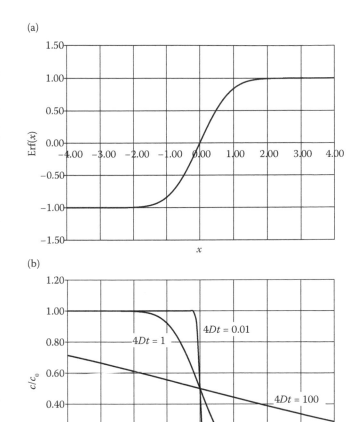

FIGURE 13.10
(a) The error function, or probability integral. The first derivative of this function is the Gaussian. (b) Solution to the diffusion equation for initial conditions $c = c_0$ for $x < 0$ and $c = 0$ for $x > 0$. As time proceeds the concentration increases to the right of $x = 0$ and decreases to the left. The derivative (gradient) of the concentration is a spreading Gaussian, $\exp - \frac{x^2}{4Dt}$.

13.5.3 Three-Dimensional Diffusion from a Point

The preceding example illustrates the random, diffusive behavior of a large number of particles confined to move along one axis. Particles move in random directions, of course, but the initial absence of particles at $x > 0$ causes more to move right than left. If a large number of particles, N_0, is placed at the origin a $t = 0$, they should likewise diffuse outward, on average. This three-dimensional diffusion example should be spherically symmetric. What do we expect the concentration distribution to look like at later times? As shown in Figure 13.11, the concentration should spread symmetrically away from the origin, conserving particles, with the concentration $c(r = 0,t)$ decreasing with time. In fact, the decrease should be quite dramatic, as the initial concentration at $r = 0$, exactly at time $t = 0$, is infinite. We have seen such time-zero

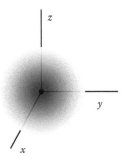

FIGURE 13.11
Radial diffusion from a point. A fixed number of particles is initially placed at the origin. They diffuse outward, preserving particle number.

infinities before; they occur when initial conditions with sharply varying concentration gradients are prepared. The characteristic infinities occurs as factors like $x^2/4Dt$. We have also seen that a time-dependent Gaussian position distributions characterize most situations when a particle moves randomly from an initial position. Before we begin the solution of the time-dependent concentration problem, it is useful to formalize our earlier average position results from the one-dimensional random walk in terms of the newly introduced diffusion constant D:

$$\text{One dimension: } \langle x^2 \rangle = 2Dt$$

$$\text{Two dimensions: } \langle x^2 + y^2 \rangle = \langle r^2 \rangle = 4Dt \tag{13.32}$$

$$\text{Three dimensions: } \langle x^2 + y^2 + z^2 \rangle = \langle r^2 \rangle = 6Dt$$

We should expect our solution to the point-diffusion case to abide by these Brownian rules. What might be a good guess for a solution? Gaussians of the form $\exp-(x^2/4Dt)$ look good for one dimension, but we must also construct a solution that becomes infinite at $t = 0$. A second issue is what the function should look like for three dimensions. The first guess would probably be $\exp-(r^2/4Dt)$, or possibly $\exp-(r^2/6Dt)$. The latter is found not to obey Equation 13.32. The conversion of Fick's law to spherical symmetry results in

$$j(r,t) = -D\frac{dc}{dr} \tag{13.33}$$

continuity becomes

$$\frac{dc(r,t)}{dt} = -\frac{dj(r,t)}{dr} \tag{13.34}$$

The diffusion equation for spherical symmetry becomes

$$\frac{dc(r,t)}{dt} = \frac{D}{r^2}\frac{d}{dr}\left(r^2\frac{dc(r,t)}{dr}\right) \tag{13.35}$$

Now we must construct a solution to these equations that becomes infinite at $t = 0$. We might first try to solve the one-dimensional case for N_0 particles initially placed at $x = 0$ and allowed to diffuse. A decent guess might be $c(x,t) = \frac{A}{t}\exp\left(-\frac{x^2}{4Dt}\right)$. As $t \to 0$ the Gaussian factor goes to zero for all x values except for $x = 0$. The $1/t$ out front then makes $c(0,0)$ infinite. It looks good but does not satisfy the diffusion equation. To be brief, successful guesses are as follows:

One-dimensional diffusion from a point:

$$c(x,t) = \frac{N_0}{\left(4\pi Dt\right)^{1/2}}\exp\left(-\frac{x^2}{4Dt}\right) \tag{13.36}$$

Two-dimensional diffusion from a point:

$$c(x,y,t) = \frac{N_0}{\left(4\pi Dt\right)}\exp\left(-\frac{x^2+y^2}{4Dt}\right) \tag{13.37}$$

Three-dimensional diffusion from a point:

$$c(r,t) = \frac{N_0}{\left(4\pi Dt\right)^{3/2}}\exp\left(-\frac{r^2}{4Dt}\right) \tag{13.38}$$

Are the units here correct? The quantity Dt has units $m^2\ s^{-1}\cdot s = m^2$, so the one-dimensional concentration has units of number/m, which is a correct one-dimensional concentration. The three-dimensional expression has units number/m^3.

EXERCISE 13.3

Integrate the concentrations given by Equations 13.36 and 13.38 over all x or space and show that particle number is conserved at all times. The integral is $\int_{-\infty}^{\infty} c(x,t)dx = \frac{N_0}{(4\pi Dt)^{1/2}} \int_{-\infty}^{\infty} \exp\left(-\frac{x^2}{4Dt}\right)dx$. This latter integral is easily found in integral tables; the value is $\sqrt{\pi 4Dt}$ and $N(t) = N_0$.

The three-dimensional case is similar, though care must be taken for the three dimensions. There is no angular dependence, so the three-dimensional integrals reduce to $\int_0^{\infty} r^2 dr \int_0^{\pi} \sin\theta \, d\theta \int_0^{2\pi} d\phi = 4\pi \int_0^{\infty} r^2 dr$. Integrating $c(r, t)$ over all space: $4\pi \int_0^{\infty} c(r,t)r^2 dr = \frac{4\pi N_0}{(4\pi Dt)^{3/2}} \int_0^{\infty} r^2 \exp\left(-\frac{r^2}{4Dt}\right)dr$. This integral is also easily found: $\int_0^{\infty} x^2 e^{r^2 x^2} dx = \frac{\sqrt{\pi}}{4r^3}$, so the value of our last integral is $\frac{\sqrt{\pi}(4Dt)^{3/2}}{4}$ and again the total number of particles is $N(t) = N_0$.

13.5.4 Radially Symmetric, Steady-State Solutions

Suppose we have a spherical cell of radius r immersed in some nutrient of initially uniform concentration c_0 (Figure 13.12). The cell consumes nutrient, reducing the nearby concentration. Call the reduced nutrient concentration c_1 at the cell surface, $r = R$. Far from the cell the concentration is still c_0. Let's identify the number of nutrient molecules the cell consumes per second as I. The diffusion equation for three-dimensional spherical symmetry in steady state is $\frac{dc(r,t)}{dt} = 0 = \frac{D}{r^2} \frac{d}{dr}\left(r^2 \frac{dc(r,t)}{dr}\right)$. This implies $r^2 \frac{dc}{dr} = A = \text{const}$ or $\frac{dc}{dr} = \frac{A}{r^2}$, or $c = \frac{-2A}{r} + B$. The boundary conditions are $c(\infty) = c_0$, so $B = c_0$, and $c(R) = c_1$, so the steady-state solution to our cell nutrient case is

$$c(r) = c_0 - \frac{(c_0 - c_1)R}{r} \tag{13.39}$$

We should be able to relate the cell's nutrient consumption rate to the reduced concentration c_1. As the cell consumes nutrient, nutrient molecules flow radially toward the cell. This flux is exactly the Fick's flux. The cell's consumption rate should equal Fick's flux integrated over the surface of the cell. According to Equation 13.33 the flux is

$$j(r) = -D\frac{dc}{dr} = -D\frac{d}{dr}\left[c_0 - \frac{(c_0 - c_1)R}{r}\right] = -\frac{D(c_0 - c_1)R}{r^2} \tag{13.40}$$

and has value $(c_0 - c_1)D/R$ at the cell radius, $r = R$. This flux is the number of nutrient molecules per second that pass through a unit area. The number per second that pass through the cell's surface area results from multiplying flux by $4\pi R^2$: $I = 4\pi R^2 \frac{(c_0 - c_1)D}{R}$. The concentration c_1 is then related to the cell's nutrient consumption rate I:

$$c_1 = c_0 - \frac{I}{4\pi RD} \tag{13.41}$$

Can the results of Equations 13.39 and 13.41 tell us anything useful about a cell, such as the rate of nutrient metabolism or the local reduction in nutrient concentration? It is not easy to measure a molecular concentration near the surface of cell with submicron precision, but perhaps we could calculate c_1 if we knew I. The first thing to notice in Equation 13.41 is that the theoretical maximum consumption rate is

$$I_{\max} = 4\pi RDc_0 \tag{13.42}$$

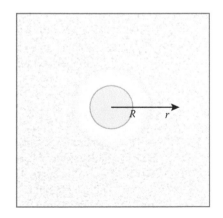

FIGURE 13.12

Cell of radius R immersed in a large bath of nutrient of concentration c_0 (far from the cell). The cell consumes nutrient, reducing the nearby nutrient concentration.

At this point the cell is reducing the nutrient concentration to zero at the cell surface. Of course, a cell cannot arbitrarily increase its consumption rate to match any given nutrient supply, but if an indicator of consumption can be found, c_0 can be varied to maximize this consumption. The concentration c_0 can be varied and the rate of cell division measured, for example. For a critical nutrient, the rate of cell division is expected to scale with consumption (with several assumptions). If the rate of cell division increases with c_0 up to some maximum value of c_0, you would have measured the cell's maximum consumption rate. For lower values of c_0 the cell would be consuming, limited by the reduction of the exterior concentration to zero.

There are even simpler, *pseudo* steady-state experiments involving cell metabolism. Suppose you place a large number N of cells in a liter of nutrient. A competent chemist could tell you the initial nutrient concentration and could measure it again after a time t perhaps 6 h. If the concentration had decreased 10% in this time, we can conclude that a *pseudo* steady state existed, at least on a time scale of a few hours. The competent biologist could examine the cell culture with a microscope before and after the 6-h period and tell you how many cells were there at both times. The competent eighth grader could then determine the average number of nutrient molecules consumed per second by one cell, as long as the nutrient concentration and number of cells did not change much, perhaps 10% or less each.

13.5.5 Diffusion Constant: Stokes and Einstein

We have seen the diffusion constant in previous sections, disguised in the more microscopic form of step probabilities, sizes, and times: $D = \frac{L_s^2}{2t_s}$, multiplied by a dimensionless constant not far from 1, depending on the detailed form of the step probability distribution. The diffusion constant can thus be traced to its microscopic origins, but step sizes and times are often of little use in the context of diffusion of many particles in a fluid. If we want to describe the random, Brownian motion—diffusion—of particles as they occur in a fluid, we need a different descriptor for the diffusion coefficient.

The most generally useful way to compute values of diffusion coefficients of particles in fluids was devised by Albert Einstein. The equation is an example of a *fluctuation-dissipation theorem*: the fluctuations of a particle are quantitatively related to the dissipation of energy by that particle. The fluctuation-dissipation relation we need is

$$\text{Einstein relation: } bD = k_{\mathrm{B}}T \tag{13.43}$$

Combined with the Stokes equation for the frictional coefficient of a spherical particle, $b = 6\pi\eta R$, we conclude

$$\text{Stokes-Einstein diffusion coefficient: } D = \frac{k_{\mathrm{B}}T}{6\pi\eta R} \tag{13.44}$$

This relation allows us to quickly and accurately compute diffusion coefficients for near-spherical particles, as well as estimate those of nonspherical. For example, the room-temperature diffusion coefficient of a sphere of radius R in nanometers can be computed as

$$D\left[\mathrm{m^2/s}\right] = \frac{2.2 \times 10^{-13}}{\eta R} \tag{13.45}$$

where η is in Pa·s. The viscosity of water is about 10^{-3} Pa·s so quick estimates of spherical-particle diffusion coefficients in water can be computed, as in Table 13.2.

It is sometimes useful to have diffusion coefficients converted to units more appropriate to cell and subcellular dimensions and times. The dimensions are, of course, about 1 μm or less, but what about the time scale? We can learn that from the value of the diffusion constant:
$D(R = 0.1\ \mu m) = 2.2 \times 10^{-12}\ \frac{\mathrm{m^2}}{\mathrm{s}} = 2.2 \times \frac{(10^{-7}\ \mathrm{m})^2}{10^{-2}\ \mathrm{s}}$, which tells us that a 0.1-μm sphere diffuse a distance corresponding to its own size in roughly 10 ms. So values of D in units of μm²/ms are useful: $D = 10^{-12}\ \frac{\mathrm{m^2}}{\mathrm{s}} = 0.001\ \frac{\mu m^2}{\mathrm{ms}}$. The millisecond time scale is also suggested in the last column of Table s 13.2: a 0.1-μm sphere diffuses about 90 nm in 1 ms.

TABLE 13.2 Approximate Diffusion Coefficients (Constants): Spheres in Water

Radius (nm)	Diffusion Coefficient D (m^2/s)	$\sqrt{\langle x^2 \rangle} = \sqrt{4Dt}$ for $t = 1\,ms$ (nm)
0.1	2.2×10^{-9}	3,000
1	2.2×10^{-10}	940
10	2.2×10^{-11}	300
100	2.2×10^{-12}	94
1,000 (1 μm)	2.2×10^{-13}	30
10,000 (10 μm)	2.2×10^{-14}	9.4

13.5.6 Reminder: Cell Interiors Are Usually Not "Bulk"

Chapter 4 introduced the idea that the interior of a cell, where the basic processes of life take place, does not consist of bulk water, though the major component is, indeed, H_2O. We expand on that fact here in a brief consideration of molecular motion inside a cell, where a cytoskeleton and many supramolecular structures exist and can interfere with the diffusive motion of molecules small and large. You may want to review Section 4.4.

The first correction we have seen to the nonbulk quality of a cell interior, and one we can apply immediately to diffusion theory, is to use an appropriate value for the viscosity of the fluid. The viscosity of pure water, about 10^{-3} Pa·s, is easy to remember, but pure water is not what cell interiors are made of. Section 12.7 notes that the viscosity of water decreases by about 30% when temperature is increased from 20°C to 35°C. This frictional decrease increases speeds and decreases frictional losses. We find in coming chapters that increasing the temperature, in general, increases biochemical reactions rates: the rule of thumb is a doubling of rate for every 10°C increase in temperature. A chemical reaction between two molecules requires that they first move together, then interact. Depending on the mechanism of the reaction, a reaction rate increase could be primarily due to viscosity decreases (faster molecular diffusion) or an increase in the fraction of molecules that have sufficient energy to approach each other closely enough. (Usually, the latter is dominant.)

Equation 12.41 suggests how to approximately correct the viscosity for dissolved solids in biosystems: $\eta - \eta_0 = 2.5\phi\eta_0$, where η_0 is the viscosity of pure fluid (water) and ϕ is the volume fraction of dissolved solids. If biophysical measurements are done in a test tube or a spectrophotometer cuvette of volume 1 cm^3 or so, we usually know what dissolved solids are in our tube. A 0.15 M NaCl solution contains $0.15 \frac{\text{millimoles}}{\text{mL}} \frac{0.058g}{\text{millimole}} \rightarrow 0.01$ vol fract of NaCl, so the viscosity correction $\eta - \eta_0 = 2.5(0.01)\eta_0 = 0.025\eta_0$ is rather small for most in vitro experiments. But what is the fraction of dissolved solids in a cell? We learned that the human body is about 70% water, which implies 30% of an "average" cell is some sort of solid, dissolved or not. Should we put in a viscosity correction of $\eta - \eta_0 = 2.5(0.3)\eta_0 = 0.75\eta_0$ for motion of small particles and molecules inside cells? The answer is "No" and the reasons are what we have just seen: (1) most particles and molecules do not move far on biologically relevant time scales, usually 1–10 nm, and (2) much of the non-H_2O components of cells are in the form of large (nanometer and larger) structures. When small molecules (O_2, Na^+Cl^-, small proteins) diffuse in three-dimensional, they diffuse through aqueous compartments that can be roughly considered water-like until they run into a large structure, where they stop, react, or change direction. The movement of small molecules is thus governed by a combination of free diffusion and geometrical crowding.

The movement of large, polymeric molecules like DNA present a special problem. Crowding effects in cells are certainly more important than corrections of bulk viscosity values. Long DNA strands presents a case of likely of crowding effects on motion inside a cell. DNA can easily be much longer than the diameter of a cell, so motion is sure to be affected at some point. Dauty and Verkman studied the case of the diffusion of DNA inside a cell as a function of length and various restricting materials inside the cell.[5] See Figure 13.13. Translational diffusion of large DNAs (>250 base pairs, length >90 nm) in cytoplasm was greatly slowed compared with that of smaller DNAs. The diffusion rate of DNAs in solutions crowded by various different materials— by Ficoll-70 (a synthetic polymer, up to 40% by weight), cytosol extracts (up to 100 mg/mL)—when normalized by the diffusion rate in an unobstructed, free solution, was approximately independent of DNA size (20–4500 bp), quite different from the strong reduction in diffusion in living cells.

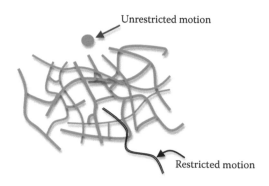

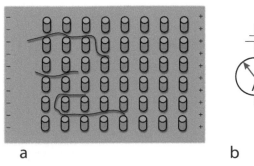

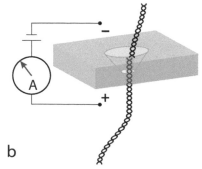

FIGURE 13.13

Restricted motion (molecular crowding) effects depend on the characteristics of the crowding agent and the diffusing molecule. Cross-linked actin matrices, like those in eukaryotic cell cytoskeletons, restrict the motion of extended DNA, but do not appreciably restrict the movement of a globular protein of mass similar to that of DNA. The extent of DNA movement-rate reduction depends on the DNA length and can range from 5 to 150 for DNA of length 20–6,000 base pairs. (After Dauty, E. and A. S. Verkman, *J. Biol. Chem.*, 280, 7823–7828, 2005.)

FIGURE 13.14

(a) Micropillar array (viewed from the top) used to separate long DNA of varying lengths. DC or AC voltages are applied along various axes at various times to entangle the DNA as it generally drifts in one direction. Longer DNA gets more entangled and is more delayed. The array can be fabricated from silicon and other materials, using modern microelectronic techniques. (Volkmuth, W. D. and R. H. Austin, *Nature* 358, 600–602, 1992; Morton, J. K. et al., *Proc. Natl. Acad. Sci.* 10, 7434–7438, 2008.) (b) Nanopore structure used to sequence DNA. See references in 6A for details of structure.

However, when the DNA was placed in a solution with a cross-linked actin network, similar to that in cells (Figures 4.7 and 4.8), the diffusion rate of DNA was reduced by factors of 5–150 for 20 to 6,000 base pair DNA. Comparable molecular-mass globular proteins did not show the movement reduction that DNA did. Furthermore, the size dependence of the diffusion rate reduction disappeared after the actin matrix cross-links were disrupted. This case suggests that a cross-linked network of fibers inside a cell is very effective in restricting the movement of long polymers but not necessarily of globular particles of similar molecular mass.

A device that uses the principle of fixed barriers organized in an array has been used to design devices to separate DNA of different lengths. The principle is similar to that of the actin cytoskeleton, which becomes an immovable set of barriers owing to cross-linking. Figure 13.14a shows a diagram of such a device, originally developed by Robert Austin's research group.[6] DNA moves, governed by diffusive, thermal motions, but an overall drift is imposed in this system by a macroscopic applied electric field. Electrical devices based on DNA motion through nanometer-scale structures have also been developed as single-molecule sequencing tools.[6A] The nanopore approach is shown in Figure 13.14b. DNA sequence is sensed by variations in the electrical current as different bases pass through the nanopore. Differences in current can be caused by different sizes and electronic properties of the bases. We now have the minimal fluid dynamics to investigate most such devices, but our electrical background is lacking. We will not pursue these "on-a-chip" devices further.

13.6 REPRISE OF PHOTOSYNTHETIC LIGHT HARVESTING

Figure 10.10 suggests that the movement of excitation energy in a photosynthetic antenna array can be described by a sort of random walk in two dimensions. Though we saw that the details of energy movement were not random at all, at least in the purple-bacterial light-harvesting (LH) system, but rather moved from one antenna complex to another and then to a reaction center (RC), LH-II → LH-I → RC, with chlorophylls arranged in symmetric arrays, we noted that there was some variability in the locations of the many primary LH-II antennae. The number of LH-II complexes closely associated with an LH-I varied, depending on conditions. There was also an associated variability in the number and arrangement of LH-II's which could transfer energy between themselves. We saw that the transfer happens rapidly, on a picosecond time scale. The overall process can then be approximately modeled as a two-dimensional random walk of energy among LH-II complexes, with energy "sinks" where LH-I complexes (with reaction center) are located, as suggested in Figure 10.10.

We now have two ways of quantitatively modeling this random motion of excitation energy in the light-harvesting array. First, we could set up a spreadsheet model, with sites for energy localization at allowed (x, y) lattice sites and sinks at particular sites, (x_m, y_m) separated by distances (ℓ_x, ℓ_y) and (L_x, L_y), respectively. Both allowed lattice sites and sink locations could be subject to geometric or distance restrictions. The distance between lattice sites would correspond to L_s, the step length, used earlier in this chapter. Single and multiple steps could be allowed with a specified probability. Step times t_s could be assigned according to those in Figure 10.19 and the latest published times. These times are all on a picosecond time scale. Even the effects of empty lattice locations, corresponding to absence of LH-II complexes, could be simulated. Likewise, the probability of transfer to an LH-I/RC sink could be assigned a relative probability. When the excitation reached the RC, it would disappear, of course. If an overall lifetime of the excitation in the lattice is assigned, perhaps 1 ns or so, corresponding to a maximum number of steps, the efficiency of transfer to a RC can be determined, as well as its dependence on several system variables. Figure 13.15 shows how such a simulation might be designed. The disadvantage of this approach is that many simulations must be run to obtain a picture of what happens, on average, to an excitation that appears randomly in the array.

How was it experimentally confirmed that absorbed photon energy actually does move in the light-harvesting array? Chapter 10 did cite indirect and spectral evidence for the antenna effect: there simply are not enough reaction centers to absorb all the photons that are, in fact, absorbed and the reaction centers have a somewhat different spectrum than that responsible for photosynthetic O_2 production. There must therefore be other chlorophylls that absorb light and transfer the energy to a reaction center. A more direct way to observe the transfer process is to fire a short laser pulse of light onto a chloroplast and observe the fluorescence spectral characteristics of the excited states. Remember, whenever a molecule absorbs light, it reemits some of the energy as fluorescence. The spectrum of the fluorescence characterizes the molecule that is excited. Such measurements lie behind the transfer times in Figure 10.19.

Another interesting type of transfer confirmation comes from the nature of a laser pulsed-excitation experiment. In order to get a measurable fluorescence signal from a chloroplast, many photons need to be packed into a short laser pulse. (Short-pulse lasers also naturally produce a large number of photons. Search "Q-switching" and "mode-locking.") The number of photons per second hitting the poor chloroplast is many, many orders of magnitude greater than the rate resulting from noon-day sunlight at the equator. This means that rather than one excitation randomly migrating in the antenna array, several or many may exist. Nothing particularly different happens until an excitation decides to hop to an already excited site. Then what? Turning back to the energy-level diagram in Figure 9.11 or 9.14, we note there are additional excited states above the first excited. The additional energy transferred from the second excitation would likely raise that state to the second excited state. Rather than two excitations on two sites, we then have a double-energy excitation on one site. What happens to doubly excited states? Usually one of the following: (1) transition back to the first excited state, with excess energy dissipated as heat, (2) transition to a triplet state, which often triggers singlet O_2 formation (not healthy), or (3) transition to the ground state, with excess energy dissipated as heat. All three of these alternatives result in fewer excited chlorophyll molecules and fewer sites that emit fluorescence: the fluorescence disappears more rapidly than when only one excitation at a time is placed on the array. This effect is termed *exciton annihilation* and has been

> An array allowing delocalization of excitation among several sites is much more difficult.

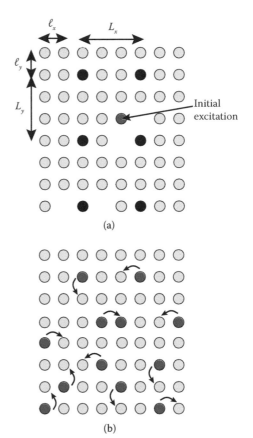

(a)

(b)

FIGURE 13.15

Array to simulate a photosynthetic light harvesting process. (a) The more numerous green sites are the primary LH-II antenna complexes, the dark sites, the LH-I complexes with reaction center (RC) excitation energy sinks. The red site can be a randomly chosen site of initial photon absorption. The sites shown form an ordered array, with several missing sites. (b) Simplified array (no sinks or missing sites) with multiple excitations caused by a high-intensity laser pulse. All excitations can move, but what happens if an excitation moves to an already excited site, as indicated near the center of this array? See text.

measured by many labs.[7] The existence of exciton annihilation in photosynthetic antenna arrays is a direct confirmation of a (likely) random walk of excitation energy in an array of chlorophyll sites.

There have been many sophisticated models for the motion of excitations in antenna arrays, but several paragraphs ago we asserted two simplistic approaches. The first was the spreadsheet simulation. A second approach could be to treat the energy transfer with a continuum diffusion model. The diffusion coefficient would approximately be $D_{ex} \approx \frac{L_s^2}{t_s}$. What do we put in for L_s? Consulting Figure 10.16, we see two choices: the distance between adjacent (bacterio) chlorophylls, about 0.9 nm, or the distance between two antenna rings, which could be as large as 3 nm. The transfer (step) time, from Figure 10.19, could be 0.7–35 ps. Some combinations of these distances and times make more sense than others, but possible values of D_{ex} range from 10^{-5} to 10^{-8} m²/s. These diffusion constants are high compared to those of particles of mass moving in water: D for water diffusing in water is about 2×10^{-9} m²/s. Homework investigates the application of a diffusion model to motion of excitation energy.

13.7 BIOPOLYMERS—RANDOM REPRISE

13.7.1 Average End-to-End Distance

We have noted that a freely jointed model for a polymer chain results in a particle of less than three apparent dimensions: $R \propto m^{1/2}$ rather than $R \propto m^{1/3}$. The fractional dimensionality is not surprising because a fully extended polymer would be one dimensional. The implications were that such a coil would be loosely packed, leaving room for solvent and other molecules—critical for DNA because even in a nonextended form enzymes and ions must gain access to specific parts of the DNA sequence for purposes of transcription and repair.

The freely jointed chain (FJC) model for a polymer is closely associated with random walks. The spreadsheet behind Figure 13.7 is capable, with minor modification, to simulate the structure of a freely jointed chain. FJC models treat polymers as small, noninteracting balls that are connected to each other by straight rods of fixed length L. There are, in fact, a variety of FJC models, but most assume that the location of each subsequent unit in the chain is found by random positioning at certain allowed orientations. A standard assumption is that if unit 1 is located at the origin, unit 2 can be located at any of the eight corners of a cube centered about the origin (Figure 13.16). Unit 3 would then be located at one of the eight corners of a cube centered at the position of unit 2. This is equivalent to a random walk of fixed length L in allowed (θ,ϕ) directions $(\pi/4, \pi/4)$, $(\pi/4, 3\pi/4)$, $(\pi/4, 5\pi/4)$, $(\pi/4, 7\pi/4)$, $(3\pi/4, \pi/4)$, $(3\pi/4, 3\pi/4)$, $(3\pi/4, 5\pi/4)$, and $(3\pi/4, 7\pi/4)$. In terms of components of steps, $(\Delta x, \Delta y, \Delta z) = \left(\pm \frac{L}{\sqrt{3}}, \pm \frac{L}{\sqrt{3}}, \pm \frac{L}{\sqrt{3}}\right)$ are allowed: each of the N steps has an equal magnitude step in the x, y, and z directions. If there are N such steps (units), the average distance from the first unit to the last is

$$R_N = \sqrt{\langle \vec{r}_N^2 \rangle} = \sqrt{\langle x_N^2 \rangle + \langle y_N^2 \rangle + \langle z_N^2 \rangle}$$

$$= \sqrt{N\left(\frac{L^2}{3}\right) + N\left(\frac{L^2}{3}\right) + N\left(\frac{L^2}{3}\right)} = \sqrt{N} L \qquad (13.46)$$

If this distance is considered as the average radius of the polymer coil, we conclude, as before, that the radius scales as square root of the mass of the polymer.

What biopolymers look like FJCs? We have seen many X-ray crystal structures of protein polymers which are globular and tightly packed: $R \propto m^{1/3}$, and the number of water molecules permeating the protein structure is small. Myosin, one of the muscle protein polymers, behaves as a loosely packed polymer when isolated and separated: the *radius of gyration* in water was measured to be about 10 times the close-packed radius (47 versus 4 nm).[8] Synthetic polystyrene also behaves as a loosely packed coil. The prime biological example of a loose polymer is DNA.

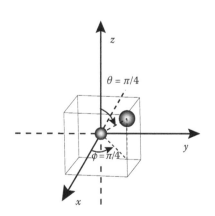

FIGURE 13.16
Random positioning of subsequent units in a freely jointed chain (FJC). The next unit in the chain is located at a corner of a sphere centered about the first unit.

13.7.2 Persistence Length, Bending, and Dynamics

The FJC model of polymers uses the distance between centers of polymer units as the primary measure of the shortest distance over which the polymer is free to bend. For DNA this presents a conundrum: the average distance between units of the DNA double strand is 0.34 nm, and we know from X-ray structures of DNA that it does not freely bend over this distance. It is, in fact, rather stiff on this distance scale: DNA structures containing 10 base pairs are mostly straight, though 20 mers often show bends. Perhaps DNA has a longer minimum length over which it can freely bend. Perhaps for hydrodynamic and average radius purposes we should consider that one DNA "unit" consists of 10 or so base pairs. The number N of hydrodynamic units would then be 10–20 times smaller than the number of base pairs, and L (Equation 13.46) 10–20 times larger. Because $R_N = \sqrt{N}\,L$ the FJC model would give a different radius of gyration, so the issue relates to measurable properties of DNA.

Contrary to the floppy image that "freely jointed" implies, DNA is a relatively rigid polymer. It is more often modeled as a wormlike chain. As noted in the previous chapter, the elastic properties of DNA involve bending, twisting, and compression. Twisting/torsional stiffness is important for the circularization of DNA, for unwinding in preparation for replication and for orientation of DNA-bound proteins relative to each other. Bending stiffness is important for DNA wrapping (e.g., around histone proteins), circularization, and protein interactions. Compression or extension of a straight DNA helix does not occur under natural cell conditions, but as the next section shows, measurement of DNA stretching directly quantifies the extension force.

We consider specific issues of DNA bending and twisting in the context of a rotary motor for DNA packing in Chapter 16, but a frequent indicator of DNA bending is the *persistence length*. We do not need to specify a detailed definition of persistence length, but it relates to the distance along the polymer axis over which the average direction of the polymer becomes uncorrelated to its initial direction. Observation of a DNA bend over a distance of 10–20 base pairs does not imply that the two ends of the sequence are uncorrelated in direction, as the bend itself may be relatively rigid. The persistence length must be longer than these typical bends observed in X-ray crystal structures. This distance corresponding to the persistence length can be qualitatively observed very readily with an atomic force (AFM, Chapter 19) or high-power fluorescence microscope, as well as by various hydrodynamic means. The persistence length of B DNA is about 50 nm, or 150 base pairs, as measured by optical measurements of DNA free in solution. Figure 13.17 shows an AFM image of DNA with increasing amount of a mitochondrial packaging protein, abf2p. This protein is responsible for packaging and compacting DNA for storage in mitochondria.[9] Before significant amounts of the packaging protein is added, the observed bending of DNA occurs over distances of somewhat less than 100 nm. Because the DNA is attached to a surface, its structure is not the average random-coil, solution structure. A uniform binding surface removes many of the bends normal to the surface, but at least some of the natural bends should be preserved. With more packaging protein, tight bends are formed. Some data indicate that the persistence length depends on sequence, but the effects are small compared to the change that occurs when the conformation changes to A DNA. Table 13.3 lists the A DNA persistence length as three times that on B DNA—A DNA is surprisingly stiffer than B DNA. Double-stranded RNA, with a 50% longer persistence length, is also somewhat stiffer than B DNA. DNA may well be stiff compared to the immensely flexible polypeptide chain, but the B form appears to give DNA more conformational adaptability than other forms.

The data on which Table 13.3 are based examined the rotational time of DNAs of lengths varying from 8 base pairs to 1.2×10^8 base pairs in solution. Hydrodynamic analysis showed that short DNA strands behaved like rigid rods in solution, while long strands formed loosely packed coil. Table 13.4 summarizes

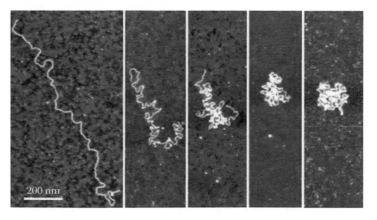

200 nm

FIGURE 13.17

Bending of DNA. AFM of linear pBR322 DNA with increasing amounts of the mitochondrial packaging protein abf2p. (Image by Raymond Friddle, Lawrence Livermore National Laboratory. See related images in Friddle, R. W. et al., *Biophys. J.*, 86, 1632–1639, 2004; Brewer, L. R. et al., *Biophys. J.*, 85, 2519–2524, 2003.)

TABLE 13.3 Bending Persistence Lengths of Double-Stranded Nucleic Acids[a]

Nucleic Acid	Persistence Length (bp)	Persistence Length (nm)
B DNA	150	50
A DNA	440	150
RNA (ds)	210	72

[a] *Source*: Serdyuk, I.N. et al., *Methods in Molecular Biophysics*. Cambridge, UK: Cambridge University Press, 425–433, 2007.

TABLE 13.4 Dependence of Overall Shape on DNA Length[a]

Number of Base Pairs	Overall Structure
20–200	Rigid rods
180–200	Begin departure from rod
200–8,000	Transition from rod to coil
>8,000	Gaussian coil

[a] *Source*: Serdyuk, I.N. et al., *Methods in Molecular Biophysics*. Cambridge, UK: Cambridge University Press, 425–433, 2007, based on molecular rotational times in solution.

the length dependence of the overall structure of double-stranded DNA. Table 13.4 tells us that we should not expect a fully "random" (Gaussian) coil structure for DNA until the number of base pairs is above 8,000.

Local DNA bending may be tailored to its function. Many measurements have indicated that bending is sequence and direction dependent (anisotropic). Because base stacking is the primary interaction governing DNA double-stranded structure, we may expect variations in DNA stacking interactions along the helix to alter bending. If less strongly stacked regions are always found on one side of the DNA helix, then the DNA preferentially bends away from that direction. There also appears to be a correlation between DNA bending and base pairing: disruption of a base pair decreases the local stiffness (increases bending).[10] There is, as well, some preferential location of A and T bases, which have weaker base pairing and more conformational freedom than G and C, in the minor grooves on the inside of bends, where tight DNA bending is induced. The winding of DNA around histone particles, as well as the packaging of DNA shown in Figure 13.17, are examples of sharp bending. Is there a connection of flexibility to biological function? The function of biomolecules, in general, is largely based on their flexibility, but Hogan and Austin found a connection between local DNA flexibility and protein binding sites, suggesting that local DNA flexibility is critical to recognition by and interaction with enzymes.[11]

13.7.2.1 Dynamics of movement

The time scale of DNA (and protein) structures and motions must also be considered. If the average radius of a loosely packed DNA coil is 200 nm, we are almost in a position to estimate the time it takes for the structure to form or un-form. Times depend on the size of the moving part of the DNA and how far it moves. Bases have ring diameters of about 0.5 nm, but the ring thickness is much smaller. If a single base were to slide out of its position in the base stack into surrounding solvent, the resistive fluid force would be small. Such motions have been observed to occur on a picosecond time scale.[12] We have also just convinced ourselves that B DNA is relatively stiff on a length scale much less than 50 nm (150 base pairs). However, biophysics tends to have researchers who devise new ways to observe biomolecular properties. These new ways sometimes do not confirm previous dogma. We should note first that Yuan et al.[13] concluded, from fluorescence and X-ray scattering measurements, that "the bending stiffness of double-stranded DNA molecules on small length scales is significantly lower than expected" and that continuum (uniform object) models for DNA, like the wormlike chain model, do not correctly predict the small-scale dynamics and flexibility of DNA. The small length scale referred to is DNA of sizes 10–90 base pairs. What does this mean for our study of biological structure? We should remind ourselves that the hallmark of biological structure is structural hierarchy and that structural features newly manifested when the length and/or time scale of observation is reduced by a factor of 5–10 should not surprise us.

13.7.2.2 Elastic forces

DNA has several levels of structure that we have studied. Its double-helical nature, with polymer backbone, base pairing and base stacking, results from strong covalent and weaker ionic and van der Waals interactions. Any distortion of this structure induces relatively strong opposing forces from the bonds. On the other hand, the loose packing of DNA and some other biopolymers into "random coils" requires no bonding at all. It would then seem that to unwind this coil would require no force and no work because there are no opposing bonds. The fallacy in this reasoning is hopefully evident: the random processes that favor a "random coil" over a straight polymer produce an entropic free energy drop. To reverse this free energy requires work. In Chapters 15 and 16 we conclude that the entropic forces are significantly smaller than those needed to stretch linear DNA, but these weaker forces are exactly those that enzymes must exert in organizing and processing DNA (along with some slightly larger forces to add somewhat tighter DNA bends), as in Figure 13.17 (right side).

13.8 POINTS TO REMEMBER

- Learn to use spreadsheet simulations of random motion.
- Remember the Einstein fluctuation-dissipation theorem and Stokes-Einstein diffusion coefficient.
- Understand what $\langle x^2 \rangle = 4Dt$ means.
- Be able to work with the diffusion "pulse" solution.
- Remember the DNA bending persistence length and how many base pairs are needed to form a "random" (Gaussian) coil.

13.9 PROBLEM-SOLVING

PROBLEM 13.1 MEAN TIME BETWEEN COLLISIONS

How rapid is the averaging process?

1. Use ideal gas theory to calculate the mean time between collisions for atmospheric oxygen gas.
2. Repeat 1, but scale the gas particle density to that corresponding to a liquid.

PROBLEM 13.2 BINOMIAL VERSUS GAUSSIAN DISTRIBUTIONS

In the following, use $p = q = 1/2$. Use Equations 13.12 and 13.15 to calculate the binomial and Gaussian probability for the following cases:

1. $P(n)$ (or $P(n)dn$) for $N = 8$ and $n = 0$, 3, and 6. For the continuous distribution state and justify the value of dn you use to compare to the discrete $P(n)$.
2. $P(n)$ (or $P(n)dn$) for $N = 30$ and $n = 0$, 5, and 15. Again, justify dn.
3. Discuss the accuracy of substituting the Gaussian for the binomial distribution for $N = 6$ and $N = 20$. How large should N be before doing this substitution?

PROBLEM 13.3 WIDTH OF A DISTRIBUTION

The Gaussian distribution in Equation 13.15 has a width characterized by $\sigma = \sqrt{Npq}$.

1. Describe precisely what this width is—full width at half maximum, half width at 10% maximum, etc.
2. Show whether a maximum or minimum width occurs when $p = q = 1/2$.
3. Use a graphing program to plot the width σ as a function of p for $N = 5{,}000$.
4. According to Equation 13.15, the most probable value of n is $n_{mp} = Np$. Find the ratio σ/n_{mp}, compute its value for $p = \frac{1}{2}$ and $N = 5 \times 10^2$, 5×10^6, and 5×10^{23}.

PROBLEM 13.4 BINOMIAL DISTRIBUTION AND PROBABILITIES OF "NET" NUMBERS

Equation 13.11 expresses the binomial probability for a net number, $m = n - (N - n)$, of particles to step to the right, or equivalently, for one stepper to end up m steps to the right after performing N steps, for the case $p = 1/2$.

1. For $N = 7$, calculate $P'(m)$ for all possible values of m. Consider negative values and discuss.
2. Repeat 1 for $N = 8$.
3. Discuss the meaning of the "missing" values of m for the two cases.
4. Discuss possible implications of the "missing" values of m when N is very large and the Gaussian distribution is substituted for the binomial distribution.

PROBLEM 13.5 $P(n)$ AND $P'(m)$

1. Convert Equation 13.15, $P(n)dn = \frac{1}{\sqrt{2\pi Npq}} \exp -\left((n - Np)^2/2Npq\right)dn$, to $P'(m)dm$ using the principle $P(n)dn = P'(m)dm$ and the relation $n = (N + m)/2$. Make sure (by inspection) that your new distribution is normalized.
2. What is the width of the $P'(m)$ distribution in terms of the width of $P(n)$? If these distributions describe steppers, what is the physical reason for the difference?
3. What is the center of the $P'(m)$ distribution? (What is m_{mp}?)

MINIPROJECT 13.6 BINOMIAL STEPPER SPREADSHEET

1. Construct a binomial spreadsheet similar to that in Figure 13.4. It is useful to have a small graph that displays directly on the spreadsheet and a graph copy on a separate sheet. Values of m should range from −100 to + 100. Make sure the spreadsheet allows computations up to at least $N = 100$ steps. Consult with a local computer/spreadsheet expert if you have difficulty. Print out graphs for $p = 0.50$, 0.75 and $N = 0$, 1, 2, 20, 60. Vertical axes usually autorange but set the m axis range so the graph is useful. Scale and paste the spreadsheet graphs onto one or two pages of a separate document (Microsoft Word, Adobe PDF, etc.).

2. Set $p = 0.70$. Determine the number of steps required for (1) first possible arrival at $m = +8$; (2) for 10% of the distribution to reach or surpass $m = +8$; (3) for 50% of the distribution to reach or surpass $m = +8$; (4) for 90% of the distribution to reach or surpass $m = +8$. Simply run the simulator many times to find the N values; no sophisticated calculation is expected.
3. Repeat 2 for $p = 0.55$. Comment where appropriate, especially in light of the expected behavior for $p = 0.50$.
4. (Graduate problem.) Design a spreadsheet calculation for <m>, the average value of m. (Do not assume the large-N, Gaussian limit for $P'(m)$.) Compare the computed values to the value of $N \cdot p$. Consider representative values of N from 1 to 100 and of p from 0.50 to 1.0.

PROBLEM 13.7 DIFFUSION OF A STEP FUNCTION (GRADUATE PROBLEM)

1. Show that $c(x,t) = \frac{c_0}{2}\left[1 - Erf\left(\frac{x}{\sqrt{4Dt}}\right)\right]$ is a solution to Fick's law and the continuity equation.
2. Use a graphing program to plot $c(x, t)$ as a function of x for $D = 3\ \mu m^2/ms = 3 \times 10^{-9}\ m^2/s$. Use a scale of x from -10 to $+10\ \mu m$ and plot for three times that "helpfully" show how the concentration distribution evolves with time. The first time should show the distribution near its initial condition. The second two should show the diffusion front proceeding toward larger values of x.

PROBLEM 13.8 DIFFUSION: STEP FUNCTION VERSUS A LOCALIZED PULSE

In the text, diffusion from an initial step-function concentration distribution resulted in a time-dependent error-function solution to the diffusion equation. (The error function is an integral of the Gaussian function.) The solution to an initial pulse of particles placed at a point in space, on the other hand, is a Gaussian. Draw diagrams of the two cases, write down the two solutions, and explain physically why the error function results for the step case. Note that the initial step function could be described by an appropriate distribution of particle "pulses" along the axis.

PROBLEM 13.9 NUTRIENT CONSUMPTION BY CELL CULTURES

You place a large number N_0 of 2.0-μm-diameter cells in a 1-L flask with an initial glucose (nutrient) concentration c_0. After a time Δt you measure the number of cells and nutrient concentration, with results, N and c, respectively.

1. Calculate, in terms of N_0, c_0, and Δt, the number of nutrient molecules consumed per second if N differs from N_0 by less than 10% and the glucose concentration decrease is 5.0%.
2. Assuming that the glucose consumption rate by the cells is the maximum possible (Equation 13.42: the glucose concentration at the cell surface is 0), calculate the product $N_0\Delta t$.

3. You plan to count cells and determine glucose concentration after 1.00 h. How many cells should you place in the flask to begin with?
4. If the initial glucose concentration were 7.0 mM (a typical concentration in mammalian blood), what is the rate of glucose consumption by cells?
5. Compare this theoretical glucose consumption rate with a "typical" cell consumption rate. You can search the Web, your favorite cell biology text, or talk to your best biologist friend. One estimate might come from the glucose consumption by the brain, 120 g per day, with 15–30 billion neurons per brain, but try to find a more appropriate number, since neurons are far from spherical.

MINIPROJECT 13.10   SPREADSHEET FOR EXCITON MOTION ON A LATTICE

Develop a spreadsheet simulator for the two-dimensional random-hopping processes illustrated in Figure 13.15a. You might begin with a 10 × 10 square array of sites, with one sink and only one-step hopping allowed. Some starting numbers—the lifetime of an excitation in the absence of sinks is 2.0 ns; the step time is 2 ps (be prepared to vary this); distance between nearest sites is 2 nm; excitation disappears when hopping to a sink occurs. Decide whether you will start excitations from particular or random lattice sites. Consider what to do if an excitation hops off the edge of your lattice: it could appear on the other side of the lattice (periodicity) or it could "reflect."

1. Determine the average lifetime of an excitation on your lattice; compare to the times given above.
2. Vary the starting position of the excitation from immediately adjacent to the sink, to several sites away. Does the average time to reach the sink depend on initial separation in a random-walk manner?
3. There are many follow-up simulations you can do once you have the basic spreadsheet created—add more sinks, remove lattice sites, put more than one excitation on the lattice. Alternatively, quit here; this has been a good deal of work.

PROBLEM 13.11   DIFFUSION OF EXCITATIONS IN 2D ARRAYS

Equation 13.37 for two-dimensional diffusion from a point can be crudely applied to a single particle in a square array of discrete sites if N_0 is set to 1 and $c(x, y,t)$ is interpreted probabilistically. For this problem, let $D = 10^{-6}$ m²/s, $d = 3.0$ nm (distance between neighboring sites), set $y = 0$ and assume an excitation starts out at the origin.

1. What are the units of $c(x, y,t)$?
2. Write down the equations describing the time dependence of the probability for the excitation to be at $x = d$ and $x = 10d$. (To account for the discrete points in space, the probability at a site $x = nd$ can be computed as the integral of the probability from $x = (n-1/2)d$ to $x = (n+1/2)d$. See *Finite-Limit Integrals of Gaussians: Error Functions* in front of book, use your favorite integration tool, or see a site such as Symbolab, https://www.symbolab.com/solver/definite-integral-calculator.)

3. Compute the times at which the probability at these points reaches a maximum.
4. Write down the equation describing the RMS distance a diffusing particle moves in two dimensions as a function of time. Do your results in 2 satisfy this equation?
5. Would your answers be appreciably different if you simply calculated $c(x,0,t)$ at the two points, $x = d$ and $x = 10d$ without the integration, since the two points are widely separated compared to d? Explain.

PROBLEM 13.12   TIME SCALE OF STRUCTURAL MOTIONS IN DNA

1. Treating a DNA base as a sphere (not a very good approximation), calculate the time required for a base to diffuse a distance equal to its own diameter in water.
2. Treat a 300-basepair length of DNA as a rod and calculate the time required to diffuse its own length in a direction perpendicular to the rod. Use information in Equations 13.43 and 12.34 to compute the rod diffusion constant.
3. Is it reasonable to treat this DNA as a rotating rigid rod? Explain.

MINIPROJECT 13.13   FREELY JOINTED CHAIN SIMULATION

Section 13.7.1 states that the spreadsheet for Figure 13.7 can be used, with minor modification, to simulate the structure of a feely jointed chain (FJC).

1. Download this spreadsheet from the text support Web site. Note that there are two sheets in the workbook: a calculation sheet and a chart sheet. If necessary, update the file for your spreadsheet program, MS Excel or other. Modify the sheet to prepare to simulate the structure of a freely jointed DNA chain. Use Tables 13.3 and 13.4 to simulate the structure for 20,000-bp B DNA. Note that your step length should correspond to the persistence length of B-DNA. State your step length (persistence length or length of the rigid segment) and total number of steps (number of segments in the entire chain).
2. Each recalculation (*Command +* in MS Excel) of the spreadsheet will simulate the structure of a FJC. Recalculate 100 times, each time recording the distance, R, of the end of the FJC from the beginning, in nm. While recalculating, copy and paste the chart for 10 of the simulations. Do not paste as a linked object that will be updated at each recalculation. Write the R value for these 10 simulations on the pasted chart.
3. Calculate the average and standard deviation of R. Compare to Equation 13.46 and to relevant published, experimental measurements. Discuss whether this distance should be considered a "diameter" or "radius".
4. Repeat for A DNA.

PROBLEM 13.14 BIOLOGICAL BINOMIAL STEPPER SPEEDS

Search your library or the Web for information on a biological molecular stepper motor. As you search, look for information on step size and time.

1. Describe this motor (where it's from, what its role is, its molecular characteristics, etc.) in about 50 words and note down the reference from which you will get information for parts 2–3 below.
2. Find the motor's step size (nm) and step time. If these quantities vary depending on conditions, record at least two pairs of numbers and describe the conditions.
3. Calculate the average speed from Equation 13.24.
4. Discuss any differences between your results for part 3 and the authors' conclusions.

REFERENCES

1. Levinthal, C. 1969. How to fold graciously. In *Mössbauer Spectroscopy in Biological Systems*, Eds. P. Debrunner, J. C. M. Tsibris, and E. Münck, 22–24. Monticello, FL: University of Illinois, Allerton House; Levinthal, C. 1968. Are there pathways for protein folding? *Journal de Chimie Physique et de Physico-Chimie Biologique.* 65:44–45. See also Zwanzig, R., A. Szabo, and B. Bagchi. 1992. Levinthal's paradox. *Proc Natl Acad Sci.* 89:20–22.
2. Einstein, A. 1905. "Über die von der molekularkinetischen Theorie der Wärme geforderte Bewegung von in ruhenden Flüssigkeiten suspendierten Teilchen," *Annalen der Physik* 17:549–560.
3. Reif, F. 1967. *Statistical Physics.* New York: McGraw-Hill, p. 73.
4. Nelson, P. 2004. *Biological Physics.* New York: W. H. Freeman, 117–118. See also "multinomial distribution," at http://en.wikipedia.org/wiki/Multinomial_distribution.
5. Dauty, E., and A. S. Verkman. 2005. Actin cytoskeleton as the principal determinant of size-dependent DNA mobility in cytoplasm. *J Biol Chem.* 280:7823–7828.
6. Volkmuth W. D., and R. H. Austin. 1992. DNA electrophoresis in microlithographic arrays. *Nature* 358:600–602; see also Morton, K. J., K. Loutherback, D. W. Inglis, O. K. Tsui, J. C. Sturm, S. Y. Chou, and R. H. Austin. 2008. Hydrodynamic metamaterials: Microfabricated arrays to steer, refract, and focus streams of biomaterials. *Proc Natl Acad Sci.* 105:7434–7438; Reisner, W., J. N. Pedersen and R. H. Austin. 2012. DNA confinement in nanochannels: Physics and biological applications. *Rep. Prog Phys.* 75(10):106601.
6A. Feng, Y., Y. Zhang, C. Ying, D. Wang, and C. Du. 2015. Nanopore-based fourth-generation DNA sequencing technology. *Genomics, Proteomics & Bioinformatics* 13(1):4–16. Lu, H., F. Giordano, and Z. Ning. 2016. Oxford nanopore MinION sequencing and genome assembly. *Genomics, Proteomics & Bioinformatics* 14(5):265–279.
7. Nordlund, T. M., and W. H. Knox. 1981. Lifetime of fluorescence from light-harvesting chlorophyll a/b proteins: Excitation intensity dependence. *Biophys. J.* 35:193–201; Greene, B. I., and R. R. Millard. 1985. Singlet-exciton fusion in molecular solids: A direct subpicosecond determination of time-dependent

annihilation rates. *Phys Rev Lett.* 55:1331–1333; Gillbro, T., Å. Sandström, M. Spangfort, V. Sundström, and R. van Grondelle. 1988. Excitation energy annihilation in aggregates of chlorophyll a/bC. *Biochim Biophys Acta.* 934:369–374.

8. Tanford, C. 1961. *Physical Chemistry of Macromolecules.* New York: John Wiley.

9. Friddle, R. W., J. E. Klare, S. S. Martin, M. Corzett, R. Balhorn, E. P. Baldwin, R. J. Baskin, and A. Noy. 2004. Mechanism of DNA compaction by yeast mitochondrial protein Abf2p. *Biophys. J.* 86:1632–1639.

10. Ramstein, J., and R. Lavery. 1988. Energetic coupling between DNA bending and base pair opening. *Proc Natl Acad Sci.* 85:7231–7235.

11. Hogan M. E., and R. H. Austin. 1987. Importance of DNA stiffness in protein-DNA binding specificity. *Nature* 329:263–266.

12. Nordlund, T. M., S. Andersson, L. Nilsson, R. Rigler, A. Graslund, and L. W. McLaughlin. 1989. Structure and dynamics of a fluorescent DNA oligomer containing the *EcoRI* recognition sequence: Fluorescence, molecular dynamics, and NMR studies. *Biochemistry* 28:9095–9103.

13. Yuan, C., H. Chen, X. W. Lou, and L. A. Archer. 2008. DNA bending stiffness on small length scales. *Phys Rev Lett.* 100:018102.

14. Brewer, L. R., et al. 2003. Packaging of single DNA molecules by the yeast mitochondrial protein Abf2p. *Biophys. J.* 85:2519–2524.

15. Friddle, R. W., et al. 2004. Mechanism of DNA compaction by yeast mitochondrial protein Abf2p. *Biophys. J.* 86:1632–1639.

CHAPTER 14

CONTENTS

Statistical Physics and Thermodynamics Primer

14

Material in this chapter generally comprises a physics, textbook-like summary of statistical physics and thermodynamics, with a clear to applications in (bio)chemistry. Readers with previous experience in intermediate or upper level undergraduate courses in those disciplines will likely find much familiar material. It would be a mistake, however, to carelessly skip this entire chapter because of presupposed facility in these areas. Sections such as 14.8 point toward novel biophysical applications of fundamental theory that have implications in current biomedical research. Persistence length, chain stiffness, entropy, and knots are governed by random processes, but the success of biological life largely rests upon the ability to manage and exploit such random events. Recent discoveries indicate that interactions modifying the random nature and interactions of the DNA "freely jointed chain" (FJC) have important effects in oncogene activation (cancer).

Flavahan et al., for example, found that

> *Reduced CTCF binding is associated with loss of insulation between topological domains and aberrant gene activation. We specifically demonstrate that loss of CTCF at a domain boundary permits a constitutive enhancer to interact aberrantly with the receptor tyrosine kinase gene PDGFRA, a prominent glioma oncogene. Treatment of IDH mutant glioma-spheres with a demethylating agent partially restores insulator function and downregulates PDGFRA. Conversely, CRISPR-mediated disruption of the CTCF motif in IDH wild-type gliomaspheres upregulates PDGFRA and increases proliferation. Our study suggests that IDH mutations promote gliomagenesis by disrupting chromosomal topology and allowing aberrant regulatory interactions that induce oncogene expression.*[1]

This quote may be largely unintelligible to readers with a mostly physics background, except for perhaps the last sentence, where the key words *chromosomal topology* indicate that the deviation of DNA folding from both random coil and the usual, controlled chromosomal folding is key to cancer gene activation. The passage illustrates the hidden connections between fundamental statistical mechanics and cell behavior. A further bit of decoding may help the physicist reader glimpse these connections:

onco-: cancer
glial cells: neural tissue cells that surround and support neurons, regulating repair, directing construction, providing nutrition and protection and other support
glioma: a broad category of brain and spinal cord tumors
glioblastoma: a very aggressive form of brain cancer
gliomasphere: a culture of glioma cells that (spontaneously) form spherical cell aggregates in suspension...

Some of the other jargon is not needed for moment, but key expressions such as *insulation between topological domains* and *domain boundaries* also intimate connections between molecular cell biology and oncology and physics. Early preparation for these relationships is a wise course of action. Overly familiar sections of this chapter should be enjoyed in their familiarity.

Physics predicts the future: Newton's and other laws describe exactly where a single particle will go if we know the exact forces and initial conditions. Even in the quantum world, the Schrödinger equation very precisely predicts what electrons and atoms will do in given circumstances, though the prediction is usually probabilistic. Statistics and probability dominate prediction of events and behavior in the micro-world. In the case of a liter of air, liquid, or solid, the first reason for using

> We remind ourselves that DNA is not simply an ideal freely jointed chain. The FJC is a model for certain characteristics of DNA, such as its tendency, in free solution, to form a Gaussian coil over length scales longer than about 8,000 base pairs.

statistics reason is practical: (1) no supercomputer could conceivably track the possible interactions and trajectories of 10^{23} particles. The corollary to reason 1 is that (2) knowing the forces on any given particle is impossible. The random, constantly vibrating and colliding environment of a microparticle or molecule makes precise prediction of the future of one particle unattainable. For cases involving large numbers of particles, $N/\sqrt{N} \gg 1$, statistical physics and its extension, thermodynamics, again allow precise prediction of the average behavior of the particles, and even of their average deviation from average behavior. The prediction gets better the larger the value of $N/\sqrt{N}$.

Biosystems are often characterized by many small particles that cannot all be tracked, but their numbers may be only $1-10^{10}$. While $10^{10}/\sqrt{10^{10}} = 100,000$ is pretty big, and a precision (average deviation) of 1 in 10^5 is pretty precise, such deviations are far larger than the precision with which the mass of the electron or the Rydberg constant have been measured. There is also the issue of observation time. The reason we can (in principle) take an instantaneous ($\Delta t \to 0$) snapshot of a liter of gas and know that the 10^{22} molecules will have a Maxwell–Boltzmann (M–B) distribution of speeds and random velocity directions is that the average collision time, given by (mean free path)/(average speed), is very short:

$$\tau = \frac{\ell}{\langle v \rangle} = \frac{\left(4\sqrt{2}\pi R^2 n\right)^{-1}}{\sqrt{8k_{B}T/\pi m}} \tag{14.1}$$

Physical equilibrium of the reactions in the sense that most particles have average properties, not that chemical equilibrium is in place.

has value about 0.5 ns for a gas at room temperature and pressure, where R and m are the radius and mass of the molecule and n is the number per unit volume.[2] Note that the quantities ℓ and $\langle v \rangle$ are individually preserved in the ratio in Equation 14.1. In the seconds, minutes, or hours before any measurement takes place, the gas sample's molecules would have undergone at least 10^9 randomizing collisions. Of course, if someone figured out a way to initially prepare the gas sample with all of the thermal kinetic energy localized on one molecule and could measure the speed distribution 10 ns later, the "average" molecule could not be treated as if it had the average M–B speed.

While a large number of randomizing collisions may not take place before every biological event, many biochemical reactions can at least be approximately described by equilibrium statistics and thermodynamics. Chapter 13 equipped us to deal with some types of nonaveraging randomness in biosystems; now we must polish up our many-particle equilibrium tools.

Statistical thermodynamics is a fearsome discipline. Before we begin we should note that most scientists have rather strong feelings about statistical thermodynamics. Some dislike and avoid it. Some believe thermodynamics is *the* most fundamental science and should overrule all other fields of science. Some believe every serious science student should know the elegant formal structure of thermodynamics, as presented by Landau and Lifshitz[3] or Walter Moore[4] or others. Some believe thermodynamics is done in the lab, using only the mathematics needed to adequately analyze and organize results. Some believe partition functions are the main thing to understand in statistical thermodynamics; some consider them normalization constants. Some consider chemical potentials central; some consider chemical potential a pretty useless idea and move on to equilibrium constants, concentrations, free energies, enthalpies, and entropies. There seem to be as many points of view as there are scientists. This chapter has a point of view of thermodynamics that is not very formal. You may well want to consult other texts with other points of view. My personal experience over four decades with six thermodynamics books spread out on my desk late at night is that the head usually drops onto the desk in despair. My advice: learn statistical thermodynamics as you need it.

14.1 IMPORTANT QUANTITIES: TEMPERATURE, PRESSURE, DENSITY, AND NUMBER

The ideal gas law,

$$PV = Nk_{B}T \tag{14.2}$$

does not accurately describe most biosystems; in fact, it hardly ever does. However, it indicates what quantities are most important to a thermodynamic description: temperature, pressure, volume, and number of particles. While P, V, and T typically do not change in biosystems, small changes in T often lie behind successful biological response to changes in the environment. The linear dependence of the energies of molecules on temperature, with resulting large changes in chemical reactivity, lies behind most biochemical reactivity. For monatomic ideal gases, recall that the translational kinetic energy, which is the only internal energy a point particle can possess, is related to temperature by

$$K_{mono} = \frac{1}{2} m \langle v^2 \rangle = \frac{1}{2} m v_{RMS}^2 = \frac{3}{2} k_B T \qquad (14.3)$$

where T is, of course, measured on the absolute Kelvin scale. As T usually varies only by about ± 20 K, RMS speeds of ideal gas molecules varies by about $\pm 3\%$.

The RMS speed of molecules depends more strongly on mass. Table 14.1 lists RMS translational speed of some atoms and molecules of various masses.

The RMS speeds above were calculated from the expression $\frac{1}{2} m v_{RMS}^2 = \frac{3}{2} k_B T$ Table 14.1 can be found in most introductory physics and chemistry textbooks, except for the entry for myoglobin (Mb). Why have we included this molecule, which is obviously not an ideal gas, with the others? Myoglobin is a globular protein molecule that normally exists in blood. Is its RMS speed really 20.8 m/s, straight from the ideal gas expression? The answer is "Yes," unless it happens to be bound to some other object (often the case), which effectively increases its mass and reduces its speed. However, the universality of the laws of statistical thermodynamics, specifically the Equipartition theorem (see Section 14.3), extends this particular molecular property, v_{RMS}, to all molecular systems, ideal gas, liquid, solid, cell. Sometimes the details are tricky, such as when a molecule is weakly bound to another, and sometimes the RMS speed is an irrelevant quantity, such as for a cell, but thermodynamics applies. This is one reason why we must learn to employ thermodynamic principles to biosystems—thermodynamics is just so darn useful.

14.2 STATISTICAL MECHANICAL VIEW AND DISTRIBUTIONS

The statistical mechanical view of any system treats its multiple particles as *well-behaved variations on a single entity.* At one extreme, particles may be truly identical in all measurable ways, such as He atoms in a gas sample. The He atoms do not all have the same velocity or kinetic energy at any given instant, but if you could observe one atom for even a few seconds, it would take on velocities and energies identical to those taken on by any other He atom in the gas. The statistical mechanical approach is to find an appropriate probability distribution to describe the variability in any property of interest, calculate average values, and average deviations from the average. The average and standard deviation of mechanical and dynamic quantities often suffices for bulk properties like pressure, but the distribution itself should be considered the primary tool for describing the system. In the case of chemical reactivity, for example, it is the atypical, high-energy molecules that are primarily responsible for chemical reactions.

Another case where distributions should be used is a protein solution, where all molecules are unlikely to be identical because of the large number of atoms, all of which can vary in position relative to nearby atoms. Protein molecules are a group of variations on one average structure. On a particular time scale, fluctuations may or may not bring a given protein molecule through all possible variations of its structure. If a reactivity variation accompanies the structural variation, then a distribution of structure carries a reactivity distribution with it. Any given molecule could, in principle, react with the structure it has at the moment, or the reaction could wait until the protein passes through a structural variation particularly favorable to the reaction. Observation time becomes a central aspect of

TABLE 14.1	Molecular Mass and RMS Speeds	
Molecule	**Molecular Mass (g/mole)**	**v_{RMS} at 20°C (m/s)**
H_2	2.02	1900
He	4.00	1350
H_2O	18.0	637
N_2	28.0	511
CO	28.0	511
O_2	32.0	478
CO_2	44.0	408
SO_2	48.0	390
Mb	16,900	20.8

distributions emerging in the binding reaction between myoglobin and oxygen, as well as other biochemical reactions.

In the statistical thermodynamic treatment of the ideal gas, the probability a molecule has a particular speed v is given by the Maxwell–Boltzmann (M–B) speed distribution:

$$P(v)dv = 4\pi \left(\frac{m}{2\pi k_B T}\right)^{3/2} \exp\left(-\frac{1/2\, mv^2}{k_B T}\right) v^2 dv \tag{14.4}$$

$$\int_0^\infty P(v)dv = 4\pi \left(\frac{m}{2\pi k_B T}\right)^{3/2} \int_0^\infty \exp\left(-\frac{1/2\, mv^2}{k_B T}\right) v^2 dv = 1 \tag{14.5}$$

This distribution should strike you as remarkably similar to those we have seen before: Gaussian exponential factor, with a T in the exponent that also appears in the normalization factor. The primary new feature of the M–B distribution is the $4\pi v^2$ term in the integrand which, as you may recall, comes from spherical geometry: larger speeds have more possible ways to have a speed between v and $v + dv$ (Figure 14.1). Figure 14.2 shows the distribution for several temperatures.

While average values are often the central quantity of interest, other values are sometimes more appropriate to use. RMS and most probable values are examples. How do we compute the most probable speed given by the M–B distribution? Like any function, the maximum values is located where the derivative equals zero:

$$\text{Most probable speed}: \frac{dP}{dv} = 0 \Rightarrow \frac{d}{dv} \exp\left(-\frac{1/2\, mv^2}{k_B T}\right) v^2 = 0 \tag{14.6}$$

The RMS speed, on the other hand, is found by finding the average value

$$\text{RMS speed}: v_{RMS}^2 = \int_0^\infty v^2 P(v)dv = 4\pi \left(\frac{m}{2\pi k_B T}\right)^{3/2} \int_0^\infty v^4 \exp\left(-\frac{1/2\, mv^2}{k_B T}\right) dv \tag{14.7}$$

and

$$\text{Average speed}: v_{avg} = \langle v \rangle = \int_0^\infty v P(v)dv = 4\pi \left(\frac{m}{2\pi k_B T}\right)^{3/2} \int_0^\infty v^3 \exp\left(-\frac{1/2\, mv^2}{k_B T}\right) dv \tag{14.8}$$

The results of these calculations are

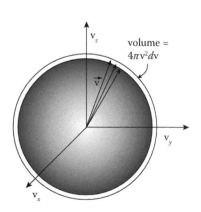

FIGURE 14.1
Density of states factor for Maxwell–Boltzmann speed distribution: more allowed velocities occupy the spherical shell of volume $4\pi v^2 dv$ the larger v is.

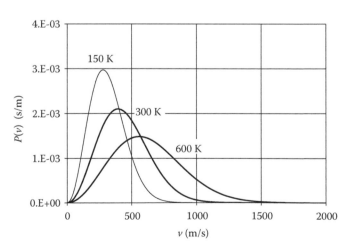

FIGURE 14.2
Maxwell–Boltzmann speed distribution for molecular mass 32 D (O_2) at three temperatures: 150, 300, and 600 K.

$$v_{mp} = \sqrt{\frac{2k_BT}{m}} = 1.41\sqrt{\frac{k_BT}{m}}$$

Ideal gas: $\quad v_{avg} = \sqrt{\frac{8k_BT}{\pi m}} = 1.60\sqrt{\frac{k_BT}{m}}$ (14.9)

$$v_{RMS} = \sqrt{\frac{3k_BT}{m}} = 1.73\sqrt{\frac{k_BT}{m}}$$

You may examine the distribution shape and locate these three characteristic speeds.

14.2.1 Boltzmann Distribution: Discrete Energies

The Maxwell–Boltzmann speed distribution is useful for ideal gases, but we seek a distribution function that is more generally useful for biomolecules in solution. That more useful distribution function is the more general and more basic function, the Boltzmann distribution. Let's start, however, with a quantity called the *Boltzmann factor*, e^{-E/k_BT}, and discrete energy values. Discrete energy values means quantum mechanics. How can that be simpler? The reason quantum systems are simpler is that if your energy resolution is sufficiently fine, you can separate and count individual energy levels; at least if there is no degeneracy. Let's assume each energy level has one unique quantum state—no degeneracy—and number the states according to energy, 1, 2, 3,..., as in Figure 14.3. The probability of a state with energy E_i is

$$P(E_i) = Ce^{-E_i/k_BT}$$ (14.10)

Note there is no dE because we have a discrete distribution and can resolve the energies of every state. This distribution has to be normalized, of course:

$$\sum_{i=1}^{\infty} P(E_i) = C\sum_{i=1}^{\infty} e^{-E_i/k_BT} = 1$$ (14.11)

but we cannot normalize the distribution unless we know what all the allowed energies are. If two or more states may have the same energy, we assign a degeneracy g_i to the state i, and that state is g_i times more probable:

$$P(E_i) = Cg_i e^{-E_i/k_BT}$$ (14.12)

with normalization

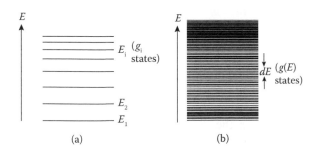

(a) (b)

FIGURE 14.3
Energy-level diagrams. (a) Discrete case: each energy level, i, has $g_i = 1, 2, 3,...$, states, where g_i is the degeneracy of the ith state. The relative probability of observing a state of energy E_i is $g_i e^{-E_i/k_BT}$. (b) If energy levels are closely spaced compared to our resolution, we assign the energy range between E and $E + dE$ a density of states factor, $g(E)$, which contains both the quantum degeneracy and the density of (unresolvable) levels in that energy range. (Consider that in (a) there are more levels per unit energy at the top of the diagram than at the bottom.) The relative probability of observing a state of energy between E and $E + dE$ is $g(E)e^{-E/k_BT}dE$.

$$\sum_{i=1}^{\infty} P(E_i) = C \sum_{i=1}^{\infty} g_i e^{-E_i/k_B T} = 1 \tag{14.13}$$

Suppose we had an ensemble of discrete (quantum) oscillators. Harmonic oscillator theory describes a wide variety of vibrations and bond rotations in biosystems, including polypeptides and polynucleotides. We know that the harmonic energy levels are given by

$$E_n = \left(n + \frac{1}{2}\right)\hbar\omega, \ n = 0,1,2,\dots \tag{14.14}$$

with $\omega = \sqrt{\frac{k}{m}}$. There is no degeneracy, so

$$P(E_n) = C \cdot 1 \cdot e^{-E_n/k_B T} = C e^{-\left(n+\frac{1}{2}\hbar\omega\right)/k_B T} \tag{14.15}$$

Where is the maximum of this distribution? Obviously, at $n = 0$, so $E_{mp} = E_{min} = \frac{\hbar\omega}{2}$ Are any of the higher states occupied with appreciable probability? The Boltzmann factor, $e^{-E/k_B T}$, says that the lowest-energy state is intrinsically the most probable and the (intrinsic) relative probability of any two states can be simply calculated by the ratio of Boltzmann factors of their energies:

> Experimentally, infrared spectroscopy uses wave number $\tilde{\nu} \equiv \frac{1}{\lambda}$ as the primary spectroscopic variable.

$$\frac{P(E_2)}{P(E_1)} = \frac{e^{-E_2/k_B T}}{e^{-E_1/k_B T}} = e^{-(E_2-E_1)/k_B T} \tag{14.16}$$

The population ratio of the $n = 1$ to the $n = 0$ state of an oscillator is $\frac{P(E_1)}{P(E_0)} = e^{-\hbar\omega/k_B T}$. Near room temperature $k_B T \sim 4.1 \times 10^{-21}$ J, so if the characteristic energy of the oscillator is $10k_B T$ or more, the first excited state ($n = 1$) is not populated with any probability. Note again the power of the exponential: $e^{-3} \sim 0.05$, but $e^{-10} \sim 5 \times 10^{-5}$. What are typical magnitudes of vibrational energies in biomolecules? The primary infrared "marker bands" of DNA—those that characterize deoxyribose, base, base pair, and phosphate bond vibrations, as they depend on overall DNA conformation—range from wave numbers of about 1,000 to 1,700 cm⁻¹. Amide bonds in polypeptides are of comparable magnitude. The vibrational energy is thus about $\hbar\omega = hf = \frac{hc}{\lambda} \cong \left(2.0\times10^{-23} \text{J}\cdot\text{cm}\right)\left(1500\text{cm}^{-1}\right) \sim 3.0\times10^{-20} \text{J} \sim 7k_B T$ and the probability that the $n = 1$ state is occupied is about 0.0006. Table 14.2 shows Boltzmann populations of the first few vibrational states for two vibrational frequencies common in biomolecules. These probabilities (populations) clearly indicate that spectroscopic measurement in the 2,000-cm⁻¹ system results from photon absorption by the ground ($n = 0$) state; in the 200-cm⁻¹ system, states from $n = 0$–$n = 3$ contribute appreciably.

Table 14.2 is graphically displayed in Figure 14.4.

If degeneracy is present, then the Boltzmann ratio is modified to

TABLE 14.2   Relative Boltzmann Populations of Oscillator States for $1/\lambda$ = 200 and 2,000 cm⁻¹

n	P_n (200 cm⁻¹, ~$k_B T$)	P_n (2,000 cm⁻¹, ~$10k_B T$)
0	1	1
1	0.379	6.1×10^{-5}
2	0.144	3.7×10^{-9}
3	0.054	2.3×10^{-13}

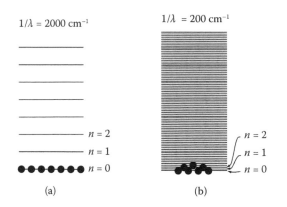

FIGURE 14.4

Boltzmann population of energy levels for level separations of 2,000 cm⁻¹ (~$10k_B T$) and 200 cm⁻¹ (~$k_B T$). (a) In the former case, only the lowest energy level is populated. (b) In the latter, three to four levels are appreciably populated. The same principle applies to a continuous set of states: only the lowest few $k_B T$ of energy states are more than a few percent populated in equilibrium. Nevertheless, a minute fraction at higher energy may be responsible for chemical reactivity.

$$\frac{P(E_2)}{P(E_1)} = \frac{g_2 e^{-E_2/k_B T}}{g_1 e^{-E_1/k_B T}} = \frac{g_2}{g_1} e^{-(E_2-E_1)/k_B T} \tag{14.17}$$

Equation 14.17 can be rewritten in a suggestive form using the substitution $g_i = e^{k_B S_i}$:

$$\frac{P(E_2)}{P(E_1)} = \frac{e^{k_B S_2} e^{-E_2/k_B T}}{e^{k_B S_1} e^{-E_1/k_B T}} = e^{[T(S_2 - S_1) - (E_2 - E_1)]/k_B T} \tag{14.18}$$

14.2.2 Boltzmann Distribution: Continuous Energies

Suppose the allowed energies formed a continuous distribution or that we had a spectrometer with too coarse a resolution to distinguish individual energy levels. We then switch our description to the number of states with energies between E and $E + dE$. In order to correctly count the states, we need to include the degeneracy factor, but we also need to know how many energy levels there are per unit energy in the range E to $E + dE$. Because we cannot even distinguish the individual energy levels, much less their degeneracy, we traditionally combine these two factors into a density of states factor, $g(E)$, and write the probability of observing a state of energy between E and $E + dE$:

$$P(E)dE = Cg(E)e^{-E/k_B T} dE \tag{14.19}$$

with

$$\int_0^\infty P(E)dE = C \int_0^\infty g(E)e^{-E/k_B T} dE \tag{14.20}$$

Let's suppose, for example, we had the monatomic ideal gas case we started with. We can derive the Boltzmann energy distribution as follows. Because we know the correct Maxwell–Boltzmann distribution as a function of speed, we can change the variable from $v \to E = \frac{1}{2}mv^2$: $P(E)dE = P(v)dv$, as we have done before. The results of this conversion are

$$P_{gas}(E)dE = \frac{2}{\sqrt{\pi}} \left(\frac{1}{k_B T}\right)^{3/2} E^{1/2} e^{-E/k_B T} dE \tag{14.21}$$

with $E = \frac{1}{2}mv^2$. For the ideal monatomic gas, which has only translational kinetic energy, the $P(v)dv$ distribution has a density of states factor proportional to v^2, which is proportional to E, but $dv/dE \propto E^{-1/2}$ must also be included in the variable conversion, resulting in the $P(E)dE$ function of Equation 14.21. For every continuous system we should expect a distribution function of Equation 14.19: density of states times Boltzmann factor. Later, we formally use the substitution $g(E) = e^{k_B S(E)}$.

> For the ideal gas, we start with the velocity distribution, find the speed distribution and then the energy distribution. We use this "backward" route because the individual states of ideal gas molecules are completely specified by the velocity vector, not by the kinetic energy. Kinetic energy does not distinguish between velocities of equal magnitudes but different directions.

14.2.3 Boltzmann, Fermi–Dirac, and Bose–Einstein Statistics

Particles are distinguished according to their intrinsic spin properties. If the particle total spin is half-integer, the particle is a fermion and the Pauli Exclusion Principle applies. If spin is zero or integral, the particle is a boson and the exclusion principle does not apply. The implications are that we should use Fermi–Dirac (F–D) statistics for fermions and Bose–Einstein (B–E) statistics for bosons, or maybe we ignore quantum mechanics and use Boltzmann statistics if we're not in a physics course or research lab. Do we actually need to decide which quantum statistics applies to our biosystems? Rarely. Quantum statistics (e.g., the exclusion principle) applies when two identical particles effectively compete to occupy the same state. This is (virtually) never the case for an ideal gas at room temperature, a solution of protein molecules, or lipids in a cell membrane. The number of quantum states (e.g., location and speeds of all atoms, vibrational states, and random changes) available to a protein molecule in solution is so vast that two molecules will not attempt to occupy the same state in the age of the universe. During our lifetime, or during the time course of a measurement, we are safe using Boltzmann and ignoring F–D and B–E. You are always welcome to use either of those latter two statistical distributions and show that they reduce to the Boltzmann distribution when applied to an ideal gas or DNA solution.

14.3 EQUIPARTITION OF ENERGY

Ideal gas theory gave the result that

$$\left\langle \frac{1}{2} mv_x^2 \right\rangle = \left\langle \frac{1}{2} mv_y^2 \right\rangle = \left\langle \frac{1}{2} mv_z^2 \right\rangle = \frac{1}{2} k_B T : \tag{14.22}$$

each dimension in which the particle is free to move results, on average, in a kinetic energy of $\frac{1}{2}k_B T$. The Equipartition theorem, which we shall not attempt to derive, states that this result extends to any distinct mode in which energy can be contained:

$$\left\langle E \text{ per mode} \right\rangle = \frac{1}{2} k_B T \tag{14.23}$$

What is a "distinct mode" (or "degree of freedom" in some books)? For our purposes a mode can contain $\frac{1}{2}k_B T$ of energy if it is a dynamic variable on which contained energy is quadratically dependent. Examples include $\frac{1}{2}mv^2$ and $\frac{1}{2}kx^2$. The Equipartition theorem applies to any system of many particles, except sometimes at low temperature, where the time to reach equilibrium may be longer than the observation time.

 An aspect of equipartition sometimes overlooked, however, is that some quantized modes are too energetic to be populated at all by thermal fluctuations (in your lifetime). How does it come about that energy gets so equally divided between the various modes? Often, as in the case of non-radiative energy dissipation in a molecule that absorbed a photon, the energy is initially deposited in a few local modes of a molecule. How does it get uniformly spread among all modes of the system? Energy equilibration among modes occurs because of interactions between modes: a particle moving in the x direction collides with another and knocks it in the y direction; a diatomic particle collides with another at an angle, transferring rotational, vibrational, and translational energy; two oscillators in a lattice exchange energy because of charge–charge interactions. This exchange of energy usually happens very rapidly: remember the 10^9 collisions per second in an ideal gas? Suppose, however, that a system consists of 10^{10} oscillators with mode spacings of $30k_B T_{300K}$. The probability the first excited state is thermally occupied is e^{-30}, or about 10^{-14}. This means 10^{-4} oscillators (i.e., none) are in the first excited state. If there is less than one particle involved in exchange of energy in one of its modes, that mode is effectively not in the energy exchange game: it cannot be treated as exchanging energy with translational kinetic energy modes, unless someone devises a measurement scheme over a very long time or raises the temperature to 3,000 K, where $k_B T_{3000K} = 4.1 \times 10^{-20}$ J. The latter is biologically irrelevant, of course. The oscillations in Figure 14.4a, for example, would not be significantly involved in equilibration of thermal energy, though they may be easily excited by infrared (IR) photon absorption.

14.4 "INTERNAL" ENERGY: KINETIC (K) AND POTENTIAL (U)

Internal energy is any form of energy that can be contained by an individual particle, that depends quadratically on a dynamic variable, and that can be exchanged with other particles. Translational kinetic energy is the simplest form of internal energy a particle can have, and it is straightforward. There are three modes (dimensions) that can contain energy, and we have seen that the Maxwell–Boltzmann distribution describes the kinetic energy "contained" in each dimension. Translational kinetic energy is also a continuous variable, so there is no minimum packet of energy that can be absorbed from the thermal environment. A free-particle collision can, in principle, transfer any fraction of kinetic energy from one particle to another.

A diatomic molecule like H_2 can contain, in addition to translational kinetic energy, vibrational kinetic and potential energy, and rotational kinetic energy. See Figure 14.5. Each vibrational "mode" includes two energy-containing modes: vibrational kinetic and potential. Rotations can be around either of two axes—$\frac{1}{2}k_B T$ thermal energy in each. Why cannot rotations be around the third axis, the axis of the H–H bond? There are two ways to understand this. First, the likelihood that a collision would be at exactly the right, glancing angle required to spin the H_2 molecule around its longitudinal axis is vanishingly small. (Remember it's the nucleus with radius about 10^{-15} m that must be given a spin.) Second, the rotational kinetic energy is $L^2/2I$, where L is the angular momentum and I is the moment of inertia about the axis in question. The moment of inertia of two spheres of radius 10^{-15} m spinning about their centers is very small. Small I means a high transferred energy, $L^2/2I$, is needed to excite this rotation.

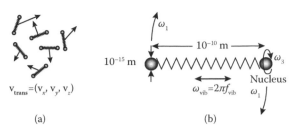

FIGURE 14.5
Various types of internal energy for a diatomic molecule—seven Boltzmann energy-containing dynamic modes. (a) Translational kinetic energy, with three modes. (b) Three rotations are possible, but the ω_3 shown does not count, because its minimum energy is too high. A rotation ω_2 that brings the atomic nuclei in and out of the paper is equivalent to ω_1. One vibration is possible, but it contains two energy modes: vibrational kinetic and vibrational potential. Vibrations may not contribute at room temperature. See text.

We have been edging ourselves further into a quantum-mechanical view of these thermodynamic internal energies. We noted that molecular vibrations are quantized, with energies better described by $E_n = \left(n + \frac{1}{2}\right)\hbar\omega$ than by $E = \frac{1}{2}kx_{max}^2$. Are molecular rotations quantized? We naturally think of rotations of a diatomic molecule as "free" rotations, implying continuously variable rotational kinetic energy. In the previous paragraph, however, we sneaked in a subtle quantum assertion when we noted that the energy needed to excite the ω_3 rotation was extremely high. You may have thought, "Wait a minute: the moment of inertia, I, may be small, but why can't the molecule rotate really slowly, making $\vec{L} = \vec{r} \times \vec{p}$ really small, and E_{rot3} reasonable?" The reason is that quantum mechanics places a minimum requirement on the magnitude of L: $L^2 = \hbar^2 \ell(\ell+1)$, where $\ell = 0, 1, 2,...$, and $E_{rot} = \frac{\hbar^2 \ell(\ell+1)}{2I}$. To go from $\ell = 0$ (no rotation) to $\ell = 1$ requires a minimum energy, which is high for ω_3.

Did we do the ideal gas all wrong? Perhaps we should have treated the ideal gas translational kinetic energy as quantized. After all, we learned that a "free particle" does not really exist; there is some distance at which some other object creates a potential energy that confines the particle.

A particle confined to a box of size L has allowed energies of $E_n = \frac{n^2 \pi^2 \hbar^2}{2mL^2}$. Perhaps the minimum energy to excite a particle from $n = 1$ to $n = 2$ is very high. It turns out we are saved by the mass of the nucleus. The small nuclear radius caused E_{rot3} to be very high; the large nuclear mass allows E_{trans} increments to be small:

$$E_n = \frac{n^2 \pi^2 \hbar^2}{2mL^2} = n^2 \frac{\pi^2 \left(1.05 \times 10^{-34}\,\text{J.}\right)\text{s}^2}{2 \cdot 1.67 \times 10^{-27}\,\text{kg} \cdot \left(0.1\text{m}\right)^2} = n^2 \cdot 3.3 \times 10^{-39}\,J \approx n^2 \cdot 10^{-18} k_B T \qquad (14.24)$$

The last expression on the right of Equation 14.24 reminds us what we compare to when we state that an energy is "large" or "small." The quantum translational E_n increments are very small compared to $k_B T$, so we are quite safe treating translational velocity and momentum as a continuous variable.

So, where were we before this pause for elementary quantum theory? Have we sidetracked ourselves into a modern physics course? Possibly so, but the facts are that vibrations and rotations of all sorts of frequencies and quantum energies occur and are important in biological macromolecules. If quantized vibrations occur in oxygen binding by hemoglobin (they do), we should be prepared to evaluate their involvement in biological reactions. Our focus right now, however, is on the ways in which energy can be contained by molecules and particles making up a biological system. The prime two molecular energy-containment tools are kinetic energy and the potential energy stored in the numerous bonds and interactions between atoms in biological macromolecules.

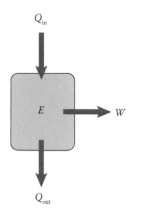

FIGURE 14.6

Conservation of energy in a bulk system. Positive net heat input into a system must result in work done by the system and/or an internal energy increase in the system. Any internal energy change is shared by the energy-containing modes of the particles making up the system, according to the Equipartition theorem.

14.5 HEAT, INTERNAL ENERGY, WORK, AND ENTHALPY

Heat is a quantity of energy associated with a macroscopic system, but heat transferred to a system can be measured by the average change in energy of those energy-exchanging modes (kinetic and potential energy) we have just discussed. When we speak of heat, we automatically enter the field formally labeled thermodynamics. Thermodynamics describes the relationships between various forms of energy and work in a bulk system and can be viewed as the "master" set of laws governing the aggregate effect of the many particles contained in the system. Though distributions of particle energies exist within the system, with more energetic particles contributing relatively more to effects such as pressure and chemical reactivity, the aggregate exchange of energy between two bulk systems is still ruled by thermodynamic laws. In spite of the fact that some biological systems cannot be classified as "bulk," their behavior is often not far from bulk behavior.

When net (positive) heat is added to a system, energy must be conserved (Figure 14.6). The system can either convert the heat into work done on the outside world or the systems internal energy, E, can increase (or both). Conservation of energy implies

$$Q_{in} - Q_{out} = \Delta E + W \tag{14.25}$$

If we integrate all of the internal energy changes over all component particles, we get the macroscopic ΔE in Equation 14.25. We can alternatively measure the temperature change of the system, which measures the average change in mode energy, $\frac{1}{2}k_B T$, multiply by the number of modes and number of particles, and get ΔE. This latter procedure has already been done for us in the form of the specific heat equation,

$$Q_{net} = mc\Delta T \tag{14.26}$$

In complex systems the specific heat may not be a constant over all temperatures. This is usually the case with biological macro-molecules. In this case, Equation 14.26 can be rewritten

$$c(T) = \frac{1}{m}\frac{dQ}{dT} \tag{14.27}$$

The total heat input into a system must then be calculated by integrating $mc(T)$ over the temperature range of interest. The work done by the system is most simply illustrated by the case of an ideal gas, where expansion at constant pressure results in $W = P\Delta V$. There is, of course, no work done if the volume is kept constant. In this case the specific heat equation can be rewritten as a molar specific heat equation at constant volume,

$$Q_{net} = nC_V\Delta T \tag{14.28}$$

where n is the number of moles of gas and C_V is its molar specific heat capacity at constant volume. For a monoatomic gas, $C_V = \frac{3}{2}R$, where $R = N_A k_B$. If, instead, the gas is held at constant pressure, C_V in Equation 14.27 is replaced by $C_P = \frac{5}{2}R$. Why is C_P greater than C_V? Because heat added at constant pressure must expand the gas, leaving less energy to raise the temperature.

However, we are interested in biological systems that are rarely similar to ideal gases. Equation 14.27 is of great importance to the biosciences. Microcalorimetry is widely used to quantify conformational changes and interactions of biomolecules in solution: small amounts of heat are added to a protein solution, for example, and the temperature rise measured and compared to that of the solvent without protein. The associated difference ΔQ is then interpreted in terms of molecular enthalpy changes. We explore this further in Chapter 15. The ideal gas example brings up an issue we must deal with carefully in biology, however. The expansion of a gas is easy to see and measure: $P\Delta V$ is macroscopic, not associated with an individual gas particle, and is not part of the internal energy of the system. In a previous chapter we noted that the molecular volume of a protein molecule can change in response to pH, binding of small molecules and other interactions. Is this molecular $P\Delta V$ part of the internal energy of the biomolecule or is it separate work, as in the ideal gas? The key question is whether this energy can exchange with other particles through the common Brownian collisions taking place in the system. It cannot, so $P\Delta V$ is not part of the internal energy of the molecule. Think of this volume change as a macroscopic volume change, $N\,P\Delta V$, where N is the number of protein molecules, but where the volume change is at the expense of the surrounding water and cannot be easily seen.

The $P\Delta V$ changes just described are not part of the internal energy, yet they are associated with individual molecules and as an energy, must be accounted for. This is nothing new to thermodynamics: a new energy quantity has been defined to include $P\Delta V$. This quantity is the *enthalpy*:

$$\text{Enthalpy: } H = E + PV \tag{14.29}$$

Because biomolecular volume changes regularly occur but are not easy to separately measure, the enthalpy largely displaces the internal energy in common usage in the biosciences. The enthalpy is then seen to be directly related to the quantities in Figure 14.6:

$$\Delta H = \Delta E + P\Delta V = Q_{net} - W + P\Delta V = Q_{net} - W' \tag{14.30}$$

where W' is any form of work done other than $P\Delta V$ work and is often assumed to be zero.

EXAMPLE 14.1

Calorimetric measurement of nucleic acid unfolding enthalpy. Raising the temperature to about 80°C causes most double-stranded DNA to come apart into single strands— the well-known DNA "melting" process, which normally takes place over about a 20°C temperature range. Measurement of the molar specific heat at constant pressure, $C_P = \frac{1}{n}\frac{\Delta Q}{\Delta T}$, over a wide temperature range (Figure 14.7) allows computation of Q_{total}, which is approximately ΔH for the conformational transition. For the 16-base pair, double-stranded DNA measured in Figure 14.7, this enthalpy is 440 kJ/mole (28 kJ/ mole of base pairs or $11k_BT$ per base pair).

Example 14.1 implies that whatever DNA internal energy is involved in the double-strand folding process, the energies are accessible to thermal fluctuations of size $\sim k_BT$ at 40°C–80°C; otherwise, they would not be measured as a calorimetric $\Delta Q/\Delta T$ change, distinct from the heat absorbed by the solvent molecules. Any elastic potential energies, for instance, could not be too high, as they typically are in diatomic molecules, where vibrations don't appreciably contribute to the specific heat until well over 1,000 K.

14.6 CONSERVATIVE AND NONCONSERVATIVE FORCES: DNA EXAMPLE

In first-year physics we learned several properties of so-called conservative forces:

- The work done by a conservative force depends only on the start and end points, not on the path taken;
- The work done by a conservative force is 0 if the start and end points are the same;
- A conservative force can be written as the derivative of a potential energy:

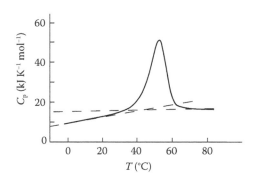

FIGURE 14.7
Microcalorimetry of 16-base pair double-stranded DNA. Integration of the curve above the dotted baselines results in an unfolding enthalpy of $\Delta H = 440$ kJ/mole. (Redrawn from Jelesarov, I. et al., *J. Mol. Biol.*, 294, 981–995, 1999.)

$$F_x^{\text{cons}} = -\frac{\partial V}{\partial x} \qquad (14.31)$$

Our prime example of a conservative force is an object of mass m initially traveling with velocity v_0 in the +y direction against the (conservative) gravitational force. The object travels to a height h above the starting height, $mgh = \frac{1}{2}mv_0^2$ and if the object is allowed to fall back, it regains all of its kinetic energy: the kinetic energy was "stored" in the gravitational field. A Hooke's spring operates the same way (Figure 14.8). Frictional force is nonconservative: kinetic energy "lost" to friction is converted not into potential energy stored in a force field, but into heat.

Here again, biology gets us into difficulty. We just convinced ourselves that heat consists of the energy contained in the accessible energy modes of the particles of a system. These modes include those vibrational potential energy modes of diatomic and multiatomic molecules. Why can't we then refer to these molecular potential energy modes as giving rise to conservative forces? Well, there is a conservative force, but that particular conservative force is coupled to kinetic energy to the particle oscillating in that mode. Of course, energy from that oscillation can be transferred to a nearby molecule through normal random collisions, but the energy is thereby randomized rather than restored to a particular molecule.

What if the molecule is a DNA molecule whose helical structure effectively forms a sort of spring? How big does a spring have to be before it may be considered as giving rise to a conservative compression/extension force? Should we treat the DNA "spring" as a macroscopic spring with restoring force accessible only to some other macroscopic agent (e.g., a micropipette or laser tweezers) as in Figure 14.9a, or should we treat the conformation as an internal energy mode, constantly exchanging $k_B T$ bits of energy with solvent molecules, allowing heat to flow, etc. (Figure 14.9c and d)? Before addressing the answer (or answers), we need to get several issues put into their proper places.

DNA can, in fact, be stretched like a spring using a micropipette and laser tweezers. A restoring force can be measured. Under appropriate conditions, this DNA stretching and relaxation *can* be repeated (reversibility), implying the force is conservative. We see this in Chapter 15. The microcalorimetry unfolding measurement described in the previous section involves structures other than simply than a stretched and relaxed spring. Such structures are illustrated in Figure 14.9b. Because only small increments of heat added to the DNA (the absolute temperature increase is not large) are responsible for the unfolding, relatively low energy vibrations and other modes must be taking up thermal energy. From 0°C to 40°C the specific heat shown in Figure 14.7 increases only gradually, indicating only a moderate number of DNA modes are taking up thermal energy. Between 40°C and 55°C, however, C_p increases rapidly: much heat is taken up while the temperature rises only a small amount. Clearly, many modes suddenly become involved in exchange of thermal energy with the environment. Why this sudden increase? Is the increase connected with any "macroscopic" elastic property of the DNA? The answers to these questions are "Cooperativity" and "Yes," but not exactly.

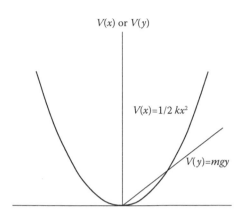

$V(x)$ or $V(y)$

$V(x) = 1/2\, kx^2$

$V(y) = mgy$

FIGURE 14.8
Potential energy derivative gives rise to force.

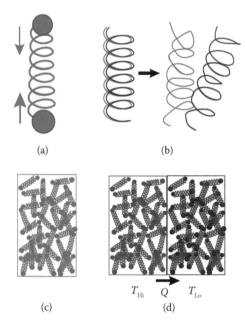

(a)

(b)

(c)

(d)

$T_{\text{Hi}} \quad Q \quad T_{\text{Lo}}$

FIGURE 14.9
Objects with spring constants. (a) A single masses-on-a-spring object. Energy $\frac{1}{2}kx^2$ can be stored and extracted from the spring without energy loss. (b) A DNA molecular "spring" can be partially dismantled by heating, separating into two single springs (approximately). (c) An ensemble of "small" springs at constant temperature has average internal energy $2 \times \frac{1}{2}k_B T$ stored in its harmonic oscillation kinetic + potential energy mode and constantly exchanges energy with other oscillators or other surrounding "small" particles. Over time, energy gains and losses average to zero, leaving thermal equilibrium. (d) Thermal conductivity. Placed in contact with a ensemble of cooler spring particles, net heat (internal energy) is transferred to the cooler springs until both ensembles are at the same temperature. When is a spring "small" enough to exchange thermal energy with the environment?

14.6.1 Cooperativity

What is a macroscopic spring? A steel spring is an appropriately shaped piece of material where deformations at one end are tightly coupled to structure at points distant from the end. This coupling of structural changes at spatially separated locations can be generically referred to as *cooperativity*. In a steel spring the (attempted) mechanical deformation of local bonds between adjacent metal atoms are spread many centimeters away by the strong interactions between neighboring atoms. If heat is added at one end of the spring, tight coupling of another sort transfers the energy rapidly along the metal to other sites. In molecules like DNA or hemoglobin, certain special structural features involve coupling of structural changes at separated sites. We deal with both cases in more detail in Chapter 15. In the case of DNA, local interacting structures with oscillation modes (e.g., base stacking and base pairing) can be excited by thermal energy, but the DNA backbone and base stack enforce a rigidity that spreads the thermal energy over many base pairs. If 50 base pairs are tightly coupled together, then roughly 50 times more energy is needed to excite the collective mode. This larger energy requires a higher temperature for thermal excitation. If the base pairs are more loosely coupled, an excitation at one site tends to be shared by other base pairs, and as temperature increases the collective mode, prevented from oscillating at lower temperature, is rapidly excited. The specific heat then increases rapidly with temperature, as in Figure 14.7. In the case of DNA, the "vibration" ultimately leads to a radically different structure, with separated base pairs and two polynucleotide strands mostly uncoupled from each other. When this happens, the excitation mode leading to the strand separation disappears and the specific heat decreases back to background levels (>60°C in Figure 14.7). Is this the same process as our purported DNA spring stretch of Figure 14.9? No, the structures and processes are not the same, but they are related. If thermal energy at 60°C can dismantle the spring, the same thermal energy should be able to excite the spring.

What is the point of these ruminations on conservative forces, site-to-site interactions and cooperativity? Biological macromolecules are poised on the borderline of two thermodynamic descriptions. On the one hand, a 1.0 μL, 10^{-9} M DNA solution can be considered as a normal thermodynamic system in which DNA molecules exchange thermal energy with solvent molecules through translational kinetic and other internal energy modes. On the other hand, a single DNA molecule can be considered as a "macroscopic" entity that has certain higher-energy modes that can be excited either by thermal energy at higher (but still moderate) temperatures or by another object (molecule, laser tweezers, AFM, etc.) that has more than a few k_BT's of energy available to it. The feature that allows this duality of biomolecular character, microscopic and macroscopic, is cooperativity.

14.7 IDEAL GAS LAW

$$PV = Nk_BT \tag{14.32}$$

The ideal gas law embodied in Equation 14.32 cannot be directly applied to many biosystems, but this law leads us to a simple expression for entropy, which we have seen is very relevant to biology. We have all performed numerous calculations of ideal gas volume and pressure changes using Equation 14.32. We do not need to repeat any of them here, but it is important to realize that the left side of this equation contains the system volume, which reflects the amount of physical space, and number of states, available to gas particles. The right side is proportional to the internal energy of the system. If pressure and particle number are kept constant, a larger volume must be accompanied by an increased internal energy content. Larger internal energy correlates with a larger number of available states for particles to be in. For a monatomic ideal gas,

$$V = \left(\frac{N}{P}\right)_{const} k_BT = \frac{2}{3}\left(\frac{N}{P}\right)_{const} E \tag{14.33}$$

Suppose incremental heat dQ is added to this ideal gas, with a fixed number of particles, but volume and pressure are allowed to vary. From the first law of thermodynamics (conservation of energy), the amount of heat added

$$dQ = dE + dW = NC_V dT + PdV \tag{14.34}$$

or, for a monatomic gas, $C_V = 3/2\ Nk_B$ and $dQ = \frac{3}{2}Nk_B dT + PdV$. From the ideal gas law we can replace P with $Nk_B T/V$, with result

$$dQ = \frac{3}{2}Nk_B dT + Nk_B T \frac{dV}{V} \tag{14.35}$$

Dividing by T,

$$\frac{dQ}{T} = \frac{3}{2}Nk_B \frac{dT}{T} + Nk_B \frac{dV}{V} \tag{14.36}$$

where $C_V = 3/2\ Nk_B$ for the monatomic ideal gas. The last equation is written in a suggestive form: an incremental change in the quantity on the left is proportional to the sum of relative changes in temperature and volume. Exercise 14.2 showed us something about ideal gas statistical mechanics, but did we learn anything about biological processes? Not really; or perhaps more carefully stated, we have seen how biological systems do *not* behave. First, biological systems are dominated by friction. Any oscillation initially set up dies quickly away. Second, thermal conductivity is high: temperature increases at a site are quickly conducted away. Third, external fields rarely play a "hands-off" role in moving objects remotely. Charges do flow across membranes and are transferred from one to another molecule, but charges are close together and do not move far. When they move, any gained electrical energy is quickly dissipated. Rather than external fields reaching into a biosystem to do work or perturb equilibrium, chemical energy is more commonly expended to initiate motion. As we have seen, the expenditure of chemical energy involves the free energy of the source and the concentrations of reactants and products. The internal energy and the entropy of the energy source and the system are intimately involved. If we encounter oscillations in a biological system, we know to look for a continual expenditure of chemical energy to maintain the oscillation and its concomitant entropy generation.

EXERCISE 14.2

Conversion of internal energy to macroscopic work. The point of this exercise is to think about how external, macroscopic work can be done using the internal energy of a system. Figure 14.10 shows a volume of a monatomic gas in an insulated container compressed to a certain volume by a frictionless piston of surface area A. The gas is compressed by a steel piston of weight mg. The height of the gas column at equilibrium is L, and the temperature is T_0. At $t = 0$, the piston and mass are lifted to a height $y_0 = L/10$ by an external magnetic field and released.

(1) Qualitatively describe what happens.
(2) Write the equation for the net force on the piston and determine the relation between m, T_0, L, and the number of gas molecules in the chamber.
(3) Calculate the work done by the net force and by the gas when the piston rises a small distance, dy.
(4) Calculate the relationship between the piston height and gas temperature, assuming (quasi) equilibrium.
(5) Calculate the position of the piston as a function of time. Use a small-change approximation, as necessary.
(6) Discuss the time dependence of important physical quantities like pressure, temperature, and entropy.

Answer

This exercise may seem somewhat intense, but the intensity is only mathematical and also illustrates the possible behavior of a simple, isolated thermodynamic system with no dissipation or external energy input.
   1. When the piston is raised, the gas pressure decreases. When dropped, the piston recompresses the gas until the pressure reverses the fall of the piston. Because

there is no frictional loss, the piston should move past its equilibrium position and then oscillate about equilibrium.

2. Two forces are on the piston: $F_{net} = F_{gas} - F_{grav} = PA - mg = \frac{Nk_BTA}{V} - mg = \frac{Nk_BT}{L+y} - mg$, where y is the height above the equilibrium height L. At equilibrium, $F_{net} = 0$, $y = 0$, $T = T_0$, and $L = \frac{Nk_BT_0}{mg}$

3. The net work done on the piston is $dW = F_{net}dy$, but the work done by the gas is $F_{gas}dy = \frac{Nk_BT}{L+y} dy$. (Gravity does the other work.)

4. Because there is no heat input into the system, $Q = 0 = \Delta E + W$, and the work done comes at the expense of the gas internal energy. Therefore we can expect a gas temperature change. Let's assemble the tools we might need to solve this problem:
   a. $PV = Nk_BT$, noting T can change.
   b. Internal energy of monatomic ideal gas: $E = 3/2\ Nk_BT$, with $dE = -dW_{gas}$.
   c. Work done by gas $F_{gas}dy = \frac{Nk_BT}{L+y} dy$.
   Equating the work done by the gas with minus the internal energy change of the gas, $dW_{gas} = \frac{Nk_BT}{L+y} dy = -dE = -\frac{3}{2} Nk_BdT$. Let's write temperature and position in terms of initial temperature and initial length of the gas column: $dW_{gas} = \frac{Nk_BT_0}{L} \left[\frac{T/T_0}{1+y/L} \right] dy = -\frac{3}{2} Nk_BdT$. Letting $u \equiv \frac{T}{T_0}$ and $x \equiv \frac{y}{L}$, the differential equation simplifies to $\left[\frac{u}{1+x} \right] dx = -\frac{3}{2} du$ or $\frac{du}{u} = -\frac{2}{3} \frac{dx}{1+x}$. We can integrate this equation directly: $\ln u = -\frac{2}{3} \ln(1+x) + c$, where c is an integration constant, or $u = (1+x)^{-2/3} C'$. The initial condition is that $u = 1$ at $x = 0$, so the relation between temperature and piston position is $u = (1+x)^{-2/3}$ or $\frac{T}{T_0} = (1+\frac{y}{L})^{-2/3}$. As y increases from 0, the gas expands and T decreases, which seems to make sense because no heat can be absorbed or emitted to the outside world.

5. Newton's law for the piston, which includes the gravitational force, says $F_{net} = m\frac{d^2y}{dt^2} = \frac{NkT_0}{L} \left[\frac{T/T_0}{1+y/L} - 1 \right]$ (after noting the relation $mg = \frac{Nk_BT_0}{L}$). The net force equation simplifies further to $\frac{d^2x}{dt^2} = \frac{g}{L} \left[\frac{(1+x)^{-2/3}}{1+x} - 1 \right]$ or $\frac{d^2x}{dt^2} = \frac{g}{L} [(1+x)^{-5/3} - 1]$. We can now see indications of oscillatory behavior in the form of g/L, which in the pendulum is the square of the angular oscillation frequency. The dependence on x, however, is nonlinear. This equation can be solved by various integration techniques, resulting in an integral expression. However, in the limit of small x, $(1+x)^{-5/3} \cong 1 - \frac{5}{3} x$, and the differential equation becomes $\frac{d^2x}{dt^2} + \frac{5g}{3L} x = 0$, with solution $x(t) \cong x_0\cos(\omega t)$, or $y(t) \cong y_0\cos(\omega t)$, with $\omega = \sqrt{\frac{5g}{3L}}$

6. This system clearly oscillates: y oscillates, which drives an oscillation in temperature because $\frac{T}{T_0} = (1+\frac{y}{L})^{-2/3} \approx 1 - \frac{2y}{3L}$, and pressure oscillates because $P = \frac{Nk_BT}{A(L+y)} = \frac{Nk_BT_0}{AL} (1+\frac{y}{L})^{-5/3} \cong \frac{Nk_BT_0}{AL} (1 - \frac{5y}{3L})$. The small-amplitude oscillation frequency is interesting. Except for the 5/3 factor, the frequency looks like that of a pendulum. Why should it appear that no dependence on the piston mass is observed? It would seem that a heavier mass should cause slower oscillations. In fact, ω does depend on m. Since $L = \frac{Nk_BT_0}{mg}$. we could rewrite ω as $\omega = \sqrt{5g/3L} = \sqrt{5mg^2/3Nk_BT_0}$. If we have 0.01 mole of gas at 300 K, then $\omega \cong 2.5 \cdot \sqrt{m}$. A 1.0-kg mass has an oscillation angular frequency of 2.5 rad/s, or a frequency of 0.4 s^{-1}. Is the value of L reasonable for these parameters? $L = \frac{Nk_BT_0}{mg} = \frac{(6.0\times10^{21})(4.1\times10^{-21} J)}{9.8N} \cong 2.5m$. Why the 5/3 factor? Those ideal gas expert readers may recognize this value as the ratio C_P/C_V, which for a monatomic ideal gas is $\left(\frac{5}{2} k_BN \right) / \frac{3}{2} k_BN$. If the gas were diatomic, the ratio would be 7/5.

What about the entropy? Clausius tells us that the entropy should remain fixed because $\Delta S = \int \frac{dQ}{T}$ and no heat is exchanged with the external world. With no friction the oscillation should last forever. Of course, in the real world this can never happen. There is always a bit of friction and thermal conductivity: the piston, walls, and gas heat up, the oscillation stops, and the entropy increases. We should also worry about our neglect of the magnetic force that initially raised the piston above its equilibrium position. Was the raising process very fast or very slow? Did the temperature change during this process? Is there any charge on the piston, which would radiate electromagnetic energy during the oscillation?

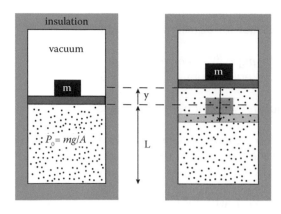

FIGURE 14.10
Expansion of a gas against the weight of a piston. Initially, the gas is at a pressure 10% higher than mg/A, where m is the total weight of the piston assembly and A is its area. When the clamping pin at the left is removed, the gas pressure pushes the piston up, giving the piston potential and kinetic energy. The piston harmonically oscillates up and down. The frictionless piston and walls of the container are perfectly insulating.

14.8 ENTROPY: GASES AND POLYMERS

The change in the macroscopic, *Clausius version of entropy* is the quantity on the left of Equation 14.37:

$$dS_{mig} \equiv \frac{dQ}{T} = \frac{3}{2} Nk_B \frac{dT}{T} + Nk_B \frac{dV}{V} \quad (14.37)$$

where the subscript "mig" stands for monatomic ideal gas ($C_V = 3/2\ Nk_B$). In order for the entropy to change, heat must flow into or out of a system. The entropy change from initial state (T_i, V_i) to (T_f, V_f) is obtained by integration:

$$\Delta S_{mig} = \left(S_f - S_i \right)_{mig} \geq \int_i^f \frac{dQ}{T} = \frac{3}{2} Nk_B \ln \frac{T_f}{T_i} + Nk_B \ln \frac{V_f}{V_i} \quad (14.38)$$

An inequality has been inserted here to reflect the fact that the equations used in the derivation assume equilibrium, and thus reversibility, is constantly maintained; this may not be the case. For nonreversible changes, the inequality holds. Increases in either the temperature or volume result in entropy increases. Equation 14.38 can also be expressed as

$$\Delta S_{mig} = \frac{3}{2} k_B N \frac{\Delta E}{E_i} + k_B N \frac{\Delta V}{V_i} \quad (14.39)$$

Because entropy is a so-called state function, depending only on the existing state of the system and not how it got there, entropy changes for a nonequilibrium process leading from state 1 to state 2 can be computed, without the inequality sign, by using an equilibrium process connecting the same two states. The classic example is the doubling of volume of an ideal gas by (nonequilibrium) expansion into an initially vacuum-filled space.

Thinking back to Exercise 14.2 we note that because ΔS in that case was zero, the internal energy and volume terms were equal and opposite.

The above equations characterize the ideal gas, of course, but a general picture of entropy emerges: increasing the temperature of a system usually (not always) increases the number of states available to the system, as does a volume increase. The latter seems natural because more physical space per particle is allowed.

We have sneaked up on what we all know by now, that another version of entropy, introduced by Boltzmann, relates to the number of states available to the system:

$$S = k_B \ln \Omega \quad (14.40)$$

where Ω is the number of distinguishable ways the system can be assembled (number of states) to give the same total system energy E. Calculation of values of Ω and S are generally difficult for real systems, but often changes can be found more readily. Entropy can thus be described by macroscopic measurables, heat added and system temperature, or by microstate descriptors of the system. In certain cases the number of states can be counted, at least approximately, from the way a system is assembled. Crystals, gases, and random structures are among these "countable" systems.

1. Entropy of an ideal gas. We present here, without proof, the translational, rotational, and vibrational entropy of an ideal gas. The interested student can consult any number of standard textbooks for derivations.
2. Translational entropy (monatomic ideal gas). Several versions of the Sackur–Tetrode formula for the translational entropy of a monatomic ideal gas are listed below.

$$S_{trans} = k_B N \ln\left[\frac{e^{5/2}V}{Nh^3}\left(2\pi m k_B T\right)^{3/2}\right]$$

$$S_{trans} = k_B N \ln\left[\left(\frac{V}{N}\right)\left(\frac{E}{N}\right)^{3/2}\right] + \frac{3}{2}k_B N\left(\frac{5}{2} + \ln\frac{4\pi m}{3h^3}\right) \quad \left(E = 3/2\,N k_B T\right) \qquad (14.41)^{5,6}$$

$$S_{trans} = k_B N \ln\left(\frac{V}{N\Lambda^3}\right) + \frac{5}{2}$$

where Λ is the thermal de Broglie wavelength $= \frac{h}{\sqrt{2\pi m k_B T}}$.

$$\text{Rotational entropy }\left(\text{diatomic ideal gas}\right): S_{rot} = k_B N + k_B N \ln\left[\frac{8\pi^2 I k_B T}{2h^2}\right] \qquad (14.42)^{7}$$

(The 2 in the denominator comes from a symmetry term when the two atoms are identical: the molecular orientation •—• cannot be distinguished from that rotated 180°.)

$$\text{Vibrational entropy }\left(\text{diatomic ideal gas}\right): S_{vib} = k_B T\left[\frac{x}{e^x - 1} - \ln\left(1 - e^x\right)\right], \quad x = \frac{hf}{k_B T} \qquad (14.43)^{7}$$

where f is the vibration frequency.

An example of magnitudes of ideal-gas entropies is shown in Table 14.3.

The vibrational entropy, not surprisingly, is small at room temperature. What is more surprising (and gratifying) is the close agreement (0.1%) of experiment with calculation for S_{tot}. The neglect of the vibrational entropy would make this agreement significantly worse. A take-home lesson from Table 14.3 is that the formation of a biopolymer from constituent atoms involves a huge translational entropy decrease, a large decrease due to arrested rotations, but an increase in vibrational entropy due to the many new vibrations possible in the polymer.

TABLE 14.3 Molar Entropy (Absolute) of F_2 Gas at T = 298 K[a]

Entropy Component	Value (J/(K·mole)	Values Used in Computation
S_{trans}	154.7	$M = 6.63 \times 10^{-26}$ kg, $V = 22.4 \times 10^{-3}$ m³
S_{rot}	48.1	$I = 32.5 \times 10^{-47}$ kg·m²
S_{vib}	0.605	$F = 2.676 \times 10^{13}$ s⁻¹, $x = 4.305$
S_{tot}	203.4	$S_{trans} + S_{rot} + S_{vib}$ (calc)
	203.2	Measured value

[a] Source: Values from Moore, W. J., *Physical Chemistry*, 4th ed., Prentice-Hall, Englewood, NJ., 195–197, 1972.

EXERCISE 14.3

The entropy of a freely jointed chain (FJC) as a system; application to DNA. (As with all "exercises" in this book, resist temptation and try this exercise before looking at the solution.) Figure 14.11 shows a model 10-mer.

1. Calculate the entropy of a polymer chain of N units, where the distance between units is L, if the polymer is forced to lie on a flat surface and the position of each subsequent unit is, randomly, $\pm L$ along the x or along the y axis. Consider the effect of a "self-avoiding" random positioning: the third unit, for example, cannot occupy the same space as the first unit.
2. Repeat part 1 if the subsequent unit can, with equal probability, be displaced at any one of n angles with respect to the preceding unit. The dimensionality is 3, and the value of n is not too large (2–20).
3. Compare to the entropy of the chain, stretched out straight.
4. Discuss the application to DNA. Is there a minimal number of DNA base pairs for this model to apply? Compare double- and single-stranded DNA.

Solution

Even though this example deals with a single molecule, we are treating it as a system in itself, with parts that can be positioned randomly or forced. In real cases the possible polymer unit positions should be weighted by a Boltzmann energy factor (some positions are energetically more favorable than others), but we assume a few possible positions are equally allowed and others are not. The structure will thus be entirely determined by entropy. We are not going to count the possible locations the first unit can be placed in, which may be anywhere within some container volume. This is not of much interest here because the changes we consider do not involve the center of mass of the polymer.

> It can also be argued that radical local steric adjustments would be needed to allow two nearby units to (nearly) occupy the same space. If unit 95 were to move one step in the $-x$ direction from the 94th, which happened to take it to the site of the 4th unit, all of the units in between 4 and 95 could, alternatively, reposition slightly to allow 95 to (nearly) move into place.

1. In this FJC, the second polymer unit can be placed in any of four positions. To avoid the first unit, the third unit can then only be placed in any of three positions relative to the four previous, or 4·3 possible positions. While a true self-avoiding random walk would ensure that an added unit does not occupy the same space as *any* previous unit (not just the first and second previous) we stick with this nearest two-neighbor scheme for simplicity. Continuing on for all N units, the number of allowed, equal-energy conformations is then

$$\text{"fourfold" FJC: } \Omega_4 = 4 \cdot 3^N \qquad (14.44)$$

The entropy is then

$$\text{"fourfold" FJC: } S_4 = k_B \ln \Omega_4 = k_B \ln(4 \cdot 3^N) = k_B[N \ln 3 + \ln 4]$$
$$= [1.10N + 1.39]k_B \qquad (14.45)$$

At temperature T the free energy corresponding to this entropy is

$$TS_4 = [1.10N + 1.39]k_B T \qquad (14.46)$$

This energy is quite large compared to $k_B T$ when N is large. There must, of course, be at least three units for this calculation to apply.

2. If each but the first two units has $n-1$ possible orientations relative to the preceding,

$$\text{"n-fold" FJC: } S_n = k_B \ln \Omega_n = k_B \ln(n \cdot (n-1)^N) = k_B[N \ln(n-1) + \ln n] \quad (14.47)$$

Note that the effect of n is weighted logarithmically. If $n = 20$, S_n is about 2.7 times S_4 for polymers with 10 or more units.

3. A stretched out chain has only one possible structure: $\Omega_{stretched} = 1$ and $S_{stretched} = 0$. Stretching restricts all units to a single position.

4. The double-stranded DNA strand is relatively stiff. For the FJC model to apply at all, we must have at least several persistence lengths of DNA—several times 50 base pairs of DNA. The stretched out length of DNA must be significantly greater than 70 nm. If experiments on such a system are to be done using an atomic force microscope, with the backbone conformation resolvable, such a DNA length is adequate. If a light microscope is used, this length should be of order 10 μm, or tens of thousands of base pairs.

14.8.1 Persistence Length and Entropy

Persistence length is not a statistical quantity that naturally emerges from a freely jointed chain model. We have been using the FJC model to this point because conformations can be easily modeled by random-walk mathematics. We have forced the persistence length into the FJC model by effectively making N, the number of units, a variable. For a given polymer of total length L_{total}, the number of units is determined indirectly by dividing by the persistence length. If something is done to change the persistence length, for example, altering the solution ionic strength, which alters the stiffness of a charged polymer, the effective number of units changes, which affects computations of numbers of equivalent-energy states and entropy. We should not accept such assertions of the model without some evaluation. What is happening to conformational states when a polymer becomes stiffer? Are allowed states disappearing? In principle, the 10th DNA base pair of a "stiffened" DNA can still be located at some coordinate in space corresponding to a too-bent conformation. Why should we no longer count that position as allowed? What has ruled that conformation out? Before we get carried away with statistical modeling, let's look at some more biology.

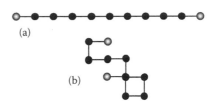

FIGURE 14.11
Ten-unit polymer in (a) a linear array and (b) random array, partially self-avoiding. The array in *a* has zero entropy, in the approximation in the text. An array of type *b* (random turns allowed at each point) is more compact and has higher entropy. See text.

14.8.2 Variable Persistence Length of DNA

Persistence lengths of double-stranded nucleic acids were discussed in the previous chapter: 50 nm for B DNA, 72 nm for double-stranded RNA, and 150 nm for A DNA. These double-stranded polymers are relatively stiff, as we have discussed, but stable in solution and easy to study. What about single-stranded chains, which might be more natural candidates for an entropic FJC model? Single strands can be obtained by raising the temperature above the "melting" point and carefully separating the single-stranded chains by chromatography or other means. Single strands can also be made in the lab by synthesizers found in most university departments of biochemistry or molecular biology, or ordered online for delivery within a week or so. Persistence lengths of such single strands have only recently been measured. Single strands are inherently unstable, preferring to form base-paired or partially based-paired structures. Most persistence-length measurements on double strands have been made using classical optical methods, e.g., electric birefringence or time-resolved rotational polarization decay of bulk solutions of DNA. These methods work well for persistence lengths (bends) above $\lambda_{photon}/10$ or so, but not for much smaller bends. Single-stranded DNA bends much more sharply. Atomic force microscopy has been employed to measure these smaller persistence lengths. Its great advantage is that individual bends are directly measured rather than inferred through average rotational times of the entire polymer. The nature of atomic force microscopy (AFM) requires adhesion of sample to a flat surface; a freely solubilized polymer cannot be imaged. Too strong adhesion distorts the structure and physical properties of the DNA; too weak adhesion allows the DNA to be pushed by the AFM tip, with no resulting image. Needless to say, these issues have mostly been solved during the past 15 years

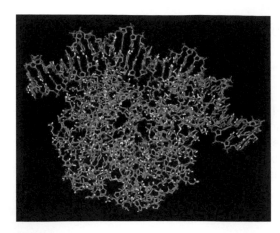

FIGURE 14.12
Sharp bend in double-stranded DNA induced by CAP protein binding. CAP (catabolite activator protein from *Escherichia coli*) is a dimeric DNA-binding enzyme that activates DNA transcription and production of enzymes that ferment catabolites such as lactose and galactose. CAP requires cAMP (cyclic AMP) to bind DNA. The cAMP is produced in *E. coli* when glucose, its favorite nutrient, is low. The bacterium then synthesizes the enzymes it needs to metabolize nonfavorite foods. This complex process saves *E. coli* the energy cost of maintaining multiple metabolic systems when there is no need—all based on a sharp bend in DNA. Protein binding increases the DNA minor-groove width significantly, which, along with other changes, allows sharper bending. The bend occurs over about 20–30 base pairs—approximately the persistence length of single-stranded DNA. (Based on Schultz, S.C. et al., *Science* 253, 1001–1007, 1991; PDB ID 1CGP, displayed using Jmol software.)

FIGURE 14.13
Deviation of the direction of a polymer unit from the preceding direction can receive an angle-dependent Boltzmann probability penalty, $\exp-(E_{i,\,i-1}/k_\text{B}T)$, in a random polymer model. See Miniproject 14.7.

for DNA imaging, mostly through developments in appropriate flat surfaces and the "tapping mode" of imaging.[8] Chapter 19 of our text addresses many of these issues. Rechendorff et al. conclude that the persistence length of single-stranded DNA depends, not surprisingly, on the ionic strength: higher NaCl counterion concentrations allow single-stranded DNA to bend more sharply, presumably because phosphate change repulsion forces are reduced. Reported persistence lengths are 4.6, 6.7, and 9.1 nm for NaCl concentrations of 10 mM, 1 mM, and "very low" ionic strength.[9] These lengths are an order of magnitude smaller than those of double-stranded nucleic acids and also illustrate that persistence length should not be treated as a materials "constant" during biological activity. Disruption of strands can allow much sharper local bends than expected for double-stranded DNA. Such partial separation regularly occurs during enzyme binding to DNA. Resulting sharp bends can be found in structures in the Protein Database. See Figure 14.12.

The FJC model has led us into an "entropic corner." In a freely jointed model there is no potential energy (enthalpy). The polymer bending that we naturally associate with a bending potential has been forced into an entropic description. The only way an FJC model has to mimic a high potential energy state is to not allow it, by not including large angular deviations as allowed states of adjacent units. We should not conclude, however, that the FJC is incorrect. The accuracy and utility of a model is determined by agreement with certain experiments, prediction of effects, and adaptability to new situations. The FJC model explains the scaling exponent (radius versus mass) of "loose" polymers such as DNA quite well and also predicts distortion forces of these structures when forces aren't too large. Some other experiments are better explained by models such as the "wormlike coil." The simplicity of the FJC model more than makes up for the limitations on its range of predictions. For a more complete treatment of simulations of polymer conformation, see works by Wall and the lucid introduction to random walks by Berg.[10,11]

Potential energies could be inserted into random-chain model simulations by assigning energetic penalties to displacements that are too far from some preferred location(s). For example, a stiffer polymer can be made by assigning a Boltzmann probability penalty to orientations of the *i*th unit that have larger angular deviation from the direction of the (*i* − 1)th unit (Figure 14.13). These potential energies could then be summed to get a total for the polymer. This is not easy in a simple spreadsheet approach, which we are taking. Miniproject 14.7, based on Figure 14.14, assigns a weighted preference for continuation of chain direction, which, in a simple spreadsheet, is much easier than directly assigning a Boltzmann energy penalty for a change of direction. Table 14.4 shows part of the Microsoft Excel spreadsheet used for Figure 14.14 and Miniproject 14.7. This rather cryptic table, as well as the next, show two of the key spreadsheet functions (RANDBETWEEN and IFERROR) and are included to facilitate work on that miniproject. (The "t" in column A in Table 14.4 refers to the polymer unit number, though it can equally well refer to the time step of a random walker.)

In Figure 14.14 and Miniproject 14.7, the direction of any step is given by the sum of a random-step vector of unit length plus a weighting factor α times the previous step vector, the vector sum then normalized to a constant step length: $\Delta\vec{r}_i = \frac{\Delta\vec{r}_\text{random} + \alpha\Delta\vec{r}_{i-1}}{|\Delta\vec{r}_\text{random} + \alpha\Delta\vec{r}_{i-1}|}L_\text{step}$. Large values of α ensure the chain maintains its direction over a distance largely compared to L_step. In Figure 14.14a a comparison is made of the *x-y* projection of a 50-unit three-dimensional polymer with a correlation factor of α = 0.5. In this case, the new, randomly chosen direction is weighted twice as much as the preceding unit direction. The net result is that the correlated-angle polymer is somewhat larger

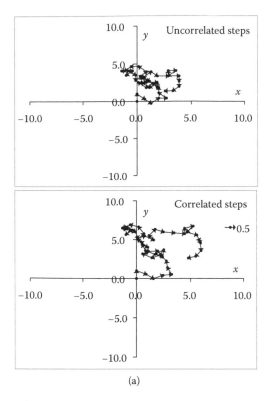

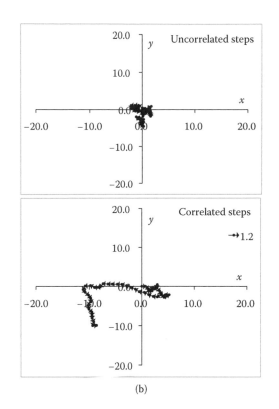

(a)   (b)

FIGURE 14.14

Random (entropic) polymer chain with (a) lower figure, weak adjacent-unit correlation ($\alpha = 0.5$), (b) lower figure, strong adjacent-unit correlation ($\alpha = 1.2$). An uncorrelated simulation is shown above each correlated simulation. The spreadsheet, described in text, simulates a three-dimensional polymer of 50 units, whose x and y coordinates are plotted. For an uncorrelated polymer, a unit i is located at the coordinates of the previous unit plus a random step of length 1, 0, or –1 in the x, y, and z directions, normalized for a total step length of 1: $(x_i, y_i, z_i) = (x_{i-1}, y_{i-1}, z_{i-1}) + (R_{x,i}, R_{y,i}, R_{z,i})/\sqrt{R_{x,i}^2 + R_{y,i}^2 + R_{z,i}^2}$. (Steps of total length 0 occur, on average, 1 in 27 steps, and are not corrected for.) To introduce correlations in a way that can easily be handled in a common spreadsheet, R_{xi} was replaced by $R_{x,i} + \alpha R_{x,i-1}$ (similarly for y and z components). Correlations tend to preserve the direction of the chain and simulate the effect of polymer stiffness (potential energy).

TABLE 14.4   Partial Spreadsheet for a Correlated Random Polymer

	A	B	E	F	I	J	M	N	Q
1	t	Random del_x	Step_size_ unnorm	x	R	Correlated del_x	Step_size_ unnorm	x	r
2	0	=RAND BETWEEN (–1,1)	=SQRT(B2^2+ C2^2+D2^2)	0	=SQRT (F2^2+ G2^2+H2^2)	=B2	=SQRT(J2^2+ K2^2+L2^2)	0	=SQRT(N2^2+ O2^2+P2^2)
3	1	=RAND BETWEEN (–1,1)	=SQRT(B3^2+ C3^2+D3^2)	=IFERROR (F2+B2/ E2,F2)	=SQRT (F3^2+ G3^2+H3^2)	=IFERROR((B3+ $R$18*J2/ $E2),B3)	=SQRT(J3^2+ K3^2+L3^2)	=IFERROR(N2+ J2/M2,N2)	=SQRT(N3^2+ O3^2+P3^2)
4	2	...	...	...	...	...	...	...	...

than the uncorrelated and most bends appear gentler. Note that in this diagram (1) both structures were computed from the same set of random vector components and (2) repetition of the random simulation leads to different structures, though general features persist. In Figure 14.14b the correlation factor has been raised to 1.2. Obligingly, the polymer smooths out, except for one or two sharp backtracks. Remember these sharp bends.

What should the average angle between subsequent units (θ in Figure 14.13) be? If units are completely uncorrelated, the angle should as often be generally backward as forward, and the average

> This average is over all possible angles, weighted by the number of possible ways each angle can be obtained; it is not the average angle for a particular simulated polymer.

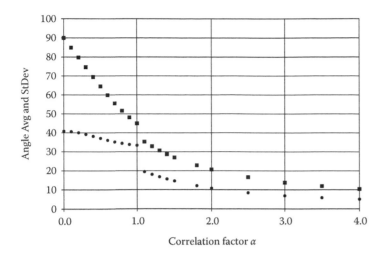

FIGURE 14.15

Average and standard deviation of allowed angles in the random, correlated polymer for values of the correlation parameter α from 0 (no correlation of the directions of the ith and $(i-1)$th units) to 4 (the direction of the previous step is weighted by a factor of 4 over a random, allowed direction). The uncorrelated polymer average angle is 90°: equally likely in forward and backward directions. Large values of α cause small angular differences between subsequent units. The discontinuity at $\alpha = 1$ is an artifact of the random simulation technique. See text.

should be 90°. Figure 14.15 shows that this is indeed the case for $\alpha = 0$. Table 14.5 was used to compute these angular averages:

The angle decreases toward 0° (straighter polymer) as α increases. If we examine the deviations of the angle from the average as a function of α, by computing the angle standard deviation, for instance, we should see that deviations decrease with increasing α. Near $\alpha = 1$, however, we see indications of interesting behavior in our spreadsheet model: there is a discontinuity in both average and standard deviation of the angle. Is this an unexpected discovery or a "glitch"? Lesson number one from experienced modelers as well as experimentalists: 99% of the time that new discovery is an "oops" rather than a breakthrough. Nevertheless, insight into the next needed step is usually present. Why does the average angle drop about 10° when α decreases from 0.99 to 1.01? Let's remember that the direction of the ith unit is determined by the direction of the vector sum $\Delta \vec{r}_{random} + \alpha \Delta \vec{r}_{i-1}$. A value of $a \sim 1$ corresponds to equal weighting of the two vectors. There are, however, nine cases in which the two vectors are in exactly opposite directions. In the case of α slightly less than 1 the resulting angle is 180°; for α slightly greater than 1 the resulting angle is 0°; the abrupt switch from 180–0 gives us our 10° discontinuity. The occasional sharp bend in Figure 14.14b (lower), with $\alpha = 1.2$, may also stem from this feature of the correlated model. Is this discontinuity physically meaningful? Is there any physical reason why a polymer unit should oscillate wildly between 0° and 180° in choosing a direction to point? Points of unstable equilibrium do occur in biology, but it is difficult to invent a set of forces that would create this situation in the case of DNA or another biopolymer. The discontinuity should be treated as a glitch to be fixed in the next, improved model. Meanwhile, there are other predictions of the model that can be looked into. For example, the ratio $\langle\theta\rangle/\Delta\theta$ seems to be about 2 for all values of α. Does this agree with experimental data on any polymers? This is a good example of a somewhat faulty theoretical model that nonetheless has testable experimental predictions. If data on the angle ratio agrees with the model, this feature of the model should be preserved in any model upgrade.

14.8.3 Entropy, Knots, and Chain Stiffness

This section is a parting thought on the polymer conformation and persistence length. There are several types of topological assumptions that go into an approximate entropy calculation such as we have done. One assumption related to self-avoidance—the presence of one part of a polymer chain

TABLE 14.5 Spreadsheet Computation of Angles in Correlated Polymer

	A	B	C	D	E	F	G	H	I	J	K	L	M	N	O
1	x1	y1	z1	r1	a	x2	y2	z2	r2	x2'	y2'	z2'	r2'	r1·r2'/ r1r2'	θ21'
2	1	0	0	=SQRT (A2^2+ B2^2+ C2^2)	0.5	1	0	0	=SQRT (F2^2+ G2^2+ H2^2)	=$E$2* A2+F2	=$E$2* B2+G2	=$E$2* C2+H2	=SQRT (J2^2+ K2^2+ L2^2)	=(A2*J2+ B2*K2+ C2*L2)/ (D2*M2)	=DEGREES (ACOS(N2))
3	1	0	0	=SQRT (A3^2+ B3^2+ C3^2)	0	1	0	=SQRT (F3^2+ G3^2+ H3^2)	=$E$2* A3+F3	=$E$2* B3+G3	=$E$2* C3+H3	=SQRT (J3^2+ K3^2+ L3^2)	=(A3*J3+ B3*K3+ C3*L3)/ (D3*M3)	=DEGREES (ACOS(N3))	

restricts other parts from occupying the same point in space. We ignored this. Another relates to experiences common to activities such as fishing, sailing, and sewing: fibers commonly get tangled. Tangling is sometimes related to that famous law of fate, "Murphy's law": *If anything unfavorable can happen, it will.* Trying to untangle strings commonly results in knots that cannot be undone. On the other hand, knots are usually *intended* in ropes and in fishing lines (at the hook), yet sailors and fishermen usually have to be trained to properly tie a knot that does not spontaneously unknot itself when stress is applied. In biology on the microscale, the presence of knots is ensured not only by Murphy's law, but also by thermodynamics, if the free energy of a knotted state is not too far above that of the unknotted state.

What is the difference between strings that naturally form irreversible knots and ropes or lines that do so only with careful topological design? Three physical factors are clearly important: string stiffness (or bending radius), self-friction, and the "deepness" of the knot. Deepness refers to how far an end of the string is from the center of the knot: a deep knot in a slippery, stiff string can still be difficult to remove. Common string has virtually zero bending stiffness: a bend can be formed with radius of curvature even smaller than the string diameter. Common string is also rough, resulting in a high coefficient of self-friction. Knots tied in string with a large force are virtually impossible to untie, unless the knot is topologically designed to do so, such as a "bow" knot. Rope and fishing line, on the other hand, commonly have minimal bending radii that are a few times the line diameter. Fishing line is also smooth, with a low coefficient of friction. Knots must be more carefully designed to avoid untying under stress.

In Exercise 14.3 we did nothing to prevent knot formation, but knots formed in DNA could be a major impediment to biological function. In our DNA confined to two dimensions a knot cannot be tied, of course, but knots can occur in three dimensions. Do they occur in biological systems? Can DNA be tied in a knot? The answer appears to be "Yes," but some important qualifications should be noted. The research group of physics Nobel Prize winner Stephen Chu, the current U.S. Secretary of Energy, was one of the first to create knots in DNA using a laser tweezers.[12] More recent video images have been recorded by Bao et al., who have tied a number of types of knots in DNA (Figure 14.16).[13] Both groups, along with many others, observed a knot behavior unlike tight string knots. Knots formed in DNA are difficult to tighten and easily slip along the length of the DNA. This behavior should now be readily explainable by the reader, in terms of basic electromechanical properties of DNA.

Nevertheless, knots in DNA do occur at a level of 10^{-2} in equilibrium systems of DNA. DNA knots have also been observed to *result* from enzyme activity.[14,15] (Remember that if an enzyme can untie a knot, it can also tie one; it's just a matter of free energy.) One of the primary tasks of the *topoisomerase* II enzyme is to remove knots and *supercoils* that naturally occur (or are created) in DNA.[16] Control of supercoiling is critical to storage of DNA, as well as unwinding of DNA during replication and transcription. We address these topics again in Chapter 17, but for a full description of the coiling/uncoiling and its relationship to DNA unwinding, see your favorite, recent biochemistry or molecular biology text. Molecular mechanical computations of energetics of certain types of DNA knots, trefoil knots, have also been done using Monte Carlo techniques.[17] Figure 14.17a shows a diagram of a trefoil knot. Figure 14.17b shows two circles that are interlinked. Both these figures illustrate structures in which a strand must be cut in order to form a structure that can be flattened on a surface (no knot). In the case of DNA, this cutting can be done by topoisomerases, which cut and reconnect the strand.

Topoisomerases enjoy a great advantage over the weekend fisherman, who cannot, except as a last resort, cut a fishing line in order to remove a knot. (Intertwined steel rings can be disentangled by magicians, of course.) Unlike the enzyme, the fisherman has no way of re-fusing the polymer fishing line once it is cut. This points to the great advantage of enzymes that control DNA structure and activity: enzymes can cut and reconnect DNA strands when necessary. Circular

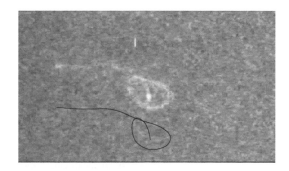

FIGURE 14.16

DNA knot tied with a laser tweezers. DNA is double stranded, with microspheres attached at either end. Spheres are moved using optical tweezers. The curved black line mimics the path of the DNA backbone. The white dot is a laser-trapped microsphere attached to one end of the DNA molecule. It is looped around and fed through the loop to form the knot, which is then tightened. See http://focus.aps.org/story/v12/st25 or original reference for video original. (Bao, X.R. et al., *Phys. Rev. Lett.* 91, 265506, 2003; erratum 95 (2005):199901; video frame from http://focus.aps.org/story/v12/st25.)

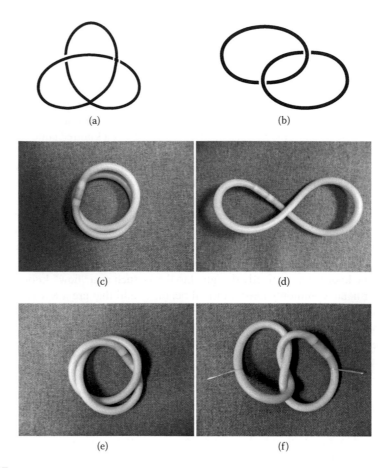

FIGURE 14.17
Trefoil knots. (a) Diagram of a trefoil knot, the simplest nontrivial topological knot, in a single, continuous strand. (b) Two intertwined circles. (c) Circle of rubber tubing with one supercoil; the two halves of the coil lie on top of each other. (d) Same coil as in c, but with two halves separated. The resulting figure eight can further be distorted back into a circle, showing no knot exists. This conformation has a slight tendency to fold back to the state in c. (e) The tubing link in c has been disconnected, one tube end placed under the second half of the coil and the link reconnected, forming a knot. (f) Unfolding the two halves of the coil, which shows the knot cannot be removed nor a single circle restored, now takes considerably more force, as the holding clips suggest.

DNA that has *supercoils*, better behaved relatives of knots, illustrated in Figure 14.17c and d, has no knots and can be reformed into a flat circle on a surface. Cutting a strand in Figure 14.17c, moving a strand end, and reforming the strand connection below another section of the coil (Figure 14.17e) forms a knot and cannot be flattened on a surface (Figure 14.17f).

14.8.4 Topoisomerases I, II, Equilibrium, and Free Energy

We refer here to topoisomerase types I and II. The literature is filled with individually numbered topoisomerases that have been numbered I III, IV, etc., but fall into one of the two type categories.

Topoisomerase I changes DNA supercoiling by cutting one DNA strand, allowing a strand to spontaneously unwrap one time about the other strand, and reconnecting the strand. No ATP energy is required for this process: the energy needed to cut the DNA is regained when the bond reforms and the *activation energy* (Chapter 15) needed to initiate the process comes from thermal energy and the binding energy of the enzyme to the DNA. Topoisomerase I then allows a solution of initially supercoiled DNA to come to equilibrium: supercoiled ⇔ relaxed. (A long time would have to pass before a solution of initially knotted or supercoiled DNA would reach equilibrium without enzymatic help, but the equilibrium is the same as with topo I.) Typically, there is a few percent supercoiled or knotted at equilibrium. Topoisomerase II is a bit different. It cuts and reconnects both strands but requires ATP → ADP hydrolysis energy. This energy requirement has long been known, but it was

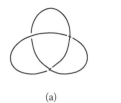
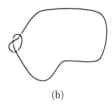

(a)   (b)

FIGURE 14.18
(a) Loose trefoil knot. (b) Tight trefoil knot. The tight knot has higher entropy: there are more ways to form the tight knot than the loose knot.

only discovered in 1997 that the consumption of energy results in a reduction of knotted or supercoiled DNA below equilibrium levels. This should not be surprising, of course, because life cannot exist solely through equilibrium processes. Vologodskii has further discussed type II topos in the topological context of how the enzymes (which only interact *locally* with their DNA binding site) "know" that cutting and passing a certain strand over another will remove a knot or relax a supercoil.[18] Type II topos do not have the advantage of "seeing" far down the DNA strands and determining which strand, cut where and moved there, will remove a knot.

The most likely explanation proposed by Vologodskii for the ability of type II topoisomerases to reduce the number of knots relies on (1) the fact that the *entropy* of the chain increases if a knot is localized in a small part of the chain[19] and (2) the enzyme's ability to recognize and bind to these hairpins. In everyday experience we note that knots that form in tangled string tend to form tighter and smaller knots as parts of the string are pulled, unless a very strategic pull is initiated. This would point toward entropy as an important factor: there are more ways to form a tight knot than a loose one. Does this hold for knots in DNA? Apparently so. Figure 14.18 shows the difference between a tight trefoil knot and a loose one. The reference just cited found that more than 55% of the simple trefoil knots formed in 500-unit FJCs were confined to 50 units or less, indicating that the chain entropy is higher for these tighter knots. In double-stranded DNA, such an entropy increase drives the chain to form these sharp bends (Figure 14.19a) but is opposed by the bending potential energy increase. The latter opposition is compensated, in the case of topoisomerase II, by the binding of the enzyme to the bend, which reduces the potential energy. Figure 14.19b shows that such sharp bends indeed characterize the binding of type II topoisomerase from yeast.

Knots also occur in proteins, but they appear to be rare. Before 2002, only two proteins were considered to have a trefoil knot. These knots were shallow, with only a few residues at one end of the chain extending through a loop exposed on the protein surface. In 2002 the first "deep" trefoil knot was found by Nureki et al., using X-ray crystallography of RNA 2'-*O*-ribose methyltransferase from *Thermus thermophilus*.[20] This knot can clearly be seen in Figure 14.20, but it is evident that even this knot could be released by an appropriate tug on the right polypeptide section in the right direction. The required distance is large, however, so random thermal motions may not be able to do this work on a reasonable time scale, especially if opposed by attractive forces (enthalpic) between chains.

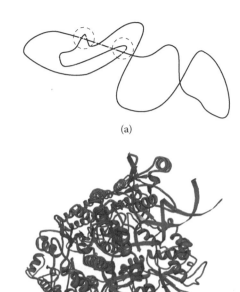

(a)

(b)

FIGURE 14.19
Knots and supercoils in DNA may be detected by sharp hairpin bends (circled in (a), to which type II topoisomerases preferentially bind. Confining the knot to a small area in the strand is entropically favored. (b) *Saccharomyces cerevisiae* Topo II and a gate-DNA segment. The DNA (blue coil) bends 150°, accompanied by large protein conformational changes, which position the DNA backbone near a reactive tyrosine and a coordinated Mg^{++} ion. (PDB ID 2RGR, by Dong, K. C., and J. M. Berger, *Nature* 450, 1201–1205, 2007.)

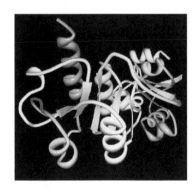

FIGURE 14.20
Trefoil knot in a *Thermus thermophilus* RNA methyltransferase enzyme. The knot is on the left side in yellow and salmon pink. (PDB ID 1IPA, by Nureki, O. et al., *Acta Cryst. D*, 58, 1129–1137, 2002.)

14.8.5 Knots, Topology, and Equilibrium

The presence of knots or supercoils in biological polymers must be governed by several considerations related to the topology. First, a knot may be prevented or facilitated by the synthesis process. If a polypeptide is synthesized from amino end first, the time-dependent formation of three-dimensional structures, assisted in some cases by *chaperonins* or other conformation-affecting factors, may prevent or promote knot formation. Second, once a knot is formed or not formed, it may be difficult (i.e., take an extremely long time) to reverse the process in the absence of enzymes that can temporarily cut a chain. This implies that there may never be any equilibrium in the absence of knot- or supercoil-removing enzymes.

14.9 FREE ENERGY (GIBBS)

The conformation of biological polymers such as nucleic acids and proteins must depend on potential energy and enthalpy, not just entropy. The limitations of the FJC model we have already encountered should convince us of this fact. Some simple measurements of conformational properties of proteins and nucleic acids also indicate this fact. Equilibrium, as well as many kinetic, processes at constant pressure are determined by the *Gibbs Free Energy*,

$$G = H - TS \qquad (14.48)$$

This simple equation shows an almost universal way to separate the contributions of entropy from enthalpy: the temperature. We have shown that the enthalpy, consisting primarily of the potential energy of the system, can be approximately equated with the heat absorbed or emitted by a system during a change. We now look at the free energy, which governs "chemical" reactions and the thermodynamic, but very real force in a system of many particles.

14.9.1 Free Energy Differences Govern Reactions

We have presented the entropy in the context of biopolymers as if its absolute numerical value is the common quantity used for analysis. We calculated the "absolute" value of entropy for a FJC model. This was, however, only part of the entropy. Suppose we were to cut a 500-unit polymer in half. According to Equation 14.45 or 14.47, the entropy would not change significantly. Yet we know that the two separated 250-unit polymers now have translational degrees of freedom that the single polymer did not have. In those formulas we did not even attempt to consider translational degrees of freedom of the overall polymer, as we did for the molecules in an ideal gas, or molecules in an ideal solution. Equations 14.45 and 14.47 really allow entropy *changes* to be computed only when conformational restrictions are applied to the intact polymer. Absolute values for biochemical entropies are rarely available, but entropy differences can easily be determined, at least experimentally. Differences of free energy are what govern reactions, so differences are what we focus on.

Reactions involving a system at constant pressure and number of particles are governed by the Gibbs free energy change. If a molecule in state A can convert to a state B (and vice versa),

$$A \rightleftharpoons B \qquad (14.49)$$

with G_A and G_B the corresponding (absolute) free energies, the free energy change on going from A to B is

$$\Delta G \equiv G_{final} - G_{initial} = G_B - G_A \qquad (14.50)$$

In physics we like to use notation ΔG_{BA} for the energy change, reflecting the ordering of the quantities on the right of Equation 14.50. In chemistry and biology, the tradition is ΔG_{AB}, reflecting the order of Equation 14.49. We use no subscripts, except when necessary for clarity. This forces us to understand the effect of a negative or positive quantity on the numbers or concentrations of states A and B—a good physical detail to focus on.

The time- and temperature-dependent concentrations relate to the free energy through a Boltzmann-like, exponential dependence:

$$\frac{P_B}{P_A} = \frac{N_B}{N_A} = \frac{c_B}{c_A} = \exp-\left(\frac{G_B - G_A}{k_B T}\right) = \exp-\left(\frac{\Delta G}{k_B T}\right) \tag{14.51}$$

where N_B (c_B) is the number of molecules in state B per unit volume (concentration of B). Where it is more fruitful to think in terms of probabilities of a single molecule, P_B is the probability the molecule is in state B. Several implications of Equation 14.51 should be considered:

1. The free energy is implicitly measured in terms of multiples of $k_B T$.
2. Read left to right, the ratio of number in state B to number in A is governed by the free energy difference.
3. Read right to left, the free energy difference between final and initial states is governed by the actual number of molecules in each state.

Implication 1 is important because of the strong mathematical dependence of the exponential function.

Before examining this strong exponential dependence, we need to remind ourselves of the two components of Gibbs free energy, enthalpy and entropy:

$$\Delta G = \Delta H - T\Delta S \tag{14.52}$$

which implies

$$\frac{P_B}{P_A} = \frac{N_B}{N_A} = \frac{c_B}{c_A} = \exp-\left(\frac{\Delta H - T\Delta S}{k_B T}\right) = e^{\Delta S/k_B} e^{-\Delta H/k_B T} \tag{14.53}$$

Table 14.6 shows this dependence for typical values of ΔH. The last column of this table reflects the pesky presence of entropy in the free energy. While the enthalpy for these types of interactions can be understood in terms of bond potential energies, which reflect the local structures directly involved in the interaction, the entropy may involve the entire biomolecule as well as solvent and ions. Bond energies can often be quantum mechanically calculated by considering only the nearest dozen or so atoms. The entropic fallout from breaking or making a bond can sometimes be estimated by crude models. For example, the cleavage of a noninteracting, 100-unit FJC polymer into two 50-unit FJCs roughly doubles the number of allowed states, but biological polymers that do not interact with solvent and solvent ions are nonexistent. Notwithstanding this undetermined term, $\exp-\Delta S/k_B$, the numerical differences between 10^{156} and 1.1 cannot be accounted for by a temperature-independent entropy factor.

Table 14.6 clearly shows why life and most chemistry that supports life occur near 300 K: equilibrium is manageable at 300 K but is not at 100 K. Except for the case of covalent bonds and (stronger) ionic interactions, there are always reasonable numbers of molecules in both states A and B at 300 K (given appropriate ΔS values), but not at 100 K. If biological processes are to manage the ratio of states A and B to adapt to changing conditions, they must have a reasonable number of molecules in each state to begin with. Stated a second way, biological processes cannot be faced with too large an energy payment to induce a favorable process. Stated a third way, the enthalpy difference ΔH should not be too many multiples of $k_B T$, so that random thermal processes have a reasonable chance of transiently promoting the system into an excited state of high enough energy to convert A to B and B to A.

TABLE 14.6 Enthalpy Differences and Temperature-Dependent Number Ratios[a]

Type of Change	Enthalpy Change, kJ/mole J/molecule	Enthalpy/$k_B T$ at 300 K at 100 K	$N_B/N_A = e^{-\Delta G/k_B T}$ at 300 K at 100 K
Covalent bond formation	−300	−120	$10^{52} \cdot e^{\Delta S/k_B}$
	-5×10^{-19}	−360	$10^{156} \cdot e^{\Delta S/k_B}$
Ionic interaction (attractive)	−80	−32	$8 \times 10^{13} \cdot e^{\Delta S/k_B}$
	-1×10^{-19}	−97	$10^{42} \cdot e^{\Delta S/k_B}$
H bond formation	−20	−8.0	$3000 \cdot e^{\Delta S/k_B}$
	-3×10^{-20}	−24	$3 \times 10^{10} e^{\Delta S/k_B}$
Dipole–dipole attraction	−9	−3.6	$37\ e^{\Delta S/k_B}$
	-1.5×10^{-20}	−11	$6 \times 10^4 e^{\Delta S/k_B}$
London (dispersion) attraction	−0.3	−0.12	$1.1\ e^{\Delta S/k_B}$
	-5×10^{-22}	−0.36	$1.4\ e^{\Delta S/k_B}$

[a] Bond energy values from Voet, D. et al., *Fundamentals of Biochemistry*, John Wiley, New York, 25, 1999.

14.9.2 Standard Free Energies: Chemical Potentials, ΔG^0 and $\Delta G^{0\prime}$

Consider a reversible reaction A + B $\rightleftharpoons$ C + D. At constant T and P, any Gibbs free-energy difference between final and initial states is defined by

$$dG = \sum_{i=A}^{D} \left(\frac{\partial G}{\partial n_i} \right)_{P,T} dn_i \equiv \sum_{i=A}^{D} \mu_i dn_i \tag{14.54}$$

> We ignore the difference between concentration and activity. When an expression like ln[C] appears, where [C] is a concentration, understand that a reference concentration, 1 M, has been removed from the logarithmic term.

Where μ_i is the *chemical potential* (or partial molar free energy) of component i. The value n_i is usually the number of moles of component i but could also be the number of molecules. The chemical potential then describes how much the free energy changes when the number of particles changes. In the case of the reaction A + B $\rightleftharpoons$ C + D, $dn_C = dn_D = -dn_A = -dn_B = dn$, so

$$dG = (\mu_C + \mu_D - \mu_A - \mu_B)dn \tag{14.55}$$

If the change in number of moles, dn, is 1 mole and the initial concentrations are 1 M,

$$\Delta G^0 = \mu_C + \mu_D - \mu_A - \mu_B \tag{14.56}$$

The chemical potentials for dilute solutions and gases are defined such that they reflect both the internal characteristics of the component molecules as well as their concentration:

$$\ln\left[C \right] = \frac{\mu_C - \mu_C^0}{RT} \tag{14.57}$$

where μ_C^0 is the chemical potential of a 1 M solution of C and R, rather than k_B, is used because μ_C^0 is a molar quantity. An alternate form of the chemical potential for dilute solutions or gases is $\mu = k_B T \ln(\frac{c}{c_0}) + \mu^0$, where c_0 is the reference concentration and energies are per molecule. (From an entropy point of view, the free energy, per mole, of an ideal solution of a molecule increases as concentration increases because the entropy decreases.) If the concentrations are not the standard 1 M, Equation 14.56 becomes

$$\Delta G = \left(\mu_C^0 + \mu_D^0 - \mu_A^0 - \mu_B^0\right) + RT \ln\left(\frac{[C][D]}{[A][B]}\right) \equiv \Delta G^0 + RT \ln\left(\frac{[C][D]}{[A][B]}\right) \qquad (14.58)$$

Some molecular species use other standard concentrations, e.g., $[H^+]$.

Recalling the expression $\Delta G = \Delta H - T\Delta S$, we recognize the concentration term as part of the entropy, which it properly is, but ΔG^0 also contains entropy related to factors other than bulk concentration. Equation 14.58 has eliminated the chemical potentials in favor of the Gibbs free energy. From this point on, we make no further use of the chemical potential.

Biology immediately causes complications for standard chemical tradition. For an important biochemical reaction like $ATP + H_2O \rightleftharpoons ADP + P_i$ the respective standard states would be the rather un-biological $[ATP] = [ADP] = [P_i] = 1$ M, while $[H_2O] = ?$ Of course water as a solvent cannot be a "standard" 1 M, but will be its own standard, 55.5 M. There are two immediate problems to be solved, one easy and one hard. First, the concentration of water virtually never changes, so why carry this constant value around? Equation 14.58 applied to ATP hydrolysis would appear as $\Delta G = \Delta G^0 + RT \ln\left(\frac{[ADP][P_i]}{[ATP][H_2O]}\right)$. Rather than constantly putting $[H_2O] = 55.5$ M into free energies, we just set $[H_2O] = 1$, put a prime on the standard free energy, and ignore the problem of units in the logarithm:

$$\Delta G = \Delta G^{0'} + RT \ln\left(\frac{[ADP][P_i]}{[ATP]}\right) \qquad (14.59)$$

So, the bottom line for the first problem is that a prime on the standard free energy indicates that at least one component of the reaction has had a special value inserted for its concentration (activity). The second problem is more serious, and we assign a new section label to it.

14.9.3 ATP: Enthalpy and Free Energy of Hydrolysis

The enthalpy and free energy of ATP is fundamental to biophysical systems. ATP is the common energy currency of life—or should we say "free-energy" currency. ATP "energy" is, to a certain extent, still an unsolved problem. The accepted numbers for ATP free energies are shown in Table 14.7.

As we can plainly see, the free energy of ATP hydrolysis is not a specific number but rather depends on concentrations existing under cellular conditions. Cells normally maintain a high level of ATP, keeping the concentration ratio $\frac{[ATP]}{[ADP][P_i]} \sim 500 M^{-1}$ or more and a larger (negative) effective free energy.[21] Not all cellular conditions are those of Table 14.7, however. From our arguments about concentrations in small cellular compartments like mitochondria, we should even question whether the bulk concentration is an appropriate number to use. Though this is a question with no definitive answer, we move on to implications of these concentrations.

14.9.4 Coupling of ATP Hydrolysis to Other Reactions

(The following is adapted from Stryer's Biochemistry textbook.[21]) Suppose a molecule A can be converted to B, but this requires a free-energy input of 20 kJ/mole:

$$A \rightleftharpoons B \qquad \Delta G^{0'} = +20 \, kJ/mole \qquad (14.60)$$

In isolation the equilibrium concentration ratio would be

$$\frac{[B]}{[A]} = e^{-\Delta G^{0'}/RT} = e^{-20,000/8.31\cdot300} = 3 \times 10^{-4} \qquad (14.61)$$

TABLE 14.7 ATP Free Energies (the Usual Values)

Reaction	$\Delta G^{0'}$ (kJ/mole)	ΔG (kJ/mole)[a]
$ATP^{4-} + H_2O \rightarrow ADP^{3-} + HPO_4^-$	−30.5	−48.1
$ATP^{4-} + H_2O \rightarrow AMP^{2-} + HPO_3PO_4^{2-}$	−32.2	

Note: $[ATP] = 3 \times 10^{-3}$ M (Voet, D. et al., *Fundamentals of Biochemistry*, John Wiley, New York, 362–363, 1999.)

[a] Values under cellular conditions: $[ADP] = 8 \times 10^{-4}$ M, $[P_i] = 4 \times 10^{-3}$ M.

Suppose, however, that the reaction were coupled to ATP hydrolysis:

$$A + ATP + H_2O \rightleftharpoons B + ADP + P_i + H^+ \tag{14.62}$$

The free energy for this coupled reaction is now $\Delta G^{0\prime} = (20-48.1)$ kJ/mole $= -28$ kJ/mole. The new equilibrium concentrations are given by $\frac{[B][ADP][P_i]}{[A][ATP]} = e^{-\Delta G^{0\prime}/RT} = e^{+28,000/8.31\cdot300} = 8\times10^4$. Putting in the concentration values from Table 14.7 $\frac{[ATP]}{[ADP][P_i]} \sim 900$, the new equilibrium concentration ratio is

$$\frac{[B]}{[A]} = 900 \cdot e^{-\Delta G^{0\prime}/RT} = 900 \cdot e^{+28,000/8.31\cdot300} = 7\times10^7 \tag{14.63}$$

> In equation 14.62 for ATP hydrolysis, we have explicitly included the proton from the water, which may be bound to Pi, depending on the pH. This foreshadows the final difficulty with ATP hydrolysis.

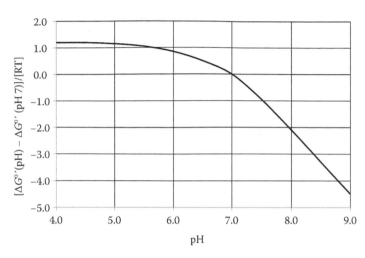

FIGURE 14.21
Dependence of Gibbs free energy for ATP hydrolysis on pH of solution. (Based on Mahler, H. R, and E. H. Cordes, *Biological Chemistry*, 2nd ed., Harper & Row, New York, 27, 1971.)

The difference between Equations 14.61 (uncoupled reaction) and 14.63 (ATP coupled) is striking, to say the least. Coupling of ATP hydrolysis to the reaction has increased the equilibrium product yield by a factor of 10^{11}. This factor comes from using most favorable values for ATP-contributed free energy. Stryer's values give a still large yield increase of $\sim 10^8$.

We have not stated exactly how ATP hydrolysis can be "coupled" to another reaction, $A \rightleftharpoons B$. We cannot simply add ATP to any reaction and expect it to dramatically shift the concentrations. It must clearly involve an enzyme that binds each of A, ATP, and possibly B and ADP: something must be done to bring A and ATP together to the same point in the cell. Enzymatic binding is the first secret to biological catalysis. Such binding should, in principle, throw into doubt all of the ideal solution theory we have been using in this section. It does, but we are still going generally in the right direction. We continue this subject in the next chapter.

We have still not broached the "final" difficulty with ATP hydrolysis. The earlier hint was that ATP hydrolysis requires the splitting of water into OH^- and H^+. The former is incorporated into the inorganic phosphate, P_i, but the latter may also be bound, depending on the pH. This innocuous "depending on pH" phrase informs us that ATP hydrolysis should also include water ionization, $H_2O \rightleftharpoons OH^- + H^+$. This further brings in all the other pH-affecting and pH-maintaining reactions of the cell. We should also consider the dependence on ion concentrations, such as $[Mg^{++}]$, but we won't. Figure 14.21 shows the dependence of $\Delta G^{0\prime}$ for ATP hydrolysis on pH. Variation of pH between 6 and 8 shifts the pH 7 value of $\Delta G^{0\prime}/k_BT$ by +1 to −2. The exponential $e^{-\Delta G^{0\prime}/k_BT}$ then changes by a multiplicative factor of about 0.36–7.4—more than a factor of 10 variation. Physiological pH changes are generally not as large as ±1 unit, but we should be wary of being too dogmatic about ATP hydrolysis energy values. Perhaps the best general lesson we should take home is to determine, as well as we can, the appropriate value of ΔG for each individual case involving ATP hydrolysis.

14.10 ENERGY DIAGRAMS

Diagrams are central to understanding most fields of physics, especially complex ones. In biological physics, complex processes and reactions can often be better understood through the use of energy diagrams. A two-state reaction process, for example, can be minimally diagrammed as in Figure 14.22a, where the two states are shown physically separated by a horizontal distance called the *reaction coordinate*, and the vertical position of the state corresponds to an "energy." This energy can be any of a number of quantities we have encountered: *E*, *V* (potential energy), *H*, *G*, *S* (or *TS*). This diagram then informs us that the change A → B brings a change of energy and coordinate. In principle, we can plot absolute values of the energies (if we know them), but in the most common case only relative energies matter. (Recall that potential energy is defined only to within an additive constant, in any case.) The equilibrium between the two states, A and B, is adequately described by the (free) energies of the two states.

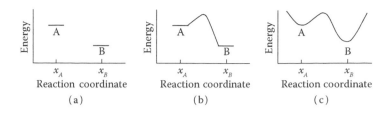

FIGURE 14.22
Energy-level diagrams for a reaction A $\rightleftharpoons$ B. The vertical axis measures any of the energies, E, V, H, TS, or G. The horizontal axis measures the "reaction coordinate," which could be a real distance (in meters), an angle, or a complicated set of coordinates involving many atoms. (a) A diagram showing only the energies of the two states may be sufficient if only equilibrium is considered. (b) Energy barriers usually exist between states. (c) The energy is, of course, a continuous function of coordinates. If the properties of the initial and final states need to be understood, the entire energy function should be drawn.

Figure 14.22a would give us no hint as to how equilibrium between the two states occurs. The time dependence of the equilibration process, whether it requires 10^{-12} s or 10^{+12} s, is determined by the energy function for values between x_A and x_B (Figure 14.22b). Most often an energy barrier exists between the two states. This energy barrier usually determines the rate at which equilibrium is reached, but not the equilibrium concentrations of A and B, which depend only on the difference between energy values at x_A and x_B. Figure 14.22c shows the energy as a continuous function of the reaction coordinate. A biomolecule in a state A or B is never, as we have seen, fixed at some particular position in space or angle. To describe the mobile behavior of the molecule in states A and B we need the shape of the energy curve for values of reaction coordinate below x_A and above x_B.

Energy diagrams may seem like a trivial matter; after all, we developed the thermodynamic equations we needed without diagrams. Diagrams have the great advantage of adding *intuitive* understanding of complex biological processes that are hard to extract from equations. We can look at an energy diagram where B is much lower in energy than A and immediately know that it would be very hard (improbable) for a molecule in state B to get to A. We can also look at the diagram in Figure 14.22c and speculate that the process of getting from A to B might depend both on the height of the barrier and its width. Guess what? We would have just discovered a reaction mechanism that none of the thermodynamics we have sketched out so far predicts: quantum-mechanical tunneling. There is, after all, no evidence of any position coordinates in the free-energy expressions for equilibrium. We will discuss this more in Chapter 15.

14.11 BOLTZMANN DISTRIBUTION IF NUMBERS VARY: GIBBS DISTRIBUTION

We have seen the Boltzmann distribution and Boltzmann factor for a system of particles of one type. The probability a particle is in a state with energy ε_i is

$$P(\varepsilon_i) = C g_i e^{-\varepsilon_i/k_B T} \tag{14.64}$$

This equation is the same as Equation 14.12, except that the replacement $E_i \rightarrow \varepsilon_i$ has been made to clarify that the energy refers to the energy of the single particle.

C is a normalization constant fixed by $\sum_i^{\text{states}} P(\varepsilon_i) = 1$, where the sum is over all the energy states.

The corresponding integral expression should be used if energies are continuous. The average number of particles in each state, $\langle n_i \rangle$, is

$$\left\langle n_i \right\rangle = C' g_i e^{-\varepsilon_i/k_B T} \tag{14.65}$$

where

$$\sum_{i=1}^{\text{states}} \left\langle n_i \right\rangle = N_{\text{part}} \tag{14.66}$$

and N_{part} is the total number of particles (assumed fixed). We know that the actual number of particles in each state is not always equal to the average. What is the probability that *any* number of particles, and not just the average number, is observed in state i? The answer is governed by the chemical potential:[22]

$$P(n_i) = C'' e^{-(\varepsilon_i - \mu)n_i / k_B T} \qquad (14.67)$$

(The independent "variable" in this distribution is the number of particles in a state, n_i, not the energy of the state. That number depends on the energy of the state, ε_i.) Equation 14.67 asserts that occupation-number probability is highest for low energies and decreases exponentially with increasing energy, with energy scale determined by the chemical potential μ. For an ideal gas or dilute solution, $\mu = k_B T \ln(c/c_0) + \mu^0$, where the reference concentration c_0 is usually 1 M, so the chemical potential should be thought of as a concentration- and temperature-dependent energy. Problem 14.10 examines these dependencies of the chemical potential for Na^+ and Cl^-.

14.12 EQUILIBRIUM CONSTANTS IN IDEAL, UNIFORM SOLUTIONS

The free-energy change of a reaction provides the drive toward equilibrium. This drive is described by Equations 14.65 through 14.67 and defines the *equilibrium constant* K_{eq}, for an ideal-solution reaction that takes place in uniform solution. Reactions that involve one, two, and three reactant molecules are called *unimolecular*, *bimolecular*, and *trimolecular*, respectively. For a unimolecular reaction,

$$A \rightleftharpoons B \qquad (14.68)$$

$$\frac{[B]}{[A]} = \exp\left[\frac{\Delta G - \Delta G^0}{RT}\right] \qquad (14.69)$$

$$\Delta G = 0 : K_{eq} \equiv \left(\frac{[B]}{[A]}\right)_{equil} = \exp\left[\frac{-\Delta G^0}{RT}\right] \qquad (14.70)$$

In this case, one type of molecule is present on the left and right sides of the reaction, the concentration ratio has no units, and the equilibrium constant (obviously not a constant because it strongly depends on temperature) is a pure number. A large value of K_{eq} means formation of the products is favored. When this simple form of the equilibrium applies, K_{eq} and the concentration ratio should not depend on initial concentrations of A.

> If two molecules of A take part in the reaction, [A] is replaced by $[A]^2$.

For a bimolecular reaction with two reactants and two products,

$$A + B \rightleftharpoons C + D \qquad (14.71)$$

$$\frac{[C][D]}{[A][B]} = \exp\left[\frac{\Delta G - \Delta G^0}{RT}\right] \qquad (14.72)$$

$$K_{eq} \equiv \left(\frac{[C][D]}{[A][B]}\right)_{equil} = \exp\left[\frac{-\Delta G^0}{RT}\right] \qquad (14.73)$$

We explore these and other equilibrium reaction equations much more in Chapter 15.

14.13 FREE ENERGY: ENTHALPY, ENTROPY, MIXING, GRADIENTS, POTENTIAL, AND ATP

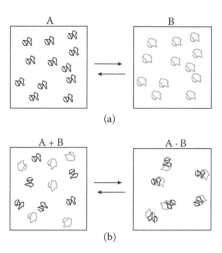

The free-energy difference "drives" a reaction. If molecules are initially present at their equilibrium concentrations, then $\Delta G = 0$ and there is no "drive": equilibrium is already present. In Figure 14.22 this would correspond to equal energies at A and B, with "energy" identified as free energy. If initial concentrations differ from equilibrium, $\Delta G \neq 0$ and the reaction proceeds until equilibrium is reached. There are two main components to this drive toward equilibrium: enthalpy and entropy, $\Delta G = \Delta H + T\Delta S$. The enthalpy is associated with the internal energy and volume of a molecule, properties local to the molecule itself. The entropy can also have a part localized to the molecule itself. We have seen the example of a double-stranded coil of DNA forming a "random" coil to maximize its molecular entropy. In the case of true unimolecular changes, all enthalpy and entropy changes are local to the molecule concerned. ("Local" can include changes to nearby solvent molecules.) Furthermore, one molecule present in the (ideal) solution is not affected by others, whether there are 1,000 or 10^{20}, as long as the solution is ideal. If a unimolecular reaction is carried out at two different molecular concentrations, the enthalpy and entropy change per molecule is the same. (Any "nonlocal" entropy associated with the concentration of the molecule does not matter, because this entropy does not change during the reaction.) Figure 14.23a diagrams a unimolecular change.

FIGURE 14.23

Entropy changes. (a) The equilibrium entropy change (per molecule) in a unimolecular reaction, e.g., a protein conformational change, does not depend on concentrations. If the initial concentration of A is doubled, no thermodynamic quantities change, as long as the solution is ideal. Such a conformational change could be initiated by addition of 30% ethanol to an aqueous solution. (b) The entropy change (per molecule) in a bimolecular reaction, e.g., protein binding to protein, A + B → A·B, can be partly due to intramolecular entropy changes—conformations available to the free protein A or B may no longer be available in the bound form A·B—but there is also a "trivial" entropy change due to concentrations. The bound forms are no longer free, for example, to occupy some of the spatial states they previously could. Furthermore, the extent of the reaction (percent of A and B reacted) depends on concentrations of both molecules. These changes are normally factored out (approximately), as in Equation 14.58.

14.13.1 Mixing Entropy

The case of bimolecular reactions is more complex. The enthalpy is still a quantity associated with the individual molecules (and their local solvent shells), but the entropy must now include a part due to the (possibly variable) concentrations. If A + B → C, the enthalpy of A and B is converted to an enthalpy associated with C. That latter enthalpy could be the same as the initial enthalpy of A + B, could be greater or lesser. The entropy is a bit different. The solution entropy (the set of accessible states) of some molecules has entirely disappeared—half, if $A_0 = B_0$ and 99+% of the molecules react. Furthermore, the number of molecules that effectively disappears from solution depends on initial concentrations of both A and B. (Consider the case where $A_0 = 2B_0$.) Looked at another way, molecule B that was previously free in solution has now been affixed to molecule A: it has lost some degrees of freedom. This solution or (un)mixing entropy change does not care what the molecules are or how they interact, as long as one type of molecule disappears. This entropy decrease always imposes certain conditions on the reaction. If the entropy decrease cannot be paid for by other energy currency, the reaction does not, on average, proceed. Often this solution ordering entropy can be partly compensated for by increased degrees of freedom of the product molecule. If, for example, H + H → H_2, the translational entropy of one hydrogen has disappeared, but the product now has rotational and (at high temperatures) vibrational degrees of freedom that the reactants did not have. One translational degree of freedom has been replaced by two internal energy-containing rotational modes. Details, such as whether or how quickly these new modes acquire energy from the initial collision or from other sources, are complicated.

The very simplest "reaction" is the mixing of two noninteracting components: an ideal solution of A is mixed with an ideal solution of B and nothing happens. The initial volumes are V_A and V_B, respectively, and the mixed volume is assumed to be $V_{mixed} = V_A + V_B$: the volume change is zero.

This would seem to be no reaction at all. Indeed, no chemical bonds form or change. However, the free energetics of the system does change because each of the molecules of types A and B has more available volume after mixing: their entropy has increased. We write this mixing entropy without derivation:

$$\Delta S_{mix} = -k_B \left[N_A \ln \frac{N_A}{N} + N_B \ln \frac{N_B}{N} \right] \tag{14.74}$$

where $N = N_A + N_B$. (See any statistical mechanics, thermodynamics, or physical chemistry book.) What do we do if we mix *solutions* of A and B? We cannot neglect the water or the solvent in the system. Suppose 0.500 L of 1.00 M A and is mixed with 0.500 L of 1.00 M B. There is 1.000 L of 55.5 M H_2O in the system, or 55.5 moles, or $55.5 N_{Avog}$ molecules of H_2O. The numbers of A and B are each $0.500 N_{Avog}$. Putting these numbers into Equation 14.74, including the water, we get

$$
\begin{aligned}
\Delta S_{mix} &= -k_B N_{Avog} \left[0.5 \ln \frac{0.5 N_{Avog}}{(56.5) N_{Avog}} + 0.5 \ln \frac{0.5 N_{Avog}}{(56.5) N_{Avog}} + 55.5 \ln \frac{55.5 N_{Avog}}{(56.5) N_{Avog}} \right] \\
&= -k_B N_{Avog} \left[-2.364 - 2.364 - 0.991 \right] \\
&= 5.72 k_B N_{Avog} = 5.72 R
\end{aligned}
$$

Not surprisingly, the entropy has increased. Again, assuming no chemical (enthalpic) interaction, the free-energy increase would be

$$\Delta G_{mix} = -T \Delta S_{mix} \tag{14.75}$$

so the free energy of the system has decreased by $-5.72 k_B T$ per molecule of (A + B). In this case, the contribution of water is clearly significant. In homework we show that when concentrations of A and B get below about 0.1 M the water term becomes negligible. The case of the entropy and free energy associated with potassium gradients across cell membranes and the entropy change when an initially localized concentration of a molecule diffuses outward is also considered.

14.13.2 Membrane Gradients

Equations 14.74 and 14.75 can be written in forms more amenable to the separation of solutions by the cell membrane. Because concentrations of a wide variety of molecules differ inside and outside the cell, a chemical entropy and free energy is stored in the gradient. For one component,

$$\Delta G = G_{in} - G_{out} = k_B T \ln \left(\frac{[A]_{in}}{[A]_{out}} \right) \tag{14.76}$$

per molecule of A. This difference is an entropic mixing free-energy difference. The potassium ion concentrations inside and outside a cell are typically about 120 and 5 mM, respectively. The free energy associated with this physical separation of potassium ions is therefore $\Delta G_K \cong k_B T \ln\left(\frac{120\,mM}{5\,mM}\right) = 3.2 k_B T$.

14.13.3 Separation of Charge

In physics we learn that if you maintain a gradient of ions across a thin membrane, you also must maintain an electrical potential difference. Though we do not explore electrical effects in cells much further in this text, we add this electrical term to Equation 14.76:

$$\Delta G = k_B T \ln \left(\frac{[A]_{in}}{[A]_{out}} \right) + Z_A e \Delta V \tag{14.77}$$

where Z_A is the number of elementary charges on A, $e = 1.602 \times 10^{-19}$ C, and $\Delta V = V_{in} - V_{out}$ is the electrical potential difference in volts/m or joules/coulomb (J/C). If a pseudoequilibrium holds for the K^+ ions, $\Delta G = 0$ and

$$\Delta V = \frac{k_B T}{Z_A e} \ln\left(\frac{[A]_{out}}{[A]_{in}}\right) \tag{14.78}$$

Equation 14.78 is often referred to as the *Nernst potential*. The value of this potential for the given K^+ concentrations is about -80 mV, with the inside negative relative to the outside. This K^+ potential is sometimes referred to as the *resting potential* of cells. Any electrical potential difference across a cell membrane can directly be used to do electrical work, so this Nernst potential, arising from entropic free-energy differences, can directly do work. We describe a case of a membrane-bound rotary motor powered by this electrical potential in Chapter 16. Of course, maintaining the potential as work is done demands compensating work be done by the cell.

14.13.4 Membrane Potentials and Work

The electrical potential can obviously perform work. Suppose the electric field transports a charge Q across a membrane. The electric potential energy change would be $Q\Delta V$, but this entire amount is not necessarily "useful" work. Frictional losses must be considered. Ions are sometime transported across membranes through water-filled channels (Figure 14.24a). If the ion has radius R and moves at speed v, there is a resistive fluid frictional force of $-bv = -6\pi\eta Rv$ and a frictional energy loss of $W_f = -6\pi\eta Rvd$. If the ion is subject to an electric force, $Q\Delta V/d$, but otherwise free in the fluid, the ion *very* quickly accelerates to a speed given by $6\pi\eta Rv = Q\Delta V/d$ and virtually all the electric work will have been converted to heat. The ion has been moved, which may have been the objective, but the work was accomplished only against frictional resistance. If the charge is coupled to a conformational change in a protein (Figure 14.24b), the electric force also does work against the opposing conformational force F_C. If the movement is slow ($6\pi\eta Rv \ll F_C$) most of the electric potential energy is converted to "useful" conformational work. The key issue, as usual in thermodynamic systems, is the rate of the energy conversion: slower is more efficient.

While the model of an external charge moving through a water channel, but coupled to a membrane-bound protein conformational change, is not likely, the case of voltage-controlled membrane protein conformational changes is central to certain types of ion gating in membranes. A voltage-induced conformational change (or aggregation) causes the protein to allow specific ions, which normally cannot pass the membrane, to flow through a channel created by the protein. An example is the KvAP protein, a voltage-gated potassium ion channel from house mouse (Figure 14.25).

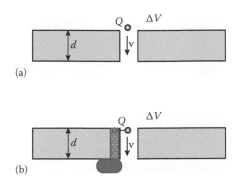

(a)

(b)

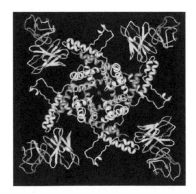

FIGURE 14.24
A gradient of concentration across a membrane can do work by driving molecules across. If the molecules are ions, a potential difference is set up. (a) Electrical work can be converted to heat and disorder, if an ion is simply moved through water. (b) If the charge is coupled to the conformation of a protein, electrical potential energy can be converted to conformational potential energy: the protein can be converted into an "active" form.

FIGURE 14.25
KvAP voltage-gated K^+ ion channel from house mouse. Structure displayed using Protein Workshop software. The channel is oriented perpendicular to the page. A row of 4 K^+ ions is located in the center of the channel. (PDB ID 2AOL by Lee, S.Y. et al., *Proc. Natl. Acad. Sci.*, 102, 15,441–15,446, 2005.)

In such realistic cases, the resistive force is often a combination of an elastic conformational force and dissipative forces resulting from motion of atoms through the lipid environment or through the protein itself. Such gating is coupled to changes in the membrane potential, which normally occur on time scales of milliseconds or longer. You can perhaps see the next step in the quantitative description of this process: the electric potential, created by ion gradients, is coupled to a conformational equilibrium of the gating protein, which then affects ionic equilibria across the membrane, which perturbs the ion gradient, etc.

14.13.5 ATP Free Energy (Again)

The expression for the free energy of an ideal solution reaction like the hydrolysis of ATP, $\Delta G = \Delta G^{0'} + RT \ln\left(\frac{[\text{ADP}][\text{P}_i]}{[\text{ATP}]}\right)$, indicates that under "equilibrium" conditions, the ATP free energy available to do work depends on concentrations. We understand, of course, that the living cell is never in equilibrium, but we would like to believe that if a steady enough state is maintained, where the cell constantly produces and consumes ATP, ADP, and P_i but maintains concentrations of each for a sufficiently long time, we can use the preceding free-energy expression to estimate how much work an ATP molecule could accomplish. This expression is still our best guess at the energy available from ATP, but we can see that there are serious, as yet unsolved, issues that need addressing.

Molecular motors use ATP energy that bind ATP and ADP (Chapter 16). The binding constants of protein for ATP and ADP almost always differ. This perturbs the local "equilibrium" of ATP and ADP. Furthermore, the proteins also bind other molecules. When they do, the binding constants for ATP and ADP usually change. This is a normal part of a molecular motor cycle. Suppose these binding energies are comparable to the ATP free energies from Table 14.7. Do we still know how much energy is provided by our ATP "fuel"? ATP in cells is not like a tank of gasoline in an automobile.

14.14 POINTS TO REMEMBER

Statistical thermodynamics is a fearsome, useful discipline. It gives target values of biosystem random-event computations, in the limit of long times and/or many particles and events and provides the framework for the remaining chapters of this book.

- The equipartition theorem: $\langle E_{\text{mode}} \rangle = \frac{1}{2} k_B T$.
- RMS molecular speeds of molecules in ideal gases range from 20 to 2,000 m/s. If you understand the equipartition theorem and the meaning of temperature, you should know if the same numbers apply to ideal solutions.
- The Maxwell–Boltzmann distribution allows for molecules with speeds $\gg v_{\text{RMS}}$.
- The Boltzmann factor and distribution: $P(E_i) = Ce^{-E_i/k_B T}$ $\frac{P(E_2)}{P(E_1)} = \frac{g_2 e^{-E_2/k_B T}}{g_1 e^{-E_1/k_B T}} = \frac{e^{k_B S_2} e^{-E_2/k_B T}}{e^{k_B S_1} e^{-E_1/k_B T}}$ $= e^{[T(S_2-S_1)-(E_2-E_1)]/k_B T}$.
- $dS_{\text{Clausius}} = \frac{dQ}{T}$.
- $S_{\text{Boltzmann}} = k_B \ln \Omega$.
- Mixing entropy.
- Approximate values of entropy types (Table 14.3).
- How to count states in an FJC polymer and use of spreadsheets to simulate random polymers.
- $G = H - TS$.
- Free energy and equilibrium constant.
- Approximate strengths of bonds (Table 14.6).
- Approximate ATP free energy: –30–50 kJ/mole.
- Energy diagrams.

14.15 PROBLEM-SOLVING

PROBLEM 14.1   DISTRIBUTION NORMALIZATION

Normalize the distribution function for the discrete harmonic oscillator (Equation 14.15), remembering you are doing the discrete case: $\sum_{n=0}^{\infty} P(E_n) = 1 = C\sum_{n=0}^{\infty} e^{-(n+\frac{1}{2})\hbar\omega/k_B T}$.
$\left(\text{Hint:} \sum_{n=0}^{\infty} e^{-\theta(2n+1)} = \frac{1}{2\sinh\theta}\right)$

PROBLEM 14.2   INCORRECT DISTRIBUTION

Suppose the Maxwell–Boltzmann speed distribution were written $P(v)dv = C\exp\left(-\frac{1/2mv^2}{k_B T}\right)dv$, i.e., someone forgot the v^2 term in front of the dv.

1. In the actual M–B distribution, the relation between the three characteristic speeds is $v_{RMS} > v_{avg} > v_{mp}$. What is the relation for this new distribution?
2. Discuss whether or not this modified speed distribution could possibly describe the speeds in some system other than an ideal gas? If yes, describe that system.

PROBLEM 14.3   FJC ENTROPY

Exercise 14.2 placed minimal self-avoidance restrictions on the conformation of polymer units. Consider the case of the simplistic three-dimensional polymer.

1. Calculate the entropic free-energy differences (versus a straight polymer, $T\Delta S$) of a 20-, 2,000-, and 20,000-unit polymer.
2. Compare to the partially self-avoiding results from Exercise 14.3.

> In 14.3 and the following problems, keep in mind the limitations of the simplified jointed chain models, as applied to DNA. Such models have associated computations that can readily be performed but are not optimized for stiff polymers like DNA. The serious student should be prepared for more complex models that can easily be found in the literature.

PROBLEM 14.4   PERSISTENCE LENGTH

DNA persistence lengths: double- versus single-stranded and entropic implications. The text discussed the persistence length of double-stranded DNA, which is about 50 nm or 150 base pairs. (See also Table 13.3.) The persistence length of single-stranded DNA is almost an order of magnitude smaller, about 4.6 nm at an NaCl concentration of 10 mM. (Refer to Rechendorff et al., *Journal of Chemical Physics* 131 (2009):095103, as necessary.) Using the crude entropy model of Exercise 14.3, take the N-unit double-stranded DNA, separate it into two single strands, and calculate the entropy of the two single strands. Compare the entropy to that of the parent double-stranded DNA. Carefully consider how to treat the shortened persistence length and the entropy of a system with two strands.

PROBLEM 14.5   DNA ENTROPY

Double-stranded DNA conformation: entropic considerations. A polymer of infinite persistence length is straight and has very low entropy. Table 13.3 shows that A DNA is stiffer than B DNA. Quantitatively compare the entropies of 10,000-base pairs A and B DNA. Use the jointed chain or other model, as necessary.

PROBLEM 14.6 DNA KNOTS

As discussed in the text, knots tied in double-stranded DNA are hard to tighten and slip easily.

1. Explain this in terms of DNA mechanical and electrical properties (~50 words).
2. Using the persistence length of DNA, estimate the minimum radius of curvature of double-stranded DNA (without applied force in excess of entropic forces).
3. Find the ratio of the radius in 2 to the radius of the double-stranded DNA. Compare to a common rope.

MINIPROJECT 14.7 ENTHALPY EFFECTS ON RANDOM WALKS: A SPREADSHEET SIMULATION (GRADUATE PROBLEM)

Figure 14.14 (upper) shows x and y positions of a 50-segment, three-dimensional polymer, where each successive segment is displaced from the preceding by one unit length, with randomly chosen components of –1, 0, or +1 in the x, y, and z directions. You may want to refer to the spreadsheet simulator in Figure 13.7, the descriptive text for Figure 14.14, and Tables 14.4 and 14.5, in this miniproject.

1. Construct a spreadsheet that simulates such a structure for a 50-segment structure. You do not have to follow Table 14.4, but it may get you started. Ignore self-avoidance effects: a unit randomly placed on top of another unit is permitted. (Assume small adjustments allow atoms to not overlap.) Also construct a graphical display of the structure, as in Figure 14.14. Run the simulator enough times to recognize "typical" structures. Copy 10 such structures and place them in a 10-part figure.
2. Copy the spreadsheet in 1, but rather than completely randomly chosen steps, let step i be given by $\Delta \vec{r_i} = \frac{\Delta \vec{r}_{\text{random}} + \alpha \Delta \vec{r}_{i-1}}{|\Delta \vec{r}_{\text{random}} + \alpha \Delta \vec{r}_{i-1}|} L_{\text{step}}$, where α is an adjustable parameter. (Again, you may find a different, better method.) In your spreadsheet the column of x, y, and z steps keeps a running tab on the displacement direction for each unit. Create "representative" 10-part figures as in part 1 for three values of α from 0 up to a value where, on average, the direction of the last of the 50 steps does not deviate from the initial direction by more than about 90° (a single, gentle curve).
3. Confirm the discontinuity in Figure 14.15 at $\alpha = 1$ by simulation polymer structures slightly above and below this value, showing that they differ more in structure than expected for a small change in α.
4. Devise a method to eliminate this discontinuity, but preserve the correlation effect in a spreadsheet polymer simulator.

PROBLEM 14.8 KNOTS IN DOUBLE- AND SINGLE-STRANDED DNA

1. Qualitatively compare the knots that could be formed in double- and single-stranded B DNA.
2. Assign some relative numbers to this comparison by estimating the minimum size of knots that could be produced by a reasonable force. State and explain the approximate magnitude of this "reasonable" force. (Hint: consider persistence lengths and entropic force.)
3. Qualitatively compare the knots that could be formed in ds A DNA to ds B DNA. See Table 13.3.

PROBLEM 14.9 TEMPERATURE DEPENDENT EQUILIBRIA

1. Calculate the derivative $\frac{d}{dT}\left(\frac{N_B}{N_A}\right)$ for each row of Table 14.6 at $T = 300$ and 30 K.
2. For a protein structural change involving the disruption of 1 average hydrogen bond, calculate the change in the percentage of H-bonded structures when the temperature is increased from 300 to 310 K.

PROBLEM 14.10 CHEMICAL POTENTIAL

1. At the Web site https://www.job-stiftung.de/index.php?data-collection the standard ($c_0 = 1$ M aqueous solution) chemical potential of Na^+ at 298 K is listed as −261.89 kJ/mole; that of Cl^- is −131.26 kJ/mole. (The chemical potential for solid NaCl is −384.03 kJ/mole.) What does a negative chemical potential (as opposed to positive) mean for solutions? Look up quantities for K^+ and KCl. Discuss differences from the Na case. You may want to look through the Web site's table to note other negative and positive cases.
2. Use Equation 14.57 (or the alternate version noted) to determine the concentration at which a molecule's chemical potential is zero.
3. Determine the concentrations at which the chemical potentials of Na^+ and Cl^- are 0. (Use $T = 298$ K.)
4. If the temperature is raised by 20°C, how are the concentrations in part 2 affected?
5. μ^0 value is noted to have a temperature coefficient −58.99 J/(mole·K) for Na^+. Calculate a corrected value of the Na^+ concentration in part 3.

PROBLEM 14.11 MIXING ENTROPY

Solution mixing entropy and free energy for small concentrations.

1. Repeat the calculation for ΔS_{mix} and ΔG_{mix} after Equation 14.74 for the mixing of two non-interacting solutions if [A] = [B] = 0.200 M before mixing. Calculate the percentage of the change due to the water in the system.
2. Repeat part 1 if [A] = [B] = 2.00 mM before mixing.

PROBLEM 14.12 MEMBRANE CONCENTRATION GRADIENTS

The K^+ concentration inside and outside a cell is typically 120 and 5 mM, respectively. Consider a spherical cell of radius 4.0 μm.

1. Calculate the entropy and free energy associated with this concentration difference for the entire cell. State and justify what your "before" and "after" assumptions are. The after condition could be, for example, excess K^+ moved from inside the cell into an equivalent volume outside the cell; or moved into a *large* outside volume, or etc.
2. Calculate the potential difference across the membrane, being careful to note the sign of the potential difference. Explain the sign of the potential difference.
3. Assume that the potential difference across the cell membrane is somehow reset by the cell to an "action" potential of +45 mV (inside positive), e.g., by adjusting other ion concentrations. Calculate the free-energy difference of the K^+ ions.

PROBLEM 14.13 MEMBRANE POTENTIALS AND EQUILIBRIUM (GRADUATE PROBLEM)

1. Discuss the issue of $\Delta G = 0$ and "equilibrium" for ion gradients across membranes. You should consider other particles. The net charge per cubic micron of volume inside and outside the membrane is zero, for instance, so there must be counterions.
2. Look up membrane potential and the Goldman Equation (Wikipedia is a good start) and quantitatively discuss the correctness of referring to the K^+ potential as the "resting potential."

PROBLEM 14.14 EQUILIBRIUM: NA⁺

The Na^+ concentrations inside and outside the cell membrane are 12 and 140 mM, respectively. Calculate the Na^+ potential, ΔV, stating any simplifying assumptions you make.

PROBLEM 14.15 ENTROPY OF BINDING

1. Derive and justify an expression for the entropy and entropic free-energy change of a reaction $A + B \rightarrow C$.
2. Calculate a value for the entropic free-energy change if 1 L of a 2 M solution of A is mixed with 1 L of a 2 M solution of B and every A and B bind together. (It cannot actually happen that every single molecule binds, except possibly at $T = 0$ K; infinite, negative ΔH.)
3. Modify Equation 14.58 to apply to the reaction $A + B \rightleftharpoons C$.

PROBLEM 14.16 UNIMOLECULAR EQUILIBRIA

Transfer of molecules from water to benzene. The molar enthalpy and free energy of transfer of CH_4 from water to benzene are +11.7 and −10.9 kJ/mole at 25°C. (Such equilibria can be measured by juxtaposing layers of the immiscible solvents.) Note that for unimolecular solution equilibria, a "standard state" is usually not needed, as long as the solutions behave as ideal.

1. Calculate the entropic free energy and entropy of transfer from water to benzene.
2. Express the entropic free energy in terms of multiples of k_BT.
3. Calculate and graph the log of the fraction of the CH_4 molecules in the benzene as a function of $1{,}000/T$, with T in K units.

PROBLEM 14.17 BIOPOLYMER STRUCTURE, ENTHALPY AND ENTROPY

Proteins and DNA are biopolymers that form more or less compact structures, both free in solution and in their normal biological habitat. Both of their structures are governed by free energies, consisting of enthalpic (potential) and entropic energies. Using what you know from Chapters 5, 6, 13, 14, and any other insights, discuss the importance of enthalpy and entropy in the various levels of biomolecular structure for both proteins and DNA. You need not (but may!) do any calculations, but be as quantitative as possible and discuss some details. For example, if the major contribution to most protein secondary structure were hydrogen bonding, state this and discuss the relative contribution of entropy and enthalpy to this H bonding. If base stacking were the major determinant of a level of DNA structure, discuss entropy and enthalpy of stacking. If a level of structure is significantly different in free solution and in the natural habitat (e.g., chromosomes), focus on the latter. If one of the biopolymers has several distinct types of structure, state so, but focus on one type. About 1500 words would be an appropriate length for these discussions.

1. Protein and DNA primary structure.
2. Protein and DNA secondary structure.
3. Protein and DNA tertiary structure.
4. Protein and DNA quaternary structure (multi-molecule aggregates).

REFERENCES

1. Flavahan, W. A., Y. Drier, B. B. Liau, S. M. Gillespie, A. S. Venteicher, A. O. Stemmer-Rachamimov, M. L. Suva and B. E. Bernstein. 2016. Insulator dysfunction and oncogene activation in IDH mutant gliomas. *Nature* 529(7584):110–114.
2. Serway, R. A. 1990. *Physics for Scientists and Engineers*. Philadelphia, PA: Saunders, pp. 574–578.
3. Landau, L. D., and E. M. Lifshitz. 1969. *Statistical Physics*. Reading, MA: Addison-Wesley.
4. Moore, W. J. 1972. *Physical Chemistry*. Englewood Cliffs, NJ: Prentice-Hall.
5. Moore, W. J. 1987. *Physical Chemistry*, Harlow, England: Longman Scientific and Technical, p. 195.

6. Wikipedia, "Sackur-Tetrode Equation," Wikipedia. http://en.wikipedia.org/wiki/Sackur-Tetrode (accessed November 19, 2009).

7. Moore, W. J. 1987. *Physical Chemistry*, Harlow, England: Longman Scientific and Technical, p. 197; Landau and Lifshitz, *Statistical Physics*, p. 132.

8. Serdyuk, I. N., N. R. Zaccai, and J. Zaccai. 2007. *Methods in Molecular Biophysics*. New York: Cambridge University Press, 641–657.

9. Rechendorff, K., G. Witz, J. Adamcik, and G. Dietler. 2009. Persistence length and scaling properties of single-stranded DNA adsorbed on modified graphite. *J. Chem. Phys.* 131:095103.

10. Wall, F. T., L. A. J. Hiller, and D. J. 1954. Wheeler. Statistical computation of mean dimensions of macromolecules. I. *J. Chem. Phys.* 22:1036–1041.

11. Berg, H. C. 1993. *Random Walks in Biology*. Princeton, NJ: Princeton University Press.

12. Perkins, T. T., D. E. Smith, and S. Chu. 1994. Direct observation of tube-like of a single polymer chain. *Science*. 264:819–822.

13. Bao, X., H. J. Lee, and S. R. Quake. 2003. Behavior of complex knots in single DNA molecules. *Phys. Rev. Lett.* 91:265506, and erratum (2005), *Phys. Rev. Lett.* 95:199901.

14. Sundin, O., and A. Varshavsky. 1981. Arrest of segregation leads to accumulation of highly intertwined catenated dimers: dissection of the final stages of SV40 DNA replication. *Cell* 25:659–669.

15. Dean, F. B., A. Stasiak, T. Koller, and N. R. Cozzarelli. 1985. Duplex DNA knots produced by *Escherichia coli* topoisomerase I. Structure and requirements for formation. *J. Biol. Chem.* 260:4975–4983.

16. Voet, D., J. G. Voet, and C. W. Pratt. 1999. *Fundamentals of Biochemistry*. New York: John Wiley, pp. 732–738.

17. Gebe, J. A., S. A. Allison, J. B. Clendenning, and J. M. Schurr. 1995. Monte Carlo simulations of supercoiling free energies for unknotted and trefoil knotted DNAs. *Biophys. J.* 68:619–633.

18. Vologodskii, A. 2009. Theoretical models of DNA topology simplification by type IIA DNA topoisomerases. *Nucleic Acids Res.* 37:3135–3133.

19. Katritch, V., W. K. Olson, A. Vologodskii, J. Dubochet, and A. Stasiak. 2000. Tightness of random knotting. *Phys. Rev. E* 61:5545–5549. One computer simulation apparently does not show a preference for tight knots: Quake, S. R. 1995. Fast Monte Carlo algorithms for knotted polymers. *Phys. Rev. E* 52:1176–1180.

20. Nureki, O., M. Shirouzu, K. Hashimoto, R. Ishitani, T. Terada, M. Tamakoshi, T. Oshima, et al. 2002. An enzyme with a deep trefoil knot for the active-site architecture, *Acta Cryst. D* 58:1129–1137.

21. Stryer, L. 1988. *Biochemistry*, 3rd ed., New York: W. H. Freeman, p. 319.

22. Landau, L. D., and E. M. Lifshitz, *Statistical Physics*, Reading, MA: Addison-Wesley, pp. 106–108.

23. Lee, S. Y., A. Lee, J. Chen, and R. MacKinnon. 2005. Structure of the KvAP voltage-dependent K+ channel and its dependence on the lipid membrane. *Proc Natl Acad Sci.* 102:15441–15446.

24. Mahler H. R., and E. H. Cordes. 1971. *Biological Chemistry*, 2nd ed., New York: Harper & Row, p. 27.

25. Nureki, O., M. Shirouzu, K. Hashimoto, R. Ishitani, T. Terada, et al., 2002. An enzyme with a deep trefoil knot for the active-site architecture. *Acta Cryst. D* 58:1129–1137.

26. Dong, K. C., and J. M. Berger, 2007. Structural basis for gate-DNA recognition and bending by type IIA topoisomerases. *Nature*. 450:1201–1205.
27. Schultz, S. C., G. C. Shields, and T. A. Steitz. 1991. Crystal structure of a CAP-DNA complex: The DNA is bent by 90 degrees. *Science*. 253:1001–1007.
28. Bao, X. R., H. J. Lee, and S. R. Quake. 2003. Behavior of complex knots in single DNA molecules. *Phys. Rev. Lett*. 91, 265506, 2003.
29. Jelesarov, I., C. Crane-Robinson, and P. L. Privalov. 1999. The energetics of HMG box interactions with DNA: thermodynamic description of the target DNA duplexes1. *J. Mol. Biol*. 294:981–995.

CONTENTS

CHAPTER 15

Reactions
Physical View

The fundamental physical factors that govern reactions are most easily comprehended through use of energy-level diagrams. Figure 15.1 shows the simplest energy diagram that describes a reaction $A \rightleftharpoons B$. Chapter 14 has taught us that three primary "energies" are relevant in biosystems: enthalpy H, consisting of internal energy and PV terms; entropy S, which includes a concentration (mixing) term if the number of independent particles changes; Gibbs free energy, $G = H - TS$, which is intrinsically temperature dependent. When applied to reactions involving proteins, such energy diagrams are often referred to as the *energy landscape*. When examining a diagram such as Figure 15.1, we must remember several important features:

- A *barrier* encountered when going from A to B implies energetic resistance (reduced Boltzmann probability) to the process for H and for G; an increase in S implies a facilitated process.
- The *reaction coordinate* can be a complex, extremely multidimensional coordinate, but in certain fundamental processes can be interpreted in terms of a real distance between two molecules or sites on molecules.
- Equilibrium is described by the local-minima energies of the two states, A and B.
- Rates of the process $A \rightarrow B$ or $B \rightarrow A$ are governed by the energy barriers encountered in the transition.
- The free-energy (G) landscape is temperature dependent and should not be thought of as a "constant" of the reaction.
- Reaction equilibrium constants and rates are all determined by energy differences in the diagram.

To a first approximation (which is about all one should try to manage in a complex biochemical reaction), H and S are temperature independent, so the physical description of a reaction is often reduced to experimentally determining the values of H, S, and G. The most direct way to determine H, S, and G is to measure the equilibrium and rates of the reaction as a function of temperature. Below, we expand on the description of reactions in terms of energies, begun in Section 14.12, because of the crucial importance of these energies and reaction coordinates.

A comprehensive treatment of biomolecular reaction theory and experiment cannot be covered in one chapter. The coverage that follows is that from a physicist's view, but should give the reader a solid foundation for more advanced or specialized applications. Those wishing coverage equivalent to a full course on reactions should consult a modern text such as that by C.R. Bagshaw (2017).[1]

15.1 ENERGY, ENTROPY, AND FREE ENERGY DIAGRAMS

15.1.1 Equilibrium

The free energy change of a reaction provides the drive toward equilibrium. For the simplest, unimolecular reaction the equilibrium was described by Equations 14.68 through 14.70 in the previous chapter. We can write these equilibrium equations in terms of enthalpy and entropy because these quantities are independent of temperature (usually):

$$A \rightleftharpoons B \tag{15.1}$$

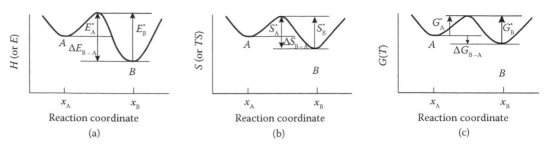

FIGURE 15.1
Energy diagrams for reactions equilibria and kinetics. (a) Enthalpy H or internal energy (or potential energy) as a function of reaction coordinate. (b) Entropy S (or TS), as a function of reaction coordinate. (c) Gibbs free energy (G) as a function of reaction coordinate. If H and S are independent of temperature, G is still temperature dependent. Two semistable states, A and B, are in equilibrium with each other, separated by energy barriers. All marked energies are energy differences, but the Δ symbol is often used only for the energy difference between the two states, A and B. Such energy-level diagrams are key to understanding biomolecular reactions.

$$\frac{[B]}{[A]} = \exp\left[\frac{\Delta G - \Delta G^0}{RT}\right] = \exp\left[\frac{\Delta H - \Delta H^0}{RT} - \frac{\Delta S - \Delta S^0}{R}\right] \tag{15.2}$$

$$\Delta G = 0 : K_{eq} \equiv \left(\frac{[B]}{[A]}\right)_{eq} = \exp\left[\frac{-\Delta G^0}{RT}\right] = \exp\left[\frac{-\Delta H^0}{RT} + \frac{\Delta S^0}{R}\right] \tag{15.3}$$

To a first approximation, reaction 15.1 has no dependence of H, S, and G on concentration, so the "0" superscripts are often dispensed with. When ΔG is not zero, a free-energy drive toward equilibrium exists, but the dynamic process that ensues cannot be described by the energy differences in Equations 15.1 through 15.3. Figure 15.2a illustrates a unimolecular reaction, a protein conformational change.

For a bimolecular reaction, the drive toward equilibrium is described by Equations 15.4 through 15.6, for an ideal solution reaction that takes place in uniform solution.

$$A + B \rightleftharpoons C + D \tag{15.4}$$

$$\frac{[C][D]}{[A][B]} = \exp\left[\frac{\Delta G - \Delta G^0}{RT}\right] = \exp\left[\frac{\Delta H - \Delta H^0}{RT} - \frac{\Delta S - \Delta S^0}{R}\right] \tag{15.5}$$

> If two molecules of A take part in the reaction, [A] is replaced by $[A]^2$.

$$\Delta G = 0 : K_{eq} \equiv \left(\frac{[C][D]}{[A][B]}\right)_{eq} = \exp\left[\frac{-\Delta H^0}{RT} + \frac{\Delta S^0}{R}\right] \tag{15.6}$$

When equal numbers of molecules are on the left and right sides of the reaction, the concentration ratio has no units and the equilibrium constant (obviously not a constant because it strongly depends on temperature) is a pure number. A large value of K_{eq} means formation of the products is favored. When this simple form of the equilibrium applies, K_{eq} and the concentration ratio should not depend on initial concentrations of any reactants. For example, if equal initial concentrations of A and B are added, equal amounts of A and B will disappear until the proper ratio of $\left(\frac{[C][D]}{[A][B]}\right)_{eq}$ is reached. Suppose, for example, that $K_{eq} = 1$. If 1 M A is mixed with 1 M B, with no initial C and D, the concentrations of A and B both decrease to 0.5 M, forming 0.5 M C and D. The latter must be equal because every reaction that produces one molecule of C must also produce one molecule of D and we have assumed no initial presence of C and D. If unequal concentrations of A and B are mixed, e.g., 1 M A and 0.1 M B, with no initial C and D, then calling the equilibrium concentration of C (=D)x, we find $K_{eq} = 1 = \left(\frac{[C][D]}{[A][B]}\right)_{eq} = \frac{x^2}{A \cdot B} = \frac{x^2}{(1M-x)(0.1M-x)}$. Solving, we get $x = 0.1/1.1 = 0.0909$ M, and the equilibrium

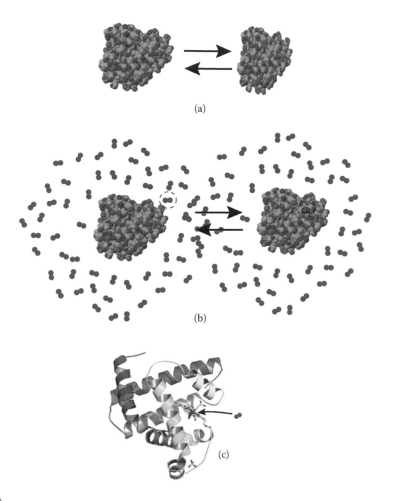

FIGURE 15.2
Spatial/physical view of two reactions. (a) Unimolecular reaction: conformational change of a protein molecule. (b) Bimolecular reaction: binding of O_2 to myoglobin (Mb). The "bimolecular" nature of the reaction is reflected by the competition of many O_2 molecules to bind to the heme site on Mb. The first part of the reaction, diffusion of O_2 molecules through solvent, is governed primarily by solvent viscosity and size of the diffusing molecules. There may also be a protein-solvent interface through which the ligand, O_2, must pass. (c) Only one O_2 can bind to the heme site. Though several oxygens may be able to diffuse in the protein structure, there is likely one preferred pathway to successful binding. Once a particular O_2 molecule has succeeded in approaching the binding site, the reaction becomes pseudo-unimolecular: the remaining rates do not depend on bulk concentration of either Mb or O_2 and are highly sensitive to details of protein structure.

concentrations are $[A]_{eq} = 0.909$ M, $[B]_{eq} = 9.09 \times 10^{-3}$ M, $[C]_{eq} = [D]_{eq} = 9.09 \times 10^{-2}$ M. Most (91%) of B has disappeared; its final concentration is even lower than those of the products. This is the effect of the large amount of molecule A initially added.

We must remind ourselves here that we are assuming ideal solutions of A, B, C, and D, and that the reaction is reversible. If, for example, A and B came from one side of a cell membrane, reacted at the membrane and released C and D on the other side, where C and D were immediately whisked away by some transport mechanism, these equilibrium expressions would not apply.

When the number of product molecules differs from the number of reactants, e.g., in the binding reaction

$$A + B \rightleftharpoons C \qquad (15.7)$$

the equilibrium expressions clearly have some mathematical issues of units:

$$\frac{[C]}{[A][B]} = \exp\left[\frac{\Delta G - \Delta G^{0'}}{RT}\right] = \exp\left[\frac{\Delta H - \Delta H^{0'}}{RT} + \frac{\Delta S - \Delta S^{0'}}{R}\right] \qquad (15.8)$$

> If $[C] \neq [D]$, the reaction is really two independent reactions and should be separated.

$$\Delta G = 0 : K_{eq} \equiv \left(\frac{[C]}{[A][B]} \right)_{eq} = \exp \left[\frac{-\Delta G^{0'}}{RT} \right] = \exp \left[\frac{-\Delta H^{0'}}{RT} + \frac{\Delta S^{0'}}{R} \right] \tag{15.9}$$

The equilibrium constant will now have units, M^{-1} in this case. The simplest way to deal with such a reaction is to choose a large value for one of the reactant concentrations, say $B \gg A$. Then $B = B_0$ is approximately a constant during the reaction and Equation 15.9 can be rewritten

$$\left[B_0 \right] K_{eq} \equiv \left(\frac{[C]}{[A]} \right)_{eq} = \left[B_0 \right] \exp \left[\frac{-\Delta G^{0'}}{RT} \right] = \left[B_0 \right] \exp \left[\frac{-\Delta H^{0'}}{RT} + \frac{\Delta S^{0'}}{R} \right] \tag{15.10}$$

This binding reaction then appears as a pseudo-unimolecular reaction, $A \rightleftharpoons C$, with a constant concentration dependence factored out of the expression. If B_0 is chosen to be large enough, B is not only approximately constant, but the equilibrium in Reaction (15.7) is shifted far to the right: almost all of A is converted to C. The physicist should still be bothered by the seeming conflict of units in Equation 15.10. Chemistry textbooks generally reassure us that all concentrations can surely be measured and that the B_0 factor represents a free-energy offset (actually an entropy offset) describing the dependence of the equilibrium concentrations of A and C on the value of B_0.

15.1.1.1 Macromolecule conformational changes: Denaturation

Because of the capability of O_2 to oxidize the heme iron atom, with resulting complex spin and oxidation-state changes, a yet simpler Mb binding reaction will be considered: Mb + CO (carbon monoxide).

The preceding discussion makes it appear that enthalpies and entropies of reaction are constants, independent of temperature. This is usually true for reactions and isomerizations of small to medium-sized molecules, as well as for binding of small molecules to proteins and DNA. After this section, we mostly assume this temperature independence. ΔG is then generally temperature dependent because $\Delta G = \Delta H - T\Delta S$. However, we have a major exception to this constancy of ΔH and ΔS in the case of protein denaturation, the unfolding of the native conformation: native $\rightleftharpoons$ denatured. Such a process involves hundreds or thousands of coupled atoms and must be complicated. While it is generally true that proteins denature at high temperature, it is sometimes true that they also denature at low temperature. Usually, ΔH^U and ΔS^U, the enthalpy and entropy of unfolding depend on temperature. Figure 15.3 illustrates the enthalpy, entropy ($\cdot T$), and Gibbs free energy of unfolding of myoglobin as a function of temperature. You should perhaps write, in the margin next to this diagram, that ΔH and ΔS for *simple* reactions are independent of temperature. Protein denaturation is not simple.[2] ΔG^U in this plot shows a maximum, indicating the point where myoglobin (Mb) is most stable against denaturation.

Physical systems are understood through diagrams. Equations quantify the understanding, make predictions, model observations, and provide the test of whether a physical model is correct or not. The equations should not be allowed to eclipse the primacy of the physical structures and good descriptive diagrams. Diagrams and real structures clue us in to alternate possibilities. Equations usually constrain our thinking to what is mathematically simplest. Mathematical simplicity is sometimes correct, sometimes not. Figure 15.2b shows what is considered the simplest bimolecular, biomolecular reaction: the binding of O_2 or CO by myoglobin (Mb). See also Figures 5.26 and 5.27a. Mb molecules placed into a solution with O_2 or CO sets up a classic bimolecular reaction. If [CO] $\gg$ Mb, the system is one where one reactant has a large, almost constant concentration. A few simple experiments with varying gas concentrations confirm this bimolecular identification. Variation of the temperature over about $\pm 10°C$

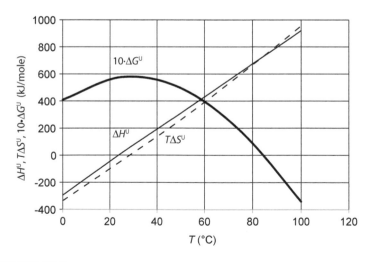

FIGURE 15.3
Protein unfolding (denaturation) often exhibits temperature-dependent enthalpies and entropies because of its complex nature. (Note the $T\Delta S^U$ line is slightly curved.) The example here is myoglobin. Note that the Gibbs free energy is multiplied by a factor of 10 in the plot, to emphasize the temperature of maximum free-energy change. (Redrawn from Privalov, P. L. and N. N. Khechinashvili, *J. Mol. Biol.*, 86, 665–684, 1974.)

then allows determination of values for ΔH, ΔS, and ΔG. So we're done, right? No, we're not done. Even the biosystem novice, looking at Figure 15.2b and 15.2c, would predict that at least two distinct phases of the binding reaction must occur. In the first phase, many O_2 or CO molecules compete to bind at the Mb heme binding site. More specifically, because the heme iron is located in the protein interior, O_2 or CO molecules randomly diffuse in the solvent until one happens to approach the heme through the "correct" trajectory. Now, must an O_2 follow a specific path from the outside of the protein to the heme "pocket," or can O_2 also diffuse through the protein matrix? This would depend on whether the Mb protein structure is loosely packed, like a freely jointed chain, or closely packed, like a three-dimensional solid. Diffusion through protein structure would be allowed in the former; preferred open channels would dominate in the latter. Mb is closer to a three-dimensional solid: it is quite closely packed, as suggested by Figure 5.26. We also learned in previous chapters, however, that proteins and other macromolecules fluctuate in structure. Fluctuations of Mb structure have, in fact, been measured, as in Figure 5.27. Such structural fluctuations could, in fact, allow a restricted diffusion. A CO molecule temporarily blocked from the heme Fe could simply "wait" until the intervening protein structure fluctuates out of the way, then move toward the Fe. With a few simple, uneducated musings based on structural diagrams, we have bumbled into a major protein discovery:

> *Proteins are usually close-packed structures, but thermal motions of atoms generate structural fluctuations that facilitate binding and other parts of the catalytic process.*

The author observed the development of this idea in the 1970s as Hans Frauenfelder, Gregorio Weber, and others argued the two extremes of the model—free ligand diffusion through the protein matrix versus channeled access to the binding site, facilitated by protein fluctuations—and the model of the fluctuating enzyme became the orthodox view.[3]

What might a model for this Mb ligand-binding process then look like? In the first phase, O_2 or CO diffuses in the bulk solvent. See Figure 15.4, numbered location 1. We know how diffusion operates in some detail from Chapter 13. What we need to add is the (average) rate of diffusion of two types of molecules present in the same solution: there is some average distance between molecules A and B, which should correspond to some average diffusion time. The inverse of this average time should be the diffusion reaction rate.

When O_2 approaches the interface between solvent and protein—the protein after all is a large structure compared to diatomic oxygen—an interfacial region is encountered. See number 2 in Figure 15.4. This interface could be described as an energy barrier O_2 must surmount to enter into the protein interior. We can also sketch an idea for modeling this process: a Boltzmann probability that the O_2 has energy at least as high as the barrier height. There is an additional geometrical probability factor, determined by the (presumed) specific location at the protein surface where the ligand must approach to be successful. (Because our model focuses on agreement with experimental data, this turns out to be an energy- and temperature-independent "frequency factor" term in the rate that cannot be separated from other such frequency factors; i.e., this is a "fudge factor," but it must be reasonable because any of you energetic readers can, with a lot of work, calculate what the value should be.) This is the approach we take. An alternative model might suppose that any energy is sufficient to pass the interface, as long as the trajectory brings the O_2 to a particular location, traveling in the particular direction, at the protein surface. We could imagine modeling this with a single Mb molecule, fixed in place, and finding the Maxwell–Boltzmann probability that an O_2 molecule has the proper velocity direction and initial location to bring O_2 along the proper trajectory. There are reaction theories for this scenario, but we do not pursue them.

When an O_2 or CO molecule successfully enters the protein, it must make its way to the binding site at the heme—location number 3 in Figure 15.4. As we saw in Chapter 5 (Figure 5.1), heme is a planar molecular group with an open binding site on one side. Clearly, the O_2

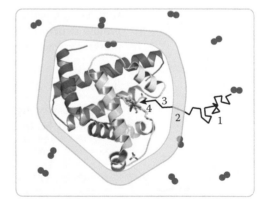

FIGURE 15.4
A "simple" biological reaction: binding of O_2 or CO to myoglobin (Mb). At least four distinct processes could be conjectured from the diagram, keeping in mind that the displayed Mb structure, though derived from X-ray crystallography, does not include space-filled structure from side chains that make the structure quite close packed. See Figure 15.2. Numbered processes: (1) diffusion in the solvent, (2) transit of a protein-solvent interface, (3) passage through obstructing protein structure, and (4) rearrangement of the heme local structure (coupled to protein) to accommodate binding. Four processes have been identified from experimental data.

or CO must approach the heme from the right direction. If an open corridor leads from the surface of the protein to the binding site (apparently *not* the case in Mb), a complicated random-walk calculation could be undertaken, including all the variable interactions with structure along the way. Such a computation is beyond the capabilities of this author, though crude simulations can be made with classical physics simulation software such as *Interactive Physics*™ (design-simulation.com). If there is no open path to the binding site, intervening protein structure must fluctuate, allowing further progress toward the heme. Computing this process in detail is also almost impossibly complex. We instead simply assign an energy barrier and an effective attempt frequency factor to this intermediate process.

When O_2 or CO finally reaches the heme site (location 4 in Figure 15.4), it finds the Fe atom, to which it must bind, retracted on the other side of the heme plane. Furthermore, the Fe is held back by a covalent bond to a protein histidine side chain, called the *proximal histidine*. This histidine group can be seen by careful examination of the bottom side of the heme group in Figure 15.4. The net result is that yet another energy barrier exists at the heme binding site. Besides being measured, this final energy barrier has provided a binding site energy barrier whose height and width has been measured.[4] More on this shortly.

15.1.1.2 Missing structural detail in early biochemical reaction data

Measurements of the binding of various ligands to Mb had been made for several decades before the fluctuating-structure, multistep binding process was discovered. Many of these experimentalists varied the temperature and ligand concentration and did rigorous mathematical analysis. Why did they miss what is obvious to us? The first mistake was an entrenched practice to vary temperature over a range of 10°C to 30°C in order to determine binding constants and rates. This temperature variation is small (3%–10%) on an absolute scale, but as we will shortly see, allows fitting of a straight line to binding data, where the slope gives the enthalpy and the intercept gives entropy. This tradition was encouraged by the rule of thumb, "Biochemical reactions double in rate with every 10°C temperature increase," and by the difficulty of measuring biochemical reactions below 0°C (solutions freeze) and above about 60°C (proteins denature). Expansion of the temperature range by employing techniques that are considerably more difficult readily shows that the data are not described by a straight line at all: the curvature indicates the presence of multiple processes. The second sort of data missing in the early 1970s was routine temperature-factor (atomic mobility) determinations for all atoms in protein structures determined by crystallographic techniques: there were no compelling data to indicate protein structures fluctuated. We can now routinely display protein structures from the PDB in terms of atomic mobility and see that a crystal of Mb is quite unlike a crystal of NaCl, in terms of atomic fluctuational motions.

> We are thinking in terms of reaction steps in *series*. There could also be *parallel* processes in binding.

15.1.1.3 Rate-determining step

The process of ligand binding to Mb, the "hydrogen atom of enzymes," is clearly complex and should be treated with some detail. Such treatment is experimentally and analytically demanding, compared to what is needed for the application of single-step reaction theory (one energy barrier). As a result, many scientists reverted to the idea of a *rate-determining step* (RDS) in the binding process, especially when faced with hundreds of samples that needed to be measured and analyzed. We employ one version of RDS analysis in the Michaelis–Menten theory. The general idea of the RDS is that one of the series of energy barriers is largest and determines what the binding rate will be (Figure 15.5). Because the binding equilibrium does not, in principle, depend on any of these intervening energy barriers or metastable states, it would seem adequate to force a single-step (two-state) theory to describe the binding.

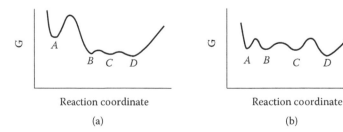

FIGURE 15.5
Energy barriers and wells describing a multistep biochemical reaction. (a) If one barrier is much larger than the others, the temptation to declare a "rate-determining step," governed by the A-B barrier is great. Analysis by a single-step theory may not lead to erroneous conclusions. (b) If barriers and energy well levels are comparable, single-step analysis will ignore details and may mislead attempts at understanding.

There are dangers in applying a rate-determining step theory to complex biological processes. First and foremost, the processes and structures are truly intricate and ingenious. Forcing a simplistic analysis prevents the elucidation of details. What is the role of the protein structure in binding rates and equilibrium constant? What prevents O_2 from oxidizing Fe in heme myoglobin and hemoglobin? Why is the Fe on the "wrong" side of the heme group before binding? How far does the Fe have to move when O_2 or Co binds? Second, while biochemists often need to measure large numbers (hundreds or even tens of thousands) of samples under a wide variety of conditions (temperature, concentrations, mutation-modified proteins) for drug discovery purposes—requiring quick, simple analysis (e.g., rate-determining-step analysis)—the role of the biophysicist is often to examine the data and biomolecules more carefully to learn structural and mechanistic details. In the end, modern computers can often process data using more complex algorithms without requiring significantly more time for the entire drug search or other process.

Search for *combinatorial chemical libraries* on the Web to explore cases where hundreds or thousands of chemical variants might need analysis.

15.1.2 Kinetics

Figure 15.1 asserts a connection between equilibrium and kinetics, at least in the case of reactions governed by equilibrium thermodynamics of a single process. We have continually noted that living systems are not in thermodynamic equilibrium, so we must take this assertion with a grain of salt. Concentrations of critical components in cells are rarely determined by only one process. Remember Figure 1.5. Nevertheless, given that changes in concentrations in cells normally occur on a time scale of 10^{-2} s or longer, an individual reaction that has a much shorter time scale can be treated as if it is in equilibrium, though that equilibrium may change during the next 10^{-2} s.

Suppose we have a single, unimolecular process, $A \underset{k_{BA}}{\overset{k_{AB}}{\rightleftharpoons}} B$, with forward and reverse rate constants k_{AB} and k_{BA}, respectively. In a bulk solution the reaction is governed by the rate equation,

$$\frac{d[A]}{dt} = -\frac{d[B]}{dt} = -k_{AB}[A] + k_{BA}[B] \tag{15.11}$$

In the steady state $\dfrac{d[A]}{dt} = -\dfrac{d[B]}{dt} = 0$, so

$$\left(\frac{[B]}{[A]}\right)_{ss} = \frac{k_{AB}}{k_{BA}} \tag{15.12}$$

We also know that if this steady state is similar to equilibrium,

$$\left(\frac{[B]}{[A]}\right)_{ss} = \frac{k_{AB}}{k_{BA}} = \left(\frac{[B]}{[A]}\right)_{eq} = K_{eq} = e^{-\Delta G^0_{B-A}/k_B T} \tag{15.13}$$

implying the kinetics-equilibrium connection. Figure 15.1c also shows that

$$\Delta G^0_{B-A} = G^*_A - G^*_B \tag{15.14}$$

where the starred quantities are the barrier heights as seen from the bottom of wells A and B. In a thermodynamic model for the reaction, the free-energy diagram demands relations (15.13) and (15.14) between equilibrium and kinetic parameters. We return to time-dependent solutions of equations like Equation 15.11 in later sections of this chapter.

The dynamics of a biochemical reaction are often governed by the energy barriers between the (quasi-) stable states of the system. The major exception to this rule is that part of molecular motion we have so far concentrated on in this book: diffusion or the random walk. While the ubiquity of diffusion suggests we might spend even more time on complex random-walk processes, it often turns out that the time needed for diffusion is insignificant in biosystems: small molecules diffuse rapidly to the active site of an enzyme, where they wait around for the more complex, slower processes of enzymatic binding or reactants and release of products (the core idea of Michaelis–Menten theory).

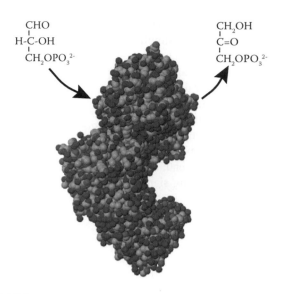

FIGURE 15.6
Triose phosphate isomerase, an "almost perfect" catalytic enzyme, delayed in its activity only by the time for substrate diffusion. The substrate shown (left) is glyceraldehyde 3-phosphate; the product, dihydroxyacetone phosphate. (PDB ID 1WYI, Kinoshita, T. et al., *Acta Crystallogr.*, 61, 346–349, 2005; NGL viewer. From Jmol: An open-source Java viewer for chemical structures in 3D. http://www.jmol.org/; Protein Workshop: Moreland, J.L. et al., *BMC Bioinformatics*, 6, 21, 2005; NGL Viewer: Rose, A.S. et al., *Bioinformatics*, 2018.)

Such a situation often holds in the case of *in vitro* experiments, where concentrations can be increased as desired, shortening the time for a reactant molecule to diffuse to its target. This does not mean that diffusion is irrelevant. In living cells (*in vivo*) the concentration may be lower and the time for diffusion longer, on average; or free diffusion may be obstructed, slowing the process. In any case, the cell must provide the concentration of reactants needed to successfully tolerate the needed diffusion time.

15.1.2.1 Enzymes and reaction kinetics: First pass

A prime goal of biochemical reaction modeling is to understand how enzymes catalyze reactions by many orders of magnitude. We have seen that even the simplest binding reaction, presumably the first step in a biochemical reaction, can be complex. A likely first step might be to determine some of the energy landscape of the enzymatic process, but let's try to foresee more of what we must quantitatively describe in real biochemical reactions. We start with an example. Suppose you are studying the enzyme triose phosphate isomerase. See Figure 15.6. According to Wikipedia,

Triose phosphate isomerase is a highly efficient enzyme, performing the reaction billions of times faster than it would occur naturally in solution. The reaction is so efficient that it is said to be catalytically perfect: it is limited only by the rate the substrate can diffuse into the enzyme's active site.[5]

The rate of this reaction is noted as 4.3×10^{-6} s^{-1} in the absence of the enzyme and 4300 s^{-1} in its presence, an acceleration of 10^9.[5] (Such measurements are done in vitro, under optimal conditions for the enzyme.) However, the meaning of these two numbers is different. We see that when we think of a real experiment we might do with an enzyme.

To carry out a real experiment, we need to decide how much "stuff" we need and how we might actually measure the rate of a reaction. First, we need to know the times needed for the nonenzymatic and (native) enzymatic reactions to occur and what that implies for measurements of rates. Second, what concentrations of substrate and enzyme should we use? Third, what experiment(s) should we do to check whether diffusion is really limiting the enzymatic reaction?

Taking the inverse of the listed rates gives some sort of reaction times. A rate of 4.3×10^{-6} s^{-1} corresponds to a time of about 2.7 days; 4300 s^{-1}, to 0.23 ms. What do these times mean? Many unimolecular reactions are described by exponential processes, $N(t) = e^{-t/\tau}$, so we expect that the reaction rate for at least the nonenzymatic reaction refers to τ^{-1}. The nonenzymatic time is quite long but can be accommodated by strategically starting your measurement at about Wednesday noon, doing several measurements before leaving for the day, doing two to three measurements throughout Thursday, taking Friday and the weekend off, and coming back Monday to do a measurement of the "long-time" numbers. Note that for the nonenzymatic reaction we were not forced to decide on concentration of substrate, glyceraldehyde 3-phosphate (GAP), because the time for 63% $(1 - e^{-1})$ of the number of molecules to spontaneously convert to product does not depend on concentration for a simple unimolecular reaction.

The mathematical expression for the "reaction time" in the case of the enzyme-catalyzed reaction is a bit different—we postpone details until the Michaelis–Menten section—but we can get a rough idea by first determining the numbers (concentrations) we are dealing with. Clearly, if we add 10^{-12} M enzyme to 1 M GAP it will take longer for 63% of the GAP to convert than if we add 0.1 M enzyme to 10^{-3} M GAP. How should concentrations be chosen and what does the enzymatic reaction rate refer to?

The substrate and/or product concentration is what needs to be measured as a function of time. There only needs to be enough enzyme to have a significant effect on the rate of the reaction. This tells us something quite important: substrate and product concentrations are normally much larger than enzyme (catalyst) concentrations. How much substrate you start with depends on how you measure concentrations. Concentrations are often measured using optical absorption (or related) techniques. In Chapter 11 we saw that light absorbance was described by Beer's law, $A = \varepsilon C L$, where A is absorbance (values between about 0.1 and 2), ε is the absorption coefficient (values ranging from 10^3 to 10^5 M^{-1} cm^{-1}), C is concentration (M), and L is the optical path length (typically 1 cm). If we assume $\varepsilon \sim 10^3$ M^{-1} cm^{-1}, then we need about 0.1 mM glyceraldehyde 3-phosphate. If we add enzyme to such a concentration of substrate, the time for 63% of the substrate to be converted depends, of course, on how much enzyme you add: more enzyme converts the substrate more quickly, so the "reaction time" does not have the same meaning as without the enzyme. As we see below, the time for an enzyme molecule to convert one substrate molecule, or the number of product molecules produced per second per enzyme molecule, is the relevant time (or rate).

> Actually, absorption will not work very well in this GAP reaction, but we will ignore this.

To analyze the reaction data, the amount of substrate or product needs to be measured and plotted as a function of time. The substrate data should then be fitted to the function $N^{\text{substrate}}(t) = e^{-t/\tau}$ and τ determined. One possible flaw in this plan is that biological systems tend not to be very stable when removed from living cells. Enzymes and biological substrate molecules in a test tube, exposed to ambient light at room temperature often photoreact or lose activity after even a few hours. Any fit to a decaying exponential function should be examined carefully.

If the enzymatic reaction obeys Michaelis–Menten kinetics, it turns out we can obtain similar information from the substrate concentration experiment, but let's wait for a bit more theory.

There are many physical/mathematical models and rate theories for reactions, but we shall focus on the simplest, which comes from the diagram of Figure 15.1 and readily provides a framework for quantitative modeling of biomolecular reactions.

15.2 RATE THEORY I: ACTIVATION-ENERGY MODEL

Reaction rate models based on equilibrium energy contours can be found in similar form in a variety of places and come from Figure 15.1 in a straightforward manner. Figure 15.7 shows a diagram of the energy contour, again simplified to a one-dimensional reaction coordinate, for two stable states, A and B, separated by an energy barrier of height E_{AB}^* above the minimum energy in well A. The energy contour can be generalized, as in Figure 15.1, to free energies. We consider the transition A $\rightarrow$ B, in which a particle (a system) must have an energy E_{AB}^* to reach the top of the barrier. We assume that the system is in quasi-thermal equilibrium, in the sense that the states in well A up to the top of the barrier are populated according to Boltzmann statistics. In order to classically pass from A to B, the particle must (1) have energy of E_{AB}^* or more and (2) travel in a "direction" toward state B.

In a true one-dimensional reaction model, a particle initially in well A that travels toward the barrier, also travels toward well B. If the particle has enough energy to get over the barrier, it will get to well B, with 100% probability. In three dimensions, having sufficient energy to surpass the barrier is still the primary requirement, but proper direction and other details generally reduce the success rate. We ignore this and simply fold this factor into activation entropy, to be determined experimentally.

> Section 15.2 is a nonrigorous derivation of the activation-energy/transition-state/thermodynamic rate constant theory. See Bowman, J. M., and A. G. Suits, 2011, Roaming reactions: The third way, *Physics Today* 64(11): 33–37, for a more general approach to rate theory.

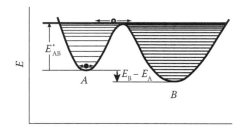

Reaction coordinate

FIGURE 15.7
Energy diagram for deriving two-state, Arrhenius-type rate constant. States A and B are each characterized by a potential well, which can be approximately described by a harmonic potential. The energy levels can be described by $E = \hbar\omega = hf$, but near the top of the well the particle is almost free, the oscillation frequency low. In thermal equilibrium $hf = k_BT$, or $f = k_BT/h$. The probability the particle has an energy equal to the barrier height, and can thus pass over the barrier, is $\frac{P(E^*)}{P(0)} = g(E^*)e^{-E^*/k_BT} = e^{s^*/k_B}e^{-E^*/k_BT} \cong e^{s^*/k_BT}e^{-H^*/k_BT} = e^{-G^*/k_BT}$, so the rate at which a particle in well A passes to state B is $K_{A \rightarrow B} = Ae^{-E^*/k_BT} = \frac{k_BT}{h}e^{S^*/k_B}e^{-E^*/k_BT}$, where f and the entropy are embedded in the *Arrhenius constant A*. (Recall for vibrational states in thermal equilibrium, P(0), the probability to be in the lowest state, is almost 1.)

(It *is* an issue of ordering.) The rate—number of unimolecular conversions per unit time—is therefore the product of an attempt frequency times a Boltzmann probability for energy E^*:

$$\text{Rate } (A \rightarrow B) \equiv k_{AB} = [\text{attempt frequency}][P(E^*)] \qquad (15.15)$$

We can quickly write the generalized Boltzmann energy probability:

$$\frac{p(E^*)}{P(0)} = \frac{g(E^*)}{g(0)} e^{-E^*/k_B T} = e^{S^*/k_B} e^{-E^*/k_B T} \cong e^{S^*/k_B} e^{-H^*/k_B T} \qquad (15.16)$$

S^* and H^* are called *activation entropy* and *activation enthalpy*, respectively, because they reflect the change in these quantities when a particle moves from the lowest-energy state in the well to the *activated state*, that at the top of the barrier. In this equation we remember that S^* and H^* are really entropy and enthalpy *differences* between activated state and lowest-energy state; we reserve ΔS and ΔH for description of differences between stable states (bottom of wells), A and B. There are several other approximations embedded in Equation 15.16, including the fact that for typical vibrations, the probability for being in the lowest energy level is virtually 1.

> The reaction must be unimolecular, of course. Bimolecular reactions like $A + B \rightleftharpoons C$ will include an entropy of mixing factor, but this often consists of simple multiplication by a concentration because entropy of random dispersal in a solvent is usually of little biological interest. See Section 15.3.

The attempt frequency is also readily approximated. The stable state, A, must have an energy minimum that can be approximated by a harmonic potential (at least near the minimum). This is illustrated by the nearly equally spaced energy levels in the bottom half of the well. Each of these energy levels has an effective frequency at which the particle oscillates back and forth. In a truly quantum model for the well, the energy levels in the bottom half of the well would be

$$E_n = (n + 1/2)\hbar\omega = (n + 1/2) \, hf \qquad (15.17)$$

with semiclassical oscillation frequency nf, for large n. Near the top of the well, the walls of the well are not so steep, corresponding to a weaker spring constant. Nevertheless, we assign an oscillation frequency f to level at the top of the barrier. Invoking thermal equilibrium in well A, we cite the equipartition theorem to assert that the vibrational energy hf has two energy modes (vibrational potential and kinetic energies), and

$$hf = k_B T \qquad (15.18)$$

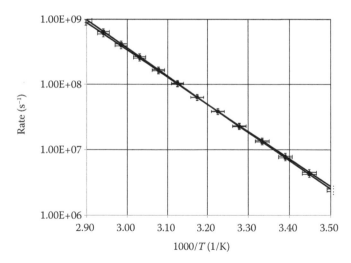

The final expression for the rate of transition from A to B is therefore

$$k_{AB} \cong \frac{k_B T}{h} e^{S^*/k_B} e^{-H^*/k_B T} \qquad (15.19)$$

Some names associated with Equation 15.19 include activated-complex theory, transition-state theory, and thermodynamic rate theory.

This version of reaction rate theory is incredibly useful. The frequency factor can be calculated without knowing any molecular details. Its value is apparently a universal constant at any given temperature. Near room temperature, its value is about

$$f = \frac{4.1 \times 10^{-21} \text{J}}{6.63 \times 10^{-34} \text{Js}} = 6.2 \times 10^{12} \text{s}^{-1} \cong 10^{13} \text{s}^{-1} \qquad (15.20)$$

FIGURE 15.8
Semilog plot of reaction rate versus $1{,}000/T$, with temperature ranging from $10°$ to $50°C$. The two lines are computed values using activated-complex theory with fixed or temperature-dependent frequency factor: $f_0 = 10^{13} \text{s}^{-1}$ or $f(T) = \frac{k_B T}{h}$. Activation enthalpies and entropies for the fixed-frequency (temperature-dependent) calculation are $H^* = 80.00$ kJ/mole (80.00 kJ/mole) and $S^*/R = 18.57$ (19.00). The goodness of fit of the two expressions cannot be distinguished for the error bars shown: $\pm 5\%$ on rates and $\pm 0.5\%$ on temperatures.

In a large majority of biological cases, this attempt frequency is simply set equal to 10^{13} s^{-1} because other theories give different temperature dependencies and because temperature variations are dominated by the exponential terms in Equation 15.19. Figure 15.8 shows that for typical rate measurements made on biosystems, the difference in frequency factor between 10^{13} s^{-1} and $k_B T/h$ cannot be resolved. In this simulated data, temperatures range from 10° to 50°C, with 0.5% uncertainty, and rates are measured with 5% precision. Two lines are computed to fit the data, using fixed frequency factor, $f_0 = 10^{13}$ s^{-1}, and temperature-dependent factor, $f(T) = k_B T/h$. The fits are equally acceptable and only the activation entropy differs: $S^*/R = 19.00$ using $f(T)$ and $S^*/R = 18.57$ for $f_0 = 10^{13}$ s^{-1}. The scatter in real rate data and the usually negligible difference between entropy values induces most researchers to use the fixed frequency factor. Homework Problem 15.3 explores this issue further.

Arrhenius Reaction Rate Theory. An earlier version of activated-complex theory was developed by Svante Arrhenius (1859–1927), a Swedish physicist/chemist, one of the founders of the science of physical chemistry. The fact that the Arrhenius equation (below), the lunar crater Arrhenius and the Arrhenius Labs at Stockholm University are named after him suggests that the contributions of this Nobel Prize winning scientist (1903, Chemistry) were quite widely appreciated. The *Arrhenius equation* is

$$k = Ae^{-E^*/k_B T} \tag{15.21}$$

The quantity A is termed the *prefactor* and E^* the *activation energy*. This equation is still used to parameterize reaction rates. Its wide utility can be understood by comparison to the activated-complex theory (Equation 15.19). Clearly, $A = fe^{S^*/k_B}$, both of which are (approximately) temperature independent, and $E^* = H^*$. The term with the activation energy (enthalpy) contains all the temperature dependence. A has the advantage of being easily determined from experimental data. Problem 15.4 demonstrates the mechanics of this.

15.3 DIFFUSION-CONTROLLED RATES (BIMOLECULAR)

> Note that if substrate concentration is much larger than enzyme concentration, the average time for an enzyme to receive a "hit" from a substrate does not depend on enzyme concentration.

Let's return to the triose isomerase catalysis example and investigate the contribution of diffusion to reaction rates. The claim is that this enzyme is so fast that it is limited only by the rate at which substrate can diffuse to it. We first ask a practical question. How do we test whether diffusion plays a role in a reaction? Perhaps the average time for molecules to diffuse together is 1 μs, while the average time to react is 10 s. There may be, for example, a large energy barrier (Figures 15.1, 15.4, and 15.5) that must be overcome for the reaction to take place. In this case many molecules would have time to diffuse together, but in only 1 in 10 million cases does the molecule have enough energy and proper directionality to surmount the barrier. (Think of the Boltzmann distribution of energies.)

Experimentally, we could ask whether we should vary the glyceraldehyde 3-phosphate concentration or the solvent viscosity to see if the triose isomerase-catalyzed reaction is limited by diffusion. The most straightforward way to test diffusion is to directly vary the molecular diffusion constants by varying the solvent viscosity. In order to do so, we add glycerol or sucrose to make the solution more viscous. This slows diffusion, but it also may affect the activity of the enzyme by altering its structure. Structural changes could be ruled out by X-ray crystallography (a big undertaking) or spectroscopy, but better ways might be (1) to create the same viscosity using two different additives (e.g., glycerol and sucrose) or (2) look for agreement of the rate's viscosity dependence with the theory below.

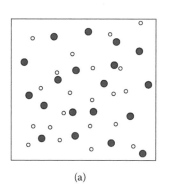

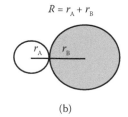

(a)   (b)

$R = r_A + r_B$

FIGURE 15.9
Diffusion-limited rate: no activation energy. Though no energy barrier exists, the rate may still be temperature dependent because η varies with T. (a) Two types of molecules, uniformly dispersed in a fluid, are separated by an average distance which requires an average time for the two molecules to diffuse together. (b) We assume no interaction between the molecules and that reaction takes place whenever the two molecules are separated by a distance $R = r_A + r_B$, where the latter are (spherical) molecular radii.

15.3.1 Diffusion-Limited Reaction Rate Theory

A *diffusion-limited* or *diffusion-controlled* rate describing a reaction of type $A + B \rightarrow C$ is one in which random thermal motion brings about encounters between two types of molecules (Figure 15.9a), where a reaction takes place as soon as the separation between the molecules approaches a distance R. In an enzyme-catalyzed reaction, diffusion of a substrate molecule to an active site on the enzyme is only the first part of the overall reaction: $A + B \rightarrow C$ would represent only the initial binding of a reactant molecule (B) to enzyme (A), forming a bound state (C = A·B). Let's simplify to the case of two molecules that diffuse together and react. In the absence of long-range forces—i.e., the molecules are not both charged or interact through dipole interactions—the distance R is usually the sum of the molecular radii (Figure 15.9b). If the molecules are both ionic or interact through a long-range potential, $V(r)$, a specific dependence on the kinetic energies of the molecules must be accounted for. (Expect another Boltzmann factor.) See Weston and Schwarz or other chemical kinetics books.[7]

Diffusion-limited reaction theory starts, not surprisingly, from Fick's law of diffusion. Generalizing to three dimensions from Equation 13.33,

$$\mathbf{j}_i = -D_i \nabla c_i \tag{15.22}$$

where $\mathbf{j}_i$ is the (vector) flux and $i = A$ or B; c_i is the concentration of species i; and D_i is its diffusion coefficient. For the case of spherical symmetry—molecule B diffuses toward molecule A until it gets to within a radial distance R (imagine for a moment that molecule A is fixed in place), where it reacts and disappears—Equation 15.22 would simplify to

$$j_B = -D_B \frac{dc_B}{dr} \tag{15.23}$$

The fact that both molecules diffuse (molecule A is *not* fixed in place) results in replacement of the diffusion constant of B with the sum of diffusion constants:

$$j_B = -\left(D_A + D_B\right)\frac{dc_B}{dr} \tag{15.24}$$

Rate constants are called *constants*, by tradition. They usually depend on temperature, pressure, solvent, and other factors. However, they should *not* depend on concentration.

(It's a good idea to think about this summation assertion. If a molecule of A moves at all, it always moves toward *some* molecule of B: A is not taking evasive action.) Let's remind ourselves of usual units: flux units molecules/(m^2 s), D in m^2/s, and c_B in molecules/m^3. If molar concentrations and/or distances in centimeters are used, appropriate conversion factors must be inserted.

In order to define a rate constant for a reaction of type $A + B \xrightarrow{k} C$, we revert to the "usual" differential equation and define the reaction rate constant, k, by

$$-\frac{d[A]}{dt} \equiv k[A][B] \tag{15.25}$$

where [A] and [B] are the (average) concentrations of A and B in the entire solution. (We must be prepared to deal with molar concentrations here.) In a pseudo steady state, the total flux toward molecules of A in a unit volume—the flux integrated over the surface of a sphere of radius $r > r_A$ times the average number of spheres per unit volume—equals the rate of reaction (Figure 15.10). The rate of the reaction is the left side of Equation 15.25. Therefore

$$-\frac{d[A]}{dt} \equiv k[A][B] = -[A]4\pi r^2 j_B \tag{15.26}$$

The reaction rate constant k is then

$$k[B] = 4\pi r^2 j_B = 4\pi r^2 \left(D_A + D_B\right)\frac{dc_B}{dr} \tag{15.27}$$

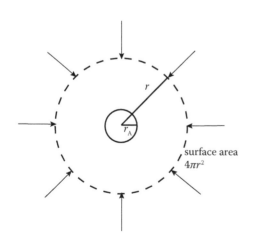

FIGURE 15.10
Inward flux toward molecule A through a spherical surface of radius r. The flux integrated over the entire surface must equal the rate of disappearance of A (or of B).

(Remember that [B] is the bulk concentration of B while c_B is its concentration as it varies with distance from an average molecule A.) Separating the derivative to the left side,

$$\frac{dc_B}{dr} = \frac{k[B]}{4\pi r^2 (D_A + D_B)} \tag{15.28}$$

and integrating from $r = R$ to $r = \infty$,

$$c_B(\infty) - c_B(R) = \frac{k[B]}{4\pi (D_A + D_B)R} \tag{15.29}$$

and yes, you should be worrying about the [B] term because the claim was that reaction rate constants should not depend on concentrations. But $c_B(\infty) = [B]$ and if the molecules react instantly when the radial distance R is reached, $c_B(R) = 0$. We immediately get

$$k_D = 4\pi(D_A + D_B)R \cong 4\pi (D_A + D_B)(r_A + r_B) \tag{15.30}$$

where k_D indicates the "diffusion" reaction rate constant, which is now independent of any concentration but depends on molecular diffusion constants and radii. The normal SI units for k_D are (molecules/m^3)$^{-1}$ s^{-1}. If k_D is expressed in chemist's usual units of M^{-1} s^{-1}, $k_D \cong 4\pi(D_A + D_B)$ $(r_A + r_B)$ 6.02 $\times$ 10^{20}, with D in cm^2/s and r in centimeters.

The diffusion coefficients can be expressed, as we know, as $D_A = \frac{k_B T}{6\pi\eta r_A}$ and similarly for D_B:

$$k_D \cong \frac{2k_B T}{3\eta}\left(\frac{1}{r_A} + \frac{1}{r_B}\right)(r_A + r_B) \tag{15.31}$$

If $r_B \gg r_A$,

$$k_D \approx \frac{2k_B T}{3\eta}\left(\frac{r_B}{r_A}\right) \tag{15.32}$$

What are typical values for this expression? Suppose we take two small molecules of the same size, $r_A = r_B = 0.2$ nm, and use Equation 15.31.

$$k_D \cong \frac{2 \cdot 4.1\times10^{-21}J}{3 \cdot 10^{-3}Pa \cdot s}\left(\frac{2}{2\times10^{-10}m}\right)\left(4\times10^{-10}m\right) = \frac{2 \cdot 4.1\times10^{-21}J}{3 \cdot 10^{-3}Pa \cdot s}(4) = 1.1\times10^{-17}m^3s^{-1} \tag{15.33}$$

In this calculation, note that the sizes actually did not matter: 20-nm spheres would have the same diffusion rate constant as 0.2 nm. The only limit is that the sphere cannot be so large that thermal motion is not the dominant motive process. Converting this rate to the usual chemist's units,

$$k_D \cong (1.1\times10^{-17}m^3s^{-1})\left(\frac{1000L}{1\,m^3}\right)\left(\frac{6.02\times10^{23}\,molecules}{mole}\right) = 6.6\times10^9 M^{-1}s^{-1} \tag{15.34}$$

How do we learn the physical meaning of this reaction rate constant? How "fast" does the reaction go? It would seem that if some concentration were 1 M, then the reaction rate would be 6.6×10^9 s^{-1}. But there are two concentrations in Equation 15.25. Which one is 1 M or must both be 1 M? The answer lies in the solution to the differential equation (later), but the simplest case to envision is if initial concentrations are $[B_0] = 1$ M and $[A_0]$ is much smaller. Throughout the reaction, $[B(t)] \approx [B_0] = 1$ M. In this case the lifetime of the small concentration of [A] in the reaction mixture is $(6.6 \times 10^9$ s$^{-1})^{-1} \approx 0.15$ ns. It doesn't matter what the concentration of A is, as long as it's much less than $[B_0]$.

EXERCISE 15.1

Explain why the lifetime of the reaction A + B → C (no back-reaction) does not depend on the concentration of A if $[B_0] \gg [A_0]$. What happens if $[A_0] \gg [B_0]$ or $[B_0] = [A_0]$?

Explanation. If there is only one molecule of A in the entire solution, at $[B_0] = 1$ M there are then 602 molecules of B in a $(10 \text{ nm})^3$ volume around A. The lifetime of A will be 0.15 ns. If there are two molecules of A, each molecule is surrounded by 602 molecules of B. The lifetime of each will be 0.15 ns. See Figure 15.11. If $[A_0]$ is 0.1 M (6 molecules per $(10 \text{ nm})^3$ volume), each A is surrounded by 602 molecules of B, except near the end of the reaction, when B has been depleted by 10%; then the reaction slows down a bit, because there are only 540 or B molecules in the $(10\text{-nm})^3$ volume. Once A is gone, the reaction stops, even though there are leftover B molecules. If $[A_0] \gg [B_0]$, then the argument holds in reverse: the reaction rate does not care which reactant is the dominant species. Even if $[A_0] = [B_0]$, the intrinsic rate of the reaction does not change. The reaction effectively slows down at long times not because the rate constant decreases but because there are very few molecules of either type left in any small volume: they must diffuse a long way to encounter each other and react.

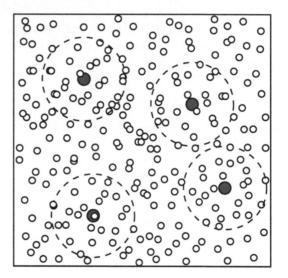

FIGURE 15.11
Bimolecular reaction with $[B_0] \gg [A_0]$. Every molecule of A in the solution has the same average number of B molecules within a "reacting" distance, so every molecule of A has the same average lifetime, independent of the concentration of A (as long as $[B_0]$ remains much larger than $[A_0]$). The decay of A appears unimolecular.

15.4 EFFECTS OF TEMPERATURE ON RATE CONSTANTS

We consider two cases here: (1) rates governed by energy barriers and (2) diffusion-controlled rates. Temperature dependence of tunneling rates is addressed in the next section.

15.4.1 Temperature Dependence of Classical "Over-the-Barrier" Rates

The system must acquire a free energy G^* in order to pass from state A to state B in Figure 15.1 or 15.7. The expression for this classical, *over-the-barrier* rate constant has a negative-exponential dependence, $e^{-H^*/k_B T}$. the activation entropy term has no temperature dependence and we neglect the possible linear temperature dependence of the frequency factor. This means that the activation enthalpy, entropy, and Gibbs free energy can be analyzed through measurements of the temperature dependence of the rate:

$$\ln\left(k_{AB}\right) \cong \ln f_0 + \frac{S^*}{k_B} - \frac{H^*}{k_B T} \tag{15.35}$$

Like the van't Hoff plot discussed in Chapter 14 for extracting equilibrium energies from experimental measurements of concentrations, a plot of the log of the rate versus absolute T^{-1} allows determination of activation enthalpy from the slope and activation entropy from the intercept at $T^{-1} = 0$. Homework problems address this procedure.

15.4.2 Measurement Biases

The strong temperature dependence of the exponential term should never be underestimated. While biological relevance would seem to restrict all measurements to a narrow (absolute) temperature range between about 0°C and 40°C, measurements over a narrow range of temperature do not allow reliable determinations of activation enthalpies and entropies, given experimental uncertainties and measurement capabilities. Besides measurements over a narrow temperature range, limited time scales impose biases on data. The shortest measurement times available in typical laboratories are often of order 10^{-4} s. See Table 15.1. If measurements are done in intervals of 10^{-4} s, digital device limitations to about 10,000 data points would allow a measurement time range of 10^{-4}–1 s. The best measurement accuracy of a signal proportional to the concentration of a reactant is typically a part in 10,000. The amplitude of a signal with a decay rate constant of 10^5 s^{-1} at time 10^{-4} s, for example, would be only about 10^{-5} times the maximal value—usually immeasurable. Problems 15.7 through 15.11 explore these limitations.

An important improvement to these experimental limitations was the development of measurement devices that measure time on a logarithmic scale.[8] Suppose, for instance, that you develop a measuring device that can take a measurement 10^{-6} s after the start of a reaction. If your digital data-storage device holds 10,000 data points, you can measure from 1 to 10,000 μs, or 10^{-6} to 10^{-2} s, and that the signal amplitude can be measured from 1,000 to 1. Suppose a reaction you are measuring decays as

$$N(t) = 1000e^{-1000t} \tag{15.36}$$

The digital values of $N(t)$ at various times are listed in Table 15.2.

While none of the measured data points could be termed unimportant, the crucial data giving information about the kinetics and the rate constant occur between about 300 and 3,000 μs—about 1 decade in time. (Exponentials decay over about 1 decade in time.) Only about 30% of the data points are useful for actual determination of rate constants, but more importantly, if a second process were also occurring with a rate constant of 10 s^{-1} or slower, measurements would miss this process entirely.

TABLE 15.1 Time Scales of Reaction Measurements

Method →	Mode-Locked Laser/ Fluorescence or Absorption	Flash Photolysis	Stop Flow	Steady-State Spectrometer	Mass Sampling of Reaction Volume
Time scale[a]	1 ps to 1 ns	10 ns to 1 s	0.1 ms^{-1} s	1 s to 6 hours	1 min to 1 year

[a] Typical values.

TABLE 15.2 Digital Signal Amplitudes N(t) from Equation 15.36

$t(\mu s)$	1	...	10	...	50	...	100	...	500	...	1,000	...	3,000	...	10,000
$N(t)$	999		990		951		905		606		368		50		0

(a)

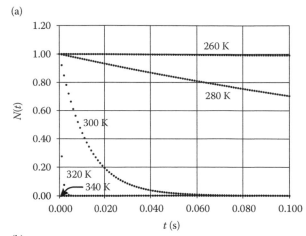

(b)

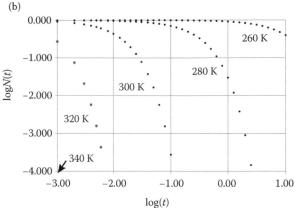

FIGURE 15.12

Temperature-dependent decay of a unimolecular reactant with an activation enthalpy of 110 kJ/mole and activation entropy of $S^*/R = 19$; $T = 260$, 280, 300, 320, and 340 K. We assume the shortest time the experimental apparatus can measure is 1.0 ms, with 0.1% resolution of $N(t)$, and 1,000 data points can be recorded and stored. (a) Linear plot of $N(t)$ decay versus t; $N(t)$ normalized to 1. The decay at $T = 340$ K is too fast to resolve, but also at 260 K no information is obtained because the decay is insignificant over the 1-ms to 1-s time period allowed by the 1,000 data points. (b) $\log N(t)$ versus $\log(t)$. Meaningful data are now recorded for four of the five temperatures. Only about 3% of the 1,000-point data memory is used.

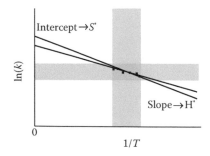

FIGURE 15.13

Correlation of H^* and S^*. Measured rate data over a small region in $1/T$ and k values, with experimental uncertainties, gives rise to correlated values of S^* and H^*.

A more sensible procedure is to space time intervals equally on a logarithmic time scale (Figure 15.12). If every decade in time had 100 time points measured, 10,000 data points would cover 100 decades in time, from 10^{-6} to 10^{94} s. The latter time is larger than the age of the universe, which really means that the measuring device need only store about 1,000 data points.

15.4.3 Compensation Effect

Experiments are subject to uncertainties and limited capabilities. The restriction of rate measurements to the temperature range 0°C–50°C or so and the typical limitation of time measurements to the range 10^{-4}–10^2 s, together with experimental error and uncertainty results in correlated values of S^* and H^*. Figure 15.13 illustrates this correlation. The data shown can be fit adequately by a range of values of H^* and S^*, but a lower value of H^* (lesser slope) correlates with a lower value of S^* (y intercept). Lowering (raising) H^* increases, while lowering S^* decreases (increases) the rates. Suppose a series of related reaction rates are measured, where a molecule is structurally modified in a series of sites. Measurements of kinetic rates of the modified molecules are done as a function of temperature and fit to an activation barrier or Arrhenius model. The results may show correlated changes in H^* and S^*. This "compensation effect" may result purely from the above mentioned experimental errors and measurement capabilities, but there may also be reasons for a real correlation effect to occur.[9] The reasons could include molecular structure of the reactant or, in the biological realm, the overall robustness of the system. The fact that life generally adapts to changes implies that many of the fundamental processes may be resistant to drastic changes in an important end result. If a structure changes for some reason and requires a larger force to break a certain bond (larger enthalpy), the overall progress of the reaction would be less affected if the organization (decreased entropy) required by the modified reactant were not so demanding on the system. One can argue that the persistence of biological life over a long period of time has ensured that important kinetic processes are, in fact, based on "compensating" reactions that tend to maintain their rates in the face of perturbations. This compensation issue has been debated for 4 decades and should be dealt with on an individual basis. In any case, the motivation for extending the range of measured temperatures is clear, which brings us to the next topic, quantum mechanical tunneling.

15.5 QUANTUM TUNNELING

Quantum-mechanical tunneling occurs when a particle crosses an energy barrier by virtually passing *through* the barrier, rather than over the barrier, as in the activated, thermodynamic processes we have dealt with so far. In classical, activated processes, a reaction occurs when a particle gains enough energy from the surrounding thermal "bath" to match or surpass the height of an energy barrier that separates initial from final states. In the simplest form of tunneling, the extension of the wave function of a particle in the initial energy well through the barrier to the final well gives the particle a finite, though normally very small, probability for appearing in the final state. We cannot say that the particle actually passes through the barrier because to demonstrate this we would have to do a measurement inside the barrier, which would disrupt the system and likely prevent any tunneling. The only measurement we can normally do is to measure the appearance of the particle in the

TABLE 15.3 Over-the-Barrier Rates at 300 and 100 K[a]

Temperature (K)	$10^{13}e^{S*/R}$ (s^{-1})	$e^{-H*/RT}$	k (s^{-1})
300	10^{13}	3.3×10^{-4}	3.3×10^9
100	10^{13}	3.5×10^{-11}	3.5×10^2
30	10^{13}	1.4×10^{-35}	1.4×10^{-22}

[a] $H* = 20$ kJ/mole, $S*/R = 0$.

final state and its disappearance from the initial state under conditions that prevent the classical, over-the-barrier process. Most often, low temperature is used to stop the classical process.

We have seen in the popular media that an ice cube can pass through the wall of a glass, a person can pass through a wall, and a camel can pass through the eye of a needle. The only question in quantum mechanics is the probability, which in these three examples is of order 10^{-10^N}, where N is perhaps 50 or so. Physically, these processes do not occur in our normal universe, though they are possible. Actually occurring biological processes that involve quantum-mechanical tunneling are surprisingly common, however. Fundamentally, the movement of any particle that can occur via classical processes—forces, collisions, passing over barriers—can also occur through quantum processes. We find that the highest probability for quantum tunneling is when the mass of the moving particle is small, so we expect that tunneling of electrons and protons would be the most common—a correct expectation. We do not have time to review all the biological tunneling processes. We shall start from the viewpoint we have just used to describe classical reactions—molecular motion or transfer—where states are separated by simple, single energy barriers.

Tunneling has been generally identified as a temperature-independent process that occurs at low temperature. The temperature independence contrasts sharply with the thermally activated, "over-the-barrier" processes we have described. A process described by a classical barrier as small as a few tens of kJ/mole virtually ceases below about 100 K. For example, if a rate is 3×10^9 s^{-1} at 300K and has an enthalpy barrier height of 20 kJ/mole, the rate at 30 K is only 10^{-22} s^{-1}, corresponding to a time of 10^{14} years, a bit longer than most organisms live (Table 15.3). In contrast, a temperature-independent tunneling rate of 10^6 s^{-1}, while insignificant at 300 K, would dominate at 150K and below.

Because 150 K is well below the freezing point of water and physiological activity has (mostly) ceased, tunneling would seem to be biologically irrelevant. Several issues signal a caution to us, however. While 20 kJ/mole ($\sim 8k_BT$) is quite small compared to most reaction energies in biophysical systems, we have seen examples where energies are even smaller. Diffusion in water is almost barrier-less. The movement of protons in water—the basis of pH control and molecular ionization in water—is ruled primarily by tunneling. Motion of electrons (charge transfer) in photosynthetic systems is also not usually faced with large energy barriers. Again, perhaps surprisingly, we find that quantum tunneling seems to be important.

Forty-five years ago, J. J. Hopfield published a seminal paper that asserts, contrary to most assumptions, that electron transfer can readily occur in biological systems by a *thermally activated quantum tunneling* process.[10] This thermal activation gives tunneling a temperature dependence and allows it to occur with significant rate at low and high temperatures. The vast majority of biochemical rates, most of which clearly involve quantum mechanics through changes in bonding, have nevertheless been described by classical reaction theory, such as we have presented so far in this chapter. Why are the descriptions classical when we have diagrammed molecules with discrete energy levels and quantized energy levels? The answer is that nothing we have done so far in reaction theory has involved the wave function or wavelength of any particle. We have only used discrete energy levels to better motivate the idea of an oscillation frequency in a well and an energy-dependent density of states, all of which could be equally well done with purely classical terminology. A computation that involves the wave function, on the other hand, has no classical counterpart. On its face, the Hopfield assertion seems to imply that all of reaction theory has been done improperly. Though quantum mechanics, in principle, should be accounted for every time a bond forms or breaks, in many cases what governs the *rate* of the reaction may be nonbonding processes or processes involving so many particles or such large quantum numbers that the overall process can be approximated classically. (Think of the harmonic oscillator with $m = 10^{-9}$ kg and

$k = 10^{-7}$ N/m, initially pulled to a displacement of 10 μm.) The case for tunneling is, however, most compelling for many processes involving the lightest biophysical particle, the electron. Electron transfer usually involves formation of an electronic excited state in a biomolecule, and motion over distances where overlap of electronic wave functions must be seriously considered. Hopfield points to electron transfer in two photosynthetic organisms and transfer distances of 0.8–1 nm. Various other authors later showed that photosynthetic electron transfer is dominated by tunneling[11] While electron transfer presents the most compelling case for tunneling in biology, we focus on tunneling of much heavier particles: protons and O_2 molecules, where we apply somewhat crude, but revealing quantum tunneling calculations.

15.5.1 O_2 and CO Binding to Myoglobin: Activated Tunneling and Structural and Energy Distributions

> Mb·O_2 binding is harder to measure, for biology-related reasons.

Carbon monoxide binding to Mb was one of the early examples of molecular tunneling in a biological context. In contrast to electron tunneling, which may be responsible for electron transfers even at physiological temperatures, CO and O_2 tunneling-mediated binding to Mb occurs well below physiological temperatures: above a temperature of about 100 K (−173°C) the activated, Arrhenius-type process is simply too fast for tunneling to compete. At low temperature, the rapid, exponential ($e^{-H^*/k_B T}$) decrease of the Arrhenius rate makes the over-the-barrier time hours or even years long. Mb·CO binding at low temperature provided the first example of the involvement of spontaneous structure fluctuations in reaction rates: individual Mb molecules have a distribution of binding rates. In addition, analysis of tunneling provided determination of not only the height of the energy barrier in Figure 15.7 but also its physical width and a correlation to a real distance on the Mb molecule.[12]

We start discussion of the Mb·CO case with the diagram in Figure 15.14a. It turns out we are being naïve about several things, but the naïve approach carries us a long way with ligand binding by myoglobin. In contrast to simple square barriers, a barrier in a real physical system has no sharp edges. This guarantees that the barrier between the two states, A and B, gets narrower near to the top, as noted in Figure 15.14b. In order to tunnel from well A to well B, the particle's wave function must have some nonzero amplitude on the other side of the barrier, in well B. As we recall from the particle-in-a-box case if the potential barrier height is infinite, the wave function's amplitude goes to zero at the barrier boundary and there is zero probability for the particle to be located on the other side. If the barrier is not infinitely high, the probability is not zero, unless the barrier is infinitely thick, which is of no interest to us. So, we expect the tunneling probability per unit time (the tunneling rate) to depend on the barrier height and width, and perhaps to a lesser extent, the shape of the barrier.

The second property of quantum tunneling we infer from Figure 15.14a and b is that a thermally activated (vibrationally excited) state in well A would have a higher probability of passing through the barrier than the ground state. This immediately violates the "temperature-independent" rumor we may have heard about tunneling, but that rumor came about largely because most tunneling measurements have been done at such low temperatures that vibrations were not thermally excited. Figure 15.14 almost guarantees that if tunneling occurs at low temperature and over-the-barrier motion occurs at high temperature, thermally activated tunneling should occur in a transition region. We find this is indeed the case with CO binding to Mb.

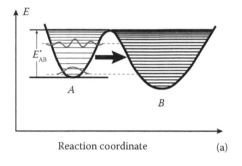

(a)

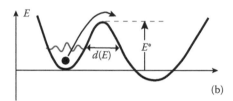

(b)

(c)

FIGURE 15.14

Quantum tunneling through a barrier. (a) The tunneling rate through the barrier is controlled primarily by the width and secondarily by the height of the barrier. A higher-energy initial state in well A has a shorter distance to tunnel, except for the unusual case of a square barrier. Higher-energy states result from thermal activation or phonon assistance. (b) The tunneling distance can be a function of the barrier height directly, if different molecules have different barriers, and indirectly, if the initial state is thermally excited. (c) The tunneling distance in myoglobin has been connected to the ~0.05-nm displacement of the heme Fe atom upon binding of CO (or O_2). (PDB ID 1MBC, Kuriyan, J. et al., J. Mol. Biol., 192, 133–154, 1986.)

A third property we can infer from Figure 15.14a comes from the discrete energy levels marked in both wells. Suppose a particle starts from an energy level in well A, tunnels through the barrier, but does not find a discrete state at exactly the same energy. Does this prevent the tunneling process, or does conservation of energy not apply here and the particle can go up or down a little in energy to find a level? Does it prefer to go down in energy? Does it go all the way to the bottom of well B? Our reflex is likely to answer "No" to each question, but they are good questions. Tunneling is, indeed, more probable when the tunneling particle finds an equal-energy state on the other side of the barrier. You *can* violate conservation of energy for a time, however, via the uncertainty principle, $\Delta E \Delta t \geq \hbar/2$. Sooner or later we have to pay for the deficit of energy or dispose of the excess energy, however. In order to do so we employ the thermal bath: we borrow or give vibrational or collisional energy from/to the surroundings. Such transfer of energy to or from the surroundings can only occur at temperatures above absolute zero and when there is some connection to the surroundings, but this is always the case in biology. The surroundings are always close by and well connected. If we go to low enough temperature, however, we lose this thermal bath of vibrations and collisions, though we can still dispose of excess energy when the particle gets to well B. The overall implication of this picture analysis is that thermally activated tunneling should occur, but as temperature is decreased a point should be reached where only temperature-independent tunneling should occur.

We have not written any equations yet, but through simple diagram analysis we have prepared ourselves for some quite advanced tunneling theory compared to what is normally encountered in even advanced quantum mechanics courses. This is what biology continually does to us: forces us to press our physics just a bit further, which is why biophysics will remain a frontier of physics for a very long time. We do have one more issue to address, however: the frequency factor. In over-the-barrier motion we identified the reaction rate frequency factor as the frequency of oscillation in well A. The particle is in well A, undergoing continual collisions with nearby atoms and gaining or losing energy. However, there is some average oscillation frequency possessed by potentially reactive particles near the top of the barrier. We identified this as the "attempt" frequency for classical motion. In the quantum case we have been taught to think of states in a potential well as "stationary": the particle wave function sits there in well A, at peace with the world. The particle would seem by this description to be isolated. This can't be in a biological system. The surrounding molecules are still there, moving and colliding. What does this do to the state in well A? Quantum students should note immediately that the vibrational states marked in well A (and in well B) result from interaction with the surroundings, or possibly by an external electric or other field. Potential wells do not occur magically. They are formed by other atoms in the neighborhood. If these atoms are at finite temperature, their interaction with our tunneling particle fluctuates and the particle can make transitions to other vibrational states in well A.

But still, what is the quantum tunneling frequency or attempt factor? How many times per second does the particle "try" to tunnel through the barrier? Is it the collisional frequency or the same frequency we used for the classical reaction rate? The question is complex enough that we often leave its answer to experimental measurement. We have neglected to seriously deal with one key quantum issue we have already mentioned, however. The states we are dealing with are not stationary states: they are time dependent. A particle placed even in a single well occupies a stationary state (no time dependence) only if we give it the exact energy of one of the quantum stationary states. If the energy does not exactly match, the particle is in a superposition of states with a time dependence given by $\psi(x,t) = a_1\varphi_1(x)e^{-iE_1 t/\hbar} + a_2\varphi_2(x)e^{-iE_2 t/\hbar} + \ldots$ We really don't want to pursue this route, however; it's too messy. Suppose we place the particle in well A of the two-well system we are describing. No matter what energy we give the particle, the particle, "placed in well A," is not in a stationary (time-independent) state. It cannot be, simply because there *is no* stationary state of the two-well system where the particle is localized in well A. The stationary states are delocalized—present in both wells at once. Therefore a particle initially placed in well A oscillates with a frequency (or distribution of frequencies) determined by the structure of the two-well system. This is a tough problem to solve, and we deal with the simpler case of two identical wells located at the same energy in the next section, proton tunneling. For now, we leave the Mb·CO frequency factor to experimental measurement.

What do measurements tell us about quantum tunneling in the binding of CO to Mb? The first observation is that if we reduce the temperature of a solution of Mb without CO to 100K and *then* allow CO access to the surface of the frozen solution, we wait months for any binding to occur.

For technical reasons, water is usually replaced by a glycerol–water mixture, which forms an almost clear glass when frozen, allowing optical measurements.

The problem is that the gas cannot travel through the frozen solution, get through the protein-solvent interface, and find the heme binding site in a reasonable time. We have to somehow produce Mb in a frozen solution with CO already inside the protein, but not already bound to the Fe atom in the heme group (the binding site). An optical technique, useful for medical treatment of cases of CO poisoning, was discovered decades ago. Absorption of a photon by Mb·CO (myoglobin with CO bound at the heme Fe) breaks the heme–CO bond with high probability, even at low temperature. After the heme–CO bond is broken, the CO is in a nearby position, cannot diffuse outside into the frozen solution, and has little choice but to rebind to the heme. This whole process can be easily monitored because the absorption spectrum of Mb and MB·CO differ significantly. A brief laser flash at one of the absorption bands of Mb (usually near 600 nm) photodissociates CO from Mb, while a continuous beam of light (usually at the heme "Soret" band near 420 nm) monitors the process of conversion from Mb·CO to Mb + CO, back to Mb·CO again. Samples can be placed in cryostats and cooled or warmed to any desired temperature.

The first observation when the temperature of the Mb sample is reduced far enough that the over-the-barrier rebinding reaction has ceased is that the decay obeys a very different mathematical form. When a collection of pairs of Mb + CO rebinds at a rate determined by a single activation energy barrier, the rebinding should obey an exponential decay law: $N(t) = N_0 e^{-kt}$, where $N(t)$ is the number (or fraction) of Mb molecules that have not yet rebound a CO. Such a decay is plotted in red in Figure 15.15a on a log–log graph ($\log(N(t))$ versus $\log(t)$). Once times comparable to k^{-1} are reached, the exponential decay drops very rapidly. On the other hand, the observed decays stretch almost linearly over many orders of magnitude.

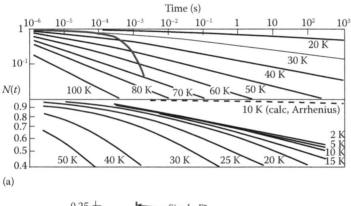

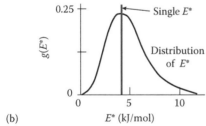

(a)

(b)

FIGURE 15.15
Rebinding of CO to Mb at low temperature after a photodissociating flash of light. (a) Rather than an exponential decay (red curve), the rebinding follows a power law, $N(t) \cong N_0(1 + t/t_0)^{-m}$, extending over many orders of magnitude in time. Below a temperature of about 20 K, the reaction occurs at rates much higher than expected from classical over-the-barrier motion. The near temperature independence below 10 K (lower panel) indicates tunneling is occurring. Above 30 K or so, the reaction occurs classically. (b) The nonexponential time dependence of the decay is readily described by a distribution of barrier heights, resulting from differing Mb conformations. An exponential decay has a single activation energy (red). At room temperature, rapid relaxation of the protein structure leads to a single, average activation energy and (approximately) exponential rebinding kinetics. Modified from original data of ref. 12.

15.5.2 Reminder: Mathematical Form of a Straight Line on a Log–Log Plot

If the form of the decay on a log–log plot is a straight line with negative slope, $\log(N(t)) = -m\log(t) = \log(t^{-m})$, then $N(t) \sim t^{-m}$. This is a power law. This functional form cannot be quite correct, of course, because $N(t) \to \infty$ as $t \to 0$. This can be fixed up by writing

$$N(t) = N_0 \left(1 + \frac{t}{t_0}\right)^{-m} \tag{15.37}$$

What do a function of the form of Equation 15.37 and the data in Figure 15.15a mean? The quick answer is that it reflects a property of macromolecules we might have expected. Proteins have hundreds or thousands of atoms that, at room temperature, are constantly fluctuating. The X-ray crystallography data (atomic mobility or temperature factor) confirm this. What happens when the protein is frozen? We expect that many, slightly differing protein conformations are "frozen in": the population of protein molecules will have a distribution of conformations. Because in Mb the protein is covalently connected to the binding site, one expects the binding site to also exhibit a distribution of structures, with a distribution of binding rates. Such a distribution of binding rates corresponds to a distribution of activation energy barriers, $g(E^*)$, notated simply $g(E)$ in the literature. The activation-energy distribution for Mb–CO is shown in Figure 15.15b. The rebinding of CO (or O_2) is now governed by a distribution of binding rates, corresponding to a weighted sum of exponentials with many different time constants k^{-1}. Why wasn't this distribution of activation energies observed at room temperature? Because the protein was not frozen, its structure rapidly fluctuates between allowed conformations, creating a time-averaged structure, which binds with an apparent single activation energy. The rebinding in Figure 15.15a from about 30 K and above occurs via a classical, over-the-barrier reaction. Below 10 K the rebinding is identified as tunneling by its temperature independence.

We are almost done, but we have not determined the tunneling equivalent of the classical Arrhenius/thermodynamic activation rate in Equation 15.19 or 15.21. We could note that k_A in Equation 15.21 or 15.19 is actually dependent on the energy barrier height, now that we know that there is a distribution of barrier energies. We likewise should write the tunneling rate as a function of barrier height and also width, but we have already convinced ourselves that any reasonable barrier width automatically increases as the height increases. We can therefore write that the tunneling rate constant $k_t(E^*)$ is a function of barrier height, E^* (or H^*). The functional dependence of a tunneling rate on the height of an energy barrier that is approximately an inverted parabola turns out to be approximately[12]

$$k_t(E^*) = A_t \exp\left(-\frac{\pi (2mE^*)^{1/2} d(E^*)}{2\hbar} \right) \tag{15.38}$$

The expression $d(E^*)$ indicates the dependence of barrier width on height. In this case the mass m is best identified with the reduced mass of the Fe–CO system.

The math we have marshaled for the Mb·CO experiment has provided a quite unique opportunity in the context of tunneling theory. Normally, the barrier width can be determined if the height is known. In the Mb case, the known functional form for the distribution of barrier heights, $g(E^*)$, determined where over-the-barrier processes dominate, allows determination of the functional dependence of barrier width on barrier height. The reference cited determined

$$d(E) \cong d_0 \left(\frac{E}{E_{peak}} \right)^{1.5}, \text{ with } d_0 = 0.05 \pm 0.01 \text{ nm} \tag{15.39}$$

What's the biological relevance here? We noted earlier that the barrier distance should describe some physical distance of motion in the Mb–CO system. This distance of 0.05 nm determined from quantum tunneling measurements of binding at very low temperatures should correspond to the physical distance the Fe–CO center of mass moves when CO binds. The value from tunneling is very close to that determined from X-ray crystallography!

15.5.3 Proton Tunneling[13]

The tunneling probability increases as the mass of the particle decreases. Next to the electron, the proton is the lowest mass particle we commonly encounter in biological systems. Tunneling has been found in a variety of biological proton-transfer processes,[14] but we focus on the single most common case, the movement of protons in water. Chapter 4 focused on water structure and properties. The structure of water is dominated by a hydrogen-bonding network, where each water molecule is hydrogen-bonded to about four neighboring molecules. We further noted that besides H bonding, a proton from a water molecule can readily detach itself, not because the energy needed to remove a proton from H_2O to a free state is so low, but that the detached H^+ readily finds a nearby H_2O site, where the energy of an H_3O^+ state is low. (There is no H^+ vacuum state in liquid water.) If a water molecule spontaneously ionizes and gives up a proton, that proton will move to another H_2O to form H_3O^+, leaving OH^-. However, the most likely occurrence is that a proton from yet another water moves over to the first water, now an OH^-. (Remember only 10^{-7} M [H^+] exists in 55 M water.) Likewise, the most likely occurrence is for the newly formed H_3O^+ to pass its extra proton further. How do these protons move? We might be quick to jump to the term *diffusion*, but this term says nothing about the mechanism of the motion. The diffusion we have treated in previous chapters implicitly assumed classical elastic collisions transferred momentum and energy from one particle to another. This cannot be easily invoked to explain the movement of a charged group from one molecule to another.

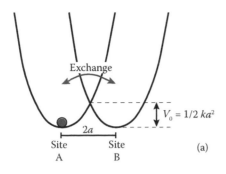

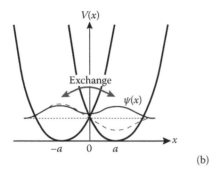

FIGURE 15.16
Model for movement of proton between two equivalent sites in water. (a) The two sites, corresponding to attachment to two adjacent water molecules, are modeled as harmonic potentials with equal-energy bottoms. The height of the barrier between the two sites is governed by the distance between sites. (b) The time-independent solution to the Schrödinger equation must have equal probabilities for occupation of either well, but if the proton initially occupies one well, the quantum probability will exchange between the two wells at a frequency $f = \omega/2\pi$. The lowest-energy symmetric and anti-symmetric wave functions are sketched.

> Why compare to a benzoic acid crystal? (1) Good data is available and (2) tunneling does not depend on fluidity, as long as the structure (the potential) is stable significantly longer than the period $2\pi/\omega$. Water structure fluctuates on a time scale of 10^{-12} s.

We should remind ourselves of the energetics of water structure before we jump into any mechanistic or tunneling description of proton movement. The H bond strength in water is about 20 kJ/mole, while London (dispersion) energies are ~0.6 kJ/mole and dipole–dipole interaction energies less than 10 kJ/ mole. The conclusion was that H bonding is the dominant interaction between water molecules. Figure 15.16a presents a simple model for the dynamics of a proton in water. The proton is bound to a site on one water molecule via a spring constant, k, but the location of a second water molecule a short distance, a, away, presents the proton with a second, equal-energy site. Those experienced in the solution to this double-well quantum problem know that the time-independent solution places equal probabilities in each well. We approach this problem with diagrams rather than detailed solution to the Schrödinger equation. Figure 15.16b shows a sketch of the lowest-energy symmetric and antisymmetric wave function solutions to the time-independent problem:

$$-\frac{\hbar^2}{2m}\frac{\partial^2\psi(x)}{\partial x^2}+\left[\frac{1}{2}k(x-a)^2+\frac{1}{2}k(x+a)^2\right]\psi(x)=E\psi(x) \quad (15.40)$$

We do not have a time-independent situation, however. We should first place the proton in well A and then ask what happens to it as time progresses. The answer is that the proton can move to well B; then it would move back. This would come from the solution to the time-dependent Schrödinger equation, and if the particle energy is less than the height of the energy barrier between the two states, $\frac{1}{2}ka^2$, the transition process would be a tunneling process. The particle would oscillate back and forth at an angular frequency ω' determined by the energy difference between the two lowest states:

$$\Delta E = 2\hbar\omega\sqrt{\frac{2V_0}{\pi\hbar\omega}}\exp\left(-\frac{2V_0}{\hbar\omega}\right) \quad (15.41)$$

(see the Merzbacher reference) and

$$\text{Tunneling (angular) frequency} \equiv \omega' = \frac{\Delta E}{\hbar} \quad (15.42)$$

where $\omega=\sqrt{\frac{k}{m}}$ and $V_0=\frac{1}{2}ka^2$. For reasonable values $V_0 = 20$ kJ/mole, $a = 0.1$ nm and $m = m_p = 1.7 \times 10^{-27}$ kg, the oscillation angular frequency is about 10^{13} Hz (rad/s). This implies a very fast proton tunneling rate, a rate that is much higher than the measured proton tunneling rate, for example, in benzoic acid crystals, 10^{10} Hz.[15]

EXERCISE 15.2

Proton tunneling and water structure fluctuation.

1. What are values for ω and ω' in the previous paragraph? What is the physical meaning of each?
2. The argument just made was that water structure and the potential are stable for about 10^{-12} s, whereas the tunneling time is much faster. What is the tunneling time (inverse of the tunneling frequency)?
3. What should be the effect of water structure fluctuation on tunneling?

Solutions

1. $V_0 = \dfrac{1}{2}ka^2 = \dfrac{1}{2}m_p\omega^2 a^2 \Rightarrow \omega = \sqrt{\dfrac{2V_0}{m_p a^2}} = \sqrt{\dfrac{2(10{,}000 \text{ J / mole}) / (6.02 \times 10^{23} \text{mole}^{-1})}{(1.67 \times 10^{-27} \text{kg})(10^{-10}\text{m})^2}}$

   $= 6.3 \times 10^{13} \text{rad/s}$

   where ω is the angular frequency of oscillation of a proton bound in either well.

   $$\begin{aligned}
   \omega' \;\; &= \;\; \frac{\Delta E}{\hbar} = 2\omega\sqrt{\frac{2V_0}{\pi \hbar \omega}}\,\exp\!\left(-\frac{2V_0}{\hbar\omega}\right) \\[2mm]
   &= \;\; 2\left(4.5 \times 10^{13} \text{rad / s}\right)\sqrt{\frac{2\left(3.32 \times 10^{-20}\,J\right)}{\pi\left(6.64 \times 10^{-20}\,J\right)}}\,\exp\!\left(-\frac{2\left(3.32 \times 10^{-20}\,J\right)}{\pi\left(6.64 \times 10^{-20}\,J\right)}\right) \\[2mm]
   &= \;\; \left(12.6 \times 10^{13} \text{rad / s}\right)\sqrt{3.18}\,\exp(-3.18) = 9.3 \times 10^{12} \text{rad / s}
   \end{aligned}$$

   ω' is the angular frequency of oscillation of a proton from one well to the other.

2. The tunneling time is $2\pi/\omega' = 6.8 \times 10^{-13}$ s. This tunneling time is faster, but not that much faster, than the water fluctuation time.

3. As water molecules randomly move around, the distance between adjacent molecules fluctuates; not much, because water is closely packed, but the critical distance is between an extra proton on one water molecule (H_2O) and the O atom on another H_2O (Figure 15.17). When a water molecule rotates, this distance changes. Fluctuations can also place a proton into a excited vibrational state in one of the wells, decreasing V_0 and the effective width of the barrier. Because a proton will not move when the distance a is large, it will tend to "wait" until a fluctuation causes the distance to decrease, then transfer.

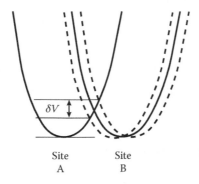

FIGURE 15.17
The fluctuating structure of water gives rise to potential wells that fluctuate in position, giving rise to a fluctuating barrier height and width, in the double-well model of Figure 15.16.

We noted that our computed tunneling rates are probably several orders of magnitude too high. While we could attribute this disagreement to the crudeness of the tunneling model we used, we should also take seriously the comments we made about the correlated movement of several protons: when a proton moves to a new molecule, a proton from that new molecule must, on average, move to yet another H_2O, where it displaces yet another, etc. While this does not imply that all 10^{25} water molecules in a liter of water must pass a proton simultaneously, there tends to be some correlation, which reduces the proton exchange rate.

15.6 A ⇌ B: UNIMOLECULAR REACTIONS

The simplest type of reaction involves only one molecule, which may have several metastable states it can exist in. These metastable states are usually local free-energy minima but generally can be any region where the slope of the free energy versus reaction coordinate is near zero. We have seen the application of unimolecular, equilibrium analysis of protein folding and microcalorimetry in the preceding chapter, with equilibrium concentrations compared explicitly to measured heats absorbed as sample temperature is raised. This section adds details of the kinetics of such unimolecular processes. We quickly find, however, that the kinetics are sensitive to the details of the reaction mechanism, such as intermediate or alternate states, that do not affect equilibrium. This has advantages and disadvantages. More sensitivity to molecular details means kinetic measurements have the potential to reveal more information about how biology operates, but these details may have to be completely worked out before any understanding is reached. The overall result in the biological sciences has been a proliferation of competing kinetic models to explain reaction mechanisms.

An alternate derivation of the activated-complex reaction rate theory makes the explicit assumption that the state at the top of the energy barrier is metastable, in the sense that it is in equilibrium with the other states in well A. (The same cannot be said of the states on the downhill side of well B, for example.) Even the Boltzmann–harmonic well model we used makes the implicit assumption that all states in the initial well A, including that at energy E^*, are in thermal equilibrium. The unimolecular reaction type A ⇌ B or A_i ⇌ A_{ii} assumes that only two metastable states are involved in the reaction, but we must be careful to separate equilibrium from kinetics here. In Figure 15.18a, two conformational states of a molecule are shown. In equilibrium, $K_{eq}^{i,ii} = \frac{[ii]_{eq}}{[i]_{eq}} = \frac{k_1}{k_{-1}}$, and the kinetics of interchange is described by the mathematics in this section. If a third state is involved (Figure 15.18b), the equilibrium between i and ii is unaffected, $K_{eq}^{i,ii} = \frac{[ii]_{eq}}{[i]_{eq}} = \frac{k_1}{k_{-1}}$, though the total number, $N_i + N_{ii}$, in the two states is reduced by displacement to state iii. Figure 15.19 shows the intermediate state in the context of a free-energy diagram, placed either in series (a) or in parallel with the process directly connecting the two states i and iii. In both cases, the equilibrium probabilities (concentrations) of i and iii are unaffected by the third state.

The differential equation describing the two-state, bulk solution reaction A ⇌ B of Figure 15.18a is

$$-\frac{d[A]}{dt} = k_1[A] - k_{-1}[B] \tag{15.43}$$

Conservation of particles, $[A] + [B] = [A]_0 + [B]_0 = const$, where subscripts indicate initial concentrations, allow solution of this differential equation. For the case

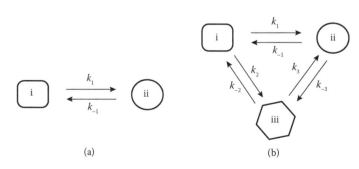

(a)

(b)

FIGURE 15.18
Unimolecular reactions and detailed balancing. (a) The equilibrium between i and ii depends only on the free-energy difference between the states i and ii. The kinetics of conversion of i to ii (or the reverse) is governed by the rates k_1 and k_{-1}. (b) If a third state (conformation) of a molecule exists, two new sets of rates are introduced, which influence the interconversion kinetics, but do not affect the equilibrium between i and ii.

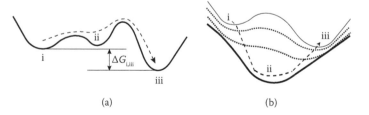

(a)

(b)

FIGURE 15.19
Intermediate states from a free energy landscape perspective. (a) Intermediate state ii between primary states i and iii. All three states are connected in series. Four states in series were invoked to explain CO binding to Mb by Austin et al., 1975. (b) Intermediate states may also be accessed along other dimensions of the reaction coordinates, effectively connecting states in parallel; see also Figure 15.18b. Four free-energy profiles (two solid and two dotted lines) are shown on a two-dimensional free-energy surface (one coordinate extends upward from page). In both cases the equilibrium between i and iii is unaffected by ii, though the kinetics of interconversion *are*.

$$B_0 = 0, B = A_0 - A : \frac{[A] - [A]_{eq}}{[A]_0 - [A]_{eq}} = e^{-(k_1 + k_{-1})t} \qquad (15.44)$$

where

$$[A]_{eq} = \frac{k_{-1}}{k_1 + k_{-1}} [A]_0 \qquad (15.45)$$

The result that the kinetics is governed by the sum of the forward and reverse rates is general. Perturbation chemistry, in which a solution at equilibrium is perturbed by suddenly changing the temperature (*T-jump*, or temperature-jump spectroscopy), pressure, or some other parameter, also measures this sum of kinetic rates.[16] Perturbation techniques have produced a large fraction of the kinetic data on solution chemical and biochemical systems.

15.7 A + B → C BINDING REACTIONS: FREE-SOLUTION REACTIONS

We have treated a detailed example of (arguably) the simplest possible biologically relevant, bimolecular reaction, the binding of O_2 or CO to the protein myoglobin. The binding process involves diffusion in water, crossing of a protein-solvent interface, passing through fluctuating protein structure, binding to a prosthetic group-bound Fe atom inside the protein, movement of the Fe, coupled to protein structure, and a distribution of activation-energy barrier heights governing the final rebinding step. We have mostly ignored the details of the intermediate steps—the multiple barriers corresponding to the solvent interface and CO movement through the protein structure. Do we assign multiple energy barriers? Must the barriers be sequentially passed or are some in parallel? How do we treat movement through a fluctuating protein structure? Simple diffusion, governed by the CO radius and the protein's average viscosity? We leave this for more advanced texts, but we should question the worth of what we are about to do next, the standard treatment of "bimolecular reactions." Such simplistic treatment is perhaps biologically questionable, but it does occur occasionally. We should be equipped to perform such unsophisticated treatments, with a mental attitude of surprise that such cases occur in biosystems.

We start where we left off in the section on diffusion-limited reaction rates. Two molecules diffuse randomly in a fluid and react when they come within a distance R. They do not have to react instantly, though the equations we developed apply to this case. We should restrict ourselves, however, to the idea of a true "bimolecular" reaction: there are many particles of each type, A and B, in the solution and no particular molecule has a corner on reaction with another molecule. The competition to react is a true competition.

> Do not confuse particle types, A and B, with states or potential wells A and B.

The simplest bimolecular reaction, A + B → C, is the dimerization of one type of molecule:

$$A + A \cdot (A)_2 \text{ or } A \cdot A \qquad (15.46)$$

(We ignore the reverse reaction for now.) The differential equation describing the simplest version of this reaction is

$$-\frac{d[A]}{dt} = 2k[A]^2 \qquad (15.47)$$

The factor of 2 is included here by tradition, in deference to the fact that both reactant molecules diffuse at the same rate. Manipulation of this equation leads to $-\frac{d[A]}{[A]^2} = 2kdt$, which can be directly integrated on both sides to yield the solution

$$\frac{1}{[A]} - \frac{1}{[A]_0} = 2kt \qquad (15.48)$$

In contrast to the exponential decay of a unimolecular reaction, bimolecular "dimerization" reactions have a slower $1/t$ dependence. There is nothing fundamental about this difference—the reaction simply slows continuously, compared to the exponential, because the concentration of the reactants decreases and diffusion over larger distances must occur.

The more general case of a bimolecular reaction is

$$A + B \rightarrow C \tag{15.49}$$

in which case the governing differential equation is

$$-\frac{d[A]}{dt} = -\frac{d[B]}{dt} = \frac{d[C]}{dt} = k[A][B] \tag{15.50}$$

Case 1: $[A]_0 = [B]_0$
If the initial concentrations of A and B are equal, then $[A] = [B]$ at all times and the mathematical description would follow that of Equation 15.47, after removing the factor of 2.

Case 2: $[B]_0 \gg [A]_0$
One reactant is present in much higher concentration. We can then approximate $[B(t)] \cong [B]_0$ and Equation 15.50 simplifies to

$$-\frac{d[A]}{dt} = \left(k[B]_0\right)[A] \tag{15.51}$$

The reaction becomes pseudo-unimolecular, with a concentration-dependent apparent rate constant, $k[B]_0$. The solution to this differential equation is the decaying exponential

$$[A(t)] = [A]_0\, e^{-k[B]_0 t} \tag{15.52}$$

When reaction mechanisms are being studied, the large excess of B makes analysis more targeted: determining if the time dependence is exponential.

Case 3: $[B]_0 \neq [A]_0$ but comparable concentrations
In this case, there will be leftover A or B, but we cannot neglect any decrease in concentration with time. The guiding differential equation is

$$-\frac{d[A]}{dt} = k[A][B] = k[A]\left([A] - [A]_0 + [B]_0\right) \tag{15.53}$$

The solution can be written

> We should be more careful here. The term *unimolecular* refers to a reaction involving only one type of molecule, while *first-order* refers to exponential reaction kinetics that depend on the concentration of only one molecule. A bimolecular reaction in the limit $[B]_0 \gg [A]_0$ is first-order, but not unimolecular.

$$\frac{1}{\Delta} \ln\left(\frac{[A]_0 [B]}{[A][B]_0}\right) = kt \tag{15.54}$$

or

$$\frac{[A]}{[B]} = \frac{[A]_0}{[B]_0} e^{-\Delta kt} \tag{15.55}$$

where $\Delta = [B]_0 - [A]_0$. Note that the ratio decreases exponentially with time, but the individual concentrations do not. In Problems 15.13 and 15.14 we examine the relation between Equations 15.48, 15.52, and 15.55. Though the more general equation, Equation 15.55, would seem to embody the approximations 15.48 and 15.52, the approximations must be made carefully. Figure 15.20 shows the difference between the decay law of Equation 15.55 and an exponential decay $e^{-\Delta kt}$.

The bimolecular reaction decay carries with it no new biophysics compared to the pseudo-unimolecular limit shown in Equation 15.52. The kinetics do not reflect, for example, any particular conformational change of a protein, tunneling versus activated reaction, etc. The kinetic difference only reflects the depletion of the free concentrations of one or both components. It is a good reminder, however, that if we try to measure an actual reaction occurring in a cell we may observe nonexponential reaction kinetics that reflect the local depletion of one reactant. In fact, this depletion is central to control of biochemical processes in the cell. Feedback to stimulate or reduce certain reactions parallels primarily the reduced concentration(s) of specific molecules and secondarily the rate of their appearance or disappearance.

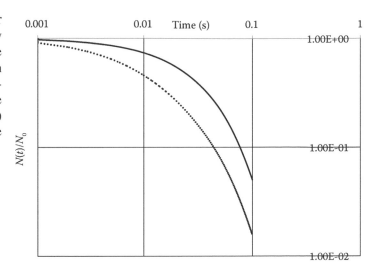

FIGURE 15.20
Bimolecular reaction kinetics: log–log plot. The reaction is A + B → C, with (lower line) $[A]_0 = 1.00 \times 10^{-3}$ M, $[B]_0 = 0.70 \times 10^{-3}$ M, and $k = 1.00 \times 10^5$ M^{-1} s^{-1}. The lower line is the (normalized) concentration of A as a function of time. The upper line is an exponential decay $e^{-([A]_0 - [B]_0)kt}$.

15.8 COMPLEX REACTIONS: RATE-DETERMINING STEPS AND MICHAELIS–MENTEN ANALYSIS

The idea of a *rate-controlling step* in a biochemical reaction stems from (1) the general complexity of biological reactions and the resulting difficult mathematical description and (2) the idea that biology takes place near 310 K, so rate measurements need be done only near that temperature. The rate-controlling step description is usually applied to a serial set of states with accompanying barriers that a set of reactants must pass through. The same principle of barriers and states that determine overall reaction rates also pertains to barriers in parallel, however. Figure 15.21a shows a case of barriers in series in which one clearly dominates any rate analysis, especially when the placement of G^* in an exponential is considered. The dominance of one barrier is not so clear in Figure 15.21b. Figure 15.21c shows the result of analysis of CO binding to Mb. There is no rate-determining step or barrier: the binding at physiological temperature is determined by a combination of many rates that cannot, in general, be expressed in analytic form. How should one determine whether the idea of one "bottom-line," rate-determining step analysis is valid? Though the temperature of immediate interest may be 310 K, measurements may have to be performed over a wide temperature range in order to interpret the contribution of various elementary steps in the reaction. It is often thought that the diffusion rate is rarely involved in biochemical reactions, because it is faster than most other processes. Yet the diffusion rate appears even in Michaelis–Menten analysis, used for virtually all enzymatic reaction analysis (below), though embedded in other parameters.

We will not spend time here to consider a complete treatment of complex reactions such as has been done in the case of rebinding of CO to Mb.[4] This analysis involved many months of work and

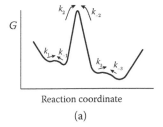
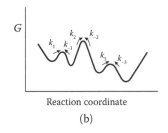
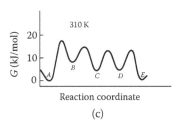

(a) (b) (c)

FIGURE 15.21
Reaction energy-barrier diagram to judge "rate-controlling" step analysis. (a) Traditional Gibbs free energy diagram showing one barrier that clearly dominates the reaction in either direction: the "rate-determining step" is over the middle barrier and the rates k_2 and k_{-2} are the only rates that matter. (Remember that barrier heights G^* go into exponential functions, amplifying differences between transition rates.) (b) A case where one rate may not adequately describe the reaction. (c) Results from the analysis of CO binding to Mb: no rate-determining step exists except at low temperature. All barriers must be taken into account at physiological temperature. Measurements over a wide temperature range are needed to determine whether rate-determining step analysis is valid.

measurements over a large temperature range, including liquid-helium temperature ranges. The question before us is whether *every* biological reaction must be analyzed in its full complexity, or whether more simplistic treatments at least point us in the right direction. General principles of uncertainty and error analysis, along with the experience of analyzing seeming enthalpy–entropy correlations caused by measurements over a small temperature range (see homework) teach us to be cautious about the physical meaning of a quick series of measurements, such as described in the following section. Nevertheless, standard measurements can at least provide a uniform way to classify enzyme systems for further investigation. Most of the time, these further investigations center on analysis of the structure of the enzyme-substrate complex, using X-ray crystallography and site-directed mutagenesis techniques to probe details of hypothesized critical reaction structures and steps.

15.8.1 Michaelis–Menten Models of Enzyme-Catalyzed Reactions

The vast majority of enzyme-catalyzed reactions have been described by the Michaelis–Menten, which is a steady-state, kinetic model based on the mechanism,

$$E + S \underset{k_{-1}}{\overset{k_1}{\rightleftharpoons}} E \cdot S \overset{k_2}{\longrightarrow} E + P \tag{15.56}$$

E is the enzyme, S its substrate, and P the product of the catalyzed reaction. The enzyme appears not to be involved in the overall reaction because it is regenerated with the product. The reaction mechanism in Equation 15.56 does not imply that the substrate and product are necessarily in thermodynamic equilibrium, since no reverse rate is included in the final reaction step. With enzyme-catalyzed reactions, the issue of equilibrium is often in doubt, as reactions without the enzyme often take hours to months to reach equilibrium—times far too long to be relevant to cell processes occurring in less than a second. Nevertheless, the usual treatment of enzymatic reactions assumes that the enzyme does not affect the free energies of the initial or final states, but only the barrier separating the two states, as shown in Figure 15.22.

The reduction of the barrier height when enzyme binds substrate (E·S) is usually caused by either a substrate conformation primed for reaction, often by placement of substrate close to a charged or polar amino acid side chain of the enzyme or the placement of substrate close to a second reactant. The latter case of a second reactant is technically not covered by reaction mechanism 15.53, but if that molecule is at high enough concentration to be constantly bound to the enzyme, its concentration may not affect the reaction under conditions of the measurement (usually bulk solution).

The differential rate equations for Equation 15.56 are

$$\frac{d[E \cdot S]}{dt} = k_1[E][S] - k_{-1}[E \cdot S] - k_2[E \cdot S] \tag{15.57}$$

$$\frac{d[E]}{dt} = -k_1[E][S] + k_{-1}[E \cdot S] + k_2[E \cdot S] \tag{15.58}$$

$$\frac{d[S]}{dt} = -k_1[E][S] + k_{-1}[E \cdot S] \tag{15.59}$$

$$\frac{d[P]}{dt} = k_2[E \cdot S] \tag{15.60}$$

The Michaelis–Menten assumption of a steady state for [E·S],

$$\frac{d[E \cdot S]}{dt} = 0 = k_1[E][S] - k_{-1}[E \cdot S] - k_2[E \cdot S] \tag{15.61}$$

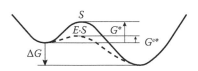

FIGURE 15.22
Reduction of activation-energy barrier height by enzymatic binding: the standard model. The barrier reduction increases the reaction rate dramatically, without affecting the equilibrium. The energy reduction is usually assumed to affect only the enthalpy or internal energy, so an equivalent diagram with G replaced by E is frequently encountered. There is no general reason to assume entropy is not affected, however.

implies $d[E]/dt = 0$ and

$$\frac{d[P]}{dt} = k_2[E \cdot S]_{ss} \tag{15.62}$$

$$[E \cdot S]_{ss} = \frac{k_1[E][S]}{k_{-1} + k_2} \tag{15.63}$$

Because the concentration $[E \cdot S]$ is not easy to measure, but $[E]_0 = [E] + [ES]$ is, it helps to rewrite equations in terms of $[E]_0$. Some algebra leads to

$$[E \cdot S]_{ss} = \frac{[E]_0[S]}{K_M + [S]} \tag{15.64}$$

with the Michaelis–Menten constant K_M defined as

$$K_M \equiv \frac{k_{-1} + k_2}{k_1} \tag{15.65}$$

If you are not puzzled over Equation 15.64, you should be. $[E \cdot S]_{ss}$ is a steady-state concentration; it should be constant. Is the right side of Equation 15.64 constant? K_M is a constant, but the concentration $[S]$ is presumably time dependent. The answer to this contradiction is that you simply do not measure for too long a time. If $[S]_0$ is chosen large compared to K_M, $[E \cdot S]$ is approximately constant. (You also must wait long enough for the complex E·S to form because initially there is none.)

The rate of appearance of the product, Equation 15.62, constitutes one of the primary measurements for enzymatic reactions in Michaelis–Menten determinations. The time dependence of concentrations of all chemical species involved in Reaction (15.56) is sketched in Figure 15.23a. The time needed for $[E \cdot S]$ to reach steady state is t_{ss0}. The concentrations of S and P change approximately linearly with time, until the substrate concentration gets too low: $[S] \sim [E \cdot S]$. The so-called initial velocity V_0 is the rate of appearance of product per unit time, usually measured after t_{ss0}, but before about 90% of the substrate has disappeared. Combining Equations (15.62) and (15.64), we obtain

$$V_0 = \frac{k_2[E]_T[S]}{K_M + [S]} = \frac{V_{max}[S]}{K_M + [S]} \tag{15.66}$$

where

$$V_{max} = k_2[E]_T \tag{15.67}$$

The turnover number or catalytic constant, k_{cat}, is defined as

$$k_{cat} = V_{max}/[E]_T \tag{15.68}$$

and measures the maximum number of product molecules per second a single enzyme molecule can produce. For the simple form of enzyme kinetics outlined above, $k_{cat} = k_2$, but this is not true for more complicated catalytic mechanisms. Values of V_{max} and K_M are generally determined from a *Lineweaver–Burk* plot of $1/V_0$ versus $1/[S]$, shown in Figure 15.23b.

(a)

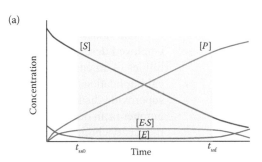

(b)

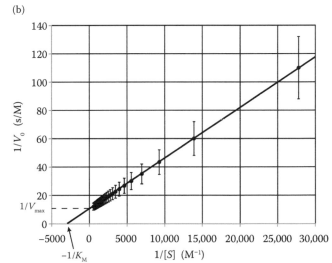

FIGURE 15.23

Michaelis–Menten (M–M) enzyme kinetics, $E + S \underset{k_{-1}}{\overset{k_1}{\rightleftharpoons}} E \cdot S \overset{k_2}{\longrightarrow} E + P$. (a) Time dependence of reaction systems that obey M–M rules. The steady-state assumption applies only for a finite time, starting at t_{ss0}, the time that the steady state between E and E·S is first reached, and ending roughly when $[S]$ approaches the steady-state $[E \cdot S]$, t_{ssf}. ($[S]_0 \gg [E]_T$. Starting at t_{ss0}, the initial velocity of the reaction, $(dP/dt)_0$, can be measured. (Adapted from Segel, I. H., *Enzyme Kinetics*, Wiley-Interscience, New York, 46, 1993. (b) Lineweaver–Burk plot used to extract V_{max} and K_M values from measurements of V_0 versus $[S]_0$. Parameter values, seen from x and y intercepts, for this calculated plot approximate those for superoxide dismutase: $[E]_{tot} = 1.0 \cdot 10^{-7}$ M, $K_M = 3.6 \times 10^{-4}$ M, $k_{cat} = 1.0 \times 10^6$ s^{-1}. The slope of the line is also K_M/V_{max}.

15.8.2 Michaelis–Menten Summary

The Michaelis–Menten treatment of enzyme kinetics can then be summarized by two constants, K_M and k_{cat}. K_M equals the substrate concentration needed to cause the enzyme to produce product at ½ of its maximal rate (Equation 15.66). A small value of this concentration reflects an enzyme that can work with very little substrate. In order to do this, it must bind the substrate tightly: the equilibrium constant $K_S = k_{-1}/k_1$ should also be small, like K_M. A high turnover number, k_{cat}, also reflects an efficient enzyme that can rapidly produce product. A combined measure of the catalytic efficiency can be obtained from the ratio

$$\frac{k_{cat}}{K_M} = \frac{k_2}{K_M} = \frac{k_1 k_2}{k_{-1} + k_2} \tag{15.69}$$

If $k_2 \gg k_{-1}$, this ratio is just k_1, the bimolecular rate constant for the reaction $E + S \rightarrow E \cdot S$. The maximum value of this rate is the diffusion-controlled rate discussed earlier in the chapter. Typical values were around 10^9 M^{-1} s^{-1}. This efficiency analysis applies, by assumption, to the case of free diffusion in bulk solution. Some values of the three parameters are shown in Table 15.4.

The values of parameters K_M and V_{max}, along with k_{cat}, the turnover number, are the most widely employed descriptors of enzyme activity. These numbers characterize, in a sense, the innate capabilities of an enzyme, but in the unnatural setting of a bulk, ideal solution. To unearth the biological implications of these numbers requires some deeper digging. You cannot say, for example, that superoxide dismutase is a "more efficient" enzyme than acetylcholinesterase because its turnover number is an order of magnitude higher, close to the theoretical maximum. You can also not assert that acetylcholinesterase, with its 25 times smaller K_M (needs less substrate to work with), is more efficient. The real question is, "What do these enzymes face in their normal (or abnormal) cellular activity?" If an enzyme must routinely deal with low substrate concentrations, but the rate of catalysis is not crucial, a small K_M value would be favored, while k_{cat} and V_{max} may be unimportant. If, like superoxide dismutase and catalase, an enzyme must rapidly eliminate reactive species that would otherwise rapidly react with critical cellular components, a large value of k_{cat} would be beneficial.

15.8.3 Control of Enzymatic Reactions: Feedback and Inhibition

Chemical reactions *in vitro* are primarily controlled by concentrations of reactants and products, and enzymes, if catalysis is involved. The most basic control of rates and equilibria we have seen so far is through temperature. Warm-blooded organisms actively control their internal temperature, with the goal of maintaining stable biological activity in the face of external temperature variation. Most of us are personally aware that mammalian temperature-control systems are complex and involve most of the organism. A single enzyme with its substrate and product, however, has no control over rates or equilibria. Such lack of control would be the death of any cell. Outside of temperature, organisms have, in fact, elaborate systems for the control of consumption and production of most chemicals. These systems can be inferred, but not understood, from diagrams like Metabolic Pathways in Figure 1.5.

TABLE 15.4   Michaelis–Menten Parameters for Several Enzymes

Enzyme	Substrate	K_M (M)	k_{cat} (s^{-1})	k_{cat}/K_M (M^{-1} s^{-1})
Catalase	H_2O_2	2.5×10^{-2}	1.0×10^7	4.0×10^8
Superoxide dismutase	O_2	3.6×10^{-4}	1.0×10^6	2.8×10^9
Acetylcholinesterase	Acetylcholine	9.5×10^{-5}	1.4×10^4	1.5×10^8

Source: Voet, D. and Voet J.G., *Biochemistry*, 3rd ed, John Wiley & Sons, New York, 480, 2004.

In contrast to test tube experiments involving one or two reactants, an enzyme, and a product, a single cell simultaneously carries out hundreds of reactions, where reactants of one reaction are products of others. Such active control over reactant and product concentrations is one key to the "livingness" of living systems. But exactly *how* do living systems manage this control? Figure 1.5 is impressive and daunting, but knowing phosphorylation of ADP controls creatine kinase activity brings understanding of neither the mechanistic details of this control nor the global importance of the reactions.

We do not address complex issues of control and organization in this book. The task is immense and would demand several, large books on enzymes, *metabolism, systems biology,* and *complex systems,* plus years of work in the lab. Preparation for such a study would be a solid biochemistry course, including a determined study of enzyme *inhibition.* Some of you may want to pause and read chapters in a biochemistry book on introductory metabolism and enzyme inhibition at this point. An alternative is a more targeted search on enzyme inhibition, competitive and noncompetitive inhibition, induction, or in the realm of genetics, and gene expression and suppression.

15.9 DRIVING FORCES

The fundamental, microscopic causes of reactions are forces between atoms and molecules. The reaction process itself must involve quantum mechanics because electron distributions around nuclei change. The interactions that govern the pre-reaction stage, when molecules diffuse in solution and approach each other, can often be satisfactorily described by classical physical forces. It is clear that on a microscopic scale, interatomic forces determine outcomes, even if they involve random forces and outcomes can only be predicted with probabilities.

Hierarchically designed living systems span the regimes of microscopic systems, where forces derived from potential energies and quantum transitions determine the end result, to clearly macroscopic systems, like interiors of large cells, where large numbers of identical molecules must diffuse, compete for reactive sites on enzymes, chemically convert to product, and diffuse, with 10^N other product molecules, to a second site. In this case, classical chemistry, with concentration-gradient (entropic) driving forces, would seem to rule. We can imagine DNA replication and conformational control might be on the borderline of these two limiting cases. Chapters 16 and 17 briefly overview aspects of such an intermediate thermodynamic case.

The general term *driving force* is used in the biochemical literature mostly in reference to Gibbs free energy differences, primarily in a nonequilibrium concentration context—the state of a biochemical system is "driven" toward equilibrium by nonzero $\Delta G = G - G^0$. In physics, we usually insist that any "force" have units of newtons. This means, of course, that any ΔG must be divided by some distance. There are usually natural distances that can be identified in cases of free-energy driving forces. We saw such an example of the electrochemical force across a membrane in Chapter 14. In this case the natural distance was the membrane thickness, but even that was an approximation that neglected any concentration gradients and nonequilibrium concentration gradients away from the membrane. The clearly entropic example of a single, freely jointed chain molecule's length and the force needed to stretch it was a Chapter 14 example of another sort, which we shall return to in the next two chapters.

15.10 REVERSIBILITY AND DETAILED BALANCE

Most biochemical reactions are reversible; if they weren't, there would be no equilibrium, and some of our analysis would technically be incorrect. (If there is no equilibrium, there can be no "drive toward equilibrium.") Michaelis–Menten kinetics, for example, does not consider conversion of product to reactant. This should not bother us, however, since from the start we knew that biology was not based on nonequilibrium processes that are, at their simplest, steady state but often not equilibrium. Reversibility is central to control of many enzymatic reactions, where a rise in concentration of product increases the back-reaction rate and reduces the net rate of product creation. This would require us to return to the M–M model and modify it to

$$E + S \underset{k_{-1}}{\overset{k_1}{\rightleftharpoons}} E \cdot S \underset{k_{-2}}{\overset{k_2}{\rightleftharpoons}} E + P \qquad (15.70)$$

Such modification is discussed in most biochemistry textbooks for particular cases, but enzymology is still ruled by the simpler M–M analysis. Michaelis–Menten is not "incorrect" in its assumption that $k_{-2} = 0$, but only applies to the biological situations where product is rapidly used and does not have time to accumulate and significantly affect the overall reaction.

We have alluded to the fact that there are other ways to control the overall state of catalyzed reactions, however. If the amount of enzyme is reduced, or if existing enzyme molecules are inhibited from reacting, the ratio of product to reactant would stay at its current value for an extended period of time, owing to the slow rate of the nonenzymatic reaction. Biology is usually practical when it comes to abiding by laws of equilibrium or drive toward equilibrium. If a delay in the thermodynamic drive toward equilibrium is less costly to the organism than shifting the equilibrium, the organism will do the former.

For systems governed strictly by random (Markov) processes, some fundamental rules apply to reaction steps, series of steps, and reaction cycles. The *Principle of Detailed Balance* states that

at equilibrium, the rates of the forward and reverse reactions are equal for elementary reactions.

Note that this principle does not state that reaction "rate constants" for the forward and reverse processes are equal ($k_1 = k_{-1}$) but the rates—rate constants multiplies by the corresponding state probabilities (or populations)—are equal ($P_1 k_1 = P_2 k_{-1}$). This principle has been the basis of the chemical reaction rate equations we have written in this chapter, as well as the Boltzmann and other distribution functions. We surmise that biology does not always pay attention to the principle of detailed balance on a cellular scale because equilibrium is often delayed indefinitely. Nevertheless, distribution functions describing speeds and energies are generally close to equilibrium distributions, near physiological temperature. Reaction rates that we have derived based on activated-complex theory implicitly assume detailed balance, in that transitions between states i and j in a well describing an initial state A are in detailed balance, $P_i k_{ij} = P_j k_{ji}$, during the reaction process A → B (Figure 15.24). If, for example, the rate of passage over the barrier is faster than the transition rates within well A, the high-energy, reactive states in well A may get depleted and the reaction may slow because of zero population in the state at E^*. The author is not aware of biological processes where this is a concern, but experiments spanning a wide range of temperature and other thermodynamic variables need to consider such a possibility.

The principle of detailed balance can be applied to a hypothesized reaction cycle of elementary steps. Suppose there are three states such as in Figure 15.25a, with corresponding forward and backward reaction rates. If equilibrium exists or can exist, then

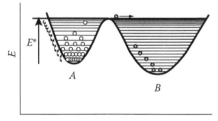

$$\frac{[B]_{eq}}{[A]_{eq}} = \frac{k_1}{k_{-1}} = K_1 \tag{15.71}$$

FIGURE 15.24
Detailed balance (or microscopic reversibility) is the basis of Boltzmann distributions. Populations in each of the microscopic states, i, in well A adjust until, at equilibrium, the transition probabilities between any two states are equal. The thermodynamic derivation of activated complex reaction-rate theory is based on assumed Boltzmann equilibrium populations, in that the probability $P(E^*)$ abides by the Boltzmann distribution. Detailed balance is also usually applied to chemical reaction steps described by activated-complex theory. Such assertions assume that the rate of passage over the barrier is much smaller than the rates at which particles make transitions between states i, j of well A.

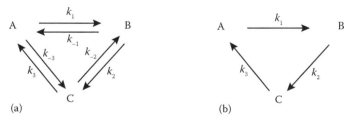

(a)   (b)

FIGURE 15.25
Detailed balance may be applied to cycles of elementary reaction steps. (a) The ratios of rates in the reversible cycle must obey the relation, $\frac{k_1}{k_{-1}} = \frac{k_{-2}k_{-3}}{k_2 k_3}$. (b) This nonreversible reaction cycle cannot exist in equilibrium.

and similarly for other reaction steps. It can then be shown that

$$\frac{k_1}{k_{-1}} = \frac{k_{-2}k_{-3}}{k_2 k_3} \tag{15.72}$$

What is the physical meaning of Equation 15.72? The equilibrium between any two states (A and B in this case) does not depend on the reaction path taken or the paths that are available. A proposed reaction cycle such as in Figure 15.25b, where the reverse reaction rate constants are zero, cannot exist in an equilibrium, stochastic system. Such irreversible cycles can and do exist in systems where frictional loss occurs, but friction does not exist for an "elementary" step that is part of a system at equilibrium. This we know from elementary physics: if frictional or other losses occur, an external source of power must be connected or else the system will come to a stop.

A more fundamental physical principle is that of *microscopic reversibility*:

at equilibrium, the rates of transition from microscopic state $i \rightarrow j$ equals the rate of transition $j \rightarrow i$.

Microscopic reversibility derives from the fact that Newton's laws and other fundamental laws of physics have the same form when time t is replaced by $-t$: $F_{nat} = ma = m\frac{d^2x}{dt^2}$. Two biophysical examples can be considered here, one of which violates reversibility and can thus not be part of an equilibrium system. These examples come from the first chapter in the book. Suppose a particle has an initial velocity $\mathbf{v}_0$, travels in a (gravitational) potential $V(y) = mgy$ or $\vec{F}(x, y, z) = -mg\vec{y}$, and no frictional losses occur. Newton's law becomes

$$F_{net,x} = F_{net,z} = 0, \quad F_{net,y} = m\frac{d^2y}{dt^2} = -mg \tag{15.73}$$

These equations are completely symmetric when t is replaced by $-t$. Therefore the process the mass m undergoes, as described by a parabolic trajectory, is reversible: the mass can retrace its path from finish to start. Suppose, on the other hand, that the mass is in a viscous fluid. If we neglect the gravitational force,

$$m\frac{d^2x}{dt^2} = -b\mathbf{v}_x, \ \mathrm{m}\frac{d^2y}{dt^2} = -b\mathbf{v}_y, \ m\frac{d^2z}{dt^2} = -b\mathbf{v}_z \tag{15.74}$$

Equation 15.74 is not symmetric to $\mathbf{v} \rightarrow -\mathbf{v}$. The left side of the equation does not change sign, while the right does. Processes involving frictional losses of the type $f = -b\mathbf{v}$ cannot be part of a system at equilibrium.

15.11 SINGLE-MOLECULE BEHAVIOR

We now touch on an area that is becoming increasingly important in biosystems: the modeling of the behavior of a single particle or small numbers of particles. We have considered the case of a single particle undergoing a single random walk. While some limits can be placed on possible outcomes of an N-step random walk, e.g., the particle cannot move farther than NL in any direction, only probabilistic statements can be made. These probabilistic statements such as the binomial probability, describe the behavior of the single particle, averaged over a *large number* of equivalent random walks, or (almost) equivalently, the average behavior of a large number of identical particles undergoing random walks. The question we posed in the beginning was whether bulk, average behavior correctly characterized what happens in cells, where small numbers of molecules or molecular interactions take place. Single-molecule measurements are designed to investigate this question. We will shortly consider two of many new techniques devised to measure the behavior of single molecules: single-molecule fluorescence spectroscopy and atomic force microscopy (AFM). The former is considered briefly, while Chapter 19 is entirely devoted to AFM. First, however, we examine issues of intracellular confinement and crowding and their effects on observed behavior of single molecules.

15.11.1   "Strange Kinetics": Biomolecular Movement in Living Cells Is Sometimes Slower

Chapter 4, Section 4.5.1, discussed random movement of molecules inside of living cells and whether crowding effects generally perturbed free diffusion of biomolecules and small molecules. In 2008, Dix and Verkman (reference 16 of Chapter 4) concluded that (1) crowding can slow the diffusion of solutes without resulting in "anomalous" diffusion, (2) large reductions in diffusion are probably indicators of interactions between solute and cellular components or of barriers to diffusion, (3) crowding reduces diffusion by only a few fold, and (4) discrepancies between simulations and experiment required further investigation. The Flavahan et al. reference 1A in Chapter 14's introduction implies a particular type of restricted diffusion, in that DNA strands in different structural domains may be positioned physically close to each other in a chromosome, allowing more efficient interactions than random positioning and movement would produce. This sort of DNA strand interaction can be crucial to an organism and its health, as disturbances can result in loss of insulation between topological domains and aberrant gene activation. These and other results seem to imply that (1) large structures such as chromosomes can facilitate or inhibit interactions (reactions) by positioning two components near each other, or that (2) large structures can slow down cellular reactions by presenting barriers to diffusion. We now explore one particular aspect of molecular diffusion in cells that sometimes results in *anomalous diffusion*.

As early as Chapter 4, and again in Chapter 13, we noted that free diffusion of a particle from an initial position resulted in motion governed by $\langle r^2 \rangle = 2Dt$, $4Dt$, or $6Dt$ in one, two, or three dimensions, respectively. In other words, the average diffusional distance scaled with the square root of time, whether the molecule was dimensionally restricted or not. While these governing laws may seem slow, they reflect the fastest a molecule can move through purely random motion. They also govern molecular encounter rates of all chemical and biochemical reactions, though we have seen that energy barriers usually regulate the details. While spatial restrictions to diffusion can be attributed to a physical barrier, they cannot be described by an *energy* barrier, with its exponential dependence on temperature, $k \mu e^{-E^*/k_B T}$.

Barkai et al., in a 2012 *Physics Today* review, focused on the observation that some observables in living cells appear to be fundamentally irreproducible and that the time averaged observation on a single molecule sometimes differs from the average on an ensemble of the same molecules. In a sense, we have been dealing with *irreproducibility* ever since Chapter 14, but the inability to exactly replicate an experiment should be rather a shock to a scientist. The shock was perhaps moderated by the presumption that while small variations in results may occur due to 10^{21} uncontrollable atoms in the surrounding, the basic experiment was good science. We must again remind ourselves that the microscopic world is not just a scaled-down version of the physics we studied through cart collisions in an 11th-grade physics class, nor is it a scaled-up version of quantum physics experiments, with its intrinsic multiple quantum states. Biologically important reactions and other events occurring inside of cells seem, at least sometimes, to have primary participants—biomolecules—that are identical in chemical makeup, yet behave somewhat differently if the reaction is repeated or another "identical" molecule in the cell replaces the first. Though the analogy is dangerous, cellular events have parallels in baseball games. Batter 1 may face Pitcher 1 four or five times in a game, but the result of the third at-bat is usually different from that of the second or fourth. While we could postulate that, averaged over a thousand encounters between Batter 1 and Pitcher 1, the average result would be the same as the average of the next one thousand encounters, a baseball game does not allow thousands of encounters. There are only nine innings and four or five at-bats for each player. When the game is over, it's over. Likewise, the batting result averaged over Batter 1's team of players will rarely be the same as the average of Batter 1 himself. We will see in Chapter 16 that molecular motor 1, carrying cargo along a track, may successfully deliver the cargo after 20 steps, while identical motor 2 may only need 12 steps. Identical motor 3 may go the wrong direction and fall off the track or lose its cargo. Nevertheless, the cell may live quite nicely, even though its health may depend on tens or hundreds of deliveries per second, rather than 10,000/s.

The violation of the ergodic theorem discussed by Barkai may (or may not) differ from that we have already encountered in Figure 15.15, where different molecules could present a different energy barrier to the reaction of two molecules. In Section 15.5.1 case CO binding to myoglobin (Mb),

---

Much higher numbers of molecules can be dealt with through signal correlation techniques. The reader is invited to do a search on *single-molecule detection* and *correlation* or *autocorrelation*.

the possible violation of ergodicity was not measured through direct comparison of ensemble average and single-molecule time average, so only an indirect test of the identicalness of myoglobin molecules could be made. The conclusion was that ergodicity or non-ergodicity was not the most helpful issue to focus on. An ensemble of myoglobin molecules at low temperature (below 120 K or so) did bind as if they were each quite different, though at high temperature (above 250 K or so), they all behaved similarly. An issue of more importance to cellular activity than ergodicity was that a myoglobin molecule could take up a variety of slightly different structures, each of which had different binding properties.

Even today, forty years after the Mb-CO experiment, the reaction of a single pair of molecules cannot be measured without considerable effort and expense. The Brownian part of a single particle's motions was tracked by Perrin as early as 1908 and by I. Nordlund in 1914, the latter on a much longer time scale.[18] The primary difference between the Perrin and Nordlund methods: the former measured short Brownian tracks of many small putty particles and averaged over the ensemble of particles, whereas Nordlund simply measured tracks of single particles over long time (a difficult task) and time averaged. Molecules are much smaller and, of course, more difficult to track individually, but sensitive new single-molecule techniques, based on molecular fluorescence and enabled by sensitive photodetectors, have made tracking of single molecules possible. See Bagshaw,[1] Chapter 9, Barbara,[19] and Weigel et al.,[20] and search for *single-molecule tracking*. Note that reference 19 makes use of emission from quantum dots, described in Section 9.6.5 of this text.

> The reader can do a computer simulation of Perrin's measurements of small putty particles by running the spreadsheet associated with Figure 13.7 for a fixed time interval, then plotting the [x,y] coordinates of the final point relative to the starting point; repeat 100 times or so and plot all final points on a coordinate chart with circles of regularly increasing radii centered about the origin. Nordlund traced trajectories of slowly dropping mercury particles in water on moving strips of film over a long timescale. The particles showed Brownian variations superimposed on uniform motion.

While some of the non-reproducible (non-ergodic) motions will resist quantitative description with equations, for a variety of reasons, some other random motions that violate a $\langle r^2 \rangle \propto Dt$ description have modeled with a small modification:

$$\langle r^2(t) \rangle = \int r^2 P(r,t) d^3r \propto D_\alpha t^\alpha \qquad (15.75)$$

Equation 15.75 refers to random motions, starting from the origin, averaged over an ensemble of particles, as a function of time. This is appropriate for the Perrin measurement. If random motion along a single particle's trajectory is measured as a function of time, the time averaged mean square change of position as a function of delay time, Δ, along the single trajectory is used:

$$\overline{\delta^2(\Delta)} = \frac{1}{t-\Delta} \int_0^{t-\Delta} \left(r(t'+\Delta) - r(t') \right)^2 dt' \propto D_\alpha \Delta^\alpha \qquad (15.76)$$

Equation 15.76 is appropriate for the Nordlund measurement. To understand equation 15.76, think of the value of r at a certain time and that at a small time Δ later, where Δ is much smaller than the total observation time, t. The quantity $\delta^2(\Delta)$ will then reflect the mean squared displacement, over a time interval, Δ, over the entire time, t. If the motion were purely linear at constant velocity, the value of α would be 2; if purely diffusive, $\alpha = 1$; if anomalously diffusive, $\alpha < 1$. In both 15.75 and 15.76, the parameter α need not be 1.0, though it has the same meaning. In 2006, Golding and Cox modeled the trajectories of single messenger RNA molecules in bacterial cells with equation 15.76. They determined the diffusion to be anomalous in two ways. First, the value of α was 0.7, meaning that RNA in a cell diffused with weaker time dependence than a RNA molecule free in ideal solution. Such weaker time dependence is sometimes referred to as *subdiffusion*. Second, the diffusion constant, D_α, showed a pronounced scatter in value for different measurements. This scatter was not a result of careless repetition of experiment; rather, the irreproducibility was reproducible. Such irreproducibility should, again, alarm the sensibilities of a scientist inexperienced with biological systems on a microscopic scale. These unexpected diffusion results turn out to be fairly common inside of living cells. Understanding such anomalous diffusion and irreproducibility results in cells is still an active research area, but some factors, such as a cell environment changing constantly with time and confinement by the cell wall, are likely involved. In Chapter 16 we discuss microtubules and tracks inside of cells, which facilitate transport of materials over long distances, in the context of free diffusion requiring an unacceptably long time. The reader should ponder whether a value of α less than 1 makes the microtubule transport less or even more necessary.

15.11.2 Fluorescence Spectroscopy of Single Molecules

Previous chapters introduced the method of optical spectroscopy to measure properties and interactions of biomolecules. A horizontal beam of light, usually of cross section 1 cm^2 or so, illuminates a rectangular cuvette, usually 1 cm × 1 cm × 3 cm (tall), as in Figure 15.26a. If a sample has concentration 1.0×10^{-5} M, the number of molecules illuminated by the beam is $10^{-5} \frac{moles}{L} \times 10^{-3} L \times 6 \times 10^{23} \frac{molecules}{mole} \approx 10^{16}$ molecules. The average fraction of the molecules that absorb a photon is given by Beer's law, $1 - 10^{-A} = 1 - 10^{-\varepsilon c L} \approx 1 - 10^{-0.05} = 0.11$. Here we have used the rule of thumb that if fluorescence is to be measured, the absorbance should be no more than 0.05 to avoid unwanted effects, like reabsorption of emitted fluorescence and nonuniform excitation through the sample. The number of molecules excited is then about 10^{15}, a large number. If a short flash of light is used, a snapshot in time of the average properties of the 10^{15} molecules will be obtained.

Suppose a laser beam is focused into a sample and that somehow signals from only the molecules within the focal volume are measured (Figures 15.26b through 15.28). How many molecules are in this volume? According to Table 15.5, the volume of the focal spot is $2\lambda^3 f^4 / n^3 (\pi R^2)^2$, where λ is the laser wavelength (in air) and f the lens focal length. For a 4-mm focal length at $\lambda = 400$ nm, this volume is $\Delta V \cong 1.4 \times 10^{-18}$ m^3. If this focal spot is in a 0.1-μM solution, the average number of molecules in the excitation volume is $N = \left(10^{-7} \, moles / L\right)\left(1000 L / m^3\right)\left(6.0 \times 10^{23} \, molec / mole\right)\left(1.4 \times 10^{-18} m^3\right) \approx 80$ molecules. Dilute by another factor of 100 and there will be about one molecule at any given time. This means that the situation in Figure 15.26b holds: the number of molecules within the excited volume changes by a large fraction as molecules randomly diffuse (Figure 15.27). The excitation volume has roughly cylindrical dimensions, transverse radius r and length Δz (Figure 15.28), which can be estimated from information in Table 15.5.

Clearly, with a laser, a good lens, and a fluorescence detection system, we have the basic components of a single-molecule detection system. The three major questions remaining are (1) can we find a low-noise detection system sufficiently sensitive to measure fluorescence from one molecule, (2) how do we get laser light in and fluorescence light out to the detector, and (3) how do we eliminate fluorescence from areas outside the focal volume? These issues are sufficiently challenging that only a few commercial manufacturers and a small number of researchers have assembled apparatus that partially or completely fulfills requirements. Zeiss and Olympus multiphoton microscopes employ nonlinear optical effects to restrict molecular excitation to the focal

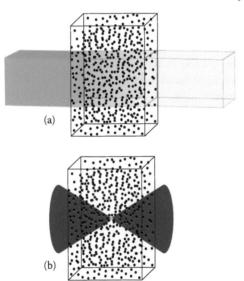

(a)

(b)

FIGURE 15.26
Example optical spectroscopic setups for observing reaction chemistry. (a) Spectrophotometer measurement of the bulk: the 1-cm^2 beam area measures behavior of the average of ~10^{16} molecules in the light beam. The light wavelength is chosen to interact directly with molecular electronic states. (b) Measurement of single-molecule behavior. The light beam is focused to a small spot (~0.5 μm diameter) with a high-quality microscope objective lens. The light wavelength is chosen such that it does not interact with the molecules, but high-intensity in the focal spot produced second-, third- or fourth-harmonic ($\lambda/2$, $\lambda/3$, $\lambda/4$) light that does interact. The sample concentration is chosen such that a few or no molecules are in the focal volume at any given time.

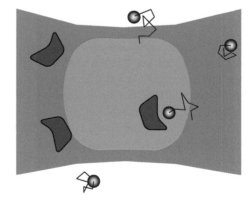

FIGURE 15.27
Expanded view of focal spot in Figure 15.26. Molecules diffuse in and out of the high-intensity region where harmonic generation occurs. If the harmonic light excites molecular fluorescence, the measured fluorescence intensity will fluctuate in time from zero to integral multiples of a standard intensity. The characteristic time scale of fluctuation is determined by diffusion. If the molecules are undergoing a reaction that affects the fluorescence intensity, additional intensity variations will occur on time scales determined by the reaction. Either or both of the two molecules shown may be fluorescent.

TABLE 15.5 Optical, laser, and lens parameters used to calculate illumination volumes (etc.) for focused molecular excitation, as in single-molecule high-intensity measurements

Quantity	Symbol	Equation	Notes
Lens focal length	f	—	in air
Initial beam radius	R	—	—
Laser wavelength	λ	$c = \lambda f$ [a]	F = light frequency, c = speed of light; in air
Focal spot radius (waist)	r	$\dfrac{\lambda f}{\pi n R}$ [b]	n = index of refraction of medium
Focal spot length (longitudinal)	Δz	$\pm\dfrac{\lambda f^2}{\pi R^2}$ [b]	Focal spot center to where beam area doubles
Focal spot volume	ΔV	$\pi r^2 \Delta z = \dfrac{2\lambda^3 f^4}{n^3\left(\pi R^2\right)^2}$ [b]	—
Instantaneous beam max power	P_m		Watts
Average beam power ("continuous wave")	P_{cw}	$P_{cw} = P_m f_{rep} t_p$	If repetitive pulse of width t_p, rep rate f_{rep}
Average beam *pulse intensity* $\left(\dfrac{photons}{s \cdot m^2}\right)$	I_p	$\dfrac{P_m n^2 \pi R^2}{\lambda f^2 hc}$ [b] $\dfrac{P_{cw} n^2 \pi R^2}{\lambda f^2 hc}\dfrac{1}{f_{rep} t_p}$ [b]	Photons s^{-1} m^{-2}

[a] Recall light frequency remains fixed when light enters a second medium.
[b] Computations are approximations of reality. Microscope objectives, complex combinations of many lenses, are used and glass/oil/glass/water interfaces affect focusing. 90% of a laser's power is often lost to multiple, imperfect reflectivities (<100%), dielectric interfaces, scattering and imperfect lens behavior. See also https://www.microscopyu.com/techniques/multi-photon/multiphoton-microscopy.

region. First, the fundamental wavelength λ of the laser is chosen in a region above the absorption bands of the molecule of interest. Typically, λ is in the near infrared region. By using a laser beam consisting of short, high-intensity pulses, two, three, or four photons can simultaneously excite a molecule located in the focal region (Figure 15.29). Outside the focus, the probability for simultaneous arrival of two photons is too low, and one photon has too little energy to promote the molecule to its first excited state. Because high-intensity, pulsed lasers are rather expensive, other, single-photon approaches have been developed to eliminate fluorescence from unwanted regions. Some of these methods involve confocal and/or time-gating techniques. Careful attention must be paid to optical layouts, lens, and detector quality. The competition to recruit experimentalists who are both qualified to assemble such a system and who understand fundamental biophysics is rather intense. We do not explore the technical issues of these fluorescence systems further.

When second or higher harmonic excitation is desired in a single-molecule or imaging measurement, the laser must be pulsed. In

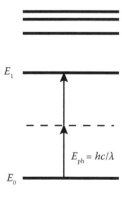

FIGURE 15.29
Multiphoton excitation. If the photon intensity is high enough, two photons can arrive at the same molecule "simultaneously," summing their energies. If twice the photon energy matches $E_1 - E_0$ of the molecule, excitation can occur, subject to slightly different selection rules than for single-photon absorption. Three- and four-photon excitation has also been successfully employed. Multiphoton effects restrict molecular excitation to the focal region of a focused, short-pulse laser beam.

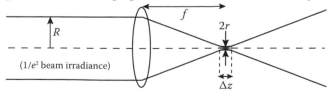

FIGURE 15.28
Parameters describing the focusing of a laser beam of Gaussian intensity profile.

Table 15.5 the pulse width is t_p and the repetition rate of the pulse is f_{rep}. If the laser is set to single-pulse mode, P_{cw} is irrelevant and the peak pulse power is used. (If pulse energy is given, the power is given by dividing by t_p, with appropriate numerical factors describing the pulse shape (time dependence.) A common source is the titanium-sapphire laser, with wavelength tunable from about 700 to 1,000 nm, pulse repetition rate 76 MHz, pulse width 0.01–1 ps, beam radius 0.4 mm, and continuous wave (cw) intensity 0.5–2 W. Area units of cm^2 are commonly used in laser contexts, by tradition and for compatibility with molecular absorption cross sections, commonly expressed in cm. Recall from earlier chapters that the relation between absorption coefficient (Beer's law) and cross section is $\sigma(cm) = 3.824 \times 10^{-21} \varepsilon$ (M^{-1} cm^{-1}).

The fundamental measurement made in a fluorescence single-molecule study is the time dependence of the fluorescence signal. In particular, the fluorescence signal fluctuates about an average. The fluctuation amplitude is determined by the fluorescent molecule and its concentration. The time periods of the fluctuations are governed, in the simplest case of a single molecule, by diffusion times in and out of the focal region (Figure 15.26b). The diffusion time is of interest, of course, but often a true binding reaction is of interest. Suppose the sample consists of two biomolecules that bind to each other (Figure 15.27). Suppose further that only one molecule fluoresces but that the fluorescence changes when the second molecule binds. Alternatively, suppose that one molecule transfers energy to the other when the two bind. (Recall energy transfer from Chapter 9.) In both these cases a second time dependence is in evidence, characterizing the molecular interaction process. If energy transfer occurs and the detection system is tuned to the wavelength of the acceptor, no signal will be observed until both molecules are within the excitation region and binding or unbinding occurs. If a second detector is tuned to the fluorescence donor wavelength, a complementary signal can be obtained. The processes of diffusion in and out of the focal region and molecular binding are disentangled using time, amplitude, and wavelength dependencies of measured signals.

The advantage of single-molecule measurements based on optical techniques is the ability to measure the signal from a true, single molecule or from large numbers of molecules in the same optical system. The latter signal would reflect the population-average properties of the molecule of interest. This is what has been described by the classical reaction theory we have mostly concentrated on in this chapter. *What new information could we get from measurements on a single molecule?* As we discussed with myoglobin, it is not always clear that the population-average properties of a molecule are appropriate for a system of a few hundred molecules in a small cell compartment. Do bulk measurements that average the signal from 10^{15} molecules over a time period of several hours properly characterize the behavior of 1,000 molecules that are activated over a period of 100 ms? The bulk measurement likely encompass the short-time behavior of those 100 molecules, but the single-molecule optical technique

EXERCISE 15.3

Optical parameters for focused laser excitation. Calculate the instantaneous, focal-spot intensity (photons/s·cm²) for the following cases of *cw mode-locked* laser excitation.

1. A cw, pulsed dye laser: $\lambda = 600$ nm, $P_{cw} = 100$ mW, $f_{rep} = 4.0$ MHz, $t_p = 1.0 \times 10^{-11}$ s, $R = 1.0$ mm, $n = 1.33$ (water), and $f = 4$ mm.

Answer

7.3×10^{33} photons/s·m², or 7.3×10^{29} photons/s·cm².

2. A cw, pulsed Ti-sapphire laser: $\lambda = 800$ nm, $P_{cw} = 1.0$ W, $f_{rep} = 76$ MHz, $t_p = 0.2 \times 10^{-12}$ s, $R = 0.4$ mm, $n = 1.33$ (water), and $f = 4$ mm.

Answer

2.3×10^{34} photons/s·m², or 2.3×10^{30} photons/s·cm².

Both of these intensities are within the range where two-photon excitation can occur. See http://www.calctool.org/CALC/chem/photochemistry/2pa, or search for *two-photon absorption calculator* for an online two-photon calculator.

has the potential for observing single (or small numbers of) molecules inside a living cell, without damaging the cell. Measurement of the "strange" biomolecular movement discussed in the previous section rely heavily on single-molecule fluorescence techniques. Problem 15.16 helps us make the connection between single-molecule measurements and the classical reaction descriptions we have presented in this chapter. Thinking of a career in lasers, imaging and cell biology? Think about single-molecule techniques.

[Original Section 15.11.2 on Atomic Force Microscopy, including Figure 15.30, has been deleted and will be reinserted in Chapter 18, where the new Chapter 19 on AFM will be introduced. A new Section 15.11.1, inserted before 15.11.2, Fluorescence Spectroscopy of Single Molecules, presents new discoveries about unusual single-molecule kinetics in cells. The figure for Problem 15.16 has been renumbered Figure 15.30.]

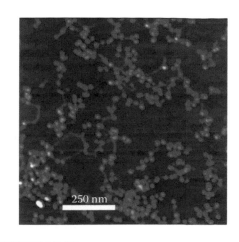

250 nm

FIGURE 15.30
(Problem 15.16) Single-molecule fluorescence of RNA hairpin. (Based on Single Molecule Biophysics Group, Taekjip Ha, University of Illinois at Urbana–Champaign.)

15.12 POINTS TO REMEMBER

- $A \rightleftharpoons B$: equilibrium. $\dfrac{[B]}{[A]} = \exp\left[\dfrac{\Delta G - \Delta G^0}{RT}\right] = \exp\left[\dfrac{\Delta H - \Delta H^0}{RT} - \dfrac{\Delta S - \Delta S^0}{R}\right]$,

$$\Delta G = 0 : K_{eq} \equiv \left(\dfrac{[B]}{[A]}\right)_{eq} = \exp\left[\dfrac{-\Delta G^0}{RT}\right]$$

- $A \rightleftharpoons B$: steady state. $\dfrac{d[A]}{dt} = -\dfrac{d[B]}{dt} = -k_{AB}[A] + k_{BA}[B]$;

steady state $\dfrac{d[A]}{dt} = -\dfrac{d[B]}{dt} = 0$, so $\left(\dfrac{[B]}{[A]}\right)_{ss} = \dfrac{k_{AB}}{k_{BA}}$

- $A \rightleftharpoons B$: time dependence. $-\dfrac{d[A]}{dt} = k_1[A] - k_2[A] - k_2[B]$, $\dfrac{[A] - [A]_{eq}}{[A]_0 - [A]_{eq}} = e^{-(k_1+k_2)t}$

- $A + B \rightleftharpoons C$: equilibrium. $\dfrac{[C]}{[A][B]} = \exp\left[\dfrac{\Delta G - \Delta G^{0'}}{RT}\right]$,

$$\Delta G = 0 : K_{eq} \equiv \left(\dfrac{[C]}{[A][B]}\right)_{eq} = \exp\left[\dfrac{-\Delta G^{0'}}{RT}\right]$$

- Remember the energy diagram that results in $k_{AB} \cong \dfrac{k_B T}{h} e^{S^*/k_B} e^{-H^*/k_B T} = A e^{e^{-E^*/k_B T}}$ (Arrhenius).
- Recognize Michaelis–Menten modeling: $E + S \underset{k_2}{\overset{k_1}{\rightleftharpoons}} E \cdot S \overset{k_2}{\longrightarrow} E + P$; V_0, V_{max}, k_{cat}
- Catalysis: Enzymes use charges and specialized groups strategically located to reduce repulsion (the energy barrier) in reactants; binding is central.
- Proteins are close-packed structures, but thermal motions of atoms generate structural fluctuations that facilitate binding and other parts of the catalytic process.

15.13 PROBLEM-SOLVING

PROBLEM 15.1 EQUILIBRIUM

An equilibrium conformational change $A \rightleftharpoons B$ of a relatively small molecule is described by enthalpy and entropy changes $\Delta H = +11.7$ kJ/mole (B is higher) and $\Delta S/R = +9.12$.

1. Compute the Gibbs free energy change at 0°C, 30°C, and 80°C.
2. If the total concentration [A] + [B] = 1.00 μM, plot a graph of the two concentrations as a function of temperature from 0°C to 100°C.

PROBLEM 15.2 PROTEIN CONFORMATIONAL CHANGES

"Thermodynamics of protein denaturation is complicated" (Mahler and Cordes, *Biological Chemistry*, 2nd ed., New York: Harper & Row, 1971, p. 182). Graphs like Figure 15.3 (as well as Figure 14.7) show that unfolding of macromolecules cannot be described by a single enthalpy and entropy change. Thermal denaturation of myoglobin (Mb) is sometimes said to involve a two-state transition, but both enthalpy and entropy differences between native and denatured structures are temperature dependent. Consider Mb denaturation between $T = 0$ and $100°C$.

1. At what temperature is Mb most stable against denaturation?
2. In terms of enthalpy and entropy (ordering), qualitatively and quantitatively describe the denaturation of Mb at $0°C$, $25°C$, $50°$, and $95°C$.
3. If $\Delta C_P = \left(\frac{\partial \Delta H}{\partial T}\right)_P$ is the specific heat difference at constant pressure, $C_{den} - C_{nat}$, between denatured and native structures, while $\Delta H = H_{den} - H_{nat}$, determine the approximate specific heat of denaturation of Mb.

PROBLEM 15.3 TEMPERATURE DEPENDENCIES IN ACTIVATED-COMPLEX THEORY

Suppose values of activation enthalpy and entropy for a reaction rate k are determined from rate data, with errors on individual rates of $\pm 5\%$ and temperature measurement errors $\pm 0.2°C$.

1. Case 1: Measurements are done at $5°C$ intervals from $20°C$ to $45°C$; assume activation enthalpies and entropies for activated-complex theory are $H^* = 45.0$ kJ/mole and $S^*/R = -12.0$. Using software, plot a graph of $\log_{10}(k)$ versus the quantity $1,000/T$ (T in Kelvin) for the temperature range $15°C–50°C$ using (1) $f_0 = 10^{13}$ s^{-1} and (2) $f(T) = k_B T/h$. (Put both plots on the same graph, as in Figure 15.7, and place f_0 in a spreadsheet cell, where it can easily be adjusted in value.) Describe the differences in the plots and the significance of using the two expressions for attempt frequency. Determine an adjusted value of S^*/R that make the f_0 plot indistinguishable from the $f(T)$ plot.
2. Case 2: Measurements are done at $10°C$ intervals from 10 to 100 K; assume activation enthalpies and entropies to be $\Delta H^* = 15.0$ kJ/mole and $\Delta S^*/R = -12.0$. Repeat the plots as in part 1. Discuss the linearity of the semilog plots. Using the software's linear-fit capability, fit a straight line to each of the two curves and determine the apparent values of H^* and S^*/R from this fit.

PROBLEM 15.4 ANALYSIS OF RATE MEASUREMENTS

The following unimolecular rate date is measured for a simple process A → B:

$T(°C)$	20	25	30	35	40	45	50
k (s^{-1})	2.5×10^3	3.5×10^3	5.0×10^3	7.1×10^3	8.8×10^3	1.0×10^4	1.4×10^4

1. Graph these data appropriately for activation-complex analysis.
2. Obtain activation enthalpies and entropies by fitting with theory. Use $f = 10^{13}$ s^{-1}.
3. Estimate uncertainties in the fit parameters, assuming temperatures are measured to $\pm 0.5°C$ precision and rates to $\pm 10\%$.

PROBLEM 15.5   DIFFUSION-CONTROLLED RATE

O_2 binding by Mb.

1. Verify that the units are correct in Equations 15.30 through 15.32.
2. Calculate the diffusion rate constant for the reaction $Mb + O_2 \rightarrow Mb{\cdot}O_2$. Express it in both SI units and in units of $M^{-1}\ s^{-1}$.
3. Search the Web or library for a measured value of the rate constant in part 2. Is oxygen binding to myoglobin diffusion limited?

PROBLEM 15.6   RESTRICTED DIFFUSION

Problem 15.6 is best assigned in conjunction with Problem 15.5.
   Consider the O_2-binding reaction in Problem 15.5.

1. By what fraction is the diffusion-controlled reaction rate reduced if O_2 can only bind when it strikes Mb at a particular 2Å × 2Å patch on the surface of Mb, with velocity vector within a cone angle of 15° of a specific direction? Make sure you draw a diagram justifying your calculation.
2. How many times, on average, will an O_2 molecule strike a Mb molecule before it successfully enters the channel to bind? (Ignore any ejection of O_2 after it enters the channel.)

PROBLEM 15.7   BINDING AND PROTEIN
CONFORMATIONAL CHANGE

Problem 15.7 is best assigned in conjunction with Problem 15.5.
   The following is a poor model of conformational change, but it provides food for future thought. Consider again the binding of O_2 by Mb. When proteins bind other molecules a conformational change generally takes place. This change may occur before, during, or after binding and takes a finite amount of time. The conformational change may be triggered by binding (or pre-binding) or may be part of an ongoing structural fluctuation process. Assume such a conformational change is an ongoing fluctuation and that a proper conformation allowing binding of O_2 can be modeled as a "gate" that opens up for time Δt at a rate $f = 1/\tau$. Treat the gate as a tunnel of length $L \sim 0.6$ nm, through which O_2 must pass before the open gate closes again.

1. Draw a diagram of this situation.
2. Passing the gate. Assume that every O_2 molecule that makes it successfully to the gate (e.g., through the exclusion process in Problem 15.6) has the proper direction to pass through the gate, but has a speed described by the Maxwell–Boltzmann distribution at $T = 300$ K (Figure 14.2): it may or may not have high enough speed to get through the gate before it closes. Write an expression for the minimum speed a molecule needs to get through an open gate.
3. Calculate the value of this minimum speed if $L = 0.60$ nm and $\Delta t = 2.0 \times 10^{-12}$ s and use Figure 14.2 to estimate what fraction of the O_2 molecules have this speed or greater.
4. If the gating frequency f is too slow, the binding rate decreases further below the (reduced) diffusion rate estimated in Problems 15.5 and 15.6. Briefly describe this possible limitation and estimate a value for f, below which the binding rate would be reduced. (If you did not do Problem 15.6, assume a reasonable value for the reduction of the diffusion rate constant in Problem 15.5 due to directionality requirements.) What fraction of the time is the gate open, $\Delta t/\tau$?

PROBLEM 15.8   REACTION GRAPHS

Using the unimolecular rate constant data from Problem 15.4, use spreadsheet/graphing software to make the following:

1. A linear plot of the population of the reactant $N(t)$ versus time at $T = 20°$ and at 45°C. Assume a normalized $N(0) = 1$ and that only 40 data points can be recorded. For data at each temperature, choose a time, Δt, between measured data points that allows recording of about 20 $N(t)$ values between $N = 1$ and $N = e^{-1} = 0.37$ and 20 values between $N = 0.37$ and 0.
2. Re-plot the data in part 1 as $\log_{10}[N(t)]$ versus t and confirm (1) that the resulting curve is a straight line and (2) that the slope of the line correctly gives the rate constants.
3. Decay data are recorded by taking measurements at time intervals defined by $t_n = \Delta t \cdot 2^n$, e.g., $t_n = \Delta t \cdot (1, 2, 4, 8,...)$ and that Δt is the shortest sampling time of a commercial analog-to-digital (A/D) converter. Look up this shortest sampling time ([sample rate]$^{-1}$) at a manufacturer's Web site, e.g., http://www.analog.com. Plot data as $\log_{10} N(t)$ versus $\log_{10}(t)$.
4. Compare the plots in parts 1 through 3, discussing their advantages and disadvantages.

PROBLEM 15.9   RATES AND DISTRIBUTIONS

Binding rates with distributions of activation energies (enthalpies).

1. Calculate classical, over-the-barrier rate constants at $T = 30°C$ corresponding to the activation-energy distribution for CO binding to myoglobin in Figure 15.15b. Choose the energies of the distribution peak and the two 1/10th-of-max points (i.e., near 2 and 9 kJ/mole). Assume the value of $S*/R$ is −9.9 and is the same for all $E*$. Use a frequency factor of 10^{13} s^{-1}.
2. Discuss the implications of this rate distribution on the binding reaction if the distribution results from differing protein conformations and these conformations are interconverting at 30°C. Be specific. (You could assume, for example, that there are actually 100 $E*$ values, corresponding to 100 conformations, and that these conformations randomly interconvert at a rate of 10^3 or 10^6, or 10^{10} s^{-1}.)

PROBLEM 15.10   BIMOLECULAR REACTIONS 1

1. Write the mathematical expression for $[B(t)]$ corresponding to case 2 of the general bimolecular reaction (Equation 15.49).
2. Plot a linear–linear graph of $[B(t)]$ for $[B]_0 = 2.00 \times 10^{-3}$ M and (1) $[A]_0 = 0.02[B]_0$ and (2) $[A]_0 = 0.2[B]_0$, with $k = 1.00 \times 10^8$ M^{-1} s^{-1}.
3. Plot a log–log graph of the functions in part 2.

PROBLEM 15.11   BIMOLECULAR REACTIONS 2

1. Assume $[B]_0 > [A]_0$ and solve Equation 15.53 for $[A(t)]$.
2. Assume $[B]_0 \gg [A]_0$ and show that Equation 15.55 or (15.54) reduces to (15.52).
3. Plot a log–log graph of $[A(t)]$ for $[B]_0 = 2.00 \times 10^{-3}$ M and (1) $[A]_0 = 0.2[B]_0$ and (2) $[A]_0 = 0.9[B]_0$, with $k = 3.00 \times 10^5$ M^{-1} s^{-1}.

PROBLEM 15.12   MICHAELIS–MENTEN KINETICS (GRADUATE PROBLEM)

1. Determine the time it takes for a Michaelis–Menten reaction to first reach steady state. You may assume, if necessary, that $k_2 \ll k_{-1}$.
2. Determine the approximate time period ($t_{ssf} - t_{ss0}$, Figure 15.23a) during which M–M conditions hold.
3. For the cases of superoxide dismutase ($K_M = 2.5 \times 10^{-2}$ M, $k_{cat} = 1.0 \times 10^7$ s^{-1}) and catalase (($K_M = 3.6 \times 10^{-4}$ M, $k_{cat} = 1.0 \times 10^6$ s^{-1}), determine these times. You may assume $k_{cat} \approx k_2$ and $[S]_0 = 3 K_M$. Superoxide dismutase catalyzes the reaction $2O_2^- + 2H^+ \rightarrow H_2O_2 + O_2$, where O_2^- is the reactive (and destructive) superoxide anion and H_2O_2 is peroxide. Catalase catalyzes the reaction $2H_2O_2 \rightarrow 2H_2O + O2$.

PROBLEM 15.13   DETAILED BALANCE AND EQUILIBRIUM

Derive Equation 15.72 and explain its meaning in more detail than given in the text.

PROBLEM 15.14   SINGLE MOLECULES

Single molecule detection by optical means (single photon excitation).

1. Estimate the dimensions (r, Δz, and ΔV) of the focus region of a 410-nm laser beam focused by a 5.0-mm focal length lens.
2. Estimate the average number, N, of myoglobin molecules in a 5.0×10^{-14} M solution that are within the focal volume.
3. The noise-to-signal ratio is given roughly by $\sqrt{N}/N$. Sketch the time dependence of a fluorescence signal from excited volume.
4. If a myoglobin molecule starts from the center of the focal region, estimate the average time before it diffuses out of the volume.

PROBLEM 15.15   MULTIPHOTON EXCITATION

Suppose you have a cw mode-locked Ti-sapphire laser, wavelength tunable from 700 to 1,000 nm, that is appropriate for multiphoton excitation.

1. Find out what *cw mode-locked* means.
2. Suppose the Soret, α, and β absorption bands of oxy-myoglobin (myoglobin with O_2 bound) are to be excited with the laser. Find the approximate wavelength regions of these absorption bands (e.g., Soret, 395–425 nm). Choose wavelengths where absorbance is above half of the peak absorbance.
3. For two-, three-, and four-photon absorption, determine the laser wavelengths that excite these absorption bands.

PROBLEM 15.16 RNA HAIRPINS

Single-molecule fluorescence and RNA hairpin folding. Figure 15.30 shows a simulated fluorescence signal from a single-molecule fluorescence resonance energy transfer (FRET) fluorescence detection system. When the RNA is in the "folded" configuration, the donor–acceptor distance is short and efficient transfer occurs. The signal is normalized such that the fluorescence of one molecule of acceptor gives a value of 1.0 intensity unit when located in the focal region of the laser beam. The signal is zero if no molecule is in the focal region and low when the hairpin is unfolded.

1. Assuming there is always an RNA molecule in the focal region, calculate the equilibrium constant, $K = $ [Folded]/[Unfolded].
2. Estimate the uncertainty in your calculation.
3. What can you say about reaction *rates* from the observed data?

PROBLEM 15.17 mRNA SUBDIFFUSION MEASURED BY SINGLE-MOLECULE FLUORESCENCE

Find the article Golding, I., and E. C. Cox, 2006, Physical nature of bacterial cytoplasm, *Phys. Rev. Lett.* 96(9): 098102, as well as the Supplemental Material, which is referred to in the Barkai et al. reference 16. Examine Figure 2(a), where anomalous diffusion results for mRNA are graphed: $\log_{10} <\delta^2>$ versus $\log_{10}\tau$. (τ is referred to as Δ in our text.)

1. Why is the graph plotted as a log–log plot?
2. What is the timescale of this data and what is the interval between measurements of position?
3. What is the stated uncertainty in measurement of the slope of the plot, which gives the value of α, as in Equations 15.75 and 15.76, and what are listed values of α for molecules in living cells and in free solution?
4. Describe the physical size of the mRNA used, as well as whether or not it should diffuse like a sphere.
5. Describe how the fluorescent RNA molecule is constructed.
6. From the Supplemental Material, obtain the video file, golding_movie_S1.avi, convert it, if necessary, to a more useful video format, and estimate the uncertainty in any determination of the position of the mRNA molecule. You should read the description of the video and sample preparation in the file, golding_suppmat_022706.doc, and also summarize their discussion of the size determination of the RNA molecules.

REFERENCES

1. Bagshaw, C. R. 2017. *Biomolecular Kinetics: A Step-by-Step Guide.* Boca Raton, FL: CRC Press.
2. Privalov, P. L., and N. N. Khechinashvili. 1974. A thermodynamic approach to the problem of stabilization of globular protein structure: A calorimetric study. *J Mol Biol* 86:665–684.
3. Welch, G. R., Ed. 1986. *The Fluctuating Enzyme.* New York: John Wiley & Sons.
4. Austin, R. H., K. W. Beeson, L. Eisenstein, H. Frauenfelder, and I. C. Gunsalus. 1975. Dynamics of ligand binding to myoglobin. *Biochemistry* 14:5355–5373.
5. Wikipedia, "Triosephosphate Isomerase," http://en.wikipedia.org/wiki/Triosephosphate_isomerase (accessed December 4, 2009).

6. Radzicka, A., and R. Wolfenden. 1995. A proficient enzyme. *Science* 267:90–93.

7. Weston Jr., R. E., and H. E. Schwarz. 1972. *Chemical Kinetics*. Englewood Cliffs, NJ: Prentice-Hall, 155–158.

8. Austin, R. H., K. W. Beeson, S. S. Chan, P. G. Debrunner, R. Downing, L. Eisenstein, H. Frauenfelder, and T. M. Nordlund. 1976. Transient analyzer with logarithmic time base. *Rev Sci Instrum* 47:445–447.

9. Wilson, M. C., and A. K. Galwey. 1973. Compensation effect in heterogeneous catalytic reactions including hydrocarbon formation on clays. *Nature* 243:402–404.

10. Hopfield, J. J. 1974. Electron transfer between biological molecules by thermally activated tunneling. *Proc Natl Acad Sci USA* 71:3640–3644.

11. DeVault, D. 1989. Tunneling enters biology. *Photosynthesis Res* 22:3–10.

12. Alberding, N., R. H. Austin, K. W. Beeson, S. S. Chan, L. Eisenstein, H. Frauenfelder, and T. M. Nordlund. 1976. Tunneling in ligand binding to heme proteins. *Science* 192:1002–1004.

13. The basics of this derivation can be found (in another context) in E. Merzbacher, Chapter 8 in *Quantum Mechanics* (New York: John Wiley & Sons, 1997).

14. Meyer, M. P., D. R. Tomchick, and J. P. Klinman. 2008. Enzyme structure and dynamics affect hydrogen tunneling: The impact of a remote side chain (I553) in soybean lipoxygenase-1. *Proc Natl Acad Sci USA* 105:1146–1151.

15. Oppenländer, A., A. C. Rambaud, H. P. Trommsdorff, and J. C. Vial. 1989. Translational tunneling of protons in benzoic-acid crystals. *Phys Rev Lett* 63:1432–1435.

16. Nobel Foundation, "The Nobel Prize in Chemistry, 1967, to M. Eigen, R. G. W. Norrish and G. Porter for Perturbation Chemistry Methodology," Nobelprize.org, http://nobelprize.org/nobel_prizes/chemistry/laureates/1967/index.html.

17. Barkai, E., Y. Garini and R. Metzler. 2012. Strange kinetics of single molecules in living cells. *Phys Today* 65(8):29–35.

18. Nordlund, I. 1914. Eine neue Bestimmung der Avogadroschen Konstante aus der Brownschen Bewegung kleiner, in Wasser suspendierten Quecksilberkügelchen. *Z Phys Chem* 87(1):40–62.

19. Barbara, P. F., Ed. 2005. Single-molecule spectroscopy. *Accounts Chem Res* 38(7):503–610.

20. Weigel, A. V., B. Simon, M. M. Tamkun, D. Krapf, and J. Lippincott-Schwartz. 2011. Ergodic and nonergodic processes coexist in the plasma membrane as observed by single-molecule tracking. *Proc Natl Acad Sci USA* 108(16):6438–6443.

21. Segel, I. H. 1993. *Enzyme Kinetics*. New York: Wiley-Interscience.

22. Voet, D., and J. G. Voet. 2004. *Biochemistry*, 3rd ed. New York: John Wiley & Sons, p. 480.

23. RCSB PDB software references:

    Jmol: An open-source Java viewer for chemical structures in 3D. http://www.jmol.org/.

    Protein Workshop: Moreland, J. L., A. Gramada, O. V. Buzko, Q. Zhang, and P. E. Bourne. 2005. The Molecular Biology Toolkit (MBT): a modular platform for developing molecular visualization applications. *BMC Bioinformatics* 6:21.

    NGL Viewer: Rose, A. S., A. R. Bradley, Y. Valasatava, J. D. Duarte, A. Prlić, and P. W. Rose. 2018. NGL viewer: Web-based molecular graphics for large complexes. *Bioinformatics*. doi:10.1093/bioinformatics/bty419.

24. Kinoshita, T., R. Maruki, M. Warizaya, H. Nakajima, and S. Nishimura. 2005. Structure of a high-resolution crystal form of human triosephosphate isomerase: Improvement of crystals using the gel-tube method. *Acta Crystallogr* 61:346–349.

25. Kuriyan, J., S. Wilz, M. Karplus, G. A. Petsko. 1986. X-ray structure and refinement of carbon-monoxy (Fe II)-myoglobin at 1.5Å resolution. *J Mol Biol* 192:133–154.

CHAPTER 16

CONTENTS

Molecular Machines
Introduction

16

Proteins move and transform matter on a molecular scale in biosystems. Proteins constitute the molecular machines that enable life processes to take place. Machines that transform matter are usually called *enzymes*, and the transport machines are called *motors*. Molecular motors comprise a class of molecular machines whose primary function is to transport objects along a track. This does not imply that enzymes that catalyze $(S + E \rightleftharpoons S \cdot E \rightarrow P + E)$-type reactions that rearrange chemical bonds do not move matter and do not work; they always do the former and often the latter.

When an enzyme binds to a substrate and reduces the height of an energy barrier, it does so by distorting bonds and moving atoms. Depending on the tissue, the majority of energy expended by the cell may be directed toward chemical synthesis rather than simple movement of molecules or packages of molecules from one site to another (as in muscle tissue and microtubule-based transport systems). Somewhere in between the "chemical reaction" enzymes and the "motor" enzymes are the enzyme complexes involved in movement of electric charge, i.e., nerve impulses. A great deal of power can be expended by these charge-moving systems. The human brain comprises 2% of the body weight but uses 20% of total body oxygen supply and 25% of total body glucose and would exhaust the local supply of ATP within a few seconds if it were not rapidly replaced. The brain is, in effect, a motor of power about 20–25 W and uses the energy mostly to move charge.[1] The remaining 100 W or so of resting power dissipation consists of (other) chemistry and physical work. In a rough way, we humans experience these approximately equivalent proportions of electrical (nervous), physical (mechanical), and chemical work done in our bodies every day—excluding, perhaps, those days spent in front of a video screen.

It would be convenient if all of the molecular machines in biology burned ATP for fuel, even if we have difficulty determining exactly how much work we get from each ATP molecule. We don't have this convenience, however. Some motor systems involve several sources of energy; some apparently involve none. Some, like the rotating flagellum, employ the charge separated across the membrane. The human body employs a huge variety of molecular motors, with different roles (chemistry, charge and physical transport, segregation) and modes of action. We can take only a brief moment to review a few important examples that have been well characterized by physical measurements or by experiments that are too beautiful to pass up. The rotary viral DNA-packaging motor is such an example. As you may guess from the chapter title, we will focus very narrowly on motors that manipulate DNA, along with single-molecule measurements quantifying the forces they exert. Typically, such measurements are made using laser tweezers or atomic force microscopy (AFM) techniques, the former for the force range 10^{-14}–10^{-10} N and the latter for 10^{-11}–10^{-8} N. Though the AFM has intrinsically smaller distance resolution, both can measure distances from a fraction of a manometer to microns.

Some of you will become fascinated by these active proteins. Additional reading can be found in many books, but the "facts" about motors change rapidly as single-molecule techniques become more reliable and sophisticated. Accepted facts about motors two years ago may no longer be facts. Books, even edited reviews of the state of the science, generally cannot keep up with research. A serious study demands regular searches of the journal literature. With these reservations in mind, some references that contain useful directions on molecular motors include

- *E. coli in Motion*, Howard C. Berg, 2004, Springer.
- *Mechanics of Motor Proteins and the Cytoskeleton*, Jonathan Howard, 2001, Sinauer Associates; and "Mechanical Signaling in Networks of Motor and Cytoskeletal Proteins," J. Howard, *Annual Review of Biophysics and Biomolecular Structure* 38 (2009):217–234.

A 2,000-calorie/day diet,

$$\left(\frac{2000 \text{ kcal}}{\text{day}}\right) \frac{4.18 \text{J}}{\text{cal}}$$

$$\frac{1 \text{ day}}{3600 \cdot 24\text{s}} \cong 100 \text{ W},$$

provides a "resting" power consumption. Of this, 20 W or so is used by the brain. An athlete can consume ~8,000 calories/day—400 W—and if the difference is caused just by a "muscle" activity increase by, say, a factor of 10, then the 100-W resting consumption is 20 W brain, 33 W muscle, and 47 W other chemistry/metabolism.

- "Understanding How the Replisome Works," K. J. Marians, *Nature* 425 no. 15 (2008): 125–127.
- *Molecular Machines & Motors*, J.-P. Sauvage, ed., 2001, Springer/Amazon.
- *Molecular Motors*, Manfred Schliwa, ed., 2003, Wiley-VCH.
- *RNA Polymerases as Molecular Motors*, Henri Buc and Terence Strick, eds., 2009, Royal Society of Chemistry, Biomolecular Sciences.
- *Motor Proteins and Molecular Motors*, Anatoly Kolomeisky, 2015, CRC Press.

16.1 BASIC CONSIDERATIONS FOR MOTORS

The variety of tasks and mechanisms makes enumeration of *basic considerations* of the motors difficult. Even among the motors specializing in nucleic acid manipulation and herding, the variety of motors is wide. The most basic task of one motor may be to collect the major molecular players—nucleic acid strand(s), ions, ATP, other proteins—in one small volume and carry out a chemical reaction, while another motor may move along a single DNA strand and prevent it from reattaching to another strand. Some characteristics of biomolecular motors are not truly "design features" in the sense of having a specific role in the biological process but are physical reflections of the microscale. These design features and characteristics include

- Randomness and directionality
- Employment of diffusion and gradients
- Characteristic step sizes and times
- Efficient use of energy

The last characteristic should not be underestimated. While precise computations of efficiencies often cannot be done under in vivo conditions, life could not rely on the efficiency of a classical Carnot heat engine,

$$\varepsilon_{\text{Carnot}} = 1 - \frac{T_{\text{Cold}}}{T_{\text{Hot}}} \tag{16.1}$$

when common thermal gradients are only a few degrees:
$\varepsilon \approx 1 - \frac{300}{310} \approx 3\%$.

16.1.1 Random Walk

The random walk enters into our consideration of DNA-manipulating molecular motors in two ways. First, most molecular motors tend to take steps but with a probability for direction and size of step (Figure 16.1a). Second, the conformation of DNA (one of the primary objects under the care of molecular motors) on a long scale (>500 nm) can be described with sufficient accuracy by a freely jointed chain (FJC) coil theory based on random walks (Figure 16.1b). We have already considered such random walks in Chapter 13. Random walks are a feature of the nanoscale, the single-molecule nature of the processes, and ubiquitous nature of thermal fluctuations. If there were no guidance to the motion—a directional drift—the motion would be a truly random walk. Even when hydrolysis of ATP is used to power a motor, the energy an ATP molecule provides is of order $20k_{\text{B}}T$, not that much larger than thermal fluctuation energies. The corresponding forces and distances are typically tens of piconewtons and a few nanometers: $20k_{\text{B}}T = (20 \text{ pN}) (4 \text{ nm})$. Table 16.1 shows examples of some ATP-driven molecular motor forces and step sizes.

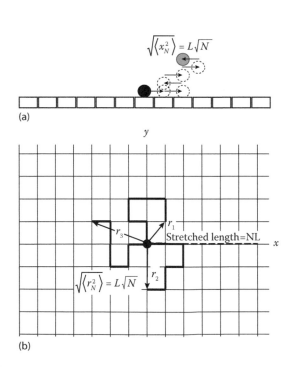

FIGURE 16.1
(a,b) A linear polymer like DNA has many more ways (higher entropy) to curl into a coil with short distance between the two ends than it has to form a straight line. Pulling the two ends of the coil therefore requires force, $F_{\text{entropic}} = -dG/dr$. The magnitude of this force for B DNA is about 0.1 pN. See text.

TABLE 16.1   Linear Molecular Motors on Tracks: Stepping Characteristics[a]

Motor/Track	Force (max) (pN)	Step Size (nm)	Speed (max)	Stays Attached?	Mechanism	Efficiency (approximate)[b]
Myosin (II)/ Actin	3–6	$5.5 \times n$	~1–8 µm/s (varies)	Rarely	Hops	20%
Kinesin/ Microtubule	5–7	8	800 nm/s	~100 steps	Walks	50%
RNA polymerase/ DNA	15–30	0.34	1–250 bases/s	Always	Crawls	20%

[a] Data abstracted from various sources.[2]
[b] Work done/ATP energy consumed.

16.1.1.1 Asymmetric steppers

The motion is not, of course, completely random under most circumstances. The protein molecular motors and the tracks or fibers to which they attach are asymmetric. Motion begins with the triggering of protein conformational changes that are not symmetric, though they may be reversible. In the cases of "advanced" motors like the family of actin/myosin muscle motors or kinesin, a well-studied and simpler motor, the asymmetric unit accomplishing directional motion is a pair of stepping feet called heads. Like our own asymmetric legs we expect the structure of the foot (head), ankle (neck), and leg (stalk) of the two-headed molecular motor to favor forward motion. This turns out to be true, but the mechanisms for directional motion in mysin/actin and kinesin are quite different. We will not pursue this fascinating topic, but the fundamental difference between these two motors is shown in Figure 16.2. A myosin motor acts as one of a large group of motors pulling on the same macroscopic object. A kinesin molecule acts alone and is entirely responsible to move its own microscopic load. This simplified structural diagram elicits many of the critical mechanistic questions about these motors. Does kinesin remain continuously attached to the track so it won't drift away, while an individual myosin molecule may not do so? Are the multiple myosin motors synchronized? If so, how? Is fluid frictional resistance significant for the kinesin load? What is the load limit for kinesin? Is the number of myosin motors on the fiber fixed? Are they uniformly spaced? The references cited at the beginning of this chapter devote many pages to addressing these questions.

> A macroscopic system with sharp edges like those shown in Figure 16.1 has issues that may allow net forward motion when random macroscopic shoves forward and backward occur. Cross-country skiers experience this phenomenon.

16.1.1.2 Thermal ratchets

The Brownian ratchet was publicized by Richard Feyman in lectures given to introductory physics classes at the California Institute of Technology. The basic idea of the Brownian or thermal ratchet is to use thermal energy—random packets of a few k_BT of energy—transferred to a ratchet mechanism that moves preferentially in one direction to produce net work. This would, of course, violate the second law of thermodynamics if the amount of net work were significantly greater than k_BT: work cannot be done in a cyclic manner from a single heat bath in an equilibrium system.

A simple version of a type of thermal ratchet is shown in Figure 16.3. A particle shown in gray sits in a saw tooth potential and experiences random, thermal collisions with other particles in the system. If this system is in equilibrium and the height of the potential rise is at least several k_BT, the particle will sit at one of the potential minima, occasionally acquiring enough energy from a collision to jump over the barrier to the left or right. What if the potential oscillates on and off with some frequency? Collisions will send the particle randomly to the right or left, but a short displacement to the right results in the particle moving to the adjacent well, where it would fall to the bottom, even further to the right (Figure 16.3d and e). An equivalent short jog

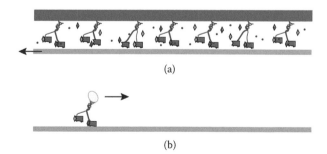

(a)

(b)

FIGURE 16.2
Diagram of molecular motors (steppers) differently tasked. (a) Many motors, coupled together on one fiber, collaborate to move another (macroscopic) fiber. Motors may not be synchronized. The myosin (motor)/actin (track) is in this category. (b) A single stepper transports a load along a track. Kinesin is the common example. Small diamonds and circles indicate that binding of ATP, ADP, Ca++, etc., is involved, but no sequential ordering is intended.

to the left would result in the particle returning to its initial position when the potential switches back on. Video simulations of this sort of Brownian motor may be found by inputting this term in any Internet search engine. If there were no thermal motions, the particle would not move, but the random thermal collisions will move the particle preferentially to the left. Net work is done on the particle, especially considering that any motion through the cloud of thermalized particles will be accompanied by an effective frictional force.

If a student is now asked, "Have random thermal motions moved the particle to the left in this asymmetric potential landscape?" she would probably reply, "I guess so... but... no, it can't be! That has to violate some law of thermodynamics." Net work extracted from a single heat bath? It is certainly good to have suspicions about this scenario, but what, exactly, is the problem or fallacy?

Before we attack the question of net work apparently done by purely randomized, microscopic thermal collisions, we should clarify what Figure 16.3 implies physically. The potential appears to be one dimensional: as the particle moves along the horizontal axis, the potential energy rises and falls in a periodic manner. On the other hand, the smaller particles responsible for the random collisions appear to be distributed vertically, as well as horizontally. If we were interested in a truly one-dimensional problem, we should restrict all particles to the z axis (longitudinal direction). But we *aren't* interested in a *truly* one-dimensional case—we are interested in biology! The "one-dimensional" examples of molecular motors such as kinesin, which travel along a linear microtubule path (Figure 16.4), involve one-dimensional transport, but three-dimensional thermal motions of molecules

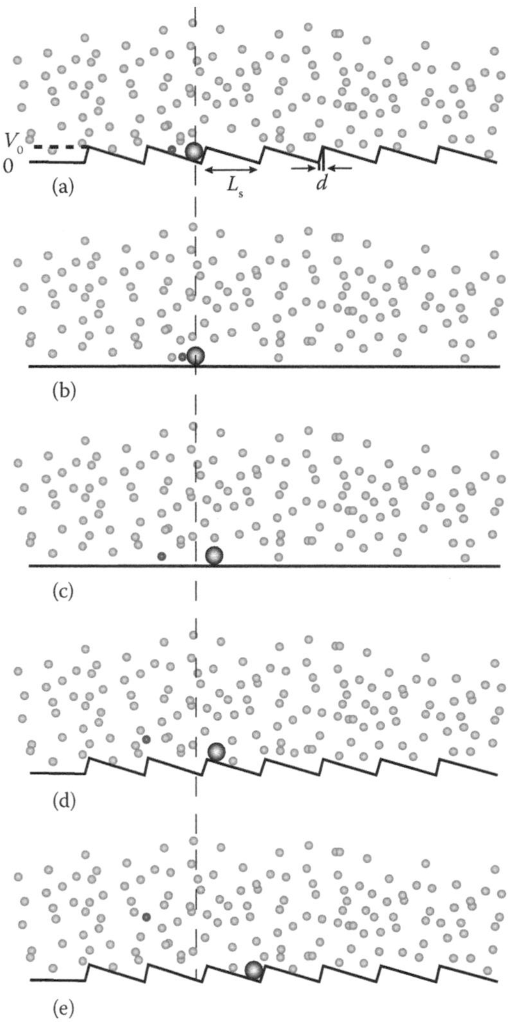

FIGURE 16.3

A (microscopic) Brownian ratchet with fluctuating potential. A particle (gray color) sits in a potential contoured like a ratchet. The potential at its lowest is zero. Particles collide with the gray particle, giving it an impulse to the right or left. As long as the system is in thermal equilibrium, the probability for net motion left or right is the same. (a) Initial state, potential in place; collisions cause little movement. (b) Potential is turned off; a random collision with the red particle sends the gray particle to the left. (c) A displacement to the left occurs. (d) The potential is turned back on, raising the particle on a left-sloping section. (e) The particle moves down the potential to the left. A random collision in b, sending the particle to the right, would likely lead to a return of the particle to its initial position in (a).

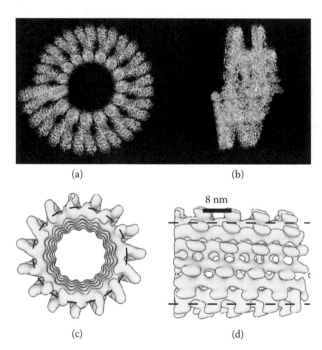

FIGURE 16.4

Structural details of a "one-dimensional" molecular motor track. (a, b) End and side view of a short section of display of kinesin13-microtubule ring complex, obtained by "docking" atomic-resolution X-ray crystal structures onto cryo-electron microscopy-derived electron densities, displayed using Protein Workshop software (PDB ID 3EDL, by Tan, D., et al., *Structure* 16, 1732–1739, 2008). (c, d) Earlier image reconstructions of microtubules decorated with monomeric and dimeric kinesins, displayed using EMViewer software. Protrusions outside of (above) dashed lines are mostly kinesin. (EMDB Entry EMD-1029, by Hoenger, A. et al., *J. Cell. Biol.*, 141, 419–430, 1998.)

that collide with the motor or diffuse from surrounding areas to the motor site. We expect any load particle moving along the horizontal axis to undergo collisions tending to push it in the $\pm z$, $\pm x$ but preferentially in the $-y$ direction, where $+y$ is upward, away from the microtubule surface. We must also assume that only the gray particle is subject to the periodic potential energy landscape, which in the case of kinesin/microtubules, would be binding to the spatially periodic microtubule, subject to the coupled, time periodic conversion of ATP to ADP. A more accurate model of the real system might then be a periodic raising of the energy of the gray particle (kinesin + load) through input of ATP energy: the particle is raised to the top of the saw tooth barrier, experiences random collisions in the longitudinal $\pm z$ direction, losing kinetic energy as it begins to move, staying at the same z position if bumped to the right and moving (downhill) to the left one step if bumped to the left.

> This downward ($-y$) osmotic pressure from particles free in the solvent would tend to keep the gray particle near its track if it were to become detached. However, as we shall see in Chapter 17, osmotic pressure is usually significant only when particles large compared to water molecules are involved.

EXERCISE 16.1   WORK DONE ON THE PARTICLE IN FIGURE 16.3

Discuss the work done on the gray particle in Figure 16.3.

> The solution to this exercise deserves a quick but thorough discussion in class. Consequently, we will not present the "solution" here.

1. Itemize the forces that do work on the gray particle from b → c in the figure. As discussed, the saw tooth potential is a one-dimensional *potential* felt by the gray particle, not a physical restriction of all particles to one dimension.
2. In Figure 16.3d the potential comes back on. The particle happens to be located near a peak in the potential. How much work is done on the particle from c to d? Does this work come from the thermal energy of the bath?
3. Between Figure 16.3d and e of the diagram the gray particle moves down the sloping potential and comes to rest in the next potential minimum. Keeping in mind the laws the rule particle motion in a liquid, what energy is converted to what form?
4. Discuss the differences if the saw tooth potential remained fixed at all times, but each time the gray particle reaches a local minimum, it remains there for a time Δt and then gains energy ΔE, which is within $k_B T$ of V_0 ($V_0 \pm k_B T$), at all times undergoing collisions with solvent molecules.
5. Discuss the magnitude of the periodic energy gain, ΔE (assume it is due to ATP hydrolysis) compared to the energy gains/losses from collisions with solvent molecules.

16.1.1.3 Coupling

In biochemical reactions in which ATP is converted to ADP, the issue of "coupling" of ATP hydrolysis to some other process is often said to be an issue. In the present context, the rigorous coupling of a single step of a motor with the hydrolysis of one ATP molecule is the question. Such coupling would fit with our human design sensibilities, but it need not be the case. The energy of one ATP hydrolysis could provide energy for two steps; or two ATPs might be needed per step. ATP hydrolysis might even produce a locally heated patch of water or a general potential gradient across a membrane (Figure 16.5). The localized hot water could constitute the "hot bath" for a classic heat engine. No such biological ATP-powered heat engine is known, but we should remember that electronic energy extracted from or transferred to any molecule is often rapidly dissipated to vibrational modes, producing, in effect, a local heat pool. It could be said that ATP is required for *every* process in *every* organism because these processes probably would not occur in a dead organism, but such a claim would certainly involve uncoupled processes. (The assertion would also be quite unilluminating.) The rotary ATP synthase of mitochondria and chloroplasts is an enzyme that uses the stored energy of a transmembrane proton gradient as a source for adding

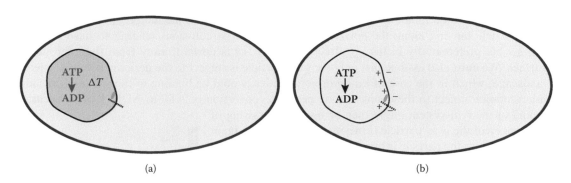

(a)   (b)

FIGURE 16.5

ATP coupling. Examples of possible motor motion that is uncoupled from ATP hydrolysis.
(a) Hydrolysis produces a locally elevated temperature, which then provides the "hot" bath for a
classic heat engine. No such biological example is known. As we saw earlier, thermal gradients
are rapidly dispersed by the high thermal conductivity of water. (b) ATP hydrolysis produces charge
separation, whose potential drives motor motion. If a motor gains the energy for a step from the
potential gradient across a membrane, the motion is uncoupled from ATP hydrolysis. If hydrolysis
of one ATP separates one charge, which is directly used to power one step, the motion is coupled
to hydrolysis. The rotary ATPase is in this category.

a phosphate group to a molecule ADP to form ATP. This enzyme works when a single proton
moves down the gradient, giving the enzyme a rotational motion. As with most enzymes, ATP
synthase can also function in reverse and use energy released by ATP hydrolysis to pump protons
against their thermodynamic gradient. The important articles by Howard Berg should be read by the
student interested in these motors and the motile mechanisms of bacteria.[3] In the case of kinesin,
the coupling of ATP hydrolysis appears to be quite direct, with ATP/ADP binding and unbinding
causing conformational changes that result in steps.[4]

16.1.1.4 Directionality

Motor transport systems involving transport along a track can be directional because of an asymmetry in the track, in the motor, or in the starting conditions (Figure 16.6). The ratchet mechanism of Exercise 16.1 includes an asymmetric structure that favors motion in one direction, but
the initial motion of the particle is completely random. Ratchets are
directional, not because forces are exerted preferentially in one direction but because structural asymmetry (in this case, asymmetry of the
track) inhibits motion in one direction. Of course, in the end a force
is exerted in the direction of motion, but this only happens when the
initial thermal motion is in the preferred direction.

Tracks, like the microtubules used by the molecular motor
kinesin, are often directionally asymmetric, with a "+" and "−"
end of the track. Unlike the model ratchet track above, however,
the microtubule is not the motile agent; the motor, kinesin, is. As
with our sidewalks and roads, it would seem that tracks need not be
directional to get directional motion. Why couldn't a motor simply
jump on a track and move in a direction determined by a concentration gradient or some other chemical signal? (We do have to
remember that molecular motors do not have eyes or other remote-
sensing devices that can signal a preferred direction.) A search of
published literature turns up cases in which the same type of motor
on the same type of track can move in either direction. If fact, it
is quite risky to list and memorize the currently known properties
of biomolecular motors, as new discoveries every few months tend
to modify or erase what was fact 6 months earlier. Some of the
contradictions between new and old measurements relate to sample
preparation or damage, removal of parts of the overall system, or

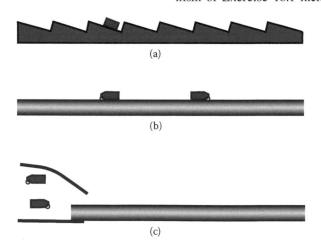

(a)

(b)

(c)

FIGURE 16.6

Directionality of a motor on a track. (a) Directionality caused
by asymmetry of the track. Nota bene: you may be misled by
this figure. See text. (b) The motor may be asymmetric and
may move in either direction, depending on its orientation
on the track. (c) If all motors attach to the track at one end
of the track, only those "pointed" in the right direction will
move. The wrongly oriented motors will immediately move
off the track and try again.

peculiarities of the measurement process itself. AFM imaging results on DNA in the early days were notoriously unreliable because of sample artifacts and too strong interactions with surface or scanning tip. Measurements on molecular motor movement involve far more complexity—motor and track removal from normal environments in cells can damage one or both; attachment to a surface or microsphere using a variety of chemical agents can introduce sample differences; the "same" motor from two species or strains may differ; chemicals and concentrations in different experiments often differ; etc.—so it is not surprising that a variety of motor behaviors may be reported. We will thus continue to focus on possible molecular motor behaviors and mechanisms, leaving the details to a study of current literature.

16.1.1.5 Flagella and direction

Rotational motors that drive whole cells through water with no tracks, like the flagella of bacteria, can also be directional. Again, we refer to the reviews of Howard Berg (and others) on the *Escherichia coli* and other flagella. At least three distinct types of flagella have been found: bacterial flagella like *E. coli*'s, eukaryotic flagella (often called *cilia* or *undulipodia*), and archaeal flagella.[5] In the case of *E. coli*, five to six flagella extending from random points on the sides of the cell body form a synchronous bundle when they rotate in one direction, driving the cell forward. When the flagella rotate in the opposite direction, they do not bundle but rotate independently, causing the cell to tumble. This tumbling changes the direction the cell is pointed, allowing a change of direction if, for example, an environment lacking in nutrients is encountered. The interested student or instructor is encouraged to explore the classic treatment by Berg on rotational diffusion and directional changes of *E. coli*.[6] Even in this more complex case of rotational motion the central role of asymmetry is clear: drive behavior is not symmetric with respect to direction.

The asymmetry in individual flagella, which rotate, and cilia, which wave, must be built into the elastic properties of the structure itself. Figure 16.7 illustrates several types of asymmetry. A cilium that waves through the solution must have different bending moduli for bending in opposite directions. The model shown in Figure 16.7a suggests a simple hinge with no bending resistance in one direction, while waving in the opposite direction results in a straightened rod that cannot bend any further. Similarly, rotation of the flagellum must produce different structural effects for opposite directions of rotation (Figure 16.7b). Figure 16.7c shows a more realistic model of a bacterial flagellum, with a helical, screw-like structure. We know that a helix has a direction, so rotation of a rigid helix with the structure of Figure 16.7c would produce torques in a different direction, depending on the direction of rotation. Because the flagellum is not infinitely rigid, the helical structure will distort differently in response to opposite torques, exhibiting different hydrodynamic behavior. Asymmetric elastic properties are commonly found in nature. Most leaves are hydrodynamically asymmetric: in a strong wind the back of a leaf will face the wind. A branch of a tree shows more displacement (lower spring constant) to an upward force than to an equivalent downward force. Even a wooden dowel,

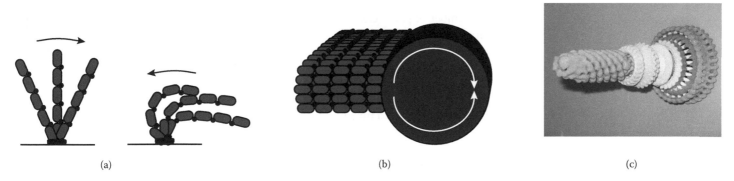

(a) (b) (c)

FIGURE 16.7
Simple examples of structural asymmetry in cilia and flagella. (a) Forces produced by the waving of a cilia through water produce different structural effects if the cilia elastic properties are asymmetric. (b) Rotation of a flagellum must result in different structural effects if the direction of rotation is reversed. (c) A more realistic physical model of a bacterial flagellum; imaged and modeled at Brandeis University in the DeRosier lab and printed at the University of Wisconsin–Madison, author Alan Wolf; fabricated on a ZCorp Z406 printer from a VRML generated at Brandeis; reproduced under the Creative Commons Attribution-Share Alike 2.0 Generic license (http://creativecommons.org/licenses/by-sa/2.0/deed.en.)

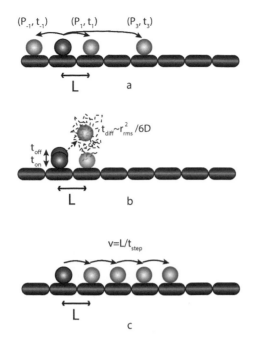

FIGURE 16.8

Relation of step length and time can be complex. (a) Linear stepper with variable step lengths, probabilities, and times. One model of such a stepper is based on the multinomial distribution described in Chapter 13. (b) A stepper can also detach from its track, with an on/off equilibrium (probability) described by an equilibrium constant. Once off, the particle can diffuse, in principle to any other site, to the original site, or away altogether. (c) The simplest model usually asserts that stepping can be described by an average velocity given by $v = L/t_{step}$.

machined into a perfect cylindrical rod, will have different bending rigidities in different directions, depending on the "grain" of the wood. This asymmetry extends, not surprisingly, to the microscopic level because biopolymer units are intrinsically asymmetric.

16.1.1.6 Step size, step time, and rate of motion

Expression of the relationship of the step size and time in terms of an average velocity can be complicated, as Figure 16.8 suggests. In the case of a perfect linear motor that never leaves the track, the motion may be described by a step probability, with corresponding step time. We saw in Chapter 13 that the binomial distribution describes the case of single steps taken to the right or left with probabilities p and q, respectively. If $p = q$, the stepper goes nowhere, on average, though there is a finite probability to be at locations $\pm NL_s$ or anywhere in between, after N steps. If there are many steppers and enough time, this may be a biologically viable mechanism for transport. More often, $p \neq q$ and the average velocity of motion is (Equation 13.24)

$$\langle v \rangle = (p-q)\frac{L_s}{t_s} = (2p-1)\frac{L_s}{t_s} \tag{16.2}$$

where L_s is the step length and t_s the step time. As we learned in Chapter 13, this average velocity, or the almost equal velocity of the peak of the probability distribution, may or may not be the biologically relevant velocity. If many equivalent steppers are moving and a cell only cares when the *first* package is delivered, the first-passage time or perhaps the arrival of a 10% occupation probability would be more meaningful. Working out such a problem requires careful review of current research. More complex kinetics are also possible, as described in Chapter 13, if there are finite probabilities for steps of lengths of multiples of L_s. If the probability for a step of length kL_s is P_k and k ranges from $-m$ to $+m$, the average velocity of the stepper will be

$$\langle v \rangle = \frac{L_s}{t_s} \sum_{k=-m}^{+m} kP_k \tag{16.3}$$

Equation 16.3 includes cases where steppers may not move ($k = 0$) during a particular step time. If the step time or length were variable, they should be moved to the right of the summation sign and given indices k also. The problem with such a stepper is the difficulty of experimentally determining such a large number of parameters, especially when we see that they all may be dependent on the load the stepper is carrying.

What should the average velocity of the stepper in Figure 16.3 be? When the potential reduces to zero, the stepper begins free diffusion. If the peak of the potential step (to the right) is a distance d from the minimum (Figure 16.3a), the stepper must diffuse a distance slightly more than d to the right. If we accept a model of one-dimensional diffusion, diffusing an average distance d in the horizontal direction requires a time given by

$$\langle d^2 \rangle = 2Dt_d \tag{16.4}$$

What does this mean? If the time that the potential stays at zero ("off") is at least $t_{off} = 2t_d = \frac{d^2}{D}$, the stepper will have time to diffuse a distance d to the right and will take a step of length L_s when the potential comes back on. The maximum average speed of the Figure 16.3 stepper for which $t_{off} = t_{on} = d^2/D$ would then be

Factor of 2 because half the time it will go to the left.

$$\langle v \rangle_{max} = \frac{L_s}{t_{off}} = \frac{DL_s}{d^2} \tag{16.5}$$

How efficient is this switching-potential, diffusion stepper? If the stepper diffuses a distance d to the right, the potential will have to do work to raise the stepper from $V = 0$ to $V = V_0$ when it comes back on. If the potential had been switched off a longer time and the stepper had diffused farther to the right, the potential would not have to raise the stepper's energy as high as V_0. If the particle diffused exactly a distance L_s to the right, the particle would already be at the potential's adjacent minimum and no work need be done. Does this mean that if the potential off time is increased so that $t_{\text{off}} = 2t_d = \frac{L_s^2}{D}$ that the particle would always have diffused to the next potential minimum and the potential would never have to do any work? Problem 16.3 explores this question, to which you I hope answered "No."

16.1.1.7 Detachment and processivity

A stepper can also, however, detach from its tracks as in Figure 16.8b. The switching-potential stepper of Figure 16.3 also implies this possibility because the particle at any time could acquire enough thermal energy to escape the local well. The detachment and reattachment can sometimes be described by an on/off equilibrium, using a free-energy difference between bound and unbound or by on and off rates:

$$\frac{\left[c_{\text{on}}\right]}{\left[c_{\text{off}}\right]} = e^{-\Delta G/k_B T} = \frac{k_{\text{on}}}{k_{\text{off}}} \tag{16.6}$$

If detachment occurs, the stepper will not be able to progress along the track, unless the motor is the switching-potential, diffusing type. Stated differently, if the drive for the stepper is built into the stepper itself, it must remain attached to the track in order to progress. For this type of self-driven motor, the fraction of time the motor remains attached to the track is termed its *processivity*. A highly processive motor remains attached most of the time and the average velocity from Equation 16.2 or 16.3 applies. If the stepper is not 100% processive, the velocity applies only when attached. If we assume that a detached motor remains near the last site of attachment the average velocity must be multiplied by the fraction of "on time":

$$\langle v \rangle = \langle v \rangle_0 \frac{k_{\text{on}}}{k_{\text{on}} + k_{\text{off}}} = \langle v \rangle_0 (1 + e^{\Delta G/k_B T}) \tag{16.7}$$

(This assumes that reattachment does not include some other issues, such as improper binding, pointing in the wrong direction, etc., which can be subsumed into the free-energy factor.) We noted that kinesin (Figure 16.9) is highly processive and remains attached to the microtubule most of the time. In contrast, the more complex, multimotor system of muscles has low processivity.

16.1.2 Friction and Dissipation

Molecular motors move loads through water, perhaps encountering other objects or media along the way. If the motor plus its load can be modeled approximately as a spherical particle, we can calculate precisely what the fluid frictional dissipation will be. The frictional force is $f = bv = \frac{k_B T}{D} v = 6\pi\eta R v$ and the power dissipated is

$$P_{\text{frict}} = f \cdot v = bv^2 = \frac{k_B T}{D} v^2 = 6\pi\eta R v^2 \tag{16.8}$$

If the fluid has an effective viscosity 100 times that of bulk water, the particle radius is 1 μm (a large load) and the stepper moves at 1 μm/s (somewhat above the maximum speed of kinesin), the frictional force is $f = 6\pi\eta R v = 6\pi(100 \cdot 10^{-3}\text{Pa} \cdot \text{s}) (10^{-6}\text{m})(10^{-6}\text{m/s}) = 2 \text{ pN}$;

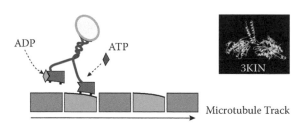

FIGURE 16.9
The two-headed (two-footed?) kinesin motor has been studied using classical Michaelis–Menten enzymology, as well as modern single-molecule force and step techniques. Type 1 kinesin steps on a microtubule track, with discreet step lengths (load displacements) of 8 nm, though each head is placed 16 nm ahead of the trailing head. Cartoon model is of complete motor with load shown (light blue). The inset shows the X-ray diffraction structure of kinesin from Norwegian rat brain. One ADP molecule is bound to each head. The two heads are not identical. Only part of the stalk region, and no load, is shown in the latter structure. (PDP ID 3KIN by Kozielski, F. et al., *Cell*, 91, 985–994, 1997, displayed using Protein Workshop.) (From Jmol: An open-source Java viewer for chemical structures in 3D. http://www.jmol.org/; Protein Workshop: Moreland, J.L. et al., *BMC Bioinformatics*, 6, 21, 2005; NGL Viewer: Rose, A.S. et al., *Bioinformatics*, 2018.)

the power dissipated is about $P_{\text{frict}} = 6\pi(100 \cdot 10^{-3}\,\text{Pa} \cdot \text{s})(10^{-6}\,\text{m})(10^{-6}\,\text{m/s})^2 \cong 2 \times 10^{-18}\,\text{W}$. Are these large or small numbers? The measured *stall force* of kinesin is about 6 pN, so the load we have constructed is about 30% of the maximum for kinesin. We could compare the power to the ATP energy consumed by kinesin. About 90 ATPs per second are converted to ADP by the two head of kinesin. If each ATP releases about 50 kJ/mole, the ATP power is $P_{\text{ATP}} \cong \frac{90\,\text{ATP}}{\text{s}}\,\frac{5\times10^4\,\text{J}}{\text{mole ATP}}\,\frac{1\,\text{mole}}{6\times10^{23}\,\text{molecules}} = 7.5\times10^{-18}\,\text{W}$. Frictional dissipation corresponds to about 25% of this power. Potentially, friction may account for a few tens of percent of the capability of a motor like kinesin. We should note that our load is larger than those commonly encountered. A 100-nm radius load moving at 200 nm/s requires a force of only 0.04 pN and power of $0.008 \times 10^{-18}\,\text{W}$.

16.1.2.1 Reynolds number and microscopic motion

We introduced the Reynolds number, $\mathcal{R} = \frac{\rho v_0 R}{\eta}$, in Chapter 12 to describe the "connectedness" or coherence of fluid molecules in the neighborhood of a moving object. The Reynolds number is unitless and depends on the fluid density and viscosity (ρ and η) and the radius and speed of a spherical particle. If $\mathcal{R} >> 1$, inertial effects and turbulence are important; if $\mathcal{R} < 1$, inertial effects, and turbulence are negligible.

Consider a 5-μm radius microorganism chasing another that it intends to eat. If they are dawdling along at about one diameter per second, $v_0 = 10$ μm/s, in somewhat "thick" pond water of viscosity $10\eta_w \sim 10^{-2}\,\text{Pa·s}$, the Reynolds number is $\mathcal{R} = \frac{(1000\,\text{kg/m}^3)(10^{-5}\,\text{m/s})(5\times10^{-6}\,\text{m})}{10^{-2}\,\text{Pa·s}} \cong 5\times10^{-6}$, which is clearly in the low-R regime. The task facing the hungry microorganism is similar to what most of us humans who cook our own breakfast encounter about one time a week.

16.1.2.2 Story of the eggshell

We decide to fry up the best scrambled eggs west of the Prime Meridian. We crack the first egg and dump it into the bowl. The second—oops, a bit of egg shell drops into the bowl. No problem: we'll get the bit of shell out with a spoon. Carefully, we position the spoon next to the shell and push to the side of the bowl. However, no matter how careful we are, the shell bit remains about 2 mm ahead of the spoon tip and, just before the bit reaches the surface, it slips off to the side and escapes. We repeat the process six times, with the same result. Some of us who have not yet had our coffee resign ourselves to crunchy eggs or head for a box of cereal. Why can't we catch the shell bit in our spoon? The reason is similar to, though not exactly the same as, the reason a microorganism has some difficulty catching its prey. The main reason the shell bit is hard to catch is that the egg whites consist of random polymers (glycoproteins) interspersed with water and many other molecules. These polymer molecules interact with each other and tend to exclude large particles from entering the polymer matrix. (Remember the DNA purification technique using PEG and magnetic microbeads?) However, the egg white also has a very large viscosity, about 160 centipoise (0.16 Pa·s), 160 times the viscosity of bulk water, making the Reynolds number smaller.[7] In small-R situations the fluid sticks together, out to a distance comparable to the size of the moving object.

16.1.2.3 Catching a runaway

In the present context of molecular motors, even the rather large flagellar motor of *E. coli* cannot hope to allow the bacterium to escape the low-$\mathcal{R}$ regime of kinematics. Even in low-viscosity bulk water, *E. coli* would have to move at 1 m/s, or 1,000,000 cell lengths per second, in order to achieve $\mathcal{R} \sim 1$. If we estimate that an organism's maximum speed is N cell lengths per second, the values of N and $\mathcal{R}$ (cell radius) needed to achieve $\mathcal{R} = 1$ are given by

$$N\mathcal{R}^2 \cong 2\times10^{-6}\,\text{m}^2 \tag{16.9}$$

in bulk water. Table 16.2 shows some typical values. Figure 16.10 shows the primary problem a small microorganism has with being a predator: the surrounding fluid, out to a distance of about an additional radius, tends to travel with the organism, pushing other particles of comparable or smaller size along with it.

TABLE 16.2 Sphere Speed and Radius Needed to Achieve $\mathcal{R} \sim 1$

Radius (µm)	Speed to Reach R>1 (diameters traveled/s)	Example Organisms
1	2,000,000	Not possible
10	20,000	Not possible
100	200	Not likely
300	20	See Figure 1.2a
1000	2	Cyclops (0.5–5 mm); see Figure 16.10b
10,000	0.02	Small fish; most aquatic organisms visible to the eye

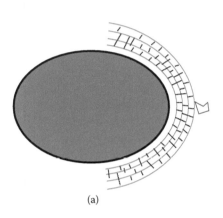

(a)

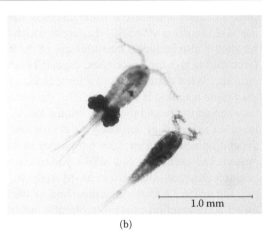

1.0 mm

(b)

FIGURE 16.10
Effect of low Reynolds number (or loose polymer fluid matrix) on pursuit in a fluid. (a) The coherence of the fluid in front of the pursuing object pushes the small object (the prey) forward. The pursuer must temporarily accelerate and approach the high-$\mathcal{R}$ regime in order to catch the prey. (b) The predatory, millimeter-sized *Cyclops bicuspidatus* (the dominant cyclopoid species in Lake Michigan) and similar organisms are just large enough to easily escape the low-$\mathcal{R}$ regime. Its prey includes ciliates, rotifers, small copepods, and fish larvae. (Public domain microphotograph by Liebig, J. NOAA GLERL, 2000.)

Large aquatic organisms (e.g., fish, submarines) have the opposite problem: their Reynolds number is so high that turbulence is difficult to avoid. Ways for minimizing high-R turbulence effects include slimy polymer coatings that minimize interactions with water.

16.1.3 Michaelis–Menten Enzyme Analysis and Molecular Motors

If the claim that enzymes are molecular motors and can be understood in terms of forces, step sizes, and efficiencies holds water, there should be a connection to classical biochemical analysis of enzymes. Chapter 15 introduced the Michaelis–Menten (M–M) description of enzyme action using the continuum model $E + S \xrightleftharpoons[k_{-1}]{k_1} E \cdot S \xrightarrow{k_2} E + P$ (Equation 15.56). M–M analysis has been performed on thousands of enzymes, mostly with success. The primary equation used in the analysis is $V_0 = \frac{k_2 [E]_T [S]}{K_M + [S]} = \frac{V_{max}[S]}{K_M + [S]}$ (Equation 15.66) with $K_M \equiv \frac{k_{-1} + k_2}{k_1}$ (Equation 15.65). V_0 and V_{max} are reaction "velocities," measured in units of concentration change per second: Ms^{-1} or molecules $m^{-3}s^{-1}$. The initial reaction velocity V_0 is plotted as function of the substrate concentration [S] (or $1/V_0$ versus $1/[S]$) in order to determine values of V_{max} and K_M. Remember V_{max} is the maximum reaction velocity attained at high substrate concentration and K_M is the substrate concentration at which the velocity is half maximum. This parameterization makes perfect sense when the main job of an enzyme is to catalyze conversion of substrate to product.

What do the Michaelis–Menten parameters mean if applied to a molecular motor that uses ATP energy and performs work, moving a load by applying a force? Is the substrate the load before it is moved and the product the transported load? This would make little sense in terms of substrate and product concentrations and reaction velocities. Perhaps the substrate is ATP and the product ADP? ATP is consumed and ADP is certainly produced, so this identification is appropriate. Describing a molecular motor in terms of an enzyme catalyzing the conversion of ATP to ADP seems a rather pointless exercise, though the M–M analysis may be straightforward. How about a hybrid M–M interpretation: the substrate concentration is [ATP] and the reaction velocity is the actual velocity of the motor, in meters per second? K_M would still have the same meaning and units as for a classical catalytic enzyme. The initial velocity V_0 will have different units than in the normal enzyme-catalysis case, but identifying the product as "moved load" seems promising, except for the fact that we have still ignored one of the primary functions of a transporting motor: the load it carries.

Transporter (motile) molecular motors move loads from one place to another. They apply a force to a load and move it some distance. Identifying the motor velocity in meters per second as the M–M reaction velocity—the rate at which product appears—is a step in the right direction, but we should also include a measurement of the force. Besides, physicists are always concerned about force during dynamic processes, especially when we expect the kinetics of the process to depend on force. What corresponds to a force in classical Michaelis–Menten reaction kinetics? Hopefully, your rapid response is something like "the activation energy," or "the slope of the free-energy barrier between substrate and product," but the fact is that Michaelis–Menten parameters do not explicitly involve the distances necessary for determination of forces. Even the basic kinetic parameters that depend on activation energies are hidden in the Michaelis–Menten analysis and are often not discussed. The exercise below allows you to expose the fundamental physical quantities—energy and position changes—that govern M–M analysis.

Figure 16.11 shows a comparison of the energy landscape faced by a transporter molecular motor and a reaction-catalysis molecular motor. The transporter motor normally makes many small, almost identical steps that require it to overcome a small energy barrier for each step (Figure 16.11a). The free energy required to move may also gradually increase, if we include the frictional (entropic) energy losses that normally occur or if the destination site has higher free energy for any reason. For example, the concentration of the load being moved may already be high at the destination and low at the origin: expenditure of energy is required to move the load against this concentration gradient. Backward motion may also occur, though attention must then be paid to frictional losses. The "catalysis" motor, on the other hand, encounters a large energy barrier along a reaction coordinate,

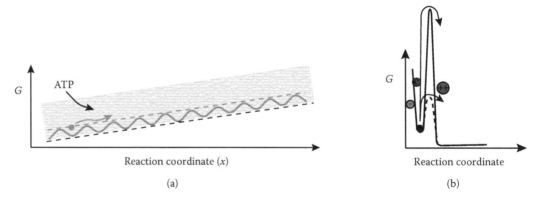

(a)   (b)

FIGURE 16.11
Energetics of enzymes: "motors" versus catalysts. (a) The task faced by a molecular motor is usually to move a particle distances of nanometer to μm over many steps, each step requiring a small force. The energy may effectively slope upward because of frictional losses or because the free energy (e.g., concentration) of the load molecules may be higher at the destination. The slope of the free energy curve reflects, roughly, the force necessary to move farther. The shading reflects obstructions the motor and load may pass through during motion. (b) An enzyme catalyzing a reaction must also move particles, but often the distance is small and only one step is needed. A noncatalytic motor would have to exert an enormous force to push the particle over the energy barrier between initial and final states. Nature is not stupid: the biological motor—the enzyme—will reduce the force necessary (reduce the energy barrier) before doing its work. In the case of the EcoRI endonuclease, the final state is lower in free energy, so no input of ATP or other energy "fuel" is needed. Lowering the energy barrier is accomplished in part through strategic placement of positive charges.

EXERCISE 16.2

Activation energies and forces in Michaelis–Menten parameters V_0, K_M, and k_{cat}.

1. Draw an energy diagram and use it to explain how activation energies and product-reactant energy differences affect the three M–M parameters.
2. Interpreting force as $\Delta E/\Delta s$, where Δs is a displacement that is accompanied by an energy change, ΔE, explain how forces affect the M–M parameters.

Solution

This is another one for the student. Draw a good diagram.

a barrier whose height the enzyme reduces by binding and association of various components in a small spatial region. (We will discuss this more in a moment.) The catalytic enzyme then simply waits for the reaction to proceed "spontaneously" over the reduced barrier.

Michaelis–Menten Analysis of Kinesin: The linear, motile molecular motor *kinesin* has been analyzed both in terms of M–M analysis and velocity-force dependence by Block and coworkers (and others).[8] Kinesin (specifically kinesin-1), sometimes referred to as the "world's tiniest biped,"[9] is a protein composed of two identical catalytic domains (heads) that each hydrolyze ATP. They can each bind to and together move along a microtubule (Figure 16.9). The two heads are attached to a ~70-nm-long common stalk, made up of α-helical, coiled-coil, and a few "hinge" regions. The kinesin heads carry out a hand-over-hand walk that moves the molecule toward the plus end of a microtubule in 8-nm steps. (Note that each head executes a 16-nm motion past its partner, but motion of the load is in 8 nm steps.) The motion is tightly coupled to the hydrolysis of a single ATP molecule. Kinesin movement is highly *processive*: the motor takes hundreds of steps before leaving its track. In the model of Figure 16.9, each head must attach and release itself from the microtubule. High processivity means that both heads rarely leave at the same time. Genetic techniques used to modify the lengths and other characteristics of kinesin's two legs have resulted in "limping," whose behavior shows that the motor moves by an asymmetric, hand-over-hand walk, where the even- and odd-numbered steps are not identical.

16.2 DNA-MANIPULATING MOTORS

There are hundreds of enzymes known to manipulate DNA. The primary role of some is to create or break covalent bonds. Others seem to interact physically with DNA without making or breaking phosphate bonds. Yet other enzymes do both. Figure 16.12 shows structures of a few enzymes—molecular motors—that manipulate DNA without breaking or making phosphate bonds in the nucleic acid. Some of these enzymes clearly have structures designed for certain purposes, like in Figure 16.12a, the DNA polymerase Pol III "sliding clamp" and *b*, the DnaB helicase. The hole in the middle of these structures would seem to be designed for a DNA strand to pass through. This is indeed the case. See figure caption. The third DNA-manipulating protein in Figure 16.12c is DNAI, whose job it is to position another protein, the helicase, onto DNA so that it can separate strands using energy derived from ATP hydrolysis.

The wise reader will study the review article by Bustamante et al., "Ten Years of Tension: Single-Molecule DNA Mechanics," 2003, *Nature* 421:423–427, as well as the volume devoted to this topic: "Single Molecule Imaging and Mechanics: Seeing and Touching Molecules of Life One at a Time," 2014, *Chemical Reviews* 114 (6)

16.2.1 Manipulating DNA

DNA in cells stores information. This "passive" molecule cannot be put away in a closet and ignored until the information is needed for cell division, however. DNA's information must constantly be accessed, transcribed, translated, checked, and corrected. DNA must, as well, be "herded" (e.g., during cell division), without perturbing the informational structure. The entities that perform these functions are generally classified as enzymes, but the requirements of these enzymes in terms of the forces

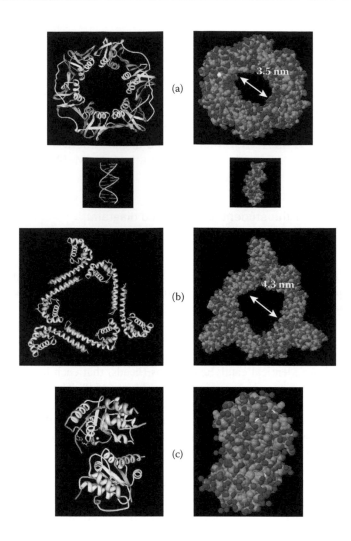

FIGURE 16.12

Structures of several "handler" proteins that manipulate DNA, without making or breaking bonds. (a) The dimeric "sliding clamp" part of the Pol III, *E. coli*'s multienzyme replicase. As you might guess, DNA passes through the central hole, which is about 3.5 nm in diameter. (PDB ID 1OK7, displayed using Protein Workshop (left) and Jmol (right) software[24]; Burnouf, D. Y., et al., *J. Mol. Biol.* 335, 1187, 2004.) The small insert between (a) and (b) is a scaled structure of BDNA (PDB ID 3BNA by Fratini, A. V. et al., *J. Biol. Chem.*, 257, 14686–14707, 1982.) (b) Hexameric ring structure of the N-terminal domain of *Mycobacterium tuberculosis* DnaB helicase. DnaB separates strands of double-stranded DNA. (PDB ID 2R5U, Biswas, T., and O. V. Tsodikov, *FEBS J.*, 275, 3064–3071, 2008.) (c) Structure of an enzyme from *Geobacillus kaustophilus*, DnaI, the helicase loader, which is required to load a hexameric helicase enzyme DnaC onto DNA at the start of replication. ADP, phosphate, and Mg++ may be identified in the structure on the left. (PDB ID 2W58, Tsai, K.L., et al., *J. Mol. Biol.*, 393, 1056, 2009.)

they must exert vary enormously. A motor that needs only to moderately straighten DNA, perhaps in preparation for cell division or segregation of the DNA in one part of the nucleus, need not exert much force. A DNA replication motor that needs to separate strands must be capable of significantly more force but usually no more than about 70 pN. At the high end of the force-exerting motors are the DNA repair and DNA-cleaving enzymes (motors), which must break covalent bonds. The description of these "chemistry" motors traditionally reverts to an enzyme framework (e.g., Michaelis–Menten) rather than a force description, but a rough estimate puts this force needed for chemistry at about 1 nN, if done in a brute force manner. We will see, however, that nature abhors brute force methods, instead relying on a variety of local molecular and ionic assistants to facilitate the reaction. In some ways this process is like chaperonin-assisted structure formation, with agents placed in the local environment to favor certain structural outcomes. Table 16.3 lists typical molecular-motor force magnitudes that have been measured or estimated for various processes involving DNA.[10]

16.2.1.1 EcoRI endonuclease and DNA cleavage

It would be a mistake to assume that the processes involving larger forces also consume more energy in the form of ATP → ADP conversion or the like. Cleavage of phosphate bonds by endonucleases often requires no fuel at all. EcoRI is a "type II" endonuclease found in the RY13 strain of *E. coli* bacteria. Its primary role is to inactivate foreign (e.g., viral) DNA by chopping it up. (*E. coli*'s own DNA is protected from cleavage by being methylated by another enzyme in the Eco family.) EcoRI specifically recognizes the DNA sequence GAATTC, catalyzes the cleavage between G and A, and requires no ATP. (Type I endonucleases do require ATP for activity.) GAATTC happens to be a self-complementary sequence—it is its own complement—so a solution of GAATTC will spontaneously form the double strand $^{G\ A\ A\ T\ T\ C}_{C\ T\ T\ A\ A\ G}$. EcoRI binds to both sides of the DNA and cleaves both strands. The structure of this enzyme-DNA complex is shown in Figure 16.13. Figure 16.13a, b and c, d reflect structure before/after cleavage.

EcoRI requires no obvious fuel source for its cleavage activity. When DNA is synthesized, the polymerases work with the high-energy form of the nucleotides, dATP, dCTP, dTTP, and dGTP, so the building blocks also provide free energy for assembly. Because double-stranded DNA is a higher free-energy form than the cleaved strands, one could claim that no energy is required for cleavage. However, before cleavage the two phosphate bonds between two sets of ribose rings were intact; after cleavage, they are not. This will not spontaneously happen, at least not on a time scale of tenths of a second. (There is some truth to the *Jurassic Park* assertion that DNA can be stable for millions of years, under the proper circumstances.)

Perhaps we should simply retreat to the position we took in Chapter 15 and say that this is "enzymatic catalysis," and all we need consider is that enzymes reduce the energy barrier and thus increase the reaction rate. This should bother, but also enlighten, the physics student: there is a connection between forces and activation energies! Roughly speaking, the slope of the free-energy versus reaction coordinate plot is the force needed to move the particle. The difference between "motile" (transporter) molecular motors and catalytic molecular motors was illustrated in Figure 16.11. The former have a series of comparatively small barriers to overcome on a generally upward sloping landscape of free energy versus position, as discussed earlier. An enzyme, on the other hand, has a large barrier to overcome along a reaction coordinate in order to get to a state that may be lower in free energy. If a molecular motor tried to pull apart two nucleotides by brute force, the required force would be enormous. EcoRI, like all biological enzymes, did not survive by being a brute. The molecular motors we call enzymes do not just force the system over the highest-energy barrier but rather employ a variety of means to reduce the barrier height.

Figure 16.13 shows two of the means EcoRI employs for this cleavage: water and doubly charged cations. First, we note that before cleavage a single phosphate group with net charge $-e$ forms the connection between two subsequent ribose rings of adjacent bases (Figures 5.11 and 5.29). An O atom from PO_4 is bound to each ribose. After cleavage, the separated ends are left with 5′-phosphate and 3′-OH. The cleaved phosphate group only has one of its O atoms bound, leaving the other free to ionize. At pH 7, a free phosphate group would be undergoing charge equilibrium $-2e \rightleftharpoons -e$ (Pk = 6.82; an H^+ from water). The addition of an extra negative charge must be paid for in stabilization energy. How can this be done? A doubly charged ion like Mg^{++} is found at the two cleavage sites in Figure 16.13c and d. A spare OH must also be found to stabilize the 3′ ribose. Where does a cell find −OH? It's all around in the water! Figure 16.13a and b shows that besides the liquid (mobile, does not show in X-ray diffraction) water bathing the entire structure, specifically bound water molecules are located around the negatively charged DNA, as well as in the vicinity of the protein. Water reduces electrostatic energies caused by either + or − charges. Careful study of the two structures shown in Figure 16.13 also suggests that

TABLE 16.3 DNA Molecular Motor Forces[a]

Process	Force
Entropic stretching, double-stranded DNA	0.1 pN
Supercoiling forces	~0.2 pN
Entropic stretching, single-stranded DNA	5 pN
Helicases (DNA zipping/unzipping)	10–15 pN
Viral DNA packaging	15–50 pN
Nucleosome removal	20 pN
DNA replication	Up to 35 pN
Transcription forces	>20 pN
Exonuclease cleavage	>40 pN
DNA overstretching	20–80 pN
Bond cleavage (brute force)	1 nN

[a] Most values from Bustamante, C., et al., *Nature* 421, 423–427, 2003.

Study of endonucleases led to the 1978 Nobel Prize for Physiology or Medicine, awarded to Daniel Nathans, Werner Arber, and Hamilton Smith, and the subsequent development of molecular cloning.

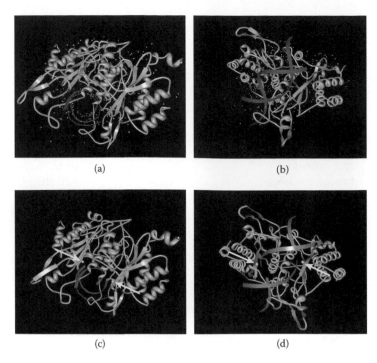

(a)

(b)

(c)

(d)

FIGURE 16.13

The enzyme/motor EcoRI is a type II endonuclease, which recognizes a specific GAATTC sequence of bases, and cleaves between G and A on both strands: $G^+ \; A \; A \; T \; T \; C$, $C \; T \; T \; A \; A \; {}_+G$. (a,b) Dimeric enzyme bound to DNA (red strands) before cleavage of DNA, also showing water molecules. In order to stabilize this structure, ions like Mg++ and Mn++ must be rigorously excluded. (PDB ID 1CKQ by Horvath, M. et al., 1999.) (c,d) EcoRI-DNA structure after DNA cleavage. Manganese ions (in place of Mg++) are shown bound near the strand breaks. Type II endonucleases leave 5'-phosphates and 3'-OH groups after cleavage. The enzymatic requirement for 2+ ions reflects the high Coulomb energy reduction requirements of the motor's actions. (PDB ID 1QPS by Horvath, M.M. and Choi, J., *Protein Sci.*, 2001.)

1. The electric dipole moment created by protein α-helical structures can interact with the charges near the cleavage site to reduce the free energy of the activated state.
2. The central base pairs in GAATTC unstack.
3. The DNA locally unwinds by 28°.
4. The major groove widens, with parts of protein helices inserted.

All of these structural changes occur with no ATP energy input, with the net effect that the free energy of the activated, transition state of the almost cleaved DNA is lowered significantly.

16.2.1.2 "Free" energy machines?

Should we conclude that the microscopic biological world is populated by thousands of catalytic "free-energy" machines, performing work at no energy cost besides the energy obtained from the thermal bath? (Exclude those enzymes that consume ATP or other high-energy compounds.) We have already answered "No" to this question by noting that the product state of the catalyzed reaction is lower in free energy than the initial state. But are we still saying that somehow biology has tricked Mother Nature into applying special energy rules for biology (i.e., reactions speed up by factors of 10^{10}) at no cost? The answer is an emphatic "No." Biology pays a price to produce these catalytic enzymes, and to set up and maintain their production facilities. Just because a particular reaction $E + S \underset{k_{-1}}{\overset{k_1}{\rightleftharpoons}} E \cdot S \overset{k_2}{\longrightarrow} E + P$ requires no direct energy input does not mean the cell expended no energy to carry it out. Consider every "E" you see in a catalyzed reaction scheme as representing a constant power overhead, or "indirect costs." Even the local concentrations of ions where cellular reactions are carried out are maintained at nonambient levels. There is a cost to this. What might this power overhead be for a typical catalyzed reaction? We won't pursue the answer here, but further discussion may be found associated with search terms such as *cellular energy budget* or *dynamic energy budget*. We noted at the start of this chapter that the human brain consumes a relatively large fraction of the body's daily energy allowance. Neural systems in general have been identified as particularly subject to energy budget limitations.[11]

16.2.1.3 Biomolecular search

We have neglected the fascinating topic of how EcoRI searches for, recognizes, and strongly and specifically binds to its double-stranded GAATTC recognition sequence. Compared to its final, catalytic task, this search process is much more like a molecular stepping (or diffusing) motor. You are prepared to simulate some of the "search-and-destroy" tasks that EcoRI has, using random-walk spreadsheets. You can, for instance, place an EcoRI molecule on a random site in a long DNA sequence and see how many steps are needed (how much time is needed) to reach a GAATTC site (Figure 16.14a). Alternatively, you could take on a more challenging task and simulate, from known Michaelis–Menten parameters for *Eco*RI, the random diffusion of the protein molecules from solution to various sites on the DNA (Figure 16.14b). We leave this large issue as a homework challenge for energetic students. Meanwhile, our conclusion for the EcoRI DNA-cleaving molecular motor is that no fuel need be consumed; rather, the enzyme recruits the assistance of water, ions, and parts of its own specific structure to reduce the high energy of the transition state, allowing the reaction to proceed on thermal energy fluctuations.

16.2.1.4 Entropic force of DNA stretching

Because DNA and its cousin, RNA, must undergo frequent manipulation by motors that pull, lift, and separate strands, the entropic contribution to the mechanical energy play a major role. We have seen that when DNA is stretched by pulling the two ends of a long DNA random coil, most of the stretching is determined by the distance-dependent entropic free energy, $TS(r)$, where r is the separation between the ends.

A FJC of N units that allows n step directions for each succeeding step has a conformational entropy of

$$S = k_B N \ \ln(n) \tag{16.10}$$

with respect to a conformation fixed along a straight line (Figure 16.1b). The average end-to-end distance of the coil is

$$r_{coil} = \sqrt{\langle r^2 \rangle} = \sqrt{NL} \tag{16.11}$$

whereas the stretched length is

$$r = NL \tag{16.12}$$

If this DNA is stretched from coil to straight, the distance change is

$$\Delta r = (N - \sqrt{n})L \tag{16.13}$$

The average force needed to stretch DNA this distance is then

$$F_{ent} = \frac{k_B T N \ \ln(n)}{\left[N - \sqrt{N} \right] L} = \frac{k_B T \ \ln(n)}{\left[1 - \sqrt{1/N} \right] L} \cong \frac{k_B T \ \ln(n)}{L} \tag{16.14}$$

where the last approximation is for N large ($N \geq 20$ ~persistence lengths or so, corresponding to a coil of several microns or more in total backbone length). If $L \sim 100$ nm and $n = 50$ possible directions,

$$F_{ent} \cong \frac{\left(4.1 \times 10^{-21} \text{J} \right)(\ln 50)}{100 \times 10^{-9} \text{nm}} = 0.1 \text{ pN} \tag{16.15}$$

This rough computation can be improved on by considering the distance dependence of the force and computing TdS/dr rather than simply $T\Delta S_{total}/\Delta r_{total}$, but our random-walk approximation does not warrant a more complex formulation. This magnitude of force is indeed approximately the observed value of the force needed to initially stretch DNA.[12] It is also well within the force capabilities of the laser-tweezers technique introduced in Chapter 9, which is how these stretching forces have been measured.

During this process of entropic stretching of DNA, the double-stranded structure—base-pairing, base-stacking, wide and narrow grooves—does not appreciably change. A force of 0.1 pN is not large enough to appreciably perturb the multiplicity of reinforcing hydrogen bonds and stacking interactions. As the stretched DNA approaches it maximum linear length, the double-stranded structure begins to be stressed. Up until a pulling force of about 65 pN, the DNA remains at its maximum length. Above this force, however, the double-stranded structure rapidly begins to distort and fall apart, and the chain extends to a length 60%–70% longer than its B contour length. Forces from 70 to nearly 100 pN do not further stretch the DNA.

Recall B DNA's persistence length is about 50 nm. More careful considerations lead to identification of L with the "Kuhn length," about twice the persistence length.

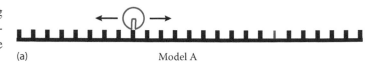

(a) Model A

(b) Model B

FIGURE 16.14
(a) Possible enzymatic searches for an error site on a DNA strand. An enzyme could attach to the DNA and carry out a "stepper" search, proceeding randomly or unidirectionally until it encounters the abnormal site. (b) Alternatively, the enzyme could diffuse from the solution until it encounters the DNA, binding more tightly when it encounters an abnormal site. In principle, the main differences between the enzymes would be the processivity—what fraction of the time the enzyme remains bound to the track—and stepping—whether the enzyme has a one-dimensional stepping mechanism.

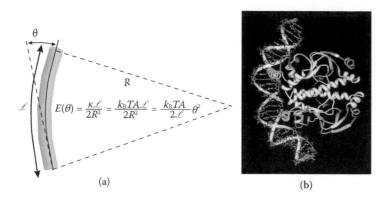

FIGURE 16.15
DNA bending can be described by worm-like coil (WLC) theory. (a) Parameters describing bending of rods. (b) Crystal structure of a CAP–DNA complex: the DNA is bent by 90 degrees. There are 30 base pairs in this strand of DNA. Two cAMP molecules, needed for binding to the DNA, are located near the center of the protein. (PDB ID 1CGP, by Schultz, S.C. et al., *Science*, 253, 1001–1007, 1991.)

16.2.1.5 DNA bending

The FJC model of DNA does not formally allow for elastic bending of the DNA helix. Each 100-nm segment is assumed to be straight, though not aligned with the previous segment, and no energy is associated with a change of direction. The wormlike chain (WLC) model of DNA allows helix bending explicitly. The model gives energy as a function of bending radius or angle:

$$E(\theta) = \frac{k\mathcal{L}}{2R^2} = \frac{k_B TA\mathcal{L}}{2R^2} = \frac{k_B TA}{2\mathcal{L}}\theta^2 \tag{16.16}$$

where L is rod length (length over which the bend is distributed), R is radius of curvature, κ is bending rigidity, and A is persistence length (~50 nm for double-stranded DNA). (See Figure 16.15.)

DNA is commonly bent when bound to enzymes. The classic example of DNA bending occurs during binding to the catabolite gene activator protein (CAP), whose job it is to "turn on" transcription (DNA → mRNA) of the *lac repressor* protein, whose job it is to repress transcription of the DNA sequence for β-galactosidase (*LacZ*), an enzyme that cleaves the disaccharide lactose into glucose and galactose. These enzymes and the DNA they interact with are part of *E. coli*'s rapid response system for changes in nutrient levels in the environment.

16.2.1.6 Lac repressor search for its DNA binding site

Before attacking the CAP–DNA bending issue, let's pause to consider the overall cellular task that is being addressed by CAP. When *E. coli* encounters increased or decreased levels of nutrient (e.g., glucose), rapid adjustment of enzyme concentrations are made. Within a few minutes of a glucose increase, *E. coli* will have eliminated production of enzymes used to cleave larger carbohydrates into smaller and increased production of other enzymes, including those needed to repress production of the aforementioned enzymes. Eukaryotic cells typically require hours to days to make similar adjustments, but eukaryotic cells enjoy a much more stable environment. A specific 26-base pair DNA sequence is recognized by the *lac repressor* protein. If lac repressor binds to this sequence, transcription of the *LacZ*-coding sequence will be inhibited. Lac repressor thus has a task search illustrated in Figure 16.14. Does lac repressor act as a linear track search agent or as a diffusion searcher, binding more strongly to the target site than to others? Direct molecular motor measurements have not answered this question, but classic chemical binding experiments strongly suggest the answer.

Consider the reaction $A + B \xrightarrow{\;k_f\;} A \cdot B$, where A is the lac repressor enzyme and B is DNA. These are both rather sizeable molecules. The maximum rate at which they find each other and bind should be less than the diffusion-limited rate discussed in Chapter 15. For small molecules diffusing in water, such rates can be of order $10^9\,\mathrm{M}^{-1}\,\mathrm{s}^{-1}$ or higher, but for a protein the size of lac repressor the rate should be about $10^7\,\mathrm{M}^{-1}\,\mathrm{s}^{-1}$. The measured rate is $k_f \sim 10^{10}\,\mathrm{M}^{-1}\,\mathrm{s}^{-1}$. This is simply not possible, so the conclusion is that protein and DNA cannot be randomly diffusing together and binding. The upper diagram in Figure 16.14 is thus more likely: lac repressor binds (weakly) to whatever random sequence it runs into on the DNA strand, then walks along the strand until the target sequence is encountered. This search mechanism, while not the only possible one, seems more efficient than the three-dimensional diffusion search. Some support for the search mechanism is also found in the binding constants to DNA. Lac repressor binds to its target sequence with a dissociation constant of $K_d \sim 10^{-13}\,\mathrm{M}$; for a random sequence, $K_d \sim 10^{-4}\,\mathrm{M}$. (A small K_d indicates strong binding. Refer to *dissociation constant* in the glossary if necessary.)

16.2.1.7 CAP and the DNA bend

"Glucose is *E. coli*'s metabolite of choice; the availability of adequate amounts of glucose prevents the full expression of >100 genes that encode proteins involved in the fermentation of numerous other catabolites, including lactose… even when these metabolites are present in high concentration."[13] When glucose is absent, *E. coli* needs to start using these other sources of energy. An early chemical signal of glucose absence is the increase in cyclic AMP (cAMP) levels in the cell. The cAMP is a *cofactor* that the CAP protein needs to bind to its DNA target (recognition) sequence: CAP + cAMP binds to DNA, but CAP does not. The structure of this CAP-cAMP complex bound to DNA is what is shown in Figure 16.15b. Note the two cAMP molecules near the middle of the protein in the figure.

When examined more closely, the 90° DNA bend is better described as two localized 40° "kinks" with an intervening, gradual 10° bend.[14] In the WLC model of Equation 16.16, this would require more bending energy, but nature is clearly telling us that at least in this case, a lower-energy CAP-induced DNA bend results from the sharper kinks. How can this be? First, we should look more closely at the structure, as described by the X-ray crystallographers who determined its structure—the group of Tom Steitz. They noted that two important sources of DNA sequence specificity in any protein-DNA complexes are direct hydrogen bonding and van der Waals interactions between protein side chains and the exposed edges of base pairs in the major groove of B-DNA and sequence-dependent bendability of DNA.[15,16] Sequence-specific bendability allows some nucleic acid sequences to adopt a particular structure required for binding to a protein at lower free energy cost than other sequences. In the case of CAP, three side chains emanating from the "recognition helix" of CAP appear to hydrogen bond directly to three base pairs in the major groove of the DNA. The tendency of adjacent AT base pairs to favor bending into the minor groove and GC base pairs favor bending into the major groove, plus several other interactions, are believed to explain why the sharp kinks occur in the CAP recognition sequence. We already noted in Chapter 13 that highly flexible DNA sequences are believed to be a characteristic of enzymatic recognition sites in general. This sequence-specific bendability requires computational techniques beyond the scope of this book.

16.2.2 Rotary Motor That Bends DNA: Bacteriophage ϕ29 Portal Motor[17]

ϕ29 (Phi 29) belongs to a family of related bacteriophages and is among the smallest known double-stranded DNA phages. Most of these phages infect *Bacillus subtilis* and, as part of the viral infection cycle, must package their newly replicated genomes for delivery to other host cells. ϕ29 packages its 6.6-μm-long double-stranded DNA into a 42 × 54 nm capsids by means of a portal motor that hydrolyses ATP. The motor must overcome entropic, electrostatic, and bending energies of the DNA to pack the DNA to near-crystalline density. The Bustamante research group used laser tweezers to pull on single DNA molecules as they were packaged, measuring the force generated by the motor. As we have already seen, the laser tweezers is designed to exert and accurately measure forces from about 0.1 to 100 pN. Figure 16.16a shows the design of the experiment.

The virus DNA is first attached by streptavidin-biotin binding techniques to a micron-sized polystyrene microsphere (upper sphere), appropriately coated for surface chemistry. The DNA is part of a stalled, partly prepackaged complex attached by means of the unpackaged end of the DNA. This microsphere is captured in the laser trap and brought into contact with a second sphere held by a micropipet. This second bead is coated with antibodies against the phage, so a stable tether is formed between the two beads (antibody/protein G attachment). When no ATP is present, the tether displays the passive stretching behavior expected for a single DNA molecule. When ATP is added, the two microspheres move closer together, indicating DNA-packaging activity. Figure 16.16b shows the packaging rate at constant applied force versus percentage of DNA packaged. The motor initially packages at a rate of nearly 100 base pairs per second, slowing to about half that rate when 60% of the genome is packed into the capsid.

The motor can work, on average, against loads of up to 57 pN, making it one of the strongest molecular motors known. Figure 16.17a shows the results of stall force measurements on ϕ29. The trap force was increased until packaging stalled or stopped. Not unexpectedly, the stall force shows a distribution of values for these single molecular motors, ranging from 40–75 pN.

Movements of over 5 μm of continuous motion were observed, indicating high processivity, though pauses and slips also occurred, particularly at higher forces. As the DNA is packed into the capsid, the internal pressure of the capsid increases. The motor must exert more force and the packing rate decreases. Figure 16.17b shows the internal force rising to over 50 pN as the capsids

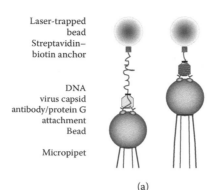
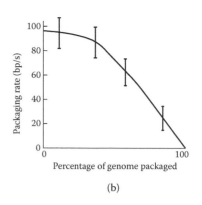

(a)   (b)

FIGURE 16.16

(a) Bustamante group experiment for force measurement of φ29–lambda phage rotary motor; 19,300 base pairs. A single φ29 packaging protein complex is attached between two microspheres. Laser tweezers trap one microsphere and measure the forces acting on it, while the other bead is held by a micropipet. (Left) start of measurement; (Right) DNA is wound into the virus capsid while a constant force is applied. (b) Packaging rate versus amount of DNA packaged, relative to the original 19,300-base pairs φ29 genome. A constant force of 5 pN was applied by the laser trap. The solid line is the average of eight measurements, smoothed using a 200-nm sliding window. The standard deviation is roughly indicated by the vertical bars. (Drawings and graph adapted by permission from Macmillan Publishers Ltd: Smith, D. E. et al., *Nature*, 413, 748–752, 2001.)

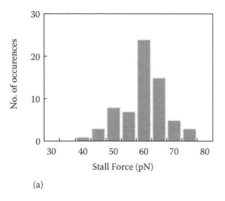
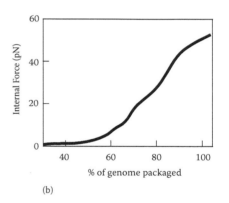

(a)   (b)

FIGURE 16.17

(a) Stall force for phage φ29 rotary motor measured for 65 individual complexes shows a distribution, with average about 57 pN. *Stall force* refers to total force—external force plus an inferred internal force of 14 pN, for the case of 2/3 packaged—needed to stop further packaging. (b) Internal capsids force versus percent of genome packaged. (Drawings and graph adapted by permission from Macmillan Publishers Ltd: Smith, D. E. et al., *Nature*, 413, 748–752, 2001.)

nears its capacity and the DNA is maximally compressed. Why does the virus not use a larger capsid, reducing the force-generating demands on the portal motor? Smith et al.[17] suggest that this internal pressure is a natural solution to the virus' next task: the ejection of the DNA from the capsid during infection of a bacterial cell. Such built-in internal pressure obviates the need for an "injection" motor.

The Bustamante experiment with the rotary-portal motor is now a classic example of the sort of measurements that are being done with molecular motors. The simplicity of the setup and interpretation of data illustrate two important facts. The large effort invested over the past 20 years into the laser trap method and the surface chemistry of nanometer- and micron-sized polymer spheres is now paying handsome dividends in the field of single-molecule motor measurements. The force measurements provide important new data to add to the more classic measurements of reaction rates (e.g., ATP consumption).

16.2.3 Direct Measurement of DNA Twist, Writhe, and Torque

DNA processing enzymes such as gyrases, RNA polymerase, topoisomerases, helicases, and the nucleosome all cause some degree of twisting or writhing of the B-DNA helix. Since any tension on or bending of DNA is coupled by the helical structure to at a least local distortion of one strand with respect to the other, all of the DNA-manipulating motors discussed in this chapter would be expected to exert not only force, but also torque on the DNA. Techniques similar to the laser tweezers have been developed to measure single-molecule twisting.[18,19] Such techniques are generically referred to as *magnetic tweezers*, since external magnets, rather than a focused laser beam, are used to both pull and twist a magnetized microbead. Figure 16.18 shows two experimental designs for applying measureable tensions and twists on DNA.

Figure 16.18a shows the conventional magnetic tweezers (MT) apparatus. DNA, typically 4–5 kb in length, is first bonded to a glass cover slip by multiple bonds that keep the DNA from rotating at the lower end. The upper end of the DNA is similarly bound to a superparamagnetic microbead, which can be pulled vertically and rotated in the horizontal plane by controlled movements of small rare earth magnets positioned above. The bead becomes magnetized in a fixed orientation, causing it to act like a compass needle. Raising (lowering) the magnets causes the force on the DNA to increase (decrease). The stretching force is determined by analyzing the Brownian motion of the bead and the end-to-end extension of the DNA. Forces from 0.01 to 100 pN can be applied and measured. The DNA is thus torsionally constrained and mechanically stretched by the soft magnetic force. The position of the bead can be measured in real time by a video microscope located below the cover glass. When the external magnet pair is rotated, the bead also rotates, introducing superhelical turns (~one negative supercoil per each clockwise rotation, one positive supercoil for each counterclockwise rotation) in the DNA. In a range of about ±3 to ±15 rotations, each rotation of the bead results in the introduction or removal of one plectonemic supercoil. The twisting will also generally cause a change in the length of the DNA, which can be measured by the video microscope. In the linear range of supercoiling, each rotation of the bead causes a ±60-nm change in DNA end-to-end extension, as it introduces or removes one supercoil. (A single plectonemic supercoil has a contour length of ~60 nm but contributes ~0 nm to the overall DNA end-to-end extension.)

To analyze measurements, the researchers in references 18 and 19 considered the degree of supercoiling, σ,

$$\sigma = \Delta Lk/Lk_0 = n/Lk_0 \tag{16.17}$$

where n is the number of rotations of the bead, Lk is the linking number, and Lk_0 is the natural linking number,

$$Lk_0 = N/h \tag{16.18}$$

where N is the number of base pairs and h is the number of base pairs per helical turn of the DNA ($h = 10.4$ for B-DNA). The linking number is equal to the sum of the number of twists and writhes,

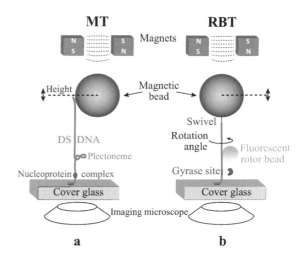

FIGURE 16.18
Two magnetic tweezers apparatuses for measuring twisting and extension or contraction of a single DNA molecule. Rather than pulling on a microbead with a focused laser beam, two aligned magnets can be adjusted vertically and rotated horizontally to pull the DNA straight and control the rotational orientation of the ~1 μm-diameter superparamagnetic bead (e.g., Invitrogen™, DYNAL™, MyOne™, Dynabeads™). In each case the height of the bead can be measured. A strand of DNA is affixed to a cover glass or the inside of a square capillary tube by multiple bonds and to a magnetic bead. The actual magnets and cover glass are much larger than the bead, which is similarly much larger than the plectoneme or protein-DNA complexes.
(a) Magnetic tweezers (MT) apparatus.[18] The magnet pair can initially be rotated to induce a specific positive or negative DNA supercoiling, which forms the looped plectoneme structure. If a nucleoprotein complex then forms or changes conformation, inducing additional twists in the DNA, the number of plectoneme loops changes. Length changes of the DNA can be directly measured, but twisting or linking number changes must be determined by a deconvolution method. In this MT experiment, multiple bonds to the bead restrains DNA twisting at the top end.
(b) Rotor bead tracking (RBT) apparatus.[19] The additional ~300 nm fluorescent micro-bead, attached at the side of the DNA, can be imaged to directly measure rotations of the DNA induced by activity at the gyrase site below it. Binding at the site, gyrase will distort the DNA, generally compacting and twisting it. A plectoneme does not form because bonding of the upper end of the DNA to the main bead acts as a swivel.

The reader should replicate the rubber-tubing experiment of Figure 14.17c and d and convince him/herself that forcing an initially supercoiled circular tube (e.g., DNA) to lie flat on a surface, with no supercoil will introduce mechanical twisting strain in the tube.

$$Lk = Tw + Wr \qquad (16.19)$$

($Wr = 0$ when $Lk = Lk_0$). The energy associated with supercoiling can be estimated[20] from

$$\frac{\Delta G}{N} = 10 k_B T \sigma^2 \qquad (16.20)$$

16.3 POINTS TO REMEMBER

- The range of typical molecular motor forces.
- No matter the motor force, statistical (random) behavior is a major part of biomolecular motor behavior.
- The range of DNA packing encountered, from entropic stretching to viral packing.
- DNA does not bend like a uniform rod when interacting with proteins and may display local, sequence-dependent, sharp bends. Local DNA flexibility is often connected with enzyme binding sites.
- Results from laser tweezers (optical traps) and/or AFM, classic reaction kinetic experiments, and X-ray crystallography measurements are usually needed to understand the mechanism of a biomolecular motor.
- Magnetic traps and other single-molecule apparatuses are rapidly being developed that can probe the responses of nucleic acids, proteins and supramolecular structures to torques and rotations, in addition to forces.

EXERCISE 16.3

Figure 16.19 shows an example MT calibration curve for the DNA extension/contraction caused by rotating the magnets (and bead) of Figure 16.19 a given number of times in either direction. Starting from DNA with no plectonemic supercoils, a small (0.3 pN) force will stretch the 4-kB DNA to a structure of (approximately) maximum extension. The magnets are then rotated, keeping the force constant (their height above bead fixed), and the height of the bead above the glass surface is measured.

1. Explain the shape of this extension vs. rotation curve.

Answer

At $n = 0$, the magnets have been raised until the DNA is approximately vertical and "gently taut." If the elasticity of the DNA (both stretching and twisting) were of single, pure types, the point of tautness would be sharp, and subsequent rotation of the magnets would immediately engage the rotational elasticity and the extension curve should look like an inverted "V." From the last several chapters, we know that macromolecules have multiple modes of flexing and moving, as well as exhibiting constant Brownian motion, which means the sharpness of the top of the curve should be smoothed. From the graph, at about three rotations in either direction, the random flexings have apparently given way to one primary rotational elastic mode, and further rotation of the magnets probes this primary mode. The linearity of extension with rotation indicates that there is a constant (likely small) resistance to this rotation.

2. Should the extension curve be symmetric about $n = 0$? Should positive and negative supercoils have the same mechanical characteristics under twisting strain? Explain.

Answer

If DNA were like the rubber tubing just discussed, rotating in either direction should produce the same results, since the rubber has uniform elastic properties. DNA does not, however, since the helical structure of DNA has a rotational handedness. Rotating in the direction of DNA's intrinsic, double-helical winding is not equivalent to the opposite direction. However, the graph shows that the two directions are at least approximately equivalent. How can this be so? The keys are the weak force applied by the magnets and the relative ease at which supercoils form, as opposed to unwinding or overwinding DNA's helix. Winding and unwinding the helix may not be equivalent, but the experiment, designed to produce supercoils, does not exert enough torque to probe this difference.

3. Calculate the extension per rotation in the approximately linear parts of the curve for both positive and negative rotations.

Answer

The slope of the graph on the left side is about $\frac{(660-20)\text{nm}}{(-5+15)\text{rotations}} \cong 64\,\text{nm/rotatior}$; on the right, $\frac{(630-40)\text{nm}}{(5-15)\text{rotations}} \cong -59\text{nm/rotation}$, where the negative sign merely indicates winding in the opposite direction. The (simulated) data does show a slight difference on the two sides, but since Figure 16.19 is not the actual data of Revyakin et al., we should defer to these researchers determination of ~60 nm/rotation in either direction.

Further questions for the reader: Is there any hysteresis in the curve? If the measurement starts from −15 rotations, then proceeds to 0, are the measured extensions the same as when measured in the opposite order? Should the extension go all the way to 0 nm for ±16 to ±17 rotations?

4. Revyakin et al., 2003[18], state that one rotation of the magnet pair quantitatively produces one plectonemic supercoil, in the linear region. Why should this be true?

Answer

One can get a feel for the behavior of supercoil formation vs. distortion of the intrinsic helical structure of dsDNA by straightening, then rotating the end of a length of soft, rubber tubing. (See Figure 14.17.) When a full twist of the tube has been made, a supercoil will have formed, if the two separated ends of the tubing are allowed to approach each other with only minimal longitudinal applied force. A second twist will create a second supercoil. Assuming the direct correlation of magnet rotation and supercoil formation, there should be little or no twisting of the DNA helix: any twisting stress is immediately converted to supercoil formation, which eliminates the stress. The researchers present other data to support the linking of rotation and supercoil formation, but the interested reader should consult the original publication. See also question 5.

5. If the calibration experiment were done with a much larger applied force, what results might be expected?

Answer

Returning to the rubber-tubing analogy, if the tubing is longitudinally pulled very hard when the twists are made, much more resistance is encountered and supercoils are inhibited. The MT apparatus is not designed to directly measure any overall rotation of the DNA helix, which will likely occur if the applied longitudinal force is larger and under other biologically relevant conditions. The RBT apparatus of Figure 16.18a and b can manage this situation.

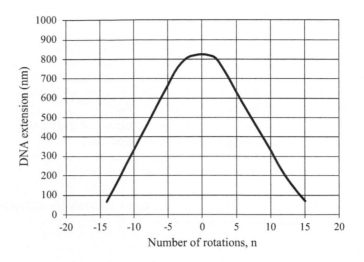

FIGURE 16.19

Sample graph showing DNA extension vs. rotation for an applied force of 0.3 pN. The number of rotations, n, is the number of turns applied to the 4-kb (1.3 µm) DNA by the magnets and the bead. Graph is patterned after Revyakin et al. (2003), Figure 1B[18] and is used for purposes of Example 16.3.

6. Calculate the degree of supercoiling and the associated free energy of supercoiling per base pair and per strand at several rotational numbers in the data of Figure 16.19, assuming 4 kb DNA.

Solution

Assuming the validity of equations 16.17, 16.18, and 16.20, the degree of supercoiling is $\sigma = \frac{nh}{N} \cong \frac{10.4n}{4,000}$ and the energy is $\frac{\Delta G}{N} = 10k_B T \sigma^2$. Below are several computed values of σ.

n	−10	−5	−2	+2	+5	+ 10
σ	−0.026	−0.013	0.0052	0.0052	0.013	0.026
$\Delta G/N$	$0.0068k_B T$	$0.0017k_B T$	$0.00027k_B T$	$0.00027k_B T$	$0.0017k_B T$	$0.0068k_B T$
ΔG	$27k_B T$	$6.8k_B T$	$1.1k_B T$	$1.1k_B T$	$6.8k_B T$	$27k_B T$

One notes that the energy per base pair associated with supercoiling is small compared to $k_B T$. Even if 100 base pairs were part of a supercoil, the overall energy would still be small. The conclusion is that the data of Figure 16.19 and reference 18 are taken under low-stress conditions.

16.4 PROBLEM-SOLVING

PROBLEM 16.1 ENZYME FORCES NEEDED FOR CHEMISTRY

1. Draw a "before" and "after" diagram of DNA cleaved by an endonuclease.
2. Estimate the force a "type I" endonuclease needs to cleave DNA. Assume that the energy of one ATP molecule is expended in the process. (How far does what structure have to move in order for the bond to break?)
3. Estimate the force for the "type II" endonuclease EcoRI.

MINIPROJECT 16.2 ENDONUCLEASE SEARCH PROCESS

Single or both parts of this miniproject may be assigned. Either part can become a sophisticated, time-consuming project that needs instructor guidance.

Two generic processes exist for an enzyme to find its recognition site on a long strand of DNA. See Figure 16.14. (1) In the first, the enzyme manages to bind at a random site on the DNA, where it binds with an energy comparable to a few $k_B T$ (compared to being free in solution). This keeps the enzyme attached to the DNA strand for a time t_0. While attached, the enzyme performs a one-dimensional random walk on the DNA. If it arrives at the recognition site, it binds strongly. (2) In a second possible search process, enzyme molecules diffuse randomly in (three-dimensional) solution and periodically collide with DNA. If the collision site is close to the recognition site, and if the enzyme orientation is correct, binding occurs.

1. Set up a quantitative model for search process 1. Your model should include, for example, the length of the DNA (number of base pairs), number of recognition sites, number of bases in the recognition site, how close the enzyme must get to the site to bind, the enzyme step size and time, probabilities for various step sizes and directions, etc. You should probably set $t_0 = \infty$ at first: the enzyme never leaves the track. Devise a spreadsheet simulator for the one-dimensional search process and determine the most important parameters and results to present. Make sure parameters are physically reasonable, if that is possible. (No guarantees either model is connected to reality.)

2. Repeat the above for the three-dimensional diffusion model. You may want to try two approaches. In both, you must try to employ crude approximations. In the first, analytically compute the average time for an enzyme molecule to diffuse to the site, starting with one linear DNA molecule surrounded by randomly placed enzyme molecules whose number density corresponds to a reasonable concentration. In the second approach, modify a three-dimensional random-walk spreadsheet (or computer program) to notify you when a randomly placed particle arrives in the target area, as specified by your model. Keep the volume of the simulation small enough so that the number of steps does not keep your computer occupied for the entire semester.

Rather than using the diffusion differential equation in three dimensions with complicated boundary conditions, for example, find the nearest 10 or so enzyme molecules to the recognition site. Then calculate the time for diffusion over the corresponding distance and the probability the diffusion is in the correct direction.

PROBLEM 16.3 STEPPER (GRADUATE PROBLEM)

Speed, work, and power of the switching-potential, diffusing stepper (Figure 16.3). Equation 16.5 and the nearby text set up the question of the stepper's speed and the power expended by the switching potential as a function of the potential switching time (frequency). Assume the potential is switched on and off in a square-wave pattern with $t_{off} = t_{on}$. If you find you need to set $d = 0$, do so.

1. Calculate the x dependence of $V(x)$ in Figure 16.3 from one minimum to the next (x ranging from 0 to L_s).
2. Assuming a stepper (particle) diffusion constant D, calculate the probability $P(x)$ that the stepper will diffuse a distance x in time t_{off}, when the potential is turned off. Think whether you need to include the negative direction.
3. When the potential is turned back on, the particle must be raised from $V = 0$ to $V(x)$. Calculate the probability $P(V(x))$ that the potential energy change (work) will be $V(x)$. Neglect diffusion steps longer than L_s and state what limits on t_{off} will ensure this.
4. Calculate the average power generated by the switching potential.

5. Diffusion less than a distance d to the right or displacement to the left results in the particle remaining in place, yet the switching potential would have done work, because the particle would need to be raised in energy. How has this work been dissipated?
6. Whenever the particle is to the right of the peak of the potential, it feels a force $-dV/dx$ pushing to the right and it will move to the next potential minimum. Calculate the average work done on the particle by this force in one cycle of t_{off}/t_{on} and the corresponding power.
7. Compare and discuss the powers in parts 4 and 6.

PROBLEM 16.4   CAP PROTEIN BENDING OF DNA

See Figure 16.15b. The original reference to this work is S. C. Schultz et al., *Science* 253 (1991):1001–1007.

1. Download the structure of DNA bound to the CAP protein. Confirm that the DNA is bent by about 90°, using molecular structure display tools.
2. Calculate the bending energy of the DNA.
3. The Schultz et al. reference describes the 90° bend in more detail as consisting of two 40° kinks, each localized to a region of a few base pairs. Estimate the elastic energy of these two kinks, assuming they each occur over a region of five base pairs.
4. Discuss the validity of the result in part 3 in view of the discussion in the text on sequence-specific bendability of DNA.

PROBLEM 16.5   CIRCLES OF WORMLIKE COILS

Consider, as a standard, a double-stranded B-DNA strand bent into a circle of radius R. Calculate the total elastic bending energy and bending energy per base pair for double-stranded DNA of 10, 100, and 5,000 base pairs.

PROBLEM 16.6   BENDING

Calculate (approximately, when necessary) the ratio of bending energies of the structures b, c, and d to structure a in Figure 16.20 below, if they all have the same length. Ignore twisting energies.

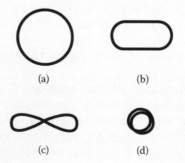

(a)   (b)

(c)   (d)

FIGURE 16.20
Double-stranded DNA shapes involving circular DNA. Bending and twisting energies are involved; we ignore that latter for Problem 16.6.

PROBLEM 16.7 Φ PORTAL MOTOR FORCE AND POWER

In Figure 16.16b, when little of the phage DNA is packaged, the portal motor packages at a rate of 90 base pairs per second against a force of 5.0 pN.

1. Calculate the power generated by the motor to package against this force.
2. Estimate the minimum number of ATPs the motor must hydrolyze per second to sustain this power output.
3. (Graduate problem) Obtain a copy of the original Smith et al. paper (see text). Compare your results to theirs; describe differences in numerical results and include their analysis of motor efficiency.

PROBLEM 16.8 Φ PORTAL MOTOR: DNA CIRCLES

Assume the 19,300 base pairs of DNA are packed inside the viral capsid, wound in a circle having a radius equal to 2/3 of the average radius of the capsid.

1. Find this radius.
2. Calculate the force needed to create this bending radius from the wormlike coil model.
3. Calculate the bending energy in the completely packed capsid.
4. *Obtain a copy of the original Smith et al. paper (see text). Compare your results to theirs and describe differences. See especially the paragraph beginning, "Figure 3c shows a histogram of the total force needed to stall the motor."

PROBLEM 16.9 MAGNETIC TWEEZERS APPLIED TO DNA

1. References 18 and 19 state that the microbeads of Figure 16.18 are magnetized or superparamagnetic. Investigate the details of the magnetic beads used in these experiments on DNA. For example, what makes the beads superparamagnetic? What materials and surface coatings are used? What size, shape, temperature, and fluid surrounds are used?
2. Refresh your knowledge of magnetic materials. What are the characteristics of ferromagnetic, paramagnetic, diamagnetic and superparamagnetic materials? Why are superparamagnetic microbeads used?
3. In your description of paramagnetism, you should have noted that alignment of magnetic moments in the material required the application of an external field. The magnetic alignment in the material is in the direction of the external field. If the field is turned off, the alignment disappears. If the field direction changes, the direction of the magnetization changes. Superparamagnetism is similar to paramagnetism, in that rotation or removal of the external field will (eventually) result in change or disappearance of the magnetic alignment in a material that is held in place. In the realm of micro- and nano-samples, superparamagnetism gets even more complex. Investigate superparamagnetic relaxation (e.g., start with https://en.wikipedia.org/

wiki/Superparamagnetic_relaxometry) on the nanoscale and discuss the limitations on the size of the microbeads and the magnetic if the desire is to rotate the micro-bead, rather than rotating the magnetic alignment inside the bead. Force and the decay time of the magnetic alignment should be considered in this discussion.

4. Comment on the training and background the bioscientists who developed the magnetic tweezers method needed.

REFERENCES

1. Abstracted from various sources, including Voet, D., and Voet, J. 2010. *Biochemistry* Hoboken, NJ: John Wiley & Sons, p. 571; see also https://www.ncbi.nlm.nih.gov/books/NBK21154/?term=brain+power and http://hypertextbook.com/facts/2001/JacquelineLing.shtml.

2. Nelson, P., 2008. *Biological Physics*. New York: W. H. Freeman, 572; Serdyuk, I. N., N. R. Zaccai, and J. Zaccai. *Methods in Molecular Biophysics*. 2007. New York: Cambridge University Press; Höök, P., X. Li, J. Sleep, S. Hughes, and L. Larsson, 1999. In vitro motility speed of slow myosin extracted from single soleus fibres from young and old rats. *J Physiol* 520:463–471.

3. Berg, H. C., 2000. Motile behavior of bacteria. *Phys. Today.* 53:24–29; Berg, H. C. 2004. *E. Coli in Motion*. New York: Springer; DiLuzio, W. R., L. Turner, M. Mayer, P. Garstecki, D. B. Weibel, H. C. Berg, and G. M. Whitesides. 2005. Escherichia coli swim on the right-hand side. *Nature* 435:1271–1274.

4. Ishii, Y., M. Nishiyama, and T. Yanagida. Mechano-chemical coupling of molecular motors revealed by single molecule measurements. 2004. *Current Protein and Peptide Science* 5(2):81–87, 5:81–87; Fehr, A. N., B. Gutiérrez-Medina, C. L. Asbury, and S. M. Block. 2009. On the origin of kinesin limping. *Biophys. J.* 97:1663–1670.

5. Ng, S. Y., B. Chaban, and K. F. Jarrell. 2006. Archaeal flagella, bacterial flagella and type IV pili: A comparison of genes and posttranslational modifications. *J Mol Microbiol Biotechnol* 11:167–191.

6. Berg, H. C. 1993. Chapter 6 in *Random Walks in Biology*. Princeton, NJ: Princeton University Press.

7. Lang, E. R., and C. Rha. 1982. Apparent shear viscosity of native egg white. *Int J Food Sci Tech*. 17:595–606.

8. Visscher, K., M. J. Schnitzer, and S. M. Block. 1999. Single kinesin molecules studied with a molecular force clamp. *Nature* 400:184–189.

9. Asbury, C. L. 2005. Kinesin: World's tiniest biped. *Curr Opin Cell Biol*. 17:89–97.

10. Bustamante, C., Z. Bryant, and S. B. Smith. 2003. Ten years of tension: single-molecule DNA mechanics. *Nature* 421:423–427.

11. Nawroth, J. C., C. A. Greer, W. R. Chen, S. B. Laughlin, and G. M. Shepherd. 2007. An energy budget for the olfactory glomerulus. *J Neurosci* 27:9790–9800; Kooijman, B. 2009. *Dynamic Energy Budget Theory for Metabolic Organisation*. New York: Cambridge University Press.

12. Smith, S. B., L. Finzi, and C. Bustamante. 1992. Direct mechanical measurements of the elasticity of single DNA molecules by using magnetic beads. *Science* 258:1122–1126.

13. Voet, D., and J. G. Voet. 2004. *Biochemistry*, 3rd ed. New York: John Wiley & Sons, 1238–1240.

14. Schultz, S. C., G. C. Shields, and T. A. Steitz. 1991. Crystal structure of a CAP–DNA complex: The DNA is bent by 90°. *Science* 253:1001–1007.

15. Hogan, M. E., and R. H. Austin. 1987. Importance of DNA stiffness in protein–DNA binding specificity. *Nature* 329:263–266.

16. Steitz, T. A. 1990. Structural studies of protein–nucleic acid interaction: The sources of sequence-specific binding. *Q Rev Biophys* 23:205–280.

17. Smith, D. E., S. J. Tans, S. B. Smith, S. Grimes, D. L. Anderson, and C. Bustamante. 2001. The bacteriophage straight ϕ29 portal motor can package DNA against a large internal force. *Nature* 413:748–752.

18. Revyakin, A., R. H. Ebright, and T. R. Strick. 2004. Promoter unwinding and promoter clearance by RNA polymerase: Detection by single-molecule DNA nanomanipulation. *Proc Natl Acad Sci USA* 101(14): 4776–4780; Revyakin, A., J. F. Allemand, V. Croquette, R. H. Ebright, and T. R. Strick. 2003. Single-Molecule DNA nanomanipulation: Detection of promoter-Unwinding events by RNA polymerase. *Methods Enzymol* 370: 577–598.

19. Basu, A., A. J. Schoeffler, J. M. Berger, and Z. Bryant. 2012. ATP binding controls distinct structural transitions of Escherichia coli DNA gyrase in complex with DNA. *Nature Struct Mol Biol* 19(5): 538–546.

20. Vologodskii A. V., A. V. Lukashin, V. V. Anshelevich, and M. D. Frank-Kamenetskii. 1979. Fluctuations in superhelical DNA. *Nucleic Acids Res* 6 (3): 967–982.

21. Burnouf, D. Y., V. Olieric, J. Wagner, S. Fujii, J. Reinbolt, R. P. Fuchs, and P. Dumas. 2004. Structural and biochemical analysis of sliding clamp/ligand interactions suggest a competition between replicative and translesion DNA polymerases. *J Mol Biol* 335:1187–1197.

22. Biswas, T., and O. V. Tsodikov. 2008. Hexameric ring structure of the N-terminal domain of Mycobacterium tuberculosis DnaB helicase. *FEBS J* 275:3064–3071.

23. Hoenger, A., S. Sack, M. Thormählen, A. Marx, J. Müller, H. Gross, and E. Mandelkow. 1998. Image reconstructions of microtubules decorated with monomeric and dimeric kinesins: Comparison with x-ray structure and implications for motility. *J Cell Biol* 141:419–430.

24. RCSB PDB software references:

    Jmol: An open-source Java viewer for chemical structures in 3D. http://www.jmol.org/.

    Protein Workshop: Moreland, J. L., A. Gramada, O. V. Buzko, Q. Zhang, and P. E. Bourne. 2005. The Molecular Biology Toolkit (MBT): a modular platform for developing molecular visualization applications. *BMC Bioinformatics* 6:21.

    NGL Viewer: Rose, A. S., A. R. Bradley, Y. Valasatava, J. D. Duarte, A. Prlić, and P. W. Rose. 2018. NGL viewer: Web-based molecular graphics for large complexes. *Bioinformatics*. doi:10.1093/bioinformatics/bty419.

25. Kozielski, F., S. Sack, A. Marx, M. Thormählen, E. Schönbrunn, V. Biou, A. Thompson, E. M. Mandelkow, and E. Mandelkow. 1997. The crystal structure of dimeric kinesin and implications for microtubule-dependent motility. *Cell* 91:985–994.

26. Fratini, A. V., M. L. Kopka, H. R. Drew, and R. E. Dickerson. 1982. Reversible bending and helix geometry in a B-DNA dodecamer: CGCGAATTBrCGCG. *J Biol Chem* 257:14686–14707.

27. Tsai, K. L., Y. H. Lo, Y. J. Sun, and C. D. Hsiao CD. 2009. Molecular interplay between the replicative helicase DnaC and its loader protein DnaI from Geobacillus kaustophilus. *J Mol Biol* 393:1056.

28. Horvath, M. M., J. Choi, Y. Kim, P. A. Wilkosz, K. Chandrasekhar, J. M. Rosenberg. 2001. The integration of recognition and cleavage: X-ray structures of pre-transition state and post-reactive DNA EcoRI endonuclease complexes. *Protein Sci*.

CHAPTER 17

CONTENTS

Assembly

There is much that is understood in detail about assembly of structures in cells, from synthesis of small molecules like ATP to proteins to almost macroscopic edifices like cell walls and cytoskeletons. The ribosome factory is now the sole subject of entire textbooks,[1–3] could easily fill the syllabus of a semester-long course, and has even become part of two music albums.[4,5] The mitochondria are also a center for assembly, in this case, of iron-sulfur (Fe-S) clusters,[6] though it has been observed that the eurkaryote *Monocercomonoides* sp. manages the assembly without a mitochondrion.[7] The clusters are usually small, two to eight Fe-S groups, carefully arranged by enzymes into geometric structures. In some cases, the clusters are assembled on a scaffold protein, e.g., the CnfU (PDB ID 2Z51) or IscA (PDB ID 1R95) proteins, which then transport the pre-built cluster to a recipient protein. Such clusters are crucial in mitochondrial respiration, in which they are required for the assembly, stability, and function of respiratory complexes, as well as for the citric acid cycle, DNA metabolism, and apoptosis (cell death). Many biochemical and genetic details of assembly are known, but mechanics and structure are not. In fact, the structural details of most biological assembly processes are unknown, as this would require a high-speed video-microscope capable of nanometer spatial resolution. Such video microscopes do not presently exist, although we will review some significant progress toward this goal at the end of this chapter.

Much of all biological construction is performed or supervised by enzymes, under the direction of the DNA of the cell's genome. The complexity of a cell's microscopic construction process has been the topic of textbooks and conferences since before the 1960s.[8] Not surprisingly, we have again used the word *complexity* in describing enzyme-mediated biological construction processes. It would not be wise to seriously start this topic near the end of a book, except from a very focused point of view. A focused view is what we pursue in this chapter, a view that contrasts with, but complements those of most biochemistry texts. We examine the construction of a few of the complex, microscopic structures in a cell, not by attacking the myriad of enzymatic processes observed in a cell's daily life but by attempting to simulate a few of the structures *without* the enzymes, i.e., self-assembly. This somewhat one-sided view of the real process seeks to determine which parts of the construction proceed in the absence of enzymatic intervention. Stated another way, how much of the structure self-assembles simply because of properties of the building block molecules and their surroundings. A spherical membrane structure can form, for example, when lipid molecules are placed in water and sonicated (subjected to vibrations driven by an ultrasonic probe). We need not look only to enzymatic construction machines for assembly of the basic cell membrane—it forms through natural processes. Insertion of specific proteins into a cell membrane may require the intervention of multiple enzymes, but even then, inserted proteins often self-assemble into active biological structures (e.g., ion channels). Biological assembly should not be equated with construction of skyscrapers, where every detail requires the action of a trained worker.

Our view of biological assembly could be described as minimalist or archaic: What sort of structures can form automatically, given the building blocks available? After familiarizing ourselves with apparently intricate structures that form from truly random processes, we proceed to consider the general topic of *entropic organization* of supramolecular structures and one supramolecular structure, the *nucleosome*, constructed of DNA and protein. Remembering the "point of view" issues discussed in Chapter 2, we expect to finish this chapter having only a little knowledge in common with those studying cell biochemistry and physiology. Such lack of overlap should motivate a good discussion—a discussion that we can be confident is important, as judged by the 2009 Nobel Prize in Medicine or the Physiology committee.[9]

> The reader may not realize it, but the path from here is toward the topics of chromosomal integrity, cell aging, and death. We do not pursue these further in this text.

17.1 OVERVIEW OF ASSEMBLY ISSUES

Biological assembly covers a wide range of phenomena. Fundamentally, assembly of biological structures is a matter of binding reactions, some of which we considered in detail in Chapter 15. The major complications in assembly are (1) the large number of molecules and binding reactions involved, (2) the complexity of the constituent molecules themselves, and (3) the number of proofing and reconstruction/repair processes that continually take place. The structures within a cell rarely have roles like those of houses or bridges, where the structure is built once, undergoes minor repairs from time to time, and is replaced at the end of a (long) life. Biological structures are constantly changing shape to perform active roles, undergo frequent damage and repair, and are duplicated to replace aging assemblies or to incorporate into new cells. Only a few of the assembly processes observed in nature can be adequately overviewed in one chapter of a book. References to review articles or books/book chapters are given in the summaries below Table 17.1. The interested student can take these references as a starting point for investigation of the primary research literature. Further careful topic limitation can lead to an interesting term paper on protein folding, aggregation, chaperonin-guided structure formation, and other topics.

Recall the frequency of UV damage to DNA in the skin.

TABLE 17.1   References on Biomolecular Folding and Assembly[a]

Year(s)	Title/Subject	Authors	Journal/Book[b]
1974	Stabilization of globular protein structure: calorimetry	Privalov & Khechinashvili	JMB 86:665–684
1988	Conformational substates in proteins	Frauenfelder et al.	ARBBC 17:451–479
1990s	Levinthal's paradox	Zwanzig et al.	PNAS 89:20–22
	Molten globular characteristics of apomyoglobin	Lin et al.	Nature Struct. Biol. 1:447–452
	Funnels, pathways, and the energy landscape	Bryngelson et al.	Proteins 21:167–195
	Calculating kinetics of protein folding from 3D structures	Muñoz & Eaton	PNAS 96: 11,311–11,316
	How RNA folds	Tinoco & Bustamante	JMB 293:271–281
2000	The structure of the large ribosomal subunit	Ban et al.	Science 289:905–920
to	Sequence evolution and the mechanism of protein folding	Ortiz & Skolnick	Biophys. J. 79:1787–1799
2005	Stability and contact order, protein folding rates	Dinner & Karplus	Nature Struct. Biol. 8:21
	Solvent effects, energy landscapes, folding kinetics	Levy et al.	PNAS 98:2188–2193
	Archaeal nucleosome stability	Bailey et al.	J. Biol. Chem. 277:9293–9301
	Free-energy surface, single-molecule fluorescence	Schuler et al.	Nature 419:743–747
	Protein folding: theory meets disease	Hunter	The Scientist 17:24–27
	Nucleosome arrays	Dorigo et al.	Science 306:1571–1573
	The energy landscape	Wales	PTRSL(A) 363:357–375
2006	Folding: single molecule spectral dynamics	Han et al.	JACS 128:672–673
to	Unfolded states of proteins, single-molecule FRET, computer simulations	Merchant et al.	PNAS 104:1528–1533
2010	Internal friction in protein folding; single molecule	Cellmer et al. Kubelka et al.	PNAS 105: 18320–18325; 18655–18662
	Protein folding reviews	(Entire issue)	NSMB 16(6):574–581
2011	The protein-folding problem, 50 years on	Dill & MacCallum	Science 338(6110): 1042–1046.
to	Macromolecular systems: multiscale and enhanced sampling techniques.	(Entire issue)	J. Phys. Chem. B 116(29)
2017	Protein folding, function and molecular machines	Whitford & Onuchic	COSB 30: 57–62
	Protein folding—how and why	Englander et al.	Ann. Rev. Biophys. 45(1): 135–152
	2D folding of polypeptides into nanostructures at surfaces	Rauschenbach et al.	ACS Nano 11(3): 2420–2427
	Programmable self-assembly of 3D nanostructures	Ong et al.	Nature 552: 72–77
	Biomolecular condensates: organizers of cellular biochemistry	Banani et al.	Nature Rev. Mol. Cell Biol. 18: 285.

[a] Selected articles; should not be taken as a list of "seminal papers."

[b] Abbreviations: ARBBC, Ann. Rev. Biophys. Biophys. Chem.; PNAS, Proc. Natl. Acad. Sci. USA; JACS, J. Amer. Chem. Soc.; JMB, J. Mol. Biol.; COSB, Curr. Opin. Struct. Biol.; COP, Curr. Opin. Pharmacol; NSMB, Nature Struct. Mol. Biol.; PTRSL(A), Philos. Trans. Roy. Soc. London., Ser. A.

17.1.1 Protein Folding

The major biochemical and structural observations concerning protein folding have already been discussed in Chapter 5. Likewise, the thermodynamics of the transition from ordered (native) to disordered (denatured) induced by temperature and/or chemical agents such as urea or guanadinium hydrochloride has been addressed. The primary interactions governing protein structure formation were stated to be peptide-bond angle restrictions (van der Waals radius overlaps), with ionic, dipole and hydrophobic interactions, and H-bonding the primarily interactions responsible. After applying Ramachandran analysis to many proteins whose structures were known, diagrams such as Figure 5.15, based on (φ, α) polypeptide angles, could be used to recognize when a new protein should have α-helical, β-sheet or other categories of structure. However, such analyses could rarely explain (in a predictive sense) why amino acid sequence A might be α-helical while a similar sequence B might be a mixture of several types of structure. The task of predicting what the overall structure of a globular protein should be is even more daunting. In some cases, protein structure is not determined solely by the amino acid sequence and how it might spontaneously form itself into an active, final structure in free solution, but also by the environment in which this protein finds itself shortly after synthesis.

17.1.2 Chaperonins: Eukaryotic Chaperonin TRiC

The geometry of the cellular environment can have a strong effect on the assembly of aggregates. Before looking at some unsophisticated models of such geometrical effects we briefly examine one case in which geometry appears to be controlled by a chaperonin structure explicitly designed for the purpose. The thermosome, an archaeal chaperonin from the microorganism *Thermoplasma acidophilum*, and its more recent relatives TRiC (or CCT) from mammals, has the job of providing a sheltered chamber in which polypeptides form their native structures. Figure 17.1 (right) shows the structure, determined from X-ray crystallography in 1998, of the microorganismal thermosome. These chaperonins are roughly spherical or barrel-shaped, hollow and have a "lid" structure, through which proteins presumably enter and leave. The lid of the *Thermoplasma* chaperonin appears to be detachable, whereas the more recent mammalian TRiC opens and closes like a sphincter.[10] In both cases the lid is closed/attached when ATP is added. The structure on the left of the figure is a simulation of the thermosome with its "lid" off. (More details and a video clip of the lid opening and closing of TRiC's lid, modeled from electron microscopy and computer simulation, can be found at https://www.youtube.com/watch?v=qkypA_7Zpgo.) Though these chaperonins may provide a relatively inert chamber in which the proper protein structures form through geometric exclusion, the consumption of ATP illustrates the fact that exclusion of unfavorable structures is an effect that must be paid for in free energy.

Chaperonins provide a fertile and fascinating field for the study of biological structure formation. Aspects ranging from genetics to crystallography are being pursued at a variety of institutes around the world (Table 17.2). The apparent involvement of such chaperonins in amyloid diseases, characterized by protein misfolding, gives such study direct health implications.

In the following sections we take a somewhat unorthodox, low-level look at assembly, starting with formation of structures from random aggregation of cubic monomers. The cubic structures are useful to consider, though not biological, because they easily allow examination of dimensionality and adjustment of overall conformation once an initial aggregate has been formed.

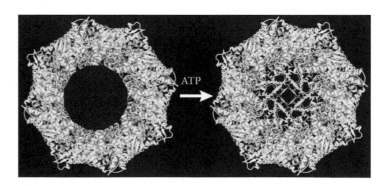

FIGURE 17.1

Chaperonin chambers from archaeal organisms. (Right) Crystal structure of the *Thermoplasma acidophilum* thermosome, an archaeal chaperonin, displayed using Protein Workshop software. (Form PDB ID 1A6D, Ditzel, L. et al., *Cell*, 93, 125–138, 1998.) The thermosome is analogous to several more recently studied chaperonins, including TRiC (or CCT) from mammals, studies on which have shown the *simulated* lid-closing transition (left to right above) upon addition of ATP. A video clip of the transition in TRiC can be found at https://www.youtube.com/watch?v=qkypA_7Zpgo.

TABLE 17.2 References on Chaperonin-Mediated Assembly

Year	Title/Subject	Authors	Journal/Book[a]
2000	*E. coli* chaperones	Gottesman & Hendrickson	COM 3(2):197–202
2002	Chaperonins and cell architecture	Csermely	*News Physiol. Sci.* 16:123–125
2003	Protein folding in the chaperonin cage	Takagi et al.	PNAS 100: 11,367
2004	Histone chaperones	Loyola & Almouzni	BBA 1677:3–11
2005	RNA and protein folding	Thirumalai & Hyeon	*Biochemistry* 44(13):4957–4970
2006	Chaperones and prions	True	*Trends Genet.* 22(2):110–117
2007	Two families of chaperonin	Horwich et al.	ARCDB 23:115–145
2008	Chaperone machines	Saibil	COSB18:35–42
2009	Chaperones in archaea	Large et al.	*Biochem. Soc. Trans.* 37: 46–51
2009	Theoretical and computational models	Lucent et al.	*Phys. Biol.* 6(1):15003
2009	Using a central cavity to kinetically assist folding	Horwich & Fenton	*Q. Rev. Biophys.* 42(2):83–116
2011	Protein folding at the exit tunnel	Fedyukina & Cavagnero	*Ann. Rev. Biophys.* 40(1):337–359
2014	Proteostasis: single-molecule methods	Bustamante et al.	*Ann. Rev. Biophys.* 43(1):119–140
2015	Molecular structure of chaperones	Bross	in *The Hsp60 Chaperone*. Springer
2016	Experimental milestones	Finka et al.	*Ann. Rev. Biochem.* 85(1): 715–742
2016	Structure and action of molecular chaperones	Gierasch et al.	*Structure and Action of Molecular Chaperones: Machines That Assist Protein Folding in the Cell*. World Scientific
2017	Prokaryotic chaperones	Kumar & Mande	*Prokaryotic Chaperones*. Springer
2017	Self-assembling protein nanomaterials	Yeates	*Ann. Rev. Biophys.* 46:23–42

[a] COM, *Curr. Opin. Microbiol.*; BBA, *Biochim. Biophys. Acta*; ARCDB, *Ann. Rev. Cell. Dev. Biol.*

17.2 KINETICS AND EQUILIBRIUM

Suppose an initially random solution of identical cubic monomers has opposite sides that are hydrophobic or "sticky," as in Figure 17.2a. Assuming the "stickiness" is enough to overcome the entropic tendency of the monomers to assume random orientations, these monomers will form linear aggregates (Figure 17.2b). If the sticky patches are located in different places on the monomer, it is easy to design structures of other aggregate form, as in Problem 17.2. There is no other ordered aggregate shape that can exist; the only question is how long the linear aggregates will be, on average, and the distribution of lengths. Because only linear arrangements are possible, any structures, e.g., two 2-mer aggregates that may form very rapidly can later convert to a 4-mer or a 3-mer plus a 1-mer. There is no energy barrier to overcome for this conversion. As long as the times for individual bonds to form or break are short enough, the final structure is the equilibrium structure: G will reach a minimum that is governed by the number of cubes available and the space in which they can self-assemble.

Suppose the monomers are sticky on all sides, as in Figure 17.3a. In this case, one-, two-, and three-dimensional aggregates can form. Figure 17.3b shows small one- and two-dimensional aggregates. Randomness (entropy) tends to favor formation of higher-dimensional structures because collisions do not have to take place strictly in one dimension. A one-dimensional dimeric, trimeric, etc. aggregate that is formed cannot be reduced without input of energy to separate to bound surfaces. It can, of course, continue to grow, though one-dimensional growth is less likely than formation of a two-dimensional structure. A two-dimensional aggregate has many equal-energy structures it can form without increasing the energy (enthalpy) of the structure

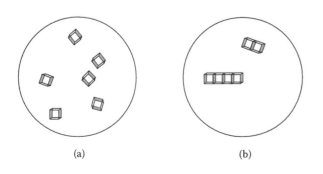

(a)   (b)

FIGURE 17.2
Aggregation of monomers with sticky opposite sides. A sticky side can only bind to another sticky side. (a) Initial random solution; (b) Aggregated structures are linear.

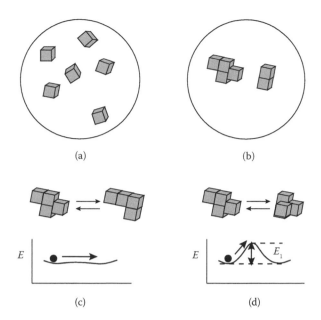

FIGURE 17.3
Aggregation of monomers that are (enthalpically) "sticky" on all sides. (a) Initial solution of monomers. (b) Two-dimensional aggregates that may form after some time. (c) Conversion of a two-dimensional aggregate to a two-dimensional does not require breakage of any bonds: there is no energy barrier to overcome. (d) Conversion from two to three dimensions requires an activation energy equivalent to (at least) one unit of aggregation energy.

(Figure 17.3c). However, conversion of a two-dimensional → three-dimensional (or vice versa) requires breakage of at least one bond (Figure 17.3d). This is equivalent to an activation energy, E_1, of one unit. The rate of such a conversion is reduced compared to that in Figure 17.3c. If $E_1 \gg k_B T$, such conversions may not occur on a biologically allowed time scale. The structures formed would then be governed by kinetics, and the lowest-energy structure may not be the one actually formed during the cell's lifetime.

EXERCISE 17.1

Rates of assembly and disassembly. (1) Estimate the rate at which dimers of the "two-sticky-sided" cubic monomers of Figure 17.2 form if a side of a cube is 1.0 nm in length. (2) Estimate the rate of disassembly of a dimer, if the separation of two monomers requires an activation energy of 100 kJ/mole. Solution. (Not exact, of course.)

1. The first issue to address is what type of rate we are dealing with. We presumably have a random solution of monomers of some concentration $c_1(0)$. Formation of a dimer likely proceeds through random collisions. This means we should look back at diffusion-controlled reaction rates in Chapter 15. The reaction is a bimolecular reaction with solution $\dfrac{1}{[A]} - \dfrac{1}{[A]_0} = 2kt$ (Equation (15.48), where $k = k_D \cong \dfrac{2k_B T}{3\eta}\left(\dfrac{1}{r_A} + \dfrac{1}{r_A}\right)(r_A + r_A) = \dfrac{8k_B T}{3\eta}$. We saw in the example given that the size of the particle actually does not matter—larger particles move more slowly but collide more frequently, so that there is one universal diffusion rate constant, approximately valid for all molecules in water at 20°C. We can thus start with the rate calculated: $k_D \cong 1.1 \times 10^{-17}$ m^3 s^{-1} = 6.6×10^9 M^{-1} s^{-1}. Now in our case collisions are only effective when the correct sides of both molecules collide. Two of the six sides of each cube are sticky, so we should (roughly) multiply

the rate constant above by $1/3 \times 1/3 = 1/9$. We get a rate constant of about $k_D \cong 1.2 \times 10^{-18}$ m^3 s^{-1} = 7.3×10^8 M^{-1} s^{-1}. This factor of 1/9 is an entropic effect. We have, of course, ignored other entropic effects, such as the entropy cost of removing water molecules from the sides of the two monomers, but we cannot afford to be too sophisticated here.

Now, if we want a rate (s^{-1}) that we can compare to the rate in part (2), we must assume a local concentration valid at some time. The rate decreases as the monomer concentration decreases, but if the initial monomer concentration is 1.0 mM, the time for the monomer concentration to reach ½ of the initial concentration can be found:

$$\frac{1}{\left[\frac{1}{2} \times 10^{-4} \text{M}\right]} - \frac{1}{\left[10^{-4}\text{M}\right]_0} = 2(7.3 \times 10^8 \text{M}^{-1}\text{s}^{-1})t_{1/2} \Rightarrow t_{1/2} = 6.8 \times 10^{-6}\text{s}, \text{ or an average}$$

rate of 1.5×10^6 s^{-1}. The time for half the monomers to dimerize is about 6.8 μs.

The subscript D stands for diffusion, but in the present context we can also think of it as dimerization.

2. Once a dimer is formed, the reaction to un-dimerize is a unimolecular reaction and does not depend on concentration. (Think this scenario through carefully: the formation of aggregates depends on concentration, the disassembly does not.) We can write an Arrhenius-type unimolecular reaction rate, $k_U = Ae^{-E_A/k_BT}$, but we must carefully think of what values we will put in. Remember the Arrhenius "prefactor," A, is related to the attempt frequency and activation entropy: $A \cong (10^{13}\text{s}^{-1})e^{s^*/k_BT}$. We are trying to avoid entropy issues in this example, so we can try "usual" values of A: 10^6 to 10^{13} s^{-1}. We are told the activation energy is 100 kJ/mole, so we can estimate the rate to un-dimerize: $k_U = Ae^{-(10^5\text{ J/mole})/[(8.31\text{ J/mole·K})(300\text{K})]} = Ae^{-40} \cong A(3.8 \times 10^{-18})$. Using the "usual" values of A gives a range of this rate of 3.8×10^{-12} s^{-1} to 3.8×10^{-5} s^{-1}. The time to un-dimerize ranges from 7 h to 8,000 years. Both of these times suggest that the reverse of the dimerization reaction does not occur during any meaningful time in the life of a cell: assembly is not an equilibrium process in this case. The assembly of the dimer and higher aggregates, on a coarse scale, is controlled by rates and probabilities (entropies). The aggregates that are most likely to be initially formed also prevail over long times. If the enthalpy of dimerization were much smaller, the un-dimerization rate could be much higher, of course.

Because we ignored so many details, we should take this conclusion with a grain of salt, but what we have done does correctly indicate that biological assembly is often not a reversible, equilibrium process. If a structure is to be adjusted after it has assembled into an incorrect configuration, an enzyme must be found to reduce the activation energy for unbinding and thus catalyze the disassembly.

An entropic "funnel" is one of the key ideas in protein folding: the protein does not explore all possible molecular conformations in order to find the lowest free-energy state. See Bryngelson et al., *Proteins* 21 (1995):167–195.

17.3 RESTRICTED SPACE FOR ASSEMBLY

Because it appears that random assembly may be importantly influenced by the initial structures formed—they may never disassemble to allow other, more favorable structures to form—we should look at how environmental structure can influence assemble. The coarse effect of a restricted space for assembly in which aggregates can form is easily seen in Figure 17.4. If the space where assembly takes place is approximately tubular, with a tube diameter not much larger than a monomer diameter, linear aggregates are the only structures that can form. The transport of monomers to the tube must be controlled, of course, so aggregation does not take place before monomers get to the assembly tube, which is a chaperonin. If, on the other hand, the space is spherical but small, three-dimensional aggregates, which have smaller maximum diameters, are favored.

Specific interactions between monomer units, such as surface patches that interact with other monomer units, also determine what structures may form. If only two patches on opposite sides of a monomer unit can bind, linear aggregates of indeterminate length will form (Figure 17.5a). If the interacting patches on each monomer are 90° apart, nonlinear structures will form (Figure 17.5b), generally growing until all interacting patches are bonded to patches on others monomers in the aggregate. If monomers are free to form aggregates in three dimensions, many different sizes and shapes are allowed. Their relative populations are determined by the free energies of the structures, compared to monomers. Figure 17.5c shows a growing aggregate formed when patches are about 110° apart on a monomer, with two types of patches, indicated by red and green. In this scenario, red patches bind to green. The aggregate again grows until all available patches are bonded, weighted by aggregate free energy. While random aggregation allows many different aggregates in three dimensions, placement of monomers in a membrane, restricting aggregates to a two-dimensional form, often selects for a unique aggregate structure. Homework problems address several of these aggregation issues.

While avoiding the further discussion of chaperonins, we have seen in the first few chapters that much of the space inside cells is crowded, at least on a scale of hundreds of nm of more, so we expect the geometric space around structures being assembled should generally be important for structures 10 nm or larger. GroEL-, TRiC-, and other chaperonin-mediated structure formation on smaller spatial scales can involve both these general geometric effects and binding of a chaperonin to specific configurations of structures during the assembly process.

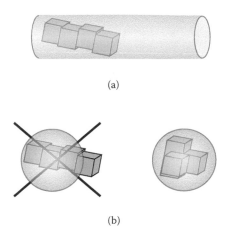

FIGURE 17.4
If aggregation takes place in a restricted space—a type of chaperoned aggregation—the shape and size of the space determines which aggregates are favored. (a) A tubular space favors one-dimensional aggregates. (b) A small spherical space favors three-dimensional aggregates, owing to their smaller maximum radius.

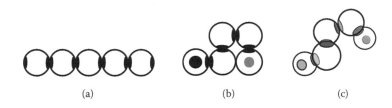

FIGURE 17.5
Aggregation of monomers with surface patches that interact. (a) Linear aggregates of indeterminate length are formed when patches are on opposite sides of monomers. (b) Two interaction patches 90° apart. Aggregates can be planar or nonplanar. (c) Patches of two different sorts (e.g., opposite charge) arranged about 110° apart on monomers. Red patches bond to green patches, but not to other red patches. Aggregates in b and c grow until all patches are bonded to other patches in the aggregate. Restriction to two dimensions often selects for one specific aggregate.

17.4 ENTROPIC DRIVE: ORDERED STRUCTURES CAN BE DRIVEN BY RANDOM PROCESSES

17.4.1 Energy Landscape and Local Minima

We have already noted that aggregate structures are more likely to be initially formed if there are more possible ways to assemble the structure. If interactions between units that approach each other are strong, these structures may also be the final structures, even if some other arrangement of monomers may have lower overall free energy. What we are discovering is that though the specific interactions that hold the aggregate together may be largely enthalpic, e.g., Coulomb attraction, the structure actually formed is largely determined by entropic considerations. A state with lower total free energy may not be accessible on any reasonable time scale, so a more easily formed, higher-energy state, may be the prevailing state *in vivo*. Figure 17.6 illustrates the energy "landscape"

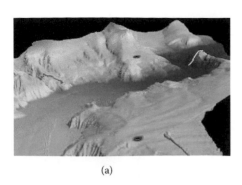

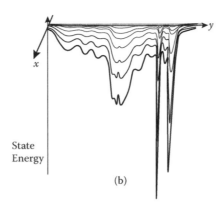

State
Energy

(a)

(b)

FIGURE 17.6

Energy landscape of structure formation. (a) The topography of the Astoria Canyon, a submarine abyss 10 km off the mouth of the Columbia River on the west coast of the United States, illustrating the energy landscape model of protein structure formation (with two small, deep wells added by the author). The depth represents the free energy, purple and black the deepest. The two small wells may represent the lowest free-energy states, but a particle falling at a random location (x,y) and rolling downhill from there would rarely roll into one of the wells. If particles are always supplied via the "riverbed," the wells would virtually never be occupied. (b) Plot of energies (enthalpies) of states versus state coordinate in two dimensions. (Image in (a) Courtesy of Lewis and Clark 2001, Office of Ocean Exploration and Research, National Oceanic and Atmospheric Administration, U.S. Department of Commerce. The yellow and fuchsia tracks represent paths of submersible explorations in 2001. http://oceanexplorer.noaa.gov/explorations/lewis_clark01/logs/jul03_astcynsummary/media/astoriacanyonwest.html).

description of such a situation. Figure 17.6a is a topographical map of the submarine Astoria Canyon off the mouth of the Columbia River between Oregon and Washington in the United States, illustrating the main idea of the landscape. The horizontal (x,y) coordinates represent the state of the (protein) structure; the depth is the energy or enthalpy of the state. Wide, flat regions indicate areas where many states have the same depth (enthalpy)—high-entropy states. To mimic the random thermal motion associated with a structure exploring its conformational space, think of a particle falling somewhere on the surface, then springing up in the air a few feet every second or so. A particle falling randomly on the map would proceed to roll/jump downhill, most likely toward and into the riverbed and down the riverbed from then on. A randomly placed particle would rarely roll into one of the small, deep holes, though they represent the lowest-energy states.

If particles are always supplied from the riverbed upstream, they would virtually never fall into a deep well, unless some active agent (an enzyme) intervened. Figure 17.6b shows a more traditional plot of the energies of the states.

These lower-energy states are sometimes accessed by more complex biological mechanisms. Polypeptide chains sometimes form temporary, "incorrect" protein structures, which are then trimmed by a *protease* enzyme, enabling the final, active structure to be formed. See Voet, D., and J. G. Voet, 2004, *Biochemistry*, New York: John Wiley & Sons, or any recent biochemistry text for more information.

17.4.2 Correlations (Cooperativity)

Suppose an N_s-segment random coil is generated by some assembly mechanism that requires enthalpy E per unit. The coil requires energy N_sE to construct. If two completely uncorrelated coils are built, the energy required is $2N_sE$. Figure 17.7 (upper) shows an example of two such seven-unit random coils. Suppose now that two coils are again constructed, but that a cooperative interaction between the two construction processes exists such that the placement of subsequent units on the two coils is correlated, as in Figure 17.7 (lower). The two coils will have identical conformations, though they may not actually be bound together. The energy to assemble these two coils is again $2N_sE$. What about the entropy of the two construction processes? The structure in the lower part of the figure is clearly more ordered than the upper and so, must have lower entropy and thus higher free energy: there is an entropic energy cost to cooperativity between the two coils. Can we calculate this entropy difference?

If the assembly of a coil is truly random and takes place on a two-dimensional lattice, as in the figure, each subsequent segment has a choice of four positions (ignoring possible problems with overlap). The N_s segments would have 4^{N_s} possible states, and the entropy of the second coil, if completely uncorrelated with the first, is $S_{uncorr} = k_B \ln(4^{N_s}) = k_B N_s \ln(4)$. (In three dimensions, the choice would be between six possible positions.) If the assembly of the second coil is completely correlated with the first, however, each subsequent segment has only one choice for its placement. The associated entropy is zero. The entropy difference between correlated and uncorrelated is

$$\Delta S = S_{corr} - S_{uncorr} = -k_B N_s \ln(4) \tag{17.1}$$

The free energy difference between these two assemblies is

$$\Delta G = G_{corr} - G_{uncorr} = \Delta H + k_B T N_s \ln(4) = k_B T N_s \ln(4) \qquad (17.2)$$

At room temperature this free energy difference is about $\Delta G \cong (14 \text{ kJ/mole})$ $N_s \cong (3.3 \text{ kcal/mole})N_s$, a nonnegligible fraction of a hydrogen-bond energy per step. There may be other energy differences—the energy needed to place two segments in correlated positions may be less than or greater than in uncorrelated positions, if frictional resistance is different, for instance—but it is useful to keep in mind that assembly of correlated structures has a free energy cost.

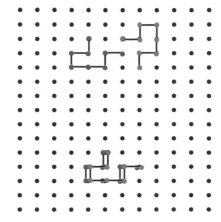

FIGURE 17.7
Entropic cost of correlation (cooperativity) in structure formation of two random 7-mers. (Upper) The two structures are each assembled randomly, with no correlation. (Lower) Placement of polymer units is correlated. The correlated structures have higher free energy construction costs because of the lower entropy.

17.4.3 Fractal Structures

Most of us are familiar with the concept of fractal structures: rough shapes that, when examined on any size scale, appear similar. The usual example is the coast of Norway; another is the high-voltage-induced dielectric breakdown in a block of Plexiglas, shown in Figure 17.8. This *Lichtenburg picture* appears similar to the branching of some trees. Another plantlike structure is *Barnsley's fern*, which again appears very ordered, but results from random processes. The self-similarity of fractal structures on all length scales reminds us of hierarchical structure in biosystems, except for one crucial difference: in biological structures, the intricate structures are *not* similar overall length scales.

Fractal structures can result from random (entropically driven) assembly of identical subunits in a relatively unrestricted environment. Figure 17.9 shows a computer simulation of the assembly of structures on a quadratic lattice as the effective interaction between adjacent particles changes. The intrinsic probability for placement next to an existing particle increases from Figure 17.9a–c. The size distribution of resulting structures depends sensitively on this probability, though the shape of the structures does not.

FIGURE 17.8
Fractal structure—a Lichtenberg figure—created by high voltage dielectric breakdown within a block of Plexiglas creates a fractal pattern similar to tree branching. The branching discharge lines are thought to extend down to the molecular level. (Courtesy of Bert Hickman, http://www.capturedlightning.com.)

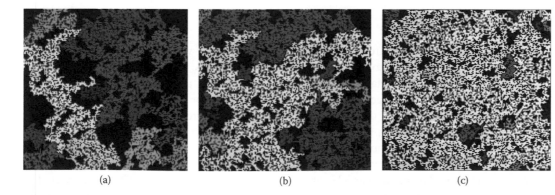

(a)   (b)   (c)

FIGURE 17.9
(a–c) Fractal structures. A two-dimensional lattice 160 × 160 places effect of increasing occupation probability on a 160 × 160 quadratic lattice (p = 0.58, 0.60, 0.62, top to bottom). Largest cluster is white; in order of decreasing size cyan, red, orange, yellow, light green, green, turquoise, and blue. Though random in construction, the final structure clearly depends on the effective interaction between particles. (From Feder, J., *Fractals*, Springer, New York, 1988, Figure C.1 reproduced by permission.)

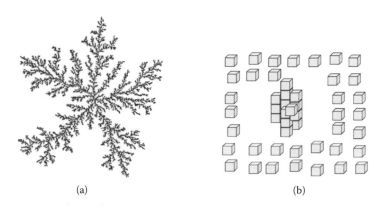

(a)   (b)

FIGURE 17.10
Snowflakes and tip growth. (a) A 50,000-particle, snowflake-like cluster produced by diffusion-limited aggregation in two dimensions. A random-walking particle starts off at random position on large circle around cluster. If the walker hits the cluster, it sticks. The cluster is initially seeded at the center of the lattice. In contrast to Figure 17.9, this cluster has obvious structure. Snowflake-like clusters can be experimentally produced by aggregation of metallic particles in solution. Program by Paul Meakin. (From Feder, J., *Fractals*, Springer, New York, 1988. Figure 3.3, by permission.) (b) Depletion of monomer stocks near large surfaces causes relatively greater growth at tips later in the growth process.

The previous random aggregates had no obvious symmetry or structure. Figure 17.10a shows a randomly assembled aggregate that shows a snowflake-like structure. The snowflake was produced on the computer by placement of an initial "seed" at the center of the coordinate system. Another particle was then randomly placed on a circle of radius much larger than the assembled structure, and then allowed to random walk on the two-dimensional lattice. If the diffusing particle collided with any part of the growing flake, it stuck. Such flake-like structures can be produced experimentally by electrodeposition of zinc at the interface between aqueous zinc sulfate and an organic liquid.[11]

Fractal structures are fascinating, but we want to keep the focus on biology. What are the biological implications of these results from random assembly? To see them, we need to probe a few more details of assembly processes that involve random walks, as in biosystems. A few basic principles of diffusion-limited assembly can be written off the cuff:

1. Monomers (single particles) diffuse the fastest.
2. Large aggregates stop moving; in order to grow further, smaller particles or smaller aggregates must move toward them.
3. The region near a large aggregate becomes depleted of monomers, if a fixed number of monomers is available. The region near a large surface will be especially depleted.

Putting principles 2 and 3 together gives us fractal growth. In the second half of the assembly process, the depletion of monomers near large surfaces means growth virtually ceases. Instead, growth continues at narrow, extended tips, where monomers are not depleted (Figure 17.10b). Voila—we have created snow-flakes! How does biology avoid creating snowflakes? First, the monomers rarely bond together at random angles in large, unobstructed areas. Instead, as in Figures 17.1 through 17.5, binding interactions occur only at specific orientations of units; when necessary, assembly takes place in chaperonin-defined spaces, in membranes, or other structures.

17.4.4 Assembly by Osmotic Pressure

Solvent and solute molecules are constantly in motion, colliding with each other and with larger structures. These collisions also take part in assembly processes, especially in restricted spaces. Consider, for example, two relatively large objects confined to a spherical volume of aqueous solution (Figure 17.11)—a large protein inside a cell space, for example. All objects undergo thermal motion, of course, but the larger (tan colored) move more slowly, constantly bombarded by the smaller, more numerous solute molecules (black). On average, the larger structures move randomly, but eventually they approach each other or the wall of the cell and begin to exclude the smaller molecules from the intervening space. As we saw earlier in the text, the average osmotic pressure exerted by the smaller particles against any surface is $P_s = c_s k_B T$. The force exerted on a side of the larger, tan particles is then $F_s = c_s k_B T A$, where A is the surface area of the larger molecule. When the distance between two apposed areas is $2r$ or less, there will be a net force keeping the two larger structures pressed together. The maximum value of this net force occurs when the two areas are entirely apposed. The free energy decrease upon association of the larger particle to the wall or to another particle is $F_s \cdot 2r = 2c_s A r (k_B T)$:

$$\Delta G_{\text{assoc}} = -2c_s A r (k_B T) \tag{17.3}$$

For the interior of a red blood cell, the hemoglobin concentration is about $c_{Hb} = 4$ mM (2.4×10^{24} m^{-3}), so for and $r_{Hb} \sim 2$ nm particles pressing larger structures of surface area $A = (20$ nm$)^2$, ΔG has a value of about $4k_B T$. What subcellular organelles might hemoglobin in RBCs organize? Mammalian RBCs are unique among the vertebrates in that they have no nuclei, mitochondria, Golgi apparatus, and endoplasmic reticulum in their mature form. These cells have nuclei during early phases but extrude them during development to provide more space for hemoglobin. Though the extrusion is not done by the hemoglobin, the presence of the large concentration of hemoglobin molecules in the cell interior makes movement of the larger structures toward the cell wall favorable. Osmotic/entropic organization effects have been observed experimentally. See Böker et al. and references therein.[12]

17.4.5 Commercial Application: Entropic Forces for DNA Purification

The human, as well as other, genome projects rely on determining the sequence of chromosomal DNA. After obtaining some appropriate cells from the organism, the first step in obtaining this sequence is to actually separate and purify the DNA from an organism. DNA purification

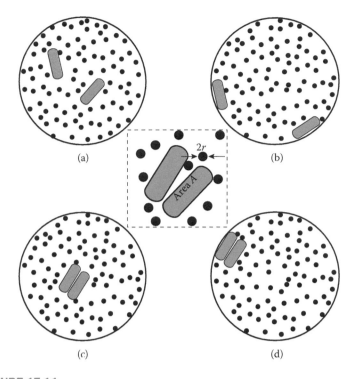

FIGURE 17.11
Osmotic assembly of structure. (a) Thermal motion of smaller particles (black) eventually force larger structures (tan) (b) to the wall (c) together, or (d) both, as long as the small particles are large enough to be excluded from the spaces between the apposed larger structures. Center of diagram: a net osmotic aggregation force occurs when smaller particles are excluded from intervening spaces.

technology has been available for decades, but genome projects require inexpensive, high-speed, massively parallel, robotic methods. Older methods relied on fracturing cells with osmotic or physical pressure jumps, followed by selective ethanol precipitation of the DNA, centrifugation, and purification using chromatography of gels (Figure 17.12a). A newer method, used for most of the Human Genome Project, employed entropic separation of DNA from other debris from other cell components, balanced by charge interactions (Figure 17.12b). Polystyrene microspheres, with negative surface charge and imbedded magnetic material, are added to the sample, along with polyethylene glycol (PEG) and salt. Because DNA and the spheres are both negatively charged, they would normally repel each other. However, PEG forms a matrix around the spheres into which the DNA cannot easily penetrate: DNA is entropically excluded from the PEG. The salt reduces the sphere-DNA repulsion, with the result that DNA binds to the sphere surface. A magnetic field then separates the spheres + DNA from the other cell components. The process can be repeated to further purify the DNA and can be carried out much faster than the centrifugation/chromatography procedure. More details on this method can be viewed at www.thermofisher.com (search DNA purification) or search on commercial producers of the magnetic spheres: BioMag® beads (Polysciences, Inc.), Estapor® superparamagnetic microspheres (Bangs Laboratories, Inc.), Sera-Mag™ magnetic microparticles (Seradyne, Inc.), Sphero™ CMX, (Spherotech, Inc.), and Dynabeads® (Dynal). The U.S. Department of Justice and other forensics services are using such technology to extract DNA from blood, semen, and other crime-scene samples.[13]

The secret to this DNA purification method is to counter enthalpic charge-repulsion forces between DNA and micro-sphere with entropic exclusion forces from PEG. Varying the ionic strength tips this balance toward the entropic or enthalpic side. This picture of the process is actually a simplification of the interactions. It has been observed, for example, that more DNA binds to positively charged spheres than is needed for charge neutralization.[14]

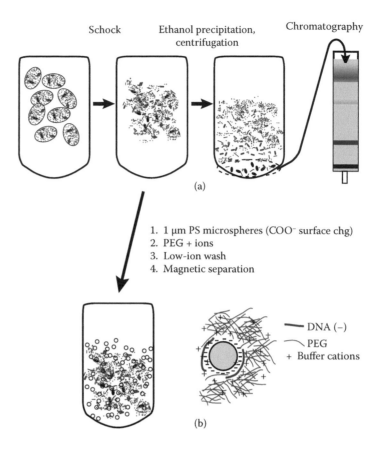

FIGURE 17.12

DNA purification for sequencing. (a) Prior method: Cells broken up by pressure shock; addition of ethanol to selectively precipitate DNA; centrifugation; chromatography or electrophoresis. (b) Magnetic microsphere/entropic separation: magnetic microspheres with negative surface charge added to cell debris; PEG + counterions added; DNA selectively binds to microspheres and are pulled to bottom of tube by a magnet. Steps in (b) can be done by robots, hundreds of samples at once.

17.4.6 Size Distributions

Random assembly processes intrinsically produce size distributions, as in Figures 17.9 and 17.2 through 17.5. Exceptions occur when binding interactions between monomers are restricted to certain surface spots, especially in two dimensions, as in Problem 17.2. Generally, however, aggregates can grow indefinitely, as long as monomers are still available. These aggregate size distributions are, of course, time dependent, and the law describing their growth reminds us of kinetics we have seen in reaction measurements in Chapter 15. The rebinding of O_2 to heme proteins was observed to follow a power-law dependence: $N(t) = (1 + t/t_0)^n$. The explanation for this behavior is that protein molecules exist in a distribution of slightly different conformations, each with its own, slightly different binding activation energy.

Figure 17.13 shows the growth of aggregates of immunoglobulin G (IgG) proteins. IgG are antibodies involved in immune response. IgG can pass through the human placenta and provide protection to the fetus *in utero*. Colostrum contains a high percentage of IgG. When IgG encounters pathogens such as viruses, bacteria, and fungi, IgG inactivates them by agglutination and immobilization (aggregation). The mathematical dependence of the aggregate size in Figure 17.13 is

$$\langle R(f) \rangle = R_0 (1 + \gamma t)^{1/D} \tag{17.4}$$

D is called the fractal dimension of the aggregate. This dimension corresponds to the description of the behavior of polymers coils in sedimentation experiments. The fractal dimension is defined by the relationship between the mass of (or number of monomers in) an aggregate and its average radius (or diameter):

$$M = \rho V = (\text{constants}) \cdot R^D \qquad (17.5)$$

A solid sphere described by $m = \rho \frac{4}{3}\pi R^3$ or cube $m = \rho a^3$ are both three-dimensional. A linear string of spheres has a fractal dimension greater than 1.0, unless the sphere radius is much smaller that the string length. A geometrically three-dimensional shell containing loosely packed monomers generally has a fractal dimension less than 3 because of spaces between spheres. We saw that the (random) conformation of a loosely packed, freely jointed chain resulted in a structure of less than three spatial dimensions, in terms of the relation between mass and average size. Now we see that such nonintegral dimensionality results from assembly processes themselves.

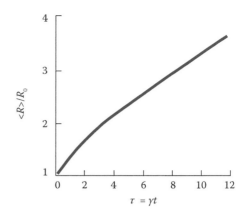

FIGURE 17.13
Growth of immunoglobulin G protein (IgG) as a function of time. Plotted is the hydrodynamic radius of the aggregates as a function of reduced time, for various temperatures and concentrations. The growth follows the law $\langle R(t) \rangle = R_0(1 + \gamma t)^{1/D}$, where D is the fractal dimension. The width of the line reflects, roughly, the scatter in the data. The cluster fractal dimension is 2.6 ± 0.3. (After Figure 3.5 in Feder, *Fractals*.)

17.4.7 Summary of "Random" Assembly (Aggregation, No Chaperonins or Assembly Apparatus)

- No ATP energy input is needed.
- Structures often have an ordered appearance.
- Fractal-like structures are common: tip growth.
- Osmotic forces tend to produce smoother structures (building blocks aligned), in the absence of strong enthalpic binding.
- "Sticky" (enthalpic) building blocks create more uneven aggregates.
- Small, specific "sticky" spots on monomers produce small, ordered aggregates if confined to two dimensions.
- Kinetics, local building-block depletion, not equilibrium, are most important.

Though biological assembly by complex cellular machinery certainly employs other modes of assembly action, we may expect to see the cell use some of the above assembly principles to its advantage, not reverse all of them. The latter would be very energy intensive.

17.5 NUCLEOSOMES AND NUCLEOSOME-LIKE STRUCTURES

Living cells have a major problem with the storage of DNA. The total length of DNA is far larger than a cell diameter and that DNA must be coiled up for inclusion in the cell's interior. Long, double-stranded DNA naturally prefers to form coils because of entropic factors, but the "random" coils we have seen would be disastrous for the cell's replication and gene expression apparatus. DNA must be stored in a compact but ordered manner, so that the cell can access specific genes when chemical signals indicate synthesis of a particular protein is needed. A random search through billions of base pairs would spell failure for cell survival. Ordered storage is needed.

> This large majority of noncoding elements are an absorbing study by themselves and reminds us that much of biology remains to be explored and understood.

Actually, this organized-storage situation is not so bad as it might seem, as far as protein coding is concerned. Sequencing of the human genome has shown rather few genes (sequences of protein-coding DNA) (e.g., ~20,000 in human, mouse, and fly, ~13,000 in roundworm, >46,000 in rice) encode all the proteins in an organism. The protein-coding sequences make up 1%–2% of the human genome. A large part of the genome appears to be little-understood introns, retrotransposons, and noncoding RNAs. Nevertheless, location of one of 20,000 coding DNA sequences among a million noncoding sequences and billions of DNA base pairs is a daunting task.

The first level of organized storage of DNA is the nucleosome. The core unit of the nucleosome is a positively charged protein (histone) octamer of about 8-nm diameter, around which the DNA wraps about 2 times, about 150 base pairs of DNA (Figure 17.14). The patches of sealing "tape" seen in the figure are histone linker proteins, H1, which lock the DNA into place and facilitate formation of higher-order structure. This structure appears to be a 10-nm fiber of beads on a string, with 10–80 base pairs of DNA separating adjacent beads. Higher order structures include the 30-nm fiber (forming an irregular zigzag) and 100-nm fiber as often observed in cells. Figure 17.15 shows a nucleosome core structure as determined from X-ray crystallography, confirming that the cartoon is not too far off.

What do we expect to happen when a negatively charged polymer is added to a suspension/solution of positively charged spheres? Certainly, our knowledge of Coulomb's law tells us to expect attractive forces. We should first get the dimensions correct, however. Spheres of diameter 8 nm are small enough to qualify as dissolved rather than suspended in water. The DNA can be microns or longer and 2-nm-thick. The spheres are therefore too large to be associated like a counterion cloud around the DNA or to insert into the double-stranded structure. The total number of amino acids carrying a positive charge on the histone core complex ranges from 160 to 220, while the number near the surface, where they can more strongly interact with DNA phosphates is 75–100.[15] The charge carried by 150 base pairs of DNA would be about $-300e$. What, then, is the net charge on the nucleosome core particle (NCP), composed of a histone core (with associated protein tails) and DNA? According to Yang et al., the NCP carries a net negative charge of about

> We will not look into the higher-order structure of chromatin. See en.wikipedia.org/wiki/ Nucleosome or Burgess, R. J., and Z. Zhang, 2013, "Histone chaperones in nucleosome assembly and human disease," *Nature Struct. Mol. Biol.* **20** (1): 14–22 for overviews.

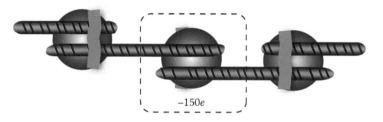

FIGURE 17.14
Cartoon diagram of nucleosome. DNA strand winds twice around each ~8-nm histone sphere and is "sealed" in place by a histone H1 protein molecule (light brown). The next larger-scale structure is formed when the spheres wrap approximately in a solenoidal structure, approximately six spheres per turn. Each nucleosome unit has a net charge of about $-150e$, excluding solvent counter-ions. DNA can slide around the histones when assisted by high enough ionic strength or enzymes.

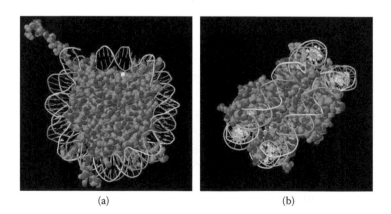

FIGURE 17.15
Complex between nucleosome (histone) core protein particle (H3,H4,H2A, H2B) and 146-bp long DNA fragment. (a) Top view, protein subunits displayed in space-filling format (CPK coloring), DNA as chain only. (b) Side view, showing two DNA windings on the histone core particle complex. (PDB ID 1AOI, displayed using JMol software, from Luger, K. et al., *Nature*, 389, 251–260, 1997.) (From Jmol: An open-source Java viewer for chemical structures in 3D. http://www.jmol.org/; Protein Workshop: Moreland, J.L. et al., *BMC Bioinformatics*, 6, 21, 2005; NGL Viewer: Rose, A.S. et al., *Bioinformatics*, 2018.)

−150*e*, consisting of a negatively charged central core (net charge about −240*e*) to which eight flexible and positively charged (net charge about +90*e*) chains are attached.[16] Then there are the counterions from the aqueous phase, but we still conclude that DNA does not simply bind to the core until the histone positive charge is neutralized. Why NCPs also aggregate to form higher-order fibers is not known in detail but is thought to be mediated by the positively charged histone tails. We discover that DNA also has a tendency to aggregate to itself, given a little help from a positively charged seed site.

The nucleosome should not, in detail, be thought of as a uniformly charged sphere wrapped with a uniformly charged, flexible string. The arrangement of charges on the histone core appears to have some correlation with the distances between phosphate charges on the DNA.[7]

> Recall from earlier that the distance-dependent dielectric constant of water, as well as counterions, shield Coulomb interactions farther than a few water layers of a charged object.

17.5.1 Nucleosomes Are Not Static

17.5.1.1 Charge switching

To keep ourselves honest, we should note that details of the charges and their locations can be a matter of survival of an organism. For example, six lysine charges on a histone "tail" region (*Tetrahymena* H2A.Z), which we are ignoring, play a key role, at least for *Tetrahymena*. These lysines are acetylated in vivo at various times, neutralizing their charges. Experiments in which the lysines were individually replaced with nonacetylable arginines showed that retention of one of the six lysines maintained cell viability. Control of the charge in specific locations, especially in regions between the spherical cores, seems to be important to survival. The authors of the study state, "It seems rather remarkable that the presence or absence of a single acetylation site in only a small fraction of the nucleosomes can result in the difference between life and death."[17] The state of acetylation of these histone regions varies rapidly, especially during transcription. Nucleosomes are assembled and remodeled during this intense activity, suggesting that charge modifications facilitate the assembly and disassembly of chromatin.[18] What a research project for a budding biophysicist: a charge-controlled mechanism for activation and remodeling of large stores of genetic information!

17.5.1.2 Nucleosome sliding

The simple picture we would like to have of negatively charged DNA wrapped around a uniformly, positively charged histone sphere would imply that sliding of DNA over the sphere would incur no free-energy penalty. This appears to be the case, under certain circumstances.[19] Specific accessory proteins, called CTCF, seem to act as nucleosome positioning anchors.[20] Although nucleosomes are intrinsically mobile, eukaryotes have developed a family of ATP-dependent chromatin remodeling enzymes to control chromatin structure. Nucleosome sliding appears to be involved in many of the enzyme-assisted processes.

17.5.2 Artificial Nucleosome-Like Structures

DNA seems to have surprising properties. DNA wraps around a positively charged sphere to a point that the structure becomes significantly negatively charged. DNA can also form a variety of helices or fiber-like structures made up of multiple individual strands of nucleic acid. We saw an extreme version of this in the method used to prepare the original sample for the Watson, Crick, Franklin, and Wilkins X-ray diffraction measurements of DNA structure (Chapter 5). Because the DNA would not crystallize, macroscopic fibers of DNA were spun from a concentrated DNA solution. The fact that large DNA fibers can be made from gentle, mechanical manipulation of a DNA solution suggests that DNA should not be thought of as a highly charged polymer that strongly repels itself. Such a picture even contradicts the double-helix model that asserts that two negatively charged strands are held together by hydrogen bonds and stacking interactions.

The nucleosome beads-on-a-string picture shows an individual DNA (double) strand wound around an 8-nm sphere. The spun X-ray diffraction sample likely had thousands of DNA strands bound together in the macroscopic DNA "rope." "Condensed" structures of DNA involving small numbers of DNA strands also form when the ionic environment is modified appropriately.[21] In all of these structures, positively charged particles—ions or histones—act to induce formation of ropes of DNA, with $1-10^4$ DNA strands bundled together. The positive particle ranges in size from 0.1–0.2 to 10 nm. What happens if we place a micron-sized sphere into a solution of DNA at normal salt concentrations (0.1–0.2 M NaCl)? Figure 17.16 shows that a beads-on-a-string structure again results, but on a scale of microns, not nanometers.[22] The left part of the figure, a scattered-light image, shows that some of the polystyrene microspheres seem to be aligned along a path. This path is also tracked by a string that is clearly visible in the image. This string must be a diameter of at least a few hundred nanometers in order to be visible in the scattered white light, indicating at least several hundred double helices bundled together. The fluorescence image on the right in Figure 17.16

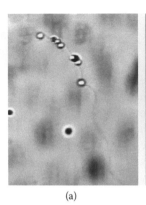 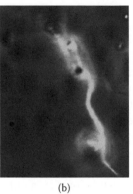

(a)   (b)

FIGURE 17.16

Microscopic images of lambda phage DNA (~16 μm long individual strands) added to 1.0-μm polystyrene microspheres with positive surface charge (pH 7.2, 0.1 M NaCl), with YOYO-1 fluorescent dye added to visualize DNA. (a) Bright-field (scattered light) image. (b) Dark-field (fluorescence) image. The fact that the DNA strand is visible in the bright field image indicates that the DNA has formed a large bundle of individual, 2-nm-diameter DNA strands. (From Krishnan, R. et al., Imaging of DNA/Nanosphere condensates, *2005 March Meeting of the American Physical Society*, American Physical Society, Los Angeles, CA, 2005.)

shows that this "string" is indeed composed of DNA, indicated by the fluorescence of YOYO-1 dye, which binds selectively to double-stranded DNA. The continuation of the DNA string tens of microns away from the last charged sphere also indicates that the bundle, once initiated by a + charged sphere, continues without significant fraying. We should also note some details in this experiment: the DNA used, from lambda phage, has a length of 16 μm. The string in Figure 17.16 is longer than this (the beads are 1 μm in diameter), showing that DNA strands added to the bundle do not need to be lined up.

The 16-μm DNA strands in Figure 17.16 are long enough to wind around the 1.0-μm spheres. How many times do they wind around? Our immediate response might be, "Two times, just like in the nucleosome." A bit of thought would make us take back this answer, however, because the physical dimensions differ by a factor of 100. We can quantify numbers of DNA strands, however, from the intensity of fluorescence. This has been done in Figure 17.17, though only relative intensities were determined. (Absolute intensity calibration, a more challenging task, would allow determination of absolute numbers of strands per bundle.) A scan of pixel regions shown in the diagram, one across a bare DNA strand, the other across a sphere wrapped with DNA, show that close to 2 times as many strands are wrapped around the

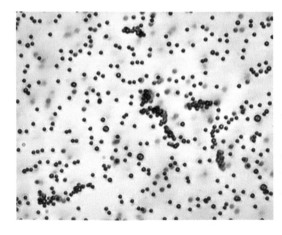 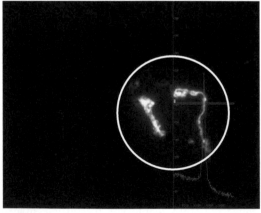

FIGURE 17.17

Beads on a DNA string. 16-μm lambda phage DNA, 1.0-mm polystyrene spheres (+ surface charge, YOYO-1 dye added, 0.1 M NaCl, pH 7.2). (a) Bright field image. (b) Fluorescence image. Only the circled region is illuminated. Computer scans over a narrow band indicated integrated the fluorescence intensity over vertical pixels, for each horizontal pixel location. The superimposed graph shows the integrated intensity versus x for a scan over a DNA strand that does not include a micro-sphere. Integrating scans were also done for bands including spheres (e.g., slightly above the indicated scan band). In this way, ratios of fluorescence including sphere to fluorescence of DNA alone could be found. The ratio is about 2. (From Krishnan, R. et al., Imaging of DNA, 2005.)

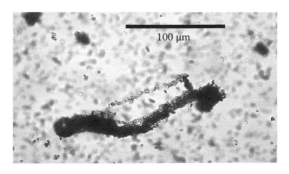

FIGURE 17.18

Model for spontaneous DNA wrapping on 1.0-µm spheres, consistent with fluorescence scan measurements. Wrapping on the spheres induces DNA to form multistrand "ropes," consisting of N DNA strands. One of these strands wraps around a sphere approximately twice.

FIGURE 17.19

A circular necklace of microspheres formed from the same microsphere DNA system of Figure 17.17. You are welcome to count the spheres and individual DNA bundles.

microsphere as are contained in the DNA bundle approaching the sphere. Surprisingly, our initial, naïve response is proven correct. Though there could be a distribution of DNA bundle sizes—we did search for a string with relatively few beads attached—the tendency of two DNA strands to wrap around spherical objects of positive surface charge seems to extend 2 orders of magnitude in size from nucleosome scale (Figure 17.18). We should note that the structures formed in these "artificial nucleosome" systems depend on ionic strength and time.

Figure 17.19 shows an impressive structure formed in the same lambda phage DNA + micron-sized sphere system. The structure is a ring of radius ~100 µm and is composed of hundreds to thousands of microspheres, probably attached to many individual bundles of DNA. Models for this bundled necklace can be found on sale in the jewelry section of any local department store.

Addition of much shorter strands of DNA to micron-sized spheres produces quite different structures. Figure 17.20 shows structures resulting from addition of 1.0-µm, positively charged spheres to herring sperm DNA fragments, of length 0.2–0.8 µm. The individual DNA strands are now smaller than a sphere diameter. DNA still binds to the spheres, but bundles of DNA do not form. Instead, DNA can be seen on the surface of the spheres but generally not in a uniform coating. Several examples of spheres with DNA located at opposite poles, like ice at the north and south poles of the Earth, are marked. If detailed analysis showed this to be the case, a DNA strand seems to prefer to attach to a site already occupied by DNA on the surface to a bare spot on the positively charged sphere.

Can the DNA, artificial polystyrene sphere system also replicate nucleosome structure on the scale occurring in nature? Polystyrene nanospheres can indeed be purchased commercially, with radius down to tens of nanometers and surface charge densities as ordered. However, it might be ill-advised to try such an experiment in a standard light microscope because it could not resolve tens of nanometers. Nevertheless, undergraduate students try many things, including the imaging experiments that lead to the data plotted in Figure 17.21. Addition of 16-µm lambda phage DNA to positively charged, 40-nm-diameter microspheres produces structures that can, indeed, be seen in a light microscope. In addition, the visible structures start forming 5–10 minutes or so after mixing of the samples but continue to grow over many hours. The structures (not shown) appear to have random, uninteresting shapes of many sizes. This is exactly what we would expect from the random aggregation discussed in this chapter. Figure 17.21 shows that this self-assembly of structure also results in a distribution of particle size. The plot of the number of observed aggregates in an image versus their average diameter shows a power-law dependence, with a slope almost equal to −1.0. Surprised?

Interpretation of "polar DNA" in the Figure 17.20 image must be done carefully, taking into account the (geometrically) larger number of fluorescent YOYO-1 molecules per pixel where the sphere surface is perpendicular to the image.

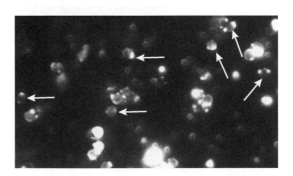

FIGURE 17.20

Short DNA (herring sperm DNA fragments, 0.2–0.6 µm) added to 1.0-µm polystyrene microspheres with positive surface charge (pH 7.2, 0.1 M NaCl), with YOYO-1 fluorescent dye added to visualize DNA. DNA does not uniformly coat the spheres, as indicated by the marked spheres, which have DNA concentrated at two poles. (From Nordlund, T., and many undergraduate students, unpublished.)

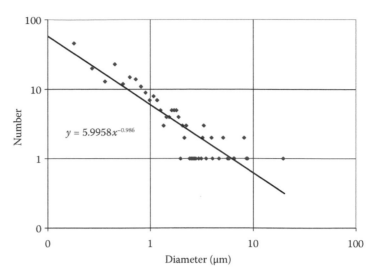

FIGURE 17.21
Aggregation of 43-nm polystyrene nanospheres (positive surface charge density) by short DNA (0.2–0.8 μm herring-sperm DNA) fragments. Aggregation is time dependent; data from an image recorded about 1 hour after addition of spheres to DNA. (From Nordlund, T. et al., unpublished.) The slope of the fit power-law line is very close to –1.

17.6 IMAGING THE ASSEMBLY OF NANOPARTICLES

Sections 17.3 and 17.4 presented diagrams and simulations asserting that nano- and microstructures can self-assemble in solution under the right conditions. The assembly of nucleosome-like structures of micron size in Section 17.5 supports these assertions, but it would be good to directly view the assembly of such structures on a nanometer scale. Scale, however, is the problem that has held such research back. While light microscopes can image structures down to about 200 nm and record video of the assembly at high speed, a true nm-scale that reflects assembly in biosystems is out of reach for optical imaging. Imaging methods such as atomic force microscopy, x-ray crystallography or electron microscopy can take over on this scale, but appropriate time resolution, 10s or shorter, has been out of reach until recently. Chapter 19 describes the time resolution afforded by AFM techniques, but AFM requires a molecule to stay in place while a probe repetitively scans over its structure: it cannot image particles freely diffusing in a liquid. In certain cases, bio-assembly takes place on surfaces, such as tyrosine kinase and other cell-signaling clusters.[23] Experimental approaches to mimic such assembly has been ongoing for about a decade. A library or Web search on terms such as "bio-assembly, self-assembly, genome packaging, genome or protein assembly" will turn up many recent publications. See, e.g., references 22.

When an electron beam enters water, *radiolysis* occurs, producing hydrated electrons (e_{aq}^-) hydrogen atoms (H), hydroxyl radicals (OH), H_3O^+, OH^-, superoxide, peroxide, and other reactive species. See Hart and Anbar, *The Hydrated Electron*, Wiley-Interscience, 1970, or Spotheim-Maurizot, M. (ed.), 2016, "Radiation Physics and Chemistry of Biomolecules: Recent Developments," *Radiation Physics and Chemistry*. Study of *radiolysis*—ionizing radiation effects on living tissue and water—grew rapidly after atomic weapons were invented, then evolved into studies of radiation treatment of cancers and materials processing.

The last decade has witnessed a significant expansion in the capabilities of transmission electron microscopy (TEM), in the areas of spatial and time resolution, and sample stabilization. The research group of Qian Chen, for example, has pioneered the technique of liquid-phase transmission electron microscopy TEM to the point where assembly of designer, nanometer-size particles in liquid solution can be observed on time scales of tens of seconds per frame.[24] Figure 17.22 shows an example of designer self-assembly of gold prisms, about 90 nm in side length and 7.5 nm in thickness. Gold material is used here because of efficient electron scattering, favorable surface-chemistry properties, and stability in solution. Like Figures 17.2, 17.3, and 17.5, these particles are designed with specific "sticky" points. Nanoprism surfaces are

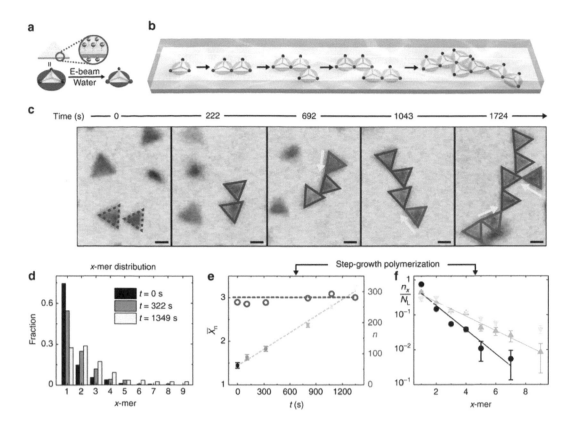

FIGURE 17.22

"Chains assembled from gold prisms via step-growth polymerization. (a) Schematics showing a single gold triangular prism coated with negatively charged thiols and labeled with spots at the three triangular tips. The spots are black when the prisms are in deionized water, meaning the repulsive cloud (red) envelopes the prism tips and renders the prisms non-reactive. Once the prism solution is illuminated by the electron beam at the appropriate dose rates (9.8–$25.8\,e^-$Å^{-2}s^{-1}) [electrons per square Angström per second], the prisms are rendered attachable at the tips (two spots changing to red) with the repulsive cloud shrunken inwards. (b) Stepwise tip-to-tip assembly schematic representing the connection scheme of the prisms highlighted in c. (c) Time-lapse liquid-phase TEM images showing the growth of a prism chain at a dose rate of $16.3\,e^-$Å^{-2}s^{-1} ... We overlay outlines on the prisms: dotted red lines for monomeric prisms before attachments, solid red lines for assembled prisms, and solid purple lines for prisms newly attached to the chain. The yellow arrows show the direction of prism attachment. (d) Distribution of x-mer, chains comprising x prisms, fraction changing over time (black: $t = 0$ s, green: $t = 322$ s, light emerald: $t = 1349$ s), which shows a shift toward higher aggregation number x. (e) The graph showing $\bar{X}_n$ (squares to the left y axis), the number-average degree of polymerization, growing linearly with time t, and the total number of prisms n (both unreacted and reacted, blue circles to the right y axis) remaining constant over time. The gray dotted line is the linear fitting of $\bar{X}_n$–t relation, while the blue dotted line is a guide to the eye. (f) A semi-log plot showing how the fraction of x-mers (n_x/N_L) is distributed at different assembly times (black circle: $t = 0$ s, light green up triangle: $t = 808$ s, emerald down triangle: $t = 1078$ s). The lines are the corresponding fit based on the Flory–Schulz distribution. Error bars denote standard deviations from counting. Scale bars: 50 nm." Qian Chen research group, University of Illinois at Urbana–Champaign. Figure 1, from Kim, J. et al., *Nature Commun.*, 8, Article no. 761, 2017; Creative Commons Open Access license http://creativecommons.org/licenses/by/4.0/.

first functionalized with -COO⁻ groups, making them electrostatically repulsive: they will avoid each other without further modifications. Placed into a liquid chamber sandwiched between carbon sheets, the prisms are free to move in two dimensions. Interactions with the chamber surface prevent true 3D diffusion, but also cause the particles to remain in the focal plane of the electron microscope. Irradiation of the sample with the electron beam allows the imaging, of course, but another effect is to monotonically increase the ionic strength, producing counterions, which screen the charge on the nanoprisms, reduce their repulsion, while interparticle van der Waals attraction remains unaffected. The net effect is to promote prism self-assembly.

In images such as Figure 17.22, the prisms are observed to effectively be reactive at the tips of the prisms (17.22a, b). The prisms appear to be divalent—two of three tips (red dots) are involved in growth into linear chains. Kim et al.[25] attribute this to steric hindrance effects between neighboring prisms. Part c of the figure shows example images of the growth of linear polymers of nanoprisms over a time of about 20 min. The time evolution of x-mers, part d, shows a progressive assembly, up to a 9-mer. The observation that smaller numbers of 2- to 8-mers exist even at time zero suggests other mechanisms for assembly. See the caption to the figure for more details.

What does such micro- and nano-structure assembly have to do with biology? While the artificial nucleosome-like structures in Section 17.5.2 seem to have some connection to biology, does assembly of irradiated gold nanoparticles? The first answer to this question is that basic principles governing nonliving nanostructures do not suddenly stop applying when "life" begins. The biosystem is indeed more complex, so additional mechanisms should be expected. Even if these components result in alternate assembly pathways, e.g., enzyme-assisted assembly, they must at least indirectly deal with the spontaneous, non-enzymatic routes. If components of an assembly spontaneously assemble themselves, perhaps in the wrong way, the bio-assisted pathway may have to outcompete the spontaneous. However, as we have seen many times in biosystems, nature does not usually block spontaneous, random, "nonliving" processes, but rather exploits, or at least tolerates, them. If light-harvesting complexes would tend to associate in membranes (Sections 10.3 and 10.4), nature will make use of this by tailoring the aggregation. Charge and light-sensitive interactions adjust the natural association tendency to the advantage of the organism: aggregation is adjusted to increase or decrease photosynthetic light harvesting.

> It is also of historical interest to note that in the 1946 publication, O. Weisz was identified as "(Miss) O. Weisz." Readers can debate the significance of this specific identification of an author as an unmarried woman. It does document the early entry of women in 20th-century chemical physics, then biophysics research.

The relevance of intelligent nanomaterial investigations to biology also relates to discussions in Chapter 1, that major advances in the understanding of biological structure and mechanism often begin with advances made by physicists (or engineers) such as Crick and Wilkens, Kendrew and Perutz, Gilbert, Delbrück, Deisenhofer, Neher and Sakman, Lauterbur and Mansfield, etc. In a 2016 Science News article,[26] reporter Meghan Rosen introduced Qian Chen as a "Taylor Swift of science," one of the year's top 10 young scientists to watch. Chen, a new (2015) assistant professor leading the research discussed in reference 23, began nanoparticle-assembly work in about 2008. According to Rosen's interview, however, her goal is "attempting to see cellular machinery with a clarity no scientist has achieved before... 'I use this as inspiration,' Chen says, grabbing her laptop and starting up a video that may well be the fantasy of anyone exploring biology's secret world. The computer animation shows molecules whizzing and whirling deep inside a cell. Gray-green blobs snap together in long chains and proteins haul giant, gelatinous bags along skinny tracks." Before Max Perutz determined the first atomic-resolution structure of a protein molecule, hemoglobin, he similarly worked on applying X-ray crystallography to molecules such as sodium chloride, then tribromo(trimethylphosphine)gold, then on to biomolecules. In 1946, it is interesting to note, Perutz and Weisz wrote, "Compounds of 2-covalent aurous gold were known to have linear structure...whereas derivatives of 4-covalent auric gold have a planar structure."[27] Now, isn't that curious? At about the same time Perutz set his sights on the x-ray crystal structure of hemoglobin he was investigating the structure of small gold-containing molecules that assembled themselves into different molecular conformations, depending on details of the atomic groups attached to the gold atom. Fourteen years later, after publishing many papers about the path to obtaining an X-ray structure, Perutz et al. published "Structure of Hæmoglobin: A Three-Dimensional Fourier Synthesis at 5.5-Å. Resolution, Obtained by X-Ray Analysis."[28] The positions of atoms in the biomolecule were then known.

The lessons here relate to the importance of fundamental physics approaches for advances in understanding the structure and function of biological systems. The reader will explore this more in homework. Much more can be said about fundamental biological self-assembly research, especially in the area of genetic, biochemical and computational approaches, but as this is a physics text, we will leave that to others.

17.7 POINTS TO REMEMBER

- Hierarchical structure requires hierarchical construction, starting from assembly of amino acid and nucleotide units to make protein and nucleic acid polymers.
- Levinthal's paradox—that there is insufficient time for biopolymers to interchange between all possible conformations—implies that biomolecular structure is not in true thermodynamic equilibrium.
- An energy landscape picture provides a coherent approach to understanding structures of individual protein molecules.
- Cooperativity, the built-in coordination of structure and dynamics through long-range coupling of structural units, is a fundamental structural property of enzymes.
- Assembly of larger intracellular structures has an important entropic contribution, enforced by osmotic and restricted-space constraints. Segregation (aggregation) of structures a few tenths of a nanometer in size is spontaneous.
- Unrestrained entropic assembly of isotropic units results in fractal-like growth and size distributions. Intrinsic asymmetry of units (e.g., "sticky ends" of peptide units) produces classes of structure (e.g., linear polymers).
- Nucleosomes—wrappings of DNA around positively charged histone-protein spheres—are the first level of structural compaction and organization of the cell nucleus. Both Coulomb and entropic forces are important contributors to nucleosome structure. "Beads on a string" is a common structure formed by DNA and positively charged spheres.
- Liquid-phase TEM, along with intelligent fluorescence microscopy, AFM and other methods, should soon allow us to see biomolecular structures assembling in space, in real time.

17.8 PROBLEM-SOLVING

PROBLEM 17.1   ENTROPIC ASSEMBLY OF CUBES

Eight cubes of side 12.0 nm form the structures as shown in Figure 17.23. In solution with the cubes is a 2.0 mM concentration of a protein of radius 1.00 nm. Consider only the entropic (osmotic) depletion forces and free energies.

1. What is the lowest free-energy structure?
2. Calculate the free energies of the structures relative to the lowest energy.

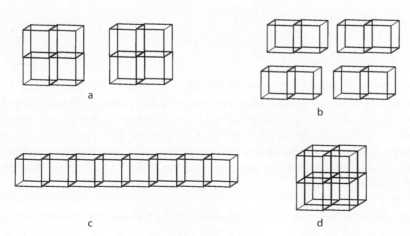

FIGURE 17.23
(a–d) Arrangement of assembled cubes for Problem 17.1.

PROBLEM 17.2 TWO-DIMENSIONAL MEMBRANE AGGREGATES

Assume a population of identical, roughly cylindrical monomers (radius R) have sticky patches located at certain points on their surfaces. Except for the sticky patches, the monomers are hydrophobic—their surfaces prefer to be surrounded by the interior portions of the membrane, away from water.

1. These monomers are components of a membrane-bound, aggregated "ion gate" that forms *symmetric*, hexameric rings in a membrane. Draw a diagram of this aggregate, preferably with your computer.
2. Determine the location of the sticky patches on the monomers.
3. Determine the approximate radius of the hole in the middle of the aggregate. (Draw a diagram to show how you define the radius.)
4. Describe three types of forces not that would make these patches stick to each other.

PROBLEM 17.3 CHARGE-COUPLED TWO-DIMENSIONAL MEMBRANE AGGREGATES

Assume that the two-dimensional, hexameric aggregates in the previous problem are protein monomers held together by electric forces and that the aggregates are located in the membrane of a roughly spherical cell, always with a specific end of the monomers at the outside surface of the membrane. (The cylinders span the membrane.)

1. Draw a diagram showing the location and sign of charges holding the aggregate together.
2. Draw another charge arrangement that would hold the aggregate together.
3. Knowing what you know about amino acids and pKs, state what amino acids might be responsible for the charges in one of your arrangements if variation of the external pH (outside the cell) in the range pH 5 to pH 9 causes the aggregate to convert between aggregated and unaggregated. State the pH at which the transition should occur. (If you have to redesign the charge arrangement in parts 1 and 2 to accommodate this scenario, do so.)

PROBLEM 17.4 THREE-DIMENSIONAL AGGREGATES

Assume a (three-dimensional) solution, initially made up of identical, cubic monomers with two adjacent sticky sides. When two sticky sides bond together, the free energy *decreases* by an amount $\Delta G_0/n$, where n is the number of monomers already in the aggregate. If a sticky side is opposed to a nonsticky, the free energy *increases* by $\alpha\Delta G_0$, where $0 < \alpha < 1$. Assume any two bonded cubes have full-face bonding. As you assemble an aggregate, discuss any issues of the order of monomer assembly on the final aggregate free energy.

1. Draw the structure of the smallest, stable structures that form over time. ("Stable" indicates that no sticky sides are still exposed to the solvent.) What is the free energy of this structure, compared to the monomeric state? (Hint: the aggregate should be planar.)

2. Draw the next larger planar aggregate. What is the aggregate's free energy? Under what conditions is this aggregate more stable than that in part 1?
Draw the next larger stable three-dimensional aggregate. What is its free energy?

PROBLEM 17.5   FRACTAL DIMENSION

Determine the fractal dimension of the following, after appropriately defining the aggregate average radius:

1. Spheres of radius r assembled in a two-dimensional hexagonal close-packed arrangement on a surface. Assemble the structure as follows: start with one sphere; pack six spheres around it; place 12 spheres around those, packing as closely to the center as possible.
2. A linear string of 10 spheres of radius r, each contacting its neighbors.
3. The freely jointed chain in three dimensions.
4. A circular ring of six spheres of radius r placed symmetrically on a circle of radius R. (Note the minimum value of the ratio R/r that allows such placement.)

PROBLEM 17.6   AGGREGATE POSSIBILITIES

Membrane adherence versus cytosolic aggregates. The entropic interactions shown in Figure 17.11 induce the larger particles to either aggregate together or move to the cell membrane, if no specific interactions are present.

1. Describe molecular interactions that would favor adherence to the cell membrane over aggregation of the large particles to each other. Be as specific as possible: if you propose Coulomb interactions, for example, state what molecular groups might be responsible.
2. Describe interactions that would favor aggregation of the large particles to each other.

MINIPROJECT 17.7   NUCLEOSOME CHARGE

Like most active areas of research in the biosciences, tomorrow's "facts" in nucleosome structure and dynamics will not be the same as today's. Investigate further the charge distribution of nucleosome core particles (NCPs). You can start with the references in the text, but find one or two others to support or correct.

1. Find the net charges on the histone core (octamer), the DNA wrapped around the histone core, and the histone "tails" associated with the NCP. Find the overall net charge.
2. Search the Protein Data Bank for X-ray data on nucleosomes that includes ions bound to the complex. Recalculate the net NCP charge, including these bound counterions. Where are these bound ions located? By what percent is the overall charge reduced?
3. Draw a surface map that shows the locations of charges that are within ± 1 nm (or your choice of distance) of the surface. Note any correlation between locations of DNA phosphate and protein side chain charges.
4. Discuss any pH effect on charge groups.

MINIPROJECT 17.8 CHARGE-CONTROLLED CHROMATIN MODIFICATION

Online searches using Google Scholar often turn up student research dissertations. These can be very useful for finding important references, but dissertations have not gone through as thorough a peer-review process as journal publications. Consult your instructor for more information.

Section 17.5.1.1 described research indicating that control of histone charge likely plays an important role in chromatin remodeling. The research in references 15 and 16 is now over a decade old, so significant advances in the understanding of this phenomenon are almost certain.

1. Locate references 15 and 16, as well as all more recent publications of these research groups on the topic of histone charge and chromatin modification. Download the reference citations, including the abstracts. Construct a mini-bibliography of these publications, marking with a "*" what you believe to be the more important papers. (Important papers can be recognized in a variety of ways: the prestige of the journal, a journal editor's comments in the same journal issue, statements of the authors, citations by other researchers, including review articles (consult results of part 2 that follows) and your own judgment.
2. Locate at least six additional significant publications on this topic by other research groups and add these to your mini-bibliography.
3. Select a review article from your bibliography. (Find one, if you don't already have one.) Summarize what the reviewer considers to be the most important results. Keep in mind that review articles often have rather vague abstracts, making it necessary to read the article more carefully.
4. List the techniques that have been applied to the problem of histone charge modification: molecular modeling, X-ray crystallography, chromatography, electrophoresis, etc.
5. Describe, in 75–100 words, what you believe to be an interesting and reasonable thesis or dissertation topic on chromatin and histone charge. State the goal of the project, the problem you intend to solve, your approach to attaining the goal and the primary research tool(s) needed. State whether you believe this to be appropriate as a senior or master's thesis or Ph.D. dissertation. Note that a review of current literature alone cannot constitute a thesis or dissertation.

PROBLEM 17.9 AGGREGATION (GRADUATE PROBLEM)

Self-aggregation of 50-nm spheres and 16-µm-long lambda phage DNA.

1. Compute the diffusion constant of 40-nm-diameter polystyrene microspheres in water at 25°C.
2. Compute the hydrodynamic (average) radius of 16-µm DNA from a freely jointed chain or other appropriate model. From this, compute the average diffusion constant of DNA in bulk water at 25°C.
3. Which particle diffuses faster?
4. Making the approximation that the slower-diffusing particle remains fixed, discuss the expected aggregation process when spheres are added to a dilute DNA solution, were DNA molecules are separated by distances >>16 µm.
5. Repeat part 4 for the case when DNA separations are comparable to DNA lengths. What bulk DNA concentration (M and particles/m^3) does this correspond to?

PROBLEM 17.10 THE ADVANCE OF SELF-ASSEMBLY RESEARCH

Investigate recent advances in the assembly work of Chen and colleagues (Section 17.6). You may want to employ bibliography software such as Endnote, Zotero, BibMe, etc.

1. Starting with reference 23, find the most recent publications of the group related to self-assembly of designed nanoparticles or biological structures. Copy into your assignment the complete reference, including publication abstract.
2. From each publication (but no more than five), select, from its own reference list, one or two of the most important publications that the Chen group cites. Find these references and copy the complete reference, including abstract.
3. Describe, in a bulleted list (10–20 words each), the advances in biological self-assembly since the 2017 Kim et al. publication. Note if an advance involves chaperonin-like structures that facilitate assembly.

PROBLEM 17.11 YOUNG SCIENTISTS TO WATCH

1. Obtain the articles in the *Science News* 2016 190(7) and 2017 192(6) (or two later years) entitled "The SN 10: Scientists to Watch." (Many college libraries subscribe to *Science News*; consider subscribing yourself for quick online access, sciencenews.org.) Make a table of the scientists' names, locations (institutions, current first), department(s), short (20-word) sketch of scientific interest(s) and a 20-word note of how physics and math play a role. For example,

Name	Location(s)	Department(s)	Scientific interests	Physics/math role
K .C. Huang	**Stanford**, MIT, Princeton	Bioengineering, Microbiol. & Immunology	Understanding the physical challenges bacteria face, photosynthetic and gut bacteria, antibiotics	Ph.D. in Physics, computer simulations, microfluidics, micromechanics.

   If you need to, search the institution's Web site to find the departments.
2. What fraction of the researchers are involved in bioscience research? In what fraction of the bioscience cases do math and physics play a significant role?

REFERENCES

1. Nierhaus, K. H., and D. N. Wilson, Eds. 2004. *Protein Synthesis and Ribosome Structure: Translating the Genome*. Weinheim, Germany: Wiley-VCH Verlag.
2. Spirin, A. S. 2000. *Ribosomes*. New York: Springer.
3. Garrett, R. A., Ed. 2000. *The Ribosome: Structure, Function, Antibiotics, and Cellular Interactions*. Washington, DC: AMS Press.
4. Cambre, K. Ribosomes, from the album *The Guy You Cheer For* (Kinder Gentler Records, 2006).
5. Stox, D. Ribosomes, from the album *Dj Tools, vol. 2*. (G Records, 2008).
6. Stiban, J., M. So, and L. S. Kaguni. 2016. Iron-sulfur clusters in mitochondrial metabolism: Multifaceted roles of a simple cofactor. *Biochemistry* (*Moscow*) 81(10):1066–1080.

7. Karnkowska, A., V. Vacek, Z. Zubáčová, S. C. Treitli, R. Petrželková, et al. 2016. A eukaryote without a mitochondrial organelle. *Current Biology* 26(10):1274–1284.

8. Hayashi, T. and A. G. Szent-Gyorgyi, Eds. 1966. *Molecular Architecture in Cell Physiology*. Englewood Cliffs, NJ: Prentice-Hall.

9. Blackburn, E. H., C. W. Greider, and J. W. Szostak. "The 2009 Nobel Prize in Physiology or Medicine: For the Discovery of "How Chromosomes Are Protected by Telomeres and the Enzyme Telomerase," http://nobelprize.org/nobel_prizes/medicine/laureates/2009/press.html.

10. Booth, C. R., A. S. Meyer, Y. Cong, M. Topf, A. Sali, S. J. Ludtke, W. Chiu, and J. Frydman. 2008. Mechanism of lid closure in the eukaryotic chaperonin TRiC/CCT. *Nat Struct Mol Biol* 15:746–753.

11. Feder, J. 1988. *Fractals*. New York: Springer.

12. Böker, A., Y. Lin, K. Chiapperini, R. Horowitz, M. Thompson, V. Carreon, T. Xu, et al. 2004. Hierarchical nanoparticle assemblies formed by decorating breath figures. *Nat Mater* 3:302–306.

13. Witt, S., J. Neumann, H. Zierdt, G. Gébel, and C. Röscheisen. 2012. Establishing a novel automated magnetic bead-based method for the extraction of DNA from a variety of forensic samples. *Forensic Sci Int: Genet* 6(5):539–547.

14. Nguyen, T. T., and B. I. Shklovskii. 2001. Complexation of DNA with positive spheres: Phase diagram of charge inversion and reentrant condensation. *J Chem Phys* 115:7298–7308.

15. Cherstvy, A. G. 2009. Positively charged residues in DNA-binding domains of structural proteins follow sequence-specific positions of DNA phosphate groups. *J Phys Chem B* 113:4242–4247.

16. Yang, Y., A. P. Lyubartsev, N. Korolev, and L. Nordenskio. 2009. Computer modeling reveals that modifications of the histone tail charges define salt-dependent interaction of the nucleosome core particles. *Biophys J* 96:2082–2094.

17. Ren, Q., and M. A. Gorovsky. 2001. Histone H2A.Z acetylation modulates an essential charge patch. *Mol Cell* 7:1329–1335.

18. Henikoff, S. 2005. Histone modifications: Combinatorial complexity or cumulative simplicity? *Proc Natl Acad Sci USA* 102:5308–5309.

19. Pennings, S., S. Muyldermans, G. Meersseman, and L. Wyns. 1989. Formation, stability and core histone positioning of nucleosomes reassembled on bent and other nucleosome-derived DNA. *J Mol Biol* 207:183–192.

20. Fu, Y., M. Sinha, C. L. Peterson, and Z. Weng. 2008. The insulator binding protein CTCF positions 20 nucleosomes around its binding sites across the human genome. *PLoS Genet* 4:e1000138.

21. Matulis, D., I. Rouzina, and V. A. Bloomfield. 2000. Thermodynamics of DNA binding and condensation: Isothermal titration calorimetry and electrostatic mechanism. *J Mol Biol* 296:1053–1063.

22. Krishnan, R., T. Jaleel, and T. Nordlund. "Imaging of DNA/Nanosphere Condensates" (paper presented at March Meeting of the American Physical Society, Los Angeles, CA, 2005).

23. Coyle, M. P., Q. Xu, S. Chiang, M. B. Francis, and J. T. Groves. 2013. DNA-mediated assembly of protein heterodimers on membrane surfaces. *J Am Chem Soc* 135(13):5012–5016; Yarden, Y., and M. X. Sliwkowski. 2001. Untangling the ErbB signalling network. *Nat Rev Mol Cell Biol* 2(2):127–137.

24. Naldi, M., E. Vasina, S. Dobroiu, L. Paraoan, D. Nicolau, and V. Andrisano. 2009. Self-assembly of biomolecules: AFM study of F-actin on unstructured and nanostructured surfaces. *Proc SPIE* 7188:Q1–Q8; Meyer, R., B. Saccà, and C. M. Niemeyer. 2015.

Site-directed, on-surface assembly of DNA nanostructures. *Angew Chem Int Ed* 54(41):12039–12043; Yeates, T. O. 2017. Geometric principles for designing highly symmetric self-assembling protein nanomaterials. *Ann Rev Biophys* 46(1):23–42.

25. Kim, J., Z. Ou, M. R. Jones, X. Song, and Q. Chen. 2017. Imaging the polymerization of multivalent nanoparticles in solution. *Nature Commun* 8(761):1–10.

26. Rosen, M. (2016). Scientists to watch: Qian chen. Making matter come alive. *Science News* 190(7):16–18.

27. Perutz, M. F., and O. Weisz. 1946. The crystal structure of tribromo(trimethylphosphine) gold. *J Chem Soc (resumed)* 0:438–442.

28. Perutz, M. F., M. G. Rossman, A. F. Cullis, H. Muirhead, G. Will, and A. C. T. North. 1960. Structure of hæmoglobin: A three-dimensional fourier synthesis at 5.5-Å. resolution, obtained by X-Ray analysis. *Nature* 185:416–422.

29. Ditzel, L., J. Löwe, D. Stock, K. O. Stetter, H. Huber, R. Huber, and S. Steinbacher. 1998. Crystal structure of the thermosome, the archaeal chaperonin and homolog of CCT. *Cell* 93(1):125–138.

30. Luger, K., A. W. Mäder, R. K. Richmond, D. F. Sargent, and T. J. Richmond. 1997. Crystal structure of the nucleosome core particle at 2.8 Å resolution. *Nature* 389(6648):251.

31. RCSB PDB software references:

    Jmol: An open-source Java viewer for chemical structures in 3D. http://www.jmol.org/.

    Protein Workshop: Moreland, J. L., A. Gramada, O. V. Buzko, Q. Zhang, and P. E. Bourne. 2005. The Molecular Biology Toolkit (MBT): a modular platform for developing molecular visualization applications. BMC Bioinformatics 6:21.

    NGL Viewer: Rose, A. S., A. R. Bradley, Y. Valasatava, J. D. Duarte, A. Prlić, and P. W. Rose. 2018. NGL viewer: Web-based molecular graphics for large complexes. Bioinformatics. doi:10.1093/bioinformatics/bty419.

CHAPTER 18

CONTENTS

Preparation for Experimental Biophysics

18

We focus in this chapter on how a new biophysicist obtains the insight to formulate an important biological hypothesis, to determine optimal experimental approaches, and to locate the needed facilities. The intent is not to expound on "the most important methods" and explain their inner workings, so that the new biophysicist can walk into a lab and attempt, for example, time-resolved electron microscopy of an H1N1 influenza virus invading a host cell. Books devoted to methods, as well as review articles and research experience courses are the proper source for such training.[1] We do delve into details and biophysical applications of atomic force microscopy, one of the twenty-first century's most revolutionary experimental methods, in Chapter 19.

Don't investigate the spherical cow. A physicist who is interested in biological things—sometimes called a *biological physicist*—might be expected to calculate and measure quantities and properties that have traditionally occupied physicists for the last century or so. While a good many biological physicists indeed measure such things such as force, velocity, kinetic energy, torque, length, volume, elasticity, etc., most of these quantities are meaningful only when the biosystem of interest has been well characterized. Calculation of a mitochondrion's volume could not be done if a mitochondrion had not yet been imaged or isolated, and a classically trained physicist would have little idea how to obtain one. Determination of the coordinates of all the atoms in DNA cannot be done before someone knows how to isolate and purify DNA. It would then appear that a biological physicist is perpetually "behind the biological curve" in comparison to biologists, biochemists, molecular biologists, and other bioscientists. We argued at the beginning of this text that such a conclusion might be warranted if the physicist did not collaborate with scientists possessing the needed biological experience. Such a collaborative effort could then be immensely fruitful, not only because the physicist brings beneficial mathematical and physical-experimental skills, but she might also have ideas that never occur to biochemist colleagues.

We are also brought back to the perennial question: What constitutes a "biosystem" that is interesting and manageable? Organisms such as bacteria or birds are biosystems, but they can be treated, physically, neither as particles nor as materials, as physicists have traditionally studied. The physicist who demands to measure or calculate a quantitative, clean physical descriptor of a biological object must have an object that is amenable to such measurement, leading to the tired biophysics joke, "Consider a spherical cow," or perhaps a model of bacteria as rockets (Figure 18.1). Up until perhaps recently, one could legitimately assert that physicists who wanted to measure physical quantities suffered from two serious weaknesses.

1. Physicists were not capable of generating or caring for a "good" biological sample on their own. As many of you know, the education of a research physicist is not easy. Perhaps until recently, a physicist was expected to have fluency in the classics: advanced mechanics and dynamics, electromagnetics, electrical circuits, sound, energetics, quantum theory, thermodynamics, fluids, optics, relativity, a bit of elementary-particle theory and cosmology, not to mention the expected advanced mathematics. The years of college coursework a research physicist needed for training in these disciplines, along with expected laboratory work, usually numbered more than nine, including the time for non-physics electives. This did not traditionally leave much time for a biological physicist to formally learn the intricacies and art of biology and biological sample cultivation, unless a research career was to be postponed until near age 30. These coursework demands are still severe, although biological physics curricula have been developed to weave in biological and biochemical training, as well as directing core physics courses toward biological applications.[2]

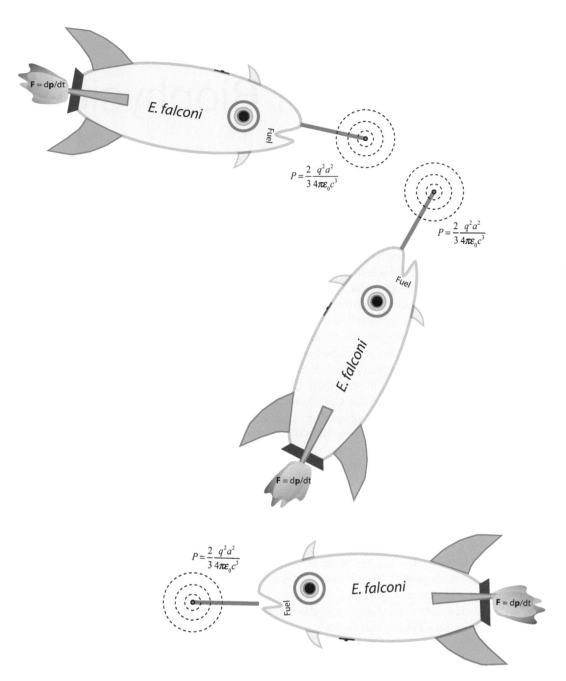

FIGURE 18.1
"Consider a spherical cow... or a rocket bacterium." Hypothetical bacteria modeled by a physicist uniformed by biological learning. Important capabilities, such as a motile, directionality, sensing, communication and nutrition are contained in the model, but it is uninformed by biological knowledge. The equations are for rocket thrust and electromagnetic transmission power.

Even great-minded physicists such as Einstein, Pauli, and Schrödinger, who each directed some thought to biology or medicine,[3] could probably not be relied upon to care for a cell culture or direct a student in a productive biophysics project.

2. Physicists did not know what the important biological questions were. Courses typically encountered in a classical physics curriculum rarely exposed students to important developments in the biosciences. Even those determined to learn some biological science could often incorporate no more than three or four semesters of biology and three or four of chemistry, leaving them at the intermediate biology/chemistry stage. Students in modern biophysics tracks, in contrast, sometimes have decent exposure to upper-level biology and chemistry, but at the expense of laser/optical, computational, and quantum physics, which we assert are critical to modern biophysical research.

> Biomedical engineering or physics curricula are often directed toward device and materials development, leaving some deficits in all the areas—biology, chemistry, laser/optical, and computational and quantum physics.

So, how can the budding biophysicist develop the competence and experience to recognize and pursue important questions in the physics of biosystems? The most effective solution is probably the same as 40 years ago: do research as soon as possible with a group that has experience in a bioscience topic that interests you. However, some preparation is in order.

18.1 WHAT DO YOU WANT TO FIND OUT, WHAT TOOLS MIGHT YOU NEED, AND WHERE MIGHT YOU FIND THESE TOOLS?

This section intends to help the reader identify what might be an interesting and fruitful foray into the field of biophysical research, including academic, industrial, computational and experimental research. A clear argument can be made that experimental (laboratory) and computational research cannot be separated, or if they are, important insights will be overlooked. If this is so, the researcher who sits in a cold room purifying biological samples on a chromatography column should also be proficient in computational/theoretical modeling or should collaborate with someone who is. The current sophistication of modeling theories and computational software, as well as the complex protocols for biological sample study makes a strong argument for collaborations.

We earlier reviewed the research effort described in A. Wagner's *Arrival of the Fittest* (2015, Penguin Group). Forty years ago, such research would be classified as evolutionary theory, because few methods were capable of experimentally addressing what has occurred in the development of even the simplest forms of life on earth.[4] A quick reading of the Wagner book will show that such efforts now employ massive supercomputer modeling (guided by theory), sophisticated generation of many thousands of organismal genotypes and phenotypes, with laboratory production and testing of some fraction of the microbial organisms. Little reliable progress could ever be made in the absence of either computation or experimentation.

18.1.1 Searching Scientific Meeting Web Sites for Experimental Methods

In the following, we illustrate how to search for important biophysical phenomena and the experimental approaches and tools the enquirer can use to productively investigate them. In most cases, such a search will reveal a combination of experimental and theoretical/computational tools.

We will begin the search by examining the American Physical Society (APS) Web site, www.aps.org, Meetings and Events, chosen because of the large number of contributors to the reported biophysics research. A similar search can be done on societies in China, Japan, Europe, India, and many other countries around the world, but APS meetings have the advantage of size and large numbers of papers from countries other than the US. For example, the search below will show 160 contributions from the Chinese Academy of Sciences (though few in the area of biophysics). One can also do a productive search on the Biophysical Society meeting Web site, www.biophysics.org/meeting-events, and similarly find many hundreds of presentations.

> In all scientific conferences, one notes a proportionally larger number of contributions from the geographical area of the meeting, because of travel expenses. Attending four days of a conference will often cost more than US$1,000, plus travel and meal expenses. Students can often attend for several hundred dollars less.

> Note that the time of your Web search is important. The APS and Biophysical Society meetings typically take place in February or March of each year. The complete meeting epitome will generally be posted sometime in December of the year prior to the meeting.

On your computer, browse to www.aps.org; select "Meetings and Events, March Meeting." A search done in May, 2018, results in

APS March Meeting 2018, Los Angeles, California, March 5–9, 2018.

The APS March Meeting 2018 brought together more than 11,000 physicists, scientists, and students from all over the world to share groundbreaking research from industry, universities, and major labs.

18.1.1.1 Explore the meeting

* View Meeting Presentations
* View the Schedule of Events
* View the Scientific Program

Click on "View the Scientific Program, "Sessions Index. Hundreds of sessions will appear. (A topical search will reduce the number.) For 2018, for example, click on Focus session A01, *Advances in Scanned Probe Microscopy I.* Notice the session sponsoring unit is "GIMS" (Group on Instrument & Measurement Science), with chair Joseph Stroscio of NIST. This already gives you useful information. The GIMS unit is one of the places to look for scanning probe microscopy (e.g., AFM) and Joseph Stroscio of NIST (National Institute of Standards and Technology) is one of the recognized experts in the area.

Click on abstract A01.00001: Data Mining in Scanning Probe Microscopy. The authors are Dusch, Banerjee, Pabbi, Snelgrove and Hudson, from the Department of Physics, Pennsylvania State University. The abstract of the presentation states, in part,

> *This necessitates the development of new instrumentation and techniques for data analysis, in order to better understand this new wealth of data. Fortunately, these developments are paralleled by the general rise of big data and the development of a number of data mining techniques which can extract hidden structure from large datasets. In this talk, I will describe the use of unsupervised learning techniques (e.g. clustering, feature extraction, and spectral mixing) that can identify patterns in data by simultaneously analyzing multiple variables... supported by the National Science Foundation under Grant No. 1229138.*

You now know the identity and location of one of the reputable scanning probe research groups and the fact that they specialize in data mining—new instrumentation and techniques for data analysis. The designation of "reputable" can be seen by the location of the research at a reputable university, as well as support from the National Science Foundation (USA), which funds only a small fraction of the research proposals submitted to it. The reader may not have realized that data mining was even a part of scanning probe microscopy (SPM), or that a person with a primary interest in computer big data analysis could be part of an SPM research effort. Our first search finding illustrates that SPM experiments in biophysics are not carried out by only persons who prepare a DNA sample and place a single molecule on a flat gold surface.

The reader should now go back to the session index and click on "Filter Sessions by Sponsor," filter by "DBIO," the Division of Biological Physics. This will restrict the many hundreds of scientific sessions down to 14 sessions and about 80 subsessions, and has the advantage of accessing most (but not all) of the biophysics research. These DBIO sessions include "Physics of Proteins, Robophysics, Physical Force Regulation of Cells, Emergent Dynamics in Neural Systems, Self Organization in the Cytoskeleton, Physics of Genome Organization, Motion and Jamming of Cells, Physics of Development and Disease, Controlling Space and Time in Biology, Self-Assembly in Liquid Crystals and other Complex Solvents, Physics of Behavior, and many others. Clicking on the session S50, "Physics of Proteins IV: Intrinsically Disordered and Aggregated States of Proteins," we see an invited presentation by Jennifer Hurley: "The Importance of Disorder in the Highly Ordered Circadian Clock." Invited speakers are generally leaders in their field of research, so the reader can read the presentation abstract and search for more information at Hurley's listed Web site and search your library for published material. Those thinking of applying to a graduate program with a research opportunity in biological order/disorder will now have identified one possibility.

In session B43: Polyelectrolyte Complexation I: Self-Assembly, Abstract B43:00004, *Structure and Properties of Complex Coacervate Core Micelles*, one sees a contribution by Heo, Kim, Lee and Choi, from Hongik and Chungnam National University, which states, "In this study, we investigate the structure of C3M as a function of molecular weight of charged block, pH, and salt concentration by dynamic light scattering (DLS) and small-angle X-ray scattering (SAXS)." The reader has now discovered a research group in Korea which uses light and X-ray scattering to investigate assembly of micelles. Therefore, light and X-ray scattering are good ways to investigate micelle structure.

Further investigation of the experimental method of interest. Suppose one decided to focus more narrowly on APS March, 2018, Meeting Abstract E01.00012: *Hybrid System of Atomic Force*

Microscopy and Optical Spectroscopy: Photo-induced Force Microscopy, authored by Bongsu Kim, Ryan Khan and Eric Potma, in Session E01: *Advances in Scanned Probe Microscopy II,* chaired by X. Chen of Tsinghua University. Such focus may be desired by one especially interested in modern microscopy, and secondarily in the biological problems themselves. The abstract of the presentation reads,

> *Photo-induced force microscopy (PiFM) is a variant of the multifrequency atomic force microscopy that measures the force between the sample and a tip induced by a laser. The PiFM technique uses a mechanical resonance of the probe to amplify the photo-induced force, enabling the detection of forces in the (sub-) picoNewton range under ambient conditions. Furthermore, by measuring a gradient of the force through detection at side-band frequencies background contributions are suppressed and sensitive measurements of the photo-induced gradient force can be made. When a femtosecond light source is used to illuminate the tip-sample junction, it is possible to record the nonlinear optical response of nanoscopic objects with PiFM. In this talk, we introduce the principles of PiFM and present recent results in ultrafast spectroscopy at the nanoscale that are enabled by force detection.*
> **This work supported by the National Science Foundation through grant CHE-1414466.*

One now has the names of three persons who may be considered possible experts in PiFM, as well as their location. More often than not, the presenter at a conference is a more junior investigator, so we will search on the last coauthor, Eric Potma. A general Web search on "Eric Potma California Irvine" brings the university's Faculty Profile System, where we note Potma's appointments and background. Research Interests reads, "molecular mobility in biological materials, vibrational coherences, ultrafast microscopy." The link to the group Web page leads to more details, including "Sum-frequency microscopy of biopolymers at laser-scanning imaging speeds," "Amplifying coherent anti-Stokes Raman scattering images by more than a million times," and "the lighter side of chemistry: General chemistry explained through cartoons," but since no force microscopy links are evident, we search on the first coauthor's name, Bongsu Kim, which leads to the Web page of the CaSTL center (Chemistry at the Space-Time limit), where Kim is listed in the Potma group as studying photo-induced force microscopy with ultrafast microscopy. A bit of poking around leads to a 2014 National Science Foundation video explaining the "Chemiscope," a version of which was apparently used in the 2018 Kim et al. APS presentation. See www.youtube.com/watch?v=nhqAc0 or www.nsf.gov/news/special_reports/science_nation/chemiscope.jsp. Figure 18.2 has two frames from this video, showing the ultrafast laser and the probe parts of the apparatus. One can presently (May 2018) download the video at the NSF site. Watching this video does not, of course, give one the training needed to actually use the Chemiscope for a biophysical project, but neither would a book chapter focused on either ultrafast laser spectroscopy or scanning probe microscopy. Further investigation of the experimental approach must be done by finding published details, which the reader can do in Problem 18.2.

> The reader should be aware that the (non-peer-reviewed) abstracts of scientific meetings sometimes over-promise the contents of the presentation a bit. Many times, the described work has not actually been completed at the time of abstract submission, some 3–4 months before the conference.

The above search procedure points the reader who already has interest in a biophysical question to the methods that are currently used for its investigation. The reader who has a specific interest in industrial research (i.e., a job in industry), can go back to the DBIO session index and Filter Sessions by Tag-"Industry," bringing up sessions including "B61: Meet Your Future: Careers in the Private Sector." Panelists for this lunch session include Director of Inertial Navigation at AOSense, quantum engineers at Rigetti Computing and Google, and Manager, Novel Computing technologies at IBM. Using the "Undergraduate" filter tag, the reader can locate the session "L16: Major Physics Organizations and Their Role in the Future of Physics." And abstract L16.00001, "Light Sources for Africa, the Americas & Middle East Project (LAAMP)." The abstract informs the reader that such light sources make crystallography possible. If, for example, one has an interest in time-resolved X-ray crystallography (short-pulse light source) of biomolecules and lives in Africa, this program should be investigated.

Finally, one should also do a session search filtered by other sponsors, such as DPOLY, the Division of Polymer Physics, DCP, the Division of Chemical Physics, DCMP, Condensed Matter Physics, DCOMP, Computational Physics, or DFP, Fluid Dynamics. A number of DPOLY sessions

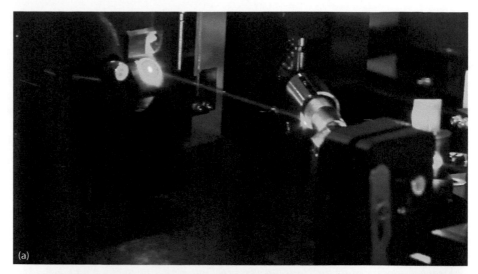

FIGURE 18.2

The Chemiscope, used in the Kim et al. presentation E01.00012: *Hybrid System of Atomic Force Microscopy and Optical Spectroscopy: Photo-Induced Force Microscopy* at the 2018 March Meeting of the American Physical Society. With support from the National Science Foundation (NSF), the Center for Chemistry at the Space-Time Limit (CaSTL at the University of California, Irvine, chemist Ara Apkarian, director. (a) The laser. (b) The scanning probe tip.

related to biophysics appear, including "Solar Energy Conversion," "Knots and Nanopores," and "Physics of Genome Organization: from DNA to Chromatin." (Some sessions are jointly sponsored by multiple sponsors, including DBIO, but some are not.)

18.1.2 Imaging: Searching a Biological Institute Website

Why is (micro)imaging so important to modern bioscience? First, human perception and understanding of the world is largely visual, based on images (movies) of events that take place. The many pictures of the structures of proteins, nucleic acids, cells and subcellular structures have hopefully given the reader considerable intuitive understanding to the words, diagrams, and equations that otherwise fill the chapters of this textbook. While geneticists and some other researchers often think of explanations for disease and organismal character and behavior in terms of genes, basic understanding of any phenomenon of life on earth must rely on the knowledge of the important physical structures that make up organisms. While large structures—bodies, bones, organs, etc.—have been observable to the naked eye for thousands of years, today's questions about living organisms largely lie at the submicron scale. Enzymes that carry out chemical reactions, as well as most other fundamental active and structural components, reveal the source of their structure and function at the nanometer scale.

Video micro-imaging of bacteria and cells has been available to bioscientists for about forty years, but until recently, the spatial resolution has been limited to about a micron and larger, and timescales of seconds or more. This has hampered study of fundamental questions in biology, since the active players are <100 nm in size and change on timescales of milliseconds or shorter. In recent years, video imaging has progressed to these scales, moving the forefront of the biosciences along with it. The biophysicist unaware of these current experimental and computational methods will be at a distinct disadvantage to those more informed.

In addition to the thousands of independent and collaborative research group research efforts across the globe, there are many institutes set up to focus on cutting-edge research in the biosciences, using the most modern methods and instrumentation. We will look at the European Molecular Biology Laboratory (EMBL), which has six locations in Europe: Heidelberg, Germany; Hamburg, Germany; Barcelona, Spain; Grenoble, France; Hinxton, U.K.; Rome, Italy. Member/Associate countries are Argentina, Australia, Austria, Belgium, the Czech Republic, Croatia, Denmark, Finland, France, Germany, Greece, Hungary, Iceland, Ireland, Israel, Italy, Luxembourg, Malta, the Netherlands, Norway, Portugal, Spain, Slovakia, Sweden, Switzerland and the United Kingdom. The Web site is www.embl.org. A useful introductory video description can be found at www.youtube.com/user/emblmedia. Especially watch the Imaging Technology Center introductory video, but there are dozens of other useful video shorts, such as, "What does a physicist do in a biology lab?" and "How to give a good scientific talk." We will concentrate in this short section on imaging, using a combination of optical fluorescence and electron microscopy. Good introductions are "Seeing is understanding" (https://www.youtube.com/watch?v=8ECHbgonbU0) and "Cell Biology and Biophysics" (https://www.youtube.com/watch?v=egfML2xmo28 or https://youtu.be/egfML2xmo28?t=12), all of which can be found at www.youtube.com/user/emblmedia/videos. The former describes the (as of September, 2017) future EMBL Imaging Technology Center, which combines light and electron microscopy. The Cell Biology and Biophysics video states some of the developments and applications of these imaging techniques to subcellular structures and events.

Optical microscopy. The standard optical microscope (Figure 18.3), consisting of a sample positioning stage with sample placed on a 1 mm-thick glass slide covered with a thin (~0.1 mm) cover slip, illumination light, objective and eyepiece lenses, and focusing mechanism, is normally covered in a first-year physics course. In this older-style microscope, illumination comes from below, through the glass slide, with imaging lenses above the sample. Such positioning allows for objects to be gently held on the sample stage by gravitational forces. Usable microscopes can be found in toy stores or ordered from scientific suppliers or manufacturers, though the price will vary by a factor of 1,000. The more expensive and capable research microscopes put the greatest effort into quality of lenses, precision adjustment mechanisms and options for introduction of (laser) illumination beams. High magnification requires a short objective focal length and is assisted by placing the objective almost in contact with the sample cover slip, with a film of oil or water between the lens and cover slip, reducing reflections and aberrations through better optical index matching. Such lenses are referred to as oil- or water-immersion lenses, the latter being quite pricey. An issue for sample integrity is the placement of sample, presumably suspended in an aqueous medium, between the glass surfaces of slide and cover slip. Depending on the amount of water, the distance between surfaces may range from 50 to 5 microns, often causing interaction between sample and glass. (Some microorganisms can be squashed.)

The actual image is usually formed by light scattered from the sample (bright-field imaging). Spectral filters can be placed in the excitation and emission beam lines, allowing a separation of emission from excitation light and imaging of sample fluorescence. As discussed in Section 9.6, synthetic fluorescent dye technology has become very sophisticated. Polarization and other filters allow emphasis and imaging of dichroism and birefringence characteristics of sample. Cameras or video cameras for imaging may take the place of the viewing eyepiece assembly or a mirror may reflect the image to a side port-mounted camera.

Optical magnification of such microscopes, approximately the product of the magnifications of objective and eyepiece lenses, has been dealt with in the appropriate science course, with result

Crystallography can have 0.1 nm resolution, but it requires the construction of a crystal, which often requires hours to months. Except for photo-initiated events within the molecules of the crystal, crystallography cannot be classified as a (video) imaging technique.

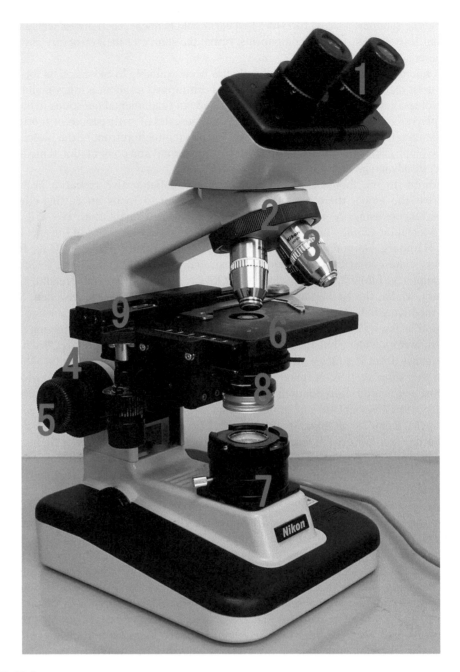

FIGURE 18.3
A standard microscope. The simplest compound microscope, using scattered light for imaging.
1: Eyepiece lens(es). 2: Turret for lens selection. 3: Objective lenses. 4–5: Focus adjustments (coarse, fine). 6: Sample translation stage. 7: Illumination light assembly. 8: Filters, polarizers, etc. for excitation light. 9: Translation stage x and y adjustments. Microscopes used for biophysical research purposes are often "inverted" and used for fluorescence, with objective lens placed beneath the sample stage. Excitation and emitted fluorescence light both pass through the objective lens and emission is separated from excitation with a dichroic mirror. Excitation may be from a laser, which is then focused to a small spot for excitation (and emission).

$$M \approx -\frac{(L - f_e)N}{f_o f_e} \tag{18.1}$$

where L = distance between eyepiece and objective lenses; N = distance between eye and the eye's near point; and f_o (f_e) = objective (eyepiece) lens focal length, and the negative sign indicates an inverted image. Research microscopes, with multiple lenses in both the objective and eyepiece assemblies, are more complicated, of course, with magnification formulas provided by the manufacturer. Usable magnifications range from about 50 to 1,500. Object sizes down to about 0.5 microns can be resolved by standard microscopes.

Fluorescence microscopes, which dominate in biological optical microscopy, are often inverted, with both excitation source and objective lenses placed below the sample positioning stage. If a standard glass slide and cover slip are used to contain the sample, the slide must be inverted, cover slip downward and in contact with the immersion objective lens. When there is a need for long *working distance* ($\geq$0.5 mm or so) between objective lens face and sample, choices of high-magnification and highly corrected objectives are limited. Long working distances might be desired, for example, if the user wants the thickness of a glass slide (~1 mm) between sample and objective lens. Plastic chambers, with cover slip bottoms, have been developed to contain liquid samples in inverted microscopes. The liquid will be held in the chamber by gravity. Free access to the chamber from above is possible and the sample volume can be 1 ml or more. The object to be imaged must, of course, be near the bottom of the chamber, or it could not be in the focal plane of a short working-distance objective. Such inverted microscopes have become quite sophisticated in the last two decades, with options for introducing multiple filters, laser beams for fluorescence excitation or even combined optical trapping and imaging.

Light sheet microscopy (multiview selective-plane illumination microscope).[5] The last 10 years has witnessed an explosion of new sorts of optical microscopes, some of which allow spatial resolution of 100 nm or better and some of which combine optical and other microscopies, such as AFM or TEM. The EMBL Imaging Technology Center specializes in this sort of microscopy. One of their recent developments has been the light sheet microscope or the multiview selective-plane illumination microscope (MuVi-SPIM), Figure 18.4. This microscope look radically different from the standard one, but one can get an idea of the function by thinking of how the image is formed in a standard optical microscope. In the simplest case, the excitation light illuminates the entire sample object, with the scattered light from the entire object collected by the objective. Light from the object side opposite the objective passes through the object and is attenuated and scattered by the object's body before being collected and relayed to the camera. One then gets an image that is a conglomeration of light directly scattered and light multiply scattered and attenuated. Usually, subcellular structures cannot be resolved. In the case of a fluorescence microscope, similar effects occur, but a fluorescence labeling molecule may be selectively bonded to a molecule or structure within the cell.

If an emission optical filter selective for the spectrum emitted by this dye is used, that object is selectively emphasized, usually by factors of thousands or more. Most images of cells that have a dark background are of this sort. Still, the image is a combination of light emitted from all parts of the object, reducing the resolution of specific parts of the object. Confocal fluorescence microscopes allow resolution of parts of small objects by focusing excitation light on a small part of the object, using either an excitation-light focusing lens and aperture or focused laser beam and a matching aperture on the emission side. (Other schemes are also used.) The emission aperture is positioned so that light from the directly illuminated spot is preferentially imaged. Fluorescence and other light scattered from other parts of the sample is mostly blocked by the aperture. By moving the illumination spot in x, y and z directions, structure throughout the object can be selectively measured. If an image of the entire object is desired, the spot must be moved (x, y, z) over the entire object, allowing a 3D structure to be assembled by the microscope-associated computer. Examples of imaging by a confocal microscope, along with detailed protocols for sample preparation, can presently be found at https://www.ncbi.nlm.nih.gov/pmc/articles/PMC3197300/.

The light sheet microscope accomplishes spatial selectivity and 3D image construction in a quite different way. Illumination is done by lasers, in which a sample is illuminated from the side with a thin light sheet (Figure 18.4d) while the emitted fluorescence is imaged by an objective lens that is perpendicular to the light sheet. The sample holder is rather more complex than a glass slide with cover slip (Figure 18.4d), since access to the sample by microscope lenses must be possible from four sides. Use of lasers, rather than light bulbs, allows use of high beam collimation and coherence to create highly tailored patterns of light. Two collinear illumination arms are used to illuminate the sample. Two detection arms on a perpendicular axis are used to collect the fluorescence generated in the light sheet and record two images with two image sensors. The MuVi-SPIM creates a virtual light sheet by scanning a laser beam through the field of view of the detection lens (Figure 18.4e). To produce focused images, the light sheet must illuminate only a thin section of the sample in the vicinity of the detection objective's focal plane. The two light sheets produced by the

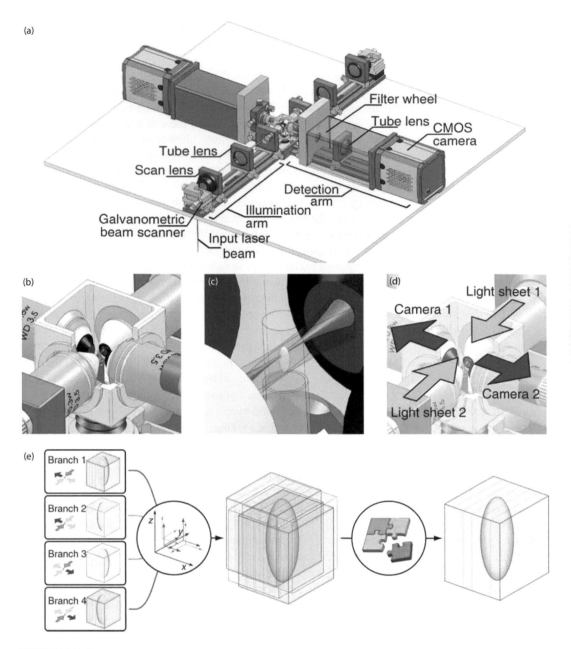

FIGURE 18.4

The Multiview selective-plane illumination microscope. "(a) MuVi-SPIM setup, consisting of two illumination and two detection arms arranged along two perpendicular axes. (b) The water-filled experimental chamber containing the agarose-mounted specimen at the intersection of the illumination and detection axes of the four objectives. (c) Specimen mounted in a cylindrical block of agarose and illuminated with a light sheet from one of the illumination arms. (d) The two illumination arms (light sheets 1 and 2) and the associated detection arms (camera 1 and 2). Orange arrows indicate the illumination direction and red arrows indicate the direction of the collected emission. (e) The four possible illumination-detection combinations (branches) for acquiring four 3D multiview images. Images are first rapidly transformed to a common coordinate system using a predetermined set of transformation parameters and then synthesized into a single high-content image." (Figure 1 from Krzic, U. et al., *Nat. Methods*, 9, 730–733, 2012. Reproduced by permission.)

two illumination arms are aligned with each other using two galvanometric scanners in such a way that they illuminate exactly the same section. The waist of the light sheet (where the light sheet is the thinnest) is positioned near the intersection of the illumination and detection axes by axially translating the two illumination objectives using two z positioners. The two detection objective lenses are then focused onto the illuminated volume using another set of z positioners. Each detection arm resembles a standard epifluorescence microscope. The addendum to this chapter describes all features of the light-sheet microscope in more detail. The reader should look again at https://www.youtube.com/watch?v=egfML2xmo28 to see examples of images produced by this microscope. An end-of-chapter problem will explore this microscope's comparative capabilities.

As one might guess from Figure 18.4, the MuVi-SPIM microscope cannot presently be purchased for the typical biology course laboratory, which is why the EMBL Imaging Technology Center supports outside users.

The main goal of Section 18.1 has been to provide the reader search tools to identify and locate the experimental tools and methods needed to investigate a wide variety of scientific issues and problems in biophysics. Our approach has been to first identify the problems of interest, then find the useful experimental and computational tools. A person who is determined to use a specific experimental tool, such as submicron optical imaging, can also search conference and major university or research institute (e.g., EMBL, NIST [U.S.], NIH [U.S.], CAS [China], IIT [India], Royal Society [U.K.], RAS [Russia], IMCB [Singapore], and more) Web sites for information. This will be left to a homework problem.

> In an epifluorescence microscope, both the illumination and emission light travels through the same objective lens.

18.2 TOOLS ALREADY DESCRIBED IN THE BOOK

Throughout this book, particular experimental methods and tools have been introduced as they were needed to understand molecular structure or the behavior of a biomolecule or biosystem. Molecular structural display software, introduced in Chapter 5, is a core analysis tool for a wide variety of researchers in the biosciences seeking to answer basic questions about the molecular basis for biological activity. This online software also constitutes an important learning tool for students entering most bioscience fields.

On the other hand, the experimental tools that provide the provide the Protein Data Bank (PDB), Nucleic Acid Data Bank (NDB) and other databases with their molecular information—X-ray crystallography, 2D- and other multidimensional nuclear magnetic resonance (NMR), electron microscopy—have been largely passed over. A practical reason for this (intentional) oversight is that most introductory (especially one-semester) biophysics courses and readers cannot afford the time to learn the rather complex devices, software, and multi-step analyses that are used today to perform NMR, crystallography, electron microscopy and other structural analysis. Undergraduate and graduate students seeking a first research experience with these investigative tools spend their first several months reading specialized books and manuals, often under the tutelage of experienced graduate students and post docs.

While this tutored introduction to experimental tools is also common to other research methods, most of those already described in this book can be learned at a functional level in days or weeks by a student with two or more years of college science coursework, especially physics and chemistry. These "simple" tools include sedimentation, chromatography, optical spectroscopy, chemical reaction analysis, and calorimetry. Methods based on the interaction of light with matter have had more extended coverage in this book because of their ubiquity in bioscience and biomedicine, because of the noninvasive nature of optical data-collection protocols, because they demand little sample, because optical methods can be simultaneously used for sample stimulation, detection and imaging, and because managing light is a daily task of many physicists. We have devoted some time to introduce atomic force microscopy (AFM), primarily because its principles come from introductory physics and important results on biosystems have emerged from AFM in the last two decades. The entire final chapter of this text describes AFM.

The locations of previous discussions of various experimental methods in this book are listed in Table 18.1, along with references to a few omitted but widely employed techniques such as electrophoresis and dialysis.

TABLE 18.1 Location of discussed experimental methods

Method	Common Applications	Location (chapter.section)
Software for macromolecular structure	Molecular display, geometric and chemical analysis, reactivity prediction	5.1
Sedimentation (centrifugation)	Molecular mass, sample purification	6.1.1, 12.3, 12.6
Chromatography	Sample purification, charge analysis, buffer exchange	6.1.2
Optical spectroscopy:		
Measuring light quantities	Sample identification, quantitation, and integrity check, imaging	9.3.1
Optical absorption	Sample identification, quantitation, and integrity check; rapid reaction sensing	6.2.2, 9.5, 9.7, 10.2.5, 11.1.2, 11.3, 11.6.2.1, 11.7.1
Light scattering	Micro-object size determination, aggregation quantitation, corneal and lens clarity measurement	9.4.1
Light refraction	Optical microscopy, index-matching	9.4.2
Fluorescence (emission spectra)	High-sensitivity molecular detection, energy and electron movement, molecular interactions, rapid reaction sensing	9.6, 9.8, 11.4, 15.11.1
Fluorescence Recovery After Photobleaching (FRAP)	Molecular diffusion in membranes	6.4.4.1
Energy Transfer (FRET)	Distance measurement (<nm), molecular binding	9.9, 10.4.7
Photon diffusion	Imaging in tissue	11.6.2.2
Optical traps (laser tweezers)	Molecular force measurement (proteins, nucleic acids, molecular motors)	9.4.3, 15.11.1, 16.2.2
Atomic Force Microscopy	Single-molecule molecular imaging and force measurement	13.7.2, 15.11.2; all of Ch 19
Calorimetry	Equilibrium and reaction energetics	14.5
Reaction analysis	Determination of rate parameters, activation and equilibrium energies	15
Dialysis	Buffer exchange, removal of small molecules	(*Wikipedia*: "Dialysis (biochemistry)"; thermofisher.com: search "pierce-protein-methods," "Nucleic Acid Purification and Analysis Support Center")
Electrophoresis	Electric charge and mobility, sample identification (e.g., DNA), separation of molecules	(*Wikipedia*: "Electrophoresis (disambiguation)"; thermofisher.com: search "pierce-protein-methods," "Nucleic Acid Purification and Analysis Support Center"

18.3 THE ROLE(S) OF A COMPUTER IN EXPERIMENTAL BIOPHYSICS

In preparing for a course or career in the experimental biosciences, a critical skill is the collection and cultivation of experimental data. Virtually all modern experimental methods involve a measuring device that produces a digital file, which must be saved. Even more critical is the experimentalist's skill at organizing and, when appropriate, archiving such data. Even in cases where the recording instrument autosaves the date and time, instrumental settings, operator's name, and other information, the user must record a variety of information special to the particular experiment. This includes things like the date of sample preparation, sample source (e.g., a commercial company or a collaborator), a reference or link to a detailed description of the sample preparation, collaborators for the experiment, any problems or issues that you note in the sample preparation or data recording (e.g., the building temperature varied, possibly affecting the sample or a device's stability; the water deionizer just had a filter change; a heavy truck drove by during data collection) locations for primary and backup data storage. As you can see, the data management problem can be daunting.

> The same data collection and organizing skills are also needed for theoretical/computational simulations of biophysical phenomena, but we will not discuss them here.

- Management of data file types. Most modern devices collect and store data in a specialized format, determined by the device manufacturer to be most convenient, efficient, and as compatible as possible. A wide variety of such formats can be found. Look on Wikipedia, for example, for *mass spectrometry data formats* and on the Protein Data Bank for biomolecule structure formats. Microscope manufacturers have specific file types for recorded images, which may be high-resolution (pixels) still-image frames, along with times between image snapshots and shutter time of snapshots, or video files compacted by various protocols. Assembly of still frames into a video and/or management of compressed video files are issues the user must take seriously and understand clearly. There are four or more "standard" formats for two-variable spectroscopic data (intensity *vs.* wavelength or other): .csv, comma-separated values; .dx, J-CAMP spectroscopy format; .spc, Galactic Industries GRAMS format; .xls, Excel.

  With this variety of data formats and the multiple files that will generally be created by the user, some rules for data management are in order. The following should be treated as minimal rules, subject to rules that may already be in place at the data-recording site.

- Rule 1. Data in its originally recorded format must be preserved. In many cases (e.g., government grant–supported, data with legal, patent, or medical implications) preservation of this data may be required, sometimes for years. File conversions to more user-friendly applications, also add a danger of data scrambling or information deletion, especially when assembling multiple data files into one, for multi-sample graphing. For example, J-CAMP files may lose date and other information on conversion to Excel. Data files in original format are the most likely to be archived in a difficult-to-access place, as it may not be in a format conducive to publication formats (graphs, images, etc.). As an aside, since the computer recording the primary data is the most important, it should be kept offline as much as possible and continuously backed up, to avoid data loss to viruses or machine malfunctions.

  Remember to convince yourself and your theoretical colleagues that the original data is holy and not to be messed with.

- Rule 2. Some type of computer spreadsheet, database, or database-management system must be developed. The spreadsheet will usually be on the user's computer or that used by multiple users. Often, this will be decided by the laboratory you are working in, but if you must develop your own, recorded information should include things like the following:

Experimenter Name, Signature	Sample Identifier(S)	Sample Prep Data	Data Recording Date	Data File Name	Computer With Original Data	Computer With Converted Data	Notes	...
T. Nordlund T. Nordlund	SDNA12.7	(link)	12oct17	FN.xlsx	(Name of computer)	(link)	(link)	

As you may expect, information like sample prep data can be fairly extended, making a simple spreadsheet more difficult to manage. A real database management system is advisable, or a linked spreadsheet. For example, for users of MS Office, Excel and Word can create links from single words to other files containing extended information. Links to separate files must be curated carefully, however, as linked files can be lost during data copy. A single folder containing all associated files is advisable. Adobe Acrobat can also do this, as well as combining primary file and linked files into one combined file. HTML is, of course, the original master of linking many subfiles to a master file. Links to data stored elsewhere on your or other computers, like converted or combined files, is also recommended.

- Rule 3. Filenames must be informative. Names are no longer limited to eight characters, with file extension, so the filename should be understandable to a collaborator with whom you have little contact. For example, DNA4.xlsx does not give the reader a clue to sample or the data, whereas Sac_nDNA_fluor_DMJ_8apr16.xlsx does: *Saccharomyces cerevisiae* yeast nuclear DNA, fluorescence recorded by D. M. Jones on 8 April, 2016. A balance must be made, of course, as such a long filename may be informative, but is also easy to mistype. From personal experience, the author (TN) can affirm that data files with uninformative filenames can result in the need to re-record data near the time of publication submission.
- Rule 4. If you collaborate with others, give them a useful filename interpreter and any needed database or spreadsheet information.
- Rule 5. Evaluate your data as you record it. If your data is being fit to a theory, fit the data as soon as possible. Have a colleague help, when necessary. Biophysics experiments usually rely on several instruments operating correctly. If the temperature-control device has gone bad, you want to know after you collect the first data file, not after 50 files and 12 hours of work. A theoretical computer data fit can be used to help run the actual experiments successfully and efficiently. See Rule 6.
- Rule 6. "Bad" data files must be dealt with carefully. The declaration that some data is bad should not be done two days after the data is collected, when you decide the data is not consistent with the last two experimental runs you made. If data processing of one recorded data file indicates disagreement with previous, the experimenter should determine the possible cause as soon as possible. For example, if the laser wavelength has possibly shifted, check it now.

18.4 THE CENTRALITY OF STRUCTURE, MOTION, AND IMAGING

At the beginning of this book, we asked the question, "Is biophysics obsolete in the age of genetics and engineering?" The answer was an unambiguous, and I hope convincing, "No," but we should remind ourselves what the central goals of the biosciences are. These goals are, in fact, a bit different than those of medicine. The prime aim of (human) medicine is to maintain and improve the health of people. The precise mechanisms of the improvement are secondary, though important, issues. Excluding a certain category of medical technicians, most of the medical personnel one encounters in a clinical setting actually do try to understand the mechanisms of the treatments they employ, but they have rarely digested the latest research. In addition, much of the research they have mastered is practical and correlative rather than mechanistic: does cause X correlate with symptom Y; does treatment A relieve disorder B. In contrast, the bioscientist seeks to understand the structures and mechanisms behind biological or biomedical phenomena.

Readers planning to seriously enter a field of the physical biosciences should spend time in the Annual Review of Biophysics every year. Biophysicists of all levels of experience could usefully spend a week studying the 2017 reviews. In 2017 (Vol. 46) the following reviews were published, many on topics we have discussed in this and previous chapters.

- *Progress and Potential of Electron Cryotomography as Illustrated by Its Application to Bacterial Chemoreceptor Arrays*, by Ariane Briegel and Grant Jensen
- *Geometric Principles for Designing Highly Symmetric Self-Assembling Protein Nanomaterials*, by Todd O. Yeates
- *Weighted Ensemble Simulation: Review of Methodology, Applications, and Software*, Daniel M. Zuckerman and Lillian T. Chong
- *Structural Insights into the Eukaryotic Transcription Initiation Machinery*, Eva Nogales, Robert K. Louder, and Yuan He
- *Biophysical Models of Protein Evolution: Understanding the Patterns of Evolutionary Sequence Divergence*, by Julian Echave and Claus O. Wilke
- *Rate Constants and Mechanisms of Protein–Ligand Binding*, by Xiaodong Pang and Huan-Xiang Zhou

- *Integration of Bacterial Small RNAs in Regulatory Networks*, by Mor Nitzan, Rotem Rehani, and Hanah Margalit
- *Recognition of Client Proteins by the Proteasome*, by Houqing Yu and Andreas Matouschek
- *What Do Structures Tell Us About Chemokine Receptor Function and Antagonism?* by Irina Kufareva, Martin Gustavsson, Yi Zheng, Bryan S. Stephens, and Tracy M. Handel
- *Progress in Human and Tetrahymena Telomerase Structure Determination*, by Henry Chan, Yaqiang Wang, and Juli Feigon
- *Theory and Modeling of RNA Structure and Interactions with Metal Ions and Small Molecules*, by Li-Zhen Sun, Dong Zhang, and Shi-Jie Chen
- *Reconstructing Ancient Proteins to Understand the Causes of Structure and Function*, by Georg K. A. Hochberg and Joseph W. Thornton
- *Imaging and Optically Manipulating Neuronal Ensembles*, Luis Carrillo-Reid, Weijian Yang, Jae-eun Kang Miller, Darcy S. Peterka, and Rafael Yuste
- *Matrix Mechanosensing: From Scaling Concepts in'Omics Data to Mechanisms in the Nucleus, Regeneration, and Cancer*, by Dennis E. Discher, Lucas Smith, Sangkyun Cho, Mark Colasurdo, Andrés J. García, and Sam Safran
- *Structures of Large Protein Complexes Determined by Nuclear Magnetic Resonance Spectroscopy*, by Chengdong Huang and Charalampos G. Kalodimos
- *How Active Mechanics and Regulatory Biochemistry Combine to Form Patterns in Development*, by Peter Gross, K. Vijay Kumar, and Stephan W. Grill
- *Single-Molecule Studies of Telomeres and Telomerase*, by Joseph W. Parks and Michael D. Stone
- *Soft Matter in Lipid–Protein Interactions*, by Michael F. Brown
- *Single-Molecule Analysis of Bacterial DNA Repair and Mutagenesis*, by Stephan Uphoff and David J. Sherratt
- *High-Dimensional Mutant and Modular Thermodynamic Cycles, Molecular Switching, and Free Energy Transduction*, by Charles W. Carter, Jr.
- *Long-Range Interactions in Riboswitch Control of Gene Expression*, by Christopher P. Jones and Adrian R. Ferré-D'Amaré
- *RNA Structure: Advances and Assessment of 3D Structure Prediction*, by Zhichao Miao and Eric Westhof
- *CRISPR–Cas9 Structures and Mechanisms*, by Fuguo Jiang and Jennifer A. Doudna
- *Predicting Binding Free Energies: Frontiers and Benchmarks*, by David L. Mobley and Michael K. Gilson

There are perhaps a few other topics or problems in biophysics which could be classified as of highest significance, but readers who have even partially grasped the above articles should be assured that the field of modern biophysics has not passed them by.

18.5 POINTS TO REMEMBER

- Think about your interests in biophysics, especially using the resources of international conferences.
- Know how to locate current research expertise in specific bioscience fields. Use resources like the American Physical Society March meeting.
- Imaging is one of the fundamental tools of the biophysicist. Know how to find the best imaging tools for your biophysical interests—confocal microscopy, light-sheet microscopy, TEM, X-ray crystallography, NMR imaging, etc.
- Develop data-management tools for collecting, organizing and archiving experimental digital data. Especially if your plans include biomedical or industrial applications, investigate legal requirements and best practices for such data management.
- Published annual reviews are good sources of current information on biophysics phenomena and experimental tools of interest. Examples are the Annual Review of: Biophysics, Biophysics and Biomolecular Structure, Biochemistry, Cell and Developmental Biology, Immunology, Microbiology, Neuroscience, Physical Chemistry, Plant Biology, and more.

18.6 PROBLEM-SOLVING

PROBLEM 18.1   STUDIES OF ASSEMBLY

Return to the aps.org "Meetings and Events" March Meeting site and locate a session on Assembly. For example, the 2018 meeting session C13 title is "Assembly and Behavior of Hierarchical Materials." (One will note a presentation of Qian Chen's work, which was discussed at the end of Chapter 17.)

1. Select an experimental presentation that has some biological relevance; copy and insert the presentation's abstract and other details into your homework assignment.
2. Locate an additional publication containing details of the instrumentation and method. Download and insert this publication into your assignment. (Hint: Some abstracts contain a published reference, which may facilitate your search.)
3. Summarize, in your own words, what the instrumentation is, how it works, and the data it records.

PROBLEM 18.2   THE CHEMISCOPE: FINDING MORE INFORMATION

Locate the sources for more information on the "Chemiscope" described in Section 18.1. Find a published description of the instrument and its capabilities, download and attach it to your homework assignment and summarize, in your own words, the instrument's primary measuring capabilities. Describe how the microscope operates to get its data.

PROBLEM 18.3   SUPER-RESOLUTION OPTICAL IMAGING

At the March APS meeting or other (bio) scientific meeting Web site, locate a presentation on very high-resolution (nanoscale) and/or single-molecule optical imaging in a biosystem or biologically relevant system. For example, session S59 of the March 2018 APS meeting, "Super Resolution Microscopy and Lithography of Polymers," contains five invited presentations, all of which are related to biological systems.

1. Copy the presentation details and insert into your assignment.
2. Summarize the methods and results in your own words. Discuss the biological importance of the described work.
3. Locate and download a published description of the method and apparatus used in the study. Insert this into your assignment. Describe how the instrumentation works. If the instrument is novel or improved, describe the innovative features.

PROBLEM 18.4 MICROSCOPE COMPARISON

(If, after an extended search, the indicated EMBL video clip [or comparable] is not available at the time of this assignment, locate another example of biological light-sheet micro-imaging, or discuss with your course instructor.)

Access the Web site https://www.youtube.com/watch?v=egfML2xmo28 again, and find, within the video, the example of light-sheet microscope's images of the drosophila embryo. The narrative notes that the microscope allows for imaging of the whole embryo in a fraction of a second, allowing a high-resolution video of development to be made for an extended period of time.

1. Download or otherwise copy this video.
2. Extract four video frames showing various stages in embryo development.
3. Determine the magnification (distance scale) of the images.
4. Locate and obtain several "brightfield" (scattered light) images of drosophila embryo development at stages comparable to those in parts 1–3 above. The samples may be stained with dyes, but only for contrast, not for fluorescence imaging. Compare and contrast to the light-sheet images.
5. Locate and obtain three or four fluorescence images (confocal or other) of drosophila embryo development at stages comparable to those in parts 1–3 above. Note that "GFP flies" have the Green Fluorescent Protein engineered into their genes, to facilitate fluorescence imaging. Compare and contrast your images to the light-sheet images.

Note that YouTube does not now allow for simple downloading of videos any longer. You may have to locate the clip of embryo development elsewhere or use a screen-capture program (e.g., Screenflick). You are not to make use of such videos outside of your own personal (homework) use, of course, except by permission of the copyright holder.

PROBLEM 18.5 RESEARCH INSTITUTE AND SOCIETY SEARCHES

Locate the Web sites of all the research institutes and societies mentioned at the end of Section 18.1. If none of these include the area in which you reside, work, or attend school, locate one within 500 miles or so of your location.

1. Locate the divisions or departments of each institute that deal with biophysical research issues. Attach those division names, along with the listed biophysics research areas, to your assignment. Briefly discuss your interpretation of the meaning of "biophysics research."
2. Find the names of two important persons associated with each division.
3. Find a recent publication from each division. Find and copy the abstract of the paper or presentation, or if none is available, summarize the research goals or findings.
4. Determine whether the institute employs or trains international workers and students.
5. If you were in a position to associate with one of these institutions or societies, which would you choose. Briefly explain.

ADDENDUM

The Multiview selective-plane illumination microscope (MuVi-SPIM)

The following detailed description of the operation of the MuVi-SPIM comes, by permission, from

Krzic, U., S. Gunther, T. E. Saunders, S. J. Streichan, and L. Hufnagel. 2012. Multiview light-sheet microscope for rapid in toto imaging. *Nature Methods* **9**: 730+.

ONLINE METHODS

Optical setup. The selective-plane illumination microscope is a fluorescence microscope with which a fluorescent specimen is illuminated from the side with a thin light sheet (Supplementary Figure 18.1) while the emitted fluorescence is imaged by an objective lens that is perpendicular to the light sheet. Multiview SPIM consists of four optical trains, or "arms," arranged along two horizontal axes that intersect at a right angle (Figure 18.1a). Each of the four arms ends with a microscope objective lens that is focused onto a specimen that lies in the intersection of the two microscope axes. Two collinear arms are used to illuminate the specimen and are therefore called illumination arms. Two arms on the axis perpendicular to the former are used to collect the fluorescence generated in the light sheet and record two images with two image sensors. They are referred to as detection arms.

MuVi-SPIM implementation creates a virtual light sheet by scanning a laser beam through the field of view of the detection lens (a method sometimes referred to as digital scanned laser beam light-sheet microscopy, or DSLM), but different methods could also be used in MuVi-SPIM: for example, a static light sheet (6) or a scanning Bessel beam (19). An illumination arm consists of a galvanometric scanner (VM500+, Cambridge Technology Inc.), scan lens (S4LFT0061/065, Sill optics GmbH and Co. KG), tube lens (TI-E 1x, Nikon Instruments Inc.) and objective lens (CFI Plan Fluor 10x/0.30W, Nikon Instruments Inc.). Each illumination arm is supplied with an illumination laser beam. A set of lasers with different wavelengths (Calypso 491 nm, Samba 532 nm, Jive 561 nm, Mambo 594 nm, all from Cobolt AB) is coupled into a single beam using a set of dichroic beam splitters (LaserMUX, Semrock Inc.) fed through an acousto-optical tunable filter (AOTFnC-400.650, A-A Optoelectronic) that controls the illumination intensity and wavelength, and is coupled into an optical fiber (kineFLEX, Qioptiq Inc.) that transports the light to the microscope. There the beam is collimated (0.7-mm beam diameter) and then diverted into the desired illumination arm using a motorized flip mirror (KSHM 40, Owis GmbH). See Supplementary Note 5 for light-sheet properties.

Each detection arm resembles a standard epifluorescence microscope and consists of an objective lens (CFI Apo 40x/0.80W or Apo LWD 25x/1.1w, both from Nikon Instruments Inc.), filter wheel (FW-1000, Applied Scientific Instrumentation) with a set of emission filters (LPD01-488RS-25 and BLP01-594R-25, both from Semrock Inc. and HQ525/30, Chroma Technology Corp.), tube lens (NT47-740, Edmund Optics Inc.), and sCMOS camera (Neo, Andor Technology plc.). A 160-mm achromatic converging lens is used as a tube lens, which results in 32x effective magnification. At this magnification, the image of a typical Drosophila embryo optimally covers the sCMOS image sensor.

Water-dipping objective lenses used in the four arms require the specimen to be immersed in an aqueous medium. This is made possible by a special experimental chamber featuring four flexible nitrile-rubber membrane seals (Supplementary Figure 18.2a). The chamber can thus be filled with a liquid while the objective lenses and the chamber remain mechanically uncoupled. This is crucial for the precise positioning of the objective lens and long-term stability of the microscope. Unlike in previous SPIM implementations, the specimen is held in the chamber from below using an original two-stage flexible seal. The latter consists of a Teflon-ring seal, which permits specimen rotation, and a flexible rubber tube, which allows translation of the specimen relative to the chamber. Coarse specimen translation is done using a manual two-axis linear stage (M-401 with two SM-13 Vernier micrometers, Newport Corp.) and motorized vertical stage (M-501, Physik Instrumente GmbH & Co. KG), whereas fine positioning is done by a two-axis piezo stage (P-628.2CL, Physik Instrumente GmbH & Co. KG). The latter is also used to move the specimen through the light sheet to produce the 3D images of the specimen.

To produce focused images, the light sheet must illuminate only a thin section of the sample in the vicinity of the detection objective's focal plane. In MuVi-SPIM, the two light sheets produced by the two illumination arms are aligned with each other using two galvanometric scanners in such a way that they illuminate exactly the same section. The waist of the light sheet (that is, the location along the light sheet where it is the thinnest) is positioned near the intersection of the illumination and detection axes by axially translating the two illumination objectives using two z positioners (G061063000, Qioptiq). The two detection objective lenses are then focused onto the illuminated volume using another set of z positioners (same as above). One of the detection objective lenses also allows for precise lateral translation along the two directions perpendicular to its optical axis (xy stage KT 65, Owis GmbH) in order to make both detection arms image precisely coincident rectangles. Lateral positioning of the illumination objective lenses is not necessary because their magnification is lower than that of the detection illumination lenses. Precise positioning of the light sheets within the field of view of the illumination objective, and their lateral size, are adjusted using the galvanometric scanners.

REFERENCES

1. Nadeau, J. L. 2017. *Introduction to Experimental Biophysics, Second Edition: Biological Methods for Physical Scientists*. CRC Press, NY.; Leake, M. C. 2016. *Biophysics: Tools and Techniques*. New York: CRC Press; Nadeau, J. 2011. *Introduction to Experimental Biophysics: Biological Methods for Physical Scientists*. New York: CRC Press; Serdyuk, I. N., N. R. Zaccai, and J. Zaccai. 2007. *Methods in Molecular Biophysics*. Cambridge, UK: Cambridge University Press; Shashkova, S., and M. C. Leake. 2017. Single-molecule fluorescence microscopy review: Shedding new light on old problems. *Bioscience Rep* 37(4):BSR20170031; Dahlberg, P. D., C. T. Boughter, N. F. Faruk, L. Hong, Y. H. Koh, M. A. Reyer, A. Shaiber, A. Sherani, J. Zhang, J. E. Jureller, and A. T. Hammond. 2016. A simple approach to spectrally resolved fluorescence and bright field microscopy over select regions of interest. *Rev Sci Instrum* 87(113704):1–6.
2. Hobbie, R. K., and B. J. Roth. 2007. *Intermediate Physics for Medicine and Biology, Fourth Edition*. New York: Springer Science + Business Media. See also your favorite university's Department of Physics academic programs. Many will have a designed biophysics or biomedical physics track.
3. van Speybroeck, L. 2009. *Exploring Pauli's (Quantum) Views on Science and Biology. Recasting Reality*. Berlin, Germany: Springer; Schrödinger, E. 1944. *What Is Life? The Physical Aspect of the Living Cell*. Cambridge, UK: Cambridge University Press; Sri Kantha, S. (1996). Einstein's medical friends and their influence on his life. *Med Hypotheses* 46(3):257–260.
4. Evolutionary theory has been around a long time, of course, and Wagner's work could be seen as descending from the work of Spiegelman, Orgel, Eigen and others. See Kettling, U., A. Koltermann, and M. Eigen. 1999. Evolutionary biotechnology— Reflections and perspectives. *Curr Topics Microbiol Immunol* 243:173–186; Oehlenschläger, F., and M. Eigen. 1997. 30 years later – a new approach to sol spiegelman's and leslie orgel's in vitro evolutionary studies, dedicated to leslie orgel on the occasion of his 70th birthday. *Orig Life Evol Biosph* 27(5–6):437–457; Lenski, R. E. 2017. Experimental evolution and the dynamics of adaptation and genome evolution in microbial populations. *ISME J* 11:2181–2194.
5. Krzic, U., S. Gunther, T. E. Saunders, S. J. Streichan, and L. Hufnagel. 2012. Multiview light-sheet microscope for rapid in toto imaging. *Nat Methods* 9:730–733.
6. Briegal, A. and G. Jensen, Eds. 2017. *Ann Rev Biophys* 46(Whole volume): 1–558. Extended reviews of many modern biophysics topics in imaging, protein structure, optical manipulation, single-molecule studies, DNA repair, CRISPR-Cas9 and imaging, binding free energies, RNA structure, molecular recognition, protein evolution, self-assembly, electron cryotomography, and other. See section 18.4 for a list of authors and titles.

CHAPTER 19

CONTENTS

Atomic Force Microscopy

19

Molecular biology requires an understanding of the physics of the nanoscale—the realm of nanometer-sized objects, such as proteins, lipids or DNA. If there is one experimental technique that is emblematic of doing science at the nanoscale, it is atomic force microscopy (AFM) and its related techniques. Since its invention in 1986, AFM has become a popular technique with a multitude of uses in biology. AFM can image and measure tiny *forces* by using a small cantilever spring with an attached sharp tip as a probe, which interacts with a sample surface. AFM is therefore a *scanning probe* technique, which means that it collects information about the sample via a probe interacting with the sample. As this probe is scanned across the sample, it generates an image line by line. AFM can achieve an imaging resolution that is only limited by the probe size, surpassing conventional optical microscopy by a factor of 100 or more. This allows the imaging of individual molecules or even atoms (Figure 19.1). AFM can also be used to measure force as a function of distance between two objects at the nanoscale, such as forces acting between two individual biomolecules.

19.1 A BRIEF HISTORY OF SCANNING PROBE MICROSCOPY

Scanning probe microscopy (SPM) is based on scanning a sharp probe across a surface to obtain a surface map. The map of the surface is based on a specific property the probe measures such as voltage, current, force, temperature or light. Scanning probes exploit the near-field effect, i.e., the fact that if we bring a probe very close to a surface it will interact across the physical dimensions of the probe only, and not through long-range fields, such as an electrical or magnetic field. Standard optical microscopy is not a near-field technique, because it works by shining light over a large area to be imaged and constructing an image from an electrical field altered by the interactions with the entire sample. Imaging resolution in optical microscopy is therefore limited by the wavelength of light, while in near-field techniques, the resolution is limited only by the probe size (Figure 19.2), which can be as small as a single atom.

The first scanning probe microscopy technique, invented in 1982, was the scanning tunneling microscope (STM). An STM uses a sharp, electrically conducting tip as a probe. To obtain an image, the tip is brought very near to the surface of a conducting sample (within less than 1 nanometer), while a voltage is applied between the tip and the sample. If the tip is close enough, electrons can "tunnel" between the sample and the tip, and a small current of about 1 nano-ampere is measured. Tunneling is a quantum-mechanical phenomenon that relies on the dual particle- and wave-nature of electrons. Although the gap between the tip and the sample should be insurmountable for the electron (electrons are held within a metal by their electrostatic attraction to the positively charged atomic nuclei), the electron "wave" leaks far enough into the outside of the metal to be picked up by another metal object, such as a tip, as long as the tip is brought close enough. Thus there is a small probability that electrons can tunnel through the gap and emerge in the tip. The tunnel current is determined by the number of electrons per unit time that can tunnel through the gap, which depends very sensitively (exponentially) on the tip-surface distance and on the electron density in the sample. The sensitivity on distance and electron density makes STM a powerful tool to image conductive samples with such high resolution that it can not only image individual atoms, but even atomic orbitals (Figure 19.3).

In the early days, STM found applications in biology, such as in imaging DNA. However, STM is not easy to use in liquids, because of the presence of electrochemical currents that create

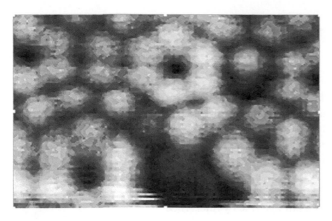

FIGURE 19.1
AFM image of a silicon crystal surface, showing individual atoms (each "blob" is an individual silicon atom). Also visible are defects (missing atoms). Such images can only be obtained in a vacuum chamber under extremely clean conditions. The width of the region imaged here is only 4 nm. Image by author (PMH).

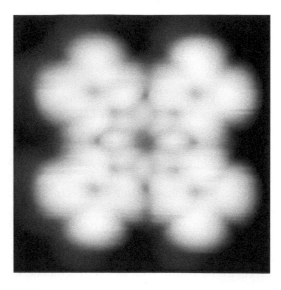

FIGURE 19.3
STM image of an individual naphthalocyanine molecule, showing molecular orbitals. Image size is 2.7 × 2.7nm. (From Gross, L. et al., *Phys. Rev. Lett.*, 107, 086101, 2011.)

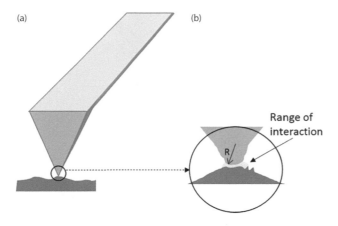

FIGURE 19.2
(a) An AFM cantilever with an attached tip interacts with a surface. (b) The magnification shows that the AFM will probe across an area that is determined by the dimensions of the probe tip, which has a nominal radius of R.

significant electrical noise. Instead, STM works best in air, or, even better, in vacuum. In addition, it requires an electrically conductive sample, while most biological systems have relatively poor electrical conductivities compared to metals. Fortunately, relatively soon after the invention of STM, researchers noticed some strange artifacts in STM images, which could be explained only by the fact that the STM tip is subject to atomic-scale forces when it is brought close to a surface. This inspired researchers to build a scanning probe microscope that measures forces instead of currents—and, in 1986, the AFM was born.

19.2 WORKING PRINCIPLES OF AFM

19.2.1 The Feedback Loop

STM can map electron densities and AFM can map forces by moving across a surface at constant tip-sample distance. However, this works only if the sample surface is very flat. Otherwise, the distance between the tip and the surface will change. Since the current or force is a function of the tip-surface separation, the image would be convolution of the signal and the changing tip-surface distance. Worse, as the tip encounters bumps on the sample surface, the tip can crash into the surface. To obtain clean data and avoid crashes, AFMs are designed to adjust the tip-surface distance during the scan by moving the tip or the sample up and down as needed using a feedback loop and an actuator (Figure 19.4). A feedback loop is an electronic device that compares a measurement to a preset desired value, called the *setpoint*. If the measurement deviates from the setpoint, the feedback loop initiates some action to bring the system closer to it. A familiar example is a house thermostat. The thermostat measures the temperature in the house, and if the temperature becomes lower than the setpoint of the thermostat, the thermostat kicks on the heater until the setpoint temperature is reached. The feedback loop in an AFM works similarly. In the simplest case, a deflection (force) setpoint is set in the computer software, and the feedback loop provides the signal to move the tip (or, alternatively, the sample) up and down to keep the cantilever deflection close to the setpoint as the tip scans across the surface.

To move the tip relative to the sample, piezoelectric actuators are usually used. Piezoelectric materials are materials with a permanent electrical dipole moment, which change shape when an electric field is applied across them.

With the feedback loop and actuator in place, scanning probe microscopes are typically used in a *constant signal mode*, i.e., they are used in a mode where the signal being measured (current or force) is held constant by controlling the probe-surface distance as the probe is scanned across the surface. The result of this approach is that AFM produces a constant force contour of the surface, much as we may obtain the contour of a surface by feeling the surface with a finger applying a constant pressure while scanning the surface. Thus AFM images represent the constant force *topography* of a surface. Figure 19.5 shows how the cantilever and tip scan across a surface to generate an image.

19.2.2 The Cantilever

To measure force, AFM uses a cantilever spring with an integrated sharp tip, which interacts with the surface. A cantilever is a straight beam of material, similar to a diving board at a local swimming pool. The cantilevers used in AFM are very small: Typical lengths range from about 50–500 μm, with widths of 20–50 μm and thicknesses of 0.4–4 μm. The tip is integrated into the free end of the cantilever and is typically 10–20 μm high and tapers to a sharp point with a radius of 2–20 nanometers. See Figure 19.6 of an AFM cantilever.

As the tip interacts with the surface, tip-surface forces lead to a bending ("deflection") of the cantilever. If the force is attractive, the cantilever will bend down towards the sample; if it is repulsive, it will bend up (Figure 19.7). If the spring constant of the cantilever, k_L, is known, the measured deflection Δd can be translated into force by Hooke's law, i.e., $F = -k_L \Delta d$.

An important question is what spring constant of the cantilever, k_L, should be chosen for different samples. Clearly, we do not want the imaging force to be so large as to damage the sample surface. This suggests that we would want to make the spring constant as small as possible. However, there is a limit to how small we can make k_L. First of all, we want to avoid sagging of the cantilever under its own weight. Secondly, a cantilever is also an oscillator (like a spring pendulum) with a natural period of $T = 2\pi \sqrt{m/k_L}$ at resonance, where m is the (reduced)

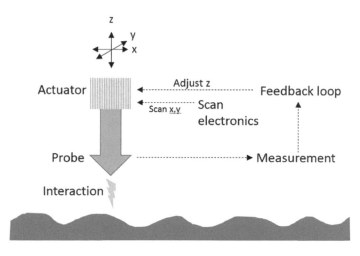

FIGURE 19.4

Principle of scanning probe microscopy: A sharp probe tip is scanned across a surface using an actuator driven by scan electronics (*x* and *y* directions). The probe measured the strength of the interaction between the tip and the surface (such as current or force). The measurement is compared to a setpoint in the feedback electronics, and the feedback directs the actuator to move the tip up or down (*z* direction) to keep the interaction strength near the setpoint.

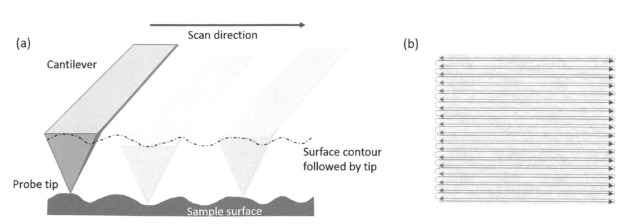

FIGURE 19.5

(a) AFM cantilever and tip scan a line across the surface of a sample. With the feedback loop keeping the cantilever deflection constant, the tip follows the contour (topography) of the sample along the scan line. (b) Top view of scan pattern across a sample surface. At each line, the tip scans forward (red) and reverse (blue) in the *x* direction (called the "fast scan direction"), and then moves a small distance in *y* ("slow scan direction") for the next scan line. A typical number of scan lines is 256 (much fewer are shown), with each scan line consisting of 256 pixels, resulting in a 256 × 256 map of the surface topography of the sample.

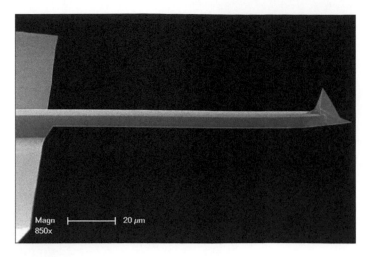

FIGURE 19.6
Scanning electron micrograph of an AFM cantilever with integrated pyramidal tip. Note that the cantilever is shown "upside-down" compared to how it would be used in an AFM. Bruker Corporation.

mass of the cantilever and T is the period. If we make k_L very small, T may become too large. Imaging requires the cantilever to react quickly to changes in force. Any change in force that occurs at smaller timescales (faster) than T has the potential to excite vibrations in the cantilever, and therefore noise. If T is a large number, one would need to scan the surface very slowly to avoid excessive noise. We therefore want T to be small. This puts a lower limit on the stiffness of the cantilever, but also tells us that we want to make the cantilever as small as possible to reduce its mass (since the formula for the period has the mass, m, in its numerator).

So, what is a good range for k_L? Imaging is a type of measurement, and measurements are comparisons. If we want to measure the length of a table, we compare the table dimensions to the dimensions of a tape measure. The tape measure has a length and accuracy that is within 1–2 orders of magnitude of the dimensions of the table—otherwise, if it were many orders of magnitude different, measurement would be difficult (for example, it would be difficult to measure the length of the Pacific coastline of the United States with a tape measure—the dimensions of the tape measure and the coastline differ too much). This suggests that if we want to image a surface at a small

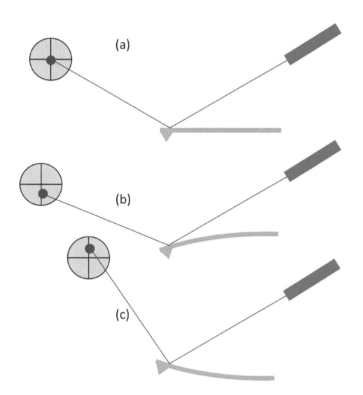

FIGURE 19.7
Cantilever bending (deflection) and resulting laser beam path (red) for three situations: (a) No force is acting on cantilever, and the cantilever is straight (no bending). (b) An attractive force is acting on the cantilever, bending the tip down (negative deflection). (c) A repulsive force is acting on the cantilever, bending the tip up (positive deflection). Note that the amount of bending is greatly exaggerated in this schematic. The laser beam hits a photodiode sensor that is sectioned into four quadrants. In all cases, the laser beam emerges from a laser diode (on right), reflects off the cantilever back, and hits a photodiode sensor that is sectioned into four quadrants. By subtracting the amount of light hitting the upper two quadrants from the light hitting the lower two quadrants, the motion of the laser beam can be tracked.

scale—ultimately the scale of atoms and molecules—we also want to match the properties of the measurement device (cantilever) with the properties of the sample. Chemical bond energies range from about 100 to 1000 kJ/mole, whereas weaker bonds, such as hydrogen bonds are in the range of 10–100 kJ/mole. Translated into energy per molecule, this gives a range of roughly $10^{-20} - 10^{-18}$ J per molecular bond. A more convenient unit for AFM is the pN nm, the work of moving 1 nm against a pN force. In these units, these molecular energies translate to a range of 10–1000 pN nm. Bonds have lengths in the range of $\Delta x = 0.1 - 0.5\,\text{nm} = 1 - 5 \times 10^{-10}$ m. With this information, we can estimate the force acting between two atoms by recognizing that the work needed to separate two atoms can be approximated by $W = F\Delta x$. Therefore $F \approx W/\Delta x$. Substituting our range of energies and an average bond length of 0.25 nm, we find bond forces ranging from about 40 to 4,000 pN. Bonds between biomolecules are typically in the 100 pN range, as they are dominated by many weak bonds, so our rough estimate is not bad. Now, using Hooke's law, we can estimate the stiffness of the sample by taking $k = F/\Delta x = 100\,\text{pN}/0.25\,\text{nm} = 400\,\text{pN/nm} = 0.4\,\text{N/m}$.

Researchers typically use cantilevers for biological imaging that are less stiff than 0.4 N/m to enhance the deflection signal. Cantilever choices range from 0.01 to 0.3 N/m, depending on the application. The unit pN/nm is a convenient unit for the stiffness of AFM cantilevers. In these units, for example, a 0.1 N/m cantilever has a stiffness of 100 pN/nm. This immediately tells us that this cantilever would deflect (bend) by 1 nm if subjected to a force of 100 pN. But can a deflection as small as 1 nanometer be measured?

EXERCISE 19.1

Choice of cantilever: You are imaging a soft material with an AFM that has a deflection sensitivity (smallest cantilever deflection it can measure) of 0.3 nm. You do not want to apply more than 500 pN of force to the sample during imaging. What is the largest cantilever stiffness that would still work for this purpose? What would be the advantages of using a less stiff cantilever? Give stiffness in units of both pN/nm and N/m.

19.2.3 The Deflection Measurement System

In the early days of AFM, physicists and engineers experimented with various approaches to measure the tiny deflections of the AFM cantilever. The first AFM used an STM tip on top of the AFM lever to measure the deflection. This proved quite cumbersome and unreliable. Eventually, the AFM community settled on using a laser deflection method as the most convenient method for measuring the bending of the cantilever. While there are some specialty AFMs that use other methods, such as interferometry or piezo-resistance, the vast majority of AFMs use laser deflection. Laser deflection works by focusing a laser beam onto the reflective back of the cantilever at an angle. The reflected laser beam moves as the cantilever bends, as seen in Figure 19.7. The position of the laser spot is measured using a four-quadrant photodiode, by subtracting the light signal falling onto the upper two quadrants from the light signal on the lower two quadrants. The deflection of the laser beam is given by $\Delta s = L\Delta\theta$, where Δs is the change in the laser spot position, L is the length of the laser path, and $\Delta\theta$ is the change in cantilever angle. If the laser path is long enough, even tiny amounts of cantilever bending can be picked up. Such a system can routinely pick up as little as 0.1 nm cantilever bending.

With a 100 pN/nm = 0.1 N/m cantilever, a 0.1 nm deflection sensitivity translates into a 10 pN force sensitivity of the AFM. Theoretically, smaller forces can be measured by using even weaker cantilevers, but weaker cantilevers are subject to more thermal noise, ultimately limiting the sensitivity. Thermal noise of the cantilever can be estimated by the equipartition theorem, which for a spring gives us $\frac{1}{2}k_B T = \frac{1}{2}k_L \langle x^2 \rangle$. Then the noise amplitude becomes $\sqrt{\langle x^2 \rangle} = \sqrt{k_B T/k_L}$. For a $k_L = 10\,\text{pN/nm} = 0.01\,\text{N/m}$ cantilever, the noise deflection amplitude is 0.64 nm, which is larger than the sensitivity of the deflection system and translates into a "noise force amplitude" of 6.4 pN as the lower limit of what can be measured above the thermal noise of the cantilever. As the cantilever stiffness is further reduced, the noise amplitude becomes so large that accurate measurements become difficult to obtain.

EXERCISE 19.2

Noise in AFM: Use a suitable program, such as Excel, to calculate the noise deflection amplitude and noise force amplitude as a function of the cantilever stiffness. Use a range of 0.1 to 100 pN/nm for the stiffness. If you want an accuracy of better than 0.5 nm for the tip position (cantilever deflection) and a force sensitivity of better then 20 pN, what cantilever stiffness would be suitable? To keep within the pN-nm unit system we use here, you can use $k_B T \approx 4.14$ pN nm at room temperature.

19.3 MEASUREMENT TECHNIQUES

AFM can perform two types of measurements: Imaging and force-distance measurements. During imaging the sample is scanned in x and y directions, while a feedback loop maintains a constant feedback signal, such as the lever deflection. In force-distance measurements, the cantilever is moved toward or away from the surface and the force is measured as a function of distance from the sample, typically while the feedback loop is switched off. In this section, we will have a closer look at different imaging techniques and methods for force-distance measurements. However, to understand these technqiues, we need to first have a closer look at the mechanics of AFM.

19.3.1 Mechanics of AFM Measurements

To understand how AFM works, it helps to have a closer look at the mechanics of the instrument. Figure 19.8a shows a diagram of the basic mechanical setup of an AFM. From this diagram, we can see that the relevant distances are related by

$$z = x + h - d \tag{19.1}$$

Here, z is the distance from the cantilever base to the sample, x is the tip-to-surface distance, h is the tip height, and d is the cantilever bending. While the tip height is fixed, the distances z, x and d can all change during a measurement. In most AFM setups, when we perform a force-distance measurement, we *control z* (by ramping it down and up) and *measure d*. There is no direct way to measure x. If we want to find the tip-surface distance x, we need to calculate it from z, h and d.

In terms of forces, neglecting friction, the force on the tip is given by:

$$ma = F_{\text{tip}} = -k_L d + F_i(x) \tag{19.2}$$

Here, m is the reduced mass of the cantilever, F_{tip} is the net force acting on the tip, k_L is the cantilever stiffness (spring constant), $-k_L d$ is the restoring force of the cantilever at a cantilever deflection of d, and F_i is the interaction force between the tip and the surface, which is a function of the tip-surface distance x. Note that we use the convention that "up" means positive. In Figure 19.8a, d is negative, because the cantilever is bending down, and the resulting restoring force of the cantilever, $-k_L d$, is positive and pointing up. This convention also explains the minus sign in front of d in Equation 19.1.

At mechanical equilibrium, when the tip is not moving, the net force on the tip is zero, and we find

$$k_L d = F_i(x) \tag{19.3}$$

This shows that at mechanical equilibrium the cantilever deflection d, when multiplied by the cantilever spring constant k_L, provides the interaction force between the tip and the sample, F_i.

When the system is not in mechanical equilibrium, i.e., when the tip is accelerating due to an oscillation of the tip, Equation 19.2 applies. If the deviation from equilibrium is small, the interaction force can be written as $F_i^{\text{eqb}} - k_i \Delta x$, and we find for the force on the tip:

$$F_{\text{tip}} = -k_L(d^{\text{eqb}} + \Delta d) + (F_i^{\text{eqb}} - k_i \Delta x) = -k_L \Delta d - k_i \Delta x \tag{19.4}$$

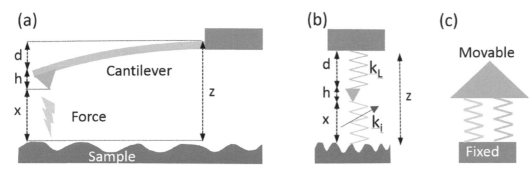

FIGURE 19.8
(a) The basic setup of the AFM. In this example, the lever experiences an attractive force, which bends the cantilever down. Here, d is the cantilever deflection, h is the tip height, x is the tip-to-surface separation, and z is the overall distance between the cantilever base and the surface of the sample. (b) Equivalent mechanical model: For small deviations from mechanical equilibrium, the change in interaction force between the tip and the sample can be represented by a spring k_i. The red arrow illustrated that this is a variable "spring," which depends on the distance x. The spring constant of the cantilever is k_L. (c) Equivalent mechanical model for the case of small motion of the tip out of equilibrium at fixed z. In this case, we can consider the springs to be in parallel with repect to the tip motion.

where we used the fact that $-k_L d^{eqb} + F_i^{eqb} = 0$. If $\Lambda z = 0$ as in a freely oscillating tip, we find from Equation 19.1 that $\Delta x = \Delta d$, and therefore

$$F_{tip} = -k_{eff}\Delta d = -(k_L + k_i)\Delta d \qquad (19.5)$$

The fact that the stiffnesses are additive ($k_{eff} = k_L + k_i$) is surprising, because we have two springs in series, as seen in Figure 19.8b. For springs in series, we add the reciprocals, i.e., $k_{eff} = \left(1/k_L + 1/k_i\right)^{-1}$. However, the springs are in series only from the point of view of the cantilever support at the top. The situation is different from the tip's point of view: If we hold z fixed, and measure the motion of the tip, we see that the tip is connected to two springs each attached to a fixed point. In our mind, we can rotate the upper surface, and see that this situation is equivalent to a parallel spring setup, as seen in Figure 19.8c—hence the additivity of the spring constants.

Equation 19.5 applies when we control z and measure d. However, during imaging and in some force-distance methods, we control d and measure z, i.e., in this case we have to adopt the point-of-view of the cantilever support. If we control d, we are, in effect, controlling the force between the tip and the surface. To control d, the feedback loop is employed. Force-distance measurements under feedback-loop control are called *force-clamp* methods. The feedback circuit controls the value of d and adjusts (and measures) z accordingly. In this case, a small change out of equilibrium changes z according to:

$$\Delta z = \Delta x - \Delta d = -\frac{\Delta F}{k_i} - \frac{\Delta F}{k_L} = -\Delta F\left(\frac{1}{k_i} + \frac{1}{k_L}\right) \qquad (19.6)$$

Comparing Equations 19.5 and 19.6, we see that in force-clamp mode, the springs act in series, while in conventional force-distance mode, the springs act in parallel. This can be important in biological measurements, if the equivalent spring constant of a biological molecule is very small. If we add a small spring constant to a larger spring constant, the small spring constant may be difficult to measure. However, if we add reciprocals, the smaller spring constant will dominate the measurement. Therefore, force-clamp methods can have advantages when performing single-molecule measurements, but they are more difficult to implement and subject to more noise.

Parallel and series springs: What is the effective spring constant of two parallel springs? When springs are parallel their deflections are equal. In other words, if I move the tip both springs experience the same magnitude change in length (see Figure 19.8c). The total force from the two springs is the sum of the forces from each individual spring. Therefore we have $F = F_1 + F_2 = k_1\Delta x + k_2\Delta x = k_{eff}\Delta x$ and consequently $k_{eff} = k_1 + k_2$.

For springs in series in mechanical equilibrium the forces in the two springs are equal and the combined change in length of the two springs is the sum of the changes in the individual springs. Therefore we find that $\Delta x = -\frac{F}{k_{eff}} = \Delta x_1 + \Delta x_2 = -\frac{F}{k_1} - \frac{F}{k_2} = -F(\frac{1}{k_1} + \frac{1}{k_1})$. Thus the effective spring constant of two springs in series is $k_{eff} = (\frac{1}{k_1} + \frac{1}{k_1})^{-1}$.

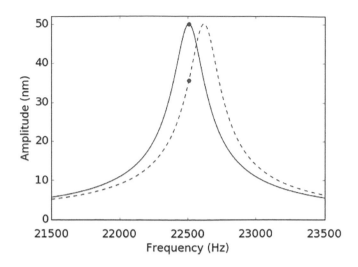

FIGURE 19.9

Solid line: Resonance curve of a 0.1 N/m cantilever with dimensions of 250 μm length, 25 μm width and 1 μm thickness. Red dot: Excitation amplitude and frequency when cantilever is unperturbed (i.e., far from the surface). Dotted line: Shifted resonance when cantilever interacts with the surface via a repulsive contact stiffness of 0.001 N/m. Blue dot: Amplitude at the same excitation frequency is reduced due to interaction with surface.

19.3.2 Imaging Modes

So far, we have discussed only the simplest way of operating an AFM for imaging: We monitor the static deflection of the cantilever as it scans across the surface and use the feedback loop to keep the deflection (and therefore the force) as constant as possible. Operating the AFM like this is called *contact mode*, because the AFM probe (or tip) is always in contact with the surface of the sample.

When AFM was first developed, all AFMs operated in contact mode. However, it was soon found that during contact mode the tip drags along the surface applying a frictional force to the surface. This can be damaging to delicate surfaces, such as biological samples. Researchers therefore searched for alternative ways to operate an AFM. They noticed that the cantilever, acting as a spring, can be driven into oscillations with a resonance frequency of $f = 1/T = \frac{1}{2\pi}\sqrt{k/m}$. Then, when the tip interacts with the surface, the effective stiffness of the system changes according to Equation 19.5 and the resonance frequency of the cantilever shifts.

This shift in resonance frequency can be exploited by monitoring the amplitude of the cantilever at resonance. If the oscillation frequency is kept constant at the *unperturbed* resonance of the cantilever, a reduction in amplitude is observed, as seen in Figure 19.9. This change in amplitude can be used as a feedback signal instead of the static deflection as in contact mode.

Using amplitude as the feedback signal is the basis of *dynamic* AFM modes. Two common dynamic modes are called *tapping mode* and *non-contact mode*. In tapping mode, the cantilever gently taps the surface, i.e., the tip contacts the surface intermittently as the cantilever oscillates up and down. In non-contact mode, the cantilever does not touch the surface, but only comes very close to interact with various longer-range forces acting between tip and surface. Non-contact mode is difficult to maintain, except in vacuum with high stiffness cantilevers, and, consequently, tapping mode is more common. Tapping mode can lead to less sample damage than contact mode, as it avoids dragging the tip, but it can lead to damage if tapping forces are not kept small.

A third dynamic mode is frequency-shift mode. In this mode, the resonance frequency itself becomes the feedback signal. A sophisticated electronic setup is used to track the resonance of the cantilever as it shifts. This mode can be very sensitive, but can be challenging to use in liquids and on biological samples.

19.3.3 Imaging Channels

During imaging, the feedback loop will do its best to maintain a constant feedback signal. The feedback signal can be cantilever deflection (in contact mode), amplitude (tapping mode), or frequency (frequency-shift mode). The feedback loop controls the cantilever-surface distance by moving the tip or the surface up and down during the scan. The voltage applied to the piezoelectric scanner to produce the up-and-down motion is recorded as the tip is scanned across the surface. Using the scanner calibration (i.e., how far the scanner moves for a given voltage), this voltage is translated into distance, and the motion of the tip is collected into a "topography" image that gives the height at each pixel of the image. Typically, topography images are presented in such a way that brighter colors indicate higher elevations of the sample, but the specific color scheme can be freely chosen by the user.

Because the feedback electronics is not perfect and cannot respond instantaneously, the feedback signal is not perfectly equal to the setpoint during the scan. When the cantilever tip encounters a bump on the surface, the feedback signal, such as the cantilever deflection in contact mode, will temporarily deviate from the setpoint until the feedback loop reacts to the change and adjusts the tip-surface distance. The feedback signal minus the setpoint can be collected into an image called

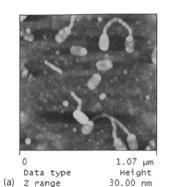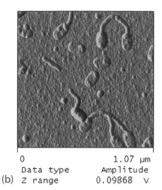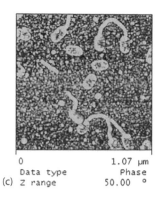

0	1.07 µm
Data type	Height
(a) Z range	30.00 nm

0	1.07 µm
Data type	Amplitude
(b) Z range	0.09868 V

0	1.07 µm
Data type	Phase
(c) Z range	50.00 °

FIGURE 19.10
Tapping mode imaging of bacteriophages with three simultaneously obtained imaging channels.
(a) Topography. (b) Error or amplitude image. (c) Phase image. (Image courtesy of Prof. Guangzhao
Mao, Wayne State University.)

the *error-signal* image (because it represents the error between the setpoint and the actual feedback
signal). The error signal is also sometimes called *deflection* (contact mode) or *amplitude* (tapping
mode). This image, while not quantitative like the topography image, emphasizes changes in topography and can therefore provide high contrast and help find small features present on a surface.

In dynamic mode imaging (such as tapping, non-contact or frequency-shift), a third channel can be
collected: the phase difference of the oscillation. The phase difference is the angular difference between
the oscillation state of the actuator that oscillates the cantilever and the actual motion of the tip. If they
move in perfect unison, the phase difference is zero, but if the tip motion lags behind the imposed
motion at the cantilever base, there will be a non-zero phase difference. At resonance, the phase difference is typically 90° or $\pi/2$ rad. However, as the tip interacts with the sample, the phase difference can
change, depending on the mechanical properties of the sample. Therefore, phase images can provide
information about the mechanical properties of a sample, but they can be difficult to interpret.

Figure 19.10 shows simultaneously acquired topography, error (amplitude) and phase images
of a sample of bacteriophages. In the topography image, the brighter regions (bacteriophages) are
higher by maximum 30 nm than the darker areas. The shading in the amplitude image (middle) is
due to the fact that the feedback signal overshoots when it approaches a bump on the surface and
undershoots when it needs to come back down from a bump. This shading enhances the contrast
of small features on the surface, but the signal measured has no quantitative meaning. The phase
image (right) shows that the bacteriophages are softer (and therefore brighter in the image) than the
substrate on which they are placed.

In contact mode, there is another image type that can be collected: As the tip is dragged across
the sample surface, it will twist sideways due to friction between the tip and the sample. This twist
can be measured by comparing the left and right quadrants of the photodiode (Figure 19.7). This
way a friction map of the surface can be obtained. This mode of imaging is known as *friction-force*
or *lateral-force microscopy*.

19.3.4 Force-Distance Measurements

Imaging is not the only measuring capability of AFM. Because AFM senses forces, it can also
measure forces as a function of distance. This is achieved by placing the tip above a fixed location
of the surface, and then approaching and retracting the tip while monitoring the tip deflection. After
converting the deflection to a force, we obtain a force-distance curve.

Force-distance curves can measure a variety of forces, including the stiffness of a cell or the
bonding strength between two protein molecules. Sometimes it is necessary to chemically modify
the cantilever tip, for example by attaching a ligand that can bind to a protein.

One complication in interpreting force-distance measurements is that the displacement measured is that of the piezoelectric actuator that moves the tip relative to the sample. This displacement is not the same as the displacement of the tip, because the cantilever also bends. The tip

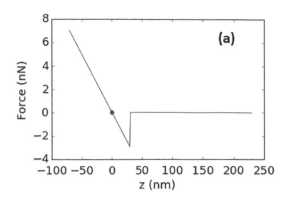 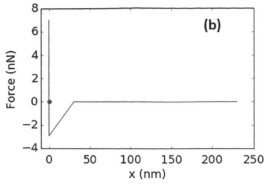

FIGURE 19.11

(a) Force-distance curve plotted versus the motion of the piezoelectric actuator, z. This example shows a retract curve on an infinitely stiff material with adhesion. Initially, the tip is pushed onto the surface with a maximum force of 7nN. As the cantilever base is retracted, the force on the tip and surface is reduced until it reaches a value of zero (red dot). This is the reference point ($x = 0$, $F = 0$) for subsequent rescaling. As the tip is further retracted, the force becomes negative because the tip adheres ("sticks") to the surface. In this regime the cantilever bending is negative. Eventually, the restoring force of the cantilever exceeds the maximum adhesion strength and the cantilever snaps off the surface. The force returns to zero. (b) Rescaled force curve to show how the force changes with actual tip-surface distance, x. As long as the tip is in contact with the sample, and assuming an infinitely stiff sample (i.e., no indentation *into* the sample), the tip-surface distance remains at zero, while the force is reduced and becomes negative in the adhesion regime. When the tip snaps off the surface, the tip-surface distance jumps. After that, the tip surface distance changes with the piezo motion.

displacement can be determined from Equation 19.1. In practice, we reference the tip-surface distance to the contact location $x = 0$. Then, as the tip moves, we find that

$$\Delta x = \Delta z + \Delta d \tag{19.7}$$

Note that we do not need to know the tip height h, because it does not change ($\Delta h = 0$).

Not all AFM data is reported with the distance axis rescaled to the actual tip motion, Δx, so it is important to read papers carefully to see how they define the distance axis in their force-distance graphs. Figure 19.11 shows the difference between plotting a force-distance curve versus piezo-motion (z) and versus tip-surface distance (x).

19.3.4.1 Types of force-distance experiments

Force-distance measurements can be separated into indentation ("pushing") and adhesion ("pulling") measurements. Further, we can distinguish between measurements using a bare, unmodified tip and those using a modified or "functionalized" tip. i.e., a tip that has been chemically altered to measure specific forces. Finally, we distinguish between conventional force-distance measurements and force-clamp measurements.

19.3.4.2 Indentation measurements and contact mechanics

Indenting the tip into a material can provide information about the elastic and plastic properties of the sample. If the sample is perfectly elastic (and the AFM performs perfectly), the approach and retract curve should overlap. In the presence of plastic deformation, there will be a mechanical hysteresis between approach and retract, because the sample deforms during compression.

To obtain quantitative information from such measurements, careful calibration of the AFM is needed:

1. The piezoelectric actuator must be calibrated, so that the voltage applied to the actuator can be converted to a distance moved. Calibration of the actuator is performed by scanning a standard sample with steps of known height. Such calibrated samples can be purchased.
2. The sensitivity of the photodiode detector must be calibrated, so that photodiode current (or voltage) can be converted to cantilever deflection. This is done by pushing the cantilever onto a hard sample. For the relatively weak cantilevers we would use for biological samples, a clean glass slide, or a metal or mica surface, will be hard enough. Assuming the sample is not

indented, the motion of the piezo will equal the motion of the tip, and this motion can then be used to calibrate the photodiode signal.

3. The cantilever stiffness (spring constant) must be calibrated, so that the cantilever deflection can be converted to force. This can be done by various methods—by measuring the cantilever dimensions and its resonance frequency or by fitting a thermal noise curve to a known Lorentzian peak shape. The latter method is more convenient and accurate and often already a part of software packages that come with commercial AFM systems.

4. The tip shape must be calibrated if parameters such as elastic moduli or adhesion energies are to be determined from the measurements. Tip shape is calibrated by either imaging it in a scanning electron microscope, or by scanning a test sample of very sharp spikes (available commercially) and using convolution methods to infer the shape of the tip.

To determine elastic parameters from indentation measurements, the simplest theory is Hertzian contact mechanics. This theory assumes that the tip is spherical and that there is no adhesion or plastic deformation. More complex theories are available in the presence of adhesion, plastic deformation, or for different tip geometries. If the elastic parameters are not to be spatially resolved or mapped, a blunt spherical tip is often used. Such a tip has the advantage that it matches the shape assumed by the theory and that the pressure it exerts on the often fragile sample (such as a biological cell) is not too large.

EXERCISE 19.3

Pressure applied by cantilever tip: Although forces used in AFM seem very small, the pressure applied to the sample by the tip can be quite high if a sharp tip is used. Assume we are applying a force of 2 nN with a tip that has a radius of 10 nm (assume it to be a circular flat with $R = 10$ nm). What is the pressure acting on the sample? Is that a lot? Why or why not?

The derivation of the Hertz equations is involved, but the resulting equations can be easily applied to AFM measurements. The contact radius, a, of a spherical tip with radius R and a flat surface is given by

$$a = \sqrt{Rx} \qquad (19.8)$$

where x is the indentation depth. Indentation depth is given by the actual tip motion (Equation 19.1). The force needed to achieve a certain indentation is given by

$$F = \frac{4}{3} E^* R^{1/2} x^{3/2} \qquad (19.9)$$

where E^* is the reduced modulus of the system, given by $\frac{1}{E^*} = \frac{1-v_1^2}{E1} + \frac{1-v_2^2}{E2}$ with $v_{1,2}$ and $E_{1,2}$ the Poisson ratios and elastic moduli of the two contacting materials (in our case, the tip and the sample). From Equations 19.8 and 19.9, we find that the contact radius is related to the indentation stiffness k by the simple equation:

$$k = \left| \frac{dF}{dx} \right| = 2E^* \sqrt{Rx} = 2E^* a \qquad (19.10)$$

EXERCISE 19.4

Indentation stiffness: A researcher indents a sample that has a known Young's modulus of 25 kPa. The radius of the spherical tip is 1 μm. If the researcher indents the sample by 50 nm, what force does she measure? What is the contact radius? What is the contact stiffness? What stiffness cantilever would you recommend for this measurement?

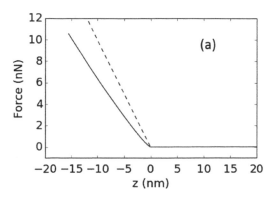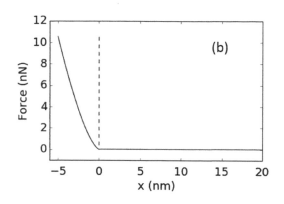

FIGURE 19.12

(a) Simulated force-distance curve (force versus piezo motion) obtained by indenting a soft material with an elastic modulus of 100 kPa with a spherical tip of radius 50 nm and a cantilever of stiffness 1 N/m. The dotted line indicates the force curve for an infinitely stiff material to show the contrast between a soft and a hard material. (b) Rescaled force curve, where the force is plotted versus the actual tip motion. The dotted curve shows the motion of the tip for an infinitely stiff material for comparison. The deviation from the vertical line is the amount of indentation into the sample.

The effective elastic modulus E^* can be determined either from the measured contact stiffness by using Equations 19.8 and 19.10, or by fitting Equation 19.9 to the measured force versus indentation curve. For soft samples, such a cells, we can assume that the elastic modulus of the tip, E_2 is much larger than the elastic modulus of the sample, E_1. This leads to $E^* \approx E_1/(1-v_1^2)$. A reasonable value for the Poisson ration of a cell is $0.3-0.4$, which gives $E^* \approx 1.1-1.2 E_1$.

Figure 19.12 shows an example of an indentation measurement into a perfectly elastic material, describable by Hertz contact mechanics.

If a sharp tip is used on a soft sample, Equation 19.9 may not provide reliable results, as the calculated indentation depth, x, may exceed the tip diameter. Sharp AFM tips are not spherical, but have on overall conical shape. Modified Hertz theory then predicts the force to be equal to

$$F = \frac{2}{\pi} E^* x^2 \tan \alpha \qquad (19.11)$$

where α is the half-opening angle of the cone, which typically ranges from $10°$ to $35°$, and is usually provided by the cantilever manufacturer.

Indentation measurements are used extensively in biological studies to determine the mechanical properties of membranes, vesicles and micelles, cells, bacteria, extracellular components (such as collagen), gels, or bone. Typical elastic moduli found for cells are in the 1–10 kPa range, while the moduli for bone are found in the 1–20 GPa range. Extracellular matrix material, such as collagen, has intermediate modulus values. Depending on the range of moduli being measured, the radius and cantilever stiffness need to be carefully chosen: weak cantilevers with dull tips work best on cells, while a sharp-tipped, high spring constant cantilever would perform better on bone.

Another parameter that can be determined from indentation measurements is the deformation work, i.e., the mechanical energy needed to indent the sample to a certain amount. This is given by

$$W = \int_0^{x_{max}} F(x)dx \qquad (19.12)$$

It is important that the integral (or sum) is done after converting the force curve to actual indentation depth, x, and not when the data is given as a function of piezo-motion z. The reason is that if we integrate the force curve $F = F(z)$, we also include the work done on the cantilever (i.e., the energy stored in the cantilever). However, we want to know the work done on the *sample*. Figure 19.13 shows an example.

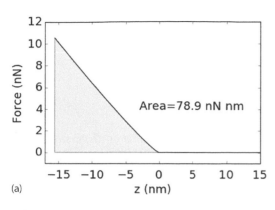
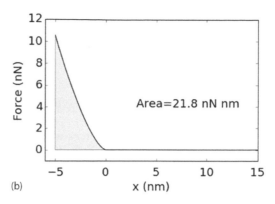

FIGURE 19.13

Calculating the work of deformation. (a) Area under force curve when force is plotted versus piezo-motion z. This includes work done on the sample *and* the cantilever. (b) Area under the force curve when force is plotted versus tip motion, x (which is equal to the indentation depth for $x < 0$). This area correctly measures the work done on the *sample*. Data parameters are the same as in Figure 19.12.

19.3.4.3 Adhesion measurements

While indentation measurements are performed during compression of the sample, i.e., during the approach of the tip towards and into the sample, adhesion measurements are performed when the tip is retracted from the surface. Adhesion measurements are used to measure the adhesion ("sticking") of cells and bacteria to various surfaces or other cells, and to measure mechanical properties of single biomolecules and their interactions. During adhesion measurements, we typically measure the maximum adhesion force, which is the maximum force that is reached before the adhesive contact breaks and the tip snaps off the surface. We can also determine the adhesion energy, which is the work done on the sample as the tip retracts.

Single molecule force measurements are a special type of adhesion measurements. Here, a ligand is chemically attached to the cantilever tip of the AFM, the tip is brought close to a surface containing target molecules, the ligand and the target molecules bind, and the tip is retracted. Then the maximum force needed to break the contact between the ligand and the target molecules (typically a protein) is measured. Another application is to measure the force to unfold a protein. Force-clamp methods are often used to measure single-molecule forces, for reasons discussed in Section 19.3.1. More about single-molecule measurements can be found in Section 19.5.2.2.

19.4 TYPES OF FORCES

There are a variety of forces that can act on the cantilever tip of an AFM. Because AFM measures the *total* force acting on the tip, we can generally not distinguish what kind of forces the AFM may register. However, depending on the situation, we can sometimes separate the various forces based on their range, the type of sample measured, or the type of tip we use.

The most common forces measured in biological applications include van der Waals, electrostatic, steric, double-layer, capillary, hydration, and hydrophobic forces. Some forces, such as chemical bonding forces, are short-range forces, i.e., they become very small after just a few nanometers of separation between the tip and the surface, while others, like electrostatic forces, can be quite long-range. In the following sections, we will discuss these force types one by one.

19.4.1 Chemical Bonding

Chemical bonding refers to covalent bonding between atoms due to the sharing of electrons. This is the main force holding molecules together. In AFM, chemical bonding between atoms in the tip and the surface of the sample can be measured, but typically only in vacuum. The reason is that in air or in a liquid, any "dangling" (unsaturated) bonds will quickly react with molecules in the air or in the liquid. Chemical bonding forces are very short-range, and typically become very weak at distances of more than 1 nanometer.

19.4.2 Van der Waals Forces

The most common attractive forces we measure in AFM are *van der Waals* or *dispersion* forces. These forces arise from a correlation between charge fluctuations in the tip with charge fluctuations in the sample or the induction of fluctuating dipoles in the tip and sample. Van der Waals forces are always present and usually, but not always, attractive. They are medium-range forces and are highly geometry dependent (i.e., they depend on the shape of the two interacting objects). Between two molecules that do not have a permanent dipole moment, the van der Waals force is given by

$$F_{vdW} = \frac{C}{d^7} \tag{19.13}$$

where C is a constant and d the distance between the molecules. For macroscopic objects, such as the AFM tip and the surface of the sample, the forces betwen constituent molecules add up. For the case of a spherical tip close to a flat surface (with separation $d \ll R$, where R is the radius of the sphere), the force becomes equal to $F_{vdW} = C'/d^2$, where C' is another constant.

A useful equation for the force between two spheres or a sphere and a flat (a flat is just a sphere of infinite radius) is given by the Derjaguin approximation, which states that

$$F(d) \approx 2\pi \frac{R_1 R_2}{R_1 + R_2} W(d) \tag{19.14}$$

Here, $W(d)$ is the interaction energy between two *flat* surfaces at a distance d, and R_1 and R_2 are the radii of the two spheres. For a sphere and a flat surface (as is usual the case in AFM), we let $R_2 \rightarrow \infty$ and obtain $F(d) \approx 2\pi R_1 W(d)$. The Derjaguin approximation is correct for *any* type of interaction between the tip and the surface and allows us to deduce the interaction energy from the measured forces. In addition, the Derjaguin approximation allows us to compare measurements obtained with different-sized tips, by reporting $F(d)/R$ instead of $F(d)$. This is because $F(d)/R = 2\pi W$ is independent of tip size.

EXERCISE 19.5

The Derjaguin approximation: A researcher measures an attractive force of 5 nN when a 50 nm radius tip is 2 nm from the surface of a sample. If she were to switch the tip for a 1 μm radius tip, how large of a force should she expect at the same tip-surface distance?

19.4.3 Electrostatic Forces

Electrostatic forces become relevant in the presence of charged surfaces. In air or vacuum, dielectric materials and semiconductors can have a surface charge, which can lead to long-range Coulomb forces. On biological samples, charges are due to ionic species, including charged macromolecules such as DNA. If the sample is in a liquid containing free ions (such as a buffer), the surface charges of the sample are screened, i.e., charges at the surface will attract oppositely charged ions, which will "neutralize" the charge over a certain distance away from the surface. This screening distance depends on the concentration of ions in solution. Forces due to a screened electrostatic interaction in solution are referred to as double-layer forces.

19.4.4 Double-Layer Forces

Double-layer forces are electrostatic forces that arise from the superposition of the surface charge of the sample and the diffuse layer of ions in solution that are screening the surface charge. Because of thermal motion, the ions in solution do not completely screen the surface charge close to the sample, but rather form a "cloud" away from the surface that screens the surface charge only over a certain thickness of the charge "double layer." As the ion concentration is increased, more ions are available to screen the charge, and the screening distance is reduced.

Double-layer forces can be repulsive (if the tip and the surface have the same charge) or attractive (if they have opposite charges), and they typically fall off exponentially with distance. The exponential term comes from the diffusion of the screening ions, which creates a Boltzmann-factor exponential concentration profile. Double layer forces and van der Waals forces are often combined to arrive at the complete force acting between particles in an electrolyte solution. The combination of these forces is known as the DLVO force, after its discoverers (Derjaguin, Landau, Verwey, and Overbeek).

19.4.5 Steric Forces

Steric forces arise when molecules approach each other so closely that they are trying to occupy the same space. When this happens, their electron clouds overlap and strongly repulse each other. Steric hindrance is the main reason for the strong repulsion experienced when the tip appraches a sample too closely. *Weak* steric forces can arise when the tip approaches an assembly of flexible molecules, such as polymers. In this case, the polymers may be moving due to thermal motion and interact with the tip on an intermittent basis. As the tip pushes deeper into a "brush" of polymers, the steric force increases. The range of these weak steric forces is of the order of the length of the polymers. Such forces can play an important role on cell surfaces, which may be covered by high densities of proteins and glycoproteins.

19.4.6 Capillary Forces

Capillary forces are very long-range forces that arise when the AFM tip approaches a surface in humid air. At finite humidity, most surfaces, especially if they are made of hydrophilic materials, are coated with a thin layer of water. When the AFM tip touches a surface coated with a water film, the water film clings to the tip and forms a meniscus. Because water has a high surface tension, the meniscus "pulls" on the tip and creates a large attractive force. Therefore, most AFM force measurements performed in air show large attractive forces, which are at least partially due to capillary forces (another important contribution can be electrostatic forces). To avoid capillary forces, which can mask other forces we would like to measure, it is best to immerse the sample completely in liquid.

19.4.7 Hydrophobic Forces

Hydrophobic forces are forces that act between two hydrophobic surfaces in an aqueous solution. The origin of hydrophobic forces is subject to much controversy, but scientists generally agree that they are an example of *entropic* forces. Entropic forces are forces that are not due to a change in a potential, but rather due to a change in average arrangement of particles, which lowers entropy. In water, nonpolar molecules cannot form hydrogen bonds with water. Because water likes to form hydrogen bonds, the presence of a nonpolar object restricts the motion of water molecules around it, decreasing its entropy. A lower entropy implies increased free energy, and therefore hydrophobic materials expel water from their surfaces. If two hydrophobic particles or surfaces approach each other in water, it is entropically favorable to bring them closer together to minimize the disruption to hydrogen bonds in the water. Therefore, an attractive force between the hydrophobic surfaces is generated. This force can be quite strong for a macroscopic object, like an AFM tip, and can make it difficult to image a hydrophobic surface in water using tapping or non-contact mode.

19.4.8 Hydration or Solvation Forces

Hydration or *solvation forces* are short-range oscillatory forces around a particle or surface in a liquid that arises from the *hydration* (in water) or *solvation* (in a general liquids) of the particle or surface. When an object (like the surface of the sample) is placed in a liquid, its presence restricts the motion of the liquid's molecules and they tend to align themselves along the surface, forming a weakly ordered layer. This first layer of ordered molecules serves as a weak barrier to molecules further away from the surface, leading to the formation of increasingly less ordered layers further away from the surface. The oscillating density of the liquid layers close to a solid surface generates an oscillating force, which can be measured by AFM. However, these forces can be quite small and typically range over only 4 to 6 molecular layers, and can therefore be difficult to detect.

19.5 BIOLOGICAL APPLICATIONS

19.5.1 Cell Biology

AFM has opened a new window into cell biology. Not only is it capable of imaging cell surfaces at high resolution, but it can also measure mechanical properties of cells, and even interactions of ligands with cell membrane components. While there were various attempts to image cells with STM, most of these failed due to the need of a conductive surface. Therefore, almost as soon as AFM was invented in 1986, researchers were excited about its possibilities in cell biology. By about 1990, the first cell images of fixed and dried cells had begun to appear. Cells that were fixed and dried proved to be much easier to image than live cells, as they were less soft and did not move.

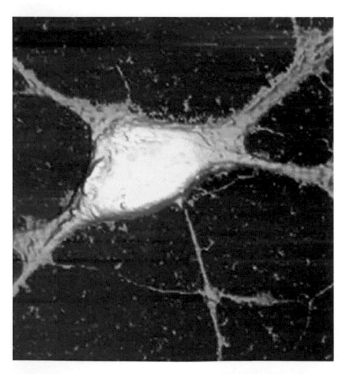

FIGURE 19.14
Topographic AFM image of a neuron (nerve cell). Image size is 30 × 30 μm and the maximum height of the cell is 2.2 μm. (From Ungureanu, A. A. et al., *Sci. Rep.*, 6, 25841, 2016.)

19.5.1.1 Cell imaging

Cell imaging has been used to measure overall morphology (Figure 19.14) and identify structures at the surface of cells (Figure 19.15). In some cases, AFM can indirectly image structures *within* the cell, as these structures (such as cytoskeleton structures) can affect the local mechanical properties of the cell and thus affect AFM imaging. Imaging *live* cells is of great interest, as it allows us to observe structures in their native condition and explore dynamic changes in living cells while they are subjected to changing conditions. However, live cell imaging of mammalian cells can be very challenging because living cells are much softer, have dynamic surfaces, may move while being imaged, and often lack good adhesion to a substrate. Consequently, imaging resolution obtained on live cells can be significantly lower than that obtained on fixed and dried cells (Figure 19.15). By fixing and drying the cells, cell movement is arrested and the cell is hardened.

Why would a soft cell be more difficult to image than a hardened cell? The reason is that even at very light imaging forces (remember that *some* force is needed for imaging), the pressures applied to a cell can be quite large. This leads to the tip penetrating into the cell, interacting with the cell over a large area. See Exercises 19.3 and 19.6 for more details.

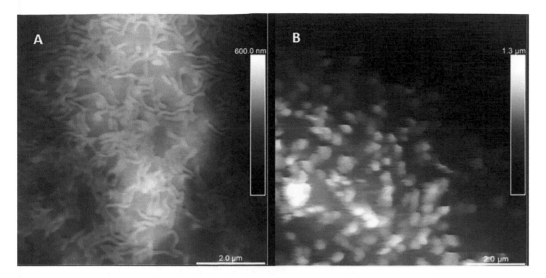

FIGURE 19.15
Topographic AFM images of microvilli on the surface of kidney cells. A: Image obtained on a fixed cell. B: Image obtained on a live cell. The imaging resolution is much better on the fixed cell. Length bars are 2 μm in both images. (From Schillers, H. et al., *J. Molec. Recog.*, 29, 95, 2016.)

EXERCISE 19.6

Imaging resolution in AFM: Imaging resolution in AFM is limited by the interaction area of the AFM tip with the surface. If the AFM tip interacts with an area of radius 100 nm, we cannot expect to clearly resolve objects of only 10 nm size. This is why it is important to image with a sharp tip. But a sharp tip may not provide high resolution if the sample is soft.

Let us assume that a live cell is imaged with a load force of 1 nN. We are using a sharp tip with a half-opening angle of 35° and the cell has an effective Young's modulus of 50 kPa. Now use Equation 19.11 to calculate the depth of indentation of the tip into the cell. Compare your result to the typical thickness of a cell membrane of about 10 nm. Assuming the tip to be a cone with a circular cross section, what is the radius of the total interaction area between the tip and the surface? What is the best image resolution we can expect?

As seen in the example, a sharp tip may push deep into a cell. This leads to reduced resolution and can even damage or rupture the cell membrane. To image small structures on cell surfaces (less than a few 10's of nanometers), cells typically need to be fixed, and imaging forces need to kept as small as possible (<0.2 nN). Imaging forces can be kept small by using a low stiffness cantilever (typically <0.3 N/m) to increase force sensitivity.

19.5.1.2 Bacteria and viruses

Bacteria are often mechanically harder due to their stiff membranes. This can make it easier to image bacteria than mammalian cells even though bacteria are much smaller in size (Figure 19.16). The same reasoning applies to viruses, which are exceedingly small, but also mechanically very hard. This makes AFM an attractive tool to image these small biological entities, which are typically too small to be imaged by optical microscopy (Figures 19.10, 19.6 and 19.17).

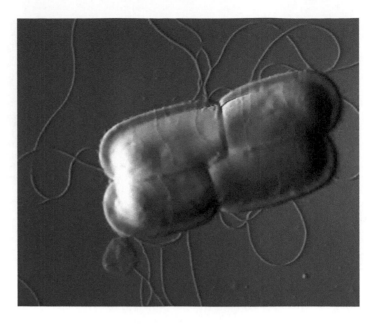

FIGURE 19.16
Deflection AFM image of *Bacteria thuringiensis* with flagella clearly visible. Image size is 8 μm.
(From Gillis, A. et al., *Nanoscale,* 4, 1585, 2012.)

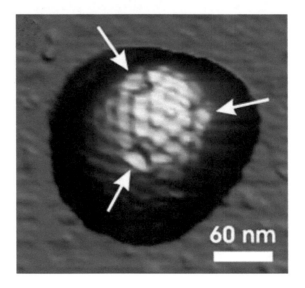

FIGURE 19.17
AFM image of a HSV1 virus capsid. Arrows point at missing capsid units, indicating 3-fold symmetry.
(From Baclayon et al., *Soft Matter,* 5273, 2010.)

19.5.1.3 Cell mechanics and mechanobiology

The ability of AFM to measure forces can be exploited to measure the mechanical properties of cells. In recent years, cell mechanics and the field of "mechanobiology" have experienced an explosive growth, when it was realized that many cellular processes, such as cell differentiation, tissue generation, and cancer metastasis are strongly affected by the mechanical properties of the extracellular matrix (ECM) and the cells themselves. AFM can be used to image the response to variations in substrate stiffness, such as changes in the cytoskeleton; to measure adhesion forces between cells, and between cells and ECM components; and to measure mechanical properties of cells, such as their effective Young's modulus.

Cell adhesion measurements are measurements of the adhesion of cells to a substrate or to other cells. These measurements can be done by coating the tip with a material that cells adhere to or by attaching a cell directly to the AFM tip. To measure the adhesion between two cells, one cell is attached to the tip and the other cell is seeded on a substrate.

Mechanical properties of cells can be measured by indenting cells. This is typically done with blunt, spherical tips as to not rupture the cell membrane. The obtained data is analyzed using contact mechanics models that were discussed in Section 19.3.4.2. These measurements provide estimates of the "stiffness" of the cell, given by the effective Young's modulus.

Force mapping is a technique in which a property such as adhesion or Young's modulus is measured across grid points to built up a map of the mechanical properties of a cell. One way to do this is to do repeated force-distance measurements across a grid. This gives quite accurate results, but can take a long time and because of time constraints is typically done only at a small number of grid points, limiting resolution. An alternative way to is to capture the tip response during imaging with a low-frequency, force-controlled tapping mode. This method can provide a high-resolution map of the mechanical properties of the surfaces, but at a possible loss of accuracy due to effects of liquid drag acting on the fast-moving tip.

19.5.1.4 Mechanical manipulation

The motion of the AFM cantilever can be directly controlled. Usually only two motions are used: scanning, combined with feedback-controlled vertical motion for imaging, and pure vertical motion for force-distance curves. However, in principle, the cantilever can be moved in any direction by the piezoelectric transducers. This direct control allows the user to use the AFM tip as a manipulation tool that can apply forces at selected locations, and move or deform micro- or nanosized objects. In cell biology, this capability has been used to observe the response of a cell to a mechanical stress or to imposed deformation.

19.5.2 Biomolecular Biology

The high imaging resolution of AFM allows the researcher to image subcellular components, even single biomolecules such as DNA or proteins. As a very sensitive force-sensing technique, AFM is also used to measure forces between and within single biomolecules.

19.5.2.1 Imaging of biomolecules

Imaging single biomolecules in live mammalian cells is extremely challenging, but such imaging has been achieved in fixed cells, as well as bacterial cells. In addition, single-molecule imaging can be done on isolated molecules, reconstituted membranes, or extracellular matrix material. Single biomolecules imaged in AFM include DNA, collagen, proteins and membrane channels. With some effort, even submolecular resolution showing the near atomic structure of large biomolecules is possible (Figure 19.18).

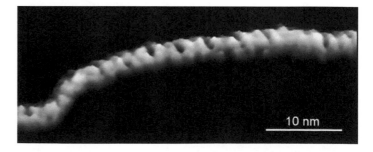

FIGURE 19.18
AFM image of DNA showing base-pair resolution. (From Pyne, A. et al., *Small*, 10, 3257, 2014.)

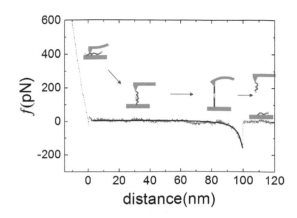

FIGURE 19.19
AFM force-distance measurement of the rupture of a protein-protein bond.

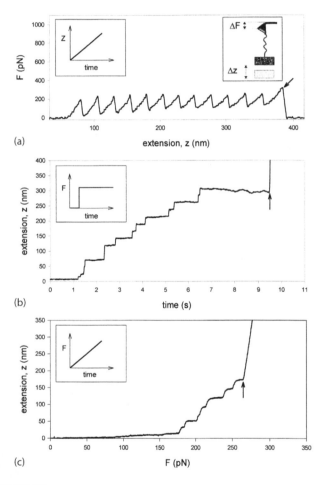

FIGURE 19.20
Force-distance ("extension") measurements of the unfolding of a titin protein. (a) Conventional force versus extension measurement, performed during a ramp of z in the absence of feedback control. Note that adhesion force is shown here as positive, unlike in previous examples. (b) Extension measured as a function of time during the application of a constant force using force clamp (force feedback). (c) Extension versus force during a ramp of the applied force using force feedback. (From Oberhauser, A. F. et al., *Proc. Nat. Aca. Sci.*, 98, 468, 2001.)

19.5.2.2 Single-molecule force measurements

With a sensitivity as small as 10 pN (10^{-11} N), AFM can directly measure interaction forces between single biomolecules and the forces needed to unfold proteins or DNA. Figure 19.19 shows a force-distance measurement of the rupture between two protein molecules. The cantilever is "retracting," i.e., it moves away from the surface. The proteins (indicated as green and yellow circles) are attached to flexible polymer linkers. At first the cantilever tip pushes on the surface, giving rise to a positive force and cantilever deflection (bending). While the tip is on the surface, the proteins have time to bond. If a bond is formed and as the tip moves away from the surface, the polymer linkers will uncoil until they are stretched, exerting a force on the protein bond. At this point, the cantilever bends downward and an adhesion force is registered. At some magnitude of the force, the protein bond breaks.

During a typical force-distance measurement, the base of the cantilever is retracted (or approached) at a constant rate, while the cantilever deflection (and therefore the force) is continuously measured. An alternative approach is the force-clamp mode, where we use a feedback loop to control the force and adjust the cantilever motion, z, to maintain a preselected force. Figure 19.20 shows an example of a force-clamp measurement of the unfolding of titin, a structural protein in our muscles.

19.5.2.3 Recognition and localization microscopy

Cells are dependent on numerous *specific* interaction between biomolecules. Enzymes interact only with specific substrates, transcription factors bind only to specific DNA sequences, and antibodies recognize a specific intruding antigen. Without such specificity, life would not be possible. This specificity also allows biomolecular researchers to recognize, tag, and image specific biomolecules in a sea of other molecules. One way to do this is to tag a target molecule with a flourescent marker attached to a ligand that specifically binds to the target molecules. Then the presence of specific molecules can be determined using fluorescence microscopy, and the molecules can even be tracked at a single-molecule level using super-resolution techniques.

AFM can also be used to recognize, localize, and track specific single molecules. This is achieved by attaching a ligand to the AFM tip that specifically binds to the target molecule. A common example for such a ligand is an antibody, but it can also be a protein, DNA or RNA sequence, or other small molecule. The tip is scanned across the surface using dynamic mode (tapping or non-contact). When the ligand attached to the tip binds to a target molecule, the tip will experience a force, and this will change the motion of the tip as it oscillates up and down. Using a variety of methods, this change in motion can be detected and compiled into a *recognition* image, which shows the locations of the targeted molecules on a surface.

19.5.3 Tissues

The nanostructure of tissues can also be measured by AFM. This can be challenging as tissues are complex, soft, and sticky—all attributes that make AFM imaging difficult. The key to imaging tissues is proper sample preparation, such as freezing the sample and then sectioning the frozen sample into thin sections. An example of tissue imaging is shown in Figure 19.21.

19.5.4 Ultrafast AFM

Typically, AFM imaging is rather slow. Scanning an image can take minutes, and it is therefore difficult to capture dynamic processes unless they are also occuring very slowly. AFM imaging is so slow because of the slow response of the feedback loop and the relatively low resonance frequency of conventional AFM cantilevers. Also, conventional piezoelectric actuators have low frequency resonances, which prevent fast scanning.

In recent years, several research groups have redesigned AFM from the ground up, including speeding up the feedback electronics, miniaturizing the cantilevers to push resonance frequencies into the Mhz range, and building scanners that can scan an entire image in as little as 20 milliseconds. These ultrafast AFM systems have been used to image dynamic processes, such as the movement of a single myosin molecule or the rotation of an ATP synthase by scanning 5–50 images per second.

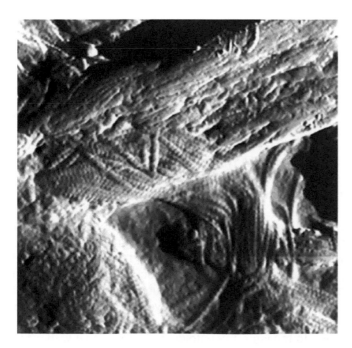

FIGURE 19.21
AFM image of extracellular matrix material of an aorta with clearly visible strands of collagen. Image size is $5 \times 5\,\mu m$. (From Graham, H.K. et al., *Matrix Biol.*, 29, 254, 2010.)

19.5.5 Other Related Techniques and Future Developments

New AFM-related and other scanning-probe measurement techniques are invented almost daily. Major research directions include combining different types of measurements, such as AFM and fluorescence microscopy, finding a more clever way to modify AFM tips to better isolate specific forces or obtain better images of a soft sample, and continuing to improve the speed and sensitivity of AFM to allow for the measurement of dynamic processes in living cells and tissues.

19.6 PROBLEM-SOLVING

PROBLEM 19.1   INDENTING A LIVING CELL

You are planning to measure the elastic modulus of a living cell. From literature, you expect that the modulus should be about $E = 100$ kPa. You want to compare data from two types of cantilevers: one with a sharp tip of cone half-angle 20°, and one with a spherical tip of radius 5μm.

1. You want the force on the cell to not exceed 100 pN. How much will each tip indent at this force?
2. If your AFM can reliably measure a lever deflection as small as 0.3 nm, what stiffness cantilever should you use in each case to keep indentation depth below 3 nm? Note that the smallest stiffness available for commercially available cantilevers is 0.01 N/m = 10 pN/nm.
3. Which of the two cantilevers should you use? What does that mean for image resolution if you wanted to create a map of the elastic modulus across the cell?

PROBLEM 19.2   MECHANICALLY DENATURING A PROTEIN

Proteins are active only in their folded form. Heat, pressure, and various chemicals can unfold and denature proteins. A typical free-energy difference between the folded and the denatured state of a protein is 50 kJ/mole. Using AFM, proteins can also be directly unfolded by applying a force, as seen in Figure 19.20.

1. How much energy is required *per molecule* to unfold a protein? Give the result in units of pN nm and k_BT (note: 1 $k_BT \approx 4.14$ pN nm at room temperature).
2. Using AFM, you measure an average force of 100 pN during unfolding of a protein. Estimate how much you stretch the molecule internally before it unfolds. Note that the distances in Figure 19.20 are much larger than the result you will obtain, because they represent the distances over which the attached polymeric linkers unfold, not the actual stretch of the protein.
3. You are trying to measure the unfolding of the protein with a force sensitivity of 10 pN. You know that your AFM cannot measure cantilever deflections smaller than 0.3 nm. What stiffness lever should you use for the measurement? Give the result in pN/nm and N/m.

PROBLEM 19.3   CALIBRATING AN AFM

Without calibration, as discussed in Section 19.3.4.2, the measurements we perform with an AFM provide meaningless results. In the absence of calibration, the photodiode signal, which represents the cantilever deflection, will be given in a unit such as mV or nA (photodiodes generate current, but these are often translated into a voltage by a converter circuit). The motion of the piezoelectric actuator will be represented by the voltage applied to the piezo. Figure 19.22 shows a measurement obtained before calibration. The goal is to find out the magnitude of the adhesion force.

1. First calibrate the z-axis. Assume you calibrated the piezoelectric actuator using a height standard and found a "piezo constant" of 22.4 nm/V. Calculate z for all cardinal points (1–5). What is the total piezo motion during the measurement (from points 1 to 5)?
2. Now that you know z, you can calibrate the photodiode to obtain the deflection of the cantilever. On a hard sample, where we expect close to no indentation, the change in cantilever deflection (Δd) should be equal the negative of the travel of the piezo ($-\Delta z$) – see Equation 19.7 with $\Delta x = 0$. In other words, the slope of the line between points 1 and 2 in the graph needs to be multiplied with a calibration factor that changes the units of the y-axis into nanometers, and makes the slope $\Delta d / \Delta z = -1$. What is value and unit of this calibration factor?
3. With the photodiode calibration in hand, what is the value of the largest deflection of the cantilever on contact (point 1). What is the value at maximum adhesion (point 3)?
4. We still do not know the force. To find the force, we need to calibrate the cantilever. In a real experiment, we would have done this before starting the experiment. From our calibration, we find that the cantilever has a stiffness of 34 pN/nm. What is the maximum force of adhesion? What is the maximum contact force?

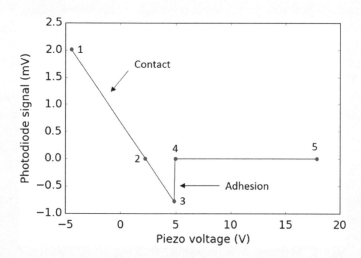

FIGURE 19.22
Uncalibrated force data taken on a very hard surface (no indentation) with adhesion present. X-axis is uncalibrated piezo voltage and y-axis is uncalibrated photodiode signal. Four cardinal points are indicated by red dots. Their coordinates are: 1: (−4.46 V, 2.01 mV), 2: (2.25 V, 0 mV), 3: (4.88 V, −0.79 mV), 4: (4.93 V, 0 mV), and 5: (17.86 V, 0 mV). Adhesion is between points 2 and 4, and repulsive contact between points 1 and 2.

Always make sure that the units work out—that will help you avoid mistakes. Think about if your results make sense based on what you learned. If distances and forces seem impossibly small or large, you probably made a mistake somewhere. Although not strictly necessary, recreating the graph in Excel or a similar program and then performing the calculation in a spreadsheet may be useful and provide you with practice in real data analysis.

REFERENCES

1. Butt, H. J., B. Cappella, and M. Kappl. 2005. Force measurements with the atomic force microscope: Technique, interpretation and applications. *Surf Sci Rep* 59:1–152.
2. Dufrêne, Y. F., T. Ando, R. Garcia, D. Alsteens, D. Martinez-Martin, A. Engel, C. Gerber, and D. J. Müller. 2017. Imaging modes of atomic force microscopy for application in molecular and cell biology. *Nat Nanotechnol* 12:295–307.
3. Jandt, K. 2001. Atomic force microscopy of biomaterials surfaces and interfaces. *Surf Sci* 491:303–332.
4. Miller, H., Z. Zhou, J. Shepherd, A. Wollman, and M. Leake. 2018. Single-molecule techniques in biophysics: A review of the progress in methods and applications. *Reports Prog Phys* 81:024601.
5. Seo, Y., and W. Jhe. 2007. Atomic force microscopy and spectroscopy. *Reports Prog Phys* 71:016101.
6. Whited, A. M., and P. S. Park. 2014. Atomic force microscopy: A multifaceted tool to study membrane proteins and their interactions with ligands. *Biochim Biophys Acta* 1838:56–58.
7. Gross, L., N. Moll, F. Mohn, A. Curioni, G. Meyer, F. Hanke, and M. Persson. 2011. High-resolution molecular orbital imaging using a p-wave STM tip. *Phys Rev Lett* 107(8):086101.
8. Ungureanu, A. A., I. Benilova, O. Krylychkina, D. Braeken, B. De Strooper, C. Van Haesendonck, C. G. Dotti, and C. Bartic. 2016. Amyloid beta oligomers induce neuronal elasticity changes in age-dependent manner: A force spectroscopy study on living hippocampal neurons. *Sci Rep* 6:25841.
9. Schillers, H., I. Medalsy, S. Hu, A. L. Slade, and J. E. Shaw. 2016. PeakForce tapping resolves individual microvilli on living cells. *J Molec Recog* 29(2):95–101.
10. Gillis, A., V. Dupres, G. Delestrait, J. Mahillon, and Y. F. Dufrêne. 2012. Nanoscale imaging of *Bacillus thuringiensis* flagella using atomic force microscopy. *Nanoscale* 4(5):1585–1591.
11. Kuznetsov, Y. G., and A. McPherson. 2011. Atomic force microscopy in imaging of viruses and virus-infected cells. *Microbiol Molec Biol Rev* 75(2):268–285.
12. Kedrov et al. 2006. *Naunyn-Schmiedeberg's Arch Pharmacol* 372:400.
13. Pyne, A., R. Thompson, C. Leung, D. Roy, and B. W. Hoogenboom. 2014. Single-molecule reconstruction of oligonucleotide secondary structure by atomic force microscopy. *Small* 10(16):3257–3261.
14. Oberhauser, A. F., P. K. Hansma, M. Carrion-Vazquez, and J. M. Fernandez. 2001. Stepwise unfolding of titin under force-clamp atomic force microscopy. *Proc Nat Acad Sci* 98(2):468–472.
15. Koehler, M., G. Macher, A. Rupprecht, R. Zhu, H. J. Gruber, E. E. Pohl, and P. Hinterdorfer. 2017. Combined recognition imaging and force spectroscopy: A new mode for mapping and studying interaction sites at low lateral density. *Sci Adv Mater* 9(1):128–134.
16. Graham, H. K., N. W. Hodson, J. A. Hoyland, S. J. Millward-Sadler, D. Garrod, A. Scothern, C. E. Griffiths, R. E. Watson, T. R. Cox, J. T. Erler, and A. W. Trafford. 2010. Tissue section AFM: In situ ultrastructural imaging of native biomolecules. *Matrix Biol* 29(4):254–260.
17. Kodera, N., D. Yamamoto, R. Ishikawa, and T. Ando. 2010. Video imaging of walking myosin V by high-speed atomic force microscopy. *Nature* 468(7320):72.

Index

Note: Page numbers in italic and bold refer to figures and tables, respectively.